The Statistical Sleuth

A Course in Methods of Data Analysis

Fred L. Ramsey
Oregon State University

Daniel W. Schafer
Oregon State University

An Alexander Kugushev Book

Duxbury Press
An Imprint of Wadsworth Publishing Company

I T P® **An International Thomson Publishing Company**

Belmont, CA • Albany, NY • Bonn • Boston • Cincinnati • Detroit • Johannesburg • London • Madrid
Melbourne • Mexico City • New York • Paris • San Francisco • Singapore • Tokyo • Toronto • Washington

Assistant Editor: Cynthia Mazow
Editorial Assistant: Martha O'Connor
Print Buyer: Barbara Britton
Permissions Editor: Peggy Meehan
Advertising Project Manager: Joseph Jodar
Production: The Book Company

Cover Designer: Harry Voigt Graphic Design
Copy Editor: Steven Gray
Accuracy Checking: Phyllis Barnidge
Compositor: Techsetters, Inc.
Printer: Quebecor Printing/Fairfield

Printed in the United States of America
1 2 3 4 5 6 7 8 9 10

For more information, contact Duxbury Press at Wadsworth Publishing Company, 10 Davis Drive,
Belmont, CA 94002, or electronically at http://www.thomson.com/wadsworth.html

International Thomson Publishing Europe
Berkshire House 168-173
High Holborn
London, WC1V7AA, England

International Thomson Editores
Campos Eliseos 385, Piso 7
Col. Polanco
11560 México D.F. México

Thomas Nelson Australia
102 Dodds Street
South Melbourne 3205
Victoria, Australia

International Thomson Publishing Asia
221 Henderson Road
#05-10 Henderson Building
Singapore 0315

Nelson Canada
1120 Birchmount Road
Scarborough, Ontario
Canada M1K 5G4

International Thomson Publishing Japan
Hirakawacho Kyowa Building, 3F
2-2-1 Hirakawacho
Chiyoda-ku, Tokyo 102, Japan

International Thomson Publishing GmbH
Königswinterer Strasse 418
53227 Bonn, Germany

International Thomson Publishing Southern Africa
Building 18, Constantia Park
240 Old Pretoria Road
Halfway House, 1685 South Africa

Library of Congress Cataloging-in-Publication Data

Ramsey, Fred L.
    The statistical sleuth : a course in methods of data analysis /
Fred L. Ramsey, Daniel W. Shafer.
            p.   cm.
    "An Alexander Kugushev publication."
    Includes bibliographical references (p.   -   ) and index.
    ISBN 0-534-25380-6
    1. Mathematical statistics.   I. Schafer, Dan.   II. Title.
    QA276.R33   1996
    519.5--dc20   96-35423

Dedications

To influential teachers Don Truax and Bob Buehler — F.L.R.

To Jeannie, Banner, and Casey — D.S.

Contents

C H A P T E R   **4**   **Alternatives to the *t*-Tools   81**

C H A P T E R   **5**   **Comparisons Among Several Samples   108**

C H A P T E R   **6**   **Linear Combinations and Multiple Comparisons of Means   142**

C H A P T E R 7   **Simple Linear Regression: A Model
                          for the Mean   167**

C H A P T E R 8   **A Closer Look at Assumptions for Simple Linear
                          Regression   198**

C H A P T E R 9   **Multiple Regression   225**

C H A P T E R 10   **Inferential Tools for Multiple Regression   255**

C H A P T E R   **11**   **Model Checking and Refinement   291**

C H A P T E R   **12**   **Strategies for Variable Selection   325**

C H A P T E R   **13**   **The Analysis of Variance for Two-way
                              Classifications   362**

C H A P T E R　18　Comparisons of Proportions or Odds　515

C H A P T E R　19　More Tools for Tables of Counts　538

C H A P T E R　20　Logistic Regression for Binary Response Variables　564

Preface

This book is written for those who need to use statistical methods to analyze data from experiments and observational studies and who need to communicate the results to others. It is intended as a text for graduate students who are preparing to design, implement, analyze, and report their research. The students must have some knowledge of basic statistical concepts such as means, standard deviations, histograms, the normal and t-distributions, but they need not be familiar with calculus or matrix algebra. All should have access to a statistical software package and a moderately powerful computer.

To the Student

Statistics is like grout—the word feels decidedly unpleasant in the mouth, but it describes something essential for holding a mosaic in place. Statistics is a common bond supporting all other sciences. It provides standards of empirical proof and a language for communicating scientific results. Statistical sleuthing is the process of using statistical tools to answer questions of interest. It includes devising experiments to unearth hidden truths, describing real data using tools based on ideal mathematical models, answering the questions of interest efficiently, verifying that the tools are appropriate, and snooping around to see if there is anything more to be learned. *The Statistical Sleuth* will show you how this all comes about.

Case studies. *The Statistical Sleuth* is organized around case studies, which begin each chapter and are used throughout to illustrate how the statistical tools operate. A small section entitled *Summary of Statistical Findings* accompanies each case study, demonstrating how to communicate statistical findings for a research publication. You should realize that the methods upon which the findings are based will be foreign to you upon first reading. After the chapter has been read, you should return to the studies and consider carefully how the chapter's methods have been used to answer the questions posed by the researchers.

Examine each case study carefully for its structural design. Ask yourself why the study was structured in the way it was. The studies will not only illustrate analytical techniques; most also present exemplary structures for your own studies.

Mathematical level.   The emphasis of this book is on the practical use of statistical methods. The correct practical use of statistical tools requires some understanding of the mathematical foundation for the tools. Sometimes algebra or elementary mathematical statistics are the best device for communicating the motivation. In general, however, the level of mathematics required to follow this book is not high.

What will you learn?   Do not expect to learn all that you will need to make you an experienced analyst. You will improve your understanding of statistical reasoning and of measures of uncertainty. You will learn how to translate mountains of computer output into short summary statements that communicate the results in a language common to all scientists. You will also learn a fairly large array of statistical tools that will be useful for a wide range of problems. But there are many more tools that are not covered in this book and many lessons that can only be gained through experience. At some point you may need to seek the help of a professional statistician. Then, at least, you will know the language, the general tools, and the spirit of statistical data analysis, which will make communication with a statistician more effective and beneficial.

To the Instructor

Level of sophistication.   The level of sophistication for this text is high when it comes to models and methods needed to analyze data and interpret results, but low when it comes to mathematics. Our foremost concern is that future researchers learn proper approaches for conducting the statistical aspects of their research. To this end, mathematics is neither sought out nor avoided.

Case studies.   Most chapters begin with two case studies for motivation and demonstration. Making these studies a central feature forces us to consider applied statistics more seriously than if we simply provided a data set to demonstrate a particular tool. It compels us to maintain a question-driven approach to the analysis of data.

The case studies are also our tool for exciting students. We cannot successfully teach them if they are not genuinely interested. We have tried to find a variety of interesting real data sets where the statistical analysis provides useful answers to questions of interest. In some cases, we found descriptions and summary statistics that made excellent examples, but we were unable to obtain the raw data. We have still used some of them, however, by generating data to match the summary statistics. To identify these cases, we use the phrase "based on a real study."

Although we have made the data problems central, limitations of space prevent us from including all the graphical displays and the different analyses we would like to present. We encourage the instructor to go into more depth in showing computer output, graphical displays, and alternative analyses.

The starting point.   At first glance, the first four chapters of *The Statistical Sleuth* appear to rehash the topics of one-sample and two-sample analysis, which are covered in introductory courses. This is not the case. These chapters are intended as a model for how topics in the rest of the book are treated. The chapters provide a detailed examination of material to which students have already been exposed, and an introduction to a philosophy of learning and using statistics.

Material covered. The Statistical Sleuth's principal tool is regression analysis. We have added several topics that are not ordinarily covered in a regression text: (1) *Generalized linear model*, including logistic and log-linear regression. This important tool enables researchers to analyze a wide range of problems that have until recently been analyzed with inappropriate tools (ANOVA) or with appropriate but difficult tools (contingency table chi-squares). We stress the parallels between generalized linear model regression and ordinary regression. With calculations provided by statistical software, this tool has become entirely understandable and extremely valuable. (2) *Repeated measures*. Whereas there is a great tendency for researchers to turn to a statistical computer packaged function that has "repeated measures" in the title, we feel there needs to be more guidance on a strategy for considering such data analysis. Chapters 16 and 17 respectively emphasize question-driven and data-driven reduction of dimensionality. (3) *Serial correlation*. Although a full treatment of time series analysis is beyond the scope of this book, by adjusting and filtering for the first-order autoregression we provide tools that expand the usefulness of regression technology to problems involving serial correlation.

Our decisions regarding coverage reflect the kinds of problems graduate researchers typically encounter. The topics chosen arise repeatedly in the campus-wide consulting service operated by Oregon State University's Statistics department for faculty and graduate students. By offering a textbook with these topics, we hope to relieve the pressure on departments to offer separate courses in categorical data analysis, in multivariate analysis, and in time series analysis for nonstatistics majors, or to enroll nonstatistics majors in classes designed primarily for statistics majors.

Possible paths through the chapters. The Statistical Sleuth was designed for a three-quarter sequence covering eight chapters each term. Typically not all the material is covered, and we have provided a number of optional topics in a section titled *Related Issues* at the end of each chapter. The book may also be used for a two-semester class in its entirety. For a one-semester or two-quarter class, we recommend the following sequence: conclusions and interpretations (1–4), several sample problems (5–6), simple linear regression (7–8), multiple regression (9–12), two-way analysis of variance (13–14), and logistic regression (20–21). There is room to mix and match to meet specific needs.

The Statistical Sleuth covers regression prior to two-way analysis of variance, in contrast with the more traditional presentation of two-way ANOVA directly after one-way ANOVA. Our reasoning here is that regression tools applied to the two-way situation are easier to interpret. They are also less subject to misunderstandings arising from imbalance in the experimental or sampling design. Experimental design chapters (23–24) appear at the end of the book. In truth, design issues are discussed throughout the text as they apply to the case studies. Topics such as replication, blocking, factorial treatment structures, and randomization appear repeatedly. So the actual chapters on design organize and summarize the issues. We also believe it is difficult to design an experiment without an understanding of the applicable analytic tools.

Statistical Computer Programs

A computer and a packaged statistical computer program are essential companions for *The Statistical Sleuth*. Many good packages are available. Unfortunately, they differ considerably in their style, language, and output. Some instruction about the particular software package must accompany your instruction for using the statistical tools in this book. The student's statistical analysis, however, should be guided by good statistical strategy, and not by the package, which is just a means for accomplishing the end. The data sets presented as case studies and as exercises are available on an enclosed disk.

Acknowledgements

We take pleasure in expressing our gratitude to those who have helped us with this project. We are especially indebted to those who taught from earlier versions of this book. Helen Berg (mayor of Corvallis), Ginny Lesser (Oregon State University), Loretta Thielman (University of Wisconsin—Stout), and Peter Thielman (University of Wisconsin—Stout) all cheerfully endured mistakes and made valuable suggestions for improvements. Thanks also to other colleagues who read drafts and provided suggestions, including Rick Rossi, Mike Scott, and Jeannie Sifneos.

We are very grateful for the use of the facilities at the Department of Statistics at Oregon State University, and the supportive cooperation of the chairman, Justus Seely, and the assistance of Genevieve Downing. Our views of statistics have benefited from fruitful discussions with our colleagues in the Statistics Department.

A portion of the early work on the book was conducted while Dan Schafer was on sabbatical leave at The University of Western Australia. The Mathematics Department there supplied facilities for which we are grateful. Special thanks go to Ian James, currently at Murdoch University in Perth.

A very enjoyable aspect of this project for us has been the discovery of modern scientific applications we have found while searching for case studies. We have gained inspiration from the many scientists who know good questions to ask in their research and whose scientific creativity is spurred by a genuine interest in finding the answers.

Very special thanks go to the students of Statistics 511–513 classes at Oregon State University. Their maturity and collective knowledge about scientific subjects have made the course a pleasure to teach. We have been driven by a desire to provide them with tools and strategies they may use to become the scientists that inspire us in the future.

We were guided through production by the staff at The Book Company, Dustine Davidson and George Calmenson. Phyllis Barnidge checked the answers to the computational exercises; her suggestions were very useful.

Finally, we wish to express our gratitude to the following reviewers: Andrew Barron, Yale University; Mauro Gasparini, Purdue University; Joseph Glaz, University of Connecticut; Mark Kaiser, Iowa State University; Frank Martin, University of Minnesota; Michael Martin, Australian National University; David Mathiason, Rochester Institute of Technology; Julia Norton, California State University—Hayward; John Rawlings, North Carolina State University; Larry Ringer, Texas A&M University; Joseph Schafer, Pennsylvania State University; Kirk Steinhorst, University of Idaho; Loretta Thielman, University of Wisconsin—Stout; Bruce Trumbo, California State University—Hayward; John Wasik, North Carolina State University.

Drawing Statistical Conclusions

Statistical sleuthing means carefully examining data to answer questions of interest. This book is about the process of statistical sleuthing, including strategies and tools for answering questions of interest and guidelines for interpreting and communicating results.

This chapter is about interpreting statistical results—a process crucially linked to study design. The setting for this and the next three chapters is the two-sample problem, where it is convenient to illustrate concepts and strategies employed in all statistical analysis.

An answer to a question is accompanied by a statistical measure of uncertainty, which is based on a probability model. When that probability model is based on a chance mechanism—like the flipping of coins to decide which subjects get which treatment or the drawing of lottery tickets to decide which members of a population get selected in a sample—measures of uncertainty and the inferences drawn from them are formally justified. Often, however, chance mechanisms are invented as conceptual frameworks for drawing statistical conclusions. For understanding and communicating statistical conclusions, it is important to understand whether a chance mechanism was used and, if so, whether it was used for sample selection, group allocation, or both.

1.1 Case Studies

1.1.1 Motivation and Creativity—A Randomized Experiment

Do grading systems promote creativity in students? Do ranking systems and incentive awards increase productivity among employees? Do rewards and praise stimulate children to learn? Although reward systems are deeply embedded in schools and in the workplace, a growing body of evidence suggests that rewards may operate in precisely the opposite way from what is intended.

A remarkable demonstration of this phenomenon was provided by psychologist Teresa Amabile in an experiment concerning the effects of intrinsic and extrinsic motivation on creativity. Subjects with considerable experience in creative writing were randomly assigned to one of two treatment groups: 24 of the subjects were placed in the "intrinsic" treatment group, and 23 in the "extrinsic" treatment group, as indicated in Display 1.1. The "intrinsic" group completed the questionnaire at the top of Display 1.2. The questionnaire, which involved ranking intrinsic reasons for writing, was intended as a device to establish a thought pattern concerning intrinsic motivation—doing something because doing it brings satisfaction. The "extrinsic" group completed the questionnaire at the bottom of Display 1.2, which was used as a device to get this group thinking about extrinsic motivation—doing something because a reward is associated with its completion.

Display 1.1   Creativity scores in two motivation groups, and their summary statistics

	Intrinsic Group		Extrinsic Group	
	12.0	20.5	5.0	17.4
	12.0	20.6	5.4	17.5
	12.9	21.3	6.1	18.5
	13.6	21.6	10.9	18.7
	16.6	22.1	11.8	18.7
	17.2	22.2	12.0	19.2
	17.5	22.6	12.3	19.5
	18.2	23.1	14.8	20.7
	19.1	24.0	15.0	21.2
	19.3	24.3	16.8	22.1
	19.8	26.7	17.2	24.0
	20.3	29.7	17.2	
Sample Size:	24		23	
Average:	19.88		15.74	
Sample Standard Deviation:	4.44		5.25	

After completing the questionnaire, all subjects were asked to write a poem in the Haiku style about "laughter." All poems were submitted to 12 poets, who evaluated them on a 40-point scale of creativity, based on their own subjective views. Judges were not told about the study's purpose. The average ratings given by the 12 judges are shown for each of the study subjects in Display 1.1. (Data based on the study in T. Amabile, "Motivation and Creativity:

Display 1.2   Questionnaires given creative writers, to rank intrinsic and extrinsic reasons for writing

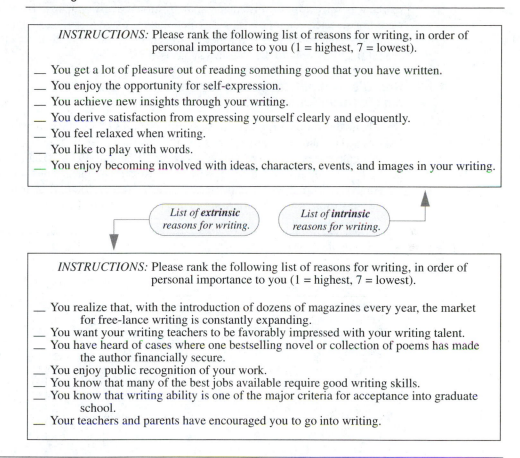

> *INSTRUCTIONS:* Please rank the following list of reasons for writing, in order of personal importance to you (1 = highest, 7 = lowest).
>
> __ You get a lot of pleasure out of reading something good that you have written.
> __ You enjoy the opportunity for self-expression.
> __ You achieve new insights through your writing.
> __ You derive satisfaction from expressing yourself clearly and eloquently.
> __ You feel relaxed when writing.
> __ You like to play with words.
> __ You enjoy becoming involved with ideas, characters, events, and images in your writing.

List of **extrinsic** reasons for writing.

List of **intrinsic** reasons for writing.

> *INSTRUCTIONS:* Please rank the following list of reasons for writing, in order of personal importance to you (1 = highest, 7 = lowest).
>
> __ You realize that, with the introduction of dozens of magazines every year, the market for free-lance writing is constantly expanding.
> __ You want your writing teachers to be favorably impressed with your writing talent.
> __ You have heard of cases where one bestselling novel or collection of poems has made the author financially secure.
> __ You enjoy public recognition of your work.
> __ You know that many of the best jobs available require good writing skills.
> __ You know that writing ability is one of the major criteria for acceptance into graduate school.
> __ Your teachers and parents have encouraged you to go into writing.

Effects of Motivational Orientation on Creative Writers," *Journal of Personality and Social Psychology* 48(2) (1985):393–99.) Is there any evidence that creativity scores tend to be affected by the type of motivation (intrinsic or extrinsic) induced by the questionnaires?

Summary of Statistical Findings

There is strong evidence that a subject would receive a lower creativity score for a poem written after the extrinsic motivation questionnaire than for one written after the intrinsic motivation questionnaire (two-sided p-value = .005, from a two-sample t-test). The estimated difference between these two scores is 4.1 points on the 0–40 point scale. A 95% confidence interval for the decrease in score due to having extrinsic motivation rather than intrinsic motivation is between 1.3 and 7.0 points. [*Note:* Details related to these statistical tools will be introduced in Chapter 2.]

Scope of Inference

Since this was a randomized experiment, one may infer that the difference in creativity scores was *caused* by the difference in motivational questionnaires. Because the subjects were not selected randomly from any population, extending this inference to any other group is speculative. This deficiency, however, is minor; the causal conclusion is strong even if it applies only to the recruited subjects.

1.1.2 Sex Discrimination in Employment— An Observational Study

Did a bank discriminatorily pay higher starting salaries to men than to women? The data in Display 1.3 are the beginning salaries for all 32 male and all 61 female skilled, entry-level clerical employees hired by the bank between 1969 and 1977. (Data from a file made public by the defense and described by H. V. Roberts, "Harris Trust and Savings Bank: An Analysis of Employee Compensation" (1979), Report 7946, Center for Mathematical Studies in Business and Economics, University of Chicago Graduate School of Business.)

Display 1.3   Starting salaries ($U.S.) for 32 male and 61 female clerical hires at a bank

Males			Females					
4,620	5,700	6,000	3,900	4,500	4,800	5,220	5,400	5,640
5,040	6,000	6,000	4,020	4,620	4,800	5,220	5,400	5,700
5,100	6,000	6,000	4,290	4,800	4,980	5,280	5,400	5,700
5,100	6,000	6,300	4,380	4,800	5,100	5,280	5,400	5,700
5,220	6,000	6,600	4,380	4,800	5,100	5,280	5,400	5,700
5,400	6,000	6,600	4,380	4,800	5,100	5,400	5,400	5,700
5,400	6,000	6,600	4,380	4,800	5,100	5,400	5,400	6,000
5,400	6,000	6,840	4,380	4,800	5,100	5,400	5,520	6,000
5,400	6,000	6,900	4,440	4,800	5,100	5,400	5,520	6,120
5,400	6,000	6,900	4,500	4,800	5,160	5,400	5,580	6,300
	6,000	8,100						6,300

Summary of Statistical Findings

A graphical summary of the distributions of male and female salaries is shown in Display 1.4. The mean starting salary for males is estimated to be $560 to $1,080 larger than the mean starting salary for females (95% confidence interval). Such a large difference is unlikely to be due to chance (one-sided p-value $< .00001$ from a two-sample t-test).

Scope of Inference

Although there is convincing evidence that the males, as a group, received larger starting salaries than the females, the statistics alone cannot address whether this difference is attributable to sex discrimination. The evidence is consistent with discrimination, but other possible explanations cannot be ruled out; for example, the males may have had more years of previous experience.

Display 1.4   Frequency histograms for male and female starting salaries

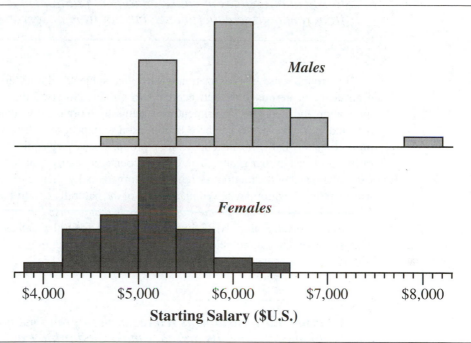

1.2   Statistical Inference and Study Design

The inferences one may draw from any study depend crucially on the study's design. Two distinct forms of inference—causal inference and inference to populations—can be justified by the proper use of random mechanisms.

1.2.1   Causal Inference

In a *randomized experiment*, the investigator controls the assignment of experimental units to groups and uses a chance mechanism (like the flip of a coin) to make the assignment. The motivation and creativity study is a randomized experiment, because a random mechanism was used to assign the study subjects to the two motivation groups. In an *observational study*, the group status of the subjects is established beyond the control of the investigator. The sex discrimination study is observational because the group status (sex of the employee) was, obviously, not decided by the investigator.

Causal Conclusions and Confounding Variables

Can statistical analysis alone be used to establish causal relationships? The answer is simple and concise:

> *Statistical inferences of cause-and-effect relationships can be drawn from randomized experiments, but not from observational studies.*

In a rough sense, randomization ensures that subjects with different and possibly relevant features are mixed up between the two groups. Surely some of the subjects in the motivation and creativity experiment were naturally more creative than others. In view of the randomization, however, there is no reason to expect that they would be placed disproportionately in the intrinsic motivation group, since every subject had the same chance of being placed in that group. There is, of course, a chance that the randomization—the particular result of the random assignment—turned out by chance in such a way that the intrinsic motivation group received many more of the naturally creative writers. *This chance, however, is incorporated into the statistical tools that are used to express uncertainty.*

In an observational study, it is impossible to draw a causal conclusion from the statistical analysis alone. One cannot rule out the possibility that confounding variables are responsible for group differences in the measured outcome.

> *A **confounding variable** is related both to group membership and to the outcome. Its presence makes it hard to establish the outcome as being a direct consequence of group membership.*

In fact, the males in the sex discrimination example generally did have more years of education than the females; and this, not sex, may have been responsible for the disparity in starting salaries. Thus, the effect of sex cannot be separated from—it is confounded with—the effect of education. Tools exist for comparing male and female salaries after accounting for the effect of education (see Chapter 9), and these help clarify matters to some extent. Other confounding variables, however, may not be recognized or measured, and these consequently cannot be accounted for in the analysis. Therefore, it may be possible to conclude that males tend to get larger starting salaries than females, even after accounting for years of education, and yet still not be possible to conclude, from the statistics alone, that this happens because they are males.

Do Observational Studies Have Value?

Is there any role at all for observational data in serious scientific inquiry? The following points indicate that the answer is yes.

1. *Establishing causation is not always the goal.* A study was conducted on 10 American men of Chinese descent and 10 American men of European descent to examine the effect of a blood pressure–reducing drug. The result—that the men of Chinese ancestry tended to exhibit a different response to the drug—does not prove that being of Chinese descent is responsible for the difference. Diets, for example, may differ in the two groups and may be responsible for the different sensitivities. Nevertheless the study does provide important information for doctors prescribing the drug to people from these populations.

2.   *Establishing causation may be done in other ways.* Although statistical methods cannot eliminate the possibility of confounding factors, there may be strong theoretical reasons for doing so. Radiation biologists counted chromosomal aberrations in a sample of Japanese atomic bomb survivors who received radiation from the blast, and compared these to counts on individuals who were far enough from the blast not to have received any radiation. Although the data are observational, the researchers are certain that higher counts in the radiation group can only be due to the radiation, and the data can therefore be used to estimate the dose–response relationship between radiation and chromosomal aberration.

3.   *Analysis of observational data may lend evidence toward causal theories and suggest the direction of future research.* Many observational studies indicated an association between smoking and lung cancer, but causation was accepted only after decades of observational studies, experimental studies on laboratory animals, and a scientific theory for the carcinogenic mechanism of smoking.

Although observational studies are undoubtedly useful, the inappropriate practice of claiming or implying cause-and-effect relationships from them—sometimes in subtle ways—is widespread.

1.2.2   Inference to Populations

A second distinction between study designs relates to how units are selected. In a *random sampling* study, units are selected by the investigator from a well-defined population. All units in the population have a chance of being selected, and the investigator employs a chance mechanism (like a lottery) to determine actual selection. Neither of the case studies in Section 1.1 used random sampling. In contrast, the study units often are *self-selected.* The subjects of the creativity study volunteered their participation. The decision to volunteer can have a strong relationship with the outcome (creativity score), and this precludes the subjects from representing a broader population.

Again, the inferential situation is straightforward:

> *Inferences to **populations** can be drawn from random sampling studies, but not otherwise.*

Random sampling ensures that all subpopulations are represented in the sample in roughly the same mix as in the overall population. Again, random selection has a chance of producing nonrepresentative samples, but *the statistical inference procedures incorporate measures of uncertainty that describe that chance.*

Simple Random Sampling

The most basic form of random sampling is simple random sampling.

> A **simple random sample** of size n *from a population is a subset of the population consisting of n members selected in such a way that every subset of size* n *is afforded the same chance of being selected.*

A typical method for choosing a simple random sample is by lottery. The names of every member of a population are placed in a box, the box is thoroughly mixed, and *n* consecutive names are drawn. This ensures that each member of the population has an equal chance of being selected, independent of which other members are selected.

1.2.3   Statistical Inference and Chance Mechanisms

Statistical analysis is used to make statements from available data in answer to questions of interest about some broader context than the study at hand. No such statement about the broader context can be made with absolute certainty, so every statistical statement includes some measure of uncertainty. In statistical analysis, uncertainty is explained by means of chance models, which relate the available data to the broader context. These ideas are expressed in the following definitions:

> An **inference** *is a conclusion that patterns in the data are present in some broader context.*
> A **statistical inference** *is an inference justified by a probability model linking the data to the broader context.*

The probability models are associated with the chance mechanisms used to select units from a population or to assign units to treatment groups. They enable the investigator to calculate measures of uncertainty to accompany inferential conclusions.

When a chance mechanism has been used in the study, uncertainty measures accompanying the researcher's inferences are backed by a *bona fide* probability structure that exactly describes the study. Often, however, units do not arise from random sampling of real populations; instead the available units are self-selected or are the result of a haphazard selection procedure. For randomized experiments, this may be no problem, since cause-and-effect conclusions can be drawn regarding the effects on the particular units selected (as in the motivation and creativity study). These conclusions can be quite strong, even if the observed pattern cannot be inferred to hold in some general population. For observational studies, the lack of truly random samples is more worrisome, because making an inference about some larger population is usually the goal. It may be possible to *pretend* that the units are as representative as a random sample, but the potential for bias from haphazard or convenience selection remains a serious concern.

Example

Researchers measured the lead content in teeth and the intelligence quotient (IQ) test scores for all 3,229 children attending first and second grades in the period between 1975 and 1978 in Chelsea and Somerville, Massachusetts. The IQ scores for those with low lead concentrations were found to be significantly higher than for those with high lead concentrations. What can be inferred? There is no random sample. In a strict sense, the statistical results apply only to the children actually measured; any extrapolation of the pattern to other children comes from supposing that the relationship between lead and IQ is similar for others. This is not necessarily a bad assumption. The point is that extending the inference to other children is surely open to question.

Statistical Inferences Based on Chance Mechanisms

Four situations are exhibited in Display 1.5, along with the type of inferences that may be drawn in each. In observational studies, obtaining random samples from the populations of

---

Display 1.5   Statistical inferences permitted by study designs

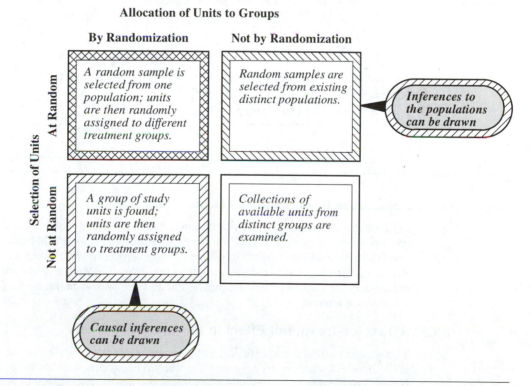

Allocation of Units to Groups

	By Randomization	Not by Randomization
At Random	A random sample is selected from one population; units are then randomly assigned to different treatment groups.	Random samples are selected from existing distinct populations.
Not at Random	A group of study units is found; units are then randomly assigned to treatment groups.	Collections of available units from distinct groups are examined.

Selection of Units

Inferences to the populations can be drawn

Causal inferences can be drawn

interest is often impractical or impossible, and inference based on assumed models may be better than no inference at all. In controlled experiments, however, there is no excuse for avoiding randomization.

1.3 Measuring Uncertainty in Randomized Experiments

The critical element in understanding probability models and the resulting statistical measures of uncertainty is visualizing *replications* of the study, exactly as it was conducted and analyzed. This section describes the probability model for randomized experiments and illustrates how a measure of uncertainty is calculated to express the evidence for a treatment difference in the creativity study.

1.3.1 A Probability Model for Randomized Experiments

A schematic diagram for the creativity study (Display 1.6) is typical of a randomized experiment. The chance mechanism for randomizing units to treatment group ensures that every subset of 24 subjects gets the same chance of becoming the intrinsic group. For example, 23 red and 24 black cards could be shuffled and dealt, one to each subject. Subjects are treated according to their assigned group, and the different treatments change their responses.

Display 1.6   Randomized experiment with two treatment groups

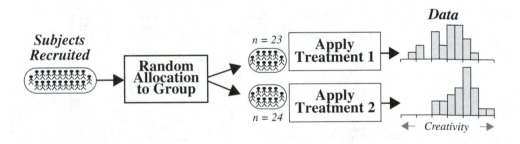

An Additive Treatment Effect Model

Let Y denote the creativity score that a subject would receive after exposure to the extrinsic questionnaire. A general model for the experiment would assert that this same subject would receive a different creativity score, Y^*, after exposure to the intrinsic questionnaire. An *additive treatment effect* model postulates that $Y^* = Y + \delta$. The treatment effect, δ, is a *parameter*—an unknown constant that describes a key feature in the model for answering questions of interest.

1.3.2 A Test for Treatment Effect in the Creativity Study

This section illustrates a statistical inference arising from the additive treatment effect probability model for the creativity study. The question concerns whether the treatment effect is real. The evidence says yes, and a *p-value* expresses the uncertainty of that inference.

Null and Alternative Hypotheses

Questions of interest are translated into questions about parameters in probability models. This usually involves some simplification—such as the assumption that the intrinsic questionnaire adds the same amount to anyone's creativity score—but the expression of a question in terms of a single parameter crystallizes the situation into one where powerful statistical tools can be applied.

In the creativity study, the question "Is there a treatment effect?" becomes a question about whether the parameter δ has a zero or a nonzero value. If $\delta = 0$, the answer to the question is no. If $\delta \neq 0$, the answer is yes. The statement that $\delta = 0$ is called the *null hypothesis*, while a proposition that $\delta \neq 0$ is an *alternative hypothesis*.

A Test Statistic

The form of the alternative hypothesis model is important, because it suggests how to summarize the data to answer the question. If $\delta \neq 0$, the best way to estimate its true value is with the difference between the average creativity score in the intrinsic group and the average creativity score in the extrinsic group. From the summary statistics in Display 1.1, $\overline{Y}_1 = 15.74$ points, $\overline{Y}_2 = 19.88$ points, and $\hat{\delta} = 4.14$ points, which is a *test statistic* for deciding the issue. ($\overline{Y}_1$ and $\overline{Y}_2$ represent the averages, also called sample means, for groups 1 and 2. The hat on δ indicates a parameter estimate.)

Randomization Distribution of the Test Statistic

If the questionnaires have no effect, the subjects would receive exactly the same creativity score regardless of the group to which they were assigned. Therefore, it is possible to determine what test statistic values would have occurred had the randomization process turned out differently. Display 1.7 has an example. The first column lists the creativity scores of 47 persons in the study. The second column lists the groups to which they were assigned. The third column lists another possible way that the group assignments could have turned out. With that grouping, the difference in averages is 2.07.

It is conceptually possible to calculate $\overline{Y}_2 - \overline{Y}_1$ for every possible outcome of the randomization process. A histogram of all these values describes the *randomization distribution* of $\overline{Y}_2 - \overline{Y}_1$ if the null hypothesis is true. The number of possible outcomes of the random assignment is prohibitively large in this example—there are 1.6×10^{13} different groupings, which would take a computer capable of one million assignments per second half a year to complete. Display 1.8, however, shows an estimate of the randomization distribution, obtained as the histogram of the values of $\overline{Y}_2 - \overline{Y}_1$ from a random subset of 1,000 groupings.

Display 1.8 suggests several things. First, it appears just as likely that test statistics will be negative as positive. Second, the majority of values fall in the range from -3.0 to $+3.0$. Third, only four of the 1,000 randomizations produced test statistics as large as 4.14. This last point indicates that 4.14 is a value corresponding to an unusually uneven randomization outcome, if the null hypothesis is correct.

The p-Value of the Test

Because the observed test statistic is improbably large under the null hypothesis, one is led to believe that the null hypothesis is wrong, that the treatment—not the randomization—caused

---

Display 1.7   An example of a different randomization for the creativity study

Creativity Score	Actual Grouping	Another Grouping		Creativity Score	Actual Grouping	Another Grouping
12.0	Intrinsic(2)	1		5.0	Extrinsic(1)	2
12.0	Intrinsic	2		5.4	Extrinsic	2
12.9	Intrinsic	1		6.1	Extrinsic	1
13.6	Intrinsic	2		10.9	Extrinsic	2
16.6	Intrinsic	2		11.8	Extrinsic	1
17.2	Intrinsic	1		12.0	Extrinsic	1
17.5	Intrinsic	2		12.3	Extrinsic	1
18.2	Intrinsic	2		14.8	Extrinsic	2
19.1	Intrinsic	1		15.0	Extrinsic	2
19.3	Intrinsic	2		16.8	Extrinsic	2
19.8	Intrinsic	2		17.2	Extrinsic	2
20.3	Intrinsic	2		17.2	Extrinsic	1
20.5	Intrinsic	1		17.4	Extrinsic	2
20.6	Intrinsic	2		17.5	Extrinsic	2
21.3	Intrinsic	1		18.5	Extrinsic	2
21.6	Intrinsic	2		18.7	Extrinsic	1
22.1	Intrinsic	1		18.7	Extrinsic	1
22.2	Intrinsic	2		19.2	Extrinsic	1
22.6	Intrinsic	1		19.5	Extrinsic	1
23.1	Intrinsic	1		20.7	Extrinsic	1
24.0	Intrinsic	1		21.2	Extrinsic	1
24.3	Intrinsic	1		22.1	Extrinsic	2
26.7	Intrinsic	1		24.0	Extrinsic	2
29.7	Intrinsic	1				

Averages from Actual Grouping

Group	Average	Difference
Intrinsic (2)	19.88	
		4.14
Extrinsic (1)	15.74	

Averages from Another Grouping

Group	Average	Difference
Group 1	18.87	
		2.07
Group 2	16.80	

---

the big difference. That conclusion could be incorrect, however, because the randomization is capable of producing such an extreme. The chance of its doing so is called the (*observed*) *p-value*. In this problem, the p-value is the probability that randomization alone leads to a test statistic as extreme or more extreme than the one observed, if the null hypothesis is true. The smaller the p-value, the more unlikely it is that chance assignment is responsible for the discrepancy between groups, and the greater the evidence that the null hypothesis is incorrect. A general definition appears in Chapter 2.

Since 4 of the 1,000 regroupings produced differences as large or larger than the observed difference, the p-value for this study is estimated from the simulation to be $4/1,000 = .004$. This is a *1-sided* p-value (for the alternative hypothesis that $\delta > 0$)

Display 1.8   Histogram of differences between group averages, from 1,000 regroupings of the creativity study data

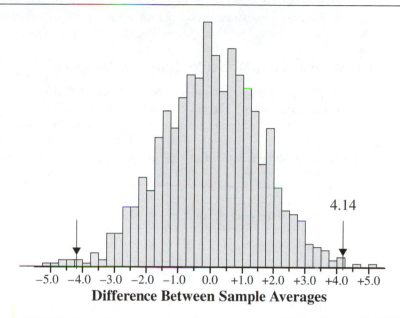

4.14

Difference Between Sample Averages

because it counted as extremes only those outcomes with test statistics as large as or larger than the observed one. Statistics that are smaller than -4.14 may provide equally strong evidence against the null hypothesis, favoring the alternative hypothesis that $\delta < 0$. If those (7 of them) are included, the result is a *2-sided p*-value of $11/1,000 = .011$, which is appropriate for the two-sided alternative hypothesis that $\delta \neq 0$.

Computing p-Values from Randomized Experiments

There are several methods for obtaining *p*-values in a randomized experiment. An enumeration of all possible regroupings of the data would represent all the ways that the data could turn out in all possible randomizations, absent any treatment effect. This would determine the answer exactly, but it is often overburdensome—as the creativity study illustrates. A second method, used earlier, is to estimate the *p*-value by simulating a large number of randomizations and finding the proportion of these that produce a test statistic at least as extreme as the observed one.

The most common method is to approximate the randomization distribution with a mathematical curve, based on certain assumptions about the distribution of the measurements and the form of the test statistic. If, for example, one assumes that creativity scores are normally distributed and that the additive model is correct, then statistical theory indicates that a normal curve provides a good approximation to the full histogram of the difference between sample averages. This approach is given formally in Chapter 2 along with a related discussion of confidence intervals for treatment effects.

1.4 Measuring Uncertainty in Observational Studies

This section describes the probability model for random sampling. Then it illustrates how an invented model provides a plausible interpretation for statistical analysis of the sex discrimination study, where no such random sampling is involved.

1.4.1 A Probability Model for Random Sampling

Display 1.9 depicts random sampling from two populations. A chance mechanism, such as a lottery, selects a subset of n_1 units from population 1 in such a way that all subsets of size n_1 have the same chance of selection. The mechanism is used again to select a random sample of size n_2 from population 2. The samples are drawn independently; that is, the particular sample drawn from one population does not influence the process of selection from the other.

Display 1.9   Random sampling study with two populations

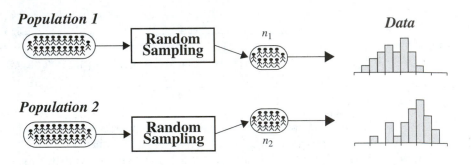

Questions of interest in sampling studies center on how features of the populations differ. In many cases, the difference between the populations can be described by the difference between their means. If the means of populations 1 and 2 are μ_1 and μ_2, respectively, then $\mu_2 - \mu_1$ becomes the single parameter for answering the questions of interest. Inferences and uncertainty measures are based on the difference, $\overline{Y}_2 - \overline{Y}_1$, between sample averages, which estimates the difference in population means.

As in Display 1.8, a *sampling distribution* for the statistic $\overline{Y}_2 - \overline{Y}_1$ is represented by a histogram of all values for the statistic from all possible samples that can be drawn from the two populations. The p-value for testing a hypothesis and the confidence intervals for estimating the parameter follow from an understanding of the sampling distribution. Full exposition appears in Chapter 2.

1.4.2 Testing for a Difference in the Sex Discrimination Study

In the sex discrimination study, there is no interest in the starting salaries of some larger population of individuals who were never hired, so a random sampling model is not relevant.

Similarly, it makes no sense to view the sex of these individuals as randomly assigned. Neither the random sampling nor the randomized experiment model applies. Any interpretation that supports a statistical analysis must be based on a fictitious chance model.

One possibility for examining the difference between the average starting salary given to males and the average starting salary given to females is a model in which the employer assigns the set of starting salaries shown in Display 1.3 to the employees *at random*. That is, the employer shuffles a set of cards, each having one of the starting salaries written on it, and deals them out. With this model, the investigator may ask whether the observed difference is a likely outcome.

Permutation Distribution of a Test Statistic

The collection of differences in averages from all possible assignments of starting salaries to individuals makes up the *permutation distribution* of the test statistic for this model. There are 8.7×10^{24} outcomes, so the million-a-second computer would now struggle along for about 275 million years. Once again, shortcuts are available for approximating the permutation distribution (see Section 4.3.1). It can be determined that an $820 difference is extremely improbable (p-value $< .00001$).

Scope of the Inferences

Inferences from fictitious models must be stated carefully. What the statistical argument shows is that the employer did not assign starting salaries at random—which was known at the outset. The strength of the argument comes from the demonstration that the actual assignment differs substantially from the expected outcome in *one model* where salary assignment is sex-blind.

1.5 Related Issues

1.5.1 Graphical Methods

This text relies heavily on the use of graphical methods for making decisions regarding the choice of statistical models. This section briefly reviews some important tools for displaying sets of numbers.

Relative Frequency Histograms

Histograms, like those in Display 1.4, are standard tools for displaying the general characteristics of a data set. Looking at the display, one should visually conclude that the central male starting salary is around $6,000, that the central female starting salary is around $5,200, that the two distributions have about the same spread—most salaries are within about $1,000 of the centers—that both shapes appear reasonably symmetric, but that there is one male starting salary that appears very high in relation to the rest of the male salaries.

A histogram is a graph where the horizontal axis displays ranges for the measurement and the vertical axis displays the relative frequency per unit of measurement. Relative frequency is therefore depicted by area.

Other features are of less concern. Choices of the number of ranges (*bins*) and their boundaries can have some influence on the final picture. However, a histogram is ordinarily used to show broad features, not exquisite detail, and the broad features will be apparent with many choices.

Stem-and-Leaf Diagrams

A *stem-and-leaf diagram* is a cross between a graph and a table. It is used to get a quick idea of the distribution of a set of measurements with pencil and paper or to present a set of numbers in a report.

Display 1.10 shows stem-and-leaf diagrams for the two sets of creativity scores. The digits in each observation are separated into a stem and a leaf. It is represented in the diagram by its leaf on the same line as its stem. All possible stem values are listed in increasing order from top to bottom, whether or not there are observations with those stems. At each stem, all corresponding leaves are listed, in increasing order. *Outliers* may require a break in the string of stems.

The stem-and-leaf diagrams show the centers, spreads, and shapes of distributions in the same way histograms do. Their advantages include exact depiction of each number, ease of

Display 1.10   Back-to-back stem-and-leaf diagrams for the creativity study data

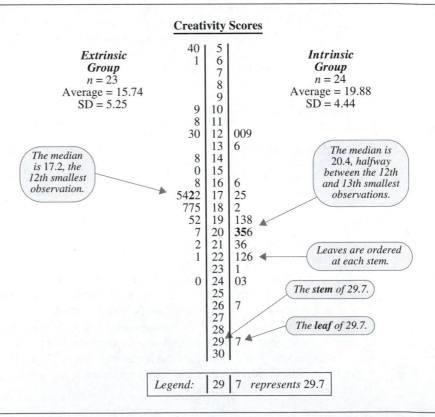

determining the median and quartiles of each set, and ease of construction. Disadvantages include difficulty in comparing distributions when the numbers of observations in data sets are very different and severe clutter with large data sets.

Finding the Median from a Stem-and-Leaf Diagram

Because stem-and-leaf diagrams show the order of numbers in a set, they offer a convenient presentation for locating the *median* of the data. At least half the values in the data are greater than or equal to the median; at least half are less than or equal to it. A median is found by the "$(n + 1)/2$" rule: calculate $k = (n + 1)/2$; if k is an integer, the median is the kth smallest observation; if k is halfway between two integers, the median is halfway between the corresponding observations (see Display 1.10).

Box Plots

A *box plot*, or box-and-whisker plot, is a graphical display that represents the middle 50% of a group of measurements by a box and highlights various features of the upper and lower 25% by other symbols. The box represents the body of the distribution of numbers, and the whiskers represent its tails. The graph gives an uncluttered view of the center, the spread, and the skewness of the distribution and indicates the presence of unusually small (or large) outlying values. Box plots are particularly useful for comparing several samples side by side. Unlike stem-and-leaf diagrams, box plots are not typically drawn by hand, but by a statistical computing program.

Most box plots use the *inter-quartile range* (IQR) to distinguish distant values from those close to the body of the distribution. The IQR is the difference between the upper and lower quartiles—the range of the middle 50% of the data.

Display 1.11 shows a box plot of per-capita incomes in 105 countries. (Data from S. Leinhardt and S. Wasserman, "Teaching Regression: An Exploratory Approach," *American Statistician* 33 (1979): 196–203.) It is easy to see that the lower quartile of incomes is about $200 per year and the upper quartile is about $1,200. The median income is marked by a line through the box, at about $400 per year. Whiskers extend from the box out through all values that are within 1.5 IQRs of the box.

Observations that are more than 1.5 IQRs away from the box are far enough from the main body of data that they are indicated separately by dots. Observations more than 3 IQRs from the box are quite distant from the main body of data and are prominently marked, as in Display 1.11, and often named.

The difference in the distributions of starting salaries for men and women is shown by placing box plots *side by side*, as in Display 1.12. One can see at a glance that the middle starting salaries for men were about $750 higher than for women, that the ranges of starting salaries were about the same for men and for women, and that the single starting salary of $8,100 is unusually large.

Notes

1. Box-plotting routines are widely available in statistical computing packages. The definitions of quartiles, extreme points, and very extreme points, however, may differ slightly among packages.

Display 1.11   Box plot of per-capita annual incomes for 105 countries, in 1974 U.S. dollars

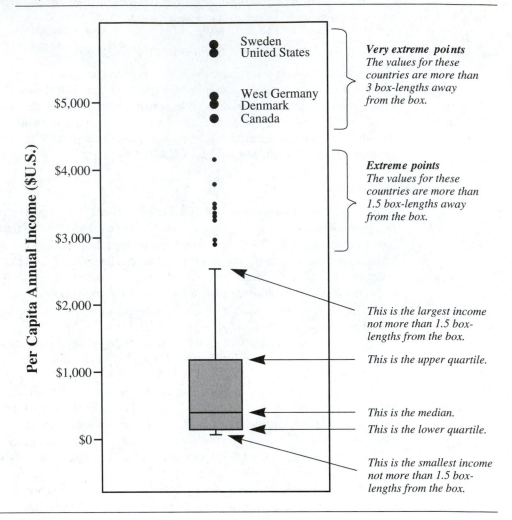

2. The choices of 1.5 and 3 IQRs are arbitrary but useful cutoffs for highlighting observations distant and quite distant from the main body.
3. The width of the box is chosen simply with a view toward making the box look nice; it does not represent any aspect of the data.
4. For presentation in a journal, it is common to define the box plot so that the whiskers extend to the 10th and 90th percentiles of the sample. This is easier to explain to readers who might be unfamiliar with box plots.
5. The data sets in Display 1.13 show a variety of different kinds of distributions. Box plots are matched with histograms of the same data sets.
6. Some statistical computer packages draw horizontal box plots.

Display 1.12 Side-by-side box plots for the starting salary data

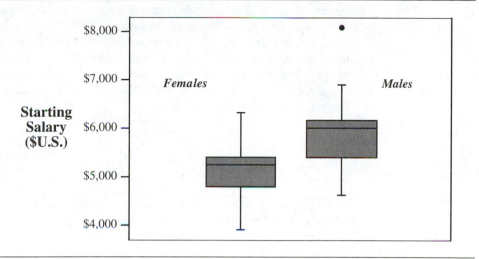

Display 1.13 Histograms and box plots for 100 observations from four distributions

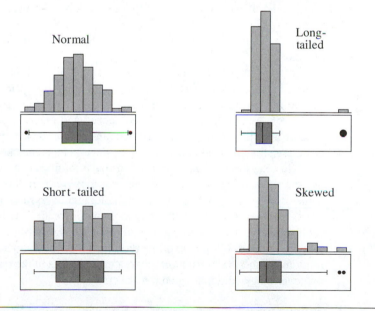

1.5.2 Standard Statistical Terminology

A *parameter* is an unknown numerical value describing a feature of a probability model. Parameters are indicated by Greek letters. A *statistic* is any quantity that can be calculated from the observed data. Statistics are represented by Roman letter symbols. An *estimate* is a statistic used as a guess for the value of a parameter. The notation for the estimate of a parameter θ is the parameter symbol with a hat on it, $\hat{\theta}$. Remember that the estimate can be calculated, but the parameter remains unknown.

The *Statistical Sleuth* uses the word *mean* when referring to an average calculated over an entire population. A mean is therefore a parameter. When referring to the average in a sample—which is both a statistic and an estimate of the population mean—the *Statistical Sleuth* uses the word *average*. Standard notation uses μ for the mean and $\overline{Y}$ for an average.

The *standard deviation* of a set of numbers $Y_1, \ldots, Y_n$ is defined as

$$\sqrt{\left(\sum_{i=1}^{n}(Y_i - \overline{Y})^2\right) \bigg/ (n-1)}.$$

The symbol for a population standard deviation is σ; for a sample standard deviation, the symbol is s. The standard deviation is a measure of spread, interpreted as the typical distance between a single number and the set's average.

1.5.3   Randomization of Experimental Units to Treatments

Experimental units are the things to which treatments are applied. They may be people, rats, samples of material, or pieces of land. Identifying the exact meaning of the experimental unit in a particular study is sometimes difficult but always important. The answer always follows the question, what are the treatments applied to? In the creativity example, the experimental units were the 47 people. The treatments applied to them were the motivation questionnaires.

In a two-treatment randomized experiment all available experimental units have the same chance of being placed in group 1. If there are 10 experimental units then a coin can be flipped for each, and the first 5 to get heads can be placed in group 1. Drawing cards from a thoroughly shuffled deck of red and black cards is an alternative. Tables of random numbers or computer generated random numbers may also be used. These have the advantage that the randomization can be documented and checked against certain patterns.

Most statistical computer programs have routines that generate random numbers. To use such a list for randomly assigning units to two experimental groups, first list the units in some order. Then generate a series of random numbers, assigning them down the list to the units. Reorder the list so that the random numbers are listed by increasing order, keeping each subject matched with its random number. Then assign the subjects with the smallest random numbers (at the top of the reordered list) to treatment 1 and those with the largest random numbers (at the bottom) to treatment 2. Ties in the random numbers can be broken by generating additional random numbers.

1.5.4   Selecting a Simple Random Sample from a Population

Every subject in the population should have the same chance of being chosen for the sample, if the sample is to be called a *simple random sample*. To accomplish this in practice requires a *frame*: a numbered list of all the subjects. If a population of 20,000 members is so listed, an integer from 1 to 20,000 is associated with each. If a random sample of 100 is desired, it is only necessary to generate 100 distinct random integers such that each of the integers 1 through 20,000 has an equal chance of being selected. This can be accomplished with a statistical computer program.

Other Random Sampling Procedures

A commonly used method for selecting a sample of 100 from the list of 20,000 is to pick a single random start from the positions 1 to 200 and then to select every 200th subject going down the frame. This is called *systematic random sampling*. Another method is to group the frame into 200 blocks of 100 consecutive subjects each and to select one of the blocks at random. This is called *random cluster sampling*. In sampling units such as lakes of different sizes, it is sometimes useful to allow larger units to have higher probabilities of being sampled than smaller units. This constitutes *variable probability sampling*. These and other random sampling schemes can be useful, but they differ from simple random sampling in fundamental ways.

1.5.5   On Being Representative

Random samples and randomized experiments are representative in the same sense that flipping a coin to see who takes out the garbage is fair. Flipping is always fair before the coin hits the table, but the outcome is always unfair to the loser. In the same way, close examination of the results of randomization or random sampling can usually expose ways in which the chosen sample is not representative. The key, however, is not to abandon the procedure when its result is suspect. Uncertainty about representativeness will be incorporated into the statistical analysis itself. If randomization were abandoned, there would be no way to express uncertainty accurately.

1.6   Summary

Cause-and-effect relationships can be inferred from randomized experiments but not from observational studies. The problem with observational studies is that confounding variables—identifiable or not—may be responsible for observed differences. Randomized experiments eliminate this problem by ensuring that differences between groups (other than those of the assigned treatments) are due to chance alone. Statistical measures of uncertainty account for this chance.

Statistically, the statements that generalize sample results to more general contexts are based on a probability model. When the model corresponds to the planned use of randomization or random sampling, it provides a firm basis for drawing inferences. The probability model may also be a fiction, created for the purposes of assessing uncertainty. Results from fictitious probability models must be viewed with skepticism.

Further Reading

Many examples of observational studies and randomized experiments are given in an interesting and easy-to-read book by Tanur (1972). More on causal conclusions and randomized experiments is available in Freedman, Pisani, Purves, and Adhikari (1991). A source of many modern graphical tools is the book by Cleveland (1993).

1.7 Exercises

Conceptual Exercises

1. Creativity Study. In the motivation and creativity experiment, the poems were given to the judges in random order. Why was that important?

2. Sex Discrimination Study. Explain why it is difficult to prove sex discrimination (that males in a company receive higher starting salaries because they are males) even if it has occurred.

3. A study found that individuals who lived in houses with more than two bathrooms tended to have higher blood pressure than individuals who lived in houses with two or fewer bathrooms. (a) Can cause-and-effect be inferred from this? (b) What confounding variables may be responsible for the difference?

4. A researcher performed a comparative experiment on laboratory rats. Rats were assigned to group 1 haphazardly by pulling them out of the cage without thinking about which one to select. Should others question the claim that this was as good as a randomized experiment?

5. In 1930 an experiment was conducted on 20,000 school children in England. Teachers were responsible for randomly assigning their students to a treatment group—to receive $\frac{3}{4}$ pint of milk each day—or to a control group—to receive no milk supplement. Weights and heights were measured before and after the four month experiment. The study found that children receiving milk gained more weight during the study period. On further investigation, it was also found that the controls were heavier and taller than the treatment group *before* the milk treatment began (more so than could be attributed to chance). What is the likely explanation and the implication concerning the validity of the experiment?

6. Ten marijuana users, aged 14 to 16, were drawn from patients enrolled in a drug abuse program and compared to nine drug-free volunteers of the same age group. Neuropsychological tests for short-term memory were given, and the marijuana group average was found to be significantly lower than the control group average. The marijuana group was held drug-free for the next six weeks, at which time a similar test was given with essentially the same result. The researchers concluded that marijuana use caused adolescents to have short-term memory deficits that continue for at least six weeks after the last use of marijuana. (a) Can a genuine causal relationship be established from this study? (b) Can the results be generalized to other 14- to 16-year-olds? (c) What are some potential confounding factors?

7. Suppose that random samples of Caucasian-American and Chinese-American individuals are obtained from patient records of doctors participating in a study to compare blood pressures of the two populations. Suppose that the individuals selected are asked whether they want to participate and that some decline. The study is conducted only on those that volunteer to participate, and a comparison of the distributions of blood pressures is conducted. Where does this study fit in Display 1.5? What assumption would be necessary to allow inferences to be made to the sampled populations?

8. More people get colds during cold weather than during warm weather. Does that prove that cold temperatures cause people to get colds? What is a potential confounding factor?

9. A study showed that children who watch more than two hours of television each day tend to have higher cholesterol levels than children who watch less than two hours of television each day. Can you think of any use for the result of this study?

10. What is the difference between a randomized experiment and a random sample?

11. A number of volunteers were randomly assigned to one of two groups, one of which received daily doses of vitamin C and one of which received daily placebos (without any active ingredient). It was found that the rate of colds was lower in the vitamin C group than in the placebo group. It became evident, however, that many of the subjects in the vitamin C group correctly guessed that they were receiving vitamin C rather than placebo, because of the taste. Can it still be said that the difference in treatments is what caused the difference in cold rates?

12. Fish Oil and Blood Pressure. Researchers used 7 red and 7 black playing cards to randomly assign 14 volunteer males with high blood pressure to one of two diets for four weeks: a fish oil diet and a standard oil diet. The reductions in diastolic blood pressure are shown in Display 1.14. (Based on a study by H. R. Knapp and G. A. FitzGerald, "The Antihypertensive Effects of Fish Oil," *New England Journal of Medicine* 320 (1989): 1037–43.) Why might the results of this study be important, even though the volunteers do not constitute a random sample from any population?

Display 1.14 Reductions in diastolic blood pressure (mm of mercury) for 14 men after 4 weeks of a special diet containing fish oil or a regular, nonfish oil

| Fish oil diet: | 8 | 12 | 10 | 14 | 2 | 0 | 0 |
| Regular oil diet: | −6 | 0 | 1 | 2 | −3 | −4 | 2 |

---

13. Why does a stem-and-leaf diagram require less space than an ordinary table?

14. What governs the *width* of a box plot?

15. What general features are evident in a box plot of data from a normal distribution? from a skewed distribution? from a short-tailed distribution? from a long-tailed distribution?

Computational Exercises

16. Planet Distances and Order from Sun (Getting Started with a Statistical Computer Package). The data in Display 1.15 are the distances from the sun (scaled so that earth = 10) and the order from the sun for the 9 planets in our solar system plus the asteroid belt (treated here as the fifth body from the sun). (a) Enter the data, including the names, into a statistical computer package. (b) Draw a scatterplot of distance (vertical axis) versus order from sun (horizontal axis). (c) Repeat (b) but plot

---

Display 1.15 Order from sun and distance from sun of the 9 planets and the asteroid belt (distance scaled so that earth's distance is 10)

Body	Order from Sun	Distance from Sun
Mercury	1	3.87
Venus	2	7.23
earth	3	10.00
Mars	4	15.24
(Asteroids)	5	29.00
Jupiter	6	52.03
Saturn	7	95.46
Uranus	8	192.00
Neptune	9	300.90
Pluto	10	395.00

the *natural logarithms* of distance from sun versus order. (d) Obtain summary statistics (average and sample standard deviation) for the distances. (e) Obtain summary statistics for the logarithms of the distances.

17.   Seven students volunteered for a comparison of study guides for an advanced course in mathematics. They were randomly assigned, 4 to study guide A and 3 to study guide B. All were instructed to study independently. Following a two-day study period, all students were given an examination about the material covered by the guides, with the following results:

Study Guide A scores: 68, 77, 82, 85

Study Guide B scores: 53, 64, 71

Perform a randomization test by listing all possible ways that these students could have been randomized to two groups. There are 35 ways. For each outcome, calculate the difference between sample averages. Finally, calculate the 2-sided *p*-value for the observed outcome.

18.   Using the creativity study data (Section 1.1.1) and a computer, assign a set of random numbers to the 47 subjects. Order the entire data set by increasing values of the random numbers, and divide the subjects into Group 1 with the 24 lowest random numbers and Group 2 with the 23 highest. Compute the difference in averages. Repeat this process five times, using different sets of random numbers. Did you get any differences larger than the one actually observed (4.14)?

19.   Write down the names and ages of 10 people. Using coin flips, divide them into 2 groups, as if for a randomized experiment. Did one group tend to get many of the older subjects? Was there any way to predict which group would have a higher average age in advance of the coin flips?

20.   Repeat Exercise 19 using a randomization mechanism that ensures that each group will end up with exactly five people.

21.   Read the methods and design sections of five published studies in your own field of specialization. (a) Categorize each according to Display 1.5. (b) Now read the conclusions sections. Determine whether inferential statements are limited to or go beyond the scope allowed in Display 1.5.

22.   Draw back-to-back stem-and-leaf diagrams of the creativity scores for the two motivation groups in Display 1.1 (by hand).

23.   Use the computer to draw side-by-side box plots for the creativity scores in Display 1.1.

24.   Using the stem-and-leaf diagrams from Exercise 22, compute the median, lower quartile, and upper quartile for each of the motivation groups. Identify these on the box plots from Exercise 23. Using a ruler, measure the length of the box (the IQR) for each group, and make horizontal lines at 1.5 IQRs above and below each box and at 3 IQRs above and below the box. Are there any *extreme points* in either group? Are there any *very extreme points* in either group?

25.   Display 1.16 exhibits box plots of the numbers of chromosome aberrations in 100 cells for 263 survivors of the Hiroshima atomic bomb who did not receive any radiation and 70 survivors who were thought to have received 1 to 100 rads of radiation from the blast. (Data from R. L. Prentice, "Binary Regression Using Extended beta-Binomial Distributions with Discussion of Correlation Induced by Covariate Measurement Error," *Journal of the American Statistical Association* 81(394) (1986): 321–27.) It is desired to know whether low levels of radiation, 1 to 100 rads, can cause chromosome aberrations. (a) What, approximately, are the median numbers of aberrations in the two groups? (b) Describe the shapes and relative spread of measurements in the two groups. (c) Although the box plots are drawn correctly, there is no middle line drawn in the box plot on the left. What is the likely reason for this? (*Note:* There are many subjects with 0 aberrations). (d) To what extent can these data be used to show that radiation causes chromosome aberrations?

Display 1.16   Chromosome aberrations per 100 cells for 333 Hiroshima A-bomb survivors

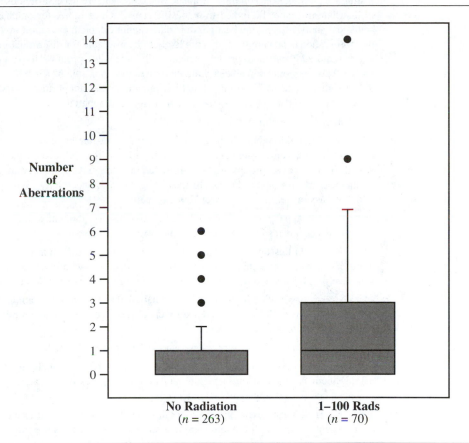

26.   The following are zinc concentrations (in mg/ml) in the blood for two groups of rats. Group A received a dietary supplement of calcium, and group B did not. Researchers are interested in variations in zinc level as a side effect of dietary supplementation of calcium.

> Group A: 1.31 1.45 1.12 1.16 1.30 1.50 1.20 1.22 1.42 1.14 1.23 1.59 1.11 1.10
> 1.53 1.52 1.17 1.49 1.62 1.29

> Group B: 1.13 1.71 1.39 1.15 1.33 1.00 1.03 1.68 1.76 1.55 1.34 1.47 1.74 1.74
> 1.19 1.15 1.20 1.59 1.47

(a) Make a back-to-back stem-and-leaf diagram (by hand) for these measurements. (b) Use the computer to draw side-by-side box plots.

Answers to Conceptual Exercises

1.   If one of the two treatment groups had their poems judged first then the effect of motivation treatment would be confounded with time effects in the judges' marking of creativity scores.

2. Statistically, the distributions of male and female starting salaries may be compared after adjusting for possible confounding variables. Since the data are necessarily observational, a difference in distribution cannot be linked to a specific cause. Once more, for emphasis: the best possible statistical analysis using the best possible data cannot establish causation in an observational study, but observational data are the only data likely to be available for discrimination cases. If, therefore, courts required plaintiffs to produce scientifically defensible proof of discrimination in order to prevail, defendants would win all discrimination cases *by definition*. As a result, some courts that wish to give weight to statistical information in discrimination cases adopt rules of evidence that allow proof to be established negatively—by the lack of an adequate rebuttal.

3. (a) No. (b) Wealth and richness of diet.

4. Yes. First of all, there is the possibility that the rats that were easier to get out of the cage are different from the others—bigger, less mobile, perhaps. Second, even if there is no obvious reason why the rats in the two groups might be different, that does not ensure they are not. Researchers must maintain skepticism about the conclusions in light of the uncertainty. With proper randomization, which is easy to carry out, there would be no doubt.

5. Teachers apparently gave the milk to the students they thought would most benefit from it. As a consequence, the results of the experiment were not valid.

6. (a) No. In this observational study, it is possible that the drug users were different than the nonusers in ways other than drug use. (b) No, because the samples are volunteers, not random samples. (c) Happiness of the child, stability of family, success of child in school.

7. It is an observational study. The population that is randomly sampled is the population of consenting subjects. To draw inference to all subjects requires the assumption that the response is unrelated to the reasons for consent.

8. No. The amount of time people spend indoors.

9. It may be possible for doctors to identify children who are likely to have high cholesterol by asking about their television watching habits. This requires no causative link. It is a minor use, however, and there is apparently no other.

10. In a randomized experiment a random mechanism is used to allocate the available subjects to treatment groups. In a random sample a random mechanism is used to select subjects from the populations of interest.

11. Yes, sort of. The treatment difference caused the different responses, but the actual "treatment" received in the vitamin C group was both a daily dose of vitamin C and knowledge that it was vitamin C. It's possible that the second aspect of this treatment is what was responsible for the difference. Researchers must make sure that the two groups are treated as similarly as possible in all respects, except for the specific agent under comparison.

12. The conclusion that the fish oil diet causes a reduction in blood pressure for these volunteers is a strong and useful one, even if it formally applies only to these particular individuals.

13. The leading digit or digits (the stems) are listed only once.

14. The width of the box is chosen to make the overall picture pleasing to the eye. It does not represent anything. For side-by-side box plots, the widths of the two boxes should be equal.

15. Normal: Median line in the middle of the box; equal whiskers; few if any extreme points; no very extreme points. Skewed: extreme points in one direction only; very extreme points possible, but few; long whisker on side of extreme points and short whisker (if any) on other side; median line closer to short whisker than to long one. Short-tailed: Like normal with no extreme points and very short whiskers. Long-tailed: Like normal (roughly symmetric) with extreme points strung out in both tails; some very extreme points possible.

Inference Using *t*-Distributions

The first of the case studies in this chapter uses the structure of two independent samples and the second uses the structure of a single sample of pairs. This chapter has the dual purposes of detailing the inferential tools for these important structures and emphasizing the conceptual basis of confidence intervals and *p*-values based on the *t*-distribution more generally. These "*t*-tools" are useful in regression and analysis of variance structures that will be taken up in subsequent chapters, but their main features can be conveyed in the relatively simple setting of a two-group comparison. The tools are derived under random sampling models when populations are normally distributed. As will be seen, the resulting tools also find application as approximations to randomization tests. Furthermore, as described in Chapter 3, they often work quite well even when the populations are not normal.

2.1 Case Studies

2.1.1 Bumpus's Data on Natural Selection—
An Observational Study

In an 1898 biology lecture at Woods Hole, Massachusetts, Hermon Bumpus reminded his audience that the process of natural selection for evolutionary change was an unproved theory: "Even if the theory of natural selection were as firmly established as Newton's theory of attraction of gravity, scientific method would still require frequent examination of its claims." As evidence in support of natural selection, he presented measurements on house sparrows brought to the Anatomical Laboratory of Brown University after an uncommonly severe winter storm. Some of these birds had survived and some had perished. Bumpus asked whether those that perished did so because they lacked physical characteristics enabling them to withstand the intensity of that particular instance of selective elimination.

Display 2.1 exhibits the humerus (arm bone) lengths for the 24 adult male sparrows that perished and for the 35 adult males that survived. Do humerus lengths tend to be different for survivors than for those that perished? If so, how large is the difference?

Display 2.1   Humerus lengths (inches) of adult male house sparrows, 24 that perished and 35 that survived in a winter storm

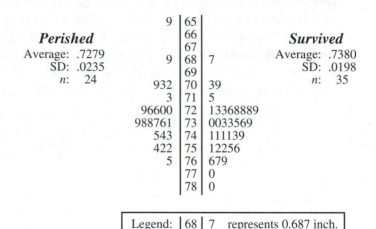

Perished		Survived
Average: .7279		Average: .7380
SD: .0235		SD: .0198
n: 24		*n*: 35

```
        9 | 65 |
          | 66 |
          | 67 |
        9 | 68 | 7
          | 69 |
      932 | 70 | 39
        3 | 71 | 5
    96600 | 72 | 13368889
   988761 | 73 | 0033569
      543 | 74 | 111139
      422 | 75 | 12256
        5 | 76 | 679
          | 77 | 0
          | 78 | 0
```

Legend: |68|7   represents 0.687 inch.

Summary of Statistical Findings

There is suggestive but inconclusive evidence of a difference between the mean humerus length of survivors and the mean humerus length of those that perished (two-sided *p*-value = .08 from a two-sample *t*-test). The mean length is estimated to be between −0.001 inch and +0.021 inch greater in those that survived than in those that perished (95% confidence interval).

Scope of Inference

First, as this was an observational study, one cannot infer a causal relationship—that the longer humerus lengths among survivors enabled them to survive. Second, the living sparrows brought to Bumpus were found in a specific area and were so stressed that they were easily collected. Inference to populations of similarly stressed sparrows is risky. Such populations are hypothetical, and there is no chance model.

2.1.2 Anatomical Abnormalities Associated with Schizophrenia—An Observational Study

Are any physiological indicators associated with schizophrenia? Early studies, based largely on postmortem analysis, suggest that the sizes of certain areas of the brain may be different in persons afflicted with schizophrenia than in others. Confounding variables in these studies, however, clouded the issue considerably. In a 1990 article, researchers reported the results of a study that controlled for genetic and socioeconomic differences by examining 15 pairs of monozygotic twins, where one of the twins was schizophrenic and the other was not. The twins were located through an intensive search throughout Canada and the United States. (Data from R. L. Suddath et al., "Anatomical Abnormalities in the Brains of Monozygotic Twins Discordant for Schizophrenia," *New England Journal of Medicine* 322(12) (1990): 789–93.)

The researchers used magnetic resonance imaging to measure the volumes (in cm^3) of several regions and subregions inside the twins' brains. Display 2.2 presents data based on the reported summary statistics from one subregion, the left hippocampus. What is the magnitude of the difference in volumes of the left hippocampus between the unaffected and the affected individuals? Can the observed difference be attributed to chance?

Display 2.2 Differences in volumes (cm^3) of left hippocampus in 15 sets of monozygotic twins where one twin is affected by schizophrenia

Pair #	Unaffected	Affected	Difference
1	1.94	1.27	0.67
2	1.44	1.63	−0.19
3	1.56	1.47	0.09
4	1.58	1.39	0.19
5	2.06	1.93	0.13
6	1.66	1.26	0.40
7	1.75	1.71	0.04
8	1.77	1.67	0.10
9	1.78	1.28	0.50
10	1.92	1.85	0.07
11	1.25	1.02	0.23
12	1.93	1.34	0.59
13	2.04	2.02	0.02
14	1.62	1.59	0.03
15	2.08	1.97	0.11

Differences

Average: 0.199
Sample SD: 0.238
n: 15

```
-2 |
-1 | 9
-0 |
 0 | 23479
 1 | 0139
 2 | 3
 3 |
 4 | 0
 5 | 09
 6 | 7
 7 |
```

Legend: | 6 | 7 represents 0.67 cm^3

Summary of Statistical Analysis

There is substantial evidence that the left hippocampus volumes are different in individuals with schizophrenia than in individuals without schizophrenia (two-sided *p*-value = .0061, from a paired *t*-test). It is estimated that the mean volume is 0.20 cm^3 smaller for those with schizophrenia (about 11% smaller). A 95% confidence interval for the difference is from 0.07 to 0.33 cm^3.

Scope of Inference

These twins were not randomly selected from general populations of schizophrenic and non-schizophrenic individuals. Tempting as it is to draw inferences to these wider populations, such inferences must be based on an assumption that these individuals are as representative as random samples are. Furthermore, the study is observational, so no causal connection between left hippocampus volume and schizophrenia can be established from the statistics alone. In fact, the researchers had no theories about whether the abnormalities preceded the disease or resulted from it.

2.2   One-Sample *t*-Tools and the Paired *t*-Test

The schizophrenia study used a *paired t*-test, in which measurements taken on paired subjects are reduced to a single set of differences for analysis. This section develops the single population methods for drawing inferences about the population mean from a random sample and at the same time introduces key concepts such as sampling distributions, standard errors, *Z*-ratios, and *t*-ratios.

2.2.1   The Sampling Distribution of a Sample Average

A random sample is drawn from a population with the objective of learning about the population's mean. Suppose the average of that sample is written on a piece of paper, which is placed in a box. Then suppose this process is repeated for every one of the equally likely samples that could be drawn. Then the distribution of all the numbers in the box is the *sampling distribution of the average*.

Display 2.3 illustrates a sampling distribution in the conceptual framework of the schizophrenia study. There is an assumed population of twins in which one of the twins has schizophrenia and the other does not. For each set of twins, Y = the difference between the left hippocampus volumes of the unaffected and the affected twin. The 15 observed differences are assumed to be a random sample from this population. To examine whether there is a structural difference between volumes, one calculates the average of the 15 measurements, $\overline{Y} = 0.20$ cm^3 as an estimate of the population mean μ.

Although only the one sample is actually taken, it is important to think about replicating the study—repeatedly collecting 15 sets of twins and repeatedly calculating the average difference. The value of the average varies from sample to sample, and a histogram of its values represents its sampling distribution.

Because in this case there is only one average with which to estimate a population mean, it would seem difficult to learn anything about the characteristics of the sampling distribution. Some illuminating facts about the sampling distribution of an average, however, come from statistical theory. If a population has mean μ and standard deviation σ,

Display 2.3   The sampling distribution of the sample average

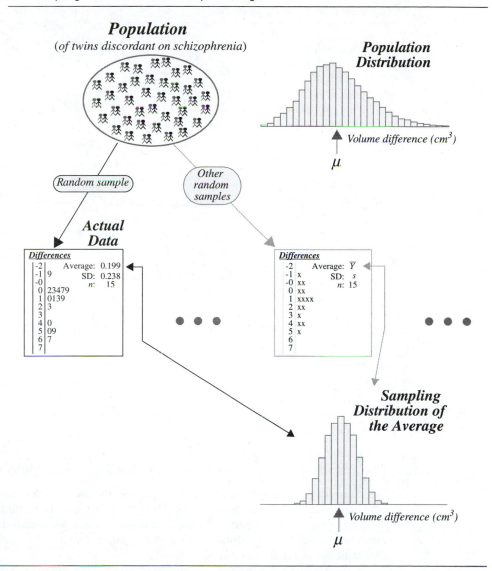

then—as illustrated in Display 2.4—the mean of the sampling distribution of the average is also μ, the standard deviation of the sampling distribution is $\sigma/\sqrt{n}$, and the shape of the sampling distribution is more nearly normal than is the shape of the population distribution. The last fact comes from the important *Central Limit Theorem*.

The standard deviation in the sampling distribution of an average, denoted by $SD(\overline{Y})$, is the typical size of $(\overline{Y} - \mu)$, the error in using $\overline{Y}$ as an estimate of μ. This error generally gets smaller as the sample size increases.

Display 2.4 Relationship between the population distribution and the sampling distribution of the average in random sampling

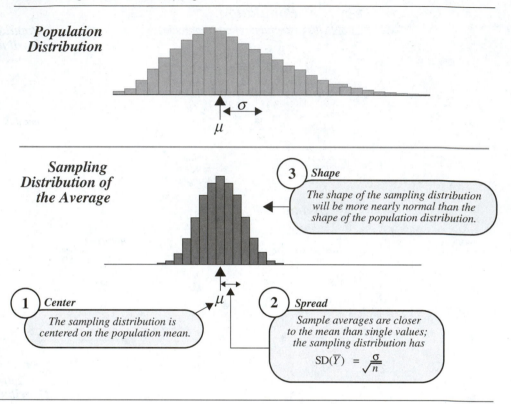

2.2.2 The Standard Error of an Average in Random Sampling

The *standard error* of any statistic is an estimate of the standard deviation in its sampling distribution. It is therefore the best guess about the likely size of the difference between a statistic used to estimate a parameter and the parameter itself. Standard errors are ordinarily calculated by substituting estimates of variability parameters into the formulas of sampling distribution standard deviations.

Associated with every standard error is a measure of the amount of information used to estimate variability, called its *degrees of freedom*, denoted d.f. Degrees of freedom are measured in units of "equivalent numbers of independent observations." Further information to explain degrees of freedom, more generally, will be provided in future chapters.

Standard Error for a Sample Average

The formula for the standard deviation of the average in a sample of size n was $\sigma/\sqrt{n}$, so if s is the sample standard deviation, the standard error for the average is

$$\mathrm{SE}(\overline{Y}) \;=\; \frac{s}{\sqrt{n}}\,, \qquad\qquad \mathrm{d.f.} = (n-1)\,.$$

The degrees of freedom in a single sample standard deviation are always one less than the sample size.

In the schizophrenia study, the average difference between volumes of unaffected and affected twins is 0.199 cm^3, and the sample standard deviation of the differences is 0.238 cm^3. The standard error of the average is therefore 0.062 cm^3, with 14 degrees of freedom. From this, one makes a preliminary judgment that the population difference is likely to be near the sample estimate, 0.199 cm^3, but that the sample estimate is likely to be somewhere in the neighborhood of 0.062 cm^3 off the mark.

2.2.3 The *t*-Ratio Based on a Sample Average

The ratio of an estimate's error to the estimate's likely size provides a convenient basis for drawing inferences about the parameter in question.

The Z-Ratio

For any parameter and its sample estimate, its *Z-ratio* is defined as $Z = $ (Estimate $-$ Parameter)/SD(Estimate). If the sampling distribution of the estimate is normal, then the sampling distribution of Z is *standard normal*, where the mean is 0 and the standard deviation is 1. The known percentiles of the standard normal distribution permit an understanding of the likely values of the Z-ratio, even though its value in any single case will not be known. From the table in Appendix A.1, for example, it is evident that for 95% of all samples the Z-ratio will fall in the interval -1.96 to 1.96. If the standard deviation of the estimate is known, this permits an understanding of the likely size of the estimation error. Consequently, useful statements can be made about the amount of uncertainty with which questions about the parameter can be resolved.

The t-Ratio

When, as is usually the case, the standard deviation of an estimate is unknown, it is natural to replace its value in the Z-ratio with the estimate's standard error. The result is the *t-ratio*,

$$t\text{-ratio} \;=\; \frac{(\text{Estimate} - \text{Parameter})}{\mathrm{SE}(\text{Estimate})}\,.$$

Associated with this *t*-ratio are the same degrees of freedom associated with the standard error of the estimate. The *t*-ratio does not have a standard normal distribution, because there is extra variability due to estimating the standard deviation. The fewer the degrees of freedom, the greater is this extra variability. Under some conditions, however, the sampling distribution of the *t*-ratio is known.

If $\overline{Y}$ is the average in a random sample of size n from a normally distributed population, the sampling distribution of its t-ratio is described by a **Student's *t*-distribution** *on n − 1 degrees of freedom.* The mathematical formula for the *t*-distributions was guessed by W. S. Gossett, a scientist who worked at the Guinness Brewery in the early 1900s and who published under the pseudonym "Student." The formula was proved to be correct by R. A. Fisher in 1925.

Histograms for *t*-distributions are symmetric about zero. For large degrees of freedom (usually 30 or more), *t*-distributions differ very little from the standard normal. For smaller degrees of freedom, they tend to have longer tails than the normal (reflecting the added uncertainty due to estimation of the standard deviation of $\overline{Y}$). Tables of selected percentiles from many *t*-distributions are provided in Appendix A.2 on page 710. Percentiles of the *t*-distribution are also available in many statistical computer programs and even in some pocket calculators.

2.2.4   Unraveling the *t*-Ratio

The average difference between hippocampus volumes for the 15 sets of twins in the schizophrenia study is 0.199 cm^3, and it standard error is 0.0615 cm^3, based on 14 degrees of freedom. The *t*-ratio is therefore $(0.199 - \mu)/0.0615$, where μ is the mean difference in the population of twins. If it can be assumed that the population distribution is normally distributed, then this *t*-ratio has a value like that typically drawn from a *t*-distribution with 14 degrees of freedom. A picture of this distribution is shown in Display 2.5.

This distribution indicates the likely values for the *t*-ratio, which in turn may be used to indicate the likely values of μ.

The Paired t-Test for No Difference

For example, consider the question "Is it plausible, based on the data, that μ could be zero?" If μ were zero, that would imply that the *t*-ratio was

$$t\text{-ratio} = (0.199 - 0)/0.0615 = 3.236.$$

Display 2.5, however, shows that 3.236 is an unusually large value to have come from the *t*-distribution. More precisely, only .003 (i.e., 0.3%) of all random samples from a population in which $\mu = 0$ lead to values of the *t*-ratio as large as or larger than 3.236. So here are your choices: (a) $\mu \neq 0$; or (b) $\mu = 0$ *and the random sampling resulted in a particularly nonrepresentative sample.* You cannot prove that $\mu \neq 0$, but you may infer it from the rarity of the converse. This type of reasoning is the conceptual basis for testing a hypothesis about a parameter, and the measure .003 is the (one-sided) *p*-value based on the *t*-distribution.

A 95% Confidence Interval for the Mean

Consider also the question "What are plausible values for μ, based on the data?" This can be answered by unraveling the *t*-ratio in a slightly different way. Display 2.5 shows that

Display 2.5   Student's *t*-distribution on 14 degrees of freedom

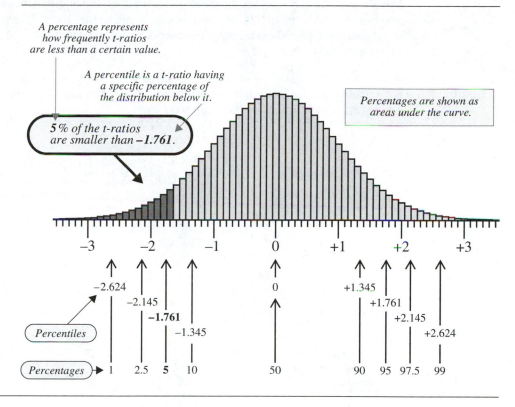

the most typical *t*-ratios are near zero, with 95% of the most likely values being between −2.145 and +2.145. If this sample produces one of these 95% most likely *t*-ratios, then

$$-2.145 < (0.199 - \mu)/0.0615 < +2.145,$$

in which case μ is between 0.067 and 0.331 cm^3. The interval from 0.067 to 0.331 is a 95% confidence interval for μ.

2.3   A *t*-Ratio for Two-sample Inference

The discussions above provide a conceptual basis for the development of inferential tools based on the *t*-distributions. This section repeats the discussions more formally, for the structure of independent samples from two normally distributed populations.

2.3.1   Sampling Distribution of the Difference Between Two Independent Sample Averages

In the Bumpus study, sparrows from the two groups were not paired. Rather, the imaginary sampling situation consists of two separate populations with samples drawn independently from each. A statistical analysis begins by calculating the difference between the averages from the group of survivors and the group of nonsurvivors. Replicating the entire process—selecting 35 survivors and 24 nonsurvivors, then calculating the difference between the sample averages—builds up the sampling distribution of the statistic $(\overline{Y}_2 - \overline{Y}_1)$. The results are summarized in Display 2.6.

Display 2.6   Facts about the sampling distribution of the difference of averages from two independent random samples (from statistical theory)

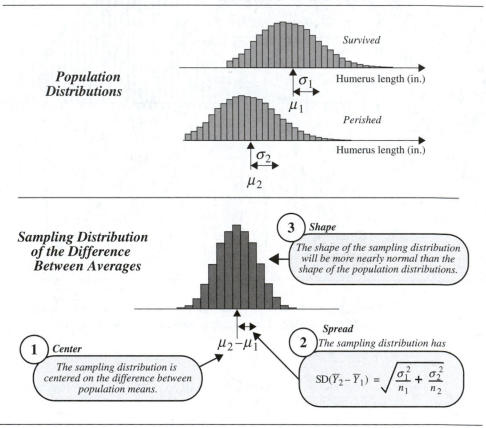

The spread of the sampling distribution would be smaller with larger sample sizes. The sampling distribution is approximately normal, and would be more so with larger sample sizes. As was the case in Section 2.2.1, the theoretical results about the sampling distribution of $\overline{Y}_2 - \overline{Y}_1$ are insufficient for making inferential statements, because $\text{SD}(\overline{Y}_1 - \overline{Y}_2)$ depends on unknown parameters.

2.3.2 Standard Error for the Difference of Two Averages

The natural estimate of the standard deviation in the sampling distribution of the difference between averages would be one obtained by replacing each of the population standard deviations by their sample versions. However, comparing averages provides a complete analysis only if all other features of the two distributions are similar. Therefore, assume in the following that the two populations have equal standard deviations: $\sigma_1 = \sigma_2 = \sigma$.

Pooled Standard Deviation for Two Independent Samples

If the two populations have the same standard deviation σ, then the sample standard deviations, s_1 and s_2, are independent estimates of it. A single estimate combines the two, and such combinations will always be formed by averaging on the variance scale. An average of the two sample variances, however, would not be quite right since the sample variance from a larger sample should be taken more seriously than a sample variance from a smaller one. A weighted average is in order, and the best single estimate of σ^2 is a weighted average in which the individual sample variances are weighted by their degrees of freedom. This is called the pooled estimate of variance, s_p^2. The square root of this is called the *pooled estimate of standard deviation, s_p*:

$$s_p = \sqrt{\frac{(n_1-1)s_1^2 + (n_2-1)s_2^2}{(n_1 + n_2 - 2)}}, \qquad \text{d.f.} = n_1 + n_2 - 2.$$

The degrees of freedom going into this estimate are the combined degrees of freedom from the individual estimates: $(n_1 - 1) + (n_2 - 1) = n_1 + n_2 - 2$.

Standard Error for the Difference

The standard deviation for the difference between averages from two independent samples is given in Display 2.6. If the two populations have equal spread, the common value of the variance factors from the two terms under the radical can be removed from under the radical as the common standard deviation. Using the pooled standard deviation as its estimate, the standard error for the difference in sample averages is:

$$\text{SE}(\overline{Y}_2 - \overline{Y}_1) = s_p \sqrt{\frac{1}{n_1} + \frac{1}{n_2}}.$$

An illustration of the computations for this standard error from the usual sample summary statistics is provided in Display 2.7 for the Bumpus data. One would estimate that the difference between sample averages (0.0101 inch) may be about 0.00567 inch off the mark, as an estimate of the difference between population means. The standard error estimate has 57 degrees of freedom.

Display 2.7   Calculation of the pooled estimate of SD and the standard error for the difference between two sample averages (Bumpus's data)

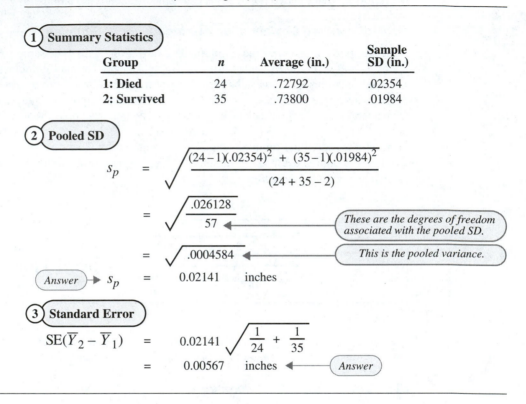

① **Summary Statistics**

Group	n	Average (in.)	Sample SD (in.)
1: Died	24	.72792	.02354
2: Survived	35	.73800	.01984

② **Pooled SD**

$$s_p = \sqrt{\frac{(24-1)(.02354)^2 + (35-1)(.01984)^2}{(24+35-2)}}$$

$$= \sqrt{\frac{.026128}{57}} \quad \longleftarrow \text{These are the degrees of freedom associated with the pooled SD.}$$

$$= \sqrt{.0004584} \quad \longleftarrow \text{This is the pooled variance.}$$

Answer $\longrightarrow$ $s_p = 0.02141$ inches

③ **Standard Error**

$$SE(\overline{Y}_2 - \overline{Y}_1) = 0.02141 \sqrt{\frac{1}{24} + \frac{1}{35}}$$

$$= 0.00567 \quad \text{inches} \longleftarrow \text{Answer}$$

2.3.3   Confidence Interval for the Difference Between Population Means

Inferences about the difference between population means arise from the consideration of a t-ratio, as in Section 2.2.3. The parameter of interest is $\mu_2 - \mu_1$. Its estimate is $\overline{Y}_2 - \overline{Y}_1$. The standard error comes from the previous section, and it has $n_1 + n_2 - 2$ degrees of freedom. In this case the t-ratio is $[(\overline{Y}_2 - \overline{Y}_1) - (\mu_2 - \mu_1)]/SE(\overline{Y}_2 - \overline{Y}_1)$. If the populations are normally distributed, this t-ratio has a t-distribution with $n_1 + n_2 - 2$ degrees of freedom.

For Bumpus's data $\overline{Y}_2 - \overline{Y}_1 = 0.01008$ inch, $SE(\overline{Y}_2 - \overline{Y}_1) = 0.00567$ inch, and the t-ratio has a sampling distribution described by a t-distribution on 57 degrees of freedom. Now a statement about the likely values for the t-ratio from this distribution can be translated into a statement about the likely values for $\mu_2 - \mu_1$.

The most likely values are near zero. A table of percentiles, such as that provided in Appendix A.2 on page 710, reveals that 95% of the most likely central values are those between -2.002 and $+2.002$. A 95% confidence interval can be obtained by supposing that the actual t-ratio is one of these 95% in the center. The extreme t-ratios, -2.002 and $+2.002$, are now used to find corresponding endpoints for an interval on the parameter. Setting

$$-2.002 < \frac{.01008 - (\mu_2 - \mu_1)}{0.00567} < 2.002$$

and solving for the parameter value yields the two interval endpoints: $-.00128 < \mu_2 - \mu_1 < .02144$.

The Mechanics of Confidence Interval Construction

A confidence interval with *confidence level* $100(1 - \alpha)\%$ is the following:

> *100(1– α)% Confidence Limits for the Difference Between Means*
>
> $$(\overline{Y}_2 - \overline{Y}_1) \; \pm \; t_{df}(1 - \alpha/2)\,\text{SE}(\overline{Y}_2 - \overline{Y}_1).$$

This formula requires that a quantity be subtracted from $\overline{Y}_2 - \overline{Y}_1$ to get the lower endpoint and that the same quantity be added to $\overline{Y}_2 - \overline{Y}_1$ to get the upper endpoint. The symbol $t_{df}(1 - \alpha/2)$ represents the $100(1 - \alpha/2)$th percentile of the *t*-distribution on d.f. degrees of freedom. For example, $t_{57}(.975)$ represents the 97.5th percentile in the *t*-distribution with 57 degrees of freedom (which is 2.002). It may seem strange that the 97.5th percentile is desired in the calculation of a 95% confidence interval, but the 2.5th and the 97.5th percentiles are the ones that divide the middle 95% of the distribution from the rest; and the 2.5th percentile is always the negative of the 97.5th. Display 2.8 summarizes the solution for the Bumpus data.

The Interpretation of a Confidence Interval

A 95% confidence interval will contain the parameter if the *t*-ratio from the observed data happens to be one of those in the middle 95% of the sampling distribution. Since 95% of all possible pairs of samples lead to such *t*-ratios, it is safe to say that the procedure of constructing a 95% confidence interval is successful (in capturing the parameter of interest) in 95% of its applications. It is impossible to say whether it is successful or not in any particular application.

The term *confidence* refers to what is known about the long-term frequency of the procedure, rather than a statement about the probability of containing the *t*-ratio for a particular sample. Many people act as if the 95% corresponds to the probability that the particular interval contains the parameter. Although this is not what the theory based on the *t*-ratio indicates, there is little harm in using this casual interpretation in practice.

Factors Affecting the Width of a Confidence Interval

There is a trade-off between the level of confidence and the width of the confidence interval. The level of confidence can be specified to be large by the user (and a high confidence level is good), but only at the expense of having a wider interval (which is bad, since the interval

Display 2.8   Construction of a 95% confidence interval for the difference between the mean
humerus lengths of sparrows that died and those that survived

Group	n	Average (in.)	SD (in.)
1: Died	24	0.72792	0.02354
2: Survived	35	0.73800	0.01984

$\overline{Y}_2 - \overline{Y}_1 = 0.73800 - 0.72792 = 0.01008$ *From Display 2.6*

$SE(\overline{Y}_2 - \overline{Y}_1) = 0.00567$ inches

Degrees of freedom $= 24 + 35 - 2 = 57$

$t_{57}(.975) = 2.002$ *From tables of the*
                                       t-distribution with
                                       57 degrees of freedom

Half-width $= (2.002)(0.00567) = 0.01136$

Lower 95% confidence limit $= 0.01008 - 0.01136 = -0.00128$ inches

Upper 95% confidence limit $= 0.01008 + 0.01136 = 0.02144$ inches

is less specific in answering the question of interest). If the consequences of not capturing the parameter are severe, then it might be wise to use 99% confidence intervals, even though they will be wider and, therefore, less informative. If the consequences of not capturing the parameter are minor, then a 90% interval might be a good choice.

The only way to decrease the interval width without decreasing the level of confidence (or similarly to increase the level of confidence without increasing width) is to get more data or, if possible, to reduce the size of σ. If a guess or initial estimate of σ is available it is easy to determine the sample size needed to get a confidence interval of a certain width. This is discussed in Chapter 23.

2.3.4   Testing a Hypothesis About the Difference Between Means

To test a hypothesized value for a parameter, a *t-statistic* is formed as the *t*-ratio that is obtained by supposing the hypothesis is true. The *t*-distribution judges whether the *t*-statistic is a likely value for a *t*-ratio and, hence, whether the hypothesis is reasonable. The *t*-statistic for the difference in means is

$$t\text{-statistic} = \frac{(\overline{Y}_2 - \overline{Y}_1) - [\text{Hypothesized value for } (\mu_2 - \mu_1)]}{SE(\overline{Y}_2 - \overline{Y}_1)}.$$

This *t*-statistic tells how many standard errors the estimate is away from the hypothesized parameter. Its sign tells whether the estimate is above the hypothesized value (+) or below it (−).

The *p-value*, used as a measure of the credibility of the hypothesis, is the proportion of all possible *t*-ratios that are farther from zero than is the *t*-statistic.

> *The* p-**value** *for a* t-*test is the probability of obtaining a* t-*ratio as extreme or more extreme than the* t-*statistic in its evidence against the null hypothesis, if the null hypothesis is correct.*

The p-value may be based on a probability model induced by random assignment in a randomized experiment (Section 1.3.2) or on a probability model induced by random sampling from populations, as here.

If the p-value is small, then either the hypothesis is correct (and the sample happened to be one of those rare ones that produce such an unusual t-ratio), or the hypothesis is incorrect. Although it is impossible to know which of these two possibilities is true, the p-value indicates the probability of the first of these results and, therefore, provides a measure of credibility for that interpretation. *The smaller the p-value, the stronger is the evidence that the hypothesis is incorrect.*

One-Sided and Two-Sided p-Values

For the Bumpus example the t-statistic for the hypothesis of "no difference" in population means is 1.778. The proportion of t-ratios that are farther from zero than 1.778 is found from the percentiles of the t_{57} distribution. The proportion of t-ratios *greater than* 1.778 is 0.040. The t-ratios *farther from zero than* 1.778 are those less than -1.778 and those greater than 1.778, and this proportion is 0.080. The proportion of t-ratios farther from zero *in one direction* is referred to as a *one-sided p-value*. The proportion of t-ratios farther from zero than the t-statistic, either positively or negatively, is a *two-sided p-value*.

The choice of one-sided or two-sided depends on how specifically the researcher wishes to declare the alternatives to the hypothesis of equal means. If Bumpus theorized that the sparrows that survived did so because their humerus tended to be larger, then he would be interested in evidence against the equal-mean hypothesis in the direction of $\mu_2 - \mu_1 > 0$, which would be manifested in t-statistics greater than zero. Bumpus did not indicate that he believed the difference to be in one particular direction, however, so a two-sided p-value is in order.

Much has been made of whether to report one-sided or two-sided p-values. There are some situations where a one-sided p-value is appropriate, some where a two-sided p-value is appropriate, and many where it is not at all clear which is appropriate. Since the two provide equivalent measures of evidence against the hypothesis of equal means (that is, the two-sided p-value is simply twice the one-sided p-value), the distinction is not terribly important. There is only one absolute when it comes to reporting: *always report whether the p-value is one- or two-sided.*

2.3.5 The Mechanics of p-Value Computation

The steps required to compute a p-value for the test of the hypothesis that $\mu_2 - \mu_1 = D$ (a specified value, like 0) are as follows.

1. Compute the estimate, $\overline{Y}_2 - \overline{Y}_1$, its standard error, and the degrees of freedom.
2. Compute the t-statistic: $t = [(\overline{Y}_2 - \overline{Y}_1) - D]/\text{SE}(\overline{Y}_2 - \overline{Y}_1)$.
3. Determine the proportion, P, of t-ratios that are less than the t-statistic, by consulting a t-distribution table with the appropriate degrees of freedom.
4. Determine the p-value based on the proportion, P, and the alternatives of interest. (a) For one-sided alternatives of the form $\mu_2 - \mu_1 > D$, t-ratios larger than t are more extreme, so the one-sided p-value is $1 - P$. (b) For the one-sided alternatives of the form $\mu_2 - \mu_1 < D$, t-ratios smaller than t are more extreme, so the one-sided p-value is P. (c) For two-sided alternatives of the form $\mu_2 - \mu_1 \neq D$, t-ratios that are larger in magnitude than t are more extreme, so the two-sided p-value is $2P$ if $P < 0.5$ or $2(1 - P)$ if $P > 0.5$.
5. Report the hypothesis, the p-value, and whether it is one- or two-sided.

An illustration of this procedure for the Bumpus data is shown in Display 2.9.

Display 2.9   The t-test for the hypothesis that the mean humerus lengths of sparrows that died is the same as the mean for sparrows that survived

---

Group	n	Average (in.)	SD (in.)
1: Died	24	0.72792	0.02354
2: Survived	35	0.73800	0.01984

$\overline{Y}_2 - \overline{Y}_1 = 0.73800 - 0.72792 = 0.01008$ *(From Display 2.7)*

$\text{SE}(\overline{Y}_2 - \overline{Y}_1) = 0.00567$ inches

Degrees of freedom $= 24 + 35 - 2 = 57$ *Hypothesized difference*

$$t\text{-statistic} = \frac{0.01008 - \mathbf{0.0}}{0.00567} = 1.778$$

$P = .960$ *From tables of the t-distribution with 57 degrees of freedom: $1.778 = t_{57}(.960)$*

One-sided p-value $= .040$ *or*   Two-sided p-value $= 2(.040) = .080$

2.4 Inferences in a Two-treatment Randomized Experiment

Probability models for randomized experiments (Section 1.3.1) are spawned by the chance mechanisms used to assign subjects to treatment groups. Probability models for random sampling (Section 1.4.1) are spawned by the chance mechanisms used to select units from real, finite populations. Chapter 2 has thus far discussed inference procedures whose motivation stems from considerations of random sampling from populations that are conceptual, infinite, and normally distributed. While there seems to be a considerable difference between the situations, it turns out that the t-distribution uncertainty measures discussed in this chapter are useful approximations to both the randomization and the random sampling uncertainty measures for a wide range of problems. The practical consequence is that t-tools are used for many situations that do not conform to the strict model upon which the t-tools are based, including data from randomized experiments. Conclusions from randomized experiments, however, are phrased in the language of treatment effects and causation, rather than differences in population means and association.

2.4.1   Approximate Uncertainty Measures for Randomized Experiments

Reconsideration of the creativity study (Section 1.1.1) illustrates the procedure just described. In the intrinsic group, the average of 24 scores is 19.88, and the SD is 4.44. In the extrinsic group, the average of 23 scores is 15.74, and the SD is 5.25. The pooled estimate of the standard deviation is 4.85, and the standard error for the difference between averages is 1.42 with 45 degrees of freedom. The difference between averages scores is 4.14 points.

Hypothesis Test of No Treatment Effect

The t-statistic, $t = (4.14 - 0)/1.42 = 2.92$, is utilized exactly as it would be for a random sampling situation. In the t-distribution with 45 degrees of freedom, 2.92 is the 99.7th percentile ($P = .997$), so the one-sided p-value for the alternative of a positive difference is .0027, and the two-sided p-value is twice that. The conclusion is phrased in terms of the additive treatment effect model.

Confidence Interval for a Treatment Effect

Construction of a confidence interval for an additive treatment effect δ is precisely the same as for the difference between population means, $\mu_2 - \mu_1$. The 97.5th percentile in the t-distribution with 45 degrees of freedom is 2.014, so the interval halfwidth is $(2.014)(1.42) = 2.85$. The interval runs from $4.14 - 2.85 = 1.29$ to $4.14 + 2.85 = 6.99$ points.

Compare this construction with one based on the randomization procedure itself (Section 1.3). The randomization distribution relies on a relationship between testing and confidence intervals: *any hypothesized parameter value should be included or excluded from a $100(1 - \alpha)\%$ confidence interval according to whether its test yields a two-sided p-value that is greater than or less than α.* Accordingly, to construct a 95% confidence interval for the treatment effect δ include only those values of δ which, when tested, have two-sided p-values greater than .05.

To determine whether $\delta = 5$ should be included in a 95% confidence interval, one must consider the randomization model for $\delta = 5$. If this is the correct value, subtracting 5 from the scores of all persons in the intrinsic group reconstructs the scores they would have had if placed in the extrinsic group. Now all 47 scores are homogeneous, so a randomization test should conclude there is no difference. Perform the randomization test. If the two-sided p-value exceeds (or equals) .05, include 5 in the interval. Otherwise, leave it out. That settles the issue for $\delta = 5$, but the limits of the interval must be found by repeating this process to find the smallest and largest δ for which the two-sided p-value is greater than or equal to .05.

Approximation of the Randomization Distribution of the t-Statistic

The *t*-based *p*-values and confidence levels are only approximations to the correct values from randomization distributions. To see how good the approximation is in the creativity study, the analysis of Chapter 1 was modified by considering the randomization distribution of the *t*-ratio rather than the difference in sample averages. A histogram of the *t*-ratios from 500 different randomizations appears in Display 2.10, along with the approximating *t*-distribution. The observed *t*-statistic (2.92) was exceeded by only one of the 500 random assignments, giving an estimated one-sided *p*-value of $1/500 = .002$, so the approximation based on the *t*-distribution (.0027) is quite good.

Display 2.10   A histogram of *t*-ratios from 500 randomizations of the creativity study data with an approximating *t*-distribution density

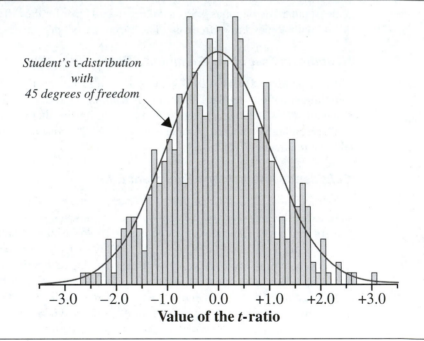

2.5   Related Issues

2.5.1   Interpretation of *p*-Values

A small *p*-value like .005 means that either the hypothesis is correct (and the randomization or random sampling led by chance to a rare outcome where the *t*-ratio is quite far from zero), or the hypothesis is incorrect. The smaller the *p*-value, the greater is the evidence that the second explanation is the correct one.

How small is small? It is difficult and unwise to decide on absolute cutoff points for believability to be applied in all situations. Display 2.11 represents a starting point for interpreting *p*-values.

Display 2.11   Interpreting the size of a *p*-value

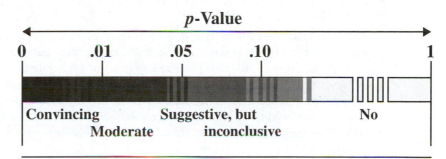

Is there evidence of a difference?

Probabilities can be comprehended by comparing them to events whose probabilities are more familiar. For example, at what point does a person flipping a series of heads begin to doubt that the coin is fair? It is not terribly unlikely to get four heads in a row. The probability of doing so with a fair coin is .0625. At five heads in a row one might start to get a bit curious. The chance of this, if the coin is fair, is .03125. When the sixth toss is also heads, one may start to question the integrity of the coin, even though it is entirely possible that a fair coin could turn up heads six times in a row (with probability .015625). Ten heads in a row is convincing evidence that the coin is not of standard issue. The probability of this event, if the chance of heads were in fact one-half, is .0009765.

The ranges of the adjectives underneath the *p*-values in Display 2.11 are flexible. For example, one might interpret six heads in a row differently depending on whether the individual flipping the coin is Mother Theresa playing a game with orphans or a character of dubious integrity making money from coin flips at a carnival. In some cases, using any such adjectives may be ill-advised because they do express a qualitative judgment over and above the statistical information.

It is tempting to think of a *p*-value as the probability of the null hypothesis being correct, but this interpretation is technically incorrect and potentially misleading. The hypothesis is or is not correct, and there is no probability associated with that. The probability arises from the uncertainty in the data. So the best technical description is the precise definition (see p. 41), clumsy as it may sound.

2.5.2   An Example of Confidence Intervals

In 1915 Albert Einstein published four papers in the proceedings of the Prussian Academy of Sciences laying the foundations of general relativity and describing some of its consequences. A paper establishing the field equation for gravity predicted that the arc of deflection of light passing the sun would be twice the angle predicted by Newton's gravity theory. Half the predicted deflection comes directly from Newton's calculations, and the other half comes from the curvature of space near the sun relative to space far away. This is represented by the equation

$$\Delta = (1/2)(1 + \gamma)\frac{1.75}{d},$$

where Δ is the deflection of light, d is the distance of the closest approach of the ray to the sun (in solar radii), and γ is the parameter describing space curvature.

The parameter γ, which is predicted by Einstein's general relativity theory to be 1 and by Newtonian physics to be 0, was estimated in 1919 by British astronomers during a total solar eclipse. Since then, measurements have been repeated many times, under various measurement conditions. The efforts are summarized in Display 2.12. (Data from C. M. Will, "General Relativity at 75: How Right Was Einstein?" *Science* 250 (November 9, 1990):770–75.)

Display 2.12 Estimates and confidence intervals for γ, the deflection of light around the sun, from 20 experiments

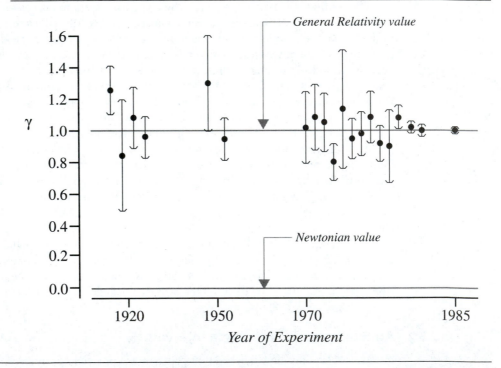

The confidence intervals around the estimates in Display 2.12 reflect uncertainty due to measurement errors. (The actual confidence levels are not given and are not important for

this illustration.) After the first relatively crude attempts to measure γ, little improvement was made until the late 1960s and the discovery of quasars. Measurements of light from quasar groups passing near the sun led to dramatic improvement in accuracy, as evident in the narrower intervals with later years.

A Note About the Cumulation of Evidence

This example shows that theories must withstand continual challenges from skeptical scientists. The essence of scientific theory is the ability to predict future outcomes. Experimental results are typically uncertain. So the fact that some intervals fail to include the value, $\gamma = 1$, is not taken to disprove general relativity, but neither would it prove general relativity right if all the intervals did include $\gamma = 1$. When a theory's predictions are consistently denied by a series of experiments—such as the Newtonian prediction of $\gamma = 0$ in this example— scientists agree that the theory is not adequate.

2.5.3 The Rejection Region Approach to Hypothesis Testing

Not long ago, statisticians took a *rejection region* approach to testing hypotheses. A *significance level* of .05, say, was selected in advance, and a p-value less than .05 called for rejecting the hypothesis at the significance level .05; otherwise the hypothesis was accepted, or more correctly, not rejected. Thus p-value $= .048$ and p-value $= .0001$ both lead to rejection at the .05 level, even though they supply vastly different degrees of evidence. On the other hand, p-value $= .049$ and p-value $= .051$ lead to different conclusions even though they provide virtually identical evidence against the hypothesis. Although important for leading to advances in the theory of statistics, more recently, in practice, the rejection region approach has largely been discarded in favor of reporting the p-values. P-values give the reader more information for judging the significance of findings.

2.6 Summary

Many research questions can be formulated as comparisons of two population distributions. Comparison of the distributions' centers effectively summarizes the difference between the parameters of interest when the populations have the same variation and general shape. This chapter concentrated on the difference in means, $\mu_2 - \mu_1$, which is estimated by the difference in sample averages.

The statistical problem is to assess the uncertainty associated with the difference between the estimate (the difference in sample averages) and the parameter (the difference in population means). The sampling distribution of an estimate is the key to understanding the uncertainty. It is represented as a histogram of values of the estimate for every possible sample that could have been selected.

Often with fairly large samples, a sampling distribution has a normal shape. A normal sampling distribution is specified by its mean and its standard deviation. When the populations have common standard deviation σ the difference in sample averages has a sampling distribution with mean $\mu_2 - \mu_1$ and standard deviation

$$\sigma \sqrt{\frac{1}{n_1} + \frac{1}{n_2}}.$$

This could be used to describe uncertainty except that it involves the unknown σ—the common standard deviation in the two populations. In practice, σ is replaced by its best estimate from the data—the pooled standard deviation, s_p, having $n_1 + n_2 - 2$ degrees of freedom. The estimated standard deviation of the sampling distribution is called the standard error.

The standard error alone, however, does not entirely describe the uncertainty in an estimate. More precise statements can be made by using the *t*-ratio, which has a Student's *t*-distribution as its sampling distribution (if the ideal normal model applies). This leads directly to confidence intervals and *p*-values as statistical tools for answering the questions of interest. The confidence interval provides a range of likely values for the parameter, and the confidence level is interpreted as the frequency with which the interval construction procedure gives the right answer. For testing whether a particular hypothesized number could be the unknown parameter, the *t*-statistic is formed by substituting the hypothesized value for the true value in the *t*-ratio. The *p*-value is the chance of getting more extreme *t*-ratios than the *t*-statistic, and it is interpreted as a measure of the credibility of the hypothesized value.

Further Reading

Elementary discussions of sampling distributions, probability, and standard errors are provided in the textbook by Freedman, Pisani, Purves, and Adhikari (1991). Elementary discussions of the standard techniques for one-sample and two-sample *t*-tests and confidence intervals are provided in all introductory statistics texts, such as Moore and McCabe (1993), Devore and Peck (1993), and Mendenhall (1993). The mathematical derivations of the properties of the sampling distribution for the average and for the difference of two independent averages are given in introductory books on mathematical statistics, such as Rice (1995). Further discussion of confidence intervals, tests of hypotheses, *p*-values, and the *t*-distribution, along with further references, may be found in the *Encyclopedia of Statistical Science* (Johnson, Kotz, and Reed, 1988). For more on the basic meaning of significance and confidence, see Mosteller and Tukey (1968).

2.7 **Exercises**

Conceptual Exercises

1. Bumpus Data. In drawing conclusions from the Bumpus data, why might it be important to know whether all the birds were measured before the ones that perished actually perished?

2. For comparing two population means when the population distributions have the same standard deviation, the standard deviation is sometimes referred to as a nuisance parameter. Explain why it might be considered a nuisance.

3. True or false? If a sample size is large, then the shape of a histogram of the sample will be approximately normal, even if the population distribution is not normal.

4. True or false? If a sample size is large then the shape of the sampling distribution of the average will be approximately normal, even if the population distribution is not normal.

5. Explain the relative merits of 90% and 99% levels of confidence.

6. What is wrong with the hypothesis that $\overline{Y}_2 - \overline{Y}_1$ is 0?

7. In a study of the effects of marijuana use during pregnancy, measurements on babies of mothers who used marijuana during pregnancy were compared to measurements on babies of mothers who did not. (Data from B. Zuckerman et al., "Effects of Maternal Marijuana and Cocaine Use on Fetal Growth," *New England Journal of Medicine* 320(12) (March 1989): 762–68.) A 95% confidence interval for the difference in mean head circumference (nonuse minus use) was 0.61 to 1.19 cm. What can be said from this statement about a p-value for the hypothesis that the mean difference is zero?

8. Suppose the following statement is made in a statistical summary: "A comparison of breathing capacities of individuals in households with low nitrogen dioxide levels and individuals in households with high nitrogen dioxide levels indicated that there is no difference in the means (two-sided p-value = .24)." What is wrong with this statement?

9. What is the difference between (a) the mean of Y and the mean of $\overline{Y}$? (b) the standard deviation of Y and the standard deviation of $\overline{Y}$? (c) the standard deviation of $\overline{Y}$ and the standard error of $\overline{Y}$? (d) a t-ratio and a t-statistic?

10. Consider blood pressure levels for populations of young women using birth control pills and young women not using birth control pills. A comparison of these two populations through an observational study might be consistent with the theory that the pill elevates blood pressure levels. What tool is appropriate for addressing whether there is a difference between these two populations? What tool is appropriate for addressing the likely size of the difference?

11. The data in Display 2.13 are survival times (in days) of guinea pigs that were randomly assigned either to a control group or to a treatment group that received a dose of tubercle bacilli. (Data from K. Doksum, "Empirical Probability Plots and Statistical Inference for Nonlinear Models in the Two-sample Case," *Annals of Statistics* 2 (1974): 267–77.) (a) Why might the additive treatment effect model be inappropriate for these data? (b) Why might the ideal normal model with equal spread be an inadequate approximation?

Display 2.13 Lifetimes of guinea pigs in two treatment groups

Control (n = 64)		Received bacilli (n = 58)
36,18	0	
91,89,87,86,52,50		76,93,97
49,20,19,18,15,14,14,08,02	1	07,08,13,14,19,36,38,39
89,78,73,67,67,66,65,60		52,54,54,60,64,64,66,68,78,79,81,81,83,85,94,98
16,12,09	2	12,13,16,20,25,25,44
92,79,78,73		53,56,59,65,68,70,83,89,91
41	3	11,15,26,26
82,80,67,55		61,73,73,76,97,98
46,32,21,21	4	06
74,63,55		59,66
46,45,05	5	
90,76,69		92,98
41,38,37,34,21,08,07,03	6	
88,85,63,50		
35,25	7	

Legend: | 5 | 98 represents 598 days

Computational Exercises

12.   Marijuana Use During Pregnancy. For the birth weights of babies in two groups, one born of mothers who used marijuana during pregnancy and the other born of mothers who did not (see Exercise 7), the difference in sample averages (nonuser mothers minus user mothers) was 280 grams, and the standard error of the difference was 46.66 grams with 1,095 degrees of freedom. From this information, provide the following: (a) a 95% confidence interval for $\mu_2 - \mu_1$, (b) a 90% confidence interval for $\mu_2 - \mu_1$, and (c) the two-sided p-value for a test of the hypothesis that $\mu_2 - \mu_1 = 0$.

13.   Fish Oil and Blood Pressure. Reconsider the changes in blood pressures for men placed on a fish oil diet and for men placed on a regular oil diet, from Exercise 12 on page 23. Do the following steps to compare the treatments:

    (a)   Compute the averages and the sample standard deviations for each group separately.

    (b)   Compute the pooled estimate of standard deviation using the formula in Section 2.3.2.

    (c)   Compute $SE(\overline{Y}_2 - \overline{Y}_1)$ using the formula in Section 2.3.2.

    (d)   What are the degrees of freedom associated with the pooled estimate of standard deviation? What is the 97.5th percentile of the t-distribution with this many degrees of freedom?

    (e)   Construct a 95% confidence interval for $\mu_2 - \mu_1$ using the formula in Section 2.3.3.

    (f)   Compute the t-statistic for testing equality as shown in Section 2.3.5.

    (g)   Find the one-sided p-value by comparing the t-statistic in (f) to the percentiles of the appropriate t-distribution (by looking up the t-statistic in a table or by reading the appropriate percentile from a computer program or calculator).

14.   Fish Oil and Blood Pressure. Find the 95% confidence interval and one-sided p-value asked for in Exercise 13(e) but use a statistical computer package to do so.

15.   Auto Exhaust and Lead Concentration in Blood. Researchers took independent random samples from two populations of policemen and measured the level of lead concentration in their blood. The sample of 126 policemen subjected to constant inhalation of automobile exhaust fumes in downtown Cairo had an average blood level concentration of 29.2 μg/dl and an SD of 7.5 μg/dl; a control sample of 50 policemen from the Cairo suburb of Abbasia, with no history of exposure, had an average blood level concentration of 18.2 μg/dl and an SD of 5.8 μg/dl. (Data from A.-A. M. Kamal, S. E. Eldamaty, and R. Faris, "Blood Lead Level of Cairo Traffic Policemen," *Science of the Total Environment* 105 (1991): 165–70.) Is there convincing evidence of a difference in the population averages?

16.   Motivation and Creativity. Verify the statements made in the summary of statistical findings for the Motivation and Creativity Data (Section 1.1.1) by analyzing the data on the computer.

17.   Sex Discrimination. Verify the statements made in the summary of statistical findings for the Sex Discrimination Data (Section 1.1.2) by analyzing the data on the computer.

18.   Bumpus's Data. Verify the computations in Display 2.8 and Display 2.9 by analyzing Bumpus's data on the computer.

19.   Fish Oil and Blood Pressure. Reconsider the fish oil and blood pressure data of Exercise 12 on page 23. Since the measurements are the reductions in blood pressure for each man, it is of interest to know whether the mean reduction is zero for each group. For the regular oil diet group do the following:

    (a)   Compute the average and the sample standard deviation. What are the degrees of freedom associated with the sample standard deviation, s_2?

    (b)   Compute the standard error for the average from this group: $SE(\overline{Y}_2) = s_2/\sqrt{n_2}$.

    (c)   Construct a 95% confidence interval for μ_2 as $\overline{Y}_2 + t_d(.975)SE(\overline{Y}_2)$, where d is the degrees of freedom associated with s_2.

(d) For the hypothesis that μ_2 is zero, construct the t-statistic $\overline{Y}_2/SE(\overline{Y}_2)$. Find the two-sided p-value as the proportion of values from a t_d distribution farther from 0 than this value.

20. Fish Oil and Blood Pressure (One-Sample Analysis). Repeat Exercise 19 for the group of men who were given the fish oil diet and then answer these questions: Is there any evidence that the mean reduction for this group is different from zero? What is the typical reduction in blood pressure expected from this type of diet (for individuals like these men)? Provide a 95% confidence interval.

Data Problems

21. Bumpus's Data: Weights of Birds that Survived and Perished. The data in Display 2.14 are weights, in grams, of the 35 male house sparrows that survived and the 24 that perished from the severe winter storm (Section 2.1.1). Analyze the data using a graphical method; then use t-tools to say whether these distributions differ and by how much. Write a short summary report, similar to the summaries in Section 2.1.

Display 2.14 Weights of male house sparrows that survived and perished (Bumpus's data)

Weights (g) of 35 Males That *Survived*

24.5 26.9 26.9 24.3 24.1 26.5 24.6 24.2 23.6 26.2 26.2 24.8 25.4 23.7 25.7
25.7 26.3 26.7 23.9 24.7 28.0 27.9 25.9 25.7 26.6 23.2 25.7 26.3 24.3 26.7
24.9 23.8 25.6 27.0 24.7

Weights (g) of 24 Males That *Perished*

26.5 26.1 25.6 25.9 25.5 27.6 25.8 24.9 26.0 26.5 26.0 27.1 25.1 26.0 25.6
25.0 24.6 25.0 26.0 28.3 24.6 27.5 31.1 28.3

---

22. Cholesterol in Urban and Rural Guatemalans. An observational study to contrast cholesterol levels in rural and urban Guatemalan Indians came up with the data in Display 2.15. (Data from M. D. Tejada et al., "The Blood Viscosity of Various Socioeconomic Groups in Guatemala," *American Journal of Clinical Nutrition* (1964):p. 303–7.) The samples are not random samples. Assess the

---

Display 2.15 Cholesterol levels in urban and rural Guatemalans

---

Serum Total Cholesterol (mg/l) Levels Among *Urban* Residents ($n = 45$)

133 134 155 170 175 179 181 184 188 189 190 196 197 199 200 200 201 201 204 205
205 205 206 214 217 222 222 227 227 228 234 234 236 239 241 242 244 249 252 273
279 284 284 284 330

Serum Total Cholesterol (mg/l) Levels Among *Rural* Residents ($n = 49$)

 95 108 108 114 115 124 129 129 131 131 135 136 136 139 140 142 142 143 143 144
144 145 145 148 152 152 155 157 158 158 162 165 166 171 172 173 174 175 180 181
189 192 194 197 204 220 223 226 231

---

difference in cholesterol distributions between the two populations. Write a summary statistical report, including a graphical display to illustrate your conclusions. What inferences are possible?

Answers to Conceptual Exercises

1. There is a potential for bias (possibly subconscious) in measuring the birds after knowing the outcome. This is especially a problem if the measurer has a particular theory in mind.

2. There is rarely any direct interest in the standard deviation, but it must be estimated in order to clear up the picture regarding means.

3. False.

4. True.

5. There is more confidence that a 99% interval contains the parameter of interest, but the extra confidence comes at the price of the interval being larger and therefore less informative about specific likely values.

6. The hypothesis must be about the population means.

7. It is less than .05.

8. The statement implies that the null hypothesis is accepted as true. It should be worded as, for example, the data are consistent with the hypothesis that there is no difference. (This issue is partly one of semantics, but it is still important to understand the distinction being made.)

9. **(a)** The mean of Y is the mean in the population of all individual measurements, and the mean of $\overline{Y}$ is the mean of the sampling distribution of the sample mean. With random sampling, the two have the same value μ.

(b) The standard deviation of Y is the standard deviation among all observations in the population, and the standard deviation of $\overline{Y}$ is the standard deviation in the sampling distribution of the sample average. The two are related, but not the same: if the standard deviation of Y is denoted by σ, then the standard deviation of $\overline{Y}$ is $\sigma/\sqrt{n}$.

(c) The standard error is an estimate of the standard deviation in the sampling distribution, obtained by replacing σ with the pooled estimate s_p.

(d) The t-ratio is the ratio of the difference between the estimate and the parameter to the standard error of the estimate. It involves the parameter, so you do not generally know what it is. The t-statistic is a trial value of the t-ratio, obtained when a hypothesized value of the parameter is used in place of the actual value.

10. A p-value. A confidence interval.

11. (a) Because the spread of the stem-and-leaf plot is larger for the control group than for the treatment group, it does not appear that the effect of treatment was simply to add a certain number of days onto the lives of every guinea pig. It may have added days for those that would not have lived long anyway, and subtracted days from those that would have lived a long time. (b) The equal variation assumption does not appear to be appropriate.

A Closer Look at Assumptions

Although statistical computer programs faithfully supply confidence intervals and p-values whenever asked, the human data analyst must consider whether the assumptions on which use of the tools is based are met, at least approximately. In this regard, an important distinction exists between the mathematical assumptions on which use of t-tools is exactly justified and the broader conditions under which such tools work quite well.

The mathematical assumptions—such as those of the model for two independent samples from normal populations with the same standard deviation—are never strictly met in practice, nor do they have to be. The two-sample t-tools are often valid even if the population distributions are nonnormal or the standard deviations are unequal. An understanding of the broader conditions, provided by advanced statistical theory and computer simulation, is needed to evaluate the appropriateness of the tools for a particular problem. After checking the actual conditions with graphical displays of the data, the data analyst may decide to use the standard tools, use them but apply the label "approximate" to the inferences, or choose an alternative approach.

An effective alternative is to apply the standard tools after transforming the data. A transformation is useful if the tools are appropriate for the conditions of the transformed data and if the questions of interest are answerable on the new scale. A particularly important transformation is the logarithm, which permits a convenient description of a multiplicative effect.

3.1 Case Studies

3.1.1 Cloud Seeding to Increase Rainfall—A Randomized Experiment

The data in Display 3.1 were collected in southern Florida between 1968 and 1972 to test a hypothesis that massive injection of silver iodide into cumulus clouds can lead to increased rainfall. (Data from J. Simpson, A. Olsen, and J. Eden, "A Bayesian Analysis of a Multiplicative Treatment Effect in Weather Modification," *Technometrics* 17 (1975): 161–66.)

Display 3.1   Rainfall (acre-feet) for days with and without cloud seeding

Rainfall from Unseeded Days ($n = 26$)

1,202.6	830.1	372.4	345.5	321.2	244.3	163.0	147.8	95.0
87.0	81.2	68.5	47.3	41.1	36.6	29.0	28.6	26.3
26.0	24.4	21.4	17.3	11.5	4.9	4.9	1.0	

Rainfall from Seeded Days ($n = 26$)

2,745.6	1,697.1	1,656.4	978.0	703.4	489.1	430.0	334.1	302.8
274.7	274.7	255.0	242.5	200.7	198.6	129.6	119.0	118.3
115.3	92.4	40.6	32.7	31.4	17.5	7.7	4.1	

On each of 52 days that were deemed suitable for cloud seeding, a random mechanism was used to decide whether to seed the target cloud on that day or to leave it unseeded as a control. An airplane flew through the cloud in both cases, since the experimenters and the pilot were themselves unaware of whether on any particular day the seeding mechanism in the plane was loaded or not (that is, they were *blind* to the treatment). Precipitation was measured as the total rain volume falling from the cloud base following the airplane seeding run, as measured by radar. Did cloud seeding have an effect on rainfall in this experiment? If so, how much?

Box plots of the data in Display 3.2(a) indicate that the rainfall tended to be larger on the seeded days. Both distributions were quite skewed, and more variability occurred in the seeded group than in the control group. The box plots in Display 3.2(b) are drawn from the same data, but on the scale of the natural logarithm of the rainfall measurements. On this scale, the measurements appear to have symmetric distributions, and the variation seems nearly the same.

Summary of Statistical Findings

It is estimated that the volume of rainfall on days when clouds were seeded was 3.1 times as large as when not seeded. A 95% confidence interval for this multiplicative effect is 1.3 times to 7.7 times. Since randomization was used to determine whether any particular suitable day was seeded or not, it is safe to interpret this as evidence that the seeding caused the larger rainfall amount.

Display 3.2 Box plots of rainfall amounts on original and transformed scales

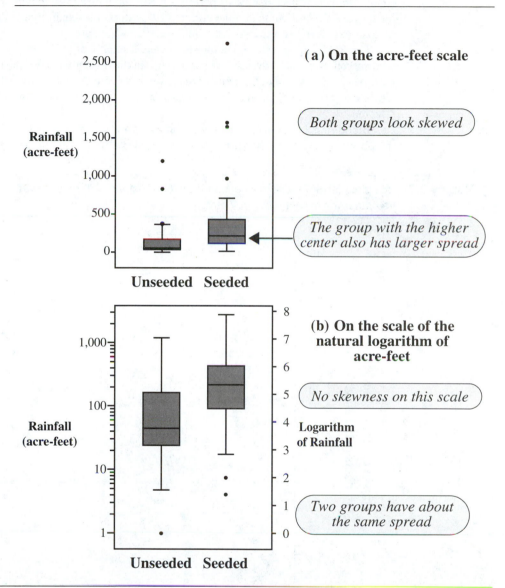

(a) **On the acre-feet scale**

Both groups look skewed

The group with the higher center also has larger spread

(b) **On the scale of the natural logarithm of acre-feet**

No skewness on this scale

Two groups have about the same spread

3.1.2 Effects of Agent Orange on Troops in Vietnam—An Observational Study

Many Vietnam veterans are concerned that their health may have been affected by exposure to Agent Orange, a herbicide sprayed in South Vietnam between 1962 and 1970. The particularly worrisome component of Agent Orange is a dioxin called TCDD, which in high doses is known to be associated with certain cancers. Recent studies have shown that high levels of this dioxin can be detected 20 or more years after heavy exposure to Agent Orange. Consequently, as part of a series of studies, researchers from the Centers for Disease Control compared the current (1987) dioxin levels in living Vietnam veterans to the dioxin levels in veterans who did not serve in Vietnam.

The 646 Vietnam veterans in the study were a sample of U.S. Army combat personnel who served in Vietnam during 1967 and 1968, in the areas that were most heavily treated with Agent Orange. The 97 non-Vietnam veterans entered the Army between 1965 and 1971 and served only in the United States or Germany. Neither sample was randomly selected.

Blood samples from each man were analyzed for the presence of dioxin. Box plots of the observed levels are shown in Display 3.3. (Data from a graphical display in Centers

Display 3.3   Box plots of 1987 dioxin concentrations in 646 Vietnam veterans and 97 veterans who did not serve in Vietnam

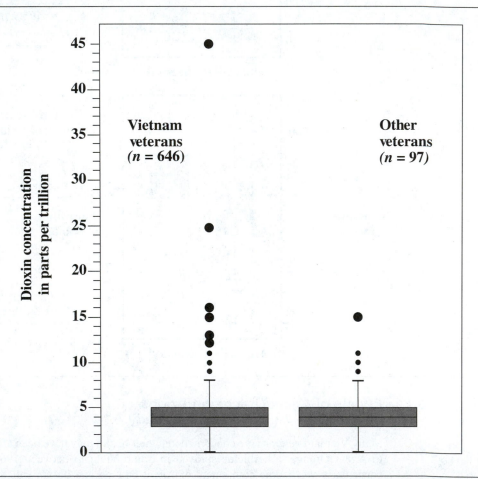

for Disease Control Veterans Health Studies, "Serum 2,3,7,8-Tetrachlorodibenzo-*p*-dioxin Levels in U.S. Army Vietnam-era Veterans," *Journal of the American Medical Association* 260 (September 2, 1988): 1249–54.) The question of interest is whether the distribution of dioxin levels tends to be higher for the Vietnam veterans than for the non-Vietnam veterans.

Summary of Statistical Findings

These data provide no evidence that the mean dioxin level in surviving Vietnam combat troops is greater than that for non-Vietnam veterans (one-sided *p*-value = .40, from a two-sample *t*-test). A 95% confidence interval for the difference in population means is −0.48 to 0.63 parts per trillion.

Scope of Inference

Since the samples were not random, inference to the populations is speculative. We must ask whether the participating veterans were representative of their respective groups. For example, nonparticipating Vietnam veterans may have failed to participate because of dioxin-related illnesses. If so, statistical statements about the populations of interest could be seriously biased.

3.2 Robustness of the Two-sample *t*-Tools

3.2.1 The Meaning of Robustness

The two-sample *t*-tools were used in the analyses of the Agent Orange study and the cloud seeding study, even though the actual conditions did not seem to match the ideal models upon which the tools are based. In the cloud seeding study, the *t*-tools were applied after taking the logarithms of the rainfalls. The *t*-tools could be used for the Agent Orange study, despite a lack of normality in the populations, because of the robustness of the *t*-tools against nonnormality.

> *A statistical procedure is **robust to departures from a particular assumption** if it is valid even when the assumption is not met.*

Valid means that the uncertainty measures—the confidence levels and the *p*-values—are very nearly equal to the stated rates. For example, a procedure for obtaining a 95% confidence interval is valid if it is 95% successful in capturing the parameter. It is robust against nonnormality if it is roughly 95% successful with nonnormal populations.

Robustness of a tool must be evaluated separately for each assumption. The following sections detail the robustness of the two-sample *t*-tools against departures from the ideal assumptions of the normal, equal standard deviation model.

3.2.2   Robustness Against Departures from Normality

The *Central Limit Theorem* asserts that averages based on large samples have approximately normal sampling distributions, regardless of the shape of the population distribution. This suggests that underlying normality is not a serious issue, as long as sample sizes are reasonably large. The theorem provides only partial reassurance of applicability with respect to *t*-tools. It states what the sampling distribution of an average should be, but it does not address the effects of estimating a population standard deviation. Many empirical investigations and related theory, however, confirm that the *t*-tools remain reasonably valid in large samples, with many nonnormal populations.

How large is large enough? That depends on how nonnormal the population distributions are. Because distributions can differ from the normal in infinitely many ways, the question of sample size is difficult to answer. Statistical theory does say something fairly general about the relative effects of skewness and long-tailedness (*kurtosis*):

1. If the two populations have the same standard deviations and approximately the same shapes, and if the sample sizes are about equal, then the validity of the *t*-tools is affected moderately by long-tailedness and very little by skewness.
2. If the two populations have the same standard deviations and approximately the same shapes, but if the sample sizes are not approximately the same, then the validity of the *t*-tools is affected moderately by long-tailedness and substantially by skewness. The adverse effects diminish, however, with increasingly large sample sizes.
3. If the skewness in the two populations differs considerably, the tools can be very misleading with small and moderate sample sizes.

Computer simulations can clarify the role of sample sizes. To further investigate the effect of nonnormality on 95% confidence intervals, a computer was instructed to generate samples from the nonnormal distributions shown in Display 3.4. For each pair of generated samples, it computed the 95% confidence interval for the difference in population means and recorded whether the interval actually captured the *true* difference in population means. The actual percentage of successful intervals from 1,000 simulations is shown in Display 3.4 for each set of conditions examined. The purpose of the simulation is to identify combinations of sample sizes and nonnormally shaped distributions for which the confidence interval procedure has a success rate of nearly 95%.

Of the five distributions examined, only the long-tailed distribution appears to have success rates that are poor enough to cause potentially misleading statements—and even those are not too bad. This distribution can be recognized in practice by the presence of outliers. For the skewed distributions, however, the normality assumption does not appear to be a major concern even for small sample sizes, at least as long as the skewness is the same in the two populations and the sample sizes are roughly equal.

3.2.3   Robustness Against Differing Standard Deviations

More serious problems may arise when the standard deviations of the two population are substantially unequal. In this case, the pooled estimate of standard deviation does not

Display 3.4   Percentage of 95% confidence intervals that are successful when the two
populations are nonnormal (but same shape and SD, and equal sample sizes)
(each percentage is based on 1,000 computer simulations)

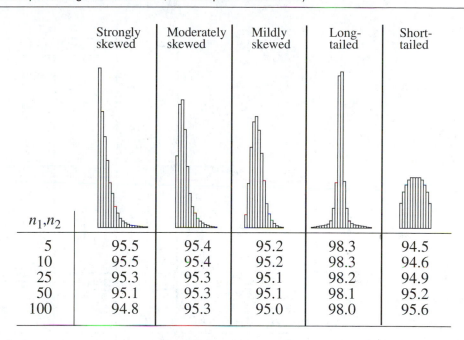

n_1, n_2	Strongly skewed	Moderately skewed	Mildly skewed	Long-tailed	Short-tailed
5	95.5	95.4	95.2	98.3	94.5
10	95.5	95.4	95.2	98.3	94.6
25	95.3	95.3	95.1	98.2	94.9
50	95.1	95.3	95.1	98.1	95.2
100	94.8	95.3	95.0	98.0	95.6

estimate any population parameter and the standard error formula, which uses the pooled
estimate of standard deviation, no longer estimates the standard deviation of the difference
between sample averages. As a result, the *t*-ratio does not have a *t*-distribution.

Theory shows that the *t*-tools remain fairly valid when the standard deviations are
unequal, as long as the sample sizes are roughly the same. For clarification, a computer
was again instructed to generate pairs of samples, this time from two normal populations
with different standard deviations, as shown in Display 3.5. It computed 95% confidence
intervals for each pair of samples and recorded whether the resulting interval successfully
captured the true difference in population means. The actual percentages successful are
displayed.

Notice that the success rates for the rows with equal sample sizes ($n_1 = n_2 = 10$ and
$n_1 = n_2 = 100$) are very nearly 95%. Thus, as suggested by theory, unequal population
standard deviations have little effect on validity if the sample sizes are equal. For substantially different σ's and different n's, however, the confidence intervals are unreliable.
The worst situation is when the ratio of standard deviations is much different from one and
the smaller sized sample is from the population with the larger standard deviation (as, for
example, when $n_1 = 100$, $n_2 = 400$, and $\sigma_1 = 4\sigma_2$).

Display 3.5   Percentage of successful 95% confidence intervals when the two populations have different standard deviations (but are normal) with possibly different sample sizes (each percentage is based on 1,000 computer simulations)

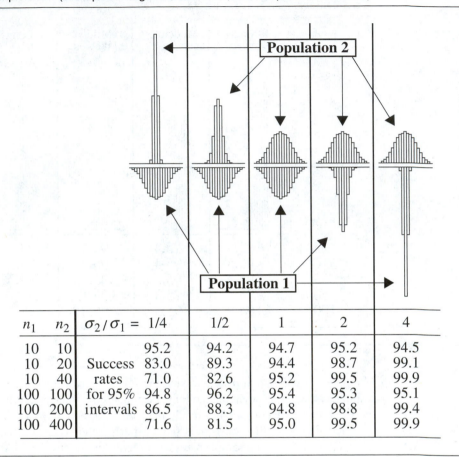

n_1	n_2	$\sigma_2 / \sigma_1 =$	1/4	1/2	1	2	4
10	10		95.2	94.2	94.7	95.2	94.5
10	20	Success	83.0	89.3	94.4	98.7	99.1
10	40	rates	71.0	82.6	95.2	99.5	99.9
100	100	for 95%	94.8	96.2	95.4	95.3	95.1
100	200	intervals	86.5	88.3	94.8	98.8	99.4
100	400		71.6	81.5	95.0	99.5	99.9

3.2.4 Robustness Against Departures from Independence

Cluster Effects and Serial Effects

Whenever knowledge that one observation is, say, above average allows an improved guess about whether another observation will be above average, *independence* is lacking. Two types of dependence (lack of independence) commonly arise in practical problems.

The first is a *cluster effect*, which sometimes occurs when the data have been collected in subgroups. For example, 50 experimental animals may have been collected from 10 litters and then randomly assigned to one of two treatment groups. Since animals from the same litter may tend to be more similar in their responses than animals from different litters, it is likely that independence is lacking.

The other type of dependence commonly encountered is caused by a *serial effect*, in which measurements are taken over time and observations close together in time tend to be more similar (or perhaps more different) than observations collected at distant time points. This can also occur if measurements are made at different locations, and measurements physically close to each other tend to be more similar than those farther apart. In the latter case the dependence pattern is called *spatial correlation*.

The Effects of Lack of Independence on the Validity of the *t*-Tools

When the independence assumptions are violated, the standard error of the difference of averages is an inappropriate estimate of the standard deviation of the difference in averages. The *t*-ratio no longer has a *t*-distribution, and the *t*-tools may give misleading results. The seriousness of the consequences depends on the seriousness of the violation. It is generally unwise to use the *t*-tools directly if cluster or serial effects are suspected. Other methods that adjust for these effects are available.

3.3   Resistance of the Two-sample *t*-Tools

Some practical suggestions will soon be provided for sizing up the actual conditions and choosing a course of action. The effect of outliers on the *t*-tools is discussed first, however, since decisions about how to deal with the outliers play an important role in the over-all strategy.

3.3.1   Outliers and Resistance

An *outlier* is an observation judged to be far from its group average. The effect of outliers on the two-sample *t*-tools has partially been addressed in the discussion of robustness. In fact, it is evident from the theoretical results and the computer simulations in Display 3.4 that the *p*-values and confidence intervals may be unreliable if the population distributions are long-tailed. Since long-tailed distributions are characterized by the presence of outliers in the sample, outliers should cause some concern.

Long-tailed population distributions are not the only explanation for outliers, however. The populations of interest may be normal but the sample may be contaminated by one or more observations that do not come from the population of interest. Often it is philosophically difficult and practically irrelevant to distinguish between a natural long-tailed distribution and one that includes outliers that result from contamination, although in some cases the identification of clear contamination may dictate an obvious course of action. For example, if it is discovered that one member of a sample from a population of 25- to 35-year old women is, in fact, over 50 years old, she should be removed from the sample.

It is useful to know how sensitive a statistical procedure may be to one or two outlying observations. The notion of resistance addresses this issue:

> *A statistical procedure is **resistant** if it does not change very much even when a small part of the data changes, perhaps drastically.*

As an example, consider the hypothetical sample: 10, 20, 30, 50, 70. The sample average is 36, and the sample median is 30. Now change the 70 to 700, and what happens? The sample average becomes 162, but the sample median remains 30. The sample average is not a resistant statistic because it can be severely influenced by a single observation. The median, however, is resistant.

Resistance is a desirable property. A resistant procedure is insensitive to outliers. A nonresistant one, on the other hand, may be greatly influenced by one or two outlying observations.

3.3.2 Resistance of *t*-Tools

Since *t*-tools are based on averages, they are not resistant. A small portion of the data can potentially have a major influence on the results. In particular, one or two outliers can affect a confidence interval or change a *p*-value enough to completely alter a conclusion.

If the outlier is due to contamination from another population, it can lead to false impressions about the population of interest. If the outlier does come from the population of interest, which happens to be long-tailed, the outcome is still undesirable for the following reason. In statistics, the goal is to describe *group* characteristics. An estimate of the center of a distribution should represent the typical value. The estimate is a good one if it represents the typical values possessed by the great majority of subjects; it is a bad one if it represents a feature unique to one or two subjects. Furthermore, a conclusion that hinges on one or two data points must be viewed as quite fragile.

3.4 Practical Strategies for the Two-sample Problem

Armed with information about the broad set of conditions under which the *t*-tools work well and the effect of outliers, the challenge to the data analyst is to size up the actual conditions using the available data and evaluate the appropriateness of the *t*-tools. This involves thinking about possible cluster and serial effects; evaluating the suitability of the *t*-tools by examining graphical displays; and considering alternatives.

In considering alternatives it is important to realize that even though the *t*-tools may still be valid when the ideal assumptions are not met, an alternative procedure that is more *efficient* (i.e., makes better use of the data) may be available. For example, another procedure may provide a narrower confidence interval.

Consider Serial and Cluster Effects

To detect lack of independence, carefully review the method by which the data were gathered. Were the subjects selected in distinct groups? Were different groups of subjects treated differently in a way that was unrelated to the primary treatment? Were different responses merely repeated measurements on the same subjects? Were observations taken at different but proximate times or locations? Affirmative answers to any of these questions suggest that independence may be lacking.

The principal remedy is to use a more sophisticated statistical tool. Identifiable clusters, which may be planned or unplanned, can be accounted for through analysis of variance (Chapters 13 and 14) or possibly through regression analysis (Chapters 9–12). Serial effects require time series analysis, the topic of Chapter 15.

Evaluate the Suitability of the t-Tools

Side-by-side histograms or box plots of the two groups of data should be examined and departures from the ideal model should be considered in light of the robustness properties of the t-tools. It is important to realize that the conditions of interest, which are those of the populations, must be investigated through graphical displays of the samples.

If the conditions do not appear suitable for use of the t-tools then some alternative is necessary. A transformation should be considered if the graphical displays of the transformed data appear to be closer to the ideal conditions. (See Section 3.5.) Alternatives tools for analyzing two independent samples are the rank-sum procedure, which is resistant and does not depend on normality (Section 4.2); other permutation tests (Section 4.3.1); and the Welch procedure for comparing normal populations that have unequal standard deviations (Section 4.3.2).

A Strategy for Dealing with Outliers

If investigation reveals that an outlying observation was recorded improperly or was the result of contamination from another population, the solution is to correct it if the right value is known or to leave it out. Often, however, there is no way to know how the outliers arose. Two statistical approaches for dealing with this situation exist. One is to employ a resistant statistical tool, in which case there is no compelling reason to ponder whether the offending observations are natural, the result of contamination, or simply blunders. (The rank-sum procedure in Section 4.2 is resistant.) The other approach is to adopt the careful examination strategy shown below. An important aspect of adopting this procedure is that an outlier does not get swept under the rug simply because it is different from the other observations. To warrant its removal, an explanation for why it is different must be established.

Outlier Strategy

1. Carry out the statistical analysis with and without the outlying observation.
2. If the statistical conclusions do not exhibit an important change when the questionable cases are removed, leave them in the data set and proceed.
3. If the conclusions do change, take further action:

   (a) Investigate the observations for possible explanations. Correct any recording errors.

   (b) If it can be established that the observations do not come from the population of interest, remove them.

Example—Agent Orange

Box plots of dioxin levels in Vietnam and non-Vietnam veterans (Display 3.3 on page 56) appear again in Display 3.6. The distributions have about the same shape and spread. Although the shape is not normal, the skewness is mild and unlikely to cause any problems with the t-test or the confidence interval. Two Vietnam veterans (#645 and #646) had considerably higher dioxin levels than the others.

Display 3.6 Outlier analysis for Agent Orange Data: effect of outliers on the p-value, for equal population means

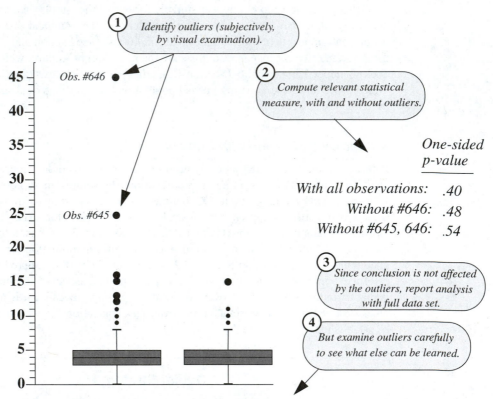

Veteran # 645: reported 180 days of indirect military exposure to herbicides.
Veteran # 646: reported no exposure (military or civilian) to herbicides.

From the results listed in Display 3.6 it is evident that the comparison of the two groups is changed very little by the removal of one or both of these outliers. Consequently, there is no need for further action. Even so, it is useful to see what else can be learned about these two, as indicated at the bottom of the display.

Notes

1. It is not useful to give a precise definition for an *outlier*. Subjective examination is the best policy. If there is any doubt about whether a particular observation deserves further examination, give it further examination.
2. It is not surprising that the outliers in the Agent Orange example have little effect, since the sample sizes are so large.
3. The apparent difference in the box plots may be due to the difference in sample sizes. If the population distributions are identical, more observations will appear in the extreme tails from a sample of size 646 than from a sample of size 97.

3.5   Transformations of the Data

3.5.1   The Logarithmic Transformation

The most useful transformation is the *logarithm* (log) for positive data. The common scale for scientific work is the *natural* logarithm, based on the number $e = 2.71828\ldots$ The logarithm of e is unity, denoted by $\log(e) = 1$. Also, the log of 1 is 0: $\log(1) = 0$. The general rule for using logarithms is that $\log(e^x) = x$. Another choice is the *common* logarithm based on the number 10, rather than e. Common logs are defined by: $\log_{10}(10^x) = x$. For statistical purposes, it does not matter which base is used. Unless otherwise stated, *log* in this book refers to the natural logarithm.

Recognizing the Need for a Log Transformation

The data themselves usually suggest the need for a log transformation. If the ratio of the largest to the smallest measurement in a group is greater than 10, then the data are probably more conveniently expressed on the log scale. Also, if the graphical displays of the two samples show them both to be skewed and if the group with the larger average also has the larger spread (see Display 3.2), the log transformation is likely to be a good choice.

In theory, the log transformation works as shown in Display 3.7. On the scale of measurement Y the two groups have skewed distributions with longer tails in the positive direction. The group with the larger center also has the larger spread. The measurements on the transformed scale have the same ordering, but small numbers get spread out more, while large numbers are squeezed more closely together. The overall result is that the two distributions on the transformed scale appear to be symmetric and have equal spread—just the right conditions for applying the t-tools.

3.5.2   Interpretation After a Log Transformation

For some measurements, the results of an analysis are appropriately presented on the transformed scale. Most users feel comfortable with the Richter scale for measuring earthquake strength, even though it is a logarithmic scale. Similarly, pH as a measure of acidity is the

Display 3.7   The logarithmic transformation used to arrive at favorable conditions for the two-sample *t*-analysis

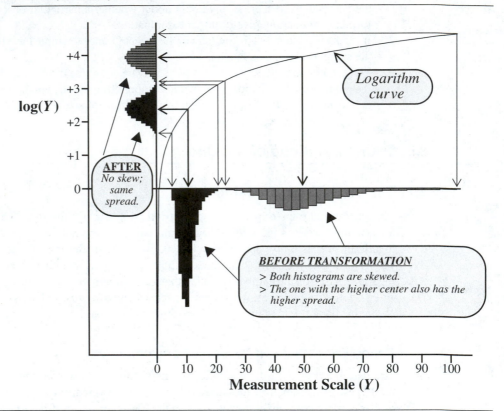

negative log of ion concentration. In other cases, however, it may be desirable to present the results on the original scale of measurement.

Randomized Experiment Model: Multiplicative Treatment Effect

If the randomized experiment model with additive treatment effect is thought to hold for the log-transformed data, then an experimental unit that would respond to treatment 1 with a logged outcome of $\log(Y)$ would respond to treatment 2 with a logged outcome of $\log(Y)+\delta$. By taking antilogarithms of these two quantities, one finds that an experimental unit that would respond to treatment 1 with an outcome of Y would respond to treatment 2 with an outcome of Ye^{δ}. Thus, e^{δ} is the *multiplicative treatment effect* on the original scale of measurement. To test whether there is any treatment effect, one performs the usual *t*-test for the hypothesis that δ is zero with the log-transformed data. To describe the multiplicative treatment effect, one back-transforms the estimate of δ and the endpoints of the confidence interval for δ.

> ### Interpretation After Log Transformation
> ### (Randomized Experiment)
>
> *It is estimated that a subject's response to treatment 2 is $\exp(\bar{Z}_2 - \bar{Z}_1)$ times as large as the subject's response to treatment 1 (where $\bar{Z}_i = $ average of $\log(Y_i)$)*

Example—Cloud Seeding

Display 3.2 shows that the log-transformed rainfalls have distributions that appear satisfactory for using the t-tools; so in Display 3.8 a full analysis is carried out on the log scale. Tests and confidence intervals are constructed in the usual way but on the transformed data. The estimate of the additive treatment effect on log rainfall is back-transformed to an estimate of the multiplicative effect of cloud seeding on rainfall.

Population Model: Estimating the Ratio of Population Medians

The t-tools applied to log-transformed data provide inferences about the difference in means of the logged measurements, which may be represented as $\text{Mean}[\log(Y_2)] - \text{Mean}[\log(Y_1)]$, where $\text{Mean}[\log(Y_2)]$ symbolizes the mean of the logged values of population 2. A problem with interpretation on the original scale arises because the mean of the logged values is not the log of the mean. Taking the antilogarithm of the estimate of the mean on the log scale does *not* give an estimate of the mean on the original scale.

If, however, the log-transformed data have symmetric distributions, the following relationships hold:

$$\text{Mean}[\log(Y)] = \text{Median}[\log(Y)]$$

(and since the log preserves ordering)

$$\text{Median}[\log(Y)] = \log[\text{Median}(Y)],$$

where $\text{Median}(Y)$ represents the *population median* (the 50th percentile of the population). In other words, the 50th percentile of the logged values is the log of the 50th percentile of the untransformed values. Putting these two equalities together, it is evident that the antilogarithm of the mean of the log values is the median on the original scale of measurements.

If $\bar{Z}_1$ and $\bar{Z}_2$ are used to represent the averages of the logged values for samples 1 and 2, then

$$\bar{Z}_2 - \bar{Z}_1 \text{ estimates } \log\left[\frac{\text{Median}(Y_2)}{\text{Median}(Y_1)}\right]$$

Display 3.8   Two-sample t-analysis and statement of conclusions after logarithmic transformation—cloud seeding example

(1) **Transform the data**

Unseeded		Seeded	
Y(acre-ft)	log(Y)	Y(acre-ft)	log(Y)
1202.6	7.092	2745.6	7.918
830.1	6.722	1697.8	7.437
372.4	5.920	1656.0	7.412
345.5	5.845	978.0	6.886
321.2	5.772	703.4	6.556
244.3	5.498	489.1	6.193
163.0	5.094	430.0	6.064
147.8	4.996	334.1	5.811
95.0	4.554	302.8	5.713
87.0	4.466	274.7	5.616
81.2	4.397	274.7	5.616
68.5	4.227	255.0	5.541
47.3	3.857	242.5	5.491
41.1	3.716	200.7	5.302
36.6	3.600	198.6	5.291
29.0	3.367	129.6	4.864
28.6	3.353	119.0	4.779
26.3	3.270	118.3	4.773
26.1	3.262	115.3	4.748
24.4	3.195	92.4	4.526
21.7	3.077	40.6	3.704
17.3	2.851	32.7	3.487
11.5	2.446	31.4	3.447
4.9	1.589	17.5	2.862
4.9	1.589	7.7	2.041
1.0	0.000	4.1	1.411

(2) **Use the two-sample t-tools on the log rainfall.**

Difference in averages = 1.1436 (SE = 0.4495).

Test of the hypothesis of no effect of cloud seeding on log rainfall: one-sided p-value from two-sample t-test = .0070 (50 df).

95% confidence interval for additive effect of cloud seeding on log rainfall: 0.2406 to 2.0467.

(3) **Back transform estimate and confidence interval.**

Estimate = $e^{1.1436}$ = 3.1382
Lower confidence limit = $e^{0.2406}$ = 1.2720.
Upper confidence limit = $e^{2.0467}$ = 7.7425.

(4) **State the conclusions on the original scale.**

Conclusion: *There is convincing evidence that seeding increased rainfall (one-sided p-value = .0070). The volume of rainfall produced by a seeded cloud is estimated to be 3.14 times as large as the volume that would have been produced in the absence of seeding (95% confidence: 1.27 to 7.74 times).*

and, therefore,

$$\exp(\overline{Z}_2 - \overline{Z}_1) \text{ estimates } \left[\frac{\text{Median}(Y_2)}{\text{Median}(Y_1)} \right].$$

The point of this is that a very useful multiplicative interpretation emerges in terms of the ratio of population medians. This is doubly important because the median is a better measure of the center of a skewed distribution than the mean. The multiplicative nature of this relationship is captured with the following wording:

> ## Interpretation After Log Transformation (Observational Study)
>
> *It is estimated that the median for population 2 is $\exp(\bar{Z}_2 - \bar{Z}_1)$ times as large as the median for population 1.*

In addition, back-transforming the ends of a confidence interval constructed on the log scale produces a confidence interval for the ratio of medians.

Example (Sex Discrimination)

Although the analysis of the sex discrimination data of Section 1.1.2, was suitable on the original scale of the untransformed salaries, graphical displays of the log-transformed salaries indicate that analysis would also be suitable on the log scale. The average male log salary minus the average female log salary is 0.147. Since $e^{0.147} = 1.16$, it is estimated that the median salary for males is 1.16 times as large as the median salary for females. Equivalently, the median salary for males is estimated to be 16% more than the median salary for females. Since a 95% confidence interval for the difference in means on the log scale is 0.100 to 0.194, a 95% confidence interval for the ratio of population median salaries is 1.11 to 1.21 ($e^{0.100}$ to $e^{0.194}$). With 95% confidence, it is estimated that the median salary for males is between 11% and 21% greater than the median salary for females.

3.5.3 Other Transformations for Positive Measurements

There are other useful transformations for positive measurements with skewed distributions where the means and standard deviations differ between groups. The *square root* transformation $\sqrt{Y}$ applies to data that are counts—counts of bacteria clusters in a dish, counts of traffic accidents on a stretch of highway, counts of red giants in a region of space—and to data that are measurements of area. The *reciprocal* transformation $1/Y$ applies to data that are waiting times—times to failure of lightbulbs, times to recurrence for cancer patients treated with radiation, reaction times to visual stimuli, and so on. The reciprocal of a time measurement can often be interpreted directly as a rate or a speed. The *arcsine square root* transformation, $\text{arcsine}(\sqrt{Y})$, and the logit transformation, $\log[Y/(1 - Y)]$, apply when the measurements are proportions between zero and one—proportions of trees infested by a wood-boring insect in experimental plots, proportions of weight lost as a side effect of leukemia therapy, proportions of winning lottery tickets in clusters of a certain size, and so forth.

Only the log transformation, however, gives such ease in converting inferences back to the original scale of measurement. One may estimate the difference in means of $\sqrt{Y_2}$ and $\sqrt{Y_1}$, but the square of this difference does not make much sense on the original scale.

Choosing a Transformation

Formal statistical methods are available for selecting a transformation. Nevertheless, it is recommended here that a trial-and-error approach, with graphical analysis, be used instead.

For positive data in need of a transformation, the logarithm should almost always be the first tried. If it is not satisfactory, the reciprocal or the square root transformations might be useful. Keep in mind that the primary goal is to establish a scale where the two groups have roughly the same spread. If several transformations are similar in their ability to accomplish this, think carefully about which one offers the most convenient interpretation.

Caveat About the Log Transformation

Situations arise where presenting results in terms of population medians is not sufficient. For example, the daily emissions of dioxin in the effluent from a paper mill have a very skewed distribution. An agency monitoring the emissions will be interested in estimating the total dioxin load released during, say, a year of operation. The total dioxin load would be the population mean times the population size, and therefore is estimated by the sample average times the population size. It cannot be estimated directly from the median, unless more specific assumptions are made.

3.6   Related Issues

3.6.1   Prefer Graphical Methods over Formal Tests for Model Adequacy

Formal tests for judging the adequacy of various assumptions exist. Tests for normality and tests for equal standard deviation are available in most statistical computer programs, as are tests that determine whether an observation is an outlier. Despite their widespread availability and ease of use, these diagnostic tests are not very helpful for model checking. They reveal little about whether the data meet the broader conditions under which the tools work well. The fact that two populations are not exactly normal, for example, is irrelevant. Furthermore, the formal tests themselves are often not very robust against their own model assumptions. Graphical displays are more informative, if less formal. They provide a good indication of whether or not the data are amenable to t-analysis and, if not, they often suggest a remedy.

3.6.2   Robustness and Transformation for Paired t-Tools

The one-sample t-test, of which the paired t-test is a special case, assumes that the observations are independent of one another and come from a normally distributed population. When cluster or serial effects are present (see Section 3.2.4), the t-tools may give misleading results. P-values and confidence intervals remain valid for moderate and large sample sizes for nonnormal distributions. For smaller sample sizes skewness can be a problem.

If the observations within each pair are positive and if the magnitudes of the differences tend to be larger for those pairs with larger average values, then a log transformation may help. The log transformation is applied *before* taking the difference. This is equivalent to forming a ratio from the measurements in each pair and performing a one-sample analysis on the logarithms of the ratios.

Suppose that there are n pairs (Y_{1i}, Y_{2i}). Let $Z_i = \log(Y_{1i}) - \log(Y_{2i})$ which is equivalent to $\log(Y_{1i}/Y_{2i})$. In an observational study, $\exp(\overline{Z})$ is an estimate of the ratio of medians of the respective populations. In a randomized, paired experiment, a model with additive treatment effect on the log scale represents a multiplicative treatment effect on the original scale. In both cases the interpretation after log transformation is the same as the interpretation for the two-sample analysis after log transformation.

Example—Schizophrenia

For the data described in Section 2.1.2, let Z_i represent the logarithm of the left hippocampus volume of the unaffected twin divided by the left hippocampus volume of the affected twin, for twin pair i. The average of the 15 Z's is 0.12849. A one-sample analysis gives a p-value of .0065 and a 95% confidence interval for the mean (of log ratios) of 0.0423 to 0.2147. By taking antilogarithms of the estimate and the endpoints of the confidence interval, the following interpretation emerges: it is estimated that the median left hippocampus volume for unaffected twins is 1.137 times as large as that for schizophrenic twins. A 95% confidence interval for the ratio of medians is from 1.043 to 1.239 times as large.

3.7 Summary

Cloud Seeding and Rainfall Study

The box plots of the rainfalls for seeded and unseeded days reveal that the two distributions of rainfall are skewed and that the distribution with the larger mean also has the larger variance. This is the situation where log-transformed data behave exactly in accordance with the ideal model. A plot of the data after transformation confirms the adequacy of the transformation. The two-sample t-test can be used as an approximation to the randomization test, and the difference in averages (of log rainfall) can be back-transformed to provide a statement about a multiplicative treatment effect: in the example, it is estimated that the rainfall is 3.1 times as much when a cloud is seeded as when it is left unseeded.

Since randomization is used, the statistical conclusion implies that the seeding causes the increase in rainfall. Since the decision about whether to seed clouds is determined (in this case) by a random mechanism, and since the airplane crew is *blind* to which treatment they are administering, human bias can have had little influence on the result.

Agent Orange Study

Graphical analysis focuses attention on the possibly undue influence of two outliers, but analyses with and without the outliers reveal no such influence, so the t-tools are used on the entire data set. The form of the sampling from the populations of living Vietnam veterans and of other veterans is a major concern in accepting the reliability of the statistical analysis. Protocols for obtaining the samples have not been discussed here, except to note that random sampling is not being used. Conclusions based on the two-sample t-test are supplied, along with the caveat that there may be biases due to the lack of random sampling.

Further Reading

Hoaglin's (1988) article on the use of transformations in everyday life accentuates the acceptability of using transformations. For more on the robustness of the two-sample *t*-tools (at a more sophisticated level), see Wetherill (1981) or Miller (1986). Cluster effects (which he calls block effects) and serial effect are discussed in the latter.

3.8 Exercises

Conceptual Exercises

1. Cloud Seeding. What is the experimental unit in the cloud seeding experiment?

2. Cloud Seeding. Randomization in the cloud seeding experiment was crucial in assessing the effect of cloud seeding on rainfall. Why?

3. Cloud Seeding. Why was it important that the airplane crew was unaware of whether seeding was conducted or not?

4. Cloud Seeding. Why would it be helpful to have the date of each observed rainfall?

5. Agent Orange. How would you respond to the comment that the box plots in Display 3.3 indicate that the dioxin levels in the Vietnam veterans tend to be larger since their values appear to be larger?

6. Agent Orange. (a) What course of action would you propose for the statistical analysis if it was learned that Vietnam veteran #646 (the largest observation in Display 3.6) worked for several years, after Vietnam, handling herbicides with dioxin? (b) What would you propose if this was learned instead for Vietnam veteran #645?

7. Agent Orange. If the statistical analysis had shown convincing evidence that the mean dioxin levels differed in Vietnam veterans and other veterans, could one conclude that serving in Vietnam was responsible for the difference?

8. Schizophrenia. In the schizophrenia study the observations in the two groups (schizophrenic and nonschizophrenic) are not independent since each subject is matched with a twin in the other group. Did the researchers make a mistake?

9. True or false? A statistical computer package will only print out a *p*-value or confidence interval if the conditions for its validity are met.

10. True or false? A sample histogram will have a normal distribution if the sample size is large enough.

11. A woman who has just moved to a new job in a new town discovers two routes to drive from her home to work. The first Monday, she flips a coin, deciding to take route A if it comes up heads and to take route B if it is tails. The following Monday, she will take the other route. The first Tuesday, she flips the coin again with the same plan. And so on for the first week. At the end of two weeks, she has traveled both routes five times and can compare their average commuting times. Why should she not use the *t*-tools for two independent samples? What should she use?

12. In which ways are the *t*-tools more robust for larger sample sizes than for smaller ones (i.e., robust with respect to normality, equal SDs, and/or independence)?

13. Fish Oil. Why is a log transformation inappropriate for the fish oil data in Display 1.14 on page 23?

14. Will an outlier from a contaminating population be more consequential in small samples or large samples?

15. What would you suggest as an alternative estimate of the standard deviation of the difference in sample averages when it is clear that the two populations have different SDs? (Check the formula for the standard deviation of the sampling distribution of the difference in averages, in Display 2.6 on page 36.)

16.   A researcher has taken tissue cultures from twenty-five subjects. Each culture is divided in half, and a treatment is applied to one of the halves chosen at random. The other half is used as a control. After determining the percent change in the sizes of all culture sections, the researcher calculates the standard error for the treatment-minus-control difference using both the paired t-analysis and the two independent sample (Chapter 2) t-analysis. Finding that the paired t-analysis gives a slightly larger standard error (and gives only half the degrees of freedom), the researcher decides to use the results from the unpaired analysis. Is this legitimate?

17.   Respiratory breathing capacity of individuals in houses with low levels of nitrogen dioxide was compared to the capacity of individuals in houses with high levels of nitrogen dioxide. From a sample of 200 houses of each type, breathing capacity was measured on 600 individuals from houses with low nitrogen dioxide and on 800 individuals from houses with high nitrogen dioxide. (a) What problem do you foresee in applying t-tools to these data? (b) Would comparing the average *household* breathing capacities avoid the problem?

18.   **Trauma and Metabolic Expenditure.** The following data are metabolic expenditures for 8 patients admitted to a hospital for reasons other than trauma and for 7 patients admitted for multiple fractures (trauma). (Data from C. L. Long, et al., "Contribution of Skeletal Muscle Protein in Elevated Rates of Whole Body Protein Catabolism in Trauma Patients," *American Journal of Clinical Nutrition* 34 (1981): 1087–93.)

Metabolic Expenditures (kcal/kg/day)

Nontrauma patients:	20.1	22.9	18.8	20.9	20.9	22.7	21.4	20.0
Trauma patients:	38.5	25.8	22.0	23.0	37.6	30.0	24.5

(a)   Is the difference in averages resistant? (*Hint:* What happens if 20.0 is replaced by 200?)

(b)   Replacing each value with its rank, from the lowest to highest, in the combined sample gives:

Metabolic Expenditures (kcal/kg/day)

Nontrauma group:	3	9	1	4.5	4.5	8	6	2
Trauma patients:	15	12	7	10	14	13	11

Consider the average of the ranks for the trauma group minus the average of the ranks for the nontrauma group. Is this statistic resistant?

19.   In each of the following data problems there is some potential violation of one of the independence assumptions. State whether there is a cluster effect or serial correlation, and whether the questionable assumption is the independence within groups or the independence between groups.

(a)   Researchers interested in learning the effects of speed limits on traffic accidents recorded the number of accidents per year for each of 10 consecutive years on roads in a state with speed limits of 90 km/h. They also recorded the number of accidents for the next 7 years on the same roads after the speed limit had been increased to 110 km/hr. The two groups of measurements are the number of accidents per year for those years under study. (Notice that there is also a potential confounding variable here!)

(b)   Researchers collected intelligence test scores on twins, one of whom was raised by the natural parents and one of whom was raised by foster parents. The data set consists of test scores for the two groups, boys raised by their natural parents and boys raised by foster parents.

(c)   Researchers interested in investigating the effect of indoor pollution on respiratory health randomly select houses in a particular city. Each house is monitored for nitrogen dioxide concentration and categorized as being either high or low on the nitrogen dioxide scale. Each member of the household is measured for respiratory health in terms of breathing capacity. The data set consists of these measures of respiratory health for all individuals from houses with low nitrogen dioxide levels and all individuals from houses with high levels.

Computational Exercises

20. Voltage and Insulating Fluid. Researchers examined the time in minutes before an insulating fluid lost its insulating property. The data below are the breakdown times for eight samples of the fluid, which had been randomly allocated to receive one of two voltages of electricity:

Times (min.) at 26 kV: 5.79   1579.52   2323.70

Times (min.) at 28 kV: 68.8   108.29   110.29   426.07   1067.60

(a)   Form two new variables by taking the logarithms of the breakdown times: $Y_1 = \log$ breakdown time at 26 kV and $Y_2 = \log$ breakdown time at 28 kV.

(b)   By hand, compute the difference in averages of the log-transformed data: $\overline{Y}_1 - \overline{Y}_2$.

(c)   Take the antilogarithm of the estimate in (b): $\exp(\overline{Y}_1 - \overline{Y}_2)$. What does this estimate? (See the interpretation for the randomized experiment model in Section 3.5.2. on p. 66.)

(d)   By hand, compute a 95% confidence interval for the difference in mean log breakdown times. Take the antilogarithms of the endpoints and express the result in a sentence.

21. Solar Radiation and Skin Cancer. The data in Display 3.9 are yearly skin cancer rates (cases per 100,000 people) in Connecticut, with a code identifying those years that came two years after higher than average sunspot activity and those years that came two years after lower than average sunspot activity. (Data from D. F. Andrews and A. M. Herzog, *Data*, New York: Springer-Verlag, 1985). (a) Is there any reason to suspect that using the two independent sample t-test to compare skin cancer rates in the two groups is inappropriate? (b) Draw scatterplots of skin cancer rates versus year, for each group separately. Are any problems indicated by this plot?

22. Sex Discrimination. With a statistical computer program, reanalyze the sex discrimination data in Display 1.3 on page 4, but use the log transformation of the salaries. (a) Draw box plots. (b) Find a p-value for comparing the distributions of salaries. (c) Find a 95% confidence interval for the ratio of population medians. Write a sentence describing the finding.

23. Agent Orange. With a statistical computer program, reanalyze the Agent Orange data of Display 3.3 with and without the two largest dioxin levels in the Vietnam veterans group. Verify the one-sided p-values in bubble 2 of Display 3.6.

24. Agent Orange. With a statistical computer package, reanalyze the Agent Orange data of Display 3.3 after taking a log transformation. Since the data set contains zeros—for which the log is undefined—try the transformation $\log(\text{dioxin} + .5)$. (a) Draw side-by-side box plots of the transformed variable. (b) Find a p-value from the t-test for comparing the two distributions. (c) Compute a 95% confidence interval for the difference in mean log measurements and interpret it on the original scale. (*Note:* Back-transforming does not provide an exact estimate of the ratio of medians since .5 was added to the dioxins; but it does provide an approximate one.)

Display 3.9 Connecticut skin cancer rates (per 100,000 people) from 1938 to 1972, with solar code (1 if there was higher than average sunspot activity and 2 if there was lower than average sunspot activity two years earlier)

Year	Rate	Code	Year	Rate	Code
1938	0.8	2	1955	2.9	2
1939	1.3	1	1956	2.5	2
1940	1.4	1	1957	2.6	2
1941	1.2	1	1958	3.2	2
1942	1.7	2	1959	3.8	1
1943	1.8	2	1960	4.2	1
1944	1.6	2	1961	3.9	1
1945	1.5	2	1962	3.7	1
1946	1.5	2	1963	3.3	2
1947	2.0	2	1964	3.7	2
1948	2.5	1	1965	3.9	2
1949	2.7	1	1966	4.1	2
1950	2.9	1	1967	3.8	2
1951	2.5	1	1968	4.7	2
1952	3.1	1	1969	4.4	2
1953	2.4	2	1970	4.8	1
1954	2.2	2	1971	4.8	1
			1972	4.8	1

25. Life Expectancy and Per Capita Income. The life expectancy and per capita income are given in Display 3.10 for 20 industrialized countries and 9 petroleum exporting countries. (Data from S. Leinhardt and S. S. Wasserman, "Teaching Regression: An Exploratory Approach," *Technometrics* 33 (1979): 196–203. Note that there is a missing value for South Africa.)

(a) (i) Draw side-by-side box plots of per capita income for industrialized countries and for petroleum exporting countries; describe the distributions and their differences. (ii) Draw side-by-side box plots for the two groups after taking logarithms. (iii) Obtain an estimate for the difference in mean log(per capita income) between the two groups. (iv) Take the antilog of the estimate in (iii) and describe the resulting estimate in a sentence. (v) Obtain a 95% confidence interval for the difference in mean log(per capita income) between the two groups. (vi) Take the antilogs of the endpoints and summarize the interval in a sentence.

(b) (i) Draw side-by-side box plots of life expectancy for the two groups; describe the distributions and their differences. (ii) Why might the variability be smaller for the industrialized countries than for the petroleum exporting countries? (iii) Can this difference in variation be remedied by a transformation? (iv) Do the *t*-tools appear appropriate for spreads that differ by this much? (Refer to Display 3.5: find the cell of the table that most closely matches the situation of these data and assess the adequacy of the *t*-tools for that situation.)

26. Pollen Removal. As part of a study to investigate reproductive strategies in plants, biologists recorded the time spent at sources of pollen and the proportions of pollen removed by bumblebee queens and honeybee workers pollinating a species of lily. (Data from L. D. Harder and J. D. Thompson, "Evolutionary Options for Maximizing Pollen Dispersal of Animal-pollinated Plants," *American Naturalist* 133 (1989): 323–44.) Their data appear in Display 3.11.

Display 3.10   Life expectancy and per capita income in petroleum exporting countries and industrialized countries (na = "not available")

| | Industrialized Country | | | Petroleum Exporting Country | |
	Life Expectancy (years)	Per Capita Income (1974 $U.S.)		Life Expectancy (years)	Per Capita Income (1974 $U.S.)
Australia	71.0	3,426	Algeria	50.7	430
Austria	70.4	3,350	Ecuador	52.3	360
Belgium	70.6	3,346	Indonesia	47.5	110
Canada	72.0	4,751	Iran	50.0	1,280
Denmark	73.3	5,029	Iraq	51.6	560
Finland	69.8	3,312	Libya	52.1	3,010
France	72.3	3,403	Nigeria	36.9	180
West Germany	70.3	5,040	Saudi Arabia	42.3	1,530
Ireland	70.7	2,009	Venezuela	66.4	1,240
Italy	70.6	2,298			
Japan	73.2	3,292			
Netherlands	73.8	4,103			
New Zealand	71.1	3,723			
Norway	73.9	4,102			
Portugal	68.1	956			
South Africa	68.2	na			
Sweden	74.7	5,596			
Switzerland	72.1	2,963			
Britain	72.0	2,503			
United States	71.3	5,523			

(a)   (i) Draw side-by-side box plots (or histograms) of the proportion of pollen removed by queens and workers. (ii) When the measurement is the proportion P of some amount, one useful transformation is $\log[P/(1 - P)]$. This is the log of the ratio of the proportion removed to the proportion not removed. Draw side-by-side box plots or histograms on this transformed scale. (iii) Test whether the distribution of proportions removed is the same or different for the two groups, using the t-test on the transformed data.

(b)   Draw side-by-side box plots of duration of visit on (i) the natural scale, (ii) the logarithmic scale, and (iii) the reciprocal scale. (iv) Which of the three scales seems most appropriate for use of the t-tools? (v) Compute a 95% confidence interval to describe the difference in means on the chosen scale. (vi) What are relative advantages of the three scales as far as interpretation goes? (vii) Based on your experience with this problem, comment on the difficulty in assessing equality of population standard deviations from small samples.

27. **Bumpus's Data.** Obtain p-values from the t-test to compare humerus lengths for sparrows that survived and those that perished (Display 2.1 on page 28), with and without the smallest length in the perished group (length = 0.659 inch). Do the conclusions depend on this one observation? What action should be taken if they do?

28. **Cloud Seeding—Multiplicative vs. Additive Effects.** On the computer, create a variable containing the rainfall amounts for only the unseeded days. (a) Create four new variables by adding 100, 200, 300, and 400 to each of the unseeded day rainfall amounts. Display a set of five box plots to illustrate what one might expect if the effect of seeding were additive. (b) Create four additional

Display 3.11 Proportions of pollen removed and visit durations by bumblebee queens and by honeybee workers

Bumblebee Queens		Honeybee Workers	
Proportion of Pollen Removed	Duration of Visit (seconds)	Proportion of Pollen Removed	Duration of Visit (seconds)
.07	2	.28	3
.10	5	.37	12
.11	7	.52	10
.12	11	.65	17
.15	12	.76	24
.19	11	.89	33
.28	9	.74	44
.31	9	.70	46
.30	16	.79	48
.34	17	.78	51
.35	12	.74	64
.39	14	.77	78
.38	23		
.40	35		
.42	21		
.40	10		
.41	9		
.42	7		
.48	11		
.48	13		
.47	14		
.49	16		
.50	14		
.51	17		
.53	22		
.58	13		
.59	13		
.65	12		
.60	19		
.60	23		
.69	21		
.70	27		
.70	28		
.51	58		
.70	15		

	Proportion of Pollen Removed	Duration of Visit (seconds)
Honeybee Workers ($n = 12$)		
Ave:	.666	35.8
SD:	.183	23.2
Bumblebee Queens ($n = 35$)		
Ave:	.426	16.2
SD:	.182	10.0

variables by multiplying each of the unseeded day rainfall amounts by 2, by 3, by 4, and by 5. Display a set of five box plots to illustrate what could be expected if the effect of seeding were multiplicative. (c) Which set of plots more closely resembles the actual data?

Data Problems

29. Iron Supplementation. A randomized experiment was performed on mice to determine whether two forms of iron, Fe^{3+} and Fe^{4+}, are retained differently. If one type is retained especially well, then it may be more useful as a dietary supplement for humans. (Data from J. Rice, *Mathematical*

Statistics and Data Analysis (1987), p. 356.) The mice were given the iron orally. The iron was radioactively labeled so that the initial amount and the amount retained after a fixed time interval could be measured. The measurements of interest (Display 3.12) are the percentages of iron retained in each mouse after the time period had elapsed. The researchers wished to know if there is higher retention for one type than for the other. If so, which type? and by how much? Use graphical and numerical methods to analyze the data. Answer the questions of interest. State your answers in one or two concise sentences that contain appropriate measures of uncertainty. Try to hold statistical jargon to a minimum.

Display 3.12   Percentages of two forms of dietary iron retained by mice

Fe^{3+} Supplement			Fe^{4+} Supplement		
0.71	2.56	4.39	2.20	4.27	6.97
1.66	2.60	4.50	2.69	4.53	6.97
2.01	3.31	5.07	3.54	5.32	7.52
2.16	3.64	5.26	3.75	6.18	8.36
2.42	3.74	8.15	3.83	6.22	11.65
2.42	3.74	8.24	4.08	6.33	12.45

30. College Tuition. Display 3.13 shows the 1993–1994 tuitions for a sample of 20 private universities and both in-state and out-of-state tuition for a sample of 20 public universities. Analyze the data to describe (a) the amount by which out-of-state tuition tends to exceed in-state tuition for

Display 3.13   Tuition in dollars for 20 private and 20 public U.S. colleges and universities

Private School	Tuition	Public School	In-state	Out-of-state
Rust College	4,000	East Carolina	764	7,248
Stephens College	13,900	CUNY — Queens College	2,650	5,250
Newberry College	10,194	Souther Connecticut State	1,842	5,962
Grinnell College	15,688	Indiana University	2,984	9,766
North Central College	11,718	Montana State	1,576	5,552
Drexel University	12,348	North Carolina, Wilmington	1,492	7,558
Bethel College	9,300	Maryland, Baltimore County	3,338	8,594
Lees-McRae College	3,807	Kansas State	1,713	6,995
Centenary College	11,400	Eastern Illinois	1,902	5,710
Muhlenberg College	16,975	Clinch Valley	3,093	7,168
Goucher College	15,588	Dakota State	1,323	3,222
Ithaca College	14,424	North Dakota	2,110	5,634
University of the Arts	12,520	Illinois, Urbana	2,760	7,560
Benedictine College	9,550	Wayne State	3,271	7,162
Rose-Hulman Inst. of Tech.	13,380	Valdosta State	1,770	4,617
Columbia College	10,425	Florida Atlantic	1,800	6,700
Liberty University	5,700	Edinboro University of Pennsylvania	3,086	7,844
Emory University	17,600	Montevallo	2,220	4,440
College of Wooster	16,240	Arkansas, Little Rock	2,136	5,352
College of Santa Fe	11,138	Arkansas, Fayetteville	1,932	5,028

public universities and (b) the amount by which private university tuition tends to exceed public university out-of-state tuition. (Data from 1995 *U.S. News and World Report's Guide to America's Best Colleges.*)

31. Brain Size and Litter Size. Display 3.14 shows relative brain weights (brain weight divided by body weight) for 51 species of mammal whose average litter size is less than 2 and for 45 species of mammal whose average litter size is greater than or equal to 2. (These are part of a larger data set considered in Section 9.1.2.) What evidence is there that brain sizes tend to be different for the two groups? How big of a difference is indicated? Include the appropriate statistical measures of uncertainty in carefully worded sentences to answer these questions.

Display 3.14 Relative brain sizes, 1,000 × (Brain weight/Body weight), for 96 species of mammals

---

1,000×(Brain weight/Body weight) for 51 species with average litter size < 2

0.42	0.86	0.88	1.11	1.34	1.38	1.42	1.47	1.63	1.73	2.17	2.42
2.48	2.74	2.74	2.79	2.90	3.12	3.18	3.27	3.30	3.61	3.63	4.13
4.40	5.00	5.20	5.59	7.04	7.15	7.25	7.75	8.00	8.84	9.30	9.68
10.32	10.41	10.48	11.29	12.30	12.53	12.69	14.14	14.15	14.27	14.56	15.84
18.55	19.73	20.00									

1,000×(Brain weight/Body weight) for 45 species with average litter size ≥ 2

0.94	1.26	1.44	1.49	1.63	1.80	2.00	2.00	2.56	2.58	3.24	3.39
3.53	3.77	4.36	4.41	4.60	4.67	5.39	6.25	7.02	7.89	7.97	8.00
8.28	8.83	8.91	8.96	9.92	11.36	12.15	14.40	16.00	18.61	18.75	19.05
21.00	21.41	23.27	24.71	25.00	28.75	30.23	35.45	36.35			

Answers to Conceptual Exercises

1. The target clouds on a day that was deemed suitable for seeding.

2. Uncontrollable confounding factors probably explain the variability in rainfall from clouds treated the same way. Randomization is needed to ensure that the confounding factors do not tend to be unevenly distributed in the two groups.

3. Blinding prevents the intentional or unintentional biases of the human investigators from having a chance to make a difference in the results.

4. There may be serial correlation. A plot of rainfall versus data could be used to check.

5. Larger values are to be expected by chance if the populations are the same, since the sample of Vietnam veterans is so much larger than the sample of non-Vietnam veterans.

6. (a) He would not be representative of the target population and should be removed from the data set for analysis. (b) Same thing.

7. No, not from the statistics alone since this is an observational study. It could be said, however, that the data are consistent with that theory.

8. No. The dependence is the result of matching and is desirable. The two-sample *t*-tools are not appropriate (but the paired *t*-tools are).

9. False.

10. False. An *average* from a sample will have a sampling distribution that will tend toward normal with large sample sizes, but the sample histogram should mirror the population distribution. As the sample size gets larger, the sample histogram should become a better approximation to the population histogram.

11. There is a cluster effect: the particular day of the week. She should use a paired-t analysis, as will be discussed in Chapter 4.

12. The t-test is robust in validity to departures from normality, especially as the sample size gets large. The robustness with respect to equal standard deviations does not depend much on what the sample sizes are, so long as they are reasonably equal. Sample size does not affect robustness with respect to independence.

13. You cannot take logarithms of negative numbers.

14. It will be more consequential in smaller samples; its effect gets washed out in large ones.

15. Replace the population SDs in the formula (Section 2.2.2) by individual *sample* SDs.

16. No. The paired analysis must be used, even though the inferences may not appear to be as precise. The unpaired analysis is inappropriate.

17. (a) Dependence of measurements on individuals in the same household (cluster effect). (b) Maybe. Getting a single measure for each household may be an easy way out of the dependence problem, but care should be used as these groups also tend to differ in the average number of persons per household.

18. Yes.

19. (a) Serial correlation both within and between groups. (Confounding variable is the time at which observations were made.) (b) Cluster effect between groups. (c) Cluster effect (members of the same household should be similar) within groups.

Alternatives to the *t*-Tools

The *t*-tools have an extremely broad range of application, extending well beyond the strict confines of the ideal model because of robustness. It extends even farther when the possibilities of transforming the data and dealing with outliers are taken into account. With little modification, the *t*-tools apply to problems in regression, to analysis of variance, and indeed to virtually every facet of statistical analysis.

Nevertheless, situations arise where the *t*-tools cannot be applied, because the model assumptions of the *t*-test are blatantly violated. For these situations, a host of other methods, based on different models, may be used. Some are presented in this chapter. Most notable are two distribution-free methods, based on models that do not specify any particular population distributions. The rank-sum test for two independent samples and the Wilcoxon signed-rank test for a sample of pairs are useful alternatives, particularly when outliers may be present or when the sample sizes are too small to permit the assessment of distributional assumptions.

4.1   Case Studies

4.1.1   Space Shuttle O-Ring Failures—An Observational Study

On January 27, 1986, the night before the space shuttle Challenger exploded, engineers at the company that built the shuttle warned National Aeronautics and Space Administration (NASA) scientists that the shuttle should not be launched because of predicted cold weather. Fuel seal problems, which had been encountered in earlier flights, were suspected of being associated with low temperatures. It was argued, however, that the evidence was inconclusive. The decision was made to launch, even though the temperature at launch time was 29°F.

The data in Display 4.1 are the numbers of O-ring incidents on previous shuttle flights, categorized into those launched at temperatures below 65°F and those launched at temperatures above 65°F. (Data from a graph in Richard P. Feynman *What Do You Care What Other People Think?* (New York: W. W. Norton, 1988).) Is there a higher risk of O-ring incidents at lower launch temperatures?

Display 4.1   Numbers of O-ring incidents on 24 space shuttle flights prior to the Challenger disaster

Launch Temperature	Number of O-Ring Incidents
Below 65° F	1 1 1 3
Above 65° F	0 0 0 0 0 0 0 0 0 0 0 0 0 0 0 0 0 1 1 2

Summary of Statistical Findings

There is strong evidence that the number of O-ring incidents was associated with launch temperature in these 24 launches. It is highly unlikely that the observed difference of the groups is due to chance (one-sided *p*-value = .0099 from a permutation test on the *t*-statistic).

Scope of Inference

These observational data cannot be used to establish causality, nor is there any broader population of which they are a sample. But the association between temperature and O-ring failure in these particular 24 launches is consistent with the theory that lower temperatures impair the functioning of the O-rings. (At one point in public hearings into the causes of the disaster, Feynman asked for a glass of ice water, placed a small O-ring in it for a time, removed it, and then proceeded to demonstrate that the rubber failed to spring back to its original form.) *Note*: Other techniques for dealing with count data are given in Chapter 22.

4.1.2 Cognitive Load Theory in Teaching— A Randomized Experiment

Consider the following problem in coordinate geometry:

> Point A has coordinates (2, 1), point B has (8, 3), and point C has (4, 6).
> What is the slope of the line that connects C to the midpoint between A and B?

Presenting the solution as a worked problem, a conventional textbook shows a picture of the layout, gives a discussion in the text, and then provides the lines of algebraic manipulation leading to the right answer. (See Display 4.2.) Recent theoretical developments in cognitive science suggest that splitting the presentation into the three distinct units of diagram, text, and algebra imposes a heavy, extraneous cognitive load on the student. The requirement that the student organize and process the separate elements constitutes a cognitive load. The load is extraneous because it is not essential to learning how to solve such problems— indeed, it impedes the learning process by placing heavy demands on cognitive resources that should be used to understand the essentials.

Display 4.2   Cognitive load experiment: conventional method of instruction (for finding the slope of the line that connects C to the midpoint between A and B)

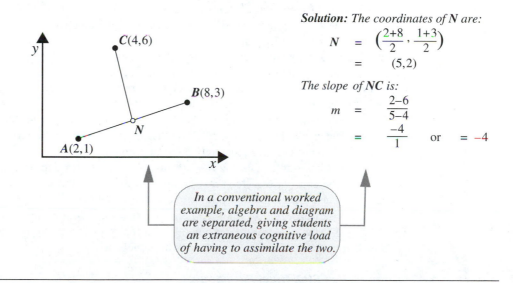

Solution: *The coordinates of N are:*

$$N = \left(\frac{2+8}{2}, \frac{1+3}{2}\right)$$
$$= (5,2)$$

The slope of NC is:

$$m = \frac{2-6}{5-4}$$
$$= \frac{-4}{1} \quad or \quad = -4$$

In a conventional worked example, algebra and diagram are separated, giving students an extraneous cognitive load of having to assimilate the two.

In a demonstration of this theory, researchers conducted a series of experiments comparing the effectiveness of conventional textbook worked examples to modified worked examples, which present the algebraic manipulations and explanation as part of the graphical display (see Display 4.3). (Data from J. Sweller, P. Chandler, P. Tierney, and

Display 4.3   Cognitive load experiment: modified method of instruction (for finding the slope of the line that connects *C* to the midpoint between *A* and *B*)

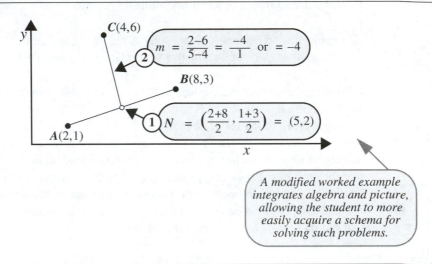

A modified worked example integrates algebra and picture, allowing the student to more easily acquire a schema for solving such problems.

M. Cooper, "Cognitive Load as a Factor in the Structuring of Technical Material," *Journal of Experimental Psychology General* 119(2) (1990): 176–92.)

They selected 28 ninth-year students in Sydney, Australia, who had no previous exposure to coordinate geometry but did have adequate mathematics to deal with the problems given. The students were randomly assigned to one of two self-study instructional groups, using conventional and modified instructional materials. The materials covered exactly the same problems, presented differently. Students were allowed as much time as they wished to study the material, but they were not allowed to ask questions. Following the instructional phase all students were tested with a common examination over three problems of different difficulty. The data in Display 4.4, based on this study, are the number of seconds required to arrive at a solution to the moderately difficult problem.

Both distributions in Display 4.4 are highly skewed. In addition, there were five students in the conventional (control) group who did not come to any solution in the 5 minutes allotted. Their solution times are considered *censored*—all that is known about them is that they exceed 300 seconds. It appears that the solution times for the "modified instructional materials" group are generally shorter than for the conventional materials group. Is there sufficient evidence to draw this conclusion?

Summary of Statistical Findings

There is convincing evidence that a student could solve the problem more quickly if taught with the modified method than if taught with the conventional method (two-sided *p*-value = .002, from the Wilcoxon rank-sum test). The sample median completion time for the 14 students taught by the modified method was 1 minute and 46 seconds, compared to 3 minutes and 55 seconds for those given the conventional method, a 55% reduction.

Display 4.4   Number of seconds to solution of a problem in coordinate geometry, for students instructed with conventional and modified materials

---

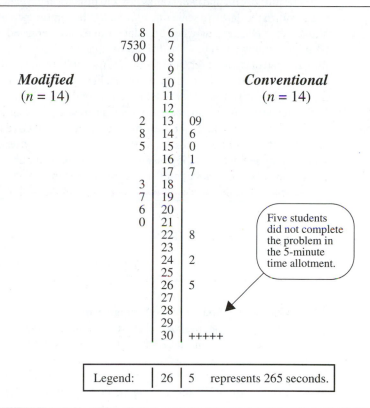

Legend: 26 | 5   represents 265 seconds.

4.2   The Rank-Sum Test

The *rank-sum test* is a resistant alternative to the two-sample *t*-test. It performs nearly as well as the *t*-test when the two populations are normal and considerably better when there are extreme outliers. Its drawbacks are that associated confidence intervals are not computed by most of the statistical computer packages and that it does not easily extend to more complicated situations.

4.2.1   The Rank Transformation

Straightforward transformations of the data were used in Chapter 3 to obtain measurements on a scale where the normality and equal spread assumptions are approximately met. The rank-sum relies on a special kind of transformation that replaces each observation by its rank in the combined sample. It should be noted that this really is a different kind of transformation for two reasons: (1) a single transformed value depends on all the data. So the transformed rank of, say, $Y = 63.925$, may be one value in one problem but a very different value in another problem. (2) There is no reverse transformation, such as the antilog, available for ranks.

The purpose behind using ranks is not to transform the data to approximate normality but to transform the data to a scale that eliminates the importance of the population distributions altogether. Notice that when the data values are replaced by their square roots or by their logarithms the sample distributions change shape dramatically. The ranks from the transformed data, however, are identical to the ranks based on the untransformed data, so the shape of the distribution of measurements has no effect on the ranks. In addition, whether the largest observation in a data set is 300 seconds or 3,000 seconds does not affect its rank. In this respect, any statistic based on ranks is resistant to outliers.

A feature of the rank-sum test that makes it an attractive choice for the cognitive load experiment is its ability to deal with *censored observations*, observations that are only known to be greater than (or possibly less than) some number. All that is known about the five students who did not complete the problem is that their solution times are greater than 300 seconds. For the rank-sum test it is enough to know that, in terms of ranks, they were tied for last.

4.2.2 The Rank-Sum Statistic

Calculation of the rank-sum test statistic for the cognitive load experiment is summarized in Display 4.5. The first four steps transform the data to their ranks in the combined sample:

1. List all observations from both samples in increasing order.
2. Identify which sample each observation came from.
3. Create a new column labeled "order," as a straight sequence of numbers from 1 to $(n_1 + n_2)$.
4. Search for ties—that is, duplicated values—in the combined data set. The ranks for tied observations are taken to be the average of the orders for those cases.

Two students tied at 80 seconds, for example. They finished sixth and seventh fastest. One cannot say which of the two deserves which order, so both are assigned the rank of $(6 + 7)/2 = 6.5$. The five students who took the full 5 minutes were the last five finishers, with orders 24, 25, 26, 27, and 28. So each is assigned rank $(24+25+26+27+28)/5 = 26$. Any observation that has a unique value gets its order as its rank.

The test statistic, T, is the sum of all the ranks in one group, called "group 1." Group 1 is conventionally the group with the smaller sample size (because that minimizes computation). The choice, however, is arbitrary.

The impressive feature of the cognitive load experiment is that the early finishers were mostly the students who studied the modified instructional material. This is reflected in the test statistic by a low rank-sum ($T = 137$) for that group.

4.2.3 Finding a *p*-Value by Normal Approximation

The rank-sum procedure is used to test the null hypothesis of no treatment effect from a two-treatment randomized experiment and also to test the null hypothesis of identical population distributions from two independent samples. If the null hypothesis is true in

Display 4.5   Rank-sum test statistic *T* for the cognitive load experiment

Y	Group	Order	Rank
68	M	1	1
70	M	2	2
73	M	3	3
75	M	4	4
77	M	5	5
80	M	6	6.5
80	M	7	6.5
130	C	8	8
132	M	9	9
139	C	10	10
146	C	11	11
148	M	12	12
150	C	13	13
155	M	14	14
161	C	15	15
177	C	16	16
183	M	17	17
197	M	18	18
206	M	19	19
210	M	20	20
228	C	21	21
242	C	22	22
265	C	23	23
300	C	24	26
300	C	25	26
300	C	26	26
300	C	27	26
300	C	28	26

1 List all observations from both samples in increasing order.

2 Identify sample membership of each value.

3 Write down an increasing sequence for list order.

4 Modify the orders, if necessary by giving the average order to tied values.

5 Add up the ranks from all observations in group 1.

$T = 137$

either case then the sample of n_1 ranks in group 1 is a random sample from the $n_1 + n_2$ available ranks. Both the randomization distribution and the sampling distribution of the rank-sum statistic, T, are represented by a histogram of the rank-sum statistics calculated for each possible regrouping of the $n_1 + n_2$ observations into samples of size n_1 and n_2.

Because conversion to ranks avoids absurd distributional anomalies, the randomization distribution of T can be approximated accurately by a normal distribution in most situations. The exceptions are when at least one sample is small—say under 5—or when large numbers of ties occur. The mean and variance of the permutation distribution, from statistical theory, are shown in Display 4.6.

These facts may be used to evaluate whether the observed rank-sum statistic is unusually small or large. If there is no difference, then the Z-statistic

$$Z\text{-stat} = [T - \text{Mean}(T)]/\text{SD}(T)$$

Display 4.6 Facts about the randomization (or sampling) distribution of the rank-sum statistic—the sum of ranks in group 1—when there is no group difference

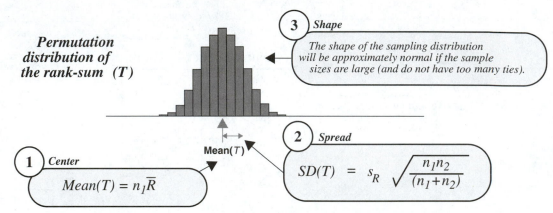

where $\bar{R}$ and s_R are the average and the sample standard deviation, respectively, for the combined set of (n_1+n_2) ranks.

should be similar to a typical value from a standard normal distribution. A *p*-value is the proportion of values from a standard normal distribution that is more extreme than the observed *Z*-statistic.

The calculations for the cognitive load data are shown in Display 4.7. The observed rank-sum statistic of 137 is smaller than the value of 203 that would be expected if there were no treatment effect. This disparity may be due to the luck involved in the random assignment, with better students placed in the modified instruction group, but the *p*-value indicates that only about one in a thousand randomizations would produce a disparity as great as or greater than the observed one, if there were no treatment effect. Such a small *p*-value provides fairly convincing evidence that a treatment effect explains the difference.

Continuity Correction

The permutation distribution of T is *discrete*, meaning T can only take on a finite set of values. A *continuity correction* is an adjustment used to improve the normal approximation to the discrete distribution of a test statistic. A continuity corrected *Z*-statistic is the same as in Step 3 of Display 4.7, but with $\frac{1}{2}$ added to or subtracted from the numerator so that it will be closer to zero. The numerator of the *Z*-statistic changes from -66 to -65.5. The resulting *Z*-statistic is $-65.5/21.7 = -3.02$ and the *p*-value is .00126. The continuity correction has little effect whenever the sample sizes are large.

Exact p-Values for the Rank-Sum Test

The normal distribution is an easy and adequate approximation to the randomization (or sampling) distribution for most problems. Troublesome situations arise where sample sizes are small or there are large numbers of ties.

Display 4.7   Finding the p-value with the normal approximation to the permutation distribution of the rank-sum statistic; calculations for the cognitive load data continued from Display 4.5

(1) *Calculate the average and sample standard deviation of the ranks from the combined sample (column 4 of Display 4.5).*

$$\overline{R} = 14.5 \qquad s_R = 8.202$$

(2) *Compute the theoretical "null hypothesis" mean and standard deviation of T, using the formulas in Display 4.6.*

$$\text{Mean}(T) = (14)(14.5) = 203; \qquad \text{SD}(T) = 8.202\sqrt{\frac{14 \times 14}{(14 + 14)}} = 21.70$$

(3) *Determine the Z-statistic.* $\longrightarrow$ $$Z = \frac{(137 - 203)}{21.70} = -3.04$$

(4) *Find the p-value from a standard normal table.* $\longrightarrow$ **One-sided p-value = .00118**

It is possible to compute the randomization distribution exactly using the technique described in Section 1.3.2. Published tables with exact percentiles exist (see, for example, Pearson and Hartley 1972, p. 227), but they are only appropriate for situations in which there are no ties. Another drawback is that only a few percentiles are published for each combination of sample sizes, so that p-values can only be bracketed.

4.2.4   A Confidence Interval Based on the Rank-Sum Test

A 95% confidence interval for a parameter can be constructed by including all hypothesized values that lead to two-sided p-values greater than or equal to .05. This relation between tests and confidence intervals can be exploited to obtain a confidence interval for the additive treatment effect δ in the cognitive load experiment.

If Y is the time to solution for a student using the conventional study materials, the additive treatment effect model says that $Y + \delta$ is the time to solution for the same student using the modified study materials. The rank-sum test in Display 4.7 indicates that 0 is not very likely as a value for δ. Could δ be, say, -50 seconds? If so, the completion times in the modified group *minus 50 seconds* should be a set of times that are similar to those for the conventional study materials group. This suggests a procedure: (1) subtract a hypothesized δ from all modified group (treatment 2) times; (2) use the rank-sum test with a criterion, two-sided p-value $> .05$, to decide whether the two resulting group differences can be explained by chance; and (3) determine—through trial and error—upper and lower limits on δ values satisfying the criterion.

Display 4.8 illustrates the process with a series of proposed values for δ. By trial and error it was determined that all hypothesized values of δ between -159 seconds and -58 lead to two-sided *p*-values greater than .05. The 95% confidence interval for the reduction in test time due to the modified instructional method is 58 to 159 seconds.

Display 4.8   Using a rank-sum test to construct a confidence interval for an additive treatment effect (cognitive load study)

Hypothesized Effect (seconds)	Two-sided *p*-Value	Confidence Interval Inclusion?	
−50	.0286	no	Try several hypothesized values for δ
−60	.0800	yes	to identify those that have two-sided *p*-
−55	.0403	no	values $\geq .05$.
−58	.0502	yes	
−150	.1227	yes	
−160	.0476	no	A 95% confidence interval is −159 sec-
−155	.0589	yes	onds to −58 seconds.
−158	.0530	yes	
−159	.0502	yes	

Notes About the Rank-Sum Procedure

1. Other names for the rank-sum test are the Wilcoxon test and the Mann–Whitney test. The different names refer to originators of different forms of the test statistic. To confuse the issue, there is also a Wilcoxon signed-rank test—which is something entirely different.

2. The rank-sum test is a *nonparametric* or *distribution-free* statistical tool, meaning there are no specific distributional assumptions required.

3. Although the *t*-test is more efficient when the populations are normal (it makes better use of the available data), the rank-sum test is not that much worse in the normal model, and is substantially better for many other situations, particularly for long-tailed distributions.

4. The theoretical mean of T in Display 4.6 can also be written as $\text{Mean}(T) = n_1(n_1 + n_2 + 1)/2$. If there are no ties, the theoretical standard deviation in the permutation distribution of T is

$$\text{SD}(T) = \sqrt{n_1 n_2 (n_1 + n_2 + 1)/12}$$

The version in bubble 2 of Display 4.6 is correct whether there are ties or not.

5. In its application to the cognitive load problem, the confidence interval method described in Display 4.8 should be modified to reflect the truncation of solution times to 5

minutes. Thus, when the modified group's times are shifted by a certain hypothesized amount, any values exceeding 300 seconds should be replaced by 300 before performing the test. Such a modification makes no change in the lower limit of the confidence interval; but the upper limit changes from 159 to 189 seconds.

4.3   Other Alternatives for Two Independent Samples

4.3.1   Permutation Tests

A *permutation test* is any test that finds a *p*-value as the proportion of regroupings—of the observed $n_1 + n_2$ numbers into two groups of size $n_1 + n_2$—that lead to test statistics as extreme as the observed one. The test statistic may be the difference in group averages, a *t*-statistic, the sum of ranks in group 1, or any other choice to represent group difference. Permutation tests were previously discussed in the context of interpretation. When used to analyze randomized experiments, for example, permutation tests are called *randomization tests* and provide statistical inferences tied to the chance mechanism in random assignment. For observational studies, they provide no inference to a broader context (except for the special case of the rank-sum test), but may nevertheless be useful for summarizing differences in the data at hand. In the cases so far discussed, the actual calculation of *p*-values and confidence intervals was based on an approximation to the permutation distribution. Sometimes there is no adequate approximation.

In the O-ring study, for example, the exact rendering of the permutation test is the only method for calculating its *p*-value. The distribution of the numbers is so nonnormal that a *t*-distribution approximation is severely inadequate. No transformation helps. The rank-sum test is inadequate because of the large number of ties: 17 of the 24 values are tied at zero. Even though the computational effort for direct calculation of the *p*-value is considerable, permutation calculations are important because they are always available and require no distributional assumptions or special conditions. The *p*-value is calculated using the following procedure:

---

1.   Decide on a test statistic, and compute its value from the two samples.
2.   List all regroupings of the $n_1 + n_2$ numbers into groups of size n_1 and n_2, and recompute the test statistic for each.
3.   Count the number of regroupings that produce test statistics at least as extreme as the observed test statistic from step 1.
4.   The *p*-value is the number found in step 3 divided by the total number of regroupings.

In problems such as the O-ring study, the procedure can be accomplished by counting with *combinatorics*, that is, by using the *combination numbers*

$$C_{n,k} = \frac{n(n-1)\cdots(n-k+1)}{k(k-1)\cdots 1}.$$

The number $C_{n,k}$—read as "n choose k"—is the number of different ways to choose k items from a list of n items. The total number of regroupings for a two-sample problem is $C_{n_1+n_2,n_1}$. Combination numbers can also be used to count the number of regroupings that lead to test statistics as extreme or more extreme than the observed one, as now shown for the O-ring data.

Step 1. The *t*-statistic was selected as the test statistic (although the difference in averages would be equally good). Its observed value is 3.888.

Step 2. It is only necessary to determine which group 1 outcomes produce *t*-statistics that are as large as or larger than the 3.888 that came from the observed outcome, $(1, 1, 1, 3)$. After some calculation, it is found that the extreme regroupings are those whose group 1 outcomes are $(1, 1, 2, 3)$, $(1, 1, 1, 3)$, or $(0, 1, 2, 3)$, with *t*-statistics 3.888, 3.888, and 5.952, respectively.

Step 3. The total number of regroupings is $C_{24,4} = 10,626$. For a one-sided *p*-value it is necessary to count the number of regroupings that produce the outcomes in Step 2. Consider the outcome $(1, 1, 2, 3)$. To get such an outcome, the single 2 and the single 3 must be selected. However, there are five 1's in the full sample, and the indicated outcome would occur with any of the $C_{5,2} = 10$ combinations of 1's. This is illustrated in Display 4.9.

Display 4.9   Counting method for regroupings that give extreme group 1 outcomes

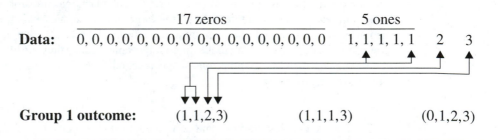

Similarly, the number of regroupings with outcome $(1, 1, 1, 3)$ is the number of ways to select three 1's from the five available in the data set, to go with the obligatory 3. This is $C_{5,3.} = 10$. Finally, a regrouping with outcome $(0, 1, 2, 3)$ can be formed by taking any one of the 17 0's in combination with any one of the five 1's, along with 2 and 3. The number of ways to do this is $C_{17,1} \times C_{5,1} = 17 \times 5 = 85$. Summing up, the number of regroupings that lead to *t*-statistics as large as or larger than the observed one is $10 + 10 + 85 = 105$.

Step 4. The one-sided *p*-value is $105/10,626 = .00988$. This result and the full permutation distribution of the *t*-statistic appear in Display 4.10.

Notes. The combinatorics method is especially useful if there are only a few regroupings that need to be counted, but it might be unmanageable if there are many. A computer can systematically enumerate the regroupings, calculate the test statistic for each, and compute the proportion that have a more extreme test statistic than the observed one. An

Display 4.10   A summary of the t-statistics calculated from all 10,626 rearrangements of the O-ring data into a Low group of size 4 and a High group of size 20

Number of Rearrangements with Identical t-Statistics	t-Statistic
2,380	−1.188
3,400	−0.463
2,040	0.231
1530	0.939
855	1.716
316	2.643
95	**3.888**
10	**5.952**

Total number of rearrangements into two groups of size 4 and 20:
10,626

Number of rearrangements with t-statistics greater than or equal to 3.888:
105

One-sided p-value from a permutation test of the t-statistic:
105/10,626 = .00988

alternative is to approximate the p-value by the proportion of a *random sample* of all possible regroupings that have a test statistic more extreme than the observed one, as discussed for the discrimination study in Section 1.1.2.

4.3.2   The Welch t-Test for Comparing Two Normal Populations with Unequal Spreads

Welch's t-test employs the individual sample standard deviations as separate estimates of their respective population standard deviations, rather than pooling to obtain a single estimate of a population standard deviation. The result is a different formula for the standard error of the difference in averages:

$$\text{SE}_W(\overline{Y}_2 - \overline{Y}_1) = \sqrt{\frac{s_1^2}{n_1} + \frac{s_2^2}{n_2}}.$$

This becomes the denominator in the t-statistic for comparing the means of populations with different spreads. Even when the populations are normal, however, the exact sampling distribution of the Welch t-ratio is unknown. It can be approximated by a t-distribution with d.f.$_W$ degrees of freedom, known as Satterthwaite's approximation:

$$\text{d.f.}_W = \frac{[\text{SE}_W(\overline{Y}_2 - \overline{Y}_1)]^4}{\dfrac{[\text{SE}(\overline{Y}_2)]^4}{(n_2 - 1)} + \dfrac{[\text{SE}(\overline{Y}_1)]^4}{(n_1 - 1)}},$$

where

$$SE(\overline{Y}_1) = \frac{s_1}{\sqrt{n_1}} \quad \text{and} \quad SE(\overline{Y}_2) = \frac{s_2}{\sqrt{n_2}}.$$

The *t*-test and confidence interval are computed exactly as with the two-sample *t*-test, except with the modified standard error and the approximate degrees of freedom (rounded down to an integer value).

A Note About the Importance of the Equal-spread Model

If the two populations have the same shape and the same spread, then the difference in means is an entirely adequate summary of their difference. Any question of interest that requires a comparison of the two distributions can be reexpressed in terms of the single parameter $\mu_2 - \mu_1$.

If, on the other hand, the populations have different means *and different standard deviations*, then $\mu_2 - \mu_1$ may be an inadequate summary. If lifetimes of brand A light bulbs have a larger mean and a larger standard deviation than lifetimes of brand B light bulbs, as shown in Display 4.11, then there is a higher proportion of long-life bulbs from brand A, but there may also be a higher proportion of short-life bulbs from brand A. A comparison of brands is not entirely resolved by a comparison of means.

Display 4.11   The conceptual difficulty with comparing population means when population spreads are not the same

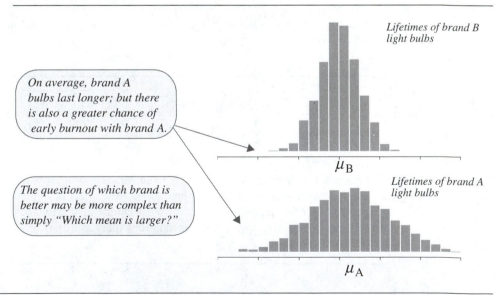

Although some statistical analysts find the unequal variance model more appealing for the two-sample problem, most of the standard methods for more complicated structures make use of a pooled estimate of variance. In that sense, the two-sample t-tools extend to more complicated situations more easily than does Welch's t-test.

4.4   Alternatives for Paired Data

Two resistant and distribution-free alternatives to the paired t-test are described here. The sign test is a tool for a quick analysis of paired data. If the sign test shows a substantial effect, that may be enough to resolve the answer to the question of interest. But it is not very efficient. If its results are inconclusive, the signed-rank test is a more powerful alternative.

4.4.1   The Sign Test

The *sign test* is a quick, distribution-free test of the hypothesis that the mean difference of a population of pairs is zero (for observational studies) or of the hypothesis that there is no treatment effect in a randomized paired experiment. It counts the number K of pairs where one group's measurement exceeds the other's. In the schizophrenia study (Section 2.1.2 on page 29), the left hippocampus volume of the unaffected twin was larger than the left hippocampus volume of the affected twin for $K = 14$ out of the $n = 15$ twin pairs. If there were no systematic difference between affected and unaffected individuals, one would expect K to be near $n/2$, as both groups share equal chances for having the larger measurement. The sign test provides evidence against the null hypothesis whenever K is far from $n/2$.

If the null hypothesis is true, the distribution of K is *binomial*, indexed by $n = 15$ trials and probability .5 of a positive difference in each trial. The chance of obtaining exactly k positive differences is given by $C_{n,k} \times (1/2)^n$. The p-value is the sum of these chances for all values of k that are as far or farther from $n/2$ than is the observed value K.

An approximation to the p-value comes from normal approximation to the binomial distribution. If the null hypothesis is true, then the Z-statistic

$$Z\text{-stat} = \frac{K - (n/2)}{\sqrt{n/4}}$$

has approximately a standard normal distribution. A p-value is found as the proportion of a standard normal distribution that is farther from zero than the observed Z-statistic. The approximation can be improved with a continuity correction, by adding $\frac{1}{2}$ to (or subtracting it from) the numerator of the Z-statistic, to make it closer to zero.

Example—Schizophrenia

Since 14 of the 15 differences are positive, the exact one-sided p-value is the probability that k is 14 plus the probability that k is 15. This is $C_{15,14} \times (1/2)^{15} + C_{15,15} \times (1/2)^{15} = 15 \times (1/32{,}768) + 1 \times (1/32{,}768) = .00049$. To illustrate the normal approximation, the Z-statistic with continuity correction is

$$Z\text{-stat} = \frac{[14 - (15/2)] - (1/2)}{\sqrt{15/4}} = 3.098,$$

which produces an approximate one-sided *p*-value of .00097.

4.4.2 The Wilcoxon Signed-Rank Test

By retaining only the signs of the difference for each pair, the sign test completely obliterates the effects of outliers, but at the expense of losing potentially useful information about the relative magnitudes of the differences. The *Wilcoxon signed-rank test* uses the ranks of the magnitudes of the differences in addition to their signs. Since ranks are used, the procedure is resistant to outliers.

Computation of the *signed-rank statistic* proceeds as follows:

---

1. Compute the difference in each of the *n* pairs.

2. Drop zeros from the list.

3. Order the *absolute differences* from smallest to largest and assign them their ranks $1, \ldots, n$ (or average rank for ties).

4. The signed-rank statistic, *S*, is the sum of the ranks from the pairs for which the difference is positive.

The computation of *S* is illustrated in the schizophrenia data in Display 4.12, for the hypothesis that the mean difference is zero.

Exact p-Value

An exact *p*-value for the signed-rank test is the proportion of all assignments of outcomes within each pair that lead to a test statistic as extreme as or more extreme than the one observed. Assignment refers to switching the group status but keeping the same observations within each pair. Within a single pair there are two possible assignments, so with *n* pairs there are a total of 2^n possible assignments. The *p*-value is therefore the number of possible assignments that provide a sum of positive ranks more extreme than the observed one, divided by 2^n.

For the schizophrenia example, there are 15 pairs, so the labels "affected" and "unaffected" can be assigned to observations within pairs in $2^{15} = 32,768$ possible ways. The assignments that produce an *S* greater than or equal to 110.5 are those where the sum of the negative ranks is less than or equal to 9.5. These are all the ways that the Rank column in Display 4.12 can be redistributed among +Rank and −Rank so that the sum of the −Rank column is 9.5 or less.

By systematically examining the possible assignments with no negative ranks, one negative rank, two negative ranks, and so on, the following assignments (indicated by the

Display 4.12 Signed-rank test statistic computations (schizophrenia study)

Pair	Unaffected	Affected	Difference	Ordered Magnitude	Order	Rank	+Ranks	−Ranks
1	1.94	1.27	.67	.02 (+)	1	1	1	
2	1.44	1.63	−.19	.03 (+)	2	2	2	
3	1.56	1.47	.09	.04 (+)	3	3	3	
4	1.58	1.39	.19	.07 (+)	4	4	4	
5	2.06	1.93	.13	.09 (+)	5	5	5	
6	1.66	1.26	.40	.10 (+)	6	6	6	
7	1.75	1.71	.04	.11 (+)	7	7	7	
8	1.77	1.67	.10	.13 (+)	8	8	8	
9	1.78	1.28	.50	.19 (+)	9	9.5	9.5	
10	1.92	1.85	.07	.19 (−)	10	9.5		9.5
11	1.25	1.02	.23	.23 (+)	11	11	11	
12	1.93	1.34	.59	.40 (+)	12	12	12	
13	2.04	2.02	.02	.50 (+)	13	13	13	
14	1.62	1.59	.03	.59 (+)	14	14	14	
15	2.08	1.97	.11	.67 (+)	15	15	15	

(1) Order the absolute differences and assign ranks to them.

= 110.5

(2) Signed-rank statistic = Sum of ranks for positive differences.

ranks associated with a negative sign) are found to have a sum of negative ranks less than 9.5: (none), (1), (2), (3), (4), (5), (6), (7), (8), (9.5), (9.5), (1, 2), (1, 3), (1, 4), (1, 5), (1, 6), (1, 7), (1, 8), (2, 3), (2, 4), (2, 5), (2, 6), (2, 7), (3, 4), (3, 5), (3, 6), (4, 5), (1, 2, 3), (1, 2, 4), (1, 2, 5), (1, 2, 6), (1, 3, 4), (1, 3, 5), and (2, 3, 4). There are 34 assignments with the sum of negative ranks less than 9.5, and hence 34 assignments with S greater than or equal to 110.5, so the exact one-sided p-value is $34/32{,}768 = .00104$.

Normal Approximation

This method of directly counting more extreme cases in order to find the p-value can be replaced by tables or by computer assisted counting. A normal approximation for convenient computation of an approximate p-value is available. The mean and standard deviation of S in its permutation distribution when the null hypothesis is true are

$$\text{Mean}(S) = n(n + 1)/4 \quad \text{and} \quad \text{SD}(S) = [n(n + 1)(2n + 1)/24]^{1/2}.$$

Comparing Z-stat $= [S - \text{Mean}(S)]/\text{SD}(S)$, with a continuity correction, to a standard normal distribution gives a good approximation to the p-value for $n \geq 20$.

For the schizophrenia study, $\text{Mean}(S) = 60$, $\text{SD}(S) = 17.61$, so that Z-stat $= 2.84$. An approximate one-sided p-value is .00226.

4.5 Related Issues

4.5.1 Practical and Statistical Significance

p-values indicate *statistical significance*, the extent to which a null hypothesis is contra-
dicted by the data. This must be distinguished from *practical significance*, which describes
the practical importance of the effect in question.

A study may find a statistically significant increase in rainfall due to cloud seeding, but
the amount of increase in rainfall, say 10%, may not be sufficient to justify the expense of
seeding. A finding that annual male salaries tend to be $16.32 larger than female salaries,
when annual incomes are on the order of $10,000, does not provide a strong indication of
persistent discrimination, even if the difference is statistically significant. A statistically
significant reduction in time to complete a test problem may not be practically relevant
if the reduction is only from 150 seconds to 147 seconds. These are issues of practical
significance. They have little to do with statistics, but they are highly relevant to the
interpretation of statistical analyses.

It is important to understand the connection between sample size and statistical significance. If there is a difference in population means (no matter how practically insignificant),
large enough samples should indicate the existence of a statistically significant difference.
On the other hand, even if there is a practically significant difference in population means
(that is, an important difference), a small sample size may fail to indicate the existence of
a statistically significant difference between means.

Three practical points deserve attention:

1. *p*-values are sample-size-dependent.
2. A result with a *p*-value of .08 can have more scientific relevance than one with a *p*-value
   of .001.
3. Tests of hypotheses by themselves rarely convey the full significance of the results.
   They should be accompanied by confidence intervals, if possible, to indicate the range
   of likely effects and to facilitate the assessment of practical significance.

4.5.2 The Presentation of Statistical Findings

The Summary of Statistical Findings sections that accompany the case studies in this book
are intended as models for communicating results. The following are some general recom-
mendations.

1. State the conclusions as they apply to the question of interest and avoid statistical jargon
   and symbols, except to include appropriate measures of statistical uncertainty and to
   state which statistical tools were used.
2. Make clear exactly what is being estimated to answer a question of interest; for example,
   the difference in population means, a ratio of medians, or an additive treatment effect.
3. Prefer confidence intervals to standard errors, particularly when the estimate does not
   have a normal sampling (or randomization) distribution.
4. Use graphical displays to help convey important features of the data.

5. Do not include unnecessarily large numbers of digits in estimates or p-values.
6. If transformations have been used, express the results on the original scale of measurement.
7. Comment on the scope of inference. Was random assignment used? Was random sampling used?

4.5.3 Levene's Test for Equality of Two Variances

Sometimes a question of interest calls for a test of equality of two population variances. The *F-test for equal variances* and its associated confidence interval are available in standard statistical computer packages, but they are not robust against departures from normality. For example, p-values can easily be off by a factor of 10 if the distributions have shorter or longer tails than the normal.

A robust alternative is *Levene's Test*. Suppose there are n_1 observations Y_{1i} from population 1 with standard deviation σ_1, and n_2 observations Y_{2i} from population 2 with standard deviation σ_2. Let Z_{1i} be the squared deviation of the ith observation in group 1 from its group average: $(Y_{1i} - \overline{Y}_1)^2$, and let Z_{2i} be the squared deviation of the ith observation in group 2 from its average: $(Y_{2i} - \overline{Y}_2)^2$. Then the mean of the Z_{1i}'s is approximately σ_1^2 and the mean of the Z_{2i}'s is approximately σ_2^2. Even though the Z's in each group are not independent of one another (since they are based on a common average) Levene's idea was to treat them as independent and conduct a two-sample t-test to compare their means. Despite problems with the exact assumptions of the t-test not being met, this test works well.

Example—Sex Discrimination

The data of Section 1.1.2 is used here to illustrate Levene's test for the hypothesis that the variance of male salaries is equal to the variance of female salaries. The test compares the 32 squared deviations of the males to the 61 squared deviations for the females. The t-statistic is 1.3787 and the two-sided p-value, by comparison to a t-distribution on 91 degrees of freedom, is .17. Thus Levene's test gives no evidence that the variances are unequal.

4.5.4 Survey Sampling

Survey sampling concerns selecting members of a specific (finite) population to be included in a survey and the subsequent analysis. One sampling method already discussed is simple random sampling, in which each member of a population has an equal chance of being included in the sample. This is appealing for its mathematical and conceptual simplicity, but is often practically unrealistic. The prohibitive part of finding a simple random sample of American voters, for example, is first finding a list of all American voters.

Instead of simple random sampling, survey organizations rely on complex sampling designs, which make use of stratification, multistage sampling, and cluster sampling. In *stratified sampling*, the population is divided into strata, based on supplementary information, and separate simple random samples are selected for each stratum. For example, samples of American voters may be taken separately for different states. *Multistage sampling*, as its name suggests, uses sampling at different stages. For example, a random sample of states is taken, then random samples of counties are taken within the selected states, and then random samples of individuals are selected from lists in the chosen counties. In *cluster sampling*, a simple random sample of clusters is selected at some stage of the sampling and all members within the cluster are included in the survey. For example, a random sample of households (the clusters) in a county may be selected and *all* voters in each household surveyed.

Standard Errors from Complex Sampling Designs

The formulas for standard errors so far presented are based on simple random sampling *with replacement*, which means that a member of a population can appear in a sample more than once. Random sampling in sample surveys, however, is conducted *without replacement* and does not permit a member to appear in a sample twice. Consequently, different standard errors must be calculated. The variance of an average from random sampling without replacement differs from that with replacement by a multiplicative factor called the *finite population correction* (FPC). If N is the population size and n the sample size, FPC $= (N-n)/N$. So the difference is small if only a small fraction of the population is sampled.

In addition, the usual standard errors are inappropriate for data from complex sampling designs. The ratio of the sampling variance of an estimate from data based on a particular sampling design to the sampling variance it would have had if it was based on a simple random sample of the same size is called the *design effect*. Design effects for estimates from complex stratified, multistage sampling designs are usually greater than 1. Therefore, application of standard statistical theory to complex sampling designs usually results in an overstatement of the degree of precision. For example, confidence intervals will tend to be narrower than they really should be. Methods for obtaining correct standard errors are discussed in texts on survey sampling.

Nonresponse

Of 979 Vietnam veterans who were deemed to be eligible for the study in Section 3.1.2, only 900 were located, only 871 of those completed the first interview, only 665 of those were available for blood samples, and only 646 of those had valid dioxin readings. Assuming the 979 represented the population of interest, is there any danger in using the available 646 (66%) who responded and provided valid results to represent the population? The answer depends on whether those who were lost along the way differed in their dioxin levels from those who were not. The question that must be asked is whether or not being unavailable or not providing valid results had anything to do with dioxin level. If not, there is little harm in ignoring the nonresponders.

In sample surveys, the issue is usually much more serious, since those individuals who tend to respond to surveys often have much different views from those who do not. As an illustration, consider the popular telephone-in surveys that news programs run after presidential debates. Viewers call in (and have to pay for their calls) to indicate who they think won. Although this is entertaining, the results give an entirely unreliable indication of what all viewers think, since those who respond tend to be a quite different crowd from those who do not.

4.6 Summary

The Space Shuttle O-Ring Study

The main feature of these data is that both the two-sample *t*-test and the rank-sum test are inappropriate. The *t*-tools are not sufficiently robust against nonnormality when the data are so discrete. The rank-sum test is not appropriate when there are so many ties. The permutation test, however, can be applied here, because it does not rely on any assumptions. The computations are difficult, and inferences apply only to the 24 launches.

It should be mentioned that some information is lost due to the form in which the data are presented. They have been rather artificially divided into cool temperature launches and warm temperature launches, whereas the actual temperature data for each launch are available. The more advanced method of Poisson log-linear regression is applied to these data in Chapter 22.

The Cognitive Load Study

These data from a randomized experiment could be suitably analyzed with any of a number of techniques if it were not for the censoring of five observations. Without having actual completion times for these five students, it is impossible to take an average. The rank-sum test is convenient in this case since for it one only needs to know that these five students were tied for last. It can be used to test for an additive treatment effect and, with some effort, to provide a confidence interval for the treatment effect.

It should be mentioned that the rank-sum test is not suitable for censored observations if the rank of the censored observation cannot be determined. Thus, it is limited to handling censoring for data problems like this one, in which a number of censored observations are tied for last.

Further Reading

A sophisticated and insightful account of one-sample and two-sample methods, model checking, and robustness is given in Chapters 1 and 2 of the book on applied statistics by Rupert Miller (1986). Further discussion, examples, and references to randomization tests, the rank-sum test, and the signed-rank test can be found in the *Encyclopedia of Statistical Science* (Kotz, Johnson, and Read 1982) under the headings randomization test, Mann–Whitney–Wilcoxon statistic, and Wilcoxon signed-rank test, respectively. See also the encyclopedia passage on survey sampling for more discussion and further references on that topic. *Sampling*, by S. K. Thompson (1990), covers biological and survey sampling.

4.7 Exercises

Conceptual Exercises

1. **Cognitive Load.** Suppose that there were two textbooks on coordinate geometry, one written with conventional worked problems and the other with modified worked problems. And suppose it is possible to identify a number of schools in which each of the textbooks is used. If you took random

samples of size 14 from schools with each text and obtained exactly the same data as in the example in Section 4.1.2, would the analysis and conclusions be the same?

2. O-Ring Data. (a) Is it appropriate to use the two-sample *t*-test to compare the number of O-ring incidents in cold and warm launches? (b) Is it appropriate to use the rank-sum test? (c) Is it appropriate to use a permutation test based on the *t*-statistic?

3. O-Ring Data. When these data were analyzed prior to the Challenger disaster it was noticed that variability was greater in the group with the larger average, so a log transformation was used. Since the log of zero does not exist, all the zeros were deleted from the data set. Does this seem like a reasonable approach?

4. O-Ring Data. Explain why the two-sided *p*-value from the permutation test applied to the O-ring data is equal to the one-sided *p*-value (see Display 4.10).

5. O-Ring Data. If you looked at the source of the O-ring data and found that temperatures for each launch were recorded in degrees F rather than as over/under 65°F, what question would that raise? Would the answer affect your conclusions about the analysis?

6. Motivation and Creativity. In what way is the *p*-value for the motivation and creativity randomized experiment (Section 1.1.1) dependent on an assumed model?

7. Are there occasions when both the two-sample *t*-test and the rank-sum test are appropriate?

8. Can the rank-sum test be used for comparing populations with unequal variances?

9. Suppose that two drugs are both effective in prolonging length of life after a heart attack. Substantial statistical evidence indicates that the mean life length for those using drug A is greater than the mean life length for those using drug B, but the variation of life lengths for drug A is substantially greater as well. Explain why it is difficult to conclude that drug A is better even though the mean life length is greater.

10. In a certain problem, the randomization test produces an exact two-sided *p*-value of .053, while the *t*-distribution approximation produces .047. One might say that since the *p*-values are on opposite sides of .05 they lead to quite different conclusions and, therefore, the approximation is not adequate. Comment on this statement.

11. What is the difference between a permutation test and a randomization test?

12. Explain what is meant by the comment that there is no single test called a randomization test.

13. What confounding factors are possible in the O-ring failure problem?

Computational Exercises

14. O-Ring Study. Find the *t*-distribution approximation to the *p*-value associated with the observed *t*-statistic. Compare this approximation to the (correct) permutation test *p*-value.

15. Considered these artificial data:

$$
\begin{array}{llll}
\text{Group 1:} & 1 & 5 \\
\text{Group 2:} & 4 & 8 & 9
\end{array}
$$

The difference in averages $\overline{Y}_1 - \overline{Y}_2$ is -4. What is a one-sided *p*-value from the permutation distribution of the *difference in averages*? (*Hint*: List the 10 possible groupings; compute the difference in averages for each of these groupings, then calculate the proportion of these less than or equal to -4.)

16. Consider these artificial data:

$$
\begin{array}{llll}
\text{Group 1:} & 5 & 7 & 12 \\
\text{Group 2:} & 4 & 6 &
\end{array}
$$

Calculate a p-value for the hypothesis of no difference, using the permutation distribution of the difference in sample averages. (You do not need to calculate the t-statistic for each grouping, only the difference in averages.)

17. **O-Ring Study.** Suppose the O-ring data had actually turned out as shown in Display 4.13.

Display 4.13   Hypothetical O-ring data

Launch Temperature	Number of O-Ring Incidents
Below 65°F	1 1 1 2
Above 65°F	0 0 0 0 0 0 0 0 0 0 0 0 0 0 0 0 0 0 0 1 1 3

These are the same 24 numbers as before, but with the 2 and 3 switched. What is the one-sided p-value from the permutation test applied to the t-statistic? (This can be answered by examining Display 4.10.)

18. Suppose that six persons have an illness. Three are randomly chosen to receive an experimental treatment, and the remaining three serve as a control group. After treatment, a physician examines all subjects and assigns ranks to the severity of their symptoms. The patient with the most severe condition has rank 1, the next most severe has rank 2, and so on up to 6. It turns out that the patients in the treatment group have ranks 3, 5, and 6. The patients in the control group have ranks 1, 2, and 4. Is there any evidence that the treatment has an effect on the severity of symptoms? Use the randomization distribution of the sum of ranks in the treatment group to obtain a p-value. (First find the sum of ranks in the treatment group. Then write down all 20 groupings of the 6 ranks; calculate the sum of ranks in the treatment group for each. What proportion of these give a rank-sum as large as or larger than the observed one?)

19. **Bumpus's Study.** Use a statistical computer program to perform the rank-sum comparison of humerus lengths in the sparrows that survived and the sparrows that perished (Display 2.1). (a) What is the two-sided p-value? (b) Does the statistical computer package report the exact p-value or the one based on the normal approximation? (c) If it reports the one using the normal approximation, does it use a continuity correction to the Z-statistic? (d) How does the p-value from the rank-sum test compare to the one from the two-sample t-test (.08; see Display 2.9 on page 42) and the one from the two-sample t-test when the smallest observation is set aside (see Exercise 27 on page 76) (e) Explain the relative merits of (i) the two-sample t-test using the strategy for dealing with outliers and (ii) the rank-sum test.

20. **Trauma and Metabolic Expenditure.** For the data in Exercise 18 on page 73 of Chapter 3: (a) Determine the rank transformations for the data. (b) Calculate the rank-sum statistic by hand (taking the trauma patients to be group 1.) (c) Mimic the procedures used in Display 4.5 and Display 4.7

to compute the Z-statistic. (d) Find the one-sided *p*-value as the proportion of a standard normal distribution larger than the observed Z-statistic.

21. Trauma and Metabolic Expenditure. Use a statistical computer package to verify the rank-sum and the Z-statistic obtained in Exercise 20. Is the *p*-value the same? (Does the statistical package use a continuity correction?)

22. Trauma and Metabolic Expenditure. Using the rank-sum procedure, find a 95% confidence interval for the difference in population medians: the median metabolic expenditure for the population of trauma patients minus the median metabolic expenditure for the population of nontrauma patients.

23. Motivation and Creativity. Use a statistical computer package to compute the randomization test's two-sided *p*-value for testing whether the treatment effect is zero for the data in Display 1.1 on page 2. How does this compare to the results from the two-sample *t*-test (which is used as an approximation to the randomization test).

24. Motivation and Creativity. Find a 95% confidence interval for the treatment effect (poem creativity score after intrinsic motivation questionnaire minus poem creativity score after extrinsic motivation questionnaire, from Display 1.1 on page 2) using the rank-sum procedure. (Use a statistical computer program.) How does this compare to the *t*-based confidence interval for the treatment effect?

25. Guinea Pig Lifetimes. Use the Welch *t*-tools to find a two-sided *p*-value and confidence interval for the effect of treatment on lifetimes of guinea pigs in Exercise 11 on page 49. Does the additive treatment effect seem like a sensible model for these data?

26. Schizophrenia Study. (a) Draw a histogram of the differences in hippocampus volumes in Display 4.12. Is there evidence that the population of differences is skewed? (b) Take the logarithms of the volumes for each of the 30 subjects, take the differences of the log volumes, and draw a histogram of these differences. Does it appear that the distribution of differences of log volumes is more nearly symmetric? (c) Carry out the paired *t*-test on the log-transformed volumes. How does the two-sided *p*-value compare with the one obtained on the untransformed data? (d) Find an estimate of and 95% confidence interval for the mean difference in log volumes. Back-transform these to get an estimate and confidence interval for the median of the population of ratios of volumes.

27. Schizophrenia Study. Find the two-sided *p*-value using the signed-rank test, as in Display 4.12, but after taking a log transformation of the hippocampus volumes. How does the *p*-value compare to the one from the untransformed data? Is it apparent from histograms that the assumptions behind the signed-rank test are more appropriate on one of these scales?

28. Darwin Data. Charles Darwin carried out an experiment to study whether seedlings from cross-fertilized plants tend to be superior to those from self-fertilized plants. He covered a number of plants with fine netting so that insects would be unable to fertilize them. He fertilized a number of flowers on each plant with their own pollen and he fertilized an equal number of flowers on the same plant with pollen from a distant plant. (He did not say how he decided which flowers received which treatments.) The seeds from the flowers were allowed to ripen and were set in wet sand to germinate. He placed two seedlings of the same age in a pot, one from a seed from a self-fertilized flower and one from a seed from a cross-fertilized flower. The data in Display 4.14 are the heights of the plants at certain points in time. (The fertilization experiments were described by Darwin in an 1878 book; these data were found in D. F. Andrews and A. M. Herzberg, *Data* (New York: Springer-Verlag, 1985), pp. 9–12.) (a) Draw a histogram of the differences. (b) Find a two-sided *p*-value for the hypothesis of no treatment effect, using the paired *t*-test. (c) Find a 95% confidence interval for the additive treatment effect. (d) Is there any indication from the plot in (a) that the paired *t*-test may be inappropriate? (e) Find a two-sided *p*-value for the hypothesis of no treatment effect for the data in Display 4.14, using the signed-rank test.

Display 4.14   Darwin's data: plant heights (inches) for 15 pairs of plants of the same age, one of which was grown from a seed from a cross-fertilized flower and the other of which was grown from a seed from a self-fertilized flower

	Plant Height (inches)	
Pair	**Cross-fertilized**	**Self-fertilized**
1	23.5	17.375
2	12	20.375
3	21	20
4	22	20
5	19.125	18.375
6	21.5	18.625
7	22.125	18.625
8	20.375	15.25
9	18.25	16.5
10	21.625	18
11	23.25	16.25
12	21	18
13	22.125	12.75
14	23	15.5
15	12	18

Data Probems

29.   Exercise and Walking Time. Can active exercise shorten the time it takes an infant to learn how to walk alone? Researchers randomly allocated 12 one-week-old, male infants from white, middle-class families to one of two treatment groups. Those in the active-exercise group received stimulation of the walking reflexes during four 3-minute sessions each day from the beginning of the second through the end of the eighth week. Those in the other group received no stimulation. (Data from P. R. Zelazo, "Walking in the Newborn," *Science* 176 (1927): 314–15.) Is there sufficient evidence to conclude that the groups differ in the typical time required to first walking? Analyze the data in Display 4.15, write a carefully worded statement of conclusion, and include some documentation of the details of the statistical analysis behind the conclusions.

Display 4.15   Ages at which infants first walked alone

Active	75,50,50,00	9	00	*No*
exercise	00	10		*exercise*
group		11	50,50	*group*
		12	00	
	00	13	00,25	

Legend: 13 | 25   represents 13.25 months

30.   Sunscreen Protection Factor. A sunscreen sunlight protection factor (SPF) of 5 means that a person who can tolerate Y minutes of sunlight without the sunscreen can tolerate $5Y$ minutes of sunlight with the sunscreen. The data in Display 4.16 are the times in minutes that 13 patients could tolerate the sun (a) before receiving treatment and (b) after receiving a particular sunscreen treatment. (Data from R. M. Fusaro and J. A. Johnson, "Sunlight Protection for Erythropoietic Protoporphyria Patients," *Journal of the American Medical Association* 229(11) (1974): 1420.) Analyze the data to estimate and provide a confidence interval for the sunlight protection factor. Treat these data as if they were from a randomized experiment (randomization was not used). Comment on whether there are any obvious potentially confounding variables in this study.

Display 4.16   Tolerance to sunlight (minutes) for 13 patients prior to treatment and after treatment with a sunscreen

---

| | *Tolerance to Sunlight (minutes)* | |
Patient	*Pretreatment*	*During Treatment*
1	30	120
2	45	240
3	180	480
4	15	150
5	200	480
6	20	270
7	15	300
8	10	180
9	20	300
10	20	240
11	60	480
12	60	300
13	120	480

---

31.   Effect of group therapy on survival of breast cancer patients. Researchers randomly assigned metastatic breast cancer patients to either a control group or a group that received weekly 90-minute sessions of group therapy and self-hypnosis, to see whether the latter treatment improved the patients' quality of life. The group therapy involved discussion and support for coping with the disease, but the patients were not led to believe that the therapy would affect the progression of their disease. Surprisingly, it was noticed in a follow-up 10 years later, that the group therapy patients appeared to have lived longer. The data on number of months of survival after beginning of the study are shown in Display 4.17 (Data from a graph in D. Spiegel, J. R. Bloom, H. C. Kraemer, and E. Gottheil, "Effect of Psychosocial Treatment on Survival of Patients with Metastatic Breast Cancer," *Lancet* (October 14, 1989): 888–91.) Notice that three of the women in the treatment group were still alive at the time of the follow-up, so their survival times are only known to be larger than 122 months. Is there indeed evidence of an effect of the group therapy treatment on survival time and, if so, how much more time can a breast cancer patient expect to live if she receives this therapy? Analyze the data as best as possible and write a brief report of the findings.

Answers to Conceptual Exercises

1.   The analysis would be the same. The conclusions would be very different. You could infer a real difference in the solution times of the two groups, but you could not attribute it to the different text types because of the possibility of a host of confounding factors.

Display 4.17 Months of survival after beginning of study for 58 breast cancer patients

Control Patients ($n = 24$)
2, 6, 8, 10, 12, 12, 14, 14, 14, 16, 16, 16, 18, 18, 18, 20, 22, 22, 26, 34, 36, 38, 40, 48

Patients Given Group Therapy for One Year ($n = 34$)
2, 2, 4, 4, 4, 6, 6, 8, 10, 10, 12, 14, 16, 16, 16, 18, 20, 22, 32, 36, 46, 46, 48, 48, 58, 58, 66, 72, 72, 82, 122, 122*, 122*, 122*

*These three patients were still alive at the end of the 122-month study period.

2. (a) No, the extent of the nonnormality in these small and unequally sized samples is more than can be tolerated by the two-sample t-test. (b) Probably not. The spreads are apparently not equal. (c) Yes. The permutation test for significance requires no model or assumptions.

3. No! Observations cannot be deleted simply because the transformation does not work on them. In this case, a major portion of the data was deleted, leaving a very misleading picture.

4. There are 105 groupings that lead to t-statistics greater than or equal to 3.888 and no groupings that lead to t-statistics less than or equal to -3.888.

5. Was the 65°F cutoff chosen to maximize the apparent difference in the two groups? If so, the p-value would not be correct. Why? Because the p-value represents the chance of getting evidence as damaging to the hypothesis when there is no difference. The "chance" is the frequency of occurrence in replications of the study, *using the same statistical analysis*. The p-value calculation assumes that the 65°F cutoff will always define the groups. If the choice of cutoff was part of the statistical analysis used on this data set, the calculation was not correct. To get a correct p-value would require that you allow for a different cutoff to be chosen in each replication. Further discussion of data snooping is given in Chapter 6.

6. The p-value is based on the two-sample t-test but it is now understood that this p-value serves as an approximation to the p-value from the exact randomization test. For this approximation to be valid, the histograms of creativity scores should be reasonably normal (which they are).

7. Yes. Since the population model for the t-tools requires that the populations be normal with equal spread they will necessarily have the same shape and spread. Therefore, the assumptions for the rank-sum test are also satisfied.

8. No. Inference to populations is based on the model in which the two populations have the same shape and spread. It is possible, however, that this model may be appropriate after some transformation of the data.

9. Generally, it is hard to make use of the difference in the centers of two distributions when the spreads are quite different. Specifically, the mean life length for drug A may be longer, but more people who use it may die sooner than for drug B.

10. The statement takes the rejection region approach too literally. There is very little difference in the degree of evidence against the null hypothesis in p-values of .053 and .047, so the approximation is pretty good.

11. A randomization test is a permutation test applied to data from a randomized experiment.

12. There is a different permutation distribution for each statistic that can be calculated from the data.

13. Perhaps workers tended to make more mistakes in cold weather or wind stress was greater on days with cold weather.

CHAPTER 5

Comparisons Among Several Samples

The issues and tools associated with the analysis of three or more independent samples (or treatment groups in a randomized experiment) are very similar to those for comparing two samples. Notable differences, however, stem from the particular kinds of questions and the greater number of them that may be asked.

An initial question, often asked in a preliminary stage of the analysis, is whether all of the means are equal. An easy-to-use F-test is available for investigating this. A typical analysis, however, goes beyond the F-test and explores linear combinations of the means to address particular questions of interest. A simple example of a linear combination of means is $\mu_3 - \mu_1$, the difference between means in populations 3 and 1. Inferences about this parameter can be made with t-tools just as for the two-independent sample problem, with the important difference that the pooled estimate of standard deviation is from all groups, not just from those being compared.

This chapter discusses the use of the pooled estimate of variance for specific linear combinations of means and the one-way analysis of variance F-test for testing the equality of several means. The next chapter looks at linear combinations more generally and the problem of compounded uncertainty from the multiple, simultaneous comparisons of means.

5.1 **Case Studies**

5.1.1 **Diet Restriction and Longevity—A Randomized Experiment**

A series of studies involving several species of animals found that restricting caloric intake can dramatically increase life expectancy. In one such study, female mice were randomly assigned to one of the following six treatment groups:

1. **NP:** Mice in this group ate as much as they pleased of a nonpurified, standard diet for laboratory mice.
2. **N/N85:** This group was fed normally both before and after weaning. (The slash distinguishes the two periods.) After weaning, the ration was controlled at 85 kcal/wk. This, rather than NP, serves as the control group because caloric intake is held reasonably constant.
3. **N/R50:** This group was fed a normal diet before weaning and a reduced-calorie diet of 50 kcal/wk after weaning.
4. **R/R50:** This group was fed a reduced-calorie diet of 50 kcal/wk both before and after weaning.
5. **N/R50 lopro:** This group was fed a normal diet before weaning, a restricted diet of 50 kcal/wk after weaning, and had dietary protein content decreased with advancing age.
6. **N/R40:** This group was fed normally before weaning and was given a severely reduced diet of 40 kcal/wk after weaning.

Display 5.1 shows side-by-side box plots for the lifetimes, measured in months, of the mice in the six groups. Summary statistics and sample sizes are reported in Display 5.2. (Data from R. Weindruch, R. L. Walford, S. Fligiel, and D. Guthrie, "The Retardation of Aging in Mice by Dietary Restriction: Longevity, Cancer, Immunity and Lifetime Energy Intake," *Journal of Nutrition* 116(4) (1986): 641–54.)

The questions of interest involve specific, pairwise comparisons of treatments as diagrammed in Display 5.3. Specifically, (a) Does lifetime on the 50 kcal/wk diet exceed the lifetime on the 85 kcal/wk diet? If so, by how much? (This calls for a comparison of the N/R50 group to the N/N85 group.) (b) Is lifetime affected by providing a reduced calorie diet before weaning, given that a 50 kcal/wk diet is provided after weaning? (This calls for a comparison of the R/R50 group to the N/R50 group.) (c) Does lifetime on the 40 kcal/wk diet exceed the lifetime on the 50 kcal/wk diet? (This calls for a comparison of the N/R40 group to the N/R50 group.) (d) Given a reduced calorie diet of 50 kcal/week, is there any additional effect on lifetime due to decreasing the protein intake? (This calls for a comparison of the N/R50 lopro diet to the N/R50 diet.) (e) Is there an effect on lifetime due to restriction at 85 kcal/week? This would indicate the extent to which the 85 kcal/wk diet served as a proper control and possibly whether there was any effect of restricting the diet, even with a standard caloric intake. (This calls for a comparison of the N/N85 group to the NP group.) Comparisons other than those indicated by the arrows are not directly meaningful because the group treatments differ in more than one way. The N/N50 lopro and the N/N40 groups, for example, differ in both the protein composition and the total calories in the diet, so a difference would be difficult to attribute to a single cause.

Display 5.1   Lifetimes of female mice fed on six different diet regimens

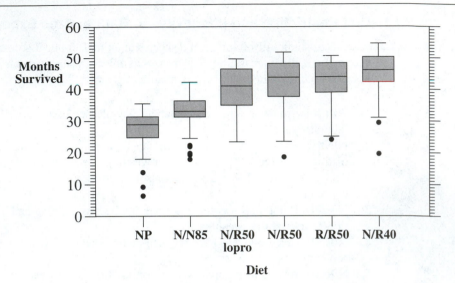

Display 5.2   Summary statistics for lifetimes of mice on six different diet regimens

Group	n	Range (months)	Average	SD	95% CI for Mean
NP	49	6.4 – 35.5	27.4	6.1	25.6 – 29.2
N/N 85	57	17.9 – 42.3	32.7	5.1	31.3 – 34.1
N/R50	71	18.6 – 51.9	42.3	7.8	40.5 – 44.1
R/R50	56	24.2 – 50.7	42.9	6.7	41.1 – 44.7
N/R50 lopro	56	23.4 – 49.7	39.7	7.0	37.8 – 41.6
N/R40	60	19.6 – 54.6	45.1	6.7	43.4 – 46.8

Summary of Statistical Findings

There is overwhelming evidence that mean lifetimes in the six groups are different (p-value $< .0001$; analysis of variance F-test). Answers to the five particular questions are indicated as follows:

(a)   There is convincing evidence that lifetime is increased as a result of restricting the diet from 85 kcal/wk to 50 kcal/wk (one-sided p-value $< .0001$; t-test). The increase is estimated to be 9.6 months (95% confidence interval: 7.3 to 11.9 months).

(b)   There is no evidence that reducing the calories before weaning increased lifetime, when the calorie intake after weaning is 50 kcal/wk (one-sided p-value $= .32$; t-test).

Display 5.3   Structure of planned comparisons among groups in the diet restriction study

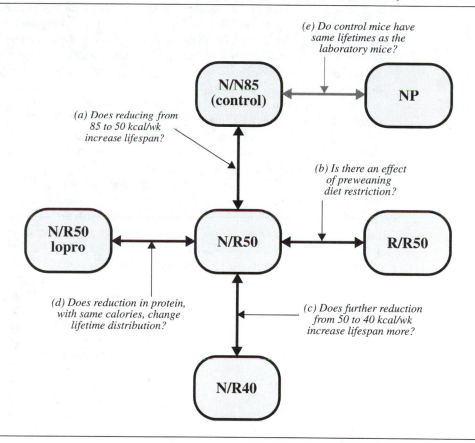

*(e) Do control mice have
same lifetimes as the
laboratory mice?*

N/N85
(control)

NP

*(a) Does reducing from
85 to 50 kcal/wk
increase lifespan?*

*(b) Is there an effect
of preweaning
diet restriction?*

N/R50
lopro

N/R50

R/R50

*(d) Does reduction in protein,
with same calories, change
lifetime distribution?*

*(c) Does further reduction
from 50 to 40 kcal/wk
increase lifespan more?*

N/R40

A 95% confidence interval for the amount by which the lifetime under the R/R50 diet exceeds the lifetime under the N/R50 diet is −1.7 to 2.9 months.

(c)   Further restriction of the diet from 50 to 40 kcal/wk tended to increase lifetime by an estimated 2.8 months (95% confidence interval: 0.5 to 5.1 months). The evidence that this effect is greater than zero is moderate (one-sided p-value = .017; t-test). (The combined effect of the reduction from 85 to 40 kcal/wk is estimated to be 12.4 months. This is a 38% increase in mean lifetime. If extended to human subjects, a 50% reduction in caloric intake might increase typical lifetimes—of 80 years—to 110 years.)

(d)   There was moderate evidence that lifetime was *decreased* by the lowering of protein in addition to the 50 kcal/wk diet (two-sided p-value = .024; t-test). The estimated decrease in lifetime is 2.6 months (95% confidence interval: 0.3 to 4.9 months).

(e)   There is convincing evidence that the control mice tended to live longer than the mice on the nonpurified diet (one-sided p-value < .0001).

5.1.2 The Spock Conspiracy Trial—An Observational Study

In 1968 Dr. Benjamin Spock was tried in United States District Court of Boston on charges of conspiring to violate the Selective Service Act by encouraging young men to resist being drafted into military service for Vietnam. The defense in that case challenged the method by which jurors were selected, claiming that women—many of whom had raised children according to popular methods developed by Dr. Spock—were underrepresented. In fact, the Spock jury had no women.

Boston juries are selected in three stages. From the City Directory, the Clerk of the Court selects at random 300 names for potential jury duty. Before a trial, a *venire* of 30 or more jurors is selected from the 300 names, again—according to the law—at random. An actual jury is selected from the venire in a nonrandom process allowing each side to exclude certain jurors for a variety of reasons.

The Spock defense pointed to the venire for their trial, which contained only one woman. That woman was released by the prosecution, making an all-male jury. Defense argued that the judge in the trial had a history of venires in which women were systematically underrepresented, contrary to the law. They compared this district judge's recent venires with the venires of six other Boston area district judges. The percents of women in those venires are presented in Display 5.4 as stem-and-leaf diagrams. (Data from H. Zeisel and H. Kalven, Jr., "Parking Tickets, and Missing Women: Statistics and the Law," in J. M. Tanur, F. Mosteller, W. H. Kruskal, R. F. Link, R. S. Pieters, and G. R. Rising, eds., *Statistics: A Guide to the Unknown*, San Francisco: Holden-Day, 1972.)

Display 5.4 Percents of women in 30-juror venires for Boston area U.S. District Court trials, grouped according to the judge presiding

	Spock Trial Judge	A	B	C	D	E	F
0	64,87						
1	33,36,50,52,77,86	68				77,97	65
2	31		70,89	10,34,75,75	43,97	15,79	07,35,64,67,95,98
3		08,36	20,27,55	05,19,25,38,38		48	19,62
4		05,89	56			02	

Legend: 4|89 represents a venire with 48.9% women.

There are two key questions: (1) Is there evidence that women are underrepresented on the Spock judge's venires compared to the venires of the other judges? and (2) Is there any evidence that there are differences in women's representation in the venires of the other six judges? The first question addresses the key issue, but the second question has considerable bearing on the interpretation of the first. If the other judges all had about the same percentage of women on their venires while the Spock judge had significantly fewer women, this would make a strong statement about that particular judge. But if the percentages of women in the venires of the other judges are all different, this would put a very different perspective on any difference that the Spock judge's venires would have from the average of the other judges' venires.

Summary of Statistical Findings

There is no evidence that the mean proportion of women on venires differed among the six other judges (p-value $= .26$; analysis of variance F-test). There is convincing evidence that the mean proportion of women in the Spock judge's venires was lower than the average of the means for the other judges (one-sided p-value $= .0000007$; t-test). The mean percent of women on the Spock judge's venires was estimated to be 15.1 percentage points less than the average of the mean percents for the other judges (95% confidence interval: 9.7% to 20.4%).

Scope of Inference

There is no indication that the venires in this observational study were randomly selected, so inference to some population is speculative. Thinking of the p-values as approximate p-values for permutation tests, however, leads one to conclude that the Spock judge tended to have a lower proportion of females on his venires than did the other judges—more so than can be explained by chance—at least for the venires included in this investigation.

5.2 Comparing Any Two of the Several Means

When subjects in a study are divided into distinct experimental or observational categories, the study itself is said to be a *several group* problem, or a *one-way classification* problem. Mice were divided into six experimental groups; samples of venires were obtained for seven judges. A typical analysis of several-group data involves graphical exploration (like side-by-side box plots), consideration of transformations, initial screening to evaluate differences between all groups, and inferential techniques chosen to specifically address the questions of interest.

In the two-sample problem the questions of interest require inference about $\mu_2 - \mu_1$. In the several-sample problem the questions of interest may involve a few pairwise differences of means, like $\mu_2 - \mu_1$ and $\mu_3 - \mu_1$; all possible pairwise differences of means; or specific linear combinations of means, like

$$[-1 \times \mu_1] + [(1/2) \times \mu_2] + [(1/2) \times \mu_3].$$

This aspect of the data structure requires careful attention to how the questions of interest can be addressed through model parameters. If there are multiple questions, as for example "Does group 1 differ from group 2, does group 1 differ from group 3, and does group 2 differ from group 3?" then attention to interpretations of multiple, simultaneous statistical inferences is important. This *multiple comparison* problem is discussed further in the next chapter. For now, the discussion focuses on the single comparison of any two means.

5.2.1   An Ideal Model for Several-sample Comparisons

The ideal population model, upon which the standard tools are derived, is a straightforward extension of the normal model for two-sample comparisons: (1) the populations have normal distributions. (2) The population standard deviations are all the same. (3) Observations within each sample are independent of each other. (4) Observations in any one sample are independent of observations in other samples.

As in the two-sample problem, the equal standard deviation model is entertained not because all data sets with several groups of measurements necessarily fit the description but because (1) it is conceptually difficult to compare populations with unequal variability, (2) it is statistically difficult as well, (3) for many problems a treatment is associated with an effect on the mean but not on the standard deviation, and (4) for many problems with unequal standard deviations it is possible to transform the data in such a way that the equal-spread model is valid on the transformed scale.

Notation

The symbols for population parameters are the same as before. A population mean is denoted by the Greek letter μ with a subscript indicating its group. The Greek letter σ is used to represent the standard deviation common to all the sampled populations. The number of samples will be represented by I, and the number of observations within the ith sample will be represented by n_i. The total number of observations from all groups combined will be $n = n_1 + n_2 + \cdots + n_I$. There are $I + 1$ parameters in the ideal model—the I means $\mu_1, \mu_2, \ldots, \mu_I$, and the single standard deviation σ.

Treatment Effects for Randomized Experiments

The discussion that follows in this chapter will continue to use the terminology of samples rather than of treatment groups, even though the methods also apply to data from randomized experiments with I treatment groups. An additive treatment effect model asserts that an experimental unit that would produce a response a of Y_1 on treatment 1 would produce a response of $Y_1 + \delta_1$ on treatment 2, $Y_1 + \delta_2$ on treatment 3, and so on. As before, exact randomization tests are available for inferences about the δ's, but approximations based on the tools developed for random samples from populations are usually adequate. The practical upshot is that data from randomized experiments will be analyzed in exactly the same way as samples from populations, but concluding statements will be worded in terms of treatment effects rather than differences in population means.

5.2.2   The Pooled Estimate of the Standard Deviation

The mean from the ith population, μ_i, is estimated by the average from the ith sample, $\overline{Y}_i$. The variance σ^2 is estimated separately by $s_i{}^2$ from each of the I samples. If all the sample sizes are equal then the best single estimate of σ^2 is their average. When the samples have different numbers of observations, a *weighted average* is more appropriate, where the estimates from larger samples are given more weight than those from smaller samples. The *pooled estimate of variance*, $s_p{}^2$, is a weighted average of sample variances in which each sample variance receives the weight of its degrees of freedom:

$$s_p^2 = \frac{(n_1-1)\,s_1^2 + (n_2-1)\,s_2^2 + \cdots + (n_I-1)s_I^2}{(n_1-1) + (n_2-1) + \cdots + (n_I-1)}.$$

Its associated degrees of freedom are the sum of degrees of freedom from all samples, $n - I$. An illustration of the calculations for the diet restriction data are shown in Display 5.5.

Display 5.5 Pooled estimate of standard deviation; diet restriction data

Group	n	Sample SD
NP	49	6.1
N/N 85	57	5.1
N/R50	71	7.8
R/R50	56	6.7
N/R50 lopro	56	7.0
N/R40	60	6.7

(1) *Calculate the pooled estimate of variance, s_p^2*

$$s_p^2 = \frac{(49-1)(6.1)^2 + (57-1)(5.1)^2 + (71-1)(7.8)^2 + (56-1)(6.7)^2 + (56-1)(7.0)^2 + (60-1)(6.7)^2}{(49-1) + (57-1) + (71-1) + (56-1) + (56-1) + (60-1)}$$

(2) *s_p is the square root.*

$$= \frac{15{,}313.90}{343} = 44.647; \quad s_p = \sqrt{44.647} = 6.68$$

$$\text{df} = 343$$

(3) *d.f. is the denominator.*

5.2.3 *t*-Tests and Confidence Intervals for Differences of Means

If $\overline{Y}_i$ is an average based on a sample from a population with mean μ_i and variance σ_i^2, and if $\overline{Y}_j$ is an average based on an independent sample from a population with mean μ_j and variance σ_j^2, then the sampling distribution of $\overline{Y}_i - \overline{Y}_j$ has variance $\sigma_i^2/n_i + \sigma_j^2/n_j$. If the two population variances are equal this reduces to $\sigma^2[(1/n_i) + (1/n_j)]$. This may be estimated by replacing the unknown σ^2 by its best estimate. The important aspect of this in the one-way classification is that if the variances from all I populations can be assumed to be equal, then the best estimate is the pooled estimate of variance from all groups. So, for example, the standard error of $\overline{Y}_3 - \overline{Y}_2$ is

$$\text{SE}(\overline{Y}_3 - \overline{Y}_2) = s_p\sqrt{\frac{1}{n_3} + \frac{1}{n_2}},$$

where s_p is the pooled estimate from *all* groups, with $(n - I)$ degrees of freedom.

The theory for confidence intervals and tests is the same as for the two-independent sample problem. A 95% confidence interval for $\mu_3 - \mu_2$ is $\overline{Y}_3 - \overline{Y}_2 \pm t_{n-1}(.975) \times \mathrm{SE}(\overline{Y}_3 - \overline{Y}_2)$ and a t-statistic for testing the hypothesis that $\mu_3 - \mu_2$ equals zero is $(\overline{Y}_3 - \overline{Y}_2)/\mathrm{SE}(\overline{Y}_3 - \overline{Y}_2)$. These are illustrated in Display 5.6.

Display 5.6 A confidence interval for $\mu_3 - \mu_2$ and a test that $\mu_3 - \mu_2 = 0$; (diet restriction data)

---

① *Get averages, sample sizes, and pooled estimate of standard deviation.*

Group	3: N/R50	2: N/N85
Sample size	71	57
Average (months)	42.3	32.7

Pooled estimate of σ: $s_p = 6.68$ months; d.f. $= 343$ (from Display 5.5)

② *Compute the estimate of $\mu_3 - \mu_2$ and its standard error.*

Estimate: $\overline{Y}_3 - \overline{Y}_2 = \quad 42.3 - 32.7 \quad = \quad 9.6$ months

$\mathrm{SE}(\overline{Y}_3 - \overline{Y}_2) = \quad 6.68 \sqrt{\dfrac{1}{71} + \dfrac{1}{57}} \quad = \quad 1.2$ months

③ *95% confidence interval for $\mu_3 - \mu_2$.*

$t_{343}(.975) \quad = \quad 1.96$

95% CI: $9.6 \pm (1.96)(1.2) \quad = \quad$
    → 7.3 months
    → 11.9 months

④ *Test the hypothesis that $\mu_3 - \mu_2 = 0$.*

$t\text{-stat} \quad = \quad \dfrac{9.6}{1.2} \quad = \quad 8.08 \quad\longrightarrow\quad$ two-sided p-value $<.0001$

5.3 The One-way Analysis of Variance *F*-Test

One question often asked, possibly in a first stage of the analysis, is "are there differences between *any* of the means?" The *analysis of variance* (ANOVA) *F*-test provides evidence in answer to this question. The term *variance* should not mislead; this is most definitely a test about *means*. It assesses mean differences by comparing the amounts of variability explained by different sources.

5.3.1   **The Extra-Sum-of-Squares Principle**

One model for the Spock data is that the percentage of women on venires for judge i comes from a normal distribution with mean μ_i and variance σ^2, for i from 1 to 7. A hypothesis for initial exploration is that all seven means are equal. That is, the null hypothesis is H: $\mu_1 = \mu_2 = \cdots = \mu_7$, and the alternative is that at least one of the means differs from the others.

Full and Reduced Models

Answering a question of interest is formulated as a problem of comparing two models for the response means. A *full model* is a general model taken as a starting point. The *reduced model* is a special case of the full model obtained by imposing the restrictions of the null hypothesis.

For comparing equality of all means in the several-sample problem, the full model is the one that has a separate mean for each group. The reduced model, obtained from the full model by supposing the hypothesis of equal means is true, specifies a single mean for all populations. For the Spock data, the means from the two models are the following:

Group:	1	2	3	4	5	6	7
Full (separate-means) model:	μ_1	μ_2	μ_3	μ_4	μ_5	μ_6	μ_7
Reduced (equal-means) model:	μ	μ	μ	μ	μ	μ	μ

The terminology of *full* and *reduced* models provides a framework for a general procedure called the *extra-sum-of-squares F*-test. For any particular application, the models will have different names that refer more specifically to the problem at hand. As a test of equality of means in one-way classification, it makes more sense to call the full model the *separate-means* model and the reduced model the *equal-means* model.

Fitting the Models

The idea behind analysis of variance is to estimate the parameters in both the full and reduced models and to see whether the variability of responses about the estimated means is comparable in the two models. The *estimated* means for each group are different for the two models:

Group:	1	2	3	4	5	6	7
Full (separate-means) model:	$\overline{Y}_1$	$\overline{Y}_2$	$\overline{Y}_3$	$\overline{Y}_4$	$\overline{Y}_5$	$\overline{Y}_6$	$\overline{Y}_7$
Reduced (equal-means) model:	$\overline{Y}$	$\overline{Y}$	$\overline{Y}$	$\overline{Y}$	$\overline{Y}$	$\overline{Y}$	$\overline{Y}$

where $\overline{Y}$ is the average of all observations, called the *grand average*.

Residuals

Associated with each observation in the data set is an estimated group mean based on the full model and a different estimated group mean based on the reduced model. Also associated with each observation is a residual for each model. A residual is the observation value minus its estimated mean. So, for observation Y_{ij} = the percentage of women on the jth venire for judge i, the residual from the full model is $Y_{ij} - \overline{Y}_i$ and the residual from the reduced model is $Y_{ij} - \overline{Y}$. Display 5.7 shows the sets of estimated means and residuals for the Spock trial data in both the full (separate-means) and the reduced (equal-means) models. Notice that the residuals tend to be larger for the equal-means model.

Display 5.7   Estimated means and residuals from two models for mean percentage of women (%W) in venires: Spock trial data

Large residuals mean that the model fits poorly.

Judge	%W	Equal Means Est.	Equal Means Res.	Separate Means Est.	Separate Means Res.
Spock	6.4	26.6	−20.2	14.6	−8.2
Spock	8.7	26.6	−17.9	14.6	−5.9
Spock	13.3	26.6	−13.3	14.6	−1.3
Spock	13.6	26.6	−13.0	14.6	−1.0
Spock	15.0	26.6	−11.6	14.6	0.4
Spock	15.2	26.6	−11.4	14.6	0.6
Spock	17.7	26.6	−8.9	14.6	3.1
Spock	18.6	26.6	−8.0	14.6	4.0
Spock	23.1	26.6	−3.5	14.6	8.5
A	16.8	26.6	−9.8	34.1	−17.3
A	30.8	26.6	4.2	34.1	−3.3
A	33.6	26.6	7.0	34.1	−0.5
A	40.5	26.6	13.9	34.1	6.4
A	48.9	26.6	22.3	34.1	14.8
B	27.0	26.6	0.4	33.6	−6.6
B	28.9	26.6	2.3	33.6	−4.7
B	32.0	26.6	5.4	33.6	−1.6
B	32.7	26.6	6.1	33.6	−0.9
B	35.5	26.6	8.9	33.6	1.9
B	45.6	26.6	19.0	33.6	12.0
C	21.0	26.6	−5.6	29.1	−8.1
C	23.4	26.6	−3.2	29.1	−5.7
C	27.5	26.6	0.9	29.1	−1.6
C	27.5	26.6	0.9	29.1	−1.6
C	30.5	26.6	3.9	29.1	1.4
C	31.9	26.6	5.3	29.1	2.8
C	32.5	26.6	5.9	29.1	3.4
C	33.8	26.6	7.2	29.1	4.7
C	33.8	26.6	7.2	29.1	4.7
D	24.3	26.6	−2.3	27.0	−2.7
D	29.7	26.6	3.1	27.0	2.7
E	17.7	26.6	−8.9	27.0	−9.3
E	19.7	26.6	−6.9	27.0	−7.3
E	21.5	26.6	−5.1	27.0	−5.5
E	27.9	26.6	1.3	27.0	0.9
E	34.8	26.6	8.2	27.0	7.8
E	40.2	26.6	13.6	27.0	13.2
F	16.5	26.6	−10.1	26.8	−10.3
F	20.7	26.6	−5.9	26.8	−6.1
F	23.5	26.6	−3.1	26.8	−3.3
F	26.4	26.6	−0.2	26.8	−0.4
F	26.7	26.6	0.1	26.8	−0.1
F	29.5	26.6	2.9	26.8	2.8
F	29.8	26.6	3.2	26.8	3.0
F	31.9	26.6	5.3	26.8	5.1
F	36.2	26.6	9.6	26.8	9.4

In the equal-means model, estimated means are equal to the grand average.

In the separate-means model, estimated means are the group averages.

If the null hypothesis is correct, then the two models should be about equal in their ability to explain the data, and the magnitudes of the residuals should be about the same. If the null hypothesis is incorrect, the magnitudes of the residuals from the equal-means model will tend to be larger. Their larger sizes will reflect model inadequacy.

Residual Sums of Squares

A single summary of the magnitude of the residuals for a particular model is the *residual sum of squares* for that model, i.e., the sum of the squared residuals. By adding the squares of the two sets of residuals in Display 5.7 separately, one finds that the residual sum of squares for the equal-means model is 3,791.53 and the sum of the squared residuals from the separate-means model is 1,864.45. Is this large difference $(3,791.53 - 1,864.45 = 1,927.08)$ due to the relatively poor fit of the equal-means model, or can it be explained by sampling variability? The *F*-test answers this question precisely.

General Form of the Extra-Sum-of-Squares F-Statistic

The *extra sum of squares* is the single number that summarizes the difference in sizes of residuals from the full and reduced models. As just calculated above,

$$\text{Extra sum of squares} =$$
$$\text{Residual sum of squares (reduced)} - \text{Residual sum of squares (full).}$$

A residual sum of squares measures the variability in the observations that remains unexplained by a model, so the extra sum of squares measures the amount of unexplained variability in the reduced model that is explained by the full model.

To determine whether there is too much variability left unexplained by the reduced model, it is necessary to compare it with the variability left unexplained by the full model, in an *F*-statistic:

$$F\text{-statistic} = \frac{(\text{Extra sum of squares})/(\text{Extra degrees of freedom})}{\hat{\sigma}^2_{\text{full}}}$$

where the *extra degrees of freedom* are the number of parameters in the mean for the full model minus the number of parameters in the mean for the reduced model, and $\hat{\sigma}^2_{\text{full}}$ is the estimate of σ^2 based on the full model. Thus, the *F*-statistic is the extra sum of squares per extra degree of freedom, scaled by the best estimate of variance.

The F-Test

Large *F*-statistics are associated with large differences in the size of residuals from the two models. They supply evidence against the hypothesis of equal means and in favor of a model with different means. The test is summarized by its *p*-value, the chance of finding an *F*-statistic as large as or larger than the observed one when all means are, in fact, equal.

F-Distributions

If all means are equal, the sampling distribution of the *F*-statistic is that of an *F-distribution*, which depends on two known parameters: the *numerator degrees of freedom* and the *denominator degrees of freedom.* For each degrees of freedom pair, there is a separate F-distribution. The F was given to this class of distributions by George Snedecor, honoring the British statistician Sir Ronald Fisher.

Theoretical histograms for four of the F-distributions are shown in Display 5.8. Notice that F values in the general range of 0.5 to 3.0 are fairly typical. An F-statistic in this range would not be considered as very strong evidence of unequal means for most degree of freedom combinations. An F-statistic in the range from 3.0 to 4.0 would be highly unlikely with large degrees of freedom in both numerator and denominator but would be only moderately suggestive of differences with smaller degrees of freedom. An F-statistic larger than 4.0 is strong evidence against equal means except for the smallest degrees of freedom, particularly in the denominator. (*Note:* All the curves have the same area below them. This means that the $F_{2,2}$ and $F_{30,2}$ curves must have considerable area, spread thinly, off the graph to the right.)

Display 5.8   Four F-distributions, having different degrees of freedom

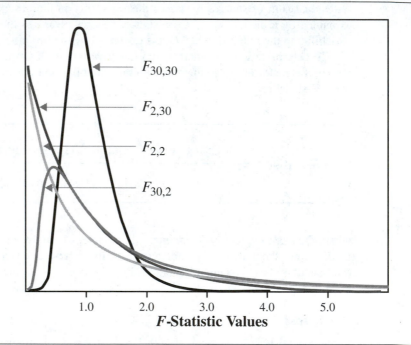

Tables of the percentage points for many F-distributions are given in Appendix A.4 on page 712. Since there are so many F-distributions, the tables do not include many percentage points. However, statistical computer packages, additional published tables, and some pocket calculators are available for more detailed computation. The p-value indicated by bubble 7 in Display 5.9, showing the F-test results for the Spock trial data, was obtained with a statistical computer package.

Display 5.9 Analysis of variance table: a test for equal mean percents of women in venires of seven judges (Spock data)

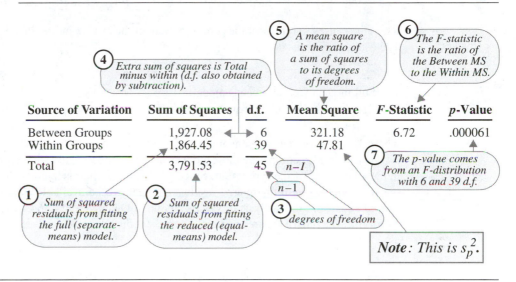

5.3.2 The Analysis of Variance Table for One-way Classification

The analysis of variance table organizes and displays the calculations used in the *F*-test. Analysis of variance tables extend to more complicated structures, where several *F*-tests simultaneously evaluate how well different models fit the data. The ANOVA table for the Spock data appears in Display 5.9.

The table is organized so that all calculations follow a fixed pattern after entry of the residual sums of squares from the full (separate-means) and reduced (equal-means) models and their degrees of freedom. The between-groups sum of squares and its degrees of freedom are found by subtraction. Each mean square equals its sum of squares divided by its degrees of freedom. The *F*-statistic equals the ratio of the between-groups mean square to the within-groups mean square; and the evidence is summarized by the *p*-value, equal to the upper tail area from the *F*-distribution, whose numerator degrees of freedom are those associated with between groups and whose denominator degrees of freedom are those associated with within groups. The *p*-value of .000061 provides convincing evidence that at least one of the judges' means differs from one of the others' means.

5.3.3 More Applications of the Extra-Sum-of-Squares *F*-Test

The one-way analysis of variance *F*-test for the Spock data provided convincing evidence that the means were not equal for all seven judges. The next step is to see whether the six other judges (not Spock's) have different mean percentages of women on their venires. To test the hypothesis $H: \mu_2 = \mu_3 = \cdots = \mu_7$ (with the Spock judge mean not necessarily equal) one may use an extra-sum-of-squares *F*-test for comparing the following full and reduced models:

Group:	1	2	3	4	5	6	7
Full model (separate-means model):	μ_1	μ_2	μ_3	μ_4	μ_5	μ_6	μ_7
Reduced model (others-equal model):	μ_1	μ_0	μ_0	μ_0	μ_0	μ_0	μ_0

where μ_0 is used to represent the common mean of the last six judges in the reduced model. The estimated means are:

Group:	1	2	3	4	5	6	7
Full model (separate-means model):	$\overline{Y}_1$	$\overline{Y}_2$	$\overline{Y}_3$	$\overline{Y}_4$	$\overline{Y}_5$	$\overline{Y}_6$	$\overline{Y}_7$
Reduced model (others-equal model):	$\overline{Y}_1$	$\overline{Y}_0$	$\overline{Y}_0$	$\overline{Y}_0$	$\overline{Y}_0$	$\overline{Y}_0$	$\overline{Y}_0$

where $\overline{Y}_0$ is the average percent among all venires for the other six judges.

The F-test for comparing these reduced and full models is not automatically computed in the analysis of variance program. Since the full model in this hypothesis is the separate-means model, its sum of squared residuals happens to be available as the within-groups sum of squares in the analysis of variance table in Display 5.9 (1,864.45 on 39 degrees of freedom). The sum of squared residuals from the reduced model can be calculated manually as

$$\sum_{j=1}^{n_1}(Y_{1j} - \overline{Y}_1)^2 + \sum_{i=2}^{7}\sum_{j=1}^{n_i}(Y_{ij} - \overline{Y}_0)^2,$$

which turns out to be 2,190.90. It could also be found as the within group sum of squares from a two-group analysis of variance, when all venires of the other judges are lumped into one group. The degrees of freedom are the sample size (46) minus the number of parameters in the mean (2), or 44.

The extra-sum-of-squares F-statistic is $[(2,190.90 - 1,864.45)/(44 - 39)]/(1,864.45/39) = 1.37$, and the p-value is the proportion of an F-distribution on 5 and 39 degrees of freedom that exceeds 1.366, which is .26. There is consequently no evidence from these data of differences in means among the six other judges.

The next step in the analysis, assuming the six other judges do have equal means, is a test of the hypothesis that the Spock judge's mean is equal to the common mean of the other six. An F-test compares a full model with two means (one for the Spock judge and one for the other six) to a reduced model with a single mean:

Group:	1	2	3	4	5	6	7
Full model (others-equal model):	μ_1	μ_0	μ_0	μ_0	μ_0	μ_0	μ_0
Reduced model (equal means model):	μ	μ	μ	μ	μ	μ	μ

Notice that the previous step in the analysis has indicated the appropriateness of the two-parameter model, with a common mean for the other judges. This then becomes the full model for the inferential question: is the Spock judge's mean different from the mean common to the six others, $H: \mu_1 = \mu_0$? The reduced model, consequently, is the equal-means model. The sum of squared residuals from this reduced model is the total sum of squares from the one-way analysis of variance table: 3,791.53 on 45 degrees of freedom. The sum of squared residuals from this full model is 2,190.90 on 44 degrees of freedom. The *F*-statistic is $[(3{,}791.53 - 2{,}190.90)/(45 - 44)]/(2{,}190.90/44)$, which is equal to 32.14. By comparison to an *F*-distribution on 1 and 44 degrees of freedom, the *p*-value is found to be 0.000001. There is convincing evidence that the Spock judge's mean differs from the others.

Summary of ANOVA Tests Involving More Than Two Models

The others-equal model (using two parameters to describe the means) lies intermediate between the equal-means model (with one parameter) and the separate-means model (with seven parameters). It is a simplification of the separate-means model, while the equal-means model is a simplification of it.

Display 5.10 illustrates the additive nature of the extra sums of squares for comparing these three nested models. The residual sum of squares in each model is viewed as a measure of the unexplained variability from that model, or as a *distance* from the model to the data. The residual sum of squares for the equal means model is 3,791.53. By including

---

Display 5.10   Residual sums of squares as distances from the data to proposed models for the means (Spock trial example)

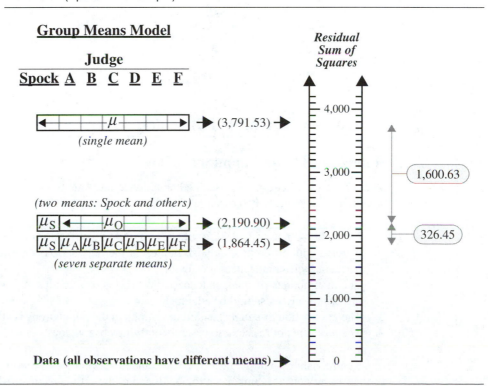

one extra parameter to allow the Spock judge to have a different mean (the others-equal model), the unexplained variability is reduced by 1,600.63. By including an additional five extra parameters (the separate-means model) the unexplained variability is further reduced by 326.45. Notice that the extra sum of squares in the comparison of the separate-means model to the equal-means model (i.e., the between-group sum of squares) is the sum of the two component sums of squares ($1,927.08 = 1,600.63 + 326.45$).

Display 5.11 suggests a detailed analysis of variance table suitable for summarizing the computations of both F-tests. The structure of the table centers on the reductions in residual sum of squares that are accomplished by the addition of parameters into the model for the means. These reductions are listed as sums of squares (sums of squared residuals further *explained* by the particular source), and the numbers of additional parameters are listed as degrees of freedom. Mean squares are the ratios of sums of squares to their degrees of freedom.

Display 5.11   Complete analysis of variance table for three tests involving the mean percents of women in venires of seven judges

Source of Variation	Sum of Squares	d.f.	Mean Square	F-Statistic	p-Value
Between groups	1,927.08	6	321.18	6.72	0.000061
Spock v. others	1,600.63	1	1,600.63	32.14	0.000001
Among others	326.45	5	65.29	1.37	0.26
Within groups	1,864.45	39	47.81		
Total	3,791.53	45			

5.4 Robustness and Model Checking

5.4.1 Robustness to Assumptions

The robustness of t-tests, F-tests, and confidence intervals to departures from model assumptions can be described essentially in the same way as in Chapter 3:

1. Normality is not critical. Extremely long-tailed distributions or skewed distribution coupled with different sample sizes present the only serious distributional problems, particularly if sample sizes are small.
2. The assumptions of independence within and across groups are critical. If lacking, different analyses should be attempted.
3. The assumption of equal standard deviations in the populations is crucial.
4. The tools are not resistant to severely outlying observations.

The robustness with respect to equal standard deviations in the populations requires further discussion. It is important to pool together the estimates of variability from all the groups in order to make the most powerful comparisons possible. If one of the populations has a very different standard deviation, however, serious problems may result, even if the comparisons do not involve the mean from that population.

A computer was used to simulate situations in which three samples from normal populations were selected, with the aim of obtaining a 95% confidence interval for the difference in means from the first two. Six different configurations of population standard deviations were used with each of four sample size combinations. Based on 2,000 simulated data sets, the results appear in Display 5.12. The procedure is robust against unequal standard deviations for those situations where success rates are approximately equal to 95%.

Display 5.12   Success rates for 95% confidence intervals for $\mu_1 - \mu_2$ from samples simulated from normal populations with possibly different SDs

			$\sigma_2=\sigma_1$			$\sigma_2=2\sigma_1$		
n_1	n_2	n_3	$\sigma_3=\sigma_1$	$\sigma_3=2\sigma_1$	$\sigma_3=4\sigma_1$	$\sigma_3=\sigma_1$	$\sigma_3=2\sigma_1$	$\sigma_3=4\sigma_1$
10	10	10	95.4	98.9	99.9	91.9	96.8	99.6
20	10	10	95.5	98.7	99.8	84.8	91.7	98.9
10	20	10	94.1	98.7	99.9	97.0	98.8	99.8
10	10	20	95.6	99.6	99.9	90.4	97.5	99.9

The simulations suggest that if σ_3 is quite different from σ_1 and σ_2, then the actual success rates of 95% confidence intervals for $\mu_1 - \mu_2$ can be quite different from 95%. Unlike the result for the two-sample tools, the effect of unequal standard deviations can be serious even if the three sample sizes are equal.

5.4.2   Diagnostics Using Residuals

Initial graphical examination of the data by stem-and-leaf diagrams, box plots, or histograms is important. If there is a large number of groups, then side-by-side box plots are particularly useful, since they eliminate much of the clutter contained in other displays. As in the two-sample problem, initial assessment helps identify (1) the centers, (2) the relative spreads, (3) the general shapes of the distributions, and (4) the presence of outliers. If the spreads are quite different, transforming the data to a different scale should be considered.

An important tool, which extends to almost all statistical methods in this book, is a *residual plot*. Because the residuals $\overline{Y}_{ij} - \overline{Y}_i$ are the original observations with their group averages subtracted out, they exhibit the variation of the observations without the visual interference caused by differences between group means. A scatterplot of these residuals versus the group averages can reveal a relationship between the spread and the group means, which may be exploited to improve the analysis.

Display 5.13 shows the residual plot from the diet restriction and longevity data. The features to look for in such a plot are (1) an increase in the spread from left to right, in a *funnel-shaped* pattern, (which would suggest the need for a log or some other transformation) or (2) seriously outlying observations. The residual plot in Display 5.13 has neither

Display 5.13   Residual plot: lifetimes of mice fed six different diets

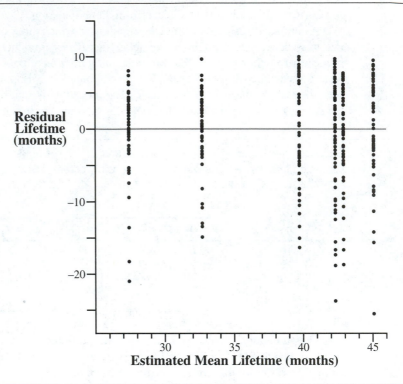

feature, suggesting that analysis on the natural scale is adequate. It is evident that the distributions of lifetimes are skewed. For these large samples, however, there should be no problem in relying on the inferential tools derived from the normal model.

Finally, if the data were collected over time, a plot of residuals versus the time or order of data collection will help to reveal any serial effects that might be present. A pattern of generally increasing or generally decreasing residuals indicates a time trend, which may be accounted for using regression techniques. A pattern in which residuals close to each other tend to be more alike (or perhaps more different) than any two arbitrarily chosen residuals may indicate serial correlation. Formal investigation into serial correlation, and methods that account for it, are provided in Chapter 15.

5.5 Related Issues

5.5.1 Further Illustration of the Different Sources of Variability

The analysis of variance is a general method of partitioning total variation into several components. This section attempts to shed further light on that partitioning.

A computer was used to generate independent random samples of size four each from nine normal populations having the same mean and the same standard deviation. The samples are illustrated in Display 5.14. The smooth curve above is the common population histogram, centered at the mean, μ. The horizontal axes below the histogram locate the separate samples. The ticks above each axis locate the four sample values, and the single arrow below each axis locates the sample's average.

Display 5.14   Three sources of variation for data simulated from the equal-means model, and mean values of averages of squares, from statistical theory

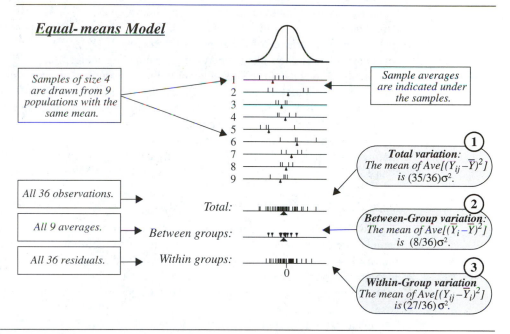

Three additional axes appear below the nine samples, labeled according to the three sources of variation in the analysis of variance table. The *Total* axis shows all 36 sample values on one common axis. Sample values are ticked on top of the axis and the grand average of all 36 is marked below the axis. On the *Between Groups* axis, bullets locate the nine sample averages on top, and their average—also the grand average—is marked by the arrowhead on the bottom. The *Within Groups* axis displays all 36 *residuals* from the separate-means model. These are the observations minus their group averages.

Variation When All Means Are Equal

Variation is different on the three different summary axes for two basic reasons: (1) *observations are always closer to their sample average than to their population mean.* This is easily visible in sample #1, where all four observations fall below the population mean and where the sample average follows them. And (2) *sample averages are less variable than individual sample values.* Reason (2) explains why the between-group variation is visibly less than the total variation. Reason (1) explains why the within-group variation is also visibly less than the total variation.

The means of the sampling distributions of the average squared distance of a tick from the axis center are listed in the three bubbles. The average of the squared residuals is $(1/36) \sum \sum (Y_{ij} - \overline{Y}_i)^2$. The mean of its sampling distribution is $(27/36)\sigma^2$. This is less than σ^2 because of reason (1) mentioned above, and illustrates the need for a degrees of freedom adjustment: the mean square of the residuals, $[1/(36 - 9)] \sum \sum (Y_{ij} - \overline{Y}_i)^2$ is an unbiased estimate of σ^2. (Its sampling distribution has mean σ^2.)

Since the populations have the same mean, the nine $\overline{Y}_i$'s are like a sample of size 9 from a normal population with mean μ and variance $\sigma^2/4$. From bubble 2 it is evident that $\sum (\overline{Y}_i - \overline{Y})^2/9$ is an unbiased estimate of $(8/9)(\sigma^2/4)$. It is not surprising then that the sample variance of the sample of averages, $\sum (\overline{Y}_i - \overline{Y})^2/8$, is an unbiased estimate of $\sigma^2/4$. It follows that the between-group mean square, which is $\sum n_i (\overline{Y}_i - \overline{Y})^2/8$, with $n_i = 4$ for all i in this case, is an unbiased estimate of σ^2. Thus, if the hypothesis of equal population means is correct, the numerator and the denominator of the F-statistic (the between-groups and within-groups mean squares) are both unbiased estimates of σ^2, so the F-statistic should be close to 1. As shown in the next display, if the population means are, in fact, unequal, then the numerator of the F-statistic is estimating something larger than σ^2 and the F-statistic will tend to be substantially larger than 1.

Variation When the Means Are Different

Display 5.15 depicts simulated samples drawn from populations with *different* means. The samples are the same as in Display 5.14, but shifted to their population means. Notice that the within-group variation is unchanged, whereas the difference in population means results in a larger between-group variation, due to the term $\sum (\mu_i - \mu)^2$, where μ is the average of the 9 population means. Hence the F-statistic should be larger than in the equal-means case.

Data that turn out like the bottom three axes in Display 5.15 show strong evidence that the between-group variability is much larger than that expected from the equal means model. The one-way analysis of variance F-test formalizes the comparison. In particular, while the denominator of the F-statistic is always an estimate of σ^2, the numerator is estimating something larger than σ^2. The amount by which it exceeds σ^2 depends on how different the means are. An F-statistic much larger than 1 provides strong evidence of differences among the population means.

5.5.2 Kruskal–Wallis Nonparametric Analysis of Variance

One method for coping with seriously outlying observations is to replace all observation values by their ranks in a single combined sample and then apply a one-way analysis of variance F-test on the rank-transformed data. The Kruskal–Wallis test, which is available in many statistical computer packages, is similar in its approach but takes advantage of the known variance of the ranks.

Display 5.15 Variations in the several group problem for data simulated from the separate-means model

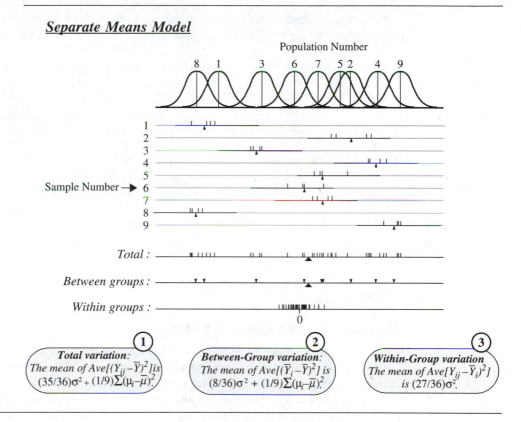

Separate Means Model

① Total variation:
The mean of $Ave[(Y_{ij}-\overline{Y})^2]$ is $(35/36)\sigma^2 + (1/9)\sum(\mu_i-\overline{\mu})^2$.

② Between-Group variation:
The mean of $Ave[(\overline{Y}_i-\overline{Y})^2]$ is $(8/36)\sigma^2 + (1/9)\sum(\mu_i-\overline{\mu})^2$.

③ Within-Group variation
The mean of $Ave[Y_{ij}-\overline{Y}_i)^2]$ is $(27/36)\sigma^2$.

The Kruskal–Wallis test statistic is

$$KW = 1/[\,\sigma_R^2\,] \times \text{Between Group Mean Square (of ranks)},$$

where σ_R^2 is the variance of all n ranks (using an $n-1$ divisor) and where n is the total number of observations in all groups. A p-value is found as the proportion of a chi-squared distribution (Appendix 4) on $(I-1)$ degrees of freedom that is larger than this test statistic.

Display 5.16 shows the rank-transformed data from the Spock trial example. Notice that Spock's judge has venires that rank quite low. The between-group sum of squares (which could be obtained from an analysis of variance table based on the data in Display

5.16) is 3,956.325. The variance of the ranks is 180.122. The value of the Kruskal–Wallis test statistic is therefore 21.96, and *p*-value is .0012 from a chi-square distribution with 6 degrees of freedom. Testing equality of the judges other than Spock's can be accomplished by performing a Kruskal–Wallis test ignoring the venires of the Spock judge. The Spock judge can be compared to the other judges combined using the rank-sum test.

Display 5.16   Spock trial data, rank-transformed

Judge	Rank of Venire from Smallest (1) to Largest (46) Percent Women								
Spock's	1	2	3	4	5	6	9.5	11	16
A	8	31	37	44	46				
B	22	26	34	36	41	45			
C	14	17	23.5	23.5	30	32.5	35	38.5	38.5
D	19	28							
E	9.5	12	15	25	40	43			
F	7	13	18	20	21	27	29	32.5	42

5.5.3 Random Effects

Rationale for the Random Effects Model

It has so far been assumed that there is direct interest in the particular groups chosen. Sometimes, however, the groups are selected as representative of some broader population, and an inference is to be drawn to that population. A distinction is made between the *fixed effects model*, in which the group means are fixed and the *random effects model*, in which the group means are a random sample from a population of means. For the case studies in this chapter there is direct interest in the particular groups, so fixed effects models were used.

To illustrate when each model is appropriate, suppose that measurements are taken on the yield of a machine operated by each of several operators. An analysis of variance may be used to compare the mean yields under the different operators. A fixed effects model would be appropriate if there was interest in only those particular operators. (They may constitute all the operators at the plant.) A random effects model would be appropriate if those operators were just a sample and if the question of interest pertained to the population of operators from which they were sampled. There may be interest, for example, in estimating the proportion of the yield variance that could be explained by between-operator variability in a plant with a large number of operators.

Is the random effects model the right one to use? There are two pertinent questions: (1) is inference desired to a larger set from which these groups are a sample, in which case one must also be concerned about? (2) are the groups (operators) truly a random sample from the larger set? A yes answer to the first question would ordinarily prompt a user to use the random effects model. Statistical inference to the larger population would only be justified, however, if there was also a yes answer to the second question. If there was no random sampling to obtain the particular operators, then the usual warnings about potential biases due to using nonrandom samples apply.

The Random Effects Model

In the fixed effects model, observed sample i is thought to be a random sample from a normal population with mean μ_i and variance σ^2. There are $I + 1$ parameters in the fixed effects model: the I means and the single variance σ^2.

In the random effects model the μ_i's themselves are thought to be a random sample from a normal population with mean μ and variance $\sigma_\mu{}^2$. The random effects model has 3 parameters: the overall mean μ, the within-group variance σ^2, and the between-group variance $\sigma_\mu{}^2$.

Analysis of the one-way classification random effects model involves a test of whether $\sigma_\mu{}^2$ is zero and an estimate of the ratio $\sigma_\mu{}^2/(\sigma_\mu{}^2 + \sigma^2)$. Notice that this ratio is between 0 and 1. It is 0 when there is no between-group variance, and it is 1 when there is no within-group variance. Since the denominator describes the total variance of the measurements, the ratio may be interpreted as the proportion of the total variance of the measurements that is explained by between-group variability. It is also called the *intraclass correlation*.

Estimation and Testing

The overall mean μ is estimated by the grand average $\overline{Y}$. The estimates of the two variances σ^2 and $\sigma_\mu{}^2$ are often found by equating the mean squares in the analysis of variance table to the means of their sampling distributions, under the random effects model. In particular, letting MS(W) and MS(B) represent the within-groups and between-groups means squares, respectively, one obtains

$$\text{Mean}\{\text{MS(W)}\} = \sigma^2$$

$$\text{Mean}\{\text{MS(B)}\} = \sigma^2 + \frac{1}{n(I-1)}\left(n^2 - \sum_{i=1}^{I} n_i^2\right)\sigma_\mu^2,$$

so the estimates of the variances are

$$\hat{\sigma}^2 = \text{MS(W)}$$

and

$$\hat{\sigma}_\mu^2 = \frac{n(I-1)[\text{MS(B)} - \text{MS(W)}]}{n^2 - \sum_{i=1}^{I} n_i^2},$$

with the modification that the latter is set to zero if the numerator turns out to be negative.

It is sometimes desired to test the hypothesis $H: \sigma_\mu{}^2 = 0$, against the alternative that it is greater than zero. It should be evident that this is analogous to the hypothesis that the means are all equal in the fixed effects model. The usual F-test for the fixed effects model (Section 5.3.2) is appropriate for this hypothesis as well.

The bottom line is that testing for between-group differences may be carried out in the usual way with an analysis of variance procedure. An additional, useful summary for the random effects model, however, is $\hat{\sigma}_\mu^2/(\hat{\sigma}_\mu^2 + \hat{\sigma}^2)$.

Example—Spock Trial Data

Although the questions of interest in the Spock example focus on the specific judges, the data set can be used as an example to demonstrate random effects, by ignoring Spock's judge and thinking of the six other judges as representative of some large population. The parameters in the random effects model are μ, the overall mean percentage of women on venires, the variance σ^2 of percentages about the judge mean, and the variance $\sigma_\mu{}^2$ of the population of judge means. A test of whether the between-judge variance is zero ($H: \sigma_\mu{}^2 = 0$) is the F-test from the standard analysis of variance. The p-value is .32, so the data are consistent with there being no between-judge variability. The estimates of $\sigma_\mu{}^2$ and σ^2 are 1.96 and 53.6, so it can be said that the proportion of variability that is due to differences between judges is $1.96/(1.96 + 53.6) = .035$. This number is also interpreted as the intraclass correlation—the correlation that percentages from two venires have if the venires come from the same judge. Of course, these inferential statements are speculative since the six judges were not, in fact, a random sample from a population of judges.

5.5.4   Separate Confidence Intervals and Significant Differences

Published research articles often present graphs of estimated means with separate confidence intervals for each. If a reader wishes to know whether two means are different, there is a fairly close—but not exact—relationship between the overlap in the confidence intervals and the result of a test of equal means. *The proper course of action for judging whether two means are equal is to carry out a t-test directly.* The comments here apply to a situation where either the reader is looking for a quick approximate answer or where the article fails to provide enough information to conduct the test.

Four categories of result are possible (Display 5.17). *Case 1*: If the intervals do not overlap, it is safe to infer that the means are different. Some readers incorrectly assume this is the only case that provides strong evidence of a difference. *Case 2*, however, also shows strong evidence. Even though the intervals overlap, the best estimate for each mean lies outside the confidence interval for the other mean. *Case 3*, where one estimate lies within the confidence interval for the other mean, but the second estimate lies outside the first interval, is difficult to judge. But in *Case 4*, where the best estimate of each mean lies inside the confidence interval for the other, there is no evidence of any difference.

Finally, it must be mentioned that the discussion of this section applies to the comparison of two confidence intervals only. If there are more than two confidence intervals then it may be quite misleading to compare the two most disparate ones, unless some adjustment for multiple comparisons is made. This topic is discussed in the next chapter.

Display 5.17   Separate confidence intervals for two group means: are the means different?

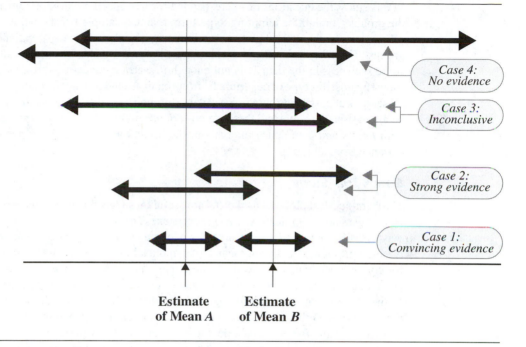

Estimate   Estimate
of Mean *A*   of Mean *B*

5.6 Summary

The term *analysis of variance* is often initially confusing as it seems to imply a comparison of variances. It is most definitely a method for comparing means, however, and the name derives from the approach for doing so—assessing variability from several sources. The analysis of variance F-test is used for assessing equality of several means.

Another point of confusion arises from the mistaken belief—due to the prevalence of the F-test in textbooks and computer programs—that the F-test necessarily plays a central role in the analysis of several samples. Usually it does not. It offers a convenient approach for detection of group differences, but it does not ordinarily provide answers to particular questions of interest. Tests and confidence intervals for pairs of means or linear combinations of means (discussed in the next chapter) provide much more specific information.

Analysis of data in several samples begins with a graphical display, like side-by-side box plots. Transformations of the data should be considered. The need for transformation and the presence of outliers is often better indicated by a residual plot—a plot of residuals versus fitted values. A funnel-shape indicates the need for a transformation like the log, for example, for positive data. The analysis of variance table provides the numerical components of the F-test for equality of means. It also contains the within-groups mean square, which exactly equals the pooled estimate of variance, the best estimate of σ^2. Confidence intervals and t-tests for pairs of means should use this pooled estimate of variance from all groups.

Diet Restriction and Longevity Study

In this study, the questions of interest called for five specific pairwise comparisons among the groups. It might be tempting to perform five two-sample t-tests, but it is a much more efficient use of the data to perform t-tests using a pooled estimate of variance from all the groups. The analysis begins with examination of side-by-side box plots. Although there is some skewness in the data, it is not enough to warrant concern—the tests are sufficiently robust against this type of departure from normality—and no transformation is suggested. A closer look at possible problems is available through a residual plot, but the spreads appear to be approximately equal and there are no serious outliers. The analysis proceeds, therefore, with t-tests and confidence intervals in the usual way, but using the pooled estimate of variance from all groups.

Spock Trial Study

The stem-and-leaf plots in Display 5.4 are useful for suggesting some answers to the question of interest and for indicating the appropriateness of the tools based on the standard one-way classification model. An analysis of variance F-test confirms the strong evidence of some differences between means. An application of the extra-sums-of-squares F-test for comparing equality of the six other judges shows no evidence of a difference in mean percentages of women on their venires. Assuming that the six other means are equal, a further F-test shows overwhelming evidence that the Spock judge mean is different from the mean of the other six. Since this is a test for equality of two means, a t-test could be used. In fact, the F-test is equivalent to a two-sided t-test when $I = 2$. The actual p-value reported in the summary of statistical findings comes from a different test, not based on the assumption of equal means among the other six judges, and is discussed in the next chapter.

Further Reading

Elementary discussions of one-way analysis of variance are found in introductory statistical texts, such as those by Moore and McCabe (1993) and by Mendenhall (1993). An easy-to-read geometrical interpretation of the analysis of variance breakdown is given by Box, Hunter, and Hunter (1978). More details on the Kruskal–Wallis test can be found in most books on nonparametric statistics, such as Sprent (1993). The extra-sum-of-squares principle is discussed again in the broader context of regression in Chapter 10. The general theory is provided in regression books, like that of Draper and Smith (1981).

5.7 Exercises

Conceptual Exercises

1.   Spock trial. Why is it important to obtain a pooled estimate of variance in the Spock trial study? Is it ever a mistake to obtain a pooled estimate of variance in a comparison involving several groups?

2.   Four methods of growing wheat are to be compared on five farms. Four plots are used on each farm and each method is applied to one of the plots. Five measurements are therefore obtained on yield per acre for each of the four growing methods. Are the methods of this chapter appropriate for analyzing the results?

3. **Diet restriction.** Is there any explanation for why the distribution of lifetimes of mice in Display 5.1 are all negatively skewed?

4. **Diet restriction.** For comparing group 3 to group 2, explain why it is better to use the t-tools presented in Section 5.2.3 (using s_p from all six groups), than to use the Chapter 2 t-tools (using s_p from only the two groups involved).

5. **Spock trial.** Should Spock's accusers question the defense on how the venires were selected for their study?

6. **Spock trial.** Why is it useful to test whether the six judges other than Spock's have equal mean percents of women on their venires?

7. What can be said about the suitability of the ideal model from the residual plots for the four-sample data sets shown in Display 5.18?

Display 5.18   Residual plots for Exercise 7 data plots

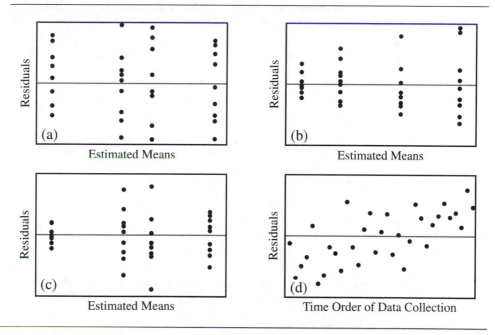

8. Why is s_p^2 not simply taken as the average of the I sample variances?

9. **Diet restriction.** If the longevity study was a planned experiment, why are the sample sizes different?

10. If s_p is zero, what must be true about the residuals?

11. Explain the role of degrees of freedom of the F-distribution associated with the F-statistic. How are degrees of freedom related to how far the F-statistic is likely to be from 1?

12. What does it mean if the F-statistic is so *small* that the chance of getting an F-statistic that small or smaller is only, say, .0001?

13. What are the most important quantities to be found in the analysis of variance table?

Computational Exercises

14. **Spock Trial.** By examining Display 5.7, answer the following:

    (a)   What is the average percent of women from all 46 venires?

    (b)   For how many of the 9 Spock judge's venires is the percentage of women less than the grand average from all 46 venires?

    (c)   For how many of the 9 Spock judge venires is the percentage of women less than the Spock judge's average?

15. **Spock trial.** Use the following summary statistics to (a) compute the pooled estimate of the standard deviation and (b) carry out a t-test for the hypothesis that the Spock judge's mean is equal to the mean for judge A.

Judge:	Spock	A	B	C	D	E	F
Average % women:	14.62	34.12	33.61	29.10	27.00	26.97	26.80
SD of % women:	5.039	11.942	6.582	4.593	3.818	9.010	5.969
Sample size:	9	5	6	9	2	6	9

16. **Spock trial.** (a) Use a calculator or statistical package to get the sample variance for the percent of women on all 46 venires treated as one sample. (b) Multiply this by 45 to get the residual sum of squares for the equal-means model. (c) Multiply s_p^2 found in Exercise 15(a) above by $(46 - 7)$ to get the residual sum of squares for the separate-means model. (d) Use these to construct an analysis of variance table, including the F-statistic for the hypothesis of equal means. Compare the result with Display 5.9.

17. **Spock trial.** Use a statistical computer package to obtain the analysis of variance table in Display 5.9.

18. Display 5.19 shows the start of an analysis of variance table. Can you fill in the whole table from what is given here? How many groups were there? Is there evidence that the group means are different?

Display 5.19 Incomplete ANOVA table for Exercise 18

Source	df	Sum of Squares	Mean Square	F-Statistic	p-Value
Between Groups	?	?	?	?	?
Within Groups	24	35,088	?		
Total	31	70,907			

19. **Fatty Acid.** The data in Display 5.20 were obtained from a randomized experiment to estimate the effect of a certain fatty acid (CPFA) on the level of a certain protein in rat livers. Only one level of the CPFA could be investigated in a day's work, so a control group (no CPFA) was investigated each day as well. (Source: Donald A. Pierce.)

    (a) Obtain estimated means for the model with six independent samples, one for each treatment. Determine the residuals and plot them versus the estimated means. Plot the residuals versus the day on which the investigation was conducted. Is there any indication that the methods of this chapter are not appropriate?

Display 5.20   Levels of protein (times 10) found in rat livers

			Treatment			
Day	**CPFA 50**	**CPFA 150**	**CPFA 300**	**CPFA 450**	**CPFA 600**	**Control**
1	154, 177, 174					157, 165, 150
2		164, 192, 159				186, 206, 195
3			157, 159, 124			192, 202, 216
4				160, 152, 141		190, 187, 160
5					147, 152, 158	191, 188, 199

(b) Obtain estimated means for the model with ten independent samples, one from each treatment-day combination. Calculate the ANOVA F-test to see whether these 10 groups have equal means.

(c) Use (a) and (b) and the methods of Section 5.3.3, to test whether the means for the control groups on different days are different. That is, compare the model with 10 different means to the model in which there are six different means.

20.   Cavity Size and Use. Biologists freely discuss the concept of competition between species, but it is difficult to measure. In one study of competition for nesting cavities in Southeast Colorado, Donald Youkey (Oregon State University Dept. of Fisheries & Wildlife) located nearly 300 cavities occupied by a variety of bird and rodent species. Display 5.21 shows box plots of the entrance area measurements from cavities chosen by nine common nesting species. The general characteristics—positive skewness, larger spreads in the groups with larger means—suggest the need for a transformation. On the logarithmic scale, the spreads are relatively uniform, and the summary statistics appear in Display 5.22. Are the species competing for the same size cavities? Or, are there differences in the cavity sizes selected by animals of different species? It appears that there are two very different sets of species here. The first six selected relatively small cavities while the last three selected larger ones. Is that the only significant difference?

(a) Compute the pooled estimate of variance.

(b) Construct an analysis of variance table to test for species differences. (The sample standard deviation of all 294 observations as one group is SD $= 0.4962$.) Perform the F-test.

(c) Verify that the analysis of variance method for calculating the between-group sum of squares yields the same answer as the formula

$$\text{Between-group SS} = \sum_{i=1}^{I} n_i \overline{Y}_i{}^2 - n\overline{Y}^2.$$

(d) Fit an intermediate model in which the first six species have one common mean and the last three species have another common mean. Construct an analysis of variance table with F-tests to compare this model with (i) the equal-means model and (ii) the separate-means model. Perform the F-test.

21.   Diet Restriction. Construct the confidence intervals for the five specific questions of interest in the longevity study. (Verify the low and high endpoints in Display 5.23.)

22.   A robust test for equality of several population variances is *Levene's test*, which was previously discussed in Section 4.5.3 for the case of two variances. This procedure carries out the usual one-way analysis of variance F-test on the squared residuals (since the mean of the squared residuals

Display 5.21 Box plots for areas of entrances to cavities used by different species

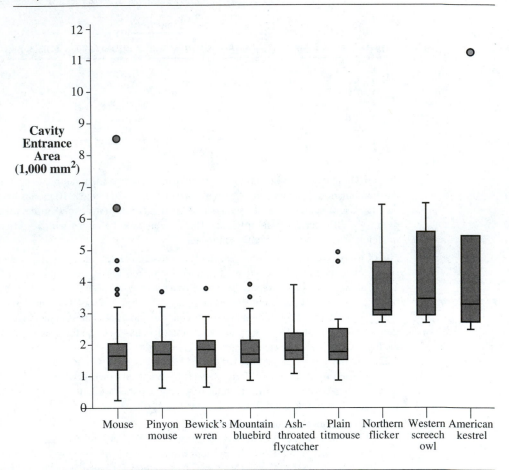

Display 5.22 Summary statistics for areas of cavity entrances (logarithmic scale)

Species	n	Mean	Sample SD
Mouse	127	7.347	.4979
Pinyon mouse	44	7.368	.4235
Bewick's wren	24	7.418	.3955
Mountain bluebird	41	7.487	.3183
Ash-throated flycatcher	18	7.563	.3111
Plain titmouse	16	7.568	.4649
Northern flicker	11	8.214	.2963
Western screech owl	7	8.272	.3242
American kestrel	6	8.297	.5842

Display 5.23 *t*-tests and 95% confidence intervals for group comparisons in the diet and lifetimes study

Comparison	Estimate	SE	95% CI Low	95% CI High	*t*-stat	Two-sided *p*-value
(a) N/R50 vs N/N85	9.6 m	1.2	7.3	11.9	8.08	< 0.0001
(b) R/R50 vs N/R50	0.6 m	1.2	−1.7	2.9	0.50	0.60
(c) N/R40 vs N/R50	2.8 m	1.2	0.5	5.1	2.39	0.017
(d) N/R50 lopro vs N/R50	−2.6 m	1.2	−4.9	−0.3	2.18	0.029
(e) N/N85 vs NP	5.3 m	1.3	2.7	7.9	4.07	< 0.0001

from each group is roughly the population variance). For practice, carry out Levene's test on the Spock data by (a) forming the residuals as the observations minus the group average, (b) computing the squared residuals, (c) obtaining an analysis of variance table for the analysis of variance of the squared residuals, (d) and finding the *p*-value for the analysis of variance *F*-test.

Data Problems

23.   Was Tyrannosaurus rex warm-blooded? Display 5.24 shows several measurements of the oxygen isotopic composition of bone phosphate in each of 12 bone specimens from a single Tyrannosaurus rex skeleton. It is known that the oxygen isotopic composition of vertebrate bone phosphate is related to the body temperature at which the bone forms. Differences in means at different bone sites would indicate nonconstant temperatures throughout the body. Minor temperature differences would be expected in warm-blooded animals. Is there evidence that the means are different for the different bones? (Data from R. E. Barrick, and W. J. Showers, "Thermophysiology of *Tyrannosaurus rex*; Evidence from Oxygen Isotopes," *Science* 265 (1994): 222–24.)

Display 5.24 Measurements of oxygen isotopic composition of vertebrate bone phosphate (per mil deviations from SMOW) in 12 bones of a single Tyrannosaurus rex specimen

Bone	Oxygen Isotopic Composition					
Rib 16	11.10	11.22	11.29	11.49		
Gastralia	11.32	11.40	11.71			
Gastralia	11.60	11.78	12.05			
Dorsal vertebra	10.61	10.88	11.12	11.24	11.43	
Dorsal vertebra	10.92	11.20	11.30	11.62	11.70	
Femur	11.70	11.79	11.91	12.15		
Tibia	11.33	11.41	11.62	12.15	12.30	
Metatarsal	11.32	11.65	11.96	12.15		
Phalange	11.54	11.89	12.04			
Proximal caudal	10.93	11.01	11.08	11.12	11.28	11.37
Mid-caudal	11.35	11.43	11.50	11.57	11.92	
Distal caudal	11.95	12.01	12.25	12.30	12.39	

24.   Vegetarians and Zinc: An Observational Study. Previous studies suggest that vegetarians may not receive enough zinc in their diets. As the zinc requirement is particularly important during pregnancy, researchers conducted a study to determine whether vegetarian pregnant women are at greater risk from low zinc levels than are nonvegetarian pregnant women. Twenty-three women were monitored: twelve vegetarians who were pregnant, six nonvegetarians who were pregnant, and five vegetarians who were not pregnant. None of these women were smokers, and none of the nonpregnant women were taking oral contraceptives. The zinc status in each woman was measured by zinc content in the blood, urine, and hair. Display 5.25 presents the zinc levels in the hair. (Data from King et al., "Effect of vegetariansim on the zinc status of pregnant women," *American Journal of Clinical Nutrition* 34 (1981): 1049–55.) What evidence is there that pregnant vegetarians tend to have lower zinc levels than pregnant nonvegetarians?

Display 5.25   Zinc levels (μg/g) in the hair of women

Pregnant Nonvegetarians	Pregnant Vegetarians	Nonpregnant Vegetarians
185	171	210
189	174	139
187	202	172
181	171	198
150	207	177
176	125	
	189	
	179	
	163	
	174	
	184	
	186	

25.   Duodenal ulcers. To clarify the importance of a certain kind of antibody activity (CCK) in gastrointestinal diseases, researchers assessed the CCK activity in the duodenal mucosa of 27 guinea pigs. Of these, 8 had gallstones, 9 had gastric ulcers, and 10 were healthy controls. The following CCK activity was determined by bioassay and measured in Ivy units per milligram of dry weight. (Data from S. Kataoka et al., "Bioassay of Cholecystokinin-Pancreozymin in Duodenal Mucosa," *Lancet* (1978): 1043.)

Controls: .11, .11, .11, .19, .21, .22, .24, .25, .31
Gallstone: .18, .27, .36, .37, .39, .47, .37, .57
Ulcer: .29, .30, .40, .45, .47, .52, .57, 1.10

Describe the difference in the distribution of CCK activity for subjects with gallstones and for healthy subjects. Describe the difference in the distribution of CCK activity for subjects with duodenal ulcers and the distribution of CCK activity for healthy subjects.

Answers to Conceptual Exercises

1.   To make comparisons, one must estimate variation. There are not many venires for any particular judge, so pooling the information gives better precision to the variance estimate. But if the groups have very different spreads, pooling is a bad idea.

2. Not appropriate. You should not expect the measurements from plots on the same farm to be independent of each other.

3. Perhaps there is something like an upper bound, a maximum possible lifetime for each group, and healthy mice all tend to get close to it. Unhealthy mice, however, die off sooner and at very different ages.

4. If the variances in all populations are equal, s_p from all groups uses much more data to estimate σ, resulting in a more precise estimator.

5. Yes. Perhaps these are just as good as random samples of all venires for each judge. If there was any bias in the selection, however—for example if the nine venires for Spock's judge were chosen because they did not have many women—the results would be misleading.

6. Spock's lawyers will have a stronger case if they can show that Spock's judge is particularly different from *all others* in having low representation of women.

7. In (a) there is no evidence of any problem with the ideal model. For (b) the spread increases with increasing estimated means. Transform the data! In (c) there are possible problems with unequal spreads. It might be wise to try a two-sample Welch confidence interval in addition to the standard analysis and see if they give very different results. If the response Y is a proportion, either of the transformations arcsine or $\sqrt{Y}$ or $\log[Y/(1-Y)]$ may help. In (d) there is a trend in the response over time, and the standard tools will not apply.

8. It is, if the sample sizes are all equal. Otherwise, it gives more weight to estimates from larger samples.

9. It is unusual for experimenters to purposefully plan on unequal sample sizes. In this study it is likely that the larger number of mice in the N/R50 group was planned, because that was the major experimental group. Inequalities in the other group sample sizes are likely the result of losing mice to factors unrelated to the experiment.

10. All the residuals would have to be identically zero for this to happen.

11. The larger the degrees of freedom in either the numerator or denominator, the less variability there is in their sampling distributions. With smaller degrees of freedom in either, sampling variability can result in an F-ratio which is considerably different from 1, even when the null hypothesis is true.

12. That would suggest that the sample averages are closer to each other than one would expect in the course of natural sampling from identical populations. You may want to check out the independence assumption.

13. There are two important quantities: (1) the within mean square is s_p^2, and (2) the p-value allows for judging group differences.

Linear Combinations and Multiple Comparisons of Means

The F-test for equality of several means gives a reliable result with any number of groups. Its weakness is that it neither tells which means are different from which others nor accounts for any structure possessed by the groups. Consequently, its role is mainly to act as an initial screening device.

If the groups have a structure, or if the research requires a specific question of interest involving several groups, a particular *linear combination* of the means may address the question of interest. This chapter shows how to make inferences about any linear combination of means and how to choose linear combinations for some important kinds of problems.

When no planned comparison is called for by the questions of interest or the group structure, one may compare all means with each other. The large number of comparisons, however, compounds the statistical uncertainty in the statements of evidence. Some methods of adjustment to account for this *multiple comparisons* problem are provided and discussed here.

6.1 Case Studies

6.1.1 Discrimination Against the Handicapped—A Randomized Experiment

The U.S. Vocational Rehabilitation Act of 1973 prohibits discrimination against people with physical disabilities. The act defines a handicapped person as any individual who has a physical or mental impairment that limits the person's major life activities. Approximately 44 million U.S. citizens fit that definition. In 1984, 9 million were in the labor force, and these individuals had an unemployment rate of 7%, compared to 4.5% in the nonimpaired labor force.

One study explored how physical handicaps affect people's perception of employment qualifications. (Data from S. J. Cesare, R. J. Tannenbaum, and A. Dalessio, "Interviewers' Decisions Related to Applicant Handicap Type and Rater Empathy," *Human Performance* 3(3)(1990):157–71.) The researchers prepared five videotaped job interviews, using the same two male actors for each. A set script was designed to reflect an interview with an applicant of average qualifications. The tapes differed only in that the applicant appeared with a different handicap. In one, he appeared in a wheelchair; in a second, he appeared on crutches; in another, his hearing was impaired; in a fourth, he appeared to have one leg amputated; and in the final tape, he appeared to have no handicap.

Seventy undergraduate students from a U.S. university were randomly assigned to view the tapes, fourteen to each tape. After viewing the tape, each subject rated the qualifications of the applicant on a 0- to 10-point applicant qualification scale. Display 6.1 shows the results. The question is, do subjects systematically evaluate qualifications differently according to the candidate's handicap? If so, which handicaps produce the different evaluations?

Display 6.1 Stem-and-leaf diagrams of applicant qualification scores given to applicants simulating five different handicap conditions

	None	Amputee	Crutches	Hearing	Wheelchair
0					
1	9	9		4	7
2	5	56		149	8
3	06	268	7	479	5
4	1 29	06	0 33	237	78
5	1 49	3589	18	589	03
6	17	1	0 234	5	1 124
7	48	2	4 45		2 46
8			5		
9					

Legend: 7 | 4 represents a score of 7.4 on the Applicant Qualification Scale.

Summary of Statistical Findings

The evidence that subjects rate qualifications differently according to handicap status is strong, but not convincing (F-test p-value $= .030$). The difference between the average qualification scores given to the *crutches* candidate and to the *hearing*-impaired candidate is difficult to attribute to chance. The difference is estimated to be 1.87 points higher for the *crutches* tape, with a 95% confidence interval from 0.14 to 3.60 points based on the Tukey–Kramer procedure. The strongest evidence supports a difference between the average scores given to the *wheelchair* and *crutches* handicaps and the average scores given to the *amputee* and *hearing* handicaps (t-statistic $= 3.19$ for a linear contrast). None of the average qualification scores from the various feigned handicaps differ significantly from the no-handicap control. (The protected least significant differences all have two-sided p-values > 0.05.)

Scope of Inference

Although the evidence suggests that differences exist among some of the handicap categories, the overall picture is made difficult by the location of the control in the middle of the groups. Any inference statements must also be qualified by the fact that the subjects used in this study may not accurately represent the population of employers making hiring decisions.

6.1.2 Preexisting Preferences of Fish—A Randomized Experiment

Charles Darwin proposed that sexual selection by females could explain the evolution of elaborate characteristics in males that appear to decrease their capacity to survive. In contrast to the usual model stressing the co-evolution of the female preference with the preferred male trait, A. L. Basolo proposed and tested a selection model in which females have a preexisting bias for a male trait, even before males of the same species possess it. She studied a Central American genus of small fish. The males in some species of the genus develop brightly-colored swordtails at sexual maturity. For her study, Basolo selected one species in the genus—the Southern Platyfish—whose males do not naturally develop the swordtails.

Six pairs of males were surgically given artificial, plastic swordtails. One male of each pair received a bright yellow sword, while the other received a transparent sword. The males in a pair were placed in closed compartments at opposite ends of a fish tank. One at a time, females were placed in a central compartment, where they could choose to engage in courtship activity with either of the males by entering a side compartment adjacent to it (see Display 6.2). Of the total time spent by each female engaged in courtship during a 20-minute observation period, the percentages of time spent with the yellow-sword male were recorded. These appear in Display 6.3. (Data from A. L. Basolo, "Female Preference

Display 6.2   Experimental tank allowing female fish to choose between males

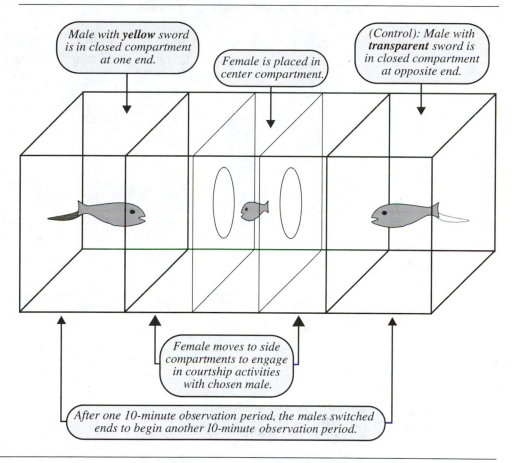

Male with **yellow** sword
is in closed compartment
at one end.

Female is placed in
center compartment.

(Control): Male with
transparent sword is
in closed compartment
at opposite end.

Female moves to side
compartments to engage
in courtship activities
with chosen male.

After one 10-minute observation period, the males switched
ends to begin another 10-minute observation period.

Predates the Evolution of the Sword in Swordtail Fish," *Science* 250(1990):808–10.) Did
these females show a preference for the males that were given yellow swordtails?

Summary of Statistical Findings

There is no evidence that the mean percentage of time with the yellow-sword male differed
from one male pair to another (p-value $= 0.56$, from an F-test). There is also no evidence
of a linear association between mean percentage of time with the yellow-sword male and
the males' body sizes (p-value $= 0.32$, from a linear contrast). The mean percentage of time
spent with the yellow-sword male is estimated to be 62.4%. A 95% confidence interval for
the mean percentage is 59.0% to 65.8%. This is convincing evidence (one-sided p-value
< 0.0001) that the mean percentage exceeds the lack-of-preference value (50%).

Display 6.3 Percentage of courtship time spent by 84 females with the yellow-sword male; body sizes of the males are shown in parentheses

	Pair 1 (35 mm)	Pair 2 (31 mm)	Pair 3 (33 mm)	Pair 4 (34 mm)	Pair 5 (28 mm)	Pair 6 (34 mm)
	43.7	52.5	91.0	72.2	78.3	33.4
	54.0	65.6	62.0	58.5	66.0	42.2
	49.8	68.5	10.0	51.0	47.7	35.6
	65.5	45.9	83.8	56.8	77.5	79.9
	53.1	80.2	91.3	92.4	58.3	59.0
	53.0	67.0	56.3	55.3	61.1	58.1
	62.3	73.0	83.6	59.3	65.1	64.2
	49.4	71.7	53.3	42.0	62.9	82.8
	45.7	55.0	36.5	68.5	61.0	75.7
	56.6	70.0	65.4	78.4		66.3
	59.0	63.2	48.1	69.6		56.3
	67.8	39.6	50.6	89.2		84.5
	73.3	41.0	40.4	67.3		61.1
	43.8	59.2	90.6	77.5		87.6
	67.4		74.9			
	58.1		56.0			
			67.5			
Average:	56.41	60.89	62.43	67.00	64.21	63.34
SD:	9.02	12.48	22.29	14.33	9.41	17.68
n:	16	14	17	14	9	14

6.2 Inferences About Linear Combinations of Group Means

6.2.1 Linear Combinations of Group Means

Questions of interest sometimes involve comparing only two group means. Each question in the diet and lifetime study (Section 5.1.1) had this feature. Examining differences between the corresponding sample averages answers such questions. But this situation is uncommon in complex studies.

More typically, questions of interest involve several group means. For example, the study of qualification scores given to handicapped applicants might focus on a comparison of two handicaps—*crutches* and *wheelchair*—with the two handicaps—*hearing* and *amputee*. If μ_1, μ_2, μ_3, μ_4, and μ_5 are the mean scores in the *none*, *amputee*, *crutches*, *hearing*, and *wheelchair* groups, respectively, that question can be explored by studying the difference between the average of mean responses, $\gamma = (\mu_3 + \mu_5)/2 - (\mu_2 + \mu_4)/2$.

The parameter γ introduced here is called a *linear combination* of the group means. Linear combinations have the form

$$\gamma = C_1\mu_1 + C_2\mu_2 + \cdots + C_I\mu_I,$$

in which the coefficients $C_1, C_2, \ldots, C_I$ are chosen by the researcher to measure specific features of interest. In the handicap example, $C_1 = 0$, $C_2 = C_4 = -1/2$, and $C_3 = C_5 = +1/2$. These particular coefficients add to zero, which gives this linear combination the special designation of being a *contrast*.

6.2.2 Inferences About Linear Combinations of Group Means

The Estimate of a Linear Combination and Its Sampling Distribution

The same linear combination of sample averages, called g, is the natural estimate of the parameter γ. It is

$$g = C_1\overline{Y}_1 + C_2\overline{Y}_2 + \cdots + C_I\overline{Y}_I.$$

The sampling distribution of this estimate has mean γ. The standard deviation in the sampling distribution is given by the formula

$$SD(g) = \sigma\sqrt{\frac{C_1{}^2}{n_1} + \frac{C_2{}^2}{n_2} + \cdots + \frac{C_I{}^2}{n_I}},$$

which depends on the nuisance parameter σ. This assumes that the equal-spread model applies. The shape of the sampling distribution is normal if the individual populations are normal, and it is approximately normal more generally.

Standard Errors for Estimated Linear Combinations

The standard error for g is obtained by substituting the pooled estimate for σ in the formula for the standard deviation of g:

$$SE(g) = s_p\sqrt{\frac{C_1{}^2}{n_1} + \frac{C_2{}^2}{n_2} + \cdots + \frac{C_I{}^2}{n_I}}.$$

Two-sample comparisons are a special case. To compare the mean score of the crutches ratings with the mean score of the no handicaps ratings, for example, the investigator chooses $C_3 = +1$ and $C_1 = -1$, with $C_2 = C_4 = C_5 = 0$. The expression under the square root in the standard error reduces to the familiar sum of the reciprocals of the two group sample sizes. The SD is estimated from information in all groups, even when only two groups are being compared.

Inferences Based on the t-Distributions

The t-ratio, $t = (g - \gamma)/\text{SE}(g)$, has an approximate Student's t-distribution with degrees of freedom equal to that of the pooled SD: d.f. $= (n_1 + n_2 + \cdots + n_I - I)$. The t-ratio may now be used as before either to construct a confidence interval or to test a hypothesized value for γ.

Example—Handicap Study

A computer can produce the averages and the pooled estimate of variability, but hand calculations are usually required from there. Display 6.4 illustrates all the steps involved in finding a confidence interval for the contrast between the *wheelchair* and *crutches* means and the average of the *amputee* and *hearing* means. A 95% confidence interval for the parameter $\gamma = (\mu_3 + \mu_5)/2 - (\mu_2 + \mu_4)/2$ extends from 0.522 to 2.264.

6.2.3 Specific Linear Combinations

Here are some examples of linear combinations that arise frequently in practical situations.

Comparing Averages of Group Means

One common problem has two *sets of groups* distinguished by a specific factor; a comparison of the two sets is of interest. The preceding comparison above is a typical example. Another example, from Section 5.1.2, involves comparing the Spock judge's mean percentage women with the average of the means from the other six judges.

If J groups in one set are to be compared to K groups in the second set, the relevant parameter is

$$\gamma = \frac{(\mu_{1,1} + \cdots + \mu_{1,J})}{J} - \frac{(\mu_{2,1} + \cdots + \mu_{2,K})}{K}.$$

The coefficients will be $+1/J$, $-1/K$, or zero, depending on whether a group is in the first set, the second set, or neither. *Important note:* One group cannot belong to both sets.

Comparing Rates

In problems like the diet restriction study in Section 5.1.1, where groups are structured according to levels of a quantitative explanatory variable, it may be desirable to report results as *rates* of increase in the mean response associated with changes in the explanatory variable. A comparison of the increase in mean lifetime associated with the reduction from

Display 6.4   Confidence interval construction for the linear combination $\gamma = (\mu_3 + \mu_5)/2 - (\mu_2 + \mu_44)/2$ in the handicap study

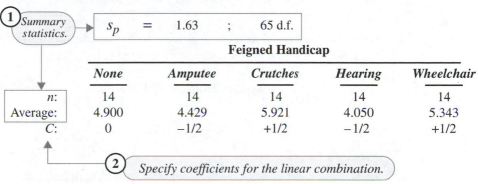

① *Summary statistics.* → S_p = 1.63 ; 65 d.f.

Feigned Handicap

	None	Amputee	Crutches	Hearing	Wheelchair
n:	14	14	14	14	14
Average:	4.900	4.429	5.921	4.050	5.343
C:	0	−1/2	+1/2	−1/2	+1/2

② *Specify coefficients for the linear combination.*

③ *Estimate the linear combination*

$$g = \frac{(\overline{Y}_3 + \overline{Y}_5)}{2} - \frac{(\overline{Y}_2 + \overline{Y}_4)}{2} = \frac{(5.921+5.343)}{2} - \frac{(4.429+4.050)}{2}$$

$$= 1.393$$

④ *Find the standard error of the estimate.*

$$SE(g) = 1.63 \sqrt{\frac{(0)^2}{14} + \frac{(-1/2)^2}{14} + \frac{(+1/2)^2}{14} + \frac{(-1/2)^2}{14} + \frac{(+1/2)^2}{14}}$$

$$= 0.436$$

⑤ *Construct the 95% confidence interval.*

$$t_{65}(.975) = 1.997 \longleftarrow \quad \text{from } t\text{-distribution table.}$$

$$1.393 \pm (1.997)(.436) \longrightarrow \text{from } 0.522 \text{ to } 2.264$$

50 to 40 kcal/wk to the increase in mean lifetime associated with the reduction from 85 to 50 kcal/wk, for example, is best made on the basis of increases associated with 1-unit changes in the caloric intake. Thus one would inquire whether the rate of increase in lifetime is the same in the reduction from 50 to 40 kcal/wk as in the reduction from 85 to 50 kcal/wk?

The rate of increase in mean lifetime associated with the reduction from 50 to 40 kcal/wk is rate2 $= (\mu_6 - \mu_3)/(50 - 40)$, which is estimated to be

$$\text{est. rate2} = \frac{(\text{Average lifetime on } N/R40) - (\text{Average lifetime on } N/R50)}{(50 - 40)}$$

$$= \frac{(45.1 - 42.3)}{10} = 0.2800 \text{ months/[kcal/wk]}.$$

Similarly, the rate of increase associated with the reduction from 85 to 50 kcal/wk is rate1 $= (\mu_3 - \mu_2)/(85 - 50)$, which is estimated to be:

$$\text{est. rate1} = \frac{(\text{Average lifetime on } N/R50) - (\text{Average lifetime on } N/N85)}{(85 - 50)}$$

$$= \frac{(42.3 - 32.7)}{35} = 0.2743 \text{ months/[kcal/wk]}.$$

Reducing caloric intake increased longevity in both instances, and the rates of increase appear to be about the same—about 0.28 months of extra lifetime for each kcal/wk reduction in caloric intake.

A formal comparison of the two rates will resolve whether the difference is real. The difference between the rates is estimated to be:

$$(\text{est. } \textit{rate1} - \text{est. } \textit{rate2}) = 0.2743 - 0/2800 = -0.0057 \text{ mo/[kcal/wk]}.$$

This is a linear combination of only three means, because μ_3 occurs in both rates. To get the correct coefficients for calculating the standard error, reduce the comparison as follows:

$$\gamma = (\textit{rate1} - \textit{rate2}) = \frac{(\mu_3 - \mu_2)}{35} - \frac{(\mu_6 - \mu_3)}{10}$$

$$= -\frac{1}{35}\mu_2 + \frac{9}{70}\mu_3 - \frac{1}{10}\mu_6.$$

This is a linear combination of three averages, so the standard error may be computed with the general formula of Section 6.2.2:

$$SE(g) = (6.68)\sqrt{\frac{\left[-\frac{1}{35}\right]^2}{57} + \frac{\left[+\frac{9}{70}\right]^2}{71} + \frac{\left[-\frac{1}{10}\right]^2}{60}}$$

$$= 0.1359 \text{ mo/[kcal/wk]}.$$

The estimate of σ, 6.68, is the pooled estimate from all six groups and has 343 degrees of freedom. The resulting t-statistic, $0.0057/0.1359 = 0.04$, provides no evidence that the two rates differ (two-sided p-value $= .97$). It might therefore be appropriate to estimate a common rate for increased lifetime over the entire 85-to-40-kcal/wk range.

Linear Trends

Sometimes the structure of group means is associated with levels of an explanatory variable. In the platyfish preference study, for example, the six groups correspond to different body sizes of the male pairs. To determine whether the group means follow a certain pattern, one hypothesizes about what the pattern is and then derives a linear combination of means to measure that specific pattern. The simplest pattern to investigate is one in which the group means fall on a straight line when plotted as a function of the explanatory variable.

Let X_i be the value of the explanatory variable associated with group i. Then the coefficient of μ_i in a linear contrast is $C_i = (X_i - \overline{X})$. Large values of the linear combination indicate strong correspondence between the means and the proposed pattern. Popular conventions multiply all the C_i's by some constant to get new coefficients that are all integers or that express the linear contrast in convenient units of measurement. Such a change is cosmetic; inferences about the contrast are unaffected.

Display 6.5 carries through these procedures to determine whether any differences exist between mean percentages of time spent with the yellow-sword male for different male pairs and to determine whether differences are related to male body sizes.

The two tests about the mean percentages in Display 6.5 differ, and both are relevant. The analysis of variance F-statistic tests the null hypothesis that $\mu_1 = \mu_2 = \cdots = \mu_I$ against the alternative that at least one mean differs from the others. The t-test for the linear contrast tests the same null hypothesis, but against the alternative that the means fall on a straight-line function of body size with nonzero slope. Although the first general conclusion might seem to imply the second, this is not the case. For instance, the linear combination might detect a significant pattern that is overlooked by the F-test. The F-test looks for departures from equal means in all possible directions, whereas the linear combination focuses on one specific direction.

Averages

Given the absence of any apparent differences between the mean percentages of times spent with the various yellow-sword males, the question of whether female platyfish prefer yellow-sword males becomes sensible globally. The average of the group means is another linear combination (but not a contrast), and the appropriate null hypothesis is that $(\mu_1 + \mu_2 + \cdots + \mu_6)/6 = 0.5$. The average of the group sample averages is 62.38%. Its standard error, 1.72%, is calculated in the same way from the general formula. This leads to the conclusive statement that the females were not dividing their time evenly between the males.

6.2.4 Caveats Concerning Linear Combinations

The overall average just described differs slightly from the grand average of all female percentages, because it gives equal weight to each male pair instead of equal weight to each female percentage. The conclusions will, in most cases, agree.

Analysis of linear combinations generally must be done manually from computer-generated summary statistics. For that reason, alternative methods for accomplishing the same thing are useful. As will be seen in the next few chapters, essentially the same analysis can be performed by computer within the context of multiple linear regression analysis.

Display 6.5   Analysis of the preexisting preference example: *F*-test for differences in mean percentage of time spent with yellow-sword male, and *t*-test for linear effect of male body size

ANOVA F-Test

Source of Variation	Sum of Squares	d.f.	Mean Square	F-Statistic	p-Value
Between male groups	938.75	5	187.75	0.786	.56
Within groups	18,636.68	78	238.93		
Total	19,575.43	83			

Conclusion: *There is no evidence that the group means are different for different pairs of males (p-value = .56, from ANOVA F-statistic).*

t-Test for Linear Effect of Body Size

Group	n	Average (%)	Standard Deviation	Male Body Size (mm)	Coefficient
Pair 1	16	56.41	9.02	35	5
Pair 2	14	60.89	12.48	31	–3
Pair 3	17	62.43	22.29	33	1 ←— *C*'s
Pair 4	14	67.00	14.33	34	3
Pair 5	9	64.21	9.41	28	–9
Pair 6	14	63.34	17.68	34	3
Pooled	84	62.13	15.46	Average = 32.5	

① *Calculate the coefficients for the linear combination.*

$$C_i = 2*(X_i - 32.5)$$

② *Calculate the effect's estimate*

$$g = (5)(56.41) + (-3)(60.89) + (1)(62.43) + (3)(67.00) + (-9)(64.21) + (3)(63.34)$$
$$= -25.06$$

and its standard error.

$$SE(g) = (15.46)\sqrt{\frac{(5)^2}{16} + \frac{(-3)^2}{14} + \frac{(1)^2}{17} + \frac{(3)^2}{14} + \frac{(-9)^2}{9} + \frac{(3)^2}{14}}$$

$$= 54.77$$

③ *Calculate the t-statistic and determine the p-value.*

$$t\text{-statistic} = \frac{-25.06}{54.77} = -0.458$$

one-sided *p*-value = .32
(from *t*-distribution with 78 d.f.)

Conclusion: *There is no evidence that the linear association between group means and male body size has a nonzero slope (one-sided p-value = .32).*

6.3 **Simultaneous Inferences**

A 95% confidence interval procedure captures its parameter 95% of the time. When several 95% confidence intervals are considered simultaneously, they constitute a *family* of confidence intervals. The frequency with which all of the intervals in a family simultaneously capture their parameters is smaller than 95%. Because this rate is often of interest, the following distinction is drawn.

> *Individual (pairwise) confidence level* is the frequency with which a single interval captures its parameter.
>
> *Overall (familywise) confidence level* is the frequency with which all intervals simultaneously capture their parameters.

If the family consists of k confidence intervals, each with pairwise confidence level 95%, the familywise confidence level can be no larger than 95% and no smaller than $100(1 - 0.05k)$%. The actual familywise confidence level depends on the degree of dependence between the intervals.

The lower limit decreases rapidly with k. With a family of 10 confidence intervals, the familywise level could be as low as 50%, indicating the strong possibility that at least one of the intervals fails to capture its parameter. In other words, one should not suppose that all the confidence intervals capture their parameters, especially when a large number of intervals are being considered simultaneously.

The issue here is *compound uncertainty*—the increased chance of making at least one mistake when drawing more than one direct inference. Compound uncertainty also arises when many tests are considered simultaneously. The greater the number of tests performed, the higher the chance that a low p-value will be found for at least one of them, even in the absence of group differences. Consequently, the researcher is likely to find group differences that are not really there.

Multiple Comparisons

Multiple comparison procedures have been developed as ways of constructing individual confidence intervals so that the familywise confidence level is controlled (at 95%, for example). The important issue for the researcher to consider is whether to control the individual confidence levels or the overall confidence level.

Planned Comparisons, Unplanned Comparisons, and Data Snooping

Consider a one-way classification with 100 groups. One researcher may be particularly interested in comparing groups 23 and 78 because the comparison answers a research question directly. The researcher knows which groups are involved before seeing the data, and the comparison will be reported regardless of its statistical and practical significance. This constitutes a *planned comparison*. The individual confidence level should be controlled for planned comparisons.

Another researcher may examine differences between all possible pairs of groups—4,950 confidence intervals in all. As a result of these efforts, the researcher finds that groups 36 and 44 and groups 27 and 90 suggest actual group differences. Only these pairs are reported as significant. They exemplify *unplanned comparisons*. The familywise confidence level should be controlled for unplanned comparisons, since the uncertainty measure must incorporate the process of searching for important comparisons.

A third researcher notices that group 36 has the largest average and group 44 has the smallest, and presents only the single confidence interval—the one comparing group 36 to group 44. This is an instance of *data snooping*, in which the particular hypothesis or comparison chosen originates from looking at the data. The overall confidence level should be controlled, for the same reason as in the second case.

The Spock trial, the diet restriction, and the platyfish preference studies all involved planned comparisons. The handicap study had no prespecified comparisons, so any comparisons reported in it should be treated as unplanned.

6.4   Some Multiple Comparison Procedures

Confidence intervals for differences between pairs of means are centered at the difference between sample averages. Interval half-widths are computed as follows:

$$\text{Interval half-width} \quad = \quad (\text{Multiplier}) \times (\text{Standard error}).$$

As usual, the standard error of the difference is the pooled standard deviation times the square root of the sum of reciprocals of sample sizes.

There are many multiple comparison procedures, and these differ in their multipliers. The two highlighted in the ensuing subsections offer strict control over the familywise confidence levels for two important families.

6.4.1   Tukey–Kramer Procedure and the Studentized Range Distributions

The Tukey–Kramer procedure utilizes the unique structure of the multiple comparisons problem by selecting a multiplier from the *studentized range distributions* rather than from the t-distributions. The idea is to incorporate the search for the two most divergent sample averages directly into the statistical procedure.

Consider the case where all group means are equal and where all sample sizes are equal. The standard errors for all comparisons are the same and are equal to SE, say. A confidence interval is successful when it includes zero, so success occurs for a particular comparison when the magnitude of the difference between sample averages, $|\overline{Y}_i - \overline{Y}_j|$, is small. All such differences are less than $M \times$ SE if $(\overline{Y}_{max} - \overline{Y}_{min})$, the range of sample averages, is less than $M \times$ SE. By selecting M in such a way that the chance of getting

$$(\overline{Y}_{max} - \overline{Y}_{min}) \leq M \times \text{SE}$$

is 95%, one guarantees that all intervals include zero 95% of the time. That is, the overall familywise confidence level is set at 95%.

A studentized range distribution describes values for the ratio of the range in I sample averages to the standard error of a single sample average, given that all samples are drawn from the same normal population. Tables of the studentized range distributions provide the $100 \times (1 - \alpha)$th percentile, $q(\alpha; I, \text{d.f.})$, depending on the number of groups (I) and the degrees of freedom (d.f.) for estimating σ. (In the one-way classification problem, d.f. $= n - I$.) The procedure originally proposed by Tukey—called Tukey's HSD, for "honest significant difference"—assumed an ideal normal model with equal spreads and also assumed equal sample sizes in all groups. The modification for unequal sample sizes provides confidence intervals with approximately the correct confidence levels, and goes by the name of the Tukey–Kramer procedure. The multiplier used in the interval half-width calculation is $q(\alpha; I, n - I)/\sqrt{2}$.

In the handicap study, $I = 5$ groups and $(n - I) = 65$ degrees of freedom. The 95th percentile in the corresponding studentized range distribution (see Appendix A.5 on page 720) is 3.975, so the multiplier for constructing confidence intervals is 2.8107.

6.4.2 Scheffé's Procedure

Scheffé proposed the multiplier

$$\sqrt{(I - 1)F_{(I-1),\text{d.f.}}(1 - \alpha)}$$

based on the F-distribution. Here, $(I-1)$ represents the between-group degrees of freedom, and d.f. is the within-group degrees of freedom. Scheffé's multiplier controls the overall confidence level for the family of parameters consisting of all possible linear contrasts among group means. When applied to the smaller family of differences between pairs of group means, the overall confidence level is *at least* $100(1 - \alpha)\%$, and generally is higher. The Scheffé method finds a more appropriate application in providing intervals for regression curves, as will be seen in later chapters.

In the handicap study, $I - 1 = 4$, d.f. $= n - I = 65$, and the 95th percentile in the F-distribution is 2.513. The resulting multiplier is 3.1705.

6.4.3 Other Multiple Comparisons Procedures

The multiple comparisons procedures in this section present a range of options for balancing pairwise and familywise confidence levels.

The LSD

The familiar choice for a multiplier is the $100(1 - \alpha)\%$ critical value in the Student's t-distribution with degrees of freedom equal to those associated with the pooled SD. The resulting interval half-width is called the *least significant difference*, or LSD. The terminology arises naturally because any difference that exceeds the LSD in size is significant in a $100\alpha\%$-level hypothesis test. The multiplier for the handicap study is $t_{65}(.975) = 1.9971$.

F-Protected Inferences

The method known as "protected LSD" is a simple and widely used alternative for testing unplanned comparisons. It is a two-step procedure, as follows:

1. Perform the ANOVA F-test.
2. (a) If the p-value from the F-test is large (> 0.05, say), do not declare any individual difference significant, even though some differences appear large enough to be declared real.
   (b) If the p-value from the F-test is small, proceed with individual comparisons, as in Section 5.2, using t-tests or confidence intervals with the t-multiplier.

It is possible to F-protect LSD tests, but there is no convenient method for F-protecting intervals. The protection for tests follows from not claiming differences when the screening F-test fails to suggest that differences exist. This reduces the chance of obtaining falsely significant findings. In the handicap study data (Display 6.6), the F-protected comparison method would proceed to step 2(b) above.

Bonferroni

If the confidence level for each of k individual comparisons is adjusted upward to $100(1 - \alpha/k)\%$, the chance that all intervals succeed simultaneously is at least $100(1 - \alpha)\%$. This result is an application of the Bonferroni inequality in probability theory. Using the Student's t-multiplier $t_{d.f.}(1 - \alpha/2k)$ allows the user to be "at least $100(1 - \alpha)\%$ confident" that all intervals succeed. In a multiple comparisons problem involving I groups, there are $k = I(I - 1)/2$ pairs of means to be compared. The exact confidence level for the Bonferroni intervals is not generally as predictable as for the Tukey–Kramer intervals in the multiple comparisons problem. Bonferroni intervals may be used in a wider range of problems, however, including some situations where the Tukey–Kramer approach is not appropriate. With the five groups in the handicap study, $k = 10$, so the t-multiplier is the 99.75th percentile in the t-distribution with 65 d.f., or 2.9060.

Others

The *Newman–Keuls* procedure also employs studentized range distributions, but with different multipliers for different ranges. *Duncan's multiple range* procedure extends the Newman–Keuls procedure, with a Bonferroni protective correction to the nominal level. *Dunnett's* method is specific to a situation (like the handicap study) where a number of treatment groups are to be compared to a single control group.

6.4.4 Multiple Comparisons in the Handicap Study

The pooled estimate of the standard deviation of the data in Display 6.1 is 1.633. All groups have the same sample size (14) so the standard error for any and all differences between sample averages is

$$SE(\overline{Y}_i - \overline{Y}_j) = 1.633\sqrt{\frac{1}{14} + \frac{1}{14}} = 0.6172.$$

The other relevant information is as follows: there are $I = 5$ groups, so there are $k = 10$ different comparisons, and the degrees of freedom for the standard error are 65.

Display 6.6   Summary of 95% confidence interval procedures for differences between treatment means in the handicap study

		Difference with ...			
Group	**Average**	*Hearing*	*Amputee*	*Control*	*Wheelchair*
Crutches	5.921	1.871	1.492	1.021	0.578
Wheelchair	5.343	1.293	0.914	0.443	
Control	4.900	0.850	0.471		
Amputee	4.429	0.379			
Hearing	4.050				

Procedure	**95% Interval Halfwidth**
LSD	1.233
Tukey–Kramer	1.735
Bonferroni	1.794
Scheffé	1.957

A confidence interval is centered at a difference with half-width given by one of the procedures.

Display 6.6 summarizes the 10 95% confidence intervals computed according to the multiple comparisons methods described earlier. Since sample sizes are all the same, confidence intervals have the same width for all comparisons under each method. The upper part of Display 6.6 shows the centers of the confidence intervals. The lower part shows the interval half-widths.

Using LSD t-multipliers leads to the conclusion that several of these differences are real. However, the Tukey–Kramer approach—the most appropriate for these data—suggests that all but one of the differences may be the result of normal sampling variation. The researcher may be discouraged from reporting any strong evidence for differences, because the *control* lies in the center of the range of averages.

Recall, however, that comparing the combined *crutches* and *wheelchair* group with the combined *amputee* and *hearing* group (Display 6.4) revealed a very clear difference. Making that comparison was suggested by an examination of the data; by analogy, the interval here should also be widened by using the Scheffé multiplier in place of the t-multiplier. When this is done, the confidence interval for the contrast is from 0.011 to 2.775, which still excludes zero. Because the Scheffé method incorporates the search among linear contrasts for the most significant, one should conclude that strong evidence exists that this difference is real.

6.4.5 Choosing a Multiple Comparisons Procedure

The LSD is the most liberal procedure (narrowest confidence intervals), and the Scheffé procedure is the most conservative (widest confidence intervals). Bonferroni and Tukey–Kramer procedures occupy intermediate positions. Aside from conducting these general comparisons, the best approach is to think carefully about whether it is desirable to control the familywise confidence level and, if so, what the appropriate family of comparisons includes. If the answer to the first question is no, then standard t-tools apply. If the family of comparisons includes pairwise differences of all group means, the Tukey–Kramer method gives precise control. If the family includes a large number of comparisons and no other method seems feasible, the Bonferroni method—although conservative—can always be applied.

6.5   Related Issues

6.5.1 Fishing Expeditions with Many Response Measurements

The multiple comparison issues discussed in this chapter relate to one response measurement. Another situation that gives rise to a multitude of research questions is one where there are many response measurements but, perhaps, only a few groups. For example, patients in two drug therapy groups may be compared on their blood pressure, lung capacity, heart rate, side effects, and so on. Since many responses are possible, the search for the most significant effects runs into similar problems of compound uncertainty and data snooping. Multiple comparisons are generally not designed for this situation, but Bonferroni methods offer reasonable protection against compounding errors.

6.5.2 Example of a Hypothesis Based on How the Data Turned Out

Although not a one-way classification problem, the following example demonstrates the need to incorporate data snooping into the assessment of uncertainty. The letters in Display 6.7 represent 2,436 mononucleotides in a DNA molecule. Mononucleotides come in four varieties—A, C, G, and T—and their sequence along the DNA strand forms a molecule's genetic code. DNA molecules break, drift for a time, and then recombine with other strands to form new molecules. The molecule shown in Display 6.7 has broken in 11 places, indicated by the dashes between two mononucleotides. Three consecutive mononucleotides form a trinucleotide, which functions as a genetic word. In this molecule, the forty occurrences of the trinucleotide TGG are shown in boldface. In line 7, a break point has appeared shortly after an occurrence of TGG, which is of interest because TGG may indicate an increased likelihood of a break to follow.

If a break occurs between any of the mononucleotides of TGG plus the four following, the break is said to be *downstream* from TGG. And in this particular molecule, six of the 11 breaks were downstream from a TGG trinucleotide. The question is, does this evidence support TGG as a precursor of breaks in the DNA?

Display 6.7   2,436 mononucleotides along a DNA molecule. All 40 occurrences of the trinucleotide TGG appear in boldface. Eleven breaks occurred in the string, at the positions indicated by dashes.

```
TAAAGAAACATAAATGCCGATATTTGTTAATACTGTGTACTGTAAGAATATATTAGCATTGT
CTATGACTAAGAATTCAAAACAATTATTGATGCTATAGGGTGGCAATATAATAGTCAATTC
TACGATATTGAAAAAGTTATCTCCTTACTTTCGCACACATTTACGTCAAAAATACACGAAA
ATAAAGATCCAGTTACTTGGGTTTGTCTAGACCTTGACATTCACAGTTTAACTTCTATAGTT
ATTTACTCGTATACTGGAAAGGTATATATAATAGTCATAACGTCGTCAATTTATTA-CGTGC
TTCTATATTAACCTCTGTAGAATTTATCATCTACACTTGTATAAACTTTATCTTACGAGATTT
TAGAAAGGAATATTGTGTCGAGTGTTACATGATGGG-TATATAATACGGACTATCCAATCTC
TTATGTCATACTAAAAACTTTATTGCCAAACACTTTTTGGAACTGGAAGATGACATCATAG
ACAATTTTGATTATCTATCTATGAAACTTATTCTAGAAAGCGATGAACTAAATGTTCCAGAT
GAGGATTATGTAGTTGATTTTGTCATTAAGTGGTATATAAAGCGAAGAAATAAATTAGGAA
ATCTGCTACTCCTTATCAAAAATGTAATCAGGTCAAATTATCTTTCTCCCAGAGGTATAAAT
AATGTAAAATGGATACTAGACTGTACCA-AAATATTTCATTGTGATAAACAACCACGCAAA
TCATACAAGTATCCATTCATAGAGTATCCTATGAACATGGATCAAATTATAGATATATCCA
TATGTGTACAAGTACTCATGTTGGAGAAGTAGTATATCTCATCGGT-GGATGGATGAACAA
TGAAATACATAACAATGCTATAGCGGTAAATTATATATCAAACAATTGGAT-TCCAATTCCT
CCGATGAATAGCCCCAGACTGTATGCTAGCGGGATACCCGCTAACA-ATAAATTATACGTAG
TAGGAGGTCTACCAAATCCCACATCTGTTGAGCGTTGGT-TCCACGGGGATGCTGCTTGGG
TTAATATGCCGAGTCTTCTGAAACCTAGATGTAATCCAGCAGTGGC-ATCCATAAACAATGT
TATATACGTAATGGGAGGACATTCTGAAACTGATACAACTACAGAATATTTGCTACCCAAT
CATGATCAGTGGCAGTTTGGACCATTCCACTTATTATCCTCATTATAAATCATGCGCGTTAG
TGTTCGGTAGAAGGTTATTCTTGGTTGGTAGAAATGCGGAATTTTATTGTGAATCCAGCAA
TACATGGCTCTGATAGATGATCCTATTTATCCGAGGGATAATCCAGAATTGATCATAGTGG
ATAATAAACTGCTATTGATAGGAGGATTTAATCGTGCATCGTATATAGATACTATAGAAGT
GTACATCATCACACTTATTCATGGAATATATGGGATGGTAAATAATTTTGAAATAAAATAT
TAGTTTTATGTTCAACATGAATATTAAC-TCACCAGTTAGATTTGTTAAGGAAACTAACAGA
GCTAAATCTCCTACTAGGCAATCACCTTACGCCGCCGGATATGATTTATATAGCGCTTACGA
TTATACTATCCCTCCAGGAGAACGACAGTTAATTAAGACAGATATTAGTATGTCCATGCCT
AAGTTCTGCTATGGTAGAATAGCTCCTAGGTCTGGTCTGTCCCTAAAAGGCATTGATATAG
GAGGCGGTGTAATAGACGAAGATTATAGGGGAAACATAGGAGTCATTCTTATTAATAATG
GAAAATGTACGTTTAATGTAAATACTGGAGATACAATAGCTCAGCTAATCTATCAACGTAT
ATATTATCCAGAACTGGAAGAAGTACAATCTCTAGATAGTACAAATAGAGGAGATCAAGG
GTTT-GGATCAACAGGACTTAGATAATAAACAATAGTATGTTGTCGATGTTTATAGTGTAAT
AATATCGTAGATTATGTAGATGATATAGATAATGGTATAGTACAGGATATAGAAGATGAG
GCTAGCAATAATGTTGATCACGACTATGTATATCCACTTCCAGAAAATATGGTATATAGAT
TTGACAAGTCCACTAACATACTCGATTATCTATCAACGGAACGGGACCCATGTAATGATGGC
TGTTCGATACTATATGAGTAAACAACGTTTAGACGACTTGTATAGACAGTTGCCCACAAAG
ACTAGATCATATATAGATATTATCAACATATATTGTGATAAAGTTAGTAATGATTATAATAG
GGACATGAATATCATGTATGA-TATGGCATCTACAAAATCATTTACAGTTTATGACATAAAT
AACGAAGTTAATACTATACTAATGGATAACAAGGGGTTGGGTGTAAGATTGGCGACAATT
TCATTTATAACCGAATTGGGTAGACGATGTATGA
```

Summary of Statistical Analysis

Two different analyses are possible, depending on whether the hypothesis that TGG was specifically indicated arose prior to viewing this molecule (planned comparison) or whether TGG was actually suggested by examination of this molecule (data snooping).

1.   Of the 2,435 possible break positions in the entire string, 235 (9.65%) are downstream from TGG trinucleotides. If breaks occurred at *random* positions, about 9.65% of the 11 breaks (that is, one) would be downstream from TGG. But in fact, six of the 11 (54.55%) occurred downstream from TGG. That would seem to be too many to have happened by chance. If breaks occurred at random positions, the possibility that six or more would occur at positions downstream from TGG trinucleotides is precisely .000243. This is very small, showing strong evidence of an association between the occurrences of breaks and TGG.

The *p*-value of .000243 assumes, however, that, in every trial determining 11 breaks, the number downstream from TGG is counted. What if the focus on TGG were the result of a search on this molecule for the trinucleotide occurring most frequently upstream of the existing breaks? *If the search for the trinucleotide was an integral part of the statistical analysis of this data set, the series of trials used to evaluate the evidence must also include the search procedure.*

2.   The computer was programmed to conduct a trial incorporating the search. It randomly selected 11 break points from the 2,435 possible, and searched upstream from all the breaks to find the trinucleotide appearing most frequently. Then it recorded the frequency of the occurrence of this trinucleotide. The computer ran 1,000 of these trials. Of course, the most frequently occurring trinucleotide differed considerably from trial to trial. The key piece of evidence is the distribution of the frequency of the most frequent upstream trinucleotide, which appears in Display 6.8. In 320 of the 1,000 trials, some trinucleotide was found upstream from the 11 breaks six times or more. In one trial, the same trinucleotide occurred upstream 11 times. Thus, having some trinucleotide upstream from many of the breaks in a molecule appears to be rather common! Consequently, if the search was conducted on this molecule, the *p*-value for an observed highest frequency of six is $p = .32$.

Display 6.8   Simulated estimate of the distribution of the highest frequency of occurrence of any trinucleotide upstream of 11 randomly selected breaks

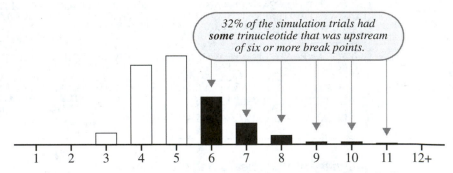

Highest Frequency of Occurrence for a Trinucleotide Upstream of the Break Points

Scope of Inference

This example illustrates the tremendous difference between evaluating the strength of evidence about a preplanned hypothesis and evaluating the strength of evidence about a hypothesis suggested by the data themselves. Correct evaluation of *post-hoc* hypotheses is possible, but it must incorporate the actual procedure by which the hypothesis was formulated. (Data from M. B. Slabaugh, N. A. Roseman, and C. K. Mathews, "Amplification of the Ribonucleotide Reductase Small Subunit Gene: Analysis of Novel Joints and the Mechanism of Gene Duplication in Vaccinia Virus," Biochemistry and Biophysics Department Report, Oregon State University, Corvallis, Oregon (1989).)

How valuable is the correct p-value? These researchers did "discover" TGG from this molecule. Had they accepted the strength of evidence .000243, they might have wasted considerable effort devising explanations for what turned out to be a nonrepeatable phenomenon.

6.5.3   Is Choosing a Transformation a Form of Data Snooping?

The Statistical Sleuth emphasizes examining graphical displays for clues to possible data transformation. When a stem-and-leaf diagram suggests a logarithmic transformation, should the resulting inferences take the data-selected transformation into account? The answer is yes. Uncertainty about which transformation (if any) is best is part of the problem, so a measure of uncertainty regarding the question of interest should incorporate uncertainty about the form of the statistical model.

In practice, however, most researchers do not incorporate model uncertainty into uncertainty measures—largely because doing so is extremely difficult. In later chapters (in particular, Chapter 12), *The Statistical Sleuth* will investigate some methods that have been devised for solving this problem. For now, the scope of inference should be limited to replicates where the transformation is selected. To caution a reader about additional uncertainty, the statistical summary should explain clearly all the steps taken to select the transformation.

6.6   Summary

Many researchers routinely examine an analysis of variance F-statistic for group differences; if it is significant, they proceed directly to multiple comparison methods to search for differences. In most cases, this is a mistake. Usually the groups have an inherent structure related to levels of specific factors set by the researcher. Linear combinations of group means can be devised to answer questions related to this structure.

A host of studies fall within the general class of one-way classifications. Yet the statistical analysis is not the same for all. A major distinction must be made between studies in which the group structure calls for specific planned comparisons and those in which the only question of interest is which means differ from which others. Selecting an appropriate statistical procedure requires the researcher to evaluate honestly whether hypotheses were clearly stated prior to data collection or whether the data themselves guided the formulation of hypotheses. In the latter case, proper statistical evaluation must acknowledge the data-snooping process.

This chapter demonstrates some statistical tools for assessing uncertainty when a family of inferences is desired. Selecting an appropriate tool requires some introspection about the nature of the family examined. This chapter also introduces a powerful tool—computer simulation—that can help evaluate evidence about more complex hypotheses suggested by the data.

Further Reading

For a synopsis of multiple comparison procedures and other references, see the listing under "multiple comparisons" in the *Encyclopedia of Statistical Science* (Kotz, Johnson, and Read 1985). For an introduction to the statistical theory underlying the Bonferroni and Tukey–Kramer methods and to linear combinations of estimates, see Rice (1995) or Miller (1986).

6.7 Exercises

Conceptual Exercises

1. Handicap Study. (a) Is it possible that the applicant's handicap in the videotape is confounded with the actor's performance? (b) Is there a way to design the study to avoid this confounding?

2. Mate Preference of Platyfish. If $\mu_1, \mu_2, \ldots, \mu_6$ represent the mean percent of time spent by females with the yellow-sword male, for the six pairs of males, (a) state the null and alternative hypotheses that are tested by (i) the analysis of variance F-test and (ii) the t-test for the hypothesis that the linear contrast (for the linear effect of male body size) is zero. (b) Say why it is possible that the second test might find evidence that the means are different even if the first does not.

3. Mate Preference of Platyfish. For the test that the mean percent of time females spent with yellow-sword males is 50%, a one-tailed p-value was reported. Why?

4. An experimenter takes twenty samples of bark from each of 10 tree species in order to estimate the differences between fuel potentials. The data give 10 species averages, the lowest being 1.6 Btu/lb and the highest 3.8 Btu/lb for a range of 2.2 Btu/lb. A colleague suggests that another species be included, so the experimenter plans to gather 20 samples from that species and calculate its average potential. Which of the following is true about the range that the 11 species averages will have when the new species is included? (a) The range will equal the old range, 2.2 Btu/lb. (b) The range will be larger than 2.2 Btu/lb. (c) The range will be smaller than 2.2 Btu/lb. (d) The range cannot be smaller than 2.2 Btu/lb. (e) The range cannot be larger than 2.2 Btu/lb. (f) It is not possible to say that any of the above options is true until the average is known.

5. O-Ring Data. The case study in Section 4.1.1 on page 82 involved the numbers of O-ring events on U.S. space shuttle flights launched at temperatures above and below 65°F. In the context of this chapter, is anything suspicious about that data? (*Hint*: Is there a possibility of data snooping?)

6. Does a confidence interval for the difference between two groups use information about variability from other groups? Why? or Why not?

7. What is the distinction between planned and unplanned comparisons?

8. Does a planned comparison always consist of estimating the difference between the means in two groups?

9. In comparing 10 groups a researcher notices that $\overline{Y}_7$ is the largest and $\overline{Y}_3$ is the smallest, and then tests the hypothesis that $\mu_7 - \mu_3 = 0$. Why should a multiple comparison procedure be used even though there is only one comparison being made?

10. If the analysis of variance screening test shows no significant evidence of any group differences, does that end the issue of there being any differences to report?

11. When choosing coefficients for a contrast, does the choice of $\{C_1, C_2, \ldots, C_I\}$ give a different t-ratio than the choice of $\{3C_1, 3C_2, \ldots, 3C_I\}$?

Computational Exercises

12. Handicap Study. Consider the groups *amputee*, *crutches*, and *wheelchair* to be handicaps of mobility and *hearing* to be a handicap affecting communication. Use the appropriate linear combination to test whether the average of the means for the mobility handicaps is equal to the mean of the communication handicap.

13. Handicap Study. Use the Bonferroni method to construct simultaneous confidence intervals for $\mu_2 - \mu_3$, $\mu_2 - \mu_5$, and $\mu_3 - \mu_5$ (to see whether there are differences in attitude towards the mobility type of handicaps).

14. Handicap Study. Examine these data with your available statistical computer package. See what multiple comparison procedures are available within the one-way analysis of variance procedure. Verify the 95% confidence interval half-widths in Display 6.6.

15. Comparison of Five Teaching Methods. An article reported the results of a planned experiment contrasting five different teaching methods. Forty-five students were randomly allocated, nine to each method. After completing the experimental course, a 1-hour examination was administered. Display 6.9 summarizes the scores on a 10-minute retention test which was given 6 weeks later. (Data from S. W. Tsai and N. F. Pohl, "Computer-Assisted Instruction Augmented with Planned Teacher/student Contacts," *Journal of Experimental Education*, 49(2)(Winter 1980–81):120–26).

Display 6.9 Test scores for the experimental CAD instruction course

Group	Logo	Teaching Method	n	Average	SD
1	L+D	Lecture and discussion	9	30.20	3.82
2	R	Programmed text	9	28.80	5.26
3	R+L	Programmed text with lectures	9	26.20	4.66
4	C	Computer instruction	9	31.10	4.91
5	C+L	Computer instruction with lectures	9	30.20	3.53

(a) Compute the pooled estimate of the standard deviation from these summary statistics.

(b) Determine a set of coefficients that will contrast the methods using programmed text as part of the method (groups 2 and 3) with those that do not use programmed text (1, 4, and 5).

(c) Estimate the contrast in (b) and compute a 95% confidence interval.

16. Adder Head Size. Red Riding Hood: "My, what big teeth you have!" Big Bad Wolf: "The better to eat you with, my dear." Are predators morphologically adapted to the size of their prey? A. Forsman studied adders on the Swedish mainland and on groups of islands in the Baltic Sea to determine if there was any relationship between their relative head lengths (RHL) and the body size of their main prey, field voles. (Data from A. Forsman, "Adaptive variation in head size in *Vipera berus* L. populations," *Biological Journal of the Linnean Society* 43(1991): 281–296.) Relative head length is head length adjusted for overall body length, determined separately for males and females. Field vole body size is a combined measure of several features, expressed on a standardized scale. The data appear in Display 6.10. The pooled estimate of standard deviation of the RHL measurements was 11.72, based on 230 degrees of freedom.

(a) Determine the half-widths of 95% confidence intervals for all 21 pairwise differences among means for the seven localities, using (i) the LSD method and (ii) the Tukey–Kramer method.

Display 6.10   Average relative head lengths of adders from seven Swedish localities with their distances to the mainland and the body sizes of prey

Locality	Sample Size	Average Relative Head Length	Distance (km) to Mainland	Field Vole Body Size
Uppsala	21	−6.98	0	− 1.75
In-Fredeln	34	−4.24	25.1	
Inre Hamnskär	20	−2.79	13.4	− 0.16
Norrpada	25	2.22	14.7	1.31
Kärringboskär	7	1.27	10.0	
Ängskär	82	1.88	22.7	1.67
Svenska Hägarna	48	4.98	39.6	2.17

   (b)   Using a linear contrast on the groups for which vole body size is available, test whether the locality means (of relative head length) are equal, against the alternative that they fall on a straight line function of vole body size, with nonzero slope.

   (c)   Repeat (b) for the pattern of distances to the mainland rather than vole body size.

17.   **Nest Cavities.** Using the nest cavity data in Display 5.20 on page 137, estimate the difference between the average of the mean entry areas for flickers, screech owls, and kestrels and the average of the mean entry areas for the other six animals (on the transformed scale). Use a contrast of means.

18.   **Diet Restriction.** For the data in Display 5.1 on page 110 (and the summary statistics in Display 5.2 on page 110), obtain a 95% confidence interval for the difference $\mu_3 - \mu_2$ using the Tukey–Kramer procedure. How does this interval differ from the LSD interval? Why is the Tukey–Kramer procedure the wrong thing to use for this problem?

Data Problems

19.   **Failure Times of Bearings.** The data in Display 6.11 are the times to fatigue failure (in units of millions of cycles) for 10 high-speed turbine engine bearings made from five different compounds. Which compounds tend to differ in their performance from the others? Analyze the data and write a brief statistical report including a summary of statistical findings, a graphical display, and a details section describing the details of the particular methods used. (Data from J. I. McCool, "Analysis of Single Classification Experiments Based on Censored Samples from the Two-parameter Weibull Distribution," *Journal of Statistical Planning and Inference*, 3(1979):39–68.

20.   **A Biological Basis for Homosexuality.** Is there a physiological basis for sexual preference? Following up on research suggesting that certain cell clusters in the brain govern sexual behavior, Simon LeVay (Data from S. LeVay, "A Difference in Hypothalamic Structure Between Heterosexual and Homosexual Men," *Science*, 253 (August 30, 1991):1034–37) measured the volumes of four cell groups in the interstitial nuclei of the anterior hypothalamus in postmortem tissue from 41 subjects at autopsy from seven metropolitan hospitals in New York and California. The volumes of one cell cluster, INAH3, are recreated in Display 6.12. The numbers are 1,000 times volumes in mm^3. Subjects are classified into five groups according to three factors: gender, sexual orientation, and cause of death. One male classified as a homosexual who died of AIDS (volume 128) was actually bisexual. LeVay used the term *presumed* heterosexual to indicate the possibility of misclassifying

Display 6.11   Failure times of bearings (millions of cycles)

Type of Compound				
I	II	III	IV	V
3.03	3.19	3.46	5.88	6.43
5.53	4.26	5.22	6.74	9.97
5.60	4.47	5.69	6.90	10.39
9.30	4.53	6.54	6.98	13.55
9.92	4.67	9.16	7.21	14.45
12.51	4.69	9.40	8.14	14.72
12.95	5.78	10.19	8.59	16.81
15.21	6.79	10.71	9.80	18.39
16.04	9.37	12.58	12.28	20.84
16.84	12.75	13.41	25.46	21.51

Display 6.12   Volumes of INAH3 (1,000 times mm^3) cell clusters from 41 human subjects at autopsy, by sex, sexual orientation, and cause of death

Males					Females	
Heterosexual		Homosexual			Heterosexual	
AIDS Death	Non-AIDS Death	AIDS Death		AIDS Death	Non-AIDS Death	
12	20	1	34	12	10	
105	37	7	39		19	
105	103	12	41		29	
118	129	15	46		105	
119	135	18	66		155	
161	140	18	86			
	161	23	128			
	175	26	142			
	179	29	193			
	209	32				

some subjects. Do heterosexual males tend to differ from homosexual males in the volume of INAH3? Do heterosexual males tend to differ from heterosexual females? Do heterosexual females tend to differ from homosexual males? Analyze the data and write a brief statistical report including a summary of statistical findings, a graphical display, and a details section describing the details of the particular methods used. Also describe the limitations of inferences that can be made. (*Hint:* One approach is to ignore cause of death, but some preliminary checking must be used to see whether it can indeed be ignored. Another approach is to use linear combinations; for example, comparing the mean of the homosexual male group to the average of the two means in the heterosexual female groups. The latter is safer since it doesn't involve an assumption about the effect of AIDS, but the former is more straightforward and possibly adequate.)

Answers to Conceptual Exercises

1. (a) Yes, it is possible that the acting performance of the actor portrayed a more competent worker in the *crutches* role, even though the script was held constant. (b) With two pairs of actors and twice as many groups, the handicap effect could be isolated from the actor effect.

2. (a) (i) Hypothesis: all means are equal; alternative: at least one is different from the others. (ii) Hypothesis: all means are equal: alternative: the means are not equal but fall on a straight line function of male body size (with nonzero slope). (b) The alternative in (ii) is more specific. For a given set of data there is more power in detecting differences if the (correct) specific alternative can be investigated. (This has previously been noticed by the fact that a one-tailed p-value is smaller that a two-tailed.)

3. The researcher had reason to believe, because of other species of fish in the same genus, that the colored tail would be more attractive to the females.

4. If the new species average is somewhere between 1.6 and 3.8 Btu/lb, the range of the set of means is unchanged. If the new species average is either less than 1.6 or greater than 3.8 Btu/lb, the range is larger. Those are the only possibilities. So the answer is (d): the range cannot be smaller than the old range (but it could be larger). The range increases as the number of groups increase.

5. Where did the 65° cutoff come from? If the analyst chose that cutoff because it produced the most dramatic difference between the two groups, the search procedure should be included in the assessment of evidence.

6. Yes, it does. It is important to pool information about variability because the population SD is difficult to estimate from small samples.

7. A planned comparison is one of a few specific comparisons that is designed to answer a question of interest. An unplanned comparison is one (of a large number) of comparisons that is suggested by the data themselves.

8. No. More complex comparisons can be made by examining linear combinations of group means.

9. The more groups there are, the larger the difference one expects between the smallest and largest averages. To incorporate the selection of this hypothesis on the basis of how the data turned out, the appropriate statistical measure of uncertainty is the same as the one that is appropriate for comparing every mean to every other mean.

10. Not necessarily. If there are no planned comparisons, it may be best to report no evidence of differences (protected LSD procedure). But the evidence about planned comparisons to answer questions of interest should be assessed on its own. It is possible that a planned comparison shows something when the F-test does not.

11. No. The parameter changes from γ to 3γ, the estimate changes from g to $3g$, and the standard error also changes from $SE(g)$ to $SE(3g) = 3SE(g)$. So the t-ratio is not changed at all. This is why one can take the convenient step of multiplying a set of coefficients by a common factor to make all coefficients into integers, if desirable.

CHAPTER 7

Simple Linear Regression: A Model for the Mean

The advice of Chapters 5 and 6 is to investigate specific questions of interest for the several-sample problem by paying attention to the structure of the grouped data. When the different groups correspond to different levels of a quantitative *explanatory* variable the idea can be extended with the simple linear regression model, in which the means fall on a straight line function of the explanatory variable.

Such a model has several advantages when it is appropriate. It offers a concise summary of the mean of the response variable as a function of the explanatory variable through two parameters: the slope and the intercept of the line. For a surprisingly large number of data problems this model is appropriate (possibly after transformation) and the questions of interest can be conveniently reworded in terms of the two parameters. The parameters are estimable even without replicate values at each distinct level of the explanatory variable.

This chapter presents the model and some associated inferential tools. The next chapter takes a closer look at the simple linear regression model assumptions and how to measure departures from them.

7.1 Case Studies

7.1.1 The Big Bang—An Observational Study

Edwin Hubble used the power of the Mount Wilson Observatory telescopes to measure features of nebulae outside the Milky Way. He was surprised to find a relationship between a nebula's distance from earth and the velocity with which it was going away from the earth. Hubble's initial data on 24 nebulae are shown as a *scatterplot* in Display 7.1. (Data from Hubble, "A Relation Between Distance and Radial Velocity Among Extra-galactic Nebulae," *Proceedings of the National Academy of Science* 15 (1929): 168–73.) The horizontal axis measures the recession velocity, in kilometers per second, which was determined with considerable accuracy by the red shift in the spectrum of light from a nebula. The vertical scale measures distance from the earth, in megaparsecs: 1 megaparsec is 1 million parsecs, and 1 parsec is about 30.9 trillion kilometers. Distances were measured by comparing mean luminosities of the nebulae to those of certain star types, a method that is not particularly accurate. The data are shown later in this chapter as Display 7.8.

Display 7.1   Scatterplot of measured distance versus velocity for 24 extra-galactic nebulae

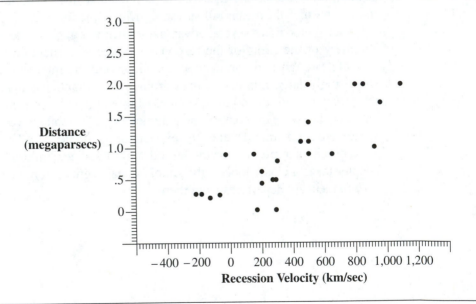

The apparent statistical relationship between distance and velocity led scientists to consider how such a relationship could arise. It was proposed that the universe came into being with a Big Bang, a long time ago. The material in the universe traveled out from the point of the Big Bang, and scattered around the surface of an expanding sphere. If the material were traveling at a constant velocity (V) from the point of the bang, then the earth and any nebulae would appear as in Display 7.2.

Display 7.2   Big Bang theory model for distance-velocity relationship of nebulae

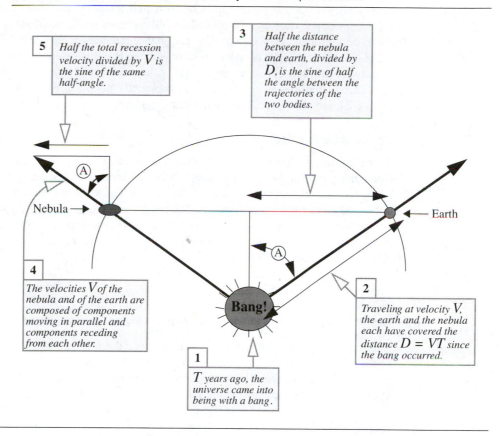

The distance (Y) between them and the velocity (X) at which they appear to be going away from each other satisfy the relationship

$$(Y/2)/VT = (X/2)/V = \sin(A),$$

where A is half the angle between them. In that case,

$$Y = TX$$

is a straight line relationship between distance and velocity. The points in Display 7.1 do not fall exactly on a straight line. It might be, however, that the *mean* of the distance measurements is TX. The slope parameter T in the equation Mean$\{Y\} = TX$ is the time elapsed since the Big Bang (that is, the age of the universe).

Several questions arise. Is the relationship between distance and velocity indeed a straight line? Is the y-intercept in the straight line equation zero, as the Big Bang theory predicts? How old is the universe?

Summary of Statistical Findings

If the theory is taken as correct, then the estimated age of the universe is .001922 megaparsecs-seconds per kilometer, or about 1.88 billion years (estimate of slope in simple linear regression through the origin). A 95% confidence interval for the age is 1.50 to 2.27 billion years. However, the data are not consistent with the Big Bang theory as proposed. Although the relationship between mean measured distance and velocity approximates a straight line, the value of the line at velocity zero is apparently not zero, as predicted by the theory (two-sided p-value $= .0028$ for a test that the intercept is zero).

Scope of Inference

These data are not a random sample, so they do not necessarily represent what would result from taking measurements from other nebulae. This analysis assumes that there is an exact linear relationship between distance and velocity, but that the measured distances do not fall exactly on a straight line because of measurement errors. The confidence interval above summarizes the uncertainty that comes from errors in distance measurements. Uncertainty due to errors in measuring velocities is not included in the p-values and confidence coefficients. Such errors are a potential source of pronounced bias. (*Note:* The Big Bang theory is still intact with estimates of the age of the universe ranging from 8 to 15 billion years, depending on what data are used.)

7.1.2 Meat Processing and pH—A Randomized Experiment

A certain kind of meat processing may begin once the pH in postmortem muscle of a steer carcass decreases to 6.0, from a pH at time of slaughter of around 7.0 to 7.2. It is not practical to monitor the pH decline for each animal, so an estimate is needed of the time after slaughter at which the pH reaches 6.0. To estimate this time, 10 steer carcasses were assigned to be measured for pH at one of five times after slaughter. The data appear in Display 7.3. (Data from J. R. Schwenke and G. A. Milliken, "On the Calibration Problem Extended to Nonlinear Models," *Biometrics* 47 (1991): 563–74.)

Display 7.3 pH of 10 steer carcasses measured at five different times after slaughter

Steer	*Time After Slaughter (hours)*	*pH*
1	1	7.02
2	1	6.93
3	2	6.42
4	2	6.51
5	4	6.07
6	4	5.99
7	6	5.59
8	6	5.80
9	8	5.51
10	8	5.36

Summary of Statistical Findings

The estimated relationship between postmortem muscle pH and time after slaughter is summarized in Display 7.4. The solid line shows the estimated mean pH as a function of the logarithm of time after slaughter. The dotted lines are the upper and lower endpoints of 95% prediction intervals for the pH of steer carcasses at times after slaughter in the range of 1 to 8 hours. It is estimated that the mean pH at 3.9 hours is 6. It is predicted that at least 95% of steer carcasses will reach a pH of 6.0 sometime between 2.94 and 5.10 hours after slaughter (95% calibration interval).

Display 7.4 Meat processing data with estimated regression line (from the simple linear regression of pH on log time after slaughter) and a 95% prediction band

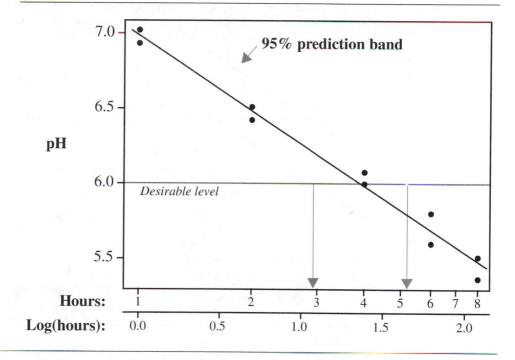

7.2 The Simple Linear Regression Model

Regression analysis describes *statistical relationships* between variables. Statistical relationships differ from mathematical relationships. If a mathematical relationship exists between two variables, knowing the value of one allows you to calculate the value of the other. In a statistical relationship, knowing the value of one variable enables you to determine certain characteristics of a possible set of values for the other, but it does not tell you the specific value.

The Big Bang example illustrates a statistical relationship. Theoretically a large number of nebulae all having the same recession velocity of 900 km/sec exist, but they have different measured distances from earth. The same is true of nebulae with recession velocities of 200 km/sec. One cannot determine the distance to a nebula knowing only that it is receding at 900 km/sec, but Display 7.1 does suggest that nebulae receding at 900 km/sec are generally farther from earth than those receding at 200 km/sec.

7.2.1 Regression Terminology

Regression analysis is used to describe the distribution of values of one variable, the *response*, as a function of other—*explanatory*—variables. This chapter deals with *simple* regression, where there is a single explanatory variable.

Think of a series of subpopulations of responses, one for each possible value of the explanatory variable. The *regression of the response variable on the explanatory variable* is a mathematical relationship between the *means* of these subpopulations and the explanatory variable. The *simple linear regression* model specifies that this relationship is a straight-line function of the explanatory variable.

Let Y and X denote respectively the response variable and the explanatory variable. The notation $\mu\{Y|X\}$ will represent the regression of Y on X, and it should be read as "the mean of Y as a function of X." When a specific value, $X = x$, is given to the explanatory variable, the same expression should be read as "the mean of Y when $X = x$." The simple linear regression model is

$$\mu\{Y \mid X\} = \beta_0 + \beta_1 X .$$

The equation involves two statistical parameters. β_0 is the *intercept* of the line, measured in the same units as the response variable. β_1 is the *slope* of the line, equal to the rate of change in the mean response per one-unit increase in the explanatory variable. That is,

$$\beta_1 = \frac{\mu\{Y \mid X = b\} - \mu\{Y \mid X = a\}}{b - a} ,$$

for any values a and b within the range of interest. The units of β_1 are the ratio of the units of the response variable to the units of the explanatory variable. Research questions frequently concern the change in mean response associated with a certain change in the explanatory variable—from a to b, say. In the simple linear regression model, the answer is

$$\mu\{Y|X = b\} - \mu\{Y|X = a\} = \beta_1(b - a).$$

Example—Big Bang

A simple linear regression model for the statistical relationship between Y = Distance and X = Velocity is:

$$\mu\{\text{Distance}|\text{Velocity}\} = \beta_0 + \beta_1 \text{Velocity},$$

in which β_0 is the mean distance, in megaparsecs, of all nebulae for which recession velocity is 0 km/sec, and β_1 is the rate of increase in the mean distance of nebulae per km/sec of recession velocity. The units of β_1 are megaparsecs/(km/sec), a unit of time.

Equal Population Standard Deviations

The standard deviations in the response distributions may also be related mathematically to the explanatory variable. The notation for the standard deviations will be $\sigma\{Y|X\}$. Ideally, these standard deviations are the same for all X. This is the equal-SD assumption: $\sigma\{Y|X\} = \sigma$, for all X.

Ideal Normal, Simple Linear Regression Model

The complete picture of a probability model emerges if, as depicted in Display 7.5, each of the subpopulations has a normal distribution. This model has only three unknown parameters—β_0, β_1, and σ. The entire statistical relationship between Y and X is described by the regression parameters, β_0 and β_1, so that most questions of interest can be phrased as questions about their values. For example, the question "is there any relationship between Y and X?" becomes the hypothesis that $\beta_1 = 0$, because $\beta_1 = 0$ in the model implies that the distributions of all subpopulations of Y are identical.

Notice that the model describes the distributions of responses, but it says nothing about the distribution of the explanatory variable. In various applications, X-values may be chosen by the researcher, or they may themselves be random.

The Independence Assumption

The independence assumption means that each response is drawn independently of all other responses from the same subpopulation and independently of all responses drawn from other subpopulations. The common types of violations are cluster effects and serial effects, as before. Suppose, for example, that pH in the meat could be measured on the same animal at several times after slaughter. The data in Display 7.4 might then arise from only two steers. This would violate the independence assumption, because if one steer's pH is above average at one time, it is more likely that it will also be above average at the next sampling time. This type of structure, in which a cluster consists of all measurements made on a particular steer, is referred to as a *repeated measure*. Repeated measures designs are important, but must be analyzed differently (Chapter 16).

7.2.2   The Scope of Inference

The simple linear regression model makes it possible to draw inferences about the mean response for any value of the explanatory variable within the observational range. Statements

Display 7.5   The ideal normal, simple linear regression model

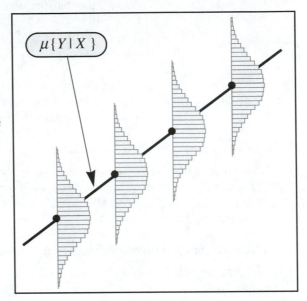

Explanatory Variable (X)

> ### *Model Assumptions*
>
> 1. **There is a normally distributed subpopulation of responses for each value of the explanatory variable.**
> 2. **The means of the subpopulations fall on a straight-line function of the explanatory variable.**
> 3. **The subpopulation standard deviations are all equal (to σ).**
> 4. **The selection of an observation from any of the subpopulations is independent of the selection of any other observation.**

about the mean at values of X not in the data set, but within the range of the observed explanatory variable values, are called *interpolations*. The ability to interpolate is a strong advantage to regression studies. Making statements for values outside of this range—*extrapolation*—is potentially dangerous, however. The straight-line model is not necessarily valid over a wider range of explanatory variable values, and the data usually provide few clues about its validity.

Although the means of responses do not necessarily fall on a straight-line function of the explanatory variable, the linear regression model is still useful, since many interesting questions can be addressed in situations where the model is believed to be only an approximation to the truth. Transformations—applied to the response, to the explanatory variable, or to both—widen the scope of application of the linear regression model. The remainder of this chapter assumes the simple regression model is appropriate for the case studies. Model checking, transformations, and alternatives will be discussed in the next chapter.

7.3   Least Squares Regression Estimation

The method of least squares is one of many procedures for choosing estimates of parameters in a statistical model. The method was described by Legendre in 1805, but the German mathematician Karl Gauss had been using it since 1795 and is generally regarded as its founder. The method provides the foundation for nearly all regression and analysis of variance procedures in modern statistical computer software.

The problem addressed by the method of least squares is that of finding the best-fitting parameter estimates, $\hat{\beta}_0$ and $\hat{\beta}_1$, to a data set consisting of n pairs of observations, (Y_i, X_i), as in Display 7.3.

7.3.1   Fitted Values and Residuals

The hat notation ($\hat{\ }$) denotes an estimate of the parameter under the hat, so $\hat{\beta}_0$ and $\hat{\beta}_1$ are symbols for estimates of β_0 and β_1, respectively. Once these estimates are determined, they combine to form an estimated mean function,

$$\hat{\mu}\{Y|X\} = \hat{\beta}_0 + \hat{\beta}_1 X.$$

This general expression may be calculated for each X_i in the data set to estimate the mean of the distribution from which the corresponding response Y_i arose. The estimated mean is called the *fitted value* (or sometimes *predicted value*) and is represented by fit$_i$. The difference between the observed response and its estimated mean is the *residual*, denoted by res$_i$:

$$\text{fit}_i = \hat{\mu}\{Y|X_i\} = \hat{\beta}_0 + \hat{\beta}_1 X_i$$

and

$$\text{res}_i = Y_i - \text{fit}_i.$$

These definitions are illustrated in Display 7.6 for a simple case.

The magnitude of a single residual measures the distance between the corresponding response and its fitted value according to the estimated mean function. A good estimate should result in a small distance. A measure of the distance between *all* responses and their fitted values is provided by the *residual sum of squares*.

Display 7.6   Illustration of the residual and fitted value for observation (X_4, Y_4) in a hypothetical data set of size 4

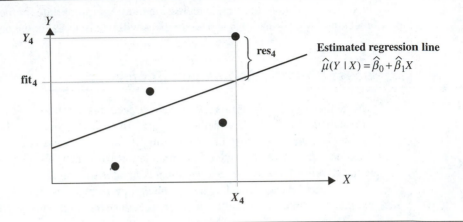

Definition of Least Squares Estimates

The least squares line is the one that has the smallest possible sum of squared residuals. Determining the solution is a problem in differential calculus. The interested reader is referred to Exercise 26. The solution is

$$
\hat{\beta}_1 = \frac{\sum\limits_{i=1}^{n} (X_i - \bar{X})(Y_i - \bar{Y})}{\sum\limits_{i=1}^{n} (X_i - \bar{X})^2}, \qquad \hat{\beta}_0 = \bar{Y} - \hat{\beta}_1 \bar{X}.
$$

A computer or calculator will compute the least squares estimates, so memorization of the formulas is not important.

For example, the least squares estimates of intercept and slope in the meat processing example are $\hat{\beta}_0 = 6.98$ pH and $\hat{\beta}_1 = -0.726$ pH/log(time), respectively.

7.3.2   Sampling Distributions of the Least Squares Estimates

The sampling distributions of $\hat{\beta}_0$ and $\hat{\beta}_1$, based on the normal simple linear regression model, are described in Display 7.7. The estimators are *unbiased*, meaning their sampling distributions are centered on the parameters they estimate. Normality and the expressions for the standard deviations of the sampling distributions lead directly to inferences based on t-ratios.

Display 7.7 Facts about the sampling distributions of the least squares estimates of slope and intercept in the ideal normal model (from statistical theory)

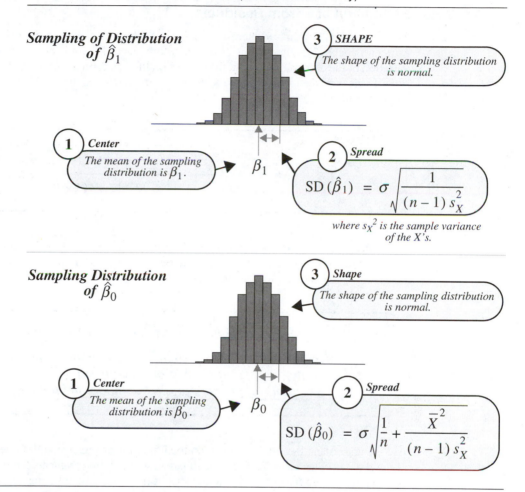

Small standard deviations in the sampling distributions indicate precise estimates. The standard deviations for both estimators depend on three quantities: the sample size n, which is known; the *sample* variation s_X^2 in the explanatory variables, which is also known; and the population standard deviation σ of the responses, which is not known. If the variations in the response and the explanatory variable are not controllable by the researcher, achieving more precision in the estimates (smaller standard deviations in their sampling distributions) can be accomplished only by increasing sample sizes. In experimental conditions, however, the researcher may specify the explanatory variable values. Smaller standard deviations in the sampling distributions can be obtained by increasing the spread of the explanatory variable values. A *caveat*, however, is that by spreading them out, one may arrive at a situation where the linear relationship no longer applies through the entire range. Reducing σ will also give smaller standard deviations in the sampling distributions, but to do so requires improved experimental technique or increased uniformity in the experimental subjects.

7.3.3 Estimation of σ from Residuals

The parameters of primary interest are β_0 and β_1, but the sampling distributions of their estimators depend on the parameter σ. Therefore, even if there is no direct interest in the variation of the observations about the regression line, an estimate of σ must be obtained in order to clear up the picture about the slope and the intercept. Residuals provide the basis for an estimate:

$$\hat{\sigma} = \sqrt{\frac{Sum\ of\ all\ squared\ residuals}{Degrees\ of\ freedom}},$$

where the degrees of freedom are the sample size minus *two*, $n - 2$. This is another instance of applying the general rule for finding the degrees of freedom associated with residuals from estimating a model for the means.

Rule for Determining Degrees of Freedom

Degrees of freedom associated with a set of residuals =
(Number of observations) −
(Number of parameters in the model for the means).

To understand the rule's application to simple linear regression, it is helpful to think about the case where $n = 2$. The least squares estimated line connects the two points, and both residuals are zero. There are no degrees of freedom for estimating variance. Only by adding a third or more observations is there any information about the variability.

Example—Big Bang

The parameter estimates for the Hubble data are: $\hat{\beta}_0 = 0.3991$ megaparsecs and $\hat{\beta}_1 = 0.001373$ megaparsecs/(km/sec). The fitted values and residuals from the least squares fit are shown in Display 7.8. The sum of the 24 squared residuals is 3.6085, with 22 degrees of freedom. The estimate of σ^2 is 0.164, and the estimate of σ is 0.405 megaparsecs.

7.3.4 Standard Errors

Standard errors for the intercept and slope estimates are obtained by replacing σ with $\hat{\sigma}$ in the formulas for their standard deviations (see Display 7.7). The degrees of freedom

Display 7.8 Fitted values and residuals for the Big Bang study

i	Nebula	Velocity X_i	Distance Y_i	fit_i	res_i
1	S. Mag.	170	0.032	0.632	−0.600
2	L. Mag.	290	0.034	0.797	−0.763
3	NGC 6822	−130	0.214	0.221	−0.007
4	NGC 598	−70	0.263	0.303	−0.040
5	NGC 221	−185	0.275	0.145	0.130
6	NGC 224	−220	0.275	0.097	0.178
7	NGC 5457	200	0.450	0.674	−0.224
8	NGC 4736	290	0.500	0.797	−0.297
9	NGC 5194	270	0.500	0.770	−0.270
10	NGC 4449	200	0.630	0.674	−0.044
11	NGC 4214	300	0.800	0.811	−0.011
12	NGC 3031	−30	0.900	0.358	0.542
13	NGC 3627	650	0.900	1.292	−0.392
14	NGC 4626	150	0.900	0.605	0.295
15	NGC 5236	500	0.900	1.086	−0.186
16	NGC 1068	920	1.000	1.662	−0.662
17	NGC 5055	450	1.100	1.017	0.083
18	NGC 7331	500	1.100	1.086	0.014
19	NGC 4258	500	1.400	1.086	0.314
20	NGC 4151	960	1.700	1.717	−0.017
21	NGC 4382	500	2.000	1.086	0.914
22	NGC 4472	850	2.000	1.566	0.434
23	NGC 4486	800	2.000	1.497	0.503
24	NGC 4649	1090	2.000	1.896	0.104

assigned to these standard errors are $n-2$, the same as those attached to the residual estimate of the standard deviation:

$$\text{SE}\,(\hat{\beta}_1) = \hat{\sigma}\sqrt{\frac{1}{(n-1)\,s_X^2}}\,, \quad \text{d.f.} = n-2.$$

and

$$\text{SE}(\hat{\beta}_0) = \hat{\sigma}\sqrt{\frac{1}{n} + \frac{\overline{X}^2}{(n-1)s_X^2}}\,, \quad \text{d.f.} = n-2.$$

Example—Big Bang

Display 7.9 shows a table that is typical of computer output for regression. Separate rows show the estimates, standard errors, and t-tests for the hypotheses that parameter values are zero. The first row corresponds to β_0, the coefficient of the oxymoronic *constant variable*. The second row corresponds to β_1, the coefficient of the variable *velocity*.

Display 7.9   Regression parameter estimates for the Big Bang study

Variable	Coefficient	Standard Error	t-Statistic	p-Value
Constant	0.3991	0.1185	3.369	.0028
Velocity	0.001373	0.000227	6.036	.0000045

Estimate of σ = 0.4050 (22 d.f.)

Ratios of coefficient estimates to their standard errors.

For hypotheses that the coefficients = 0.

Another recommended way to report the results of regression analysis is to present the estimated equation with standard errors in parentheses below the corresponding parameter estimates:

Estimated Mean Distance = 0.3991 + 0.001373 (Velocity)
                              (0.1185) (0.000227)
Estimated SD of Distances = 0.405 megaparsecs (22 d.f.)

7.4 Inferential Tools

This section puts the theory to work, answering a variety of questions that arise in the regression framework.

7.4.1 Tests and Confidence Intervals for Slope and Intercept

Inferences are based on the t-ratios, $(\hat{\beta}_0 - \beta_0)/\mathrm{SE}(\hat{\beta}_0)$ and $(\hat{\beta}_1 - \beta_1)/\mathrm{SE}(\hat{\beta}_1)$, which have t-distributions with $n - 2$ degrees of freedom under the conditions of the normal simple linear regression model. As usual, *tests* are appropriate when the question is whether the data are consistent with some particular hypothesized value for the parameter, as in "Is 7 a possible value for β_0?" *Confidence intervals* are appropriate for presenting values of a parameter that are consistent with the data, as in "What are the possible values of β_0?"

The p-values for the tests that β_0 and β_1 are individually equal to zero are included as standard output from statistical computer packages, as shown in Display 7.9. Although the test that β_1 is zero is often very important—so as to judge whether the mean of the response is unrelated to the explanatory variable—neither of the reported tests necessarily answers a relevant question. To obtain p-values for testing other hypothesized values or to construct confidence intervals, the user must perform other calculations based on the estimates and their standard errors. Some examples follow.

Does Hubble's Data Support the Big Bang Theory? (Test That β_0 Is 0.)

According to the theory discussed in Section 7.1.1 the true distance of the nebulae from earth is a constant times the recession velocity, so, if Y is measured distance, $\mu\{Y|X\} = \beta_1 X$. Thus, if the simple theory is right, the linear regression of measured distance on velocity will have intercept zero. Since the relationship does indeed appear to be a straight line, a check on the theory is accomplished by a test that β_0 is 0. From Display 7.9 the two-sided p-value for this test is reported as .0028, providing convincing evidence that these data do not support the theory as stated.

What is the Age of the Universe According to the Big Bang Theory? (Confidence Interval for β_1)

According to the Big Bang theory, $\mu\{Y|X\} = \beta_1 X$, and β_1 is the age of the universe. To estimate β_1 with the value (0.001373) reported in Display 7.9 would be wrong, because that estimate refers to a different model—one with an intercept parameter—in which the slope parameter has a somewhat different meaning. A least squares fit to the Big Bang model—without the intercept parameter—gives an estimate for β_1 of 0.001922 megaparsec-second per km, with a standard error of 0.000191 based on 23 degrees of freedom. Since $t_{23}(.975) = 2.069$, the confidence interval for β_1 is $.001922 \pm (2.069)(.000191)$, or from .001527 to .002317, or 1.50 to 2.27 billion years. The best estimate is 1.88 billion years. (One megaparsec-second per kilometer is about 979.8 billion years.)

7.4.2 Describing the Distribution of the Response Variable at Some Value of the Explanatory Variable

At some specified value, $X = X_0$, for the explanatory variable, the response distribution has mean $\beta_0 + \beta_1 X_0$ and standard deviation σ. For example, the distribution of pH levels of steers four hours after slaughter will have mean $\beta_0 + \beta_1 \log(4)$. Not all steers will have this pH; the variability of individual pH's around this mean is represented by the standard deviation σ. With least squares estimates, $\hat{\beta}_0 = 6.98$, $\hat{\beta}_1 = -.7257$, and $\hat{\sigma} = .0823$, it can be estimated that the distribution of pH's at 4 hours has a mean of $6.98 - .7257 \log(4) = 5.98$ and an SD of 0.0823.

The Standard Error for the Estimated Mean

The standard error of $\hat{\beta}_0 + \hat{\beta}_1 X_0$, for some specific number X_0, is:

$$
\text{SE}[\hat{\mu}\{Y|X_0\}] \;=\; \hat{\sigma}\sqrt{\frac{1}{n} + \frac{(X_0 - \overline{X})^2}{(n-1)s_X^2}}, \quad \text{d.f.} = n-2.
$$

Once this standard error is computed, a confidence interval or a test for the mean of Y at X_0 follows according to standard t-tools. In addition to $\hat{\sigma}$, the formula requires the average and the sample variance of the explanatory variable values in the data set. The calculations, as demonstrated on the meat processing example in Display 7.10, are straightforward, but inconvenient to do by hand.

Display 7.10   95% confidence interval for the estimated mean pH of steers 4 hours after slaughter (from the estimated regression of pH on log(time) after slaughter for the meat processing data)

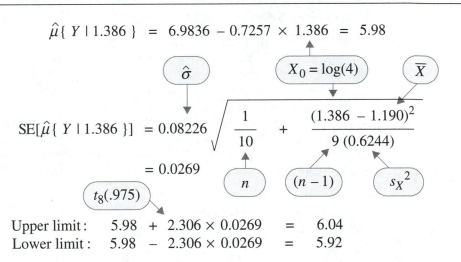

$$
\hat{\mu}\{\,Y \mid 1.386\,\} \;=\; 6.9836 - 0.7257 \times 1.386 \;=\; 5.98
$$

$$
\text{SE}[\hat{\mu}\{\,Y \mid 1.386\,\}] \;=\; 0.08226\sqrt{\frac{1}{10} + \frac{(1.386 - 1.190)^2}{9\,(0.6244)}}
$$

$$
= 0.0269
$$

Upper limit : $5.98 + 2.306 \times 0.0269 \;=\; 6.04$
Lower limit : $5.98 - 2.306 \times 0.0269 \;=\; 5.92$

Computer Centering Trick

A computer trick can be used to bypass the hand calculations. First, create an artificial explanatory variable, $X^* = X - X_0$, by subtracting X_0 from all explanatory variable values. This centers X at X_0, in the sense that the value $X = X_0$ becomes the value

$X^* = 0$. Next, fit the simple linear regression of Y on X^*. The *intercept* in the linear regression model $\mu\{Y|X^*\}$ is the mean of Y at $X^* = 0$, which is exactly the mean of Y at $X = X_0$. Therefore, you may now read the standard error for the estimated mean directly off the computer output as the standard error for the intercept parameter. This trick extends to many more complicated situations.

Additional Notes About the Estimated Mean at Some Value of X

1.   *SE for estimated intercept.* Notice that for $X_0 = 0$, the formula for the standard error of the estimated mean, above, reduces to the formula for the standard error of the estimated intercept.

2.   *Precision in estimating $\mu\{Y|X\}$ is not constant for all values of X.* The formula for the standard error indicates that the precision of the estimated mean response decreases as X_0 gets farther from the sample average of X's.

3.   *Compound uncertainty in estimating several means simultaneously.* This section has addressed the situation where there is one single value of X_0 at which the mean response is desired. The construction of a 95% confidence interval is such that 95% of repetitions of the sampling process result in intervals that include the correct mean response *at that specified* X_0. But if one constructs different 95% confidence intervals for the mean responses at two values of X, the proportion of repetitions that result in both intervals correctly covering the respective means is something less than 95%. This is another case of compound uncertainty. To account for compound uncertainty in estimating mean responses at k different values of X, the Bonferroni procedure (Section 6.4.3 on page 156) can be used. If k is very large, or if it is desired to obtain a procedure that protects against an unlimited number of comparisons, the Scheffé method can be used to construct a confidence band, as discussed next.

4.   *Where is the regression line? (The Workman–Hotelling procedure)* Suppose the question is, "what is the mean response at all X-values within the observational range?" That is, where is the entire line, as opposed to where is one specific value on the line? By the device of replacing the t-multiplier in confidence intervals with a Scheffé multiplier based on an F-percentile, the confidence interval for the mean response can be converted to a *confidence band*, having the property that at least 95% of the repetitions mentioned above produce bands that include the correct mean response *everywhere*. If the 95th percentile in the F-distribution based on 2 and $(n - 2)$ degrees of freedom is $F_{2,n-2}(.95)$, then the 95% confidence band has upper and lower bounds at each X given by

$$\hat{\mu}\{Y \mid X\} \ \pm \ \sqrt{2 \times F_{2,n-2}(.95)} \times SE[\hat{\mu}\{Y \mid X\}]$$

A 95% confidence band for the regression line in the Hubble example uses $F_{2,22}(.95) = 3.443$, so the multiplier for the standard error is 2.624 instead of the t-multiplier of 2.074.

Band limits are calculated for different values of X and the confidence band is obtained by connecting all the upper endpoints and all lower endpoints, as shown in Display 7.11. Notice that the confidence band is not substantially wider than the intervals for individual means. The *prediction band* in the display is discussed in Section 7.4.3.

Display 7.11   The 95% confidence band on the population regression line, the 95% confidence interval band for single mean estimates, and a 95% prediction interval band for the Big Bang example

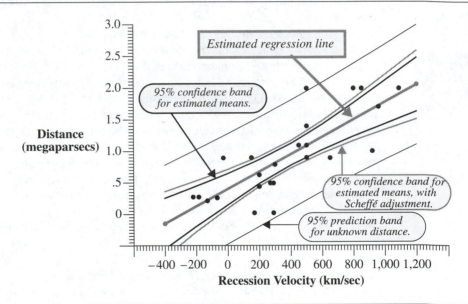

7.4.3 Prediction of a Future Response

Confidence intervals indicate likely values for parameters. Another inferential tool is a *prediction interval*, which indicates likely values for a future value of a response variable at some specified value of the explanatory variable.

Notice the difference between these two questions: (1) What is the *mean pH* 4 hours after slaughter? (2) What will be the *pH of a particular steer carcass* 4 hours after slaughter? The first question is about a parameter and the second is about an individual within the population. Even if the mean and standard deviation of pH at 4 hours after slaughter are known exactly, it is impossible to predict exactly the pH of an individual steer.

The best single prediction of a future response at X_0, denoted by $\text{Pred}\{Y|X_0\}$, is the estimated mean response at X_0:

$$\text{Pred}\{Y|X_0\} = \hat{\mu}\{Y|X_0\} = \hat{\beta}_0 + \hat{\beta}_1 X_0.$$

A *prediction interval* is an interval of likely values, along with a measure of the likelihood that the interval will include the future response value. This measure must account for two independent sources of uncertainty: uncertainty about the location of the subpopulation mean, and uncertainty about where the future value will be in relation to its mean. The error in using the predictor above is

$$
\begin{aligned}
Y - \mathrm{Pred}\{Y|X_0\} &= Y - \hat{\mu}\{Y|X_0\} \\
&= [Y - \mu\{Y|X_0\}] - [\hat{\mu}\{Y|X_0\} - \mu\{Y|X_0\}].
\end{aligned}
$$

$$
\boxed{\text{Prediction error}} = \boxed{\text{Random sampling error}} + \boxed{\text{Estimation error}}
$$

The variance of the random sampling error, as evident from the definition of the simple linear regression model, is σ^2. The variance of the estimation error is the variance of the sampling distribution of $\hat{\beta}_0 + \hat{\beta}_1 X_0$. The standard error of prediction combines the estimated variances as

$$
\mathrm{SE}[\mathrm{Pred}\{Y|X_0\}] = \sqrt{\hat{\sigma}^2 + \mathrm{SE}[\hat{\mu}\{Y|X_0\}]^2}.
$$

Some statistical programs provide standard errors of prediction at requested values. If not, both quantities under the square root sign will be available from computer output, if the computer centering trick of the previous section is invoked. As a last resort, the formula for hand calculation is

$$
\mathrm{SE}[\mathrm{Pred}\{Y|X_0\}] = \hat{\sigma}\sqrt{1 + \frac{1}{n} + \frac{(X_0 - \overline{X})^2}{(n-1)s_X{}^2}}.
$$

Prediction intervals may be obtained from this standard error and a multiplier from the t-distribution with $n-2$ degrees of freedom. Display 7.12 shows the construction of a 95% prediction interval for the pH's of steers at 4 hours after slaughter. The 95% prediction interval succeeds in capturing the future Y in 95% of its applications.

The prediction limits for the Big Bang data in Display 7.11 and for the meat processing data in Display 7.4 are for single predictions. If prediction intervals are desired for multiple new responses at many different values of the explanatory variable (in the range covered by the data), compound uncertainty will arise. For k prediction intervals with approximately 95% familywise inclusion chances, replace the t-multiplier by the Scheffé multiplier M, where $M^2 = k \times F_{k,n-2}(.95)$.

Display 7.12   95% prediction interval for the pH of a steer carcass 4 hours after slaughter (from the estimated regression of pH on log time after slaughter)

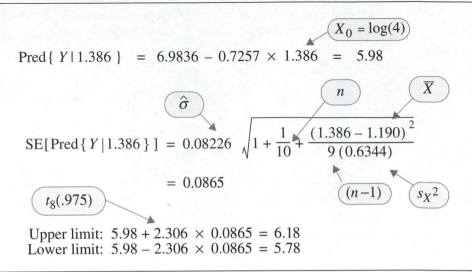

$$\text{Pred}\{\,Y\mid 1.386\,\} \;=\; 6.9836 - 0.7257 \times 1.386 \;=\; 5.98$$

$$X_0 = \log(4)$$

$$\text{SE}[\text{Pred}\{\,Y\mid 1.386\,\}] \;=\; 0.08226\,\sqrt{1 + \frac{1}{10} + \frac{(1.386 - 1.190)^2}{9\,(0.6344)}}$$

$$= 0.0865$$

Upper limit: $5.98 + 2.306 \times 0.0865 = 6.18$
Lower limit: $5.98 - 2.306 \times 0.0865 = 5.78$

7.4.4   Calibration: Guessing the *X* That Results in *Y* = *Y*₀

Sometimes it is desired to guess the X that produces a specific response Y. This happens to be the case for the meat processing example: "At what time after slaughter will the pH in any particular steer be 6?" This is a *calibration* problem, also known as *inverse prediction*.

In a typical calibration problem, an experiment is used to estimate the regression of an imprecise measurement (Y) of some quantity on its actual value (X), as determined by a more precise, but more expensive measuring instrument. The regression provides a calibration equation, so that the accurate measurement can be predicted from the cheaper, inaccurate one. It may seem appropriate to call the precise value Y, to use the imprecise value as X, and to employ prediction as in the previous section. This is not possible, however, when the precise values are controlled by the researcher, because the distributional model makes no sense.

The simplest method for calibration is to invert the prediction relationship. If $\text{Pred}\{Y|X_0\}$ is taken to be $\hat{\beta}_0 + \hat{\beta}_1 X_0$, then $\text{Pred}\{X|Y_0\}$ is $(Y_0 - \hat{\beta}_0)/\hat{\beta}_1$. This is the value of X at which the predicted Y is Y_0. A graphical method for obtaining a calibration interval (inverse prediction interval) for the X at some Y_0 is to plot prediction bands (the ones for predicting Y from X), draw a horizontal line at Y_0, then draw vertical lines down to the X-axis from the points where the horizontal line intersects the upper and lower prediction limits. These two points are the limits of the calibration interval. This procedure is shown for the meat processing data in Display 7.4.

The uncertainty in a calibration interval has to do with the prediction of new Y's. For the meat processing data: at 2.94 hours after slaughter (i.e., when log hours = 1.08), it is predicted that 95% of steer carcasses will have a pH between 6.0 and 6.3. At 5.10 hours (when log hours $= 1.63$) it is predicted that 95% of steer carcasses will have a pH between 5.7 and 6.0. According to these statements, only 2.5% of carcasses will have a pH less than 6.0 at time 2.94 hours, and only 2.5% of carcasses will have a pH above 6.0 at time 5.10 hours.

A similar method directly estimates (or predicts) $\hat{X} = (Y_0 - \hat{\beta}_0)/\hat{\beta}_1$ with an approximate standard error of either

$$SE(\hat{X}) = \frac{SE(\hat{\mu}\{Y|\hat{X}\})}{|\hat{\beta}_1|}$$

(to estimate the X at which the mean of Y is Y_0) or

$$SE(\hat{X}) = \frac{SE(Pred\{Y|\hat{X}\})}{|\hat{\beta}_1|}$$

(to find the X at which the value of Y is Y_0). An interval is centered at $\hat{X}$, with a half-width equal to a t-multiplier (d.f. $= n - 2$) times the standard error. (*Note:* When the regression line is too flat, these methods are not satisfactory. See the references at the end of this chapter for further discussion.)

Example—Meat Processing

The log time at which the mean pH has decreased to 6.0 is estimated to be $\hat{X} = 1.3554$. Its standard error is 0.0367, so a 95% confidence interval is from 1.2708 to 1.4400. Hence the estimate of time required is 3.88 hours, with an approximate 95% confidence interval from 3.56 to 4.2 hours. To predict the time when the pH of a single steer might decline to 6.0, the standard error must account for the pH not being at its average. The second standard error is 0.1191, so the interval for $\log(\hat{X})$ is from 1.0806 to 1.6301. The estimate of the time is the same, 3.88 hours, but the approximate 95% confidence interval is from 2.95 to 5.10 hours. The interval is nearly identical to the one determined graphically in Display 7.4.

7.5 **Related Issues**

7.5.1 **Historical Notes About Regression**

Regression analysis was first used by Sir Francis Galton, a nineteenth-century English scientist studying genetic similarities in parents and offspring. In describing the mean of sons' heights as a function of their fathers' heights, Galton found that sons of tall fathers tended to be tall, but on average not as tall as their fathers. Similarly, sons of short fathers tended to be short, but not on average as short as their fathers. He described this as "regression towards mediocrity." Others using his method to analyze statistical relationships referred to the method as Galton's "regression line," and the term *regression* stuck.

The Regression Effect

Galton's phenomenon of regression towards the mean appears in any test–retest situation. Subjects who score high on the first test will, as a group, score closer to the average (lower) on the second test, while subjects who score low on the first test will, as a group, also score closer to the average (higher) on the second test. The reason is illustrated in Display 7.13. Each point shows the scores of a single subject on the two tests. The horizontal axis contains the first test score, and a vertical box isolates a set of subjects who scored about one standard deviation above average on the first test. Subjects in the set are there (1) because their overall skill is about 1 SD above average, and they performed at their skill level; (2) because their overall skill level was lower than 1 SD above average, and they performed somewhat above their overall skill level on this test; or (3) because their overall skill level was higher than 1 SD above average, and they performed somewhat below their overall skill level on this test. Performances on single tests that differ from one's overall skill level might be called luck or chance.

Display 7.13 Test–retest scores, illustrating the regression effect

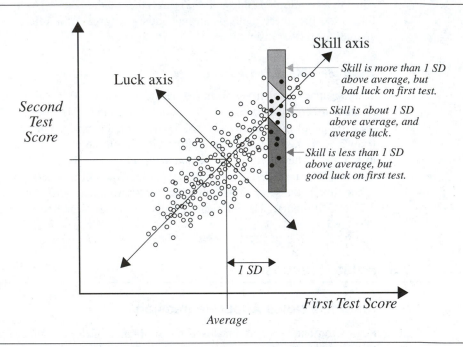

As is apparent in Display 7.13, more subjects fall into category (2) than into category (3), and they bring the average of the second test scores to a level lower than 1 SD above average. This is the *regression effect*, and it is real. The reason for the regression effect is that there are many more individuals in the whole set with skill levels near average than there are with skill levels farther than one standard deviation from it. So, by chance, more appear in the strip whose skill level is closer to the average (and their skill level tends to catch up with them on the second test).

The regression effect is therefore a natural phenomenon in any test–retest situation. It is often mistakenly interpreted as something more, suggesting some theory about the attitude toward the second test of the test takers based on their first test scores. The term *regression fallacy* is used to describe the mistake of attaching some special meaning to the regression effect.

7.5.2 Differing Terminology

The terms *dependent variable* and *independent variable* have been used in the past for response and explanatory variables, respectively. *The Statistical Sleuth* uses *response* and *explanatory* to avoid confusion with the notion of statistical independence, which is entirely different. (Economists often use *endogenous* and *exogenous*, respectively, to distinguish between variables that are determined within the system under study and those determined externally to it. However, the merits of this terminology have not led to its adoption by a wider circle of users.)

7.5.3 Causation

There is a tendency to think of the explanatory variable as a causative agent and the response as an effect, even with observational studies. But the same conditions apply here as before. With randomized experiments, cause-and-effect terminology is appropriate; with observational studies, it is not. With data from observational studies, the most satisfactory terminology describes an *association* between the mean response and the value of the explanatory variable.

7.5.4 Correlation

The *sample correlation coefficient* for describing the degree of linear association between any two variables X and Y is

$$r_{XY} = \frac{\sum_{i=1}^{n} (X_i - \bar{X})(Y_i - \bar{Y}) / (n-1)}{s_X s_Y},$$

where s_X and s_Y are the sample standard deviations. Unlike the estimated slope in regression, the sample correlation coefficient is symmetric in X and Y, so it does not depend on labeling one of them the response and one of them the explanatory variable.

A correlation is dimension-free. It falls between -1 and $+1$, with the extremes corresponding to the situations where all the points in a scatterplot fall exactly on a straight line with negative and positive slopes, respectively. Correlation of zero corresponds to the situation where there is no linear association.

Although the sample correlation coefficient is always a useful summary measure, inferences based on it only make sense when the pairs (X, Y) are selected randomly from a population. This is more restrictive than the conditions for inference from the regression model, where X need not be random. It is a common mistake to base conclusions on correlations when the X's are not random.

If the data are randomly selected pairs, then the sample correlation coefficient estimates a population correlation, $\rho_{XY} = \text{Mean}\{(X - \mu_X)(Y - \mu_Y)\}/(\sigma_X\sigma_Y)$. Conveniently, a test of the hypothesis that $\rho_{XY} = 0$ turns out to be identical to the test of the hypothesis that the slope in the linear regression of Y on X is zero. Tests for nonzero hypothesized values are also available, but they are quite different and are not pursued here.

Correlation only measures the degree of *linear* association. It is possible for there to be an exact relationship between X and Y and yet the sample correlation coefficient is zero. This would be the case, for example, if there was a parabolic relationship with no linear component.

7.5.5 Planning an Experiment: Replication

Replication means applying exactly the same treatment to more than one experimental unit. With the 10 steers available for the meat processing experiment, the researchers decided to administer five treatment levels—the five different times after slaughter—with two replicate observations at each level. Although they could have investigated 10 different times after slaughter, the benefits of replication can outweigh the extra information that would come from more explanatory variable values. The main benefit is that replication allows for an estimate of σ^2 that does not depend on any model being correct (the pooled estimate of the sample variances). This is sometimes referred to as the *pure error* estimate of σ^2. This, in turn, permits an assessment of model adequacy that would not otherwise be available. (The details of a formal test for model adequacy are shown in Section 8.5.3 on page 211.) A secondary benefit is a clearer picture of the relationship between the variance and the mean, which may permit an easier assessment of the need for and the type of transformation for the response variable.

Even though the benefits of replication are many, the practical decision about allocating a fixed number of experimental units to treatment levels is difficult. The researcher must weigh the benefits of replication against the benefits of further treatment levels. It would not be good in the meat processing study, for example, to place all 10 steers in only two treatment levels unless the researchers were certain that the regression was a straight line.

7.6 Summary

In the simple linear regression model the mean of a response variable is a straight-line function of an explanatory variable. The slope and intercept of the regression line can be estimated by the method of least squares. The variance about the regression is estimated by the sum of squared residuals divided by the degrees of freedom $n - 2$. Inferential tools discussed in this chapter are t-tests and confidence intervals for slope and intercept, *t*-tests and confidence intervals for the mean of the response at any particular value of the explanatory variable, prediction intervals for a future response at any particular value of the explanatory variable, and calibration intervals for predicting the X associated with a specified Y.

Big Bang Analysis

The analysis begins with a scatterplot of distance versus velocity. It is apparent from this (Display 7.1) that the regression of distance on velocity is at least approximately a straight line, as predicted by the Big Bang theory. The least squares method provides a way to estimate the slope and intercept in this regression. Two questions are asked: is the intercept zero, as predicted by the theory, and what is the slope; which, according to the theory, is the age of the universe? The evidence in answer to the first question is expressed through a *p*-value for the test that β_0 is zero. The evidence in answer to the second is expressed through a confidence interval for β_1.

Meat Processing Analysis

Even though the question of interest calls for finding the value of time after slaughter at which pH is 6, it is necessary to specify pH to be the response and time (log hours after slaughter) to be the explanatory variable, since the latter was controlled by the researchers. The regression of pH on log time is well-approximated by a straight line, at least for times up to 8 hours. The calibration-like question of interest requires the data analyst to find the value of X (log time) at which Y (pH) is predicted to be 6.

Further Reading

For inferences about the population correlation, and for further details of calibration (inverse prediction) intervals, see Draper and Smith (1981) or Neter, Wasserman, and Kutner (1989). For an illustration of the troubles with inferences from correlation when the X's are controlled, see Weisberg (1985, Section 3.2).

7.7 Exercises

Conceptual Exercises

1. Big Bang Data. Can the estimated regression equation be used to make inferences about (a) the mean distance of nebulae whose recession velocities are zero? (b) the mean distance of nebulae whose recession velocities are 1,000 km/sec? (c) the mean distance of nebulae whose velocities are 2,000 km/sec? (See Display 7.1.)

2. Big Bang Data. Explain why improved measurement of distance would lead to more precise estimates of the regression coefficients.

3. Meat Processing Data. By inspecting Display 7.4, describe the distribution of pH's for steer carcasses 1.65 hours after slaughter (where $X = \log(1.65) = 0.5$).

4. What is wrong with this formulation of the regression model: $Y = \beta_0 + \beta_1 X$?

5.   Is it possible to fit a separate-means (one-way analysis of variance) model to (a) the Big Bang data? (b) the meat processing data?

6.   Consider a regression of weight (kg) on height (cm) for a sample of adult males. What are the units of measurement of (a) the intercept? (b) the slope? (c) the SD? (d) the correlation?

7.   Explain the differences between the following terms: regression, regression model, and simple linear regression model.

8.   What assumptions are made about the distribution of the explanatory variable in the normal simple linear regression model?

9.   A group of children were all given a dexterity test in their fifth grade physical education class. The teacher noted a small group who performed exceptionally well, and she informed the grade six teacher as to which children were in the group. When the same children were given a similar dexterity test the next year, they performed reasonably well, but not as well as the sixth grade teacher had expected. What might have caused this?

10.   Consider the regression of weight on height for a sample of adult males. Suppose the intercept is 5 kg. (a) Does this imply that males of height 0 weigh 5 kg, on average? (b) Would this imply that the simple linear regression model is meaningless?

11.   At what value of X will there be the most precise estimate of the mean of Y? (b) At what value of X will there be the most precise prediction of a future Y?

12.   What is the standard error of prediction as the sample size approaches infinity?

Computational Exercises

13.   Suppose that the fit to the simple linear regression of Y on X from 6 observations produces the following residuals: $-3.3, 2.1, -4.0, -1.5, 5.1, 1.6$. (a) What is the estimate of σ^2? (b) What is the estimate of σ? (c) What are the degrees of freedom?

14.   **Big Bang Data.** Using the results displayed in Display 7.9, find a 95% confidence interval for the intercept in the regression of measured distance on recession velocity.

15.   **Planet Distance and Order from Sun.** Reconsider the planet order and distances in Display 1.15 on page 23. Recall that a straight line model appears to be appropriate for describing the mean log distance as a function of order from the sun. Using a statistical computer package, (a) obtain the least squares estimates of slope and intercept, (b) compute the fitted values for each of the 10 observations, and (c) compute the residuals. (d) Compute the estimate of σ, manually with calculator, from the residuals.

16.   **Planet Distance and Bode's Law.** Bode's Law is an empirical rule relating the distances of the planets in the solar system to their order. It was discovered by Titius of Wittenberg in 1766 and later expounded by Johannes Bode. The statement of Bode's Law is that the mean distance from the sun roughly doubles with each successive planet, so the distance of the ith planet is approximately $d_i = \alpha 2^i$, where α is some constant. By taking logarithms of both sides, $\log(d_i) = \log(\alpha) + i \log(2)$. (a) Using a statistical computer package, fit the simple linear regression of log distance on order from the sun ($\log(d_i)$ on i) with the data in Display 1.15 on page 23; obtain the estimates of slope and intercept, and their standard errors. (b) Check Bode's law by finding a two-sided p-value for the test that β_1 is $\log(2)$. (c) Obtain a 95% confidence interval for β_1.

17.   **Life Expectancy and Per Capita Income.** Reconsider the life expectancy data in Display 3.10 on page 76. Is there any association between life expectancy and per capita income? (a) Draw a scatterplot of life expectancy versus per capita income for the industrialized countries (but excluding South Africa, for which the income is not available). (b) Fit the regression of life expectancy on per

capita income. Find a one-sided *p*-value for testing whether the mean life expectancy is not associated with per capita income, against the alternative that the mean life expectancy increases with increasing per capita income. (c) Find the sample correlation coefficient for the degree of linear association between life expectancy and per capita income for these countries.

18. **Pollen Removal.** Reconsider the data in Display 3.11 on page 78. (a) Draw a scatterplot of proportion of pollen removed versus duration of visit, for the bumblebee queens. (b) Fit the simple linear regression of proportion of pollen removed on duration of visit. Draw the estimated regression line on the scatterplot. (Do this with the computer, if possible; otherwise draw the line with pencil and ruler.) Is there any indication of a problem with this fit? (This problem will be continued in the next chapter.)

19. Most computational procedures utilize the following identities:

$$\sum_{i=1}^{n}(X_i - \overline{X})^2 = \sum_{i=1}^{n} X_i^2 - \frac{1}{n}\left(\sum_{i=1}^{n} X_i\right)^2$$

and

$$\sum_{i=1}^{n}(X_i - \overline{X})(Y_i - \overline{Y}) = \sum_{i=1}^{n} X_i Y_i - \frac{1}{n}\left(\sum_{i=1}^{n} X_i\right)\left(\sum_{i=1}^{n} Y_i\right).$$

The left-hand expressions are needed for calculating $\hat{\beta}_1$, but the right-hand equivalents require less effort. Using the meat processing data, calculate the left-hand versions. Then calculate the right-hand versions. Did you get the same answers? Which version requires fewer steps? Which version requires less storage in a computer?

20. **Meat Processing.** (a) Enter the data from Display 7.3 into a computer and find the least squares estimates for the simple linear regression of pH on log hours. (b) Noting that the average and sample standard deviation of X are provided in Display 7.12, calculate the estimated mean pH at 5 hours after slaughter, and the standard error of the estimated mean (using the formula in Section 7.4.2). (c) Find the standard error of the estimated mean pH at 5 hours after slaughter using the computer trick described in Section 7.4.2, on page 183.

21. **Meat Processing.** (a) Find the standard error of prediction for the prediction of pH at 5 hours after slaughter. (b) Construct a 95% prediction interval at 5 hours after slaughter.

22. **Meat Processing.** Compute a 95% calibration interval (using the graphical approach) for the time at which pH of steer carcasses should be 7.

23. **Meat Processing (Sample Size Determination).** The standard error of the estimated slope based on the 10 data points is 0.0344. Using the formula for SE in Section 7.3.4, and supposing that the spread of the X's and the estimate of σ will be about the same in a future study, calculate how large the sample size would have to be in order for the SE of the estimated slope to be 0.01.

24. **Old Faithful.** Old Faithful Geyser in Yellowstone National Park, Wyoming, derives its name and its considerable fame from the regularity (and beauty) of its eruptions. As they do with most geysers in the park, rangers post the predicted times of eruptions on signs nearby, and people gather beforehand to witness the show. R. A. Hutchinson, a park geologist, collected measurements of the eruption durations (X, in minutes) and the subsequent intervals before the next eruption (Y, in minutes) over an 8-day period. These data appear in Display 7.14. (Data from S. Weisberg, *Applied*

Linear Regression (New York: John Wiley, 1985), p. 231.) Use them to obtain a method for predicting the interval between eruptions from the duration of the previous one. If possible with your statistical computer program, construct a 95% prediction band on a scatterplot of duration versus interval. Or construct 95% prediction intervals at each of several values for durations between 2 and 5 minutes.

Display 7.14 Durations (*X*, in minutes) of Old Faithful eruptions and intervals (*Y*, in minutes) until subsequent eruption, from August 1 to August 8, 1978

Date	X	Y	Date	X	Y	Date	X	Y	Date	X	Y
1	4.4	78	2	4.3	80	3	4.5	76	4	4.0	75
1	3.9	74	2	1.7	56	3	3.9	82	4	3.7	73
1	4.0	68	2	3.9	80	3	4.3	84	4	3.7	67
1	4.0	76	2	3.7	69	3	2.3	53	4	4.3	68
1	3.5	80	2	3.1	57	3	3.8	86	4	3.6	86
1	4.1	84	2	4.0	90	3	1.9	51	4	3.8	72
1	2.3	50	2	1.8	42	3	4.6	85	4	3.8	75
1	4.7	93	2	4.1	91	3	1.8	45	4	3.8	75
1	1.7	55	2	1.8	51	3	4.7	88	4	2.5	66
1	4.9	76	2	3.2	79	3	1.8	51	4	4.5	84
1	1.7	58	2	1.9	53	3	4.6	80	4	4.1	70
1	4.6	74	2	4.6	82	3	1.9	49	4	3.7	79
1	3.4	75	2	2.0	51	3	3.5	82	4	3.8	60
									4	3.4	86
5	4.0	71	6	1.8	55	7	3.5	81	8	4.2	77
5	2.3	67	6	4.6	75	7	2.0	53	8	4.4	73
5	4.4	81	6	3.5	73	7	4.3	89	8	4.1	70
5	4.1	76	6	4.0	70	7	1.8	44	8	4.1	88
5	4.3	83	6	3.7	83	7	4.1	78	8	4.0	75
5	3.3	76	6	1.7	50	7	1.8	61	8	4.1	83
5	2.0	55	6	4.6	95	7	4.7	73	8	2.7	61
5	4.3	73	6	1.7	51	7	4.2	75	8	4.6	78
5	2.9	56	6	4.0	82	7	3.9	73	8	1.9	61
5	4.6	83	6	1.8	54	7	4.3	76	8	4.5	81
5	1.9	57	6	4.4	83	7	1.8	55	8	2.0	51
5	3.6	71	6	1.9	51	7	4.5	86	8	4.8	80
5	3.7	72	6	4.6	80	7	2.0	48	8	4.1	79
5	3.7	77	6	2.9	78						

25.   Crab Claw Size and Force. As part of a study of the effects of predatory intertidal crab species on snail populations, researchers measured the mean closing forces and the propodus heights of the claws on several crabs of three species. Their data (read from their Figure 7) appear in Display 7.15. (Data from S. B. Yamada and E. G. Boulding, "Shell-breaking Efficiency of Predatory Crabs Influences the Distribution of an Intertidal Snail," Technical Report, Zoology Department, Oregon State University 1992.)

(a)   Estimate the slope in the simple linear regression of log force on log height, separately for each crab species. Obtain the standard errors of the estimated slopes.

(b)   Use a *t*-test to compare the slopes for *C. productus* and *L. bellus*. Then compare the slopes for *C. productus* and *H. nudus*. The standard error for the difference in two slope estimates from independent samples is the following:

$$SE[\hat{\beta}_{1(1)} - \hat{\beta}_{1(2)}] = \sqrt{[SE(\hat{\beta}_{1(1)})]^2 + [SE(\hat{\beta}_{1(2)})]^2},$$

where $\hat{\beta}_{1(j)}$ represents the estimate of slope from sample j. Use t-tests with the sum of the degrees of freedom associated with the two standard errors. What do you conclude? (*Note:* A better way to perform this test, using multiple regression, is described in Chapter 9.)

Display 7.15 Closing strengths and propodus heights in three predatory crab species

Hemigrapsus nudus ($n = 14$)		*Lophopanopeus bellus* ($n = 12$)		*Cancer productus* ($n = 12$)	
Force	**Height**	**Force**	**Height**	**Force**	**Height**
3.2	5.0	2.1	5.1	5.0	6.7
6.4	6.0	8.7	5.9	7.8	7.1
2.0	6.4	2.9	6.6	14.6	11.2
2.0	6.5	6.9	7.2	16.8	11.4
4.9	6.6	8.7	8.6	17.7	9.4
3.0	7.0	15.1	7.9	19.8	10.7
2.9	7.9	14.6	8.1	19.6	13.1
9.5	7.9	17.6	9.6	22.5	9.4
4.0	8.0	20.6	10.2	23.6	11.6
3.4	8.2	19.6	10.5	24.4	10.2
7.4	8.3	27.4	8.2	26.0	12.5
2.4	8.8	29.4	11.0	29.4	11.8
4.0	12.1				
5.2	12.2				

---

26. **For those literate in calculus and linear algebra.** The least squares problem is that of finding estimates of β_0 and β_1 that minimize the sum of squares,

$$SS(\beta_0, \beta_1) = \sum_{i=1}^{n}(Y_i - \beta_0 - \beta_1 X_i)^2.$$

(a) Setting the partial derivatives of $SS(\beta_0, \beta_1)$ with respect to each parameter equal to zero, show that β_0 and β_1 must satisfy the *normal equations*:

$$\beta_0 n + \beta_1 \sum_{i=1}^{n} X_i = \sum_{i=1}^{n} Y_i$$

$$\beta_0 \sum_{i=1}^{n} X_i + \beta_1 \sum_{i=1}^{n} X_i^2 = \sum_{i=1}^{n} X_i Y_i.$$

(b) Show that $\hat{\beta}_0$ and $\hat{\beta}_1$ given in Section 7.3.1 on page 175 satisfy the normal equations. (c) Verify that the solutions provide a minimum to the sum of squares.

Data Problems

27. Big Bang II. The data in Display 7.16 are measured distances and recession velocities for 10 clusters of nebulae, much farther from earth than the nebulae reported in Section 7.1.1. (Data from E. Hubble and M. Humason, "The Velocity-Distance Relation Among Extra-galactic Nebulae," *Astrophysics Journal* 74 (1931): 43–50.) If Hubble's theory is correct then the mean of the measured distance, as a function of velocity, should be β_1 Velocity, and β_1 is the age of the universe. Are the data consistent with the theory (that the intercept is zero)? What is the estimated age of the universe? (*Note:* The slope here is in units of megaparsecs-seconds per kilometer. Multiply by 979.8 to get an answer in billions of years. You should find out how to fit simple linear regression through the origin—that is, how to drop the intercept term—with your statistical computer package.) To what extent is the relationship shown by these far-away nebulae clusters similar to and different from the relationship indicated in Case Study 7.1.1? (Analyze the data and write a brief statistical report including a summary of statistical findings, a graphical display, and a details section describing the details of the particular methods used.)

Display 7.16   Measured distance (million parsecs) and recession velocity (km/sec) for 10 clusters of nebulae

Cluster	Distance	Velocity
Virgo	1.8	890
Pegasus	7.25	3810
Pisces	7.00	4638
Cancer	9.00	4820
Perseus	11.00	5230
Coma	13.80	7500
Ursa Major	22.00	11,800
Leo	32.00	19,600
Isolated nebulae I	4.20	2350
Isolated nebulae II	2.15	630

28. Number of Stories and Building Height. The *1994 World Almanac* (Mahwah, N.J.: Funk & Wagnalls, 1993) reports heights and number of stories for notable tall buildings in North America. The list in Display 7.17 is a random sample of size 60 of those for which dates of completion were available. What is the distribution of number of stories as a function of height? What is the mean height per story? Are there any buildings that have particularly fewer stories than expected for their height? Are there any buildings which have particularly more stories than expected for their height? Is there any indication that the number of stories per height has changed over the time period represented here? (Analyze the data and write a brief statistical report including a summary of statistical findings, a graphical display, and a details section describing the details of the particular methods used.)

Answers to Conceptual Exercises

1.   (a) Yes. (b) Yes. (c) Such an extrapolation would be risky.

2.   The standard deviation σ about the regression reflects measurement error variation. Making this smaller will cause the standard deviations of the sampling distributions of the least squares estimates to be smaller (see Display 7.7).

Display 7.17   Year of completion, height, and number of stories for 60 North American buildings

Building	Year	Height	Stories	Building	Year	Height	Stories
191 Peachtree Plaza (Atlanta)	1990	770	54	Celanese Bldg. (New York)	1973	592	45
Southern Bell Telephone (Atlanta)	1980	677	47	520 Madison Ave. (New York)	1983	577	42
Bell South Enterprises (Atlanta)	1990	428	28	Liberty Tower (Oklahoma City)	1971	500	36
Club Towers Apts. (Atlanta)	1989	410	38	Woodmen Tower (Omaha)	1969	469	30
One Georgia Center (Atlanta)	1966	371	29	First Natl. Center (Omaha)	1971	320	22
Scotia Centre (Calgary)	1976	504	38	Sun Bank Center (Orlando)	1988	441	31
Amoco (Chicago)	1974	1136	80	Two Liberty Place (Philadelphia)	1989	845	52
IBM Plaza (Chicago)	1991	695	52	2000 Market St. (Philadelphia)	1973	435	29
Madison Plaza (Chicago)	1982	551	45	Two Logan Square (Philadelphia)	1987	435	34
190 Lasalle (Chicago)	1986	550	40	Penn Mutual Life (Philadelphia)	1931	375	20
Carew Tower (Cincinnati)	1931	568	49	Medical Tower (Philadelphia)	1931	364	33
Central Trust Tower (Cincinnati)	1979	504	33	Inquirer Bldg. (Philadelphia)	1924	340	18
Cityplace Center (Dallas)	1988	560	50	Nix Professional Bldg. (St. Paul)	1931	375	23
2001 Bryan (Dallas)	1973	512	40	Emerald-Shapery Cntr. (San Diego)	1991	450	30
Skyway Tower (Dallas)	1981	448	31	First & Market (San Francisco)	1973	529	38
Burnett Plaza (Fort Worth)	1983	538	40	Embarcadero Cntr. (San Francisco)	1976	412	31
Houston Light. & Pwr. (Houston)	1968	410	27	AT&T Gateway Tower (Seattle)	1990	722	62
Niels Esperson (Houston)	1927	409	31	First Interstate Center (Seattle)	1983	574	48
Indiana Natl. Bank (Indianapolis)	1969	504	35	520 Pike Tower (Seattle)	1984	498	29
Norwest (Minneapolis)	1988	777	57	Key Tower (Seattle)	1986	493	40
Lincoln Centre (Minneapolis)	1987	496	81	Century Square (Seattle)	1986	379	30
1st Bank Place West (Minneapolis)	1960	386	26	100 N. Tampa (Tampa)	1992	579	42
Energy Centre (New Orleans)	1984	530	39	First Financial Tower (Tampa)	1973	458	36
Hyatt Regency (New Orleans)	1976	360	25	NCNB Plaza (Tampa)	1988	454	33
Hibernia Bank (New Orleans)	1920	355	23	First Canadian Place (Toronto)	1979	952	72
Empire State (New York)	1931	1250	102	Commerce Court W. (Toronto)	1972	784	57
Cityspire (New York)	1989	802	72	Commerce Court N. (Toronto)	1930	476	34
Citibank (New York)	1907	741	57	Palace Pier (Toronto)	1978	453	46
World Finance. Center (New York)	1988	739	54	T-D Bank Tower (Vancouver)	1978	440	30
712 5th Ave. (New York)	1990	650	56	Harbour Centre (Vancouver)	1977	428	21

3. The model says that the distribution is normal. The estimated mean pH is about 6.6. The prediction limits will be about 2 SDs up and down, so the SD is about 0.1 pH units.

4. This implies an exact relationship between Y and X. The model should be for the *mean* of Y as a function of X.

5. (a) No. There are not replicate observations at several groups. (b) Yes.

6. (a) kg; (b) kg/cm; (c) kg; (d) none.

7. Regression refers to the mean of a response variable as a function of an explanatory variable. A regression model is a function used to describe the regression. The simple linear regression model is a particular regression model in which the regression is a straight-line function of a single explanatory variable.

8. None.

9. This is the regression effect. It is exactly what you can expect to happen.

10. (a) No. Height = 0 is outside the range of observed values, so the model may not extend to that situation. (b) No. It may be useful for answering questions pertaining to the regression of weight on height for heights in a certain range.

11. (a) At the sample average of the X's used in the estimation. (b) Same as (a).

12. σ.

A Closer Look at Assumptions for Simple Linear Regression

The inferential tools of the previous chapter are based on the normal simple linear regression model with constant variance. Since real data do not necessarily conform to this model, the data analyst must size up the situation and choose a course of action based on an understanding of the robustness of the tools to model violations.

This chapter presents some informal graphical tools and a formal test for assessing the lack of fit. As before, the graphical procedures are used to find suitable transformations to scales where the simple linear regression model seems appropriate. The lack-of-fit test, on the other hand, looks specifically at the issue of whether the straight-line assumption is plausible.

8.1 Case Studies

8.1.1 Island Area and Number of Species—An Observational Study

Biologists have noticed a consistent relation between the area of islands and the number of animal and plant species living on them. If S is the number of species and A is the area, then $S = CA^\gamma$ (roughly), where C is a constant and γ is a biologically meaningful parameter that depends on the group of organisms (birds, reptiles, or grasses, for example). Estimates of this relationship are useful in conservation biology for predicting species extinction rates due to diminishing habitat.

The data in Display 8.1 are the numbers of reptile and amphibian species and the island areas for seven islands in the West Indies. (Data on species from E. O. Wilson, *The Diversity of Life* (New York: W. W. Norton, 1992); areas from *The 1994 World Almanac* (Mahwah, N.J.: Funk & Wagnalls, 1993).) These are typical of data used to estimate the area effect. It is of interest to estimate γ in the species-area equation for this group of animals and to summarize the effect of area on the typical number of species.

Display 8.1 Island area and number of reptile and amphibian species for seven islands in the West Indies

Island	Area (square miles)	Number of Species
Cuba	44,218	100
Hispaniola	29,371	108
Jamaica	4,244	45
Puerto Rico	3,435	53
Montserrat	32	16
Saba	5	11
Redonda	1	7

Summary of Statistical Findings

A second graph in Display 8.2 is a log-log scatterplot of number of species versus island area, along with the estimated line for the regression of log number of species on log area. The parameter γ in the species-area relation, Median $\{S|A\} = CA^\gamma$, is estimated to be .250 (a 95% confidence interval is .219 to .281). It is estimated that the median number of species for islands of area $2A$ will be 19% greater than the median number of species for islands of area A.

Scope of Inference

The statistical association from these observational data cannot be used to establish a causal connection. Furthermore, any generalization of these results to islands in other parts of the world or to a wider population like "islands of rain forest," is purely speculative. Never-

Display 8.2   Scatterplot and log-log-scatterplot of number of reptile and amphibian species versus area for seven islands in the West Indies

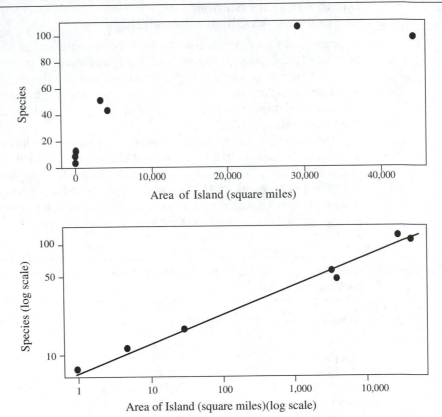

theless, the results from these data are consistent with results for other islands and with small-scale randomized experiments. In summarizing the studies, Wilson offered a conservatively optimistic guess that the current rate of environmental destruction amounts to 27,000 species vanishing each year and that twenty percent of all plant and animal species currently on earth will be extinct by the year 2022.

8.1.2 Breakdown Times for Insulating Fluid Under Different Voltages—A Controlled Experiment

In an industrial laboratory, under uniform conditions, batches of electrical insulating fluid were subjected to constant voltages until the insulating property of the fluids broke down. Seven different voltage levels, spaced two kilovolts (kV) apart from 26 to 38 kV, were studied. The measured responses were the times, in minutes, until breakdown, as listed in Display 8.3). (Data from W. B. Nelson, Schenectady, N.Y.: GE Co. Technical Report 71-C-011 (1970), as discussed in J. F. Lawless, *Statistical Models and Methods for Lifetime Data*

Display 8.3 Times (in minutes) to breakdown of 76 samples of an insulating fluid subjected to different constant voltages

Group #:	1	2	3	4	5	6	7
Voltage(kV):	26	28	30	32	34	36	38
Sample size:	3	5	11	15	19	15	8
Times (min):	5.79	68.85	7.74	0.27	0.19	0.35	0.09
	1579.52	108.29	17.05	0.40	0.78	0.59	0.39
	2323.70	110.29	20.46	0.69	0.96	0.96	0.47
		426.07	21.02	0.79	1.31	0.99	0.73
		1067.60	22.66	2.75	2.78	1.69	0.74
			43.40	3.91	3.16	1.97	1.13
			47.30	9.88	4.15	2.07	1.40
			139.07	13.95	4.67	2.58	2.38
			144.12	15.93	4.85	2.71	
			175.88	27.80	6.50	2.90	
			194.90	53.24	7.35	3.67	
				82.85	8.01	3.99	
				89.29	8.27	5.35	
				100.59	12.06	13.77	
				215.10	31.75	25.50	
					32.52		
					33.91		
					36.71		
					72.89		

(New York: John Wiley & Sons, 1982), chap. 4.) How does the distribution of breakdown time depend on voltage?

Summary of Statistical Findings

Display 8.4 shows a scatterplot of the responses on a logarithmic scale plotted against the voltage levels, along with a summary of the parameter estimates in the simple linear regression of log breakdown time on voltage. Each one kV increase in voltage decreases the median breakdown time by an estimated 60%, for voltages in the range of 26 to 37 kV. A 95% confidence interval for this factor is from 54% to 68%.

Scope of Inference

The laboratory setting for this experiment allowed the experimenter to hold all factors constant except the voltage level, which was assigned at different levels to different batches. Therefore it seems reasonable to infer that the different voltage levels must be directly responsible for the observed differences in time to breakdown. It can be inferred that other batches in the same laboratory setting would follow the same pattern so long as the voltage level is within the experimental range. Inference to voltage levels outside the experimental range and to performance under nonlaboratory conditions cannot be made from these data. If such inference is required, further testing or stronger assumptions must be invoked.

Display 8.4   Scatterplot of breakdown times (natural logarithm scale) versus voltage levels and summary of estimated simple linear regression

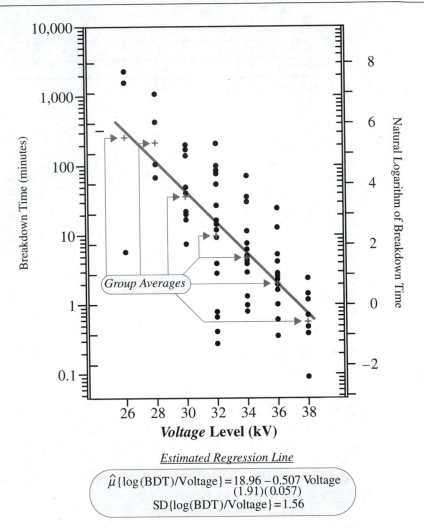

Estimated Regression Line

$$\hat{\mu}\{\log(BDT)/\text{Voltage}\} = 18.96 - 0.507\,\text{Voltage}$$
$$(1.91)(0.057)$$
$$SD\{\log(BDT)/\text{Voltage}\} = 1.56$$

8.2   Robustness of Least Squares Inferences

Exact justification of the statistical statements—tests, confidence intervals, and prediction intervals—based on least squares estimates depends on these features of the regression model:

1. **Linearity.** The plot of response means against the explanatory variable is a straight line.

2. **Constant variance.** The spread of the responses around the straight line is the same at all levels of the explanatory variable.
3. **Normality.** The subpopulations of responses at the different values of the explanatory variable all have normal distributions.
4. **Independence.** The location of any response in relation to its mean cannot be predicted, either fully or partially, from knowledge of where other responses are in relation to their means. (Furthermore, the location of any response in relation to its mean cannot be predicted from knowledge of the explanatory variable values.)

The Linearity Assumption

Two violations of the first assumption may occur: a straight line may be an inadequate model for the regression (the regression might contain some curvature, for example); or a straight line may be appropriate for most of the data, but contamination from one or several outliers from different populations may render it inapplicable to the entire set. Both of these violations can cause the least squares estimates to give misleading answers to the questions of interest. Estimated means and predictions can be biased—they systematically under- or overestimate the intended quantity—and tests and confidence intervals may inaccurately reflect uncertainty. Although the severity of the consequences is always related to the severity of the violation, it is undesirable to use simple linear regression when the linearity assumption is not met. Remedies are available for dealing with this situation.

The Equal Spread Assumption

The consequences for violating this assumption are the same as for one-way analysis of variance. Although the least squares estimates are still unbiased even if the variance is nonconstant, the standard errors inaccurately describe the uncertainty in the estimates. Tests and confidence intervals can be misleading.

The Normality Assumption

Estimates of the coefficients and their standard errors are robust to nonnormal distributions. Although the tests and confidence intervals originate from normal distributions, the consequences of violating this assumption are usually minor. The only situation of substantial concern is when the distributions have long tails (outliers are present) and sample sizes are moderate to small.

If prediction intervals are used, on the other hand, departures from normality become important. This is because the prediction intervals are based directly on the normality of the *population distributions* whereas tests and confidence intervals are based on the normality of the *sampling distributions of the estimates* (which may be approximately normal even when the population distributions are not).

The Independence Assumption

Lack of independence causes no bias in least squares estimates of the coefficients, but standard errors are seriously affected. As before, cluster and serial effects, if suspected, should be incorporated with more sophisticated models. See Chapters 13 and 14 for incorporating cluster effects using multiple regression and Chapter 15 for ways to incorporate serial effects into regression models.

8.3 Graphical Tools for Model Assessment

The principal tools for model assessment are scatterplots of the response variable versus the explanatory variable and of the residuals versus the fitted values. An initial scatterplot of species number versus island area, on the top of Display 8.2, immediately indicates problems with a straight line model for the means. Similarly, a scatterplot of breakdown time versus voltage, in Display 8.5, shows problems of nonlinearity and nonconstant variance.

Display 8.5   Scatterplot of breakdown times versus voltage

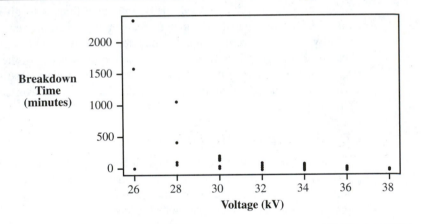

8.3.1 Scatterplot of the Response Variable Versus the Explanatory Variable

Scatterplots with some commonly occurring patterns are shown in Display 8.6. The eye should group points into vertical strips, as shown in (a), to explore the characteristics of subpopulation distributions. Lines connecting strip averages roughly exhibit the relationship between the mean of the responses and the explanatory variable. Vertical lines show the strip standard deviations, which should be examined for patterns in variability. As discussed for each of the plots, the particular patterns for the mean and the variability often suggest a course of action.

(a)   This is the ideal situation. The regression is a straight line and the variability is about the same at all locations along the line. Use simple linear regression.

(b)   The regression is not a straight line, but it is monotonic (the mean of Y either strictly increases or strictly decreases in X), and the variability is about the same at all values

Display 8.6 Some hypothetical scatterplots of response versus explanatory variable with suggested courses of action; (a) is ideal

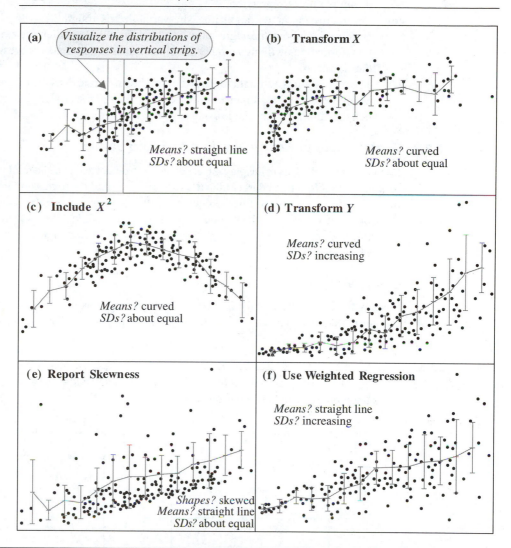

of X. Try transforming X to a new scale where the straight line assumption appears defensible. Then use simple linear regression.

(c) The regression is not a straight line and is not monotonic, and the variability is about the same at all values of X. No transformation of X can yield a straight line relationship. Try *quadratic regression* ($\mu\{Y|X\} = \beta_0 + \beta_1 X + \beta_2 X^2$), which is discussed in Chapter 9.

(d) The regression is not a straight line, and the variability increases as the mean of Y increases. Try a transformation of Y, such as log, reciprocal, or square root, followed by simple linear regression.

(e) The regression is a straight line, the variability is roughly constant, but the distribution of Y about the regression line is skewed. Remedies are unnecessary, and transformations will create other problems. Use simple linear regression, but report the skewness.

(f) The regression is a straight line but the variability increases as the mean of Y increases. Simple linear regression gives unbiased estimates of the straight line relationship, but better estimates are available using *weighted regression*, as discussed in Section 11.6.1.

8.3.2 Scatterplots of Residuals Versus Fitted Values

Sometimes the patterns in Display 8.6 are difficult to detect because the total variability of the response variable is much larger than the variability around the regression line. Scatterplots of the residuals versus the fitted values are better for finding patterns because the linear component of variation in the responses has been removed, leaving a clearer picture about curvature and spread. The residual plot alerts the user to *nonlinearity, nonconstant variance, and the presence of outliers*.

Example

Display 8.5 is a scatterplot of the breakdown time versus voltage. This scatterplot resembles the pattern in part (d) of Display 8.6, suggesting the need for transformation. Display 8.7

Display 8.7   Scatterplot of the square root of breakdown time versus voltage and a residual plot based on the simple linear regression fit

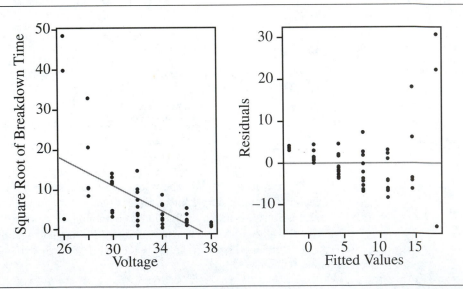

shows the scatterplot and residual plot after applying a square root transformation to the response. The residual plot is more informative than the scatterplot. It has a classic *horn-shaped* pattern, showing a combination of a poor fit to the subpopulation averages and increasing variability. It is evident that the square root transformation has not resolved the problem. A log transformation, however, works well (see Display 8.4). *Note*: The horn-shaped pattern is always a key indicator of the need to transform, whether or not there are replicate samples at specific values of the explanatory variable.

Transformations Indicated by Horn-shaped Residual Plots

A horn-shaped pattern in the residual plot suggests a response transformation like the square root, the logarithm, or the reciprocal. The logarithm is the easiest to interpret. The reciprocal, $1/Y$, works better when the nonconstant spread is more severe, and the square root works better when the nonconstant spread is less severe. Judging severity of nonconstant variance in the horn-shaped residual plot is difficult and unnecessary, however. Try one of the transformations, re-fit the regression model on the transformed data, and redraw the residual plot to see if the transformation has worked. If not, then one of the others can be attempted.

8.4   Interpretation After Log Transformations

A data analyst must interpret regression results in a way that makes sense to the intended audience. The appropriate wording for inferential statements after logarithmic transformation depends on whether the transformation was applied to the response, to the explanatory variable, or to both.

When the Response Variable Is Logged

If $\mu\{\log(Y)|X\} = \beta_0 + \beta_1 X$, and if the distribution of the transformed responses about the regression is symmetric, then

$$\text{Median}\{Y|X\} = \exp(\beta_0)\exp(\beta_1 X).$$

Consequently,

$$\text{Median}\{Y|(X+1)\}/\text{Median}\{Y|X\} = \exp(\beta_1),$$

so an increase in X of 1 unit is associated with a multiplicative change of $\exp(\beta_1)$ in $\text{Median}\{Y|X\}$.

   For example, the estimated relationship between the breakdown time (BDT) of insulating fluid and voltage is $\hat{\mu}\{\log(\text{BDT}) \mid \text{Voltage}\} = 19.0 - 0.51\,\text{Voltage}$. A one kV increase in voltage is associated with a multiplicative change in median BDT of $\exp(-.51)$, or .60. So, the median breakdown time at 28 kV is 60% of what it is at 27 kV; the median breakdown time at 29 kV is 60% of what it is at 28 kV, and so on. Since a 95% confidence interval for β_1 is $-.62$ to $-.39$, a 95% confidence interval for $\exp(\beta_1)$ is $\exp(-.62)$ to $\exp(-.39)$, or 0.54 to 0.68.

When the Explanatory Variable Is Logged

The relationship $\mu\{Y|\log(X)\} = \beta_0 + \beta_1\log(X)$ can be described in terms of multiplicative changes in X, either as a change in the mean of Y for each doubling of X or a change in the mean of Y for each ten-fold increase in X. The chosen multiple should be consistent with the range of X's in the data set.

Notice that

$$\mu\{Y|\log(2X)\} - \mu\{Y|\log(X)\} = \beta_1\log(2),$$

so a doubling of X is associated with a $\beta_1\log(2)$ change in the mean of Y. Similarly, a ten-fold increase in X is associated with a $\beta_1\log(10)$ change in the mean of Y.

For the meat processing data of Section 7.1.2, $\hat{\mu}\{pH\ |\log(Time)\} = 6.98 - 0.726$ log(Time), so a doubling of time after slaughter is associated with a $\log(2)(-.726) = -0.503$ unit change in pH. Since a 95% confidence interval for β_1 is from $-.805$ to $-.646$, a 95% CI for $\log(2)\beta_1$ is from $-.558$ to $-.448$. In words: it is estimated that the mean pH is reduced by .503 for each doubling of time after slaughter (95% confidence interval .448 to .558).

When Both the Response and Explanatory Variable Are Logged

The interpretation is a combination of the previous two. If $\mu\{\log(Y)|\log(X)\} = \beta_0 + \beta_1\log(X)$, then $\text{Median}\{Y|X\} = \exp(\beta_0)X^{\beta_1}$. A doubling of X is associated with a multiplicative change of 2^{β_1} in the median of Y. Or, a ten-fold increase in X is associated with a 10^{β_1}-fold change in the median of Y.

For the island size and number of species data, $\hat{\mu}\{\log(\text{species})|\log(\text{area})\} = 1.94 + 0.250\log(\text{area})$. Thus, an island area of 2A is estimated to have a median number of species that is $2^{0.250}$-fold (or 1.19-fold) larger than the median number of species for an island of area A. Since a 95% confidence interval for β_1 is 0.219 to 0.281, a 95% confidence interval for the multiplicative factor in the median is $2^{0.219}$ to $2^{0.281}$, or 1.16 to 1.22.

The Need for Interpretation

As demonstrated, convenient ways to interpret regression results with log-transformed variables are available. This may also be the case for other transformations in certain instances. For example, the square root of the cross-sectional area of a tree may be reexpressed as the diameter; the reciprocal of the time to complete a race may be interpreted as the speed. In general, however, interpretation for other transformations may be awkward.

For two types of questions, interpretation is not critical. If the regression is used only to assess whether the distribution of the response is *associated with* the explanatory variable, it is sufficient to test the hypothesis that the slope in the regression of Y on X is zero, where Y or X are transformed variables. If the distribution of the transformed response is associated with X, the distribution of the response is associated with X, even though that association may be difficult to describe. Secondly, if the purpose is prediction, no interpretation of the regression coefficients is needed. It is only necessary to express the prediction on the original scale, regardless of the expression used to make the prediction.

8.5   Assessment of Fit Using the Analysis of Variance

An analysis of variance can be used to compare several models. When there are replicate response variables at several explanatory variable values, an analysis of variance F-test for comparing the simple linear regression model to the separate-means (one-way analysis of variance) model supplies a formal assessment of the goodness of fit of simple linear regression. This is called the *lack-of-fit F-test*.

8.5.1   Three Models for the Population Means

In the following three model descriptions for means, the subscript i refers to the group number (e.g., different voltage levels).

1. Separate-means model: $\mu\{Y|X_i\} = \mu_i$, for $i = 1, \ldots, I$.
2. Simple linear regression model: $\mu\{Y|X_i\} = \beta_0 + \beta_1 X_i$, for $i = 1, \ldots, I$.
3. Equal-means model: $\mu\{Y|X_i\} = \mu$, for $i = 1, \ldots, I$.

The separate-means model has no restriction on the values of any of the means. It has I different parameters—the individual group means. The simple linear regression model has two parameters, the slope and the intercept. The equal-means model has a single parameter.

These models form a *hierarchical* set. The equal-means model is a special case of the simple linear regression model, which in turn is a special case of the separate-means model. Viewed conversely, the separate-means model is a generalization of the simple linear regression model, which in turn is a generalization of the equal-means model. A generalization of one model is another model that contains the same features but uses additional parameters to describe a more complex structure.

8.5.2   The Analysis of Variance Table Associated with Simple Regression

Display 8.8 contains two different analysis of variance tables. Table (b) comes from a one-way analysis of variance (Chapter 5), showing the details of an F-test that compares the separate-means model to the equal-means model. Table (a) comes from a simple linear regression analysis showing the details of an F-test that compares the simple linear regression model to the equal-means model.

If β_1 is zero, the simple linear regression model reduces to the equal-means model, $\mu\{Y|X\} = \beta_0$. The hypothesis that $\beta_1 = 0$ can therefore be tested with a comparison of the sizes of the residuals from fits to these two models. An analysis of variance table associated with simple linear regression displays the components of the F-statistic for this test (Display 8.8(a)). Its two main features are the p-value is the same as the two-sided p-value for the t-test of H_0: $\beta_1 = 0$ (the F-statistic is the square of the t-statistic); and the residual mean square is the estimate of variance, $\hat{\sigma}^2$.

The sums of squares column contains the important working pieces. In Table (a), the *residual sum of squares* is the sum of squares of residuals from the regression model, while the *total sum of squares* is the sum of squares of residuals from the equal-means model. The *regression sum of squares* is their difference; i.e., it is the amount by which the residual

Display 8.8   Analysis of variances tables for the insulating fluid data from a simple linear regression analysis and from a separate-means (one-way ANOVA) analysis

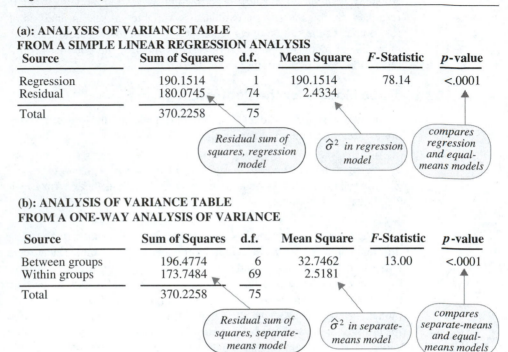

(a): ANALYSIS OF VARIANCE TABLE
FROM A SIMPLE LINEAR REGRESSION ANALYSIS

Source	Sum of Squares	d.f.	Mean Square	F-Statistic	p-value
Regression	190.1514	1	190.1514	78.14	<.0001
Residual	180.0745	74	2.4334		
Total	370.2258	75			

Residual sum of squares, regression model $\hat{\sigma}^2$ *in regression model*   *compares regression and equal-means models*

(b): ANALYSIS OF VARIANCE TABLE
FROM A ONE-WAY ANALYSIS OF VARIANCE

Source	Sum of Squares	d.f.	Mean Square	F-Statistic	p-value
Between groups	196.4774	6	32.7462	13.00	<.0001
Within groups	173.7484	69	2.5181		
Total	370.2258	75			

Residual sum of squares, separate-means model $\hat{\sigma}^2$ *in separate-means model*   *compares separate-means and equal-means models*

sum of squares decreases when the model for the mean of Y is generalized by adding $\beta_1 X$. The generalization adds one parameter to the model, and that is the number of degrees of freedom associated with the regression sum of squares.

Except for differences in terminology, this is completely analogous to the sums of squares column in Table (b), where the *between sum of squares* is the amount by which the residual sum of squares is reduced by the generalization from the equal-means to the separate-means model, involving the change from one to seven parameters (difference $= 6$ parameters added).

8.5.3   The Lack-of-Fit *F*-Test

When replicate values of the response occur at some or all of the explanatory variable values, a formal test of the adequacy of the straight-line regression model is available. It compares the regression model to the more general separate-means (one-way analysis of variance) model. The extra sum of squares upon which the F-statistic is based (Section 5.3 on page 117) is the sum of squared residuals from the reduced (simple linear regression) model minus the sum of squared residuals from the full (separate-means) model.

Specifically, the lack-of-fit F-statistic is where

$$F\text{stat} = \frac{[\text{SSRes}_{LR} - \text{SSRes}_{SM}]/[\text{d.f.}_{LR} - \text{d.f.}_{SM}]}{\hat{\sigma}_{SM}^2},$$

where SSRes_{LR} and SSRes_{SM} are the sums of squares of residuals from the simple linear regression ("Residual" in Display 8.8(a)) and separate-means models ("Within groups" in Display 8.8(b)), respectively; and d.f._{LR} and d.f._{SM} are the degrees of freedom associated with these residual sums of squares. The denominator of the F-statistic is the estimate of σ^2 from the separate-means model. The p-value, for a test of the null hypothesis that the simple linear regression model fits, is found as the proportion of values from an F distribution that exceed the F-statistic. The numerator degrees of freedom are $\text{d.f.}_{LR} - \text{d.f.}_{SM}$, and the denominator degrees of freedom are d.f._{SM}.

Test Computation

Some statistical computer packages will automatically compute the lack-of-fit F-test statistic within the simple regression procedure, as long as there are indeed replicates. Others might perform this test if requested. If neither of these options is available, the user must obtain the analysis of variance tables from both the regression fit and the one-way analysis of variance, as in Display 8.8, extract the necessary information, and compute the F-statistic manually.

Example—Insulating Fluid

From Display 8.8, $\text{SSRes}_{LR} = 180.0745$, $\text{SSRes}_{SM} = 173.7484$, $\text{d.f.}_{LR} = 74$, and $\text{d.f.}_{SM} = 69$. So the F-statistic is $[(180.0745 - 173.7484)/(74 - 69)]/2.5181 = 0.502$. The proportion of values from an F-distribution on 5 and 69 degrees of freedom that exceed 0.502 is .78. This large p-value provides no evidence of lack-of-fit to the simple linear regression model. A small p-value would have suggested that the variability between group means cannot be explained by a simple linear regression model.

8.5.4   A Composite Analysis of Variance Table

Since the simple linear regression model is intermediate between the equal-means and the separate-means models, the reduction in residual sum of squares (196.4774) associated with generalizing from the equal-means model to the separate-means model can be broken up into contributions associated with an initial generalization to the simple linear regression model (190.1514) and with a further generalization from there to the separate-means model. See Display 8.9. The amount associated with the latter step is the difference between the two, $196.4774 - 190.1514 = 6.3260$. Therefore, a useful composite of the analysis of variance tables (a) and (b) in Display 8.8 is given in Display 8.10.

Essentially this is the same table as the analysis of variance table for the separate-means model, Display 8.8 (b), except that the "Between group" sum of squares has been broken into two components—one that measures the pattern in the means that follows a straight line, and one that measures any patterns that fail to follow the straight-line model. The latter

Display 8.9   Reductions in sums of squared residuals in hierarchical models for mean responses in the insulating fluid study

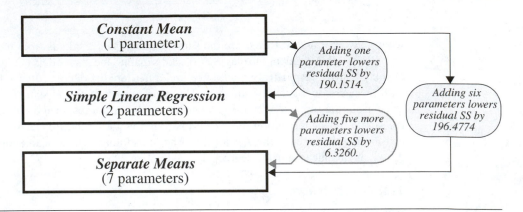

Display 8.10   Composite analysis of variance table with *F*-test for lack-of-fit

Source of Variation	Sum of Squares	d.f.	Mean Square	*F*-Statistic	*p*-Value
Between Groups	*196.4774*	*6*	*32.7462*	*13.00*	*<.0001*
Regression	190.1514	1	190.1514	75.51	<.0001
Lack of Fit	**6.3260**	**5**	**1.2652**	**0.50**	**.78**
Within Groups	*173.7484*	*69*	*2.5181*		
Total	370.2258	75			

By subtraction

LEGEND
Normal type items come from Regression Analysis (a).
Italicized items come from separate-means Analysis (b).
Bold face items are new and calculated here.

---

is called the *lack-of-fit* component. The mean squares in each row are the sum of squares divided by degrees of freedom.

Any residual from the regression model can be decomposed into two parts:

$$Y_{ij} - (\hat{\beta}_0 + \hat{\beta}_1 X_{ij}) = [Y_{ij} - \overline{Y}_i] + [\overline{Y}_i - (\hat{\beta}_0 + \hat{\beta}_1 X_{ij})]$$

Residual from simple linear regression model   Sampling error   Lack of fit

The first part is the difference between the observation and its group average, and the second part is the difference between the group average and the regression fitted value. The variability of the first set of components simply reflects the variability of observations about their group mean (sampling variability) and is unrelated to the adequacy of the regression model. The variability of the second set of components reflects the difference between the group means and the means predicted by the regression model; it will be large if the regression model is inadequate. The sum of squared residuals is equal to the sum of squares of the first part (the within groups sum of squares) *plus* the sum of squares of the second part (the lack-of-fit sum of squares).

8.6 Related Issues

8.6.1 *R*-Squared: The Proportion of Variation Explained

The *R-squared* statistic, or *coefficient of determination*, is the percentage of the total response variation explained by the explanatory variable. Referring to the analysis of variance table (a) in Display 8.8, the residual sum of squares for the equal-means model, $\mu\{Y\} = \beta_0$, is 370.2258. This measures total response variation. The residual sum of squares for the simple linear regression model $\mu\{Y|X\} = \beta_0 + \beta_1 X$ is 180.0745. This residual sum of squares measures the response variation that remains unexplained after inclusion of the $\beta_1 X$ term in the model. Including the explanatory variable therefore reduced the variability by 190.1513. *R*-Squared expresses this reduction as a percentage of total variation:

$$R^2 = 100 \left(\frac{\text{Total sum of squares} - \text{Residual sum of squares}}{\text{Total sum of squares}} \right) \%$$

For the insulating fluids example,

$$R^2 = 100 \left(\frac{370.2258 - 180.0745}{370.2258} \right) \% = 51.4\%.$$

This should be read as "Fifty-one percent of the variation in log breakdown times was explained by the linear regression on voltage." The statement is phrased in the past tense because R^2 describes what happened in the analyzed data set.

Notice that if the residuals are all zero (a perfect fit) then R^2 is 100%. At the other extreme, if the best fitting regression line has slope 0 (and therefore intercept $\overline{Y}$) the residuals will be exactly equal to $Y_i - \overline{Y}$, the residual sum of squares will be exactly equal to the total sum of squares, and R^2 will be zero.

A judgment about what constitutes "good" values for *R*-squared depends on the context of the study. In precise laboratory work, *R*-squared values under 90% may be low enough to require refinements in technique or the inclusion of other explanatory information. In some social science contexts, however, where a single variable rarely explains a great deal of the variation in a response, R-squared values of 50% may be considered remarkably good.

For simple linear regression, R-squared is identical to the square of the sample correlation coefficient for the response and the explanatory variable. As with the correlation, R-squared only estimates some population quantity if the (X, Y) pairs are randomly drawn from a population of pairs. It should not be used for inference, and it should never be used to assess the adequacy of the straight line model, because R^2 can be quite large even when the simple linear regression model is inadequate.

8.6.2 Simple Linear Regression or One-way Analysis of Variance?

When the data are arranged in groups that correspond to different levels of an explanatory variable (as in the meat processing and insulating fluid studies), the statistical analysis may be based on either simple linear regression or one-way analysis of variance. The choice between these two techniques is straightforward: *if the simple linear regression model fits* (possibly after transformation) then it is preferred. The regression approach accomplishes three things: it allows for interpolation; it gives more degrees of freedom for error estimation; and it gives smaller standard errors for estimates of the mean responses.

As an illustration of the increased precision from using regression, Display 8.11 shows estimated means in the insulating fluid example from the separate-means model (i.e., the group averages), and the simple linear regression model (i.e., the fitted values). It also shows three standard errors for the estimated means: the internal standard error of the mean, which is the one-sample standard error (equal to the sample SD over the square root of the sample size); the pooled standard error of the mean, which is the one-way

---

Display 8.11   Estimates of group means and standard errors using different approaches

Means estimated by:
(1) sample averages
(2) regression model

Voltage Level (kV)	n	Average log(BDT)	Internal SE	Pooled SE	Regression Estimate	SE
26	3	5.6240	1.9371	0.9162	5.7640	0.4467
28	5	5.3295	0.5119	0.7907	4.7492	0.3446
30	11	3.8220	0.3350	0.4785	3.7345	0.2536
32	15	2.2285	0.5675	0.4097	2.7198	0.1904
34	19	1.7864	0.3499	0.3640	1.7050	0.1858
36	15	0.9022	0.2866	0.4097	0.6903	0.2432
38	8	−0.4243	0.3506	0.5610	−0.3244	0.3318

Accuracy estimated by:
(i) sample SD/$\sqrt{n}$
(ii) pooled SD/$\sqrt{n}$
(iii) regression method

analysis of variance standard error (equal to the pooled SD over the square root of the sample size); and the standard error of the estimated mean response from the regression model (Section 7.4.2). The two nonregression standard errors are comparable in size (but, of course, confidence intervals will be narrower in the second case because of the larger degrees of freedom attached to the pooled SD). The standard errors from the regression model are uniformly smaller, however. Since the mean (at $X = 26$, say) is postulated to be $\beta_0 + \beta_1 26$, all 76 observations are used to estimate this, not just those from group 1.

As a general rule, it is appropriate to find the simplest model—the one with the fewest parameters—that adequately fits the data. This ensures both the most precise answers to the questions of interest and the most straightforward interpretation.

8.6.3   Other Residual Plots for Special Situations

Residuals Versus Time Order

If the data are collected over time (or space), serial correlation may occur. A plot of the residuals versus the time order of data collection may be examined for patterns. Four possibilities are shown in Display 8.12. In part (a) no pattern emerges, and no serial correlation is indicated. In part (b) a linear trend over time can be observed. It may be possible to include time as an additional explanatory variable (using multiple linear regression, as in Chapter 9). The pattern in part (c) shows a positive serial correlation, in which residuals tend to be followed in time by residuals of the same sign and of about the same size. The pattern in part (d) shows a negative serial correlation, where residuals of one sign tend to be followed by residuals of the opposite sign. The situations in parts (c) and (d) require time series techniques (Chapter 15).

Normal Probability Plots

The *normal probability plot* is a scatterplot involving the ordered residuals and a set of expected values of an ordered sample of the same size from a standard normal distribution. These expected values are available from statistical theory and are not described further here. Some statistical packages plot the expected values along the y-axis against the ordered residuals along the x-axis (as *The Statistical Sleuth* does), while other packages reverse the axes. Assessing normality visually from a normal plot is easier than from a histogram.

Several normal probability plots for hypothetical data are shown in Display 8.13. When subpopulations have normal distributions, the normal plot is approximately a straight line (a). In a long-tailed distribution (b), residuals in the tails have wider gaps than expected in a normal distribution, with the gaps increasing as one gets further into the tails. A skewed distribution (c) will show wider gaps in one tail and shorter gaps in the other, with a reasonably smooth progression of increase in the longer tail. This is in contrast to a situation with outliers (d), where the bulk of the distribution appears normal with one or a few exceptional cases.

Normal probability plots for the residuals from the meat processing data and the insulating fluid data are shown in Display 8.14. The patterns are close enough to a straight

Display 8.12   Possible patterns in plots of residuals versus time order of data collection

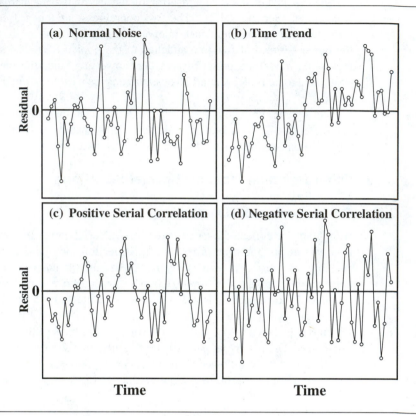

Display 8.13   Normal probability plots illustrating four distributional patterns

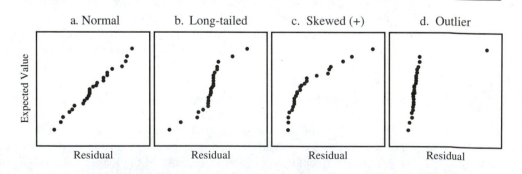

line that the prediction intervals based on the normal assumption should be adequate. (The normal distribution of observations about the regression line is important for the validity of prediction intervals, but not for the validity of estimates, tests, and confidence intervals.)

Display 8.14 Normal probability plots of residuals from simple linear regression fits to the meat processing data and the insulating fluid data

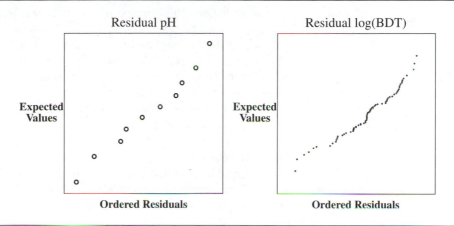

8.6.4 Planning an Experiment: Balance

Balance means having the same number of experimental units in each treatment group. The insulating fluid data are *unbalanced*, because there are three observations in the first treatment group, five in the next, and so on. For the several-treatment experiment, balance is generally desirable in providing equal accuracy for all treatment comparisons, but it is not essential. The voltage experiment was designed as unbalanced presumably because of the much greater waiting time for breakdowns at low voltages and the primary interest in voltages between 30 and 36 kV. Balance will play a more important role when the data are cross-classified according to two factors since some simplifying formulas are only appropriate for balanced data and, more importantly, it allows unambiguous decomposition in the analysis of variance.

8.7 Summary

Exploring statistical relationships begins with viewing scatterplots. Nonlinear regressions, nonconstant spread, and outliers can often be identified at this stage. In cases where problems are less apparent, a simple linear regression can be fit tentatively, and the decision about its appropriateness can be based on the residual plot.

When replicate response variables occur at some of the explanatory variable values, it is possible to conduct a formal lack-of-fit F-test. The test is a special case of the extra-sum-of-squares F-test for comparing two models. The models involved are the simple linear regression (reduced) model and the separate-means (full) model.

Insulating Fluid and Area–Species Studies

Scatterplots and residual plots for the insulating fluid data and for the area–species data reveal nonconstant spread and nonlinear regressions, suggesting transformation of the response variable. In both cases, the spread increases as the mean level increases, indicating a logarithmic, square root, or reciprocal transformation. The scatterplot does not indicate which transformation is best. Several may be tried, with the final choice depending on what is appropriate to the scientific context of the study and to the statistical model assumptions.

After a logarithmic transformation of the times to breakdown, a simple linear regression model fits the insulating fluid data well. No evidence (from a residual plot and a lack-of-fit test) indicates lack of fit or (from the normal plot nonnormality) anything but a normal distribution of the residuals. Mean estimation and prediction can proceed from that model, with results back-transformed to the original scale. Other approaches are possible—another sensible analysis of these data assumes the Weibull distribution on the original scale and gives similar results.

To estimate the parameters in the species–area study, both response and explanatory variables are log-transformed. Here the logarithmic transformations to a simple linear regression model are indicated by theoretical model considerations. Weak evidence remains of increasing variability in the residual plot and of long-tailedness in the normal plot. These data should not, however, be used for predictions. With this small sample size, no further action is required, but confidence limits should be described as approximate.

Further Reading

Residual analysis is covered in standard textbooks on regression, such as those mentioned in Chapter 7. A number of training plots for gathering experience with normal probability plots are given by Daniel and Wood (1980). See also the passage on residuals in *The Encyclopedia of Statistical Science* (Kotz, Johnson, and Read, 1985).

8.8 Exercises

Conceptual Exercises

1.   Island Area and Species Count. The estimated regression line for the data of Section 8.1.1 is $\hat{\mu}\{$log species | log area$\} = 1.94 + .250$ log(area). Show how this estimates that islands of area $.5A$ have a median number of species that is 16% lower than the median number of species for islands of area A.

2.   Insulating Fluid. For the insulating fluid data of Section 8.1.2 explain why the regression analysis allows for statements about the distribution of breakdown times at 27 kV while the one-way analysis of variance does not.

3.   Big Bang. In the data set of Section 7.1.1 on page 168 multiple distances were associated with a few recession velocities. Would it be possible to perform the lack-of-fit test for these data?

4.   Insulating Fluid. If the sample correlation coefficient between the square root of breakdown time and voltage is $-.648$, what is R^2 for the regression of square root of breakdown time on voltage?

**5.   Why can an R^2 close to 1 not be used as evidence that the simple linear regression model is appropriate?

**6.   A study is made of the stress response exhibited by a sample of 45 adults to rock music played at nine different volume levels (5 adults at each level). What is the difference between using the volume as an explanatory variable in a simple linear regression model and using the volume level as a group designator in a one-way classification model?

7. In a study where four levels of a magnetic resonance imaging (MRI) agent are each given to three cancer patients (so there are 12 patients in all), the response is a measure of the degree of seizure activity (an unpleasant side effect). The F-test for lack of fit to the simple linear regression model with $X =$ agent level has a p-value of .0082. The t-tools estimate that the effect of increasing the level of the MRI agent by 1 mg/cm^2 is to increase the level of seizure activity by 2.6 units (95% confidence interval from 1.8 to 3.4 units). (a) How should the latter inference be interpreted? (b) How many degrees of freedom are there for (i) the within-groups variation? (ii) the lack-of-fit variation?

8. Insulating Fluid. Why would it be of interest to know whether batches of insulating fluid were *randomly* assigned to the different voltage levels?

9. Suppose the (Y, X) pairs are: (5,1), (3,2), (4,3), (2,4), (3,5), and (1,6). Would the least squares fit to these data be much different from the least squares fit to the same data with the first pair replaced by (15,1)?

10. (a) What assumptions are used for exact justification of tests and confidence intervals for the slope and intercept in simple regression? (b) Are any of these assumptions relatively unimportant?

11. Suppose you had data on pairs (Y, X) which gave the scatterplot shown in Display 8.15. How would you approach the analysis?

Display 8.15 Scatterplot for Exercise II

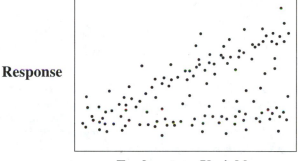

Response

Explanatory Variable

12. What is the technical difficulty with using the separate-means model as a basis for the lack-of-fit F-test when there are no replicate responses?

13. Researchers at a university wish to estimate the effect of class size on course comprehension. An intermediate course in statistics can be taught to classes of any size between 25 and 185 students, and four instructors are available. Suppose the researchers truly believe that the average course comprehension, measured by the average of student scores on a standardized test, is indeed a straight line in class sizes over the range from 25 to 185. What four class sizes should be used in the experiment? Why?

14. Insulating Fluid. Which would you use to predict the log breakdown time for a batch of insulating fluid which is to be put on test at 30 kV: the regression estimate of the mean at 30 kV or the average from the batches that were tested at 30 kV? Why?

Computational Exercises

15. Island Size and Species. (a) Draw a scatterplot of the (untransformed) number of species on the (untransformed) area of the island (Display 8.2, top). (b) Fit the simple linear regression of number of species on area and obtain a residual plot. (c) What features in the two plots indicate a need for transformation?

16. Meat Processing. The data in Display 7.3 on page 170 are a subset of the complete data on postmortum pH in 12 steer carcasses. (Data from J. R. Schwenke and G. A. Milliken, "On the Calibration Problem Extended to Nonlinear Models," *Biometrics* 47(2) (1991):563–74). Once again, the purpose is to determine how much time after slaughter is needed to ensure that the pH reaches 6.0. In Chapter 7 the simple linear regression of pH on log(Hour) was fit to the first 10 carcasses only. Refit the model with all 12 carcasses (data given in Display 8.16). (a) Assess lack-of-fit using a residual plot. (b) Assess lack-of-fit using the lack-of-fit F-test. (c) The inappropriateness of the simple linear regression model can be remedied by dropping the last two carcasses. Is there any justification for doing so? (*Hint:* In order to answer the question of interest, what range of X's appears to be important?)

Display 8.16   pH of steer carcasses 1 to 24 hours after slaughter

Animal Number:	1	2	3	4	5	6	7	8	9	10	11	12
Processing Hour:	1	1	2	2	4	4	6	6	8	8	24	24
pH:	7.02	6.93	6.42	6.51	6.07	5.99	5.59	5.80	5.51	5.36	5.30	5.47

---

17. Biological Pest Control. In a study of the effectiveness of biological control of the exotic weed tansy ragwort, researchers manipulated the exposure to the ragwort flea beetle on 15 plots that had been planted with a high density of ragwort. Harvesting the plots the next season, they measured the average dry mass of ragwort remaining (grams/plant) and the flea beetle load (beetles/gram of ragwort dry mass) to see if the ragwort plants in plots with high flea beetle loads were smaller as a result of herbivory by the beetles. (Data from P. McEvoy and C. Cox, "Successful Biological Control of Ragwort, *Senecio jacobaea*, by introduced insects in Oregon," *Ecological Applications* 1(4) (1991):430–42. The data in Display 8.17 were read from McEvoy and Cox, Figure #2.)

Display 8.17   Dry mass of ragwort weed on 15 plots exposed to flea beetles

Plot#:	1	2	3	4	5	6	7	8	9	10	11	12	13	14	15
Flea beetleload:	12.2	14.6	15.8	25.3	38.6	76.4	163	182	415	446	628	377	770	1,446	1,012
Ragwort mass:	18.2	17.5	7.22	30.6	6.66	6.14	5.21	.502	.611	.630	.427	.011	.012	.006	.002

(a) Use scatterplots of the raw data, along with trial and error, to determine transformations of $Y =$ Ragwort dry mass and of $X =$ Flea beetle load that will produce an approximate linear relationship.

(b) Fit a linear regression model on the transformed scale; calculate residuals and fitted values.

(c) Look at the residual plot. Do you want to try other transformations? What do you suggest?

18. **Chernobyl Fallout.** One of the most dangerous contaminants deposited over European countries following the Chernobyl accident of April 1987 was radioactive cesium. To study cesium transfer from contaminated soil to plants, researchers collected soil samples and samples of mushroom mycelia from 17 wooded locations in Umbria, Central Italy, from August 1986 to November 1989. Measured concentrations (Bq/kg) of cesium in the soil and in the mushrooms are listed in Display 8.18. (Data from R. Borio et al., "Uptake of Radiocesium by Mushrooms," *Science of the Total Environment* 106 (1991):183–90.) (a) Construct a scatterplot using $Y =$ concentration in mushrooms and $X =$ concentration in soil. (b) Fit a simple linear regression model. (c) Fit a simple linear regression model excluding sample number 17. (d) What are your conclusions?

Display 8.18 Cesium (^{134}Cs + ^{137}Cs) concentrations (in Bq/kg) in soil and mushrooms after Chernobyl accident

Sample #:	1	2	3	4	5	6	7	8	9	10	11	12	13	14	15	16	17
Mushrooms:	1	9	14	17	20	17	14	15	34	41	46	49	53	60	79	99	190
Soils:	33	55	138	319	415	425	442	475	279	329	82	86	55	60	144	292	1310

19. **Pollen Removal.** Reconsider the pollen removal data of Display 3.12 and the regression of pollen removed on time spent on flower, for the bumblebee queens only. (a) What problems are evident in the residual plot? (b) Do log transformations of Y or X help any? (c) Try fitting the regression only for those times less than 31 seconds (i.e., excluding the two longest times). Does this fit better? (*Note*: If the linear regression fits for a restricted range of the X's, it is acceptable to fit the model with all the other X's excluded and to report the range of X's for which the model holds.)

20. **Election Fraud.** In a special election to fill a Pennsylvania State Senate seat in 1993, the Democrat, William Stinson, received 19,127 machine-counted votes and the Republican, Bruce Marks, received 19,691. In addition, however, there were 1,391 absentee ballots for Stinson and 366 absentee ballots for Marks, so that the total tally showed Stinson the winner by 461 votes. The large disparity between the machine-counted and absentee votes, and the resulting reversal of the outcome due to the absentee ballots caused some concern about possible illegal influence on the absentee votes. To see whether the discrepancy in absentee votes was larger than could be explained by chance, an econometrician considered the data shown in Display 8.19 (read from a graph in *The New York Times*, 11 April 1994).

(a) Fit the regression of absentee count on machine count. Draw a 95% prediction band. Where does the 1993 result (i.e., absentee $= 1,025$, machine $= -564$) fall with respect to the 95% prediction band? Does this provide some information about the unusualness of the discrepancy between the two?

(b) Is there any problem with using the normal simple linear regression model for the purpose intended? (Does the model fit? Are there possible block or serial effects? Is the normality assumption that is used for the prediction bands reasonable?)

Display 8.19   Democratic minus Republican vote counts, by (a) absentee ballot and (b) voting machine, for 21 elections in Philadelphia's senatorial districts over the last 10 years

	Absentees	Machine		Absentees	Machine
1	−841	−1375	11	167	2573
2	−436	−599	12	364	2620
3	−224	−129	13	384	3596
4	160	47	14	320	3619
5	597	7132	15	240	4195
6	−356	1116	16	1365	4395
7	300	940	17	392	4512
8	288	987	18	404	5088
9	609	1927	19	408	5570
10	412	2150	20	693	6286
			21	400	6533

Data Problems

21.   Ecosystem Decay. As an introduction to their study on the effect of Amazon forest clearing (Data from T. E. Lovejoy, J. M. Rankin, R. O. Bierregaard, Jr., K. S. Brown, Jr., L. H. Emmons, and M. E. Van der Woot, "Ecosystem Decay of Amazon Forest Remnants," in M. H. Nitecki, ed., *Extinctions* (Chicago: University of Chicago Press, 1984) the researchers stated: "fragmentation of once continuous wild areas is a major way in which people are altering the landscape and biology of the planet." Their study takes advantage of a Brazilian requirement that 50% of the land in any development project remain in forest and tree cover. As a consequence of this requirement "islands" of forest of various sizes remain in otherwise cleared areas. The data in Display 8.20 are the number of butterfly species in 16 such islands. Summarize the role of area in the distribution of number of butterfly species. Write a brief statistical report including a summary of statistical findings, a graphical display, and a section detailing the methods used to answer the questions of interest.

Display 8.20   Forest patch area (hectares) and number of butterfly species found

Reserve	Area	Species	Reserve	Area	Species
1	1	14	9	10	33
2	1	50	10	10	53
3	1	55	11	10	50
4	1	34	12	100	110
5	1	40	13	100	70
6	1	57	14	100	119
7	10	43	15	100	60
8	10	103	16	1000	145

22.   Wine Consumption and Heart Disease. The data in Display 8.21 are the average wine consumption rates (in liters per person) and number of ischemic heart disease deaths (per 1,000 men aged 55 to 64 years old) for 18 industrialized countries. (Data from A. S. St. Leger, A. L. Cochrane, and

F. Moore, "Factors Associated with Cardiac Mortality in Developed Countries with Particular Reference to the Consumption of Wine," *Lancet* (June 16, 1979): 1017–20.) Do these data suggest that the heart disease death rate is associated with average wine consumption? If so, how can that relationship be described? Do any countries have significantly higher or lower death rates than others with similar wine consumption rates? Analyze the data and write a brief statistical report that includes a summary of statistical findings, a graphical display, and a section detailing the methods used to answer the questions of interest.

Display 8.21 Wine consumption (liters per person per year) and heart disease mortality rates (deaths per 1,000) in 18 countries

Country	Wine Consumption	Heart Disease Mortality
Norway	2.8	6.2
Scotland	3.2	9.0
England	3.2	7.1
Ireland	3.4	6.8
Finland	4.3	10.2
Canada	4.9	7.8
United States	5.1	9.3
Netherlands	5.2	5.9
New Zealand	5.9	8.9
Denmark	5.9	5.5
Sweden	6.6	7.1
Australia	8.3	9.1
Belgium	12.6	5.1
Germany	15.1	4.7
Austria	25.1	4.7
Switzerland	33.1	3.1
Italy	75.9	3.2
France	75.9	2.1

Answers to Conceputal Exercises

1. Median $\{$species$|$area$\}$ = $\exp(1.94)$area$^{0.250}$. So Median$\{$species $|0.5$ area$\}$/Median$\{$species $|$ area$\}$ = $.5^{.250}$ = $.84$. Thus Median$\{$species $|0.5$ area$\}$ = 0.84 Median$\{$species $|$ area$\}$ and, finally, [Median$\{$species $|$ area$\}$ − Median$\{$species $| 0.5$ area$\}$]/Median$\{$species $|$ area$\}$ = $1 − 0.84 = 0.16$.

2. The ANOVA model states that there are 7 means, one for each voltage level tested, but does not describe any relation between mean and voltage. The regression model establishes a pattern between mean log breakdown time and voltage, for all voltages in the range of 26 to 38 kV.

3. Yes. (*Note:* The multiple observations occurred because of rounding, so they do not represent repeated draws from the same distribution. Nevertheless, they are near replicates, at least, and can be used as such for the lack-of-fit test.)

4. R^2 = square of correlation coefficient = $(−.648)^2 = .420$.

5. Although a high R^2 reflects a strong degree of linear association, this linear association may well be accompanied by curvature (and by nonconstant variance).

6. In the simple linear regression model, the nine mean stress levels lie on a straight line against volume. In the one-way classification (the separate-means model) the mean stress levels may or may not lie on the straight line—their values are not restricted.

7.   (a) If the data do not fit the model, then the parameters of the model are not adequate descriptive summaries. No inference should be drawn, at least until a better model is found. (b) (i) 8; (ii) 2.

8.   Even in laboratory circumstances, confounding variables are possible. If, for example, batches are spooned from a large container that has density stratification, then assigning consecutive batches to the lowest voltage, then the next lowest, and so on, will confound fluid density with voltage. Randomization never hurts, and usually helps.

9.   Yes, very much so. The least squares method is not resistant to the effects of outliers.

10.   (a) linearity, constant variance, normality, independence. (b) normality.

11.   It may appear that this is a case where the spread of Y increases as the mean of Y does, so a simple transformation may be in order. This will not produce good results. Notice that the average Y at a large X lies in a no-man's land with few observations. Looking at the distributions in strips, you will see two separate groups in each distribution. Look for some important characteristic that separates the data into two groups, then build a separate regression model for each group.

12.   In fitting the separate-means model, all residuals are zero. The denominator of the F-statistic is zero, so the F-statistic is not defined.

13.   Put two classes at 25 students and two at 185 students. This makes the standard deviation in the sampling distribution of the slope parameter as small as possible, given the constraints of the situation. (But it gives no information on lack of fit; check the degrees of freedom).

14.   The regression estimate. It is more precise (see Display 8.11).

Multiple Regression

Multiple regression analysis is one of the most widely used statistical tools, and for good reason: it is remarkably effective for answering questions involving many variables. Although more difficult to visualize than simple regression, multiple regression is a straightforward extension. It models the mean of a response variable as a function of *several* explanatory variables.

Many issues, tools, and strategies are associated with multiple regression analysis, as discussed in this and the next three chapters. This chapter focuses on the meaning of the regression model and strategies for analysis. The details of estimation and inferential tools are deliberately postponed until the next chapter so that the student may concentrate first on understanding regression coefficients and the types of data structures that may be analyzed with multiple regression analysis.

9.1 Case Studies

9.1.1 Effects of Light on Meadowfoam Flowering—A Randomized Experiment

Meadowfoam (*Limnanthes alba*) is a small plant found growing in moist meadows of the U.S. Pacific Northwest. It has been domesticated at Oregon State University for its seed oil, which is unique among vegetable oils for its long carbon strings. Like the oil from sperm whales, it is nongreasy and highly stable.

Researchers reported the results from one study in a series designed to find out how to elevate meadowfoam production to a profitable crop. In a controlled growth chamber, they focused on the effects of two light-related factors: light intensity, at the six levels of 150, 300, 450, 600, 750, and 900 μmol/m^2/sec; and the timing of the onset of the light treatment, either at photoperiodic floral induction (PFI)—the time at which the photoperiod was increased from 8 to 16 hours per day to induce flowering—or 24 days before PFI. The experimental design is depicted in Display 9.1. (Data from M. Seddigh and

Display 9.1   Time line for light variation experiment on meadowfoam

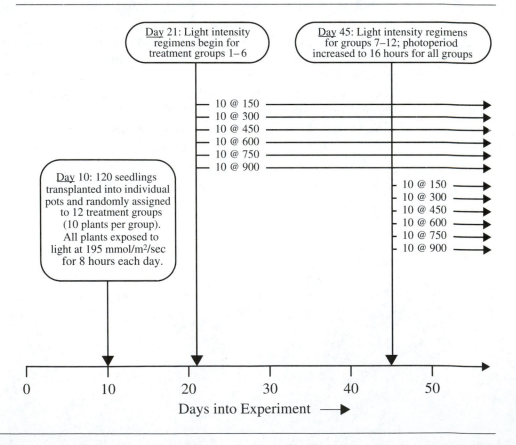

G. D. Jolliff, "Light Intensity Effects on Meadowfoam Growth and Flowering," *Crop Science* 34 (1994): 497–503.)

The design consists of 12 treatment groups—the six light intensities at each of the two timing levels. Ten seedlings were randomly assigned to each treatment group. The number of flowers per plant is the primary measure of production, and it was measured by averaging the numbers of flowers produced by the 10 seedlings in each group.

The entire experiment was replicated twice, yielding the data in Display 9.2. (Technically, the separate experiments are different blocks. Since no difference was found between these blocks, the data are presented here as if the two responses at each treatment level are replicates.) What are the effects of differing light intensity levels? What is the effect of the timing? Does the effect of intensity depend on the timing?

Display 9.2   Average number of flowers per meadowfoam plant, in 12 treatment groups

		light intensity (μmol/m^2/sec)					
		150	300	450	600	750	900
timing	at PFI	62.3 77.4	55.3 54.2	49.6 61.9	39.4 45.7	31.3 44.9	36.8 41.9
	24 days before PFI	77.8 75.6	69.1 78.0	57.0 71.1	62.9 52.2	60.3 45.6	52.6 44.4

Summary of Statistical Findings

Display 9.3 shows the fit of a multiple linear regression model that specifies parallel regression lines for the mean number of flowers as functions of light intensity. Increasing light intensity decreased the mean number of flowers per plant by an estimated 4.0 flowers per plant per 100 μmol/m^2/sec (95% confidence interval from 3.0 to 5.1). Beginning the light treatments 24 days prior to PFI increased the mean numbers of flowers by an estimated 12.2 flowers per plant (95% confidence interval from 6.7 to 17.6). There is no evidence that the effect of light intensity depends on the timing of its initiation (two-sided p-value = .91, from a t-test for interaction, 5 degrees of freedom).

Scope of Inference

The researchers can infer that the effects above were caused by the light intensity and timing manipulations, because this was a randomized experiment.

9.1.2   Why Do Some Mammals Have Large Brains for Their Size?—An Observational Study

Evolutionary biologists are keenly interested in the characteristics that enable a species to withstand the selective mechanisms of evolution. An interesting variable in this respect

---

Display 9.3   Summary of relationship of flowers produced per plant with increasing light
intensities, at and 24 days prior to photoperiodic floral induction (PFI)

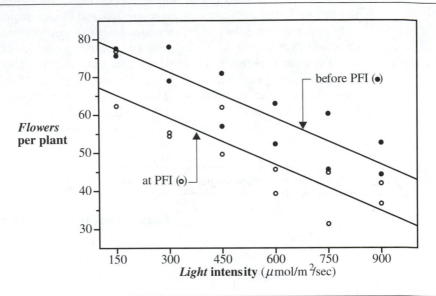

---

is brain size. One might expect that bigger brains are better, but certain penalties seem
to be associated with large brains, such as the need for longer pregnancies and fewer
offspring. Although the individual members of the large-brained species may have more
chance of surviving, the benefits for the species must be good enough to compensate for
these penalties. To shed some light on this issue, it is helpful to determine exactly which
characteristics are associated with large brains, after getting the effect of body size out of
the way.

The data in Display 9.4 are the average values of brain weight, body weight, gestation
lengths (length of pregnancy), and litter size for 96 species of mammals. (Data from
G. A. Sacher and E. F. Staffeldt, "Relation of Gestation Time to Brain Weight for Placental
Mammals; Implications for the Theory of Vertebrate Growth," *American Naturalist* 108
(1974): 593–613. The common names for the species correspond to the Latin names given
in the original paper, and those followed by a Roman numeral indicate subspecies.) Since
brain size is obviously related to body size, the question of interest is this: which, if any,
variables are associated with brain size, after accounting for body size?

Summary of Statistical Findings

There was convincing evidence that brain weight was associated with either gestation length
or litter size, even after accounting for the effect of body weight (p-value $< .0001$; extra
sum of squares F-test). There was strong evidence that litter size was associated with brain
weight after accounting for body weight and gestation (two-sided p-value $= .0089$) and
that gestation period was associated with brain weight after accounting for body weight and
litter size (two-sided p-value $= .0038$).

Display 9.4 Average values of brain weight, body weight, gestation length, and litter size in 96 species of mammal

Species	Brain Weight (grams)	Body Weight (kilograms)	Gestation Period (days)	Litter Size	Species	Brain Weight (grams)	Body Weight (kilograms)	Gestation Period (days)	Litter Size
Quokka	17.5	3.5	26	1.0	Acouchis	9.9	0.78	98	1.2
Hedgehog	3.50	0.93	34	4.6	Chinchilla	5.25	0.43	110	2.0
Tree shrew	3.15	0.15	46	3.0	Nutria	23.	5.0	132	5.5
Elephant shrew I	1.14	0.049	51	1.5	Dolphin	1,600.	160.	360	1.0
Elephant shrew II	1.37	0.064	46	1.5	Porpoise	537.	56.	270	1.0
Lemur	22.	2.1	135	1.0	Dog	70.2	8.5	63	4.0
Slow loris	12.8	1.2	90	1.2	Red fox	48.	6.0	52	4.0
Bush baby	9.9	0.7	135	1.0	Gray fox	37.3	3.8	63	3.7
Howler monkey	54.	7.7	139	1.0	Bat-eared fox	28.5	3.2	65	4.0
Ring-tail monkey	73.	3.7	180	1.0	Grizzly bear	400.	250.	219	2.3
Spider monkey I	114.	9.1	140	1.0	Beaked whale	500.	250.	240	1.8
Spider monkey II	109.	7.7	140	1.0	Raccoon	41.6	5.3	63	3.5
Gentle lemur	7.8	0.22	145	2.0	Kinkajou	31.2	2.0	77	1.1
Rhesus monkey I	84.6	6.0	175	1.0	Badger	53.	6.0	60	2.2
Rhesus monkey II	107.	8.7	165	1.1	Domestic cat	28.4	2.5	63	4.0
Hamadryas baboon	183.	21.	180	1.0	Lynx	75.	12.	60	2.5
Western baboon	179.	32.	180	1.0	Leopard	157.	46.	92	2.5
Vervet guenon	67.	4.6	195	1.0	Lion	260.	180.	108	3.0
Leaf monkey	65.5	5.8	168	1.0	Tiger	302.	210.	104	3.0
White-handed gibbon	102.	5.5	210	1.0	Fur seal	355.	250.	254	1.0
Orangutan	343.	37.	270	1.0	Sea lion	363.	100.	343	1.0
Chimpanzee	360.	45.	230	1.0	Harp seal	442.	110.	240	1.0
Gorilla	406.	140.	265	1.0	Weddell seal	550.	400.	310	1.0
Human being	1,300.	65.	270	1.0	African Elephant	4,480.	2,800.	655	1.0
Long-nosed armadillo	12.	3.7	120	4.0	Hyrax	20.5	3.8	225	2.4
Aardvark	9.6	2.2	31	5.0	Horse	712.	480.	330	1.0
Jack rabbit	13.3	2.9	41	2.5	Tapir	250.	230.	390	1.0
Tree squirrel	6.23	0.33	38	3.0	Wild boar	185.	150.	120	4.0
Flying squirrel	1.89	0.052	40	3.1	Domestic pig	180.	190.	115	8.0
Canadian beaver	40.	20.	128	2.9	Hippopotamus	590.	1,400.	240	1.0
Beaver	45.	25.	128	4.0	Pygmy hippopotamus	260.	150.	205	1.0
Deer mouse I	0.68	0.027	23	3.7	Llama	225.	93.	330	1.0
Deer mouse II	0.63	0.026	23	5.0	Vicuna	198.	45.	300	1.1
Deer mouse III	0.52	0.017	24	5.0	Barking deer	124.	16.	183	1.1
Deer mouse IV	0.69	0.024	24	5.0	Fallow deer	223.	80.	240	1.0
Hamster I	0.67	0.036	21	4.6	Axis deer	219.	89.	218	1.0
Hamster II	1.12	0.13	16	6.3	Red deer	435.	200.	255	1.0
Pygmy gerbil	1.04	0.065	21	4.0	Elk	365.	120.	235	1.0
Rat I	0.72	0.05	23	7.3	Sambar	383.	120.	246	1.1
Rat II	2.38	0.34	21	8.0	Caribou	288.	110.	225	1.0
House mouse	0.45	0.024	19	5.0	Eland	480.	560.	255	1.0
Hopping mouse	1.18	0.15	27	5.6	Yak	334.	250.	255	1.0
Porcupine I	37.	11.	112	1.2	Cattle	456.	520.	280	1.0
Porcupine II	37.	14.	112	1.2	Duiker	93.	13.	120	1.0
Porcupine III	24.	6.6	113	1.0	Blackbuck Antelope	200.	39.	180	1.0
Guinea pig	4.28	0.97	67	2.6	Barbary sheep	210.	66.	158	1.2
Capybara	76.	30.	123	3.0	Domestic sheep	125.	49.	150	2.4
Agoutis	20.3	2.8	104	1.3	Domestic goat	106.	30.	151	2.0

Scope of Inference

As suggestive as the findings may be, inferences that go beyond these data are unwise. The data were summarized from available studies and cannot be representative of any wider population. As usual, no causal interpretation can be made from these observational data.

9.2   Regression Coefficients

9.2.1   The Multiple Linear Regression Model

Regression

The *regression of Y on X_1 and X_2* is a rule, such as an equation, that describes the mean of the distribution of a response variable (Y) for particular values of explanatory variables (X_1 and X_2, say). For example, the regression of the response variable $Y = flowers$ on $X_1 = light$ intensity and $X_2 = time$ of light manipulation specifies how the mean number of flowers per plant depends on the levels of light intensity and timing. (The italicized words specify variable names.)

The data in Display 9.2 are samples of size 2 from subpopulations of measurements corresponding to the 12 combinations of *light* and *time* in the study. Notice how the distribution of *flowers* tends to change with different levels of these explanatory variables. Although the spreads of the distributions are roughly equal (see Display 9.3), the centers tend to decrease with increasing light intensity and to change with the timing. It is natural, therefore, to focus on the means of the response distributions as a function of the explanatory variables.

The symbol for the regression is $\mu\{flowers \mid light, time\}$, which is read as the "mean number of flowers, as a function of light intensity and timing." In terms of a generic response variable, Y, and two explanatory variables, X_1 and X_2:

$$\text{The regression of } Y \text{ on } X_1 \text{ and } X_2 \text{ is } \mu\{Y \mid X_1, X_2\}.$$

With specific values for X_1 and X_2 inserted, the same expression may be read somewhat differently. For example, $\mu\{flowers \mid light = 300, time = 24\}$ (or $\mu\{flowers \mid 300, 24\}$ if it is clear which variables the numerical values refer to) is read as, "the mean number of flowers when light intensity is 300 μmol/m^2/sec and time is 24 days prior to PFI."

Multiple Regression Models

In *multiple regression* there is a single response variable and *multiple explanatory variables*. The regression rule giving the mean response for each combination of explanatory variables will not be very helpful if it is a huge list or a complicated function. Furthermore, it is usually unwise to think that there is some exact, discoverable regression equation. Many possible *models* are available, however, for describing the regression. Although they should not be thought of as the truth, one or two models may adequately approximate the mean of the response as a function of the explanatory variables, and conveniently allow for the questions of interest to be investigated.

Multiple Linear Regression Model

One family of models is particularly easy to deal with and works for the majority of regression problems—the family of linear regression models. The term *linear* refers to the linearity of the regression's *parameters*:

> *Multiple Linear Regression Model (for two explanatory variables):*
> $$\mu\{Y \mid X_1, X_2\} = \beta_0 f_0(X_1, X_2) + \beta_1 f_1(X_1, X_2) + \beta_2 f_2(X_1, X_2) + \cdots,$$
> *where the $f_j(X_1, X_2)$'s are known functions of the explanatory variables.*

The β's are unknown model parameters, referred to as *regression coefficients*. The functions $f_j(X_1, X_2)$ are often simple, as in the following examples:

> *Examples of Multiple Linear Regression Models*
> $$\mu\{Y \mid X_1, X_2\} = \beta_0 + \beta_1 X_1 + \beta_2 X_2,$$
> $$\mu\{Y \mid X_1\} = \beta_0 + \beta_1 X_1 + \beta_2 X_1^2,$$
> $$\mu\{Y \mid X_1, X_2\} = \beta_0 + \beta_1 X_1 + \beta_2 X_2 + \beta_3 X_1 X_2,$$
> $$\mu\{Y \mid X_1, X_2\} = \beta_0 + \beta_1 \log(X_1) + \beta_2 \log(X_2).$$

The extension to more than two explanatory variables is straightforward. Notice that the first term in the examples is β_0 (which multiplies the trivial function, 1). This *constant term* appears in multiple linear regression models unless a specific reason for excluding it exists.

Multiple Regression Model with Constant Variance

The ideal regression model includes an assumption of constant variation. For the meadowfoam example, the assumption states that

$$\text{Var}\{\textit{flowers} \mid \textit{light, time}\} = \sigma^2.$$

The notation Var{*flowers* | *light, time*} is read as "the variance of the numbers of flowers, as a function of light and time." The right-hand side of the equation asserts that this variance is constant—the same for all values of light and time.

The interpretation of regression models is most straightforward when the response variance is constant, because the mean structure describes the entire relationship between the response and the explanatory variables. Standard inferential tools for regression (discussed in the next chapter) assume constant variance.

9.2.2 Interpretation of Regression Coefficients

The purposes of regression analysis are to find a good fitting model for the response mean, to word the questions of interest in terms of model parameters (the regression coefficients), to estimate the parameters with the available data, and to employ appropriate inferential tools for answering the questions of interest and for expressing the uncertainty in the answers. Before turning to issues of estimation and inference, it is important to discuss the meaning of regression coefficients and what questions can be answered through them.

Regression Surfaces

The multiple linear regression model, which includes a separate term for each individual variable,

$$\mu\{flowers \mid light, time\} = \beta_0 + \beta_1 light + \beta_2 time,$$

describes the regression surface as a *plane*. The parameter β_0 is the height of the plane when both *light* and *time* equal zero; the parameter β_1 is the slope of the plane as a function of *light* for any fixed value of *time*, and the parameter β_2 is the slope of the plane as a function of *time* for any fixed value of *light*. This model is represented in Display 9.5. By contrast the model that contains terms for both *light* and the square of *light*,

$$\mu\{flowers \mid light, time\} = \beta_0 + \beta_1 light + \beta_2 time + \beta_3 light^2,$$

describes a curved surface for the mean response over the explanatory variable values.

When more than two explanatory variables are present, it is difficult and not always useful to consider the geometry of regression surfaces. Instead, the regression coefficients are interpreted in terms of the effects that the selected explanatory variables have on the mean of the response when other explanatory variables are also included in the model.

Effects of Explanatory Variables

The *effect* of an explanatory variable is the change in the mean response that is associated with a one-unit increase in that variable while holding all other explanatory variables fixed. In a regression for the meadowfoam study,

$$light \text{ effect} = \mu\{flowers \mid light + 1, time\} - \mu\{flowers \mid light, time\},$$
$$time \text{ effect} = \mu\{flowers \mid light, time + 1\} - \mu\{flowers \mid light, time\}.$$

In some regression models, the effects might be different when the variables are incremented from different starting levels.

Display 9.5   Model for the regression surface of flowers per plant under 12 treatment levels
as a regression plane

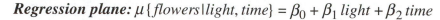

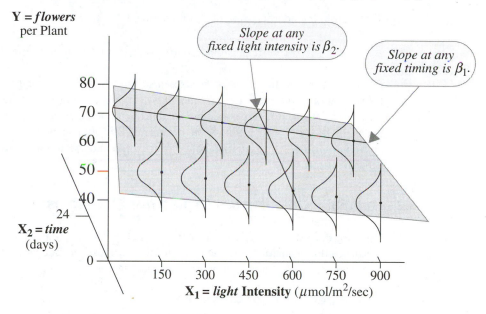

In the planar model, however, effects are the same at all levels of the explanatory variables. The change in mean *flowers* associated with increasing *light* from 150 to 151 μmol/m^2/sec with *time* at 24 days, for example, is the same as the change associated with increasing *light* from 600 to 601 μmol/m^2/sec with *time* at 0 days. The precise value of the effect is found by performing the subtraction with the specified form of the model:

$$light \text{ effect} = \mu\{flowers \mid light + 1, time\} - \mu\{flowers \mid light, time\}$$
$$= [\beta_0 + \beta_1(light + 1) + \beta_2 time] - [\beta_0 + \beta_1 light + \beta_2 time] = \beta_1.$$

In this model, the coefficient of one explanatory variable measures the effect of that variable at fixed values of the other.

When regression analysis is applied to data from a randomized experiment, the effects of design variables can be interpreted as causal effects: "A one-μmol/m^2/sec increase in light intensity causes the mean number of flowers to change by β_1-units." When applied to observational studies, the term "association" avoids any statistical implication of causation: "A one-day difference in gestation period in mammals of the same body weight and the same litter size is associated with a difference in brain weight of β_2 grams."

Interpretation Depends on What Other X's Are Included

For the mammal brain weight data, for example, the interpretation of β_1 in the model

$$\mu\{brain \mid gestation\} = \beta_0 + \beta_1 gestation$$

differs from the interpretation of β_1 in the model

$$\mu\{brain \mid gestation, body\} = \beta_0 + \beta_1 gestation + \beta_2 body.$$

Since mammals with larger body sizes also tend to have longer gestation lengths, the coefficient of *gestation* in the first model will indirectly reflect the effect of body size on brain size. When body size is also included in the model, the coefficient of *gestation* reflects the association between *gestation* and mean brain size after accounting for the effect of body size. It describes the association between *gestation* and mean brain size for animals of the same body size.

9.3   Specially Constructed Explanatory Variables

One can dramatically expand the scope of multiple linear regression by using specially constructed explanatory variables. With them, regression models can exhibit curvature, interactions, and features relating to categorical factors.

9.3.1   A Squared Term for Curvature

Display 9.6 is a scatterplot of the corn yield versus rainfall in six U.S. corn-producing states (Iowa, Nebraska, Illinois, Indiana, Missouri, and Ohio), recorded for each year from 1890 to 1927 (see also Exercise 15 on page 252). A straight-line regression model is not adequate. Although increasing rainfall is associated with higher mean yields for rainfalls up to 12 inches, increasing rainfall at higher levels is associated with no change or perhaps a decrease in mean yield. In short, the rainfall effect depends on the rainfall level.

One model for incorporating curvature includes squared rainfall as an additional explanatory variable:

$$\mu\{corn \mid rain\} = \beta_0 + \beta_1 rain + \beta_2 rain^2.$$

It is not necessary that this model corresponds to any natural law (but it could). It is, however, a convenient way to incorporate curvature in the regression of *corn* on *rain*. (Remember that *linear* regression means linear in β's, not linear in X, so quadratic regression is a special case of multiple linear regression.)

The model incorporates curvature by having the effect of rainfall be different at different levels of rainfall:

$$\begin{aligned}
\mu\{corn \mid rain + 1\} - \mu\{corn \mid rain\} &= [\beta_0 + \beta_1(rain + 1) + \beta_2(rain + 1)^2] \\
&\quad - [\beta_0 + \beta_1 rain + \beta_2 rain^2] \\
&= \beta_1 + \beta_2[(2 \times rain) + 1].
\end{aligned}$$

Display 9.6   Yearly corn yield versus rainfall (1890–1927) in six U.S. states

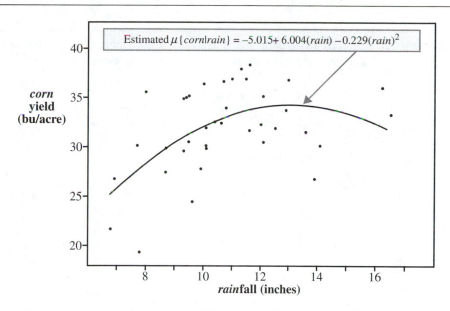

Estimated $\mu\{corn|rain\} = -5.015 + 6.004(rain) - 0.229(rain)^2$

From the fitted model shown in Display 9.6, the effect of a unit increase in rainfall from 8 to 9 inches is estimated to be an increase in mean yield of about 2.1 bushels of corn per acre. But the effect of a unit increase in rainfall from 14 to 15 inches is estimated to be a *decrease* in mean yield of about 0.6 bu/acre.

Attempting to interpret the individual coefficients in this example is difficult and unnecessary. The statistical significance of the squared rainfall coefficient is that it highlights the inadequacy of the straight line regression model and suggests that increasing yield is associated with increasing rainfall only up to a point. In many applications squared terms are useful for incorporating slight curvature and the purpose of the analysis does not require that the coefficient of the squared term be interpreted. In specialized situations, higher-order polynomial terms may also be included as explanatory variables.

9.3.2   An Indicator Variable to Distinguish Between Two Groups

Display 9.3 shows a scatterplot of the *flower* numbers versus *light* intensity, with *time* = 0 days coded differently from *time* = 24 days. The lines on this scatterplot do not represent simple linear regression equations for *time* = 0 and *time* = 24 separately. They are the result of a multiple linear regression model that incorporates an *indicator variable* to represent the two levels of the timing variable. The model requires that the lines have equal slopes.

An *indicator variable* (or dummy variable) takes on one of two values: "1" (one) indicates that an attribute is present, and "0" (zero) indicates that the attribute is absent. The variable *day24*, for example, is set equal to 1 for units where the timing was 24 days prior to PFI and is set equal to 0 for units where light intensity was not varied prior to

PFI. The indicator variable *day24* (along with variables to be introduced shortly) appears in Display 9.7.

Display 9.7 Original variables, indicator variables, and an interaction variable for use in analyzing the meadowfoam data with multiple linear regression

Original Variables			Timing Indicators		Light Level Dedicators						Interaction
flowers	*light*	*time*	*day24*	*day0*	*L150*	*L300*	*L450*	*L600*	*L750*	*L900*	*light×day24*
62.3	150	0	0	1	1	0	0	0	0	0	0
77.4	150	0	0	1	1	0	0	0	0	0	0
77.8	150	24	1	0	1	0	0	0	0	0	150
75.6	150	24	1	0	1	0	0	0	0	0	150
55.3	300	0	0	1	0	1	0	0	0	0	0
54.2	300	0	0	1	0	1	0	0	0	0	0
69.1	300	24	1	0	0	1	0	0	0	0	300
78.0	300	24	1	0	0	1	0	0	0	0	300
49.6	450	0	0	1	0	0	1	0	0	0	0
61.9	450	0	0	1	0	0	1	0	0	0	0
57.0	450	24	1	0	0	0	1	0	0	0	450
71.1	450	24	1	0	0	0	1	0	0	0	450
39.4	600	0	0	1	0	0	0	1	0	0	0
45.7	600	0	0	1	0	0	0	1	0	0	0
62.9	600	24	1	0	0	0	0	1	0	0	600
52.2	600	24	1	0	0	0	0	1	0	0	600
31.3	750	0	0	1	0	0	0	0	1	0	0
44.9	750	0	0	1	0	0	0	0	1	0	0
60.3	750	24	1	0	0	0	0	0	1	0	750
45.6	750	24	1	0	0	0	0	0	1	0	750
36.8	900	0	0	1	0	0	0	0	0	1	0
41.9	900	0	0	1	0	0	0	0	0	1	0
52.6	900	24	1	0	0	0	0	0	0	1	900
44.4	900	24	1	0	0	0	0	0	0	1	900

The regression model is

$$\mu\{\,flowers \mid light, day24\} = \beta_0 + \beta_1 light + \beta_2 day24.$$

If *time* = 0, then *day24* = 0, and the regression line is

$$\mu\{\,flowers \mid light, 0\} = \beta_0 + \beta_1 light.$$

If *time* = 24, then *day24* = 1, and the regression line is

$$\mu\{\,flowers \mid light, 1\} = \beta_0 + \beta_1 light + \beta_2.$$

This multiple regression model states that the mean number of flowers is a straight-line function of light intensity for both levels of timing. The slope of both lines is β_1; the intercept for units with timing at PFI is β_0; and the intercept for units with timing 24 days prior to PFI is $\beta_0 + \beta_2$. Because the slopes are the same, the model is called the *parallel lines* regression model. Furthermore, the coefficient of the indicator variable, β_2, is the amount by which the mean number of flowers with prior timing at 24 days exceeds that with no prior timing, after accounting for the effect of light intensity differences. In Display 9.3, the lines showing (estimates for) the mean numbers of flowers versus intensity are separated by a constant vertical difference of β_2 units.

The indicator variable can be defined for either group without affecting the model. If, instead of the indicator variable *day24*, the regression had been based on an indicator variable *day0*, taking 1 for *time* = 0 and 0 for *time* = 24, then the resulting fit would be exactly the same; it would produce exactly the same lines on Display 9.3. The only difference would be that the intercept for units with timing prior to PFI is β_0 and the intercept for units with timing at PFI is $\beta_0 + \beta_2$. This β_2 is the negative of the β_2 in the previous version of the model.

An indicator variable may be included in a regression model just as any other explanatory variable. The coefficient of the indicator variable is the difference between the mean response for the indicated category (= 1) and the mean response for the other category (= 0), at fixed values of the other explanatory variables.

9.3.3 Sets of Indicator Variables for Categorical Explanatory Variables with More Than Two Categories

A categorical explanatory variable with more than two categories, like the type of training an individual has had (clerical, managerial, operations research, computer analyst) or a fertilizer application (none, fertilizer A, fertilizer B), can also be incorporated into a regression model. When a categorical variable is used in regression it is called a *factor* and the individual categories are called the *levels* of the factor. If there are k levels then $k - 1$ indicator variables are needed as explanatory variables.

The meadowfoam study can be used to demonstrate. Let *L150* be equal to 1 for all experimental units that received light intensity of 150 μmol/m^2/sec and equal to 0 for all other units. Let *L300* similarly indicate the units that received light intensity of 300 μmol/m^2/sec; and so on, to *L900* indicating the units that received light intensity of 900 μmol/m^2/sec. This creates an indicator variable for every level of intensity (Display 9.7). To treat light intensity as a factor, all but one of these are included as explanatory variables in the regression model. The level whose indicator variable is not used in the set is called the *reference level* for that factor.

Intensity has six levels, so five indicator variables will represent all levels. Selecting the first level, 150 μmol/m^2/sec intensity, as the reference level, the multiple linear regression model is

$$\mu\{\text{flowers} \mid \text{light}, \text{day24}\} =$$
$$\beta_0 + \beta_1 L300 + \beta_2 L450 + \beta_3 L600 + \beta_4 L750 + \beta_5 L900 + \beta_6 \text{day24}.$$

The coefficients of the indicator variables in this model allow the mean number of flowers to vary arbitrarily with light intensity.

The questions of interest in the meadowfoam study, however, are more easily addressed with the simpler model that uses *light* as a numerical explanatory variable. So further discussion of using a set of indicator variables for representing a factor with many levels is postponed to later chapters.

9.3.4   A Product Term for Interaction

Two explanatory variables are said to *interact* if the effect that one of them has on the mean response depends on the value of the other. A secondary question of interest in the meadowfoam study was one of interaction: does the effect of light intensity on mean number of flowers depend on the timing of the light regime?

An Interaction Model for the Meadowfoam Data

In multiple regression, an explanatory variable for interaction can be constructed as the product of the two explanatory variables that are thought to interact. For the meadowfoam example with an indicator variable for timing, the product variable *light* × *day24* (see Display 9.7) introduces an interaction between light intensity and timing. Consider the model

$$\mu\{flowers \mid light, \ day24\} = \beta_0 + \beta_1 light + \beta_2 day24 + \beta_3 (light \times day24).$$

This is a way of expressing the regression of *flowers* on *light*, for different levels of *day24*. When *day24* = 0, the regression equation is a straight-line function of intensity with intercept β_0 and slope β_1. When *day24* = 1, the mean number of flowers is also a straight-line function of intensity, but with intercept $\beta_0 + \beta_2$ and slope $\beta_1 + \beta_3$. Rearranging the model as

$$\mu\{flowers \mid light, \ day24\} = (\beta_0 + \beta_2 day24) + (\beta_1 + \beta_3 day24) light$$

shows how both the intercept and slope in the regression of *flowers* on *light* depend on the timing. The light intensity effect in the interaction model is $(\beta_1 + \beta_3 day24)$; while the timing effect is $(\beta_2 + \beta_3 light)$. The effect of the light intensity depends on the timing; the effect of the timing depends on the light intensity. This representation of the regression leads directly to a graph that shows the mean of flowers as a function of light intensity, with different lines corresponding to different levels of timing, as shown in the top panel of Display 9.8.

As with squared terms in models for curvature, it is often difficult to interpret individual coefficients in an interaction model. The coefficient, β_1, of *light* has changed from being a global slope to being the slope when *time* = 0. The coefficient, β_3, of the product term is the difference between the slope of the regression line on *light* when *time* = 24 and the slope when *time* = 0. If it is only necessary to test whether interaction is present, no lengthy interpretation is needed. The best method of communicating findings about the presence of significant interaction may be to present a table or graph of estimated means at various combinations of the interacting variables.

When to Include Interaction Terms

Interaction terms are not routinely included in regression models. Inclusion is indicated in three situations: when a question of interest pertains to interaction (as in the meadowfoam

Display 9.8 Regression models for separate lines, parallel lines, and equal lines in two groups—meadowfoam study

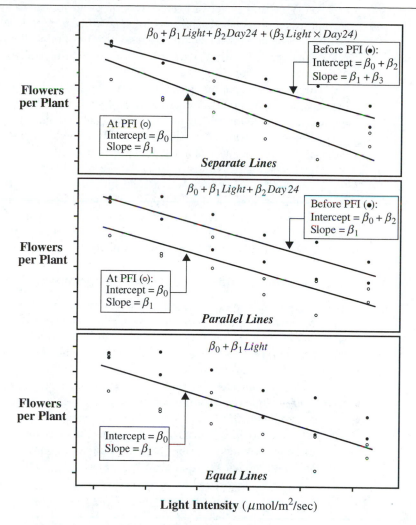

study); when good reason exists to suspect interaction; or when interactions are proposed as a more general model for the purpose of examining the goodness of fit of a model without interaction.

Except in special circumstances, a model including a product term for interaction between two explanatory variables should also include terms with each of the explanatory variables individually, even though their coefficients may not be significantly different from zero. Following this rule avoids the logical inconsistency of saying that the effect of X_1 depends on the level of X_2 but that there is no effect of X_1.

A Separate Regression Lines Model

A common situation is one in which the analyst wishes to fit simple linear regressions of Y on X separately for different levels of a categorical factor. This analysis can be accomplished by repeated application of simple linear regression for each level; but it can also be accomplished with multiple regression. The multiple linear regression model has as its explanatory variables: X, the $k - 1$ indicators to distinguish the k levels, and all products of the indicators with X.

The multiple regression approach has advantages. Questions about similarities between the separate regressions are expressible as hypotheses about specific parameters in the multiple regression model: are the slopes all equal? are the intercepts equal? Furthermore, the multiple regression approach conveniently provides a single, combined estimate of variance, which is equivalent to the pooled estimate from the separate simple regressions.

Display 9.8 depicts three models for the meadowfoam study, each specifying that the mean number of flowers is a straight-line function of light intensity. The separate regression lines model in the top panel is the most general of the three. If the coefficient of the interaction term, β_3, is zero, then the separate regression lines model reduces to the parallel regression lines model, which indicates no interaction of *light* and *time*. Similarly, if β_2 is zero in the parallel regression lines model, then it reduces to the equal lines model.

9.3.5   A Shorthand Notation for Model Description

The notation for regression models is sometimes unnecessarily bulky when there are a large number of indicator variables or interaction terms. An abbreviation for specifying a categorical explanatory variable is to list the categorical variable name in uppercase letters to represent the entire set of indicator variables used to model it. For example, $\mu\{flowers \mid light, TIME\}$ indicates that the variable *TIME* is a factor, whose effects will be modeled by indicator variables. In this example, since *TIME* has only two levels, this shorthand notation does not provide any savings over $\mu\{flowers \mid light, day24\}$. If *TIME* were a factor with 10 levels, however, this notation would avoid the need for specifying the nine indicator variables in the list of the explanatory variables.

Additional savings are achieved by omitting the parameters when specifying the terms included in the model. For example,

$$\mu\{flowers \mid light, TIME\} = light + TIME$$

describes the parallel lines model with *light* treated as a numerical explanatory variable and *TIME* as a factor. Similarly,

$$\mu\{flowers \mid light, TIME\} = light + TIME + (light \times TIME)$$

specifies the separate lines model with the effects of the numerical explanatory variable *light*, the factor *TIME*, and the interaction terms formed as the products of *light* with the indicator variable(s) for the factor *TIME*. Extensions of this notation conveniently describe models with any number of numerical and categorical variables.

9.4   **A Strategy for Data Analysis**

Display 9.9 shows a general scheme for using statistical models to analyze data. Data analysis involves more than inserting data into a computer and pressing the right button. Substantial exploratory analysis may be required at the outset, and several dead ends may be encountered in the pursuit of suitable models.

Display 9.9   A strategy for data analysis using statistical models

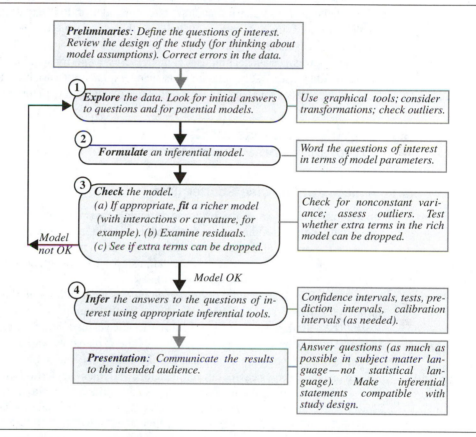

Preliminaries: Define the questions of interest. Review the design of the study (for thinking about model assumptions). Correct errors in the data.

1 *Explore* the data. Look for initial answers to questions and for potential models.

Use graphical tools; consider transformations; check outliers.

2 *Formulate* an inferential model.

Word the questions of interest in terms of model parameters.

3 *Check* the model.
(a) If appropriate, *fit* a richer model (with interactions or curvature, for example). (b) Examine residuals.
(c) See if extra terms can be dropped.

Check for nonconstant variance; assess outliers. Test whether extra terms in the rich model can be dropped.

Model not OK

Model OK

4 *Infer* the answers to the questions of interest using appropriate inferential tools.

Confidence intervals, tests, prediction intervals, calibration intervals (as needed).

Presentation: Communicate the results to the intended audience.

Answer questions (as much as possible in subject matter language—not statistical language). Make inferential statements compatible with study design.

The strategy for data analysis centers on the development of an *inferential model* where answers to the questions of interest can be found by drawing inferences about key parameters. The choice of an inferential model may be guided by the graphical displays. Further checking on model adequacy is accomplished by residual analysis and by informal testing of terms in a richer model that contains additional parameters representing possible curvature, interactions, or more complex features.

The adequacy of a model may be clear from the graphical displays and residual analysis alone. When it is not, informal testing is an important part of finding an adequate model. In particular, if replicates of the response occur at each combination of explanatory variables (as in the meadowfoam study), it is always possible to compare the model that is convenient for analysis to the separate-means model (the one with a separate response mean at each combination of the explanatory variables). Moreover, suspicions of nonlinearities and interactions can be investigated by testing the appropriate additional terms that model these inadequacies. If a particular, convenient inferential model is found to be inappropriate, the analyst has two options: renew the search for a better inferential model, or offer a more complicated explanation of the study's conclusions in terms of the richer model.

Ensuing chapters provide examples of this strategy. Readers should refer to Display 9.9 often, as the analysis patterns develop. The remainder of this chapter deals with graphical exploratory techniques (for steps 1 and 2 of this strategy). The next chapter details the estimation of parameters and the main inferential tools (for steps 3 and 4). Chapter 11 addresses new issues and techniques for model assessment (for step 3). Chapter 12 discusses the difficulties with a set containing too many explanatory variables and some computer assisted techniques for reducing the set to a reasonable number (for steps 1 and 2).

9.5   Graphical Methods for Data Exploration and Presentation

9.5.1   A Matrix of Pairwise Scatterplots

Individual scatterplots of the response variable versus each of the explanatory variables are usually helpful. Although the observed relationships in these plots will not necessarily indicate the effects of interest in a *multiple* regression model, the plots are still useful for observing the marginal (one at a time) relationships, drawing attention to interesting points, and suggesting the need for transformations.

A *matrix of scatterplots* (or *draftsman plot*) is a consolidation of all possible pairwise scatterplots from a set of variables, as shown for the mammal brain weight data in Display 9.10. The bottom row shows the response variable versus each of the explanatory variables and the higher rows show the scatterplots of the explanatory variables versus each other. There is a lot to look at in such a display. Typically, one would investigate the scatterplots of the response versus each of the explanatory variables (the bottom row) first.

Different statistical computer packages have different display formats. Instead of the lower left triangular arrangement shown in Display 9.10, some packages display a corresponding upper right triangle (in addition to the lower left one) that contains the same plots but with axes reversed. This does not contain any more information, although it may provide a different visual perspective. The resolution with which the plots can be drawn may also be an issue. It becomes increasingly difficult to read a matrix of scatterplots as the number of variables is increased. For data sets with many potential explanatory variables, it may be necessary to select subsets for display.

Notes About the Scatterplots for the Brain Weight Data

Consider the bottom row of Display 9.10 first. The scatterplot of brain weight versus body weight is not very helpful since most of the data points are clustered on top of each other in

Display 9.10 Matrix of scatterplots for brain weight data

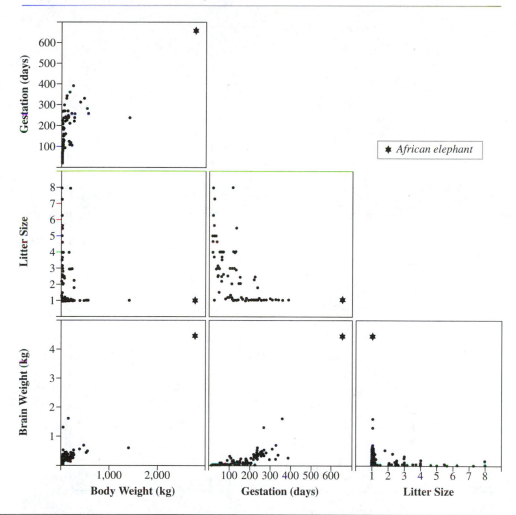

the bottom left-hand corner. The upper bounds for the X and Y axes are determined largely by one point (the African elephant). This scatterplot indicates what should have been evident from the start—that the mammals in this list differ in size by orders of magnitude. The display should be redrawn after brain weight and body weight are transformed to their logarithms.

Gestation values also appear to be quite skewed. A trial plot using the log of this variable shows that it too should be transformed. Display 9.11 shows a revised version of the matrix of scatterplots where all variables have been placed on their natural logarithmic scales.

The bottom row of Display 9.11 shows less overlap and a more distinct pattern. Notice the pronounced relationship between log brain weight and each of the explanatory variables.

Display 9.11 Matrix of scatterplots for brain weight data, after log transformations

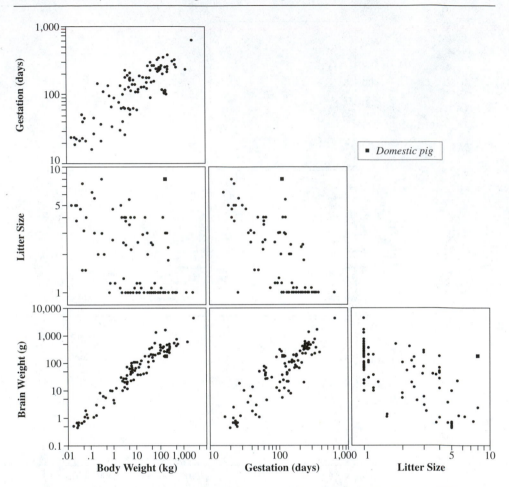

Notice also, from the upper plots in the first column, that gestation and litter size are also related to body weight. Therefore the question of whether there is an association between gestation and brain weight after accounting for the effect of body weight, and the question of whether there is an association between litter size and brain weight after accounting for the effect of body weight, are not resolved by these plots. Nevertheless, they suggest the next course of action, which is to fit a regression model for log brain weight on log body weight, log gestation, and log litter size.

In the scatterplot of log brain weight versus log litter size there is one data point that stands out—a mammal with littersize eight whose brain size is quite a bit larger than those for other mammals whose litter size is eight. The analyst should determine which mammal this is. It may be influential in the analysis, and it may provide useful information or additional questions for research. Modern statistical computer packages with interactive labeling features make the identification of such points on a graph fairly easy.

9.5.2   Coded Scatterplots

Display 9.3 is an example of a *coded scatterplot*—a scatterplot with different symbols or letters used as plotting marks to distinguish two or more groups. In the example, the groups correspond to different levels of timing. This is a great way of observing the joint effects of one numerical and one categorical explanatory variable. For observing the joint effects of two numerical explanatory variables, it is often very helpful to plot the response versus one of them, and to use a plotting code to represent groups defined by ranges of the other (like educational level less than 8 years, between 9 and 12 years, or greater than 12 years).

9.5.3   Jittered Scatterplots

The scatterplot of log brain weight versus log litter size (Display 9.11) contains many points that overlap because many species of mammals have the same litter size. This makes it difficult to judge relative frequency. When a variable has only a few distinct values it may be possible to improve the visual information of the plot by *jittering* that variable—adding small computer-generated random numbers to it before plotting. In Display 9.12, log brain weight is plotted against a jittered version of litter size. A jittered variable is only used in

Display 9.12   Jittered scatterplot: log brain weight versus litter size (jittered)

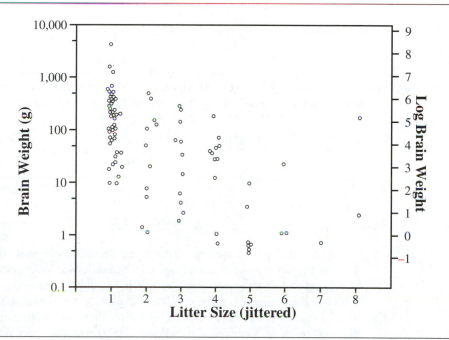

the scatterplot; subsequent numerical analyses must be based on the unjittered version. The choice as to the size of the random numbers to add in order to make each point visually distinct requires trial and error.

9.6   Related Issues

9.6.1   Computer Output

Regression output from statistical computer packages can be intimidating for the simple reason that it tends to include everything that the user could possibly want and more. One important component, which all packages display, is the list of estimates and their standard errors. Display 9.13 and Display 9.14 show the least squares estimates of the coefficients in specified models for the two examples of this chapter, their standard errors, and two-sided p-values from t-tests for the hypotheses that each β is zero in the specified model.

Display 9.13   Estimates of regression coefficients in the multiple regression of *flowers* on *light*, *day24*, and *light* × *day24*—the meadowfoam study

Variable	Coefficient	Standard Error	t-Statistic	p-Value
Constant	71.6233	4.3433	16.4905	<.0001
light	− 0.0411	0.0074	5.5247	<.0001
day24	11.5233	6.1424	1.8760	.0753
light × *day 24*	0.0012	0.0105	0.1150	.9096

Display 9.14   Estimates of regression coefficients in the multiple regression of log brain weight on log body weight, log gestation, and log litter size—brain weight data

Variable	Coefficient	Standard Error	t-Statistic	p-Value
Constant	0.8548	0.6617	1.2919	.1996
lbody	0.5751	0.0326	17.6468	<.0001
lgest	0.4179	0.1408	2.9687	.0038
llitter	−0.3101	0.1159	2.6747	.0089

Although the least squares estimates and the inferential tools are discussed further in the next chapter, it is not difficult to see which questions are addressed by the p-values. The p-value of .9096 for the interaction term from Display 9.13 indicates that zero is a plausible value for the coefficient of the interaction explanatory variable; so there is no evidence of an interaction. The p-value of .0089 for the coefficient of *llitter* indicates strong evidence that the litter size is associated with brain weight, even after accounting for body weight.

9.6.2 Factorial Treatment Arrangement

The meadowfoam study investigated two treatment factors applied jointly to each experimental unit: the light intensity (at one of six levels) and the timing (at one of two levels). All 12 combinations of these two treatment factors were applied to some experimental units. The term *factorial treatment arrangement* describes the formation of experimental treatments from all combinations of the levels of two or more distinct, individual treatment factors. This study was a 6×2 factorial arrangement, meaning all six levels of one factor appear in combination with both levels of the other.

Three benefits of testing multiple factors all at once rather than through one-at-a-time experiments are: interactive effects can be explored; there is more efficient use of the available experimental units with the multifactor arrangement (resulting in smaller standard errors for parameters of interest); the results for the multifactor experiment are more general, since each treatment is investigated at several levels of the other treatment.

9.7 Summary

Multiple regression analysis refers to a large set of tools associated with developing and using regression models for answering questions of interest. Multiple regression models describe the mean of a single response variable as a function of several explanatory variables. Transformations, indicator variables for grouped data (factors), squared explanatory variables for curvature, and product terms for interaction greatly enhance the usefulness of this model.

Substantial exploratory analysis is recommended for gaining initial insight into what the data have to say in answer to the questions of interest and for suggesting possible regression models for answering them more formally. Some standard graphical procedures are presented here. The data analyst should be prepared to be creative in using the available computer tools to best display the data, while keeping in mind the questions of interest and the statistical tools that might be useful for answering them.

Brain Weight Study

Is brain weight associated with gestation period and/or litter size after accounting for the effect of body weight? This is exactly the type of question for which multiple regression is useful. The regression coefficient of gestation in the regression of brain weight on body weight and gestation describes the effect of gestation for species of roughly the same body weight. With multiple linear regression, the coefficients can be estimated from all the animals without a need for grouping into subsets of similar body weight. Initial scatterplots (or initial inspection of the data) indicate that the regression model should be formed after transforming all the variables to their logarithms.

Meadowfoam Study

A starting point for the analysis is the coded scatterplot of number of flowers per plant versus light intensity, with different codes to represent the two levels of the timing factor (Display 9.3). The plot suggests that the mean number of flowers per plant decreases with increasing light intensity, that the rate of decrease does not depend on timing, and that (for any light intensity value) a larger mean number of flowers is associated with the "before PFI" level of the timing factor. Using an indicator variable for one of the timing levels permits fitting the parallel regression lines model. Further inclusion of the interaction of timing and intensity produces a model that fits separate regression lines for each level of timing. This model permits a check on whether the regression lines are indeed parallel.

Further Reading

There are numerous books on multiple regression. *Applied Regression Analysis* by Draper and Smith (1981) is an authoritative reference. A complete reference is *Applied Linear Statistical Models* by Neter, Wasserman, and Kutner (1990). For an encyclopedic passage with references, see the section on linear regression in *The Encyclopedia of Statistical Science* (Johnson, Kotz, and Reed 1985). The book *Data Analysis and Regression* (Mosteller and Tukey 1977) contains some solid warnings about the meaning and interpretation of regression coefficients. There are many modern books on graphical analysis. One with modern techniques for regression situations is *Visualizing Data* by W. S. Cleveland (1993).

9.8 Exercises

Conceptual Exercises

1. **Meadowfoam.** (a) Write down a multiple regression model with parallel regression lines of *flowers* on *light* for the two separate levels of *time* (using an indicator variable). (b) Add a term to the model in (a) so that the regression lines are not parallel.

2. **Meadowfoam.** A model (without interaction) for the mean *flowers* is estimated to be $71.3058 - .0405light + 12.1583day24$. For a fixed level of timing, what is the estimated difference between the mean *flowers* at 600 and 300 μmol/m^2/sec of *light* intensity?

3. **Meadowfoam.** (a) Why were the numbers of flowers from 10 plants averaged to make a response, rather than representing them as 10 different responses? (b) What assumption is assisted by averaging the numbers from the 10 plants?

4. **Mammal Brain Weights.** The three-toed sloth has a gestation period of 165 days. The Indian fruit bat has a gestation period of 145 days. From Display 9.14 the estimated model for the mean of log brain weight is $.8548 + .5751lbody + .4179lgest - .3101llitter$. Since *lgest* for the sloth is .1292 more than *lgest* for the fruit bat, does this imply that an estimate of the mean log brain weight for the sloth is $(.4179)(.1292)$ more than the mean log brain weight for the bat (i.e., the median is 5.5% higher)? Why? Why not?

5. **Insulating Fluid** (Section 8.1.2). Would it be possible to test for lack of fit to the straight-line model for the regression of log breakdown time on voltage by including a voltage-squared term in the model, and testing whether the coefficient of the squared term is zero?

6. **Island Area and Species.** For the island area and number of species data in Section 8.1.1, would it be possible to test for lack of fit to the straight-line model for the regression of log number of species on log island area by including the square of log area in the model and testing whether its coefficient is zero?

7. Which of the following regression models are *linear*?

   (a) $\mu\{Y|X\} = \beta_0 + \beta_1 X + \beta_2 X^2 + \beta_3 X^3$
   (b) $\mu\{Y|X\} = \beta_0 + \beta_1 10^X$
   (c) $\mu\{Y|X\} = (\beta_0 + \beta_1 X)/(\beta_0 + \beta_2 X)$
   (d) $\mu\{Y|X\} = \beta_0 \exp(\beta_1 X)$.

8. Describe what σ measures in the meadowfoam problem and in the brain weight problem.

9. **Pollen Removal.** Reconsider the data on proportion of pollen removed and duration of visit to the flower for bumblebee queens and honeybee workers, in Display 3.12. (a) Write down a model that describes the mean proportion of pollen removed as a straight-line function of duration of visit, with separate intercepts and separate slopes for bumblebee queens and honeybee workers. (b) How would you test whether the effect of duration of visit on proportion removed is the same for queens as for workers?

10. **Breast Milk and IQ.** In a study, Intelligence Quotient (IQ) test scores were obtained for 300 eight-year-old children who had been part of a study of premature babies in the early 1980s. Because they were premature, all the babies were fed milk by a tube. Some of them received breast milk entirely, some received a prepared formula entirely, and some received some combination of breast milk and formula. The proportion of breast milk in the diet depended on whether the mother elected to provide breast milk and to what extent she was successful in expressing any, or enough, for the baby's diet. The researchers reported the results of the regression of the response variable—IQ at age 8—on social class (ordered from 1, the highest, to 5), mother's education (ordered from 1, the lowest, to 5), an indicator variable taking the value 1 if the child was female and 0 if male, the number of days of ventilation of the baby after birth, and an indicator variable taking the value 1 if there was any breast milk in the baby's diet and 0 if there was none. The estimates are reported in Display 9.15 along with the p-values for the tests that each coefficient is zero. (Data from A. Lucas, et al. "Breast Milk and Subsequent Intelligence Quotient in Children Born Preterm," *Lancet* 339 (1992): 261–64).

Display 9.15 Breast milk and intelligence data

Explanatory Variable	Estimated Coefficient	p-Value
Social class	−3.5	.0004
Mother's education	2.0	.01
Female indicator	4.2	.01
Days of ventilation	−2.6	.02
Breast milk indicator	8.3	<.0001

(a) After accounting for the effects of social class, mother's education, whether the child was a female, and days after birth of ventilation, how much higher is the estimated mean IQ for those children who received breast milk than for those who did not?

(b) Is it appropriate to use the variables "Social class" and "Mother's education" in the regression even though in both instances the numbers 1 to 5 do not correspond to anything real but are merely ordered categories?

(c) Does it seem appropriate for the authors to simply report $< .0001$ for the p-value of the breast milk coefficient rather than the actual p-value?

(d)   Previous studies on breast milk and intelligence could not separate out the effects of breast milk and the act of breast feeding (the bonding from which might encourage intellectual development of the child). How is the important confounding variable of whether a child is breast fed dealt with in this study?

(e)   Why is it important to have social class and mother's education as explanatory variables?

(f)   In a subsidiary analysis the researchers fit the same regression model as above except with the indicator variable for whether the child received breast milk replaced by the percentage of breast milk in the diet (between 0 and 100%). The coefficient of that variable turned out to be .09. (i) From this model, how much larger is the estimated mean IQ for children who received 100% breast milk than for those who received 50% breast milk, after accounting for the other explanatory variables? (ii) What is the importance of the percentage of breast milk variable in dealing with confounding variables?

11.   **Glasgow Graveyards.** Do persons of higher socioeconomic standing tend to live longer? This was addressed by George Davey Smith and colleagues through the relationship of the heights of commemoration obelisks and the life lengths of the corresponding grave site occupants. In burial grounds in Glasgow a certain design of obelisk is quite prevalent, but the heights vary greatly. Since the height would influence the cost of the obelisk, it is reasonable to believe that height is related to socioeconomic status. The researchers recorded obelisk height, year of death, age at death, and gender for 1,349 individuals who died prior to 1921. Although they were interested in the relationship between mean life length and obelisk height, it is important that they included year of construction as an explanatory variable since life lengths tended to increase over the years represented (1801 to 1920). For males, they fit the regression of life length on obelisk height (in meters) and year of obelisk construction and found the coefficient of obelisk height to be 1.93. For females they fit the same regression and found the coefficient of obelisk height to be 2.92. (Data from Smith et al., "Socioeconomic Differentials in Mortality: Evidence from Glasgow Graveyards," *British Medical Journal* 305 (1992): 1557–60.)

(a)   After accounting for year of obelisk construction, each extra meter in obelisk height is associated with Z extra years in mean life time. What is the estimated Z for males? What is the estimated Z for females?

(b)   Since the coefficients differ significantly from zero, would it be wise for an individual to build an extremely tall obelisk, to ensure a long life time?

(c)   The data were collected from eight different graveyards in Glasgow. Since there is a potential blocking effect due to the different graveyards, it might be appropriate to include a graveyard effect in the model. How can this be done?

Computational Exercises

12.   **Mammal Brain Weights.** (a) Draw a matrix of scatterplots for the mammal brain weight data (Display 9.4) with all variables transformed to their logarithms (to reproduce Display 9.11). (b) Fit the multiple linear regression of log brain weight on log body weight, log gestation, and log litter size, to confirm the estimates in Display 9.14. (c) Draw a matrix of scatterplots as in (a) but with litter size on its natural scale (untransformed). Does the relationship between log brain weight and litter size appear to be any better or any worse (more like a straight line) than the relationship between log brain weight and log litter size?

13.   **Meat Processing.** One way to check on the adequacy of a linear regression is to try to include an X-squared term in the model to see if there is significant curvature. Use this technique on the meat processing data of Section 7.1.2 on page 170. (a) Fit the multiple regression of pH on hour and hour-squared. Is the coefficient of hour-squared significantly different from zero? What is the p-value?

(b) Fit the multiple regression of pH on log(hour) and the square of log(hour). Is the coefficient of the squared-term significantly different from zero? What is the *p*-value? (c) Does this exercise suggest a potential way of checking the appropriateness of taking the logarithm of *X* or of leaving it untransformed?

14. **Pace of Life and Heart Disease.** Some believe that individuals with a constant sense of time urgency (often called type-A behavior) are more susceptible to heart disease than are more relaxed individuals. Although most studies of this issue have focused on individuals, some psychologists have investigated geographical areas. They considered the relationship of city-wide heart disease rates and general measures of the pace of life in the city.

For each region of the United States (Northeast, Midwest, South, and West) they selected three large metropolitan areas, three medium-size cities, and three smaller cities. In each city they measured three indicators of the pace of life. The variable *walk* is the walking speed of pedestrians over a distance of 60 feet during business hours on a clear summer day along a main downtown street. *Bank* is the average time a sample of bank clerks take to make change for two $20 bills or to give $20 bills for change. The variable *talk* was obtained by recording responses of postal clerks explaining the difference between regular, certified, and insured mail and by dividing the total number of syllables by the time of their response. The researchers also obtained the age-adjusted death rates from ischemic heart disease (a decreased flow of blood to the heart) for each city (*heart*). The data in Display 9.16 were read from a graph in the published paper. (Data from R. V. Levine, "The Pace of Life," *American Scientist* 78 (1990): 450–59.) The variables have been standardized, so there are no units of measurement involved.

Display 9.16 Pace of life data: bank clerk speed, pedestrian walking speed, postal clerk talking speed, and age-adjusted death rates due to heart disease, in 36 cities

City	*bank*	*walk*	*talk*	*heart*	City	*bank*	*walk*	*talk*	*heart*
Boston, MA	31	28	24	24	Chicago, IL	24	23	25	20
Buffalo, NY	30	23	23	29	Philadelphia, PA	31	12	19	18
New York, NY	29	24	18	31	Louisville, KY	27	23	17	16
Salt Lake City, UT	28	28	23	26	Canton, OH	28	20	18	19
Columbus, OH	27	22	30	26	Knoxville, TN	21	20	17	23
Worcester, MA	26	25	24	20	San Francisco, CA	19	22	18	11
Providence, RI	30	26	24	17	Chattanooga, TN	34	14	22	27
Springfield, MA	28	30	21	19	Dallas, TX	24	20	23	18
Rochester, NY	33	22	18	26	Oxnard, CA	25	17	19	15
Kansas City, MO	33	22	22	24	Nashville, TN	25	26	19	20
St. Louis, MO	22	23	23	26	San Diego, CA	20	19	22	18
Houston, TX	30	25	20	25	East Lansing, MI	22	23	23	21
Paterson, NJ	32	23	23	14	Fresno, CA	26	13	22	11
Bakersfield, CA	29	18	25	11	Memphis, TN	29	16	21	14
Atlanta, GA	25	27	27	19	San Jose, CA	25	17	18	19
Detroit, MI	24	22	14	24	Shreveport, LA	22	17	15	15
Youngstown, OH	27	23	24	20	Sacramento, CA	24	16	10	18
Indianapolis, IN	26	22	24	13	Los Angeles, CA	13	20	12	16

(a) Draw a matrix of scatterplots of the four variables. Construct it so that the bottom row of plots all have *heart* on the vertical axis. If you do not have this facility, draw scatterplots of *heart* versus each of the other variables individually.

(b) Obtain the least squares fit to the linear regression of *heart* on *bank*, *walk*, and *talk*.

(c) Plot the residuals versus the fitted values. Is there evidence that the variance of the residuals increases with increasing fitted values or that there are any outliers?

(d) Report a summary of the least squares fit. Write down the estimated equation with standard errors below each estimated coefficient.

15. **Rainfall and Corn Yield.** The data on corn yields and rainfall, discussed in Section 9.3.1, appear in Display 9.17. (Data from M. Ezekiel and K. A. Fox, *Methods of Correlation and Regression Analysis* (New York: John Wiley & Sons, 1959); originally from E. G. Misner, "Studies of the Relationship of Weather to the Production and Price of Farm Products, I. Corn" [mimeographed publication, Cornell University, March 1928].)

(a) Plot corn yield versus rainfall.

(b) Fit the multiple regression of corn yield on *rain* and *rain²*.

(c) Plot the residuals versus year. Is there any pattern evident in this plot? What does it mean? (Anything to do, possibly, with advances in technology?)

(d) Fit the multiple regression of corn yield on *rain*, *rain²*, and *year*. Write the estimated model and report standard errors, in parentheses, below estimated coefficients. How do the coefficients of *rain* and *rain²* differ from those in the estimated model in (b)? How does the estimate of σ differ? (larger or smaller?) How do the standard errors of the coefficients differ? (larger or smaller?) Describe the effect of an increase of one inch of rainfall on the mean yield over the range of rainfalls and years.

(e) Fit the multiple regression of corn yield on *rain*, *rain²*, *year*, and *year × rain*. Is the coefficient of the interaction term significantly different from zero? Could this term be used to say something about technological improvements regarding irrigation?

Display 9.17   Average corn yield and rainfall in six U.S. states (1890–1927)

Corn Yield (bu/acre)
Rainfall (in/year)

Year								
1890	24.5	9.6	1903	30.2	14.1	1916	29.7	9.3
1891	33.7	12.9	1904	32.4	10.6	1917	35.0	9.4
1892	27.9	9.9	1905	36.4	10.0	1918	29.9	8.7
1893	27.5	8.7	1906	36.9	11.5	1919	35.2	9.5
1894	21.7	6.8	1907	31.5	13.6	1920	38.3	11.6
1895	31.9	12.5	1908	30.5	12.1	1921	35.2	12.1
1896	36.8	13.0	1909	32.3	12.0	1922	35.5	8.0
1897	29.9	10.1	1910	34.9	9.3	1923	36.7	10.7
1898	30.2	10.1	1911	30.1	7.7	1924	26.8	13.9
1899	32.0	10.1	1912	36.9	11.0	1925	38.0	11.3
1900	34.0	10.8	1913	26.8	6.9	1926	31.7	11.6
1901	19.4	7.8	1914	30.5	9.5	1927	32.6	10.4
1902	36.0	16.2	1915	33.3	16.5			

---

16. **Pollen Removal.** The data in Display 3.11 are the proportions of pollen removed and the duration of visits on a flower for 35 bumblebee queens and 12 honeybee workers. It is of interest to understand the relationship between the proportion removed and duration and the relative pollen removal efficiency of queens and workers. (a) Draw a coded scatterplot of proportion of pollen removed versus duration of visit; use different symbols or letters as the plotting codes for queens and workers. Does it appear that the relationship between proportion removed and duration is a straight line? (b) The logit transformation is often useful for proportions between 0 and 1. If p is the proportion then the logit is $\log[p/(1 - p)]$. This is the log of the ratio of the amount of pollen

removed to the amount not removed. Draw a coded scatterplot of the logit versus duration. (c) Draw a coded scatterplot of the logit versus log duration. From the three plots, which transformations appear to be worthy of pursuing with a regression model? (d) Fit the multiple linear regression of the logit of the proportion of pollen removed on (i) log duration, (ii) an indicator variable for whether the bee is a queen or a worker, and (iii) a product term for the interaction of the first two explanatory variables. By examining the p-value of the interaction term, determine whether there is any evidence that the proportion of pollen depends on duration of visit differently for queens than for workers. (e) Refit the multiple regression but without the interaction term. Is there evidence that, after accounting for the amount of time on the flower, queens tend to remove a smaller proportion of pollen than workers? Why is the p-value for the significance of the indicator variable so different in this model than in the one with the interaction term?

17. Old Faithful. With the Old Faithful data from Display 7.16, (a) draw a coded scatterplot of interval versus duration, with different codes for the different days; and (b) using 7 indicator variables for the 8 days, fit the multiple regression of interval on both duration and the factor day. Write the estimated model and show standard errors in parentheses below the estimated coefficients.

Answers to Conceptual Exercises

1. (a) Let $day24 = 1$ if $time = 24$ and 0 if $time = 0$. Then $\mu\{flowers \mid light, day24\} = \beta_0 + \beta_1 light + \beta_2 day24$. (b) $\beta_0 + \beta_1 light + \beta_2 day24 + \beta_3 (light \times day24)$.

2. The difference is 300 $\mu mol/m^2/sec$ times the coefficient of *light*, or about -12.15 flowers.

3. (a) The principal reason is that the 10 plants were all treated together and grown together in the same chamber. The experimental unit is always defined as the unit that receives the treatment, here, plants in the same chamber. (b) The assumption of normality is assisted. Averages tend to have normal distributions, so the averaging may alleviate some distributional problems that could arise from looking at separate numbers of flowers.

4. No. The difficulty with interpreting regression coefficients individually, as in a controlled experiment, is that explanatory variables cannot be manipulated individually. In this instance, the sloth and the fruit bat also have different body weights—the sloth weighs 50 times what the fruit bat weighs. (The full model estimates the brain weight of the fruit bat to be only about 35% of the brain weight of the sloth.) One might attempt to envision a fruit bat having the same weight (0.9 kg) as the sloth and the same litter size (1.0), but having a gestation period of 165 instead of 145 days. This approach, however, is generally unsatisfactory because it extrapolates beyond the experience of the data set (resulting in animals like a fish-eating kangaroo with wings).

5. Yes. A common way to explore lack of fit is to introduce curvature and interaction terms to see if measured effects change as the configuration of explanatory variables changes.

6. Yes.

7. Keep your eye on the parameters. If the mean is linear in the parameters, the model is *linear*.

    (a) Yes, even though it is not linear in X.
    (b) Yes.
    (c) No. Both numerator and denominator are linear in parameters and X, but the whole is not.
    (d) No. But this is a very useful model.

8. In both, σ is a measure of the magnitude of the difference between a response and the mean in the population from which the response was drawn. In the meadowfoam problem, σ measures the typical size of differences between seedling flowers (averaged from 10 plants) and the mean seedling flowers (averaged from 10 plants) treated similarly (same intensity and timing potential). In the brain weight problem, it is more difficult to describe what σ measures because the theoretical model invents a hypothetical subpopulation of animal species all having the same body weight, gestation, and litter size.

9.   (a) $\mu\{\text{pollen} \mid \text{duration, queen}\} = \beta_0 + \beta_1\text{duration} + \beta_2\text{queen} + (\beta_3\text{duration} \times \text{queen})$, where queen is 1 for queens and 0 for workers (or the other way around). (b) $H_0: \beta_3 = 0$.

10.   (a) 8.3 points. (b) As long as the effects are indeed linear in these coded variables it is a useful way to include them in the multiple regression (and more powerful than considering them to be factors). (c) Yes. The evidence is overwhelming that this coefficient is not zero. (d) The use of premature babies who all had to be fed by tube makes it so that some babies received breast milk but all babies were administered their milk in exactly the same way. (e) It is possible that the decision to provide breast milk and the ability to express breast milk are related to social class and mother's education, which are likely to be related to child's IQ score. It is desired to examine the effect of breast milk that is separate from the association with these potentially confounding variables. (f) (i) 4.5 points. (ii) An important confounding variable in the previous result is the mother's decision to provide breast milk, which may be associated with good mothering skills, which may be associated with better intelligence development in the child. Using the proportion of breast milk as an explanatory variable allows the dose-response assessment of breast milk, which indicates that children of mothers who provided breast milk for 100% of the diet tended to score higher on the IQ test than children of mothers who also decided to provide breast milk but were only capable of supplying a smaller proportion of the diet.

11.   (a) 1.93 years. 2.92 years. (b) No. No cause-and-effect is implied by the analysis of these observational data. (c) Seven indicator variables can be included to distinguish the 8 graveyards.

C H A P T E R   10

Inferential Tools for Multiple Regression

Data analysis involves finding a good-fitting model whose parameters relate to the questions of interest. Once the model has been established, the questions of interest can be investigated through the parameter estimates, with uncertainty expressed through p-values, confidence intervals, or prediction intervals, depending on the nature of the questions asked.

The primary inferential tools associated with regression analysis—t-tests and confidence intervals for single coefficients and linear combinations of coefficients, F-tests for several coefficients, and prediction intervals—are described in this chapter. As they are for other procedures in the statistical tool kit, the numerical calculations for these tools are done with the help of a computer. The most difficult parts of the task are knowing what model and what inferential tool best suit the need and knowing how to interpret and communicate the results.

The inferential tools are illustrated in this chapter on models that incorporate special explanatory variables from the previous chapter: indicator variables, quadratic terms, and interaction terms. The tests and confidence statements found in the examples, and their interpretations, are typical of the ones used in many fields and for many different kinds of data.

10.1   Case Studies

10.1.1   Galileo's Data on the Motion of Falling Bodies—A Controlled Experiment

In 1609 Galileo proved mathematically that the trajectory of a body falling with a horizontal velocity component is a parabola. His discovery of this result, which preceded the mathematical proof by a year, was the result of empirical findings in an experiment conducted for another purpose.

Galileo's search for an experimental setting in which horizontal motion was not affected appreciably by friction (to study inertia) led him to construct an apparatus like the one shown in Display 10.1. He placed a grooved, inclined plane on a table, released an ink-covered bronze ball in the groove at one of several heights above the table, and measured the horizontal distance between the table and the resulting ink spot on the floor. The data from one experiment are shown in Display 10.1 in units of *punti* (points). One *punto* is 169/180 millimeters. (Data from S. Drake and J. MacLachlan, "Galileo's Discovery of the Parabolic Trajectory," *Scientific American* 232(1975): 102–10.)

Display 10.1   Galileo's experimental results

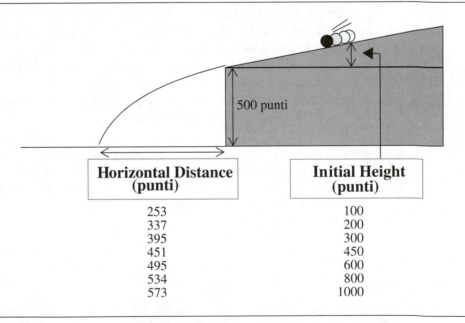

Horizontal Distance (punti)	Initial Height (punti)
253	100
337	200
395	300
451	450
495	600
534	800
573	1000

Galileo conducted this experiment to determine whether, in the absence of any appreciable resistance, the horizontal velocity of a moving object is constant. While sketching the paths of the trajectories in his notebook, he apparently came to believe that the trajectory was a parabola. Once the idea of a parabola suggested itself to Galileo, he found that proving it mathematically was straightforward. Although Galileo's experiment preceded Gauss's invention of least squares and Galton's empirical fitting of a regression line by more than 200 years, it is interesting to use regression here to see what form of trajectory the data support.

Summary of Statistical Findings

As shown in Display 10.2, a quadratic curve for the regression of horizontal distance on height fits quite well. There is strong evidence that the coefficient of a cubic term differs from zero (two-sided p-value $= .007$). Nonetheless, the quadratic model accounts for 99.03% of the variation in measured horizontal distances, and the cubic term explains only an additional 0.91% of the variation. (The significance of the cubic term can be explained by the effect of resistance.)

Display 10.2   Scatterplot of Galileo's horizontal distances versus initial heights, with estimated quadratic regression model (with standard errors in parentheses)

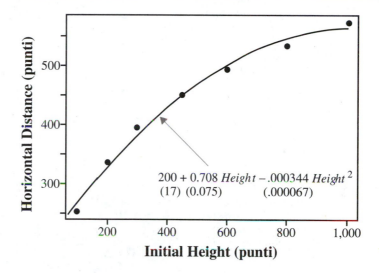

$$200 + 0.708 \, Height - .000344 \, Height^2$$
$$(17) \quad (0.075) \qquad\qquad (.000067)$$

10.1.2 The Energy Costs of Echolocation by Bats—An Observational Study

To orient themselves with respect to their surroundings, some bats use echolocation. They send out pulses and then read the echoes that are bounced back from surrounding objects. Such a trait has evolved in very few animal species, perhaps because of the high energy costs involved in producing pulses. Because flight also requires a great deal of energy, zoologists wondered whether the combined energy costs of echolocation and flight in bats was the sum of the flight energy costs and the at-rest echolocation energy costs, or whether the bats had developed a means of echolocation in flight that made the combined energy cost less than the sum.

They considered the data in Display 10.3 on in-flight energy expenditure and body mass from 20 energy studies on three types of flying vertebrates: echolocating bats, non-echolocating bats, and non-echolocating birds. They believed that, if the combined energy expenditure for flight and echolocation were additive, the amount of energy expenditure (after accounting for body size) would be greater for echolocating bats than for non-echolocating bats and non-echolocating birds. Display 10.4 shows a log–log scatterplot of in-flight energy expenditure versus body mass for the three types. (Data from J. R. Speakman and P. A. Racey, "No Cost of Echolocation for Bats in Flight," *Nature* 350(1991): 421–23.)

Display 10.3   Mass and in-flight energy expenditure for 4 non-echolocating bats (Type = 1), 12 non-echolocating birds (Type = 2), and 4 echolocating bats (Type = 3)

Species	Mass (g)	Type	Flight Energy Expenditure (W)
Pteropus gouldii	779	1	43.7
Pteropus poliocephalus	628	1	34.8
Hypsignathus monstrosus	258	1	23.3
Eidolon helvum	315	1	22.4
Meliphaga virescens	24.3	2	2.46
Melipsittacus undulatus	35	2	3.93
Sturnus vulgaris	72.8	2	9.15
Falco spaverius	120	2	13.8
Falco tinnunculus	213	2	14.6
Corvus ossifragus	275	2	22.8
Larus atricilla	370	2	26.2
Columba livia	384	2	25.9
Columba livia	442	2	29.5
Columba livia	412	2	43.7
Columba livia	330	2	34.0
Corvus crytoleucos	480	2	27.8
Phyllostomas hastatus	93	3	8.83
Plecotus auritus	8	3	1.35
Pipistrellus pipistrellus	6.7	3	1.12
Plecotus auritus	7.7	3	1.02

Summary of Statistical Findings

There is no evidence of differences in median in-flight energy expenditure for non-echolocating birds, non-echolocating bats, and echolocating bats, after accounting for body size (p-value $= .66$; extra-sums-of-squares F-test). The median in-flight energy expenditure for echolocating bats is estimated to be about 1.08 times as large as the median in-flight energy expenditure for non-echolocating bats, after accounting for body mass. A 95% confidence interval for the ratio of the echolocating median to the non-echolocating median, accounting for body mass, is 0.70 to 1.66.

Display 10.4   Log–log scatterplot of in-flight energy expenditure versus body mass for 4
non-echolocating bats, 12 non-echolocating birds, and 4 echolocating bats

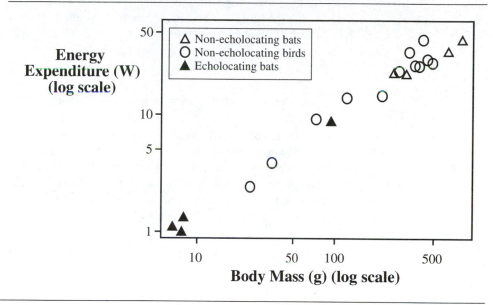

Scope of Inference

The species used in the statistical analysis were those for which relevant data were available;
any inference to a larger population of species is speculative. The statistical inferences must
also be interpreted in light of the likely violation of the independence assumption, due to
treating separate studies on the same species as independent observations.

The scientific results are premised on several facts. Bats emit the echolocation pulses
once on each wing beat, starting in the latter phase of the up-stroke, coinciding with the mo-
ment that air is expelled from the lungs. The coupling of these three processes apparently accounts for the relative energy economy. The energy used to expel air is also used to send out
the pulses, and these events occur just before the greatest demand is made on the wing beat.

10.2   Inferences About Regression Coefficients

Whether the energy expenditure is the same for echolocating bats as for non-echolocating
bats, after accounting for body mass, can be investigated by a test of the coefficient of the
indicator variable for echolocating bats in the parallel regression lines model. The simplicity
with which the question of interest can be investigated makes this the natural choice for an
inferential model. Many regression problems share this feature: questions of interest can
be answered through tests or confidence intervals for single coefficients in the model.

10.2.1   Least Squares Estimates and Standard Errors

The least squares estimates are the β values that minimize the sum of squared residuals. Formulas for the estimates come from calculus and are best expressed in matrix notation (see Exercises 10.20 and 10.21). A statistical computer program can do these computations. Users can perform all the functions of regression analysis on their data without knowing the relevant formulas. As an example of computer-provided estimates, consider the parallel regression lines model used for the bat echolocation data:

$$\mu\{lenergy \mid lmass, TYPE\} = \beta_0 + \beta_1 lmass + \beta_2 bird + \beta_3 ebat,$$

where *lenergy* is the log of the in-flight energy, *lmass* is the log of the body mass, and *TYPE* is the three-level factor represented by the indicator variables *bird* (which takes on a value of 1 if the type of species is a bird, and 0 if not) and *ebat* (which takes on a value of 1 if the species is an echolocating bat, and 0 if not). The third level of *TYPE*, non-echolocating bats, is treated here as the reference level, so its indicator variable does not appear in the regression model. The model is sketched in Display 10.5.

Display 10.5   The parallel regression lines model for the bat echolocation data

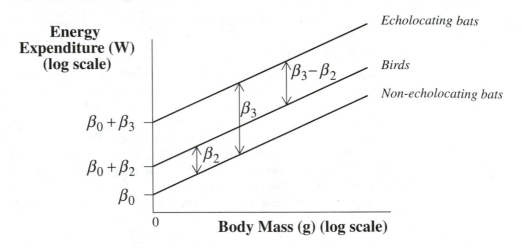

A portion of the output is shown in Display 10.6. The least squares estimates of the β's are shown in the column labeled "Coefficient." The estimate of σ^2 is the sum of squared residuals divided by the degrees of freedom associated with the residuals. The usual rule for finding degrees of freedom applies: (Sample size) minus (Number of unknown regression coefficients in the model for the mean). Since the sample size is 20 and there are 4 coefficients here—β_0, β_1, β_2, and β_3—the value for degrees of freedom is 16. The square root of the estimate of variance is the estimated standard deviation about the regression, which is 0.1860 in this case. This goes by several names on computer output, including *residual SD*, *residual SE*, and *root mean squared error*.

Display 10.6   Partial summary of the least squares fit to the regression of log energy expenditure on log body mass, an indicator variable for bird, and an indicator variable for echolocating bat

Variable	Coefficient	Standard Error	t-Statistic	Two-Sided p-Value
CONSTANT	−1.5764	0.2872	5.4880	<0.0001
lmass	0.8150	0.0445	18.2966	<0.0001
bird	0.1023	0.1142	0.8956	0.3837
ebat	0.0787	0.2027	0.3881	0.7030

Estimate of $\sigma = 0.1860$, **df** $= 16$

The column labeled "Standard Error" contains the estimated standard deviations of the sampling distributions of the least squares estimates. As in the case of previous estimates, the formulas for the standard deviations are available from statistical theory. In this instance, the formulas can be conveniently expressed only by using matrix notation, as shown in Exercise 10.22. Again, as with simple regression, they depend on the known values of the explanatory variables in the data set and on the unknown value of σ^2. The standard errors are the values of these standard deviations when σ^2 is replaced by its estimated value. Consequently, the degrees of freedom associated with a standard error are the same as the degrees of freedom associated with the estimate of σ (the sample size minus the number of β's).

Let $\hat{\beta}_j$ ("beta-hat j") represent the estimate of the jth coefficient ($j = 0, 1, 2,$ or 3 in the current example). If the distribution of the response for each combination of the explanatory variables is normal with constant variance, then the t-ratio,

$$t - \text{ratio} = (\hat{\beta}_j - \beta_j)/\text{SE}(\hat{\beta}_j),$$

has a sampling distribution described by the t-distribution with degrees of freedom equal to the degrees of freedom associated with the residuals. This theory leads directly to tests and confidence intervals for individual regression coefficients, in the familiar way.

10.2.2   Tests and Confidence Intervals for Single Coefficients

Do Non-echolocating Bats Differ from Echolocating Bats? Test for $\beta_3 = 0$

In the parallel regression lines model for log energy on log mass, the mean log energy for non-echolocating bats is $\beta_0 + \beta_1 lmass$. The mean log energy for echolocating bats is $\beta_0 + \beta_1 lmass + \beta_3$. Thus, for any given mass, the mean log energy expenditure for echolocating bats is β_3 units more than the mean log energy for non-echolocating bats (see Display 10.5). The question of whether the mean log energy expenditure for echolocating bats is the same as the mean log energy expenditure for non-echolocating bats of similar size may be examined through a test of the hypothesis that β_3 equals 0. From Display 10.6, it is evident that the two-sided p-value is .7030, providing no reason to doubt that β_3 is zero. The p-value was obtained by the computer as the proportion of values in a t-distribution on 16 degrees of freedom that are farther from 0 than 3.881 (the t-statistic for the hypothesis that β_3 is 0).

Tests of hypotheses $H: \beta_j = c$ for values of c other than zero are occasionally of interest. For these, the user must construct the t-statistic $(\hat{\beta}_j - c)/\text{SE}(\hat{\beta}_j)$ manually and find the p-value by reference to the appropriate t-percentiles.

How Much More Flight Energy Is Expended by Echolocating Bats? Confidence Interval for β_3

Although a test has revealed no reason to doubt that β_3 is zero, neither has it proved that β_3 is zero. Echolocating bats might have a higher energy expenditure, but the available study may not be powerful enough to detect this difference. As is usual for tests of hypotheses, a confidence interval should be reported in addition to the p-value, to emphasize this possibility and to provide an entire set of likely values for β_3.

Since the estimate of β_3 is 0.0787, with a standard error of 0.2027, and since the 97.5th percentile of a t-distribution with 16 degrees of freedom is 2.120, the 95% confidence interval for β_3 is

$$0.0787 - 2.12(0.2027) \qquad \text{to} \qquad 0.0787 + 2.12(0.2027),$$

or -0.351 to $+0.508$.

Coefficients after log transformations are interpreted in the same way as are coefficients described for simple regression in Section 8.4. If the parallel regression lines model is correct, the median in-flight energy expenditure is $\exp(\beta_3)$ times as great for echolocating bats as it is for non-echolocating bats of similar body mass. A 95% confidence interval for $\exp(\beta_3)$ is obtained by taking the anti-logs of the endpoints: $\exp(-0.351) = .70$ to $\exp(0.508) = 1.66$. The result can be communicated as follows: it is estimated that the median in-flight energy expenditure for echolocating bats is 1.08 times as great as the median expenditure for non-echolocating bats of similar body size. A 95% confidence interval for this multiplicative effect, accounting for body size, is 0.70 to 1.66.

Significance Depends on What Other Explanatory Variables Are Included

The meaning of the coefficient of an explanatory variable depends on what other explanatory variables are included in the regression. The p-value for the test of whether a coefficient is zero must also be interpreted in light of what other variables are included.

Consider three multiple linear regression models for the echolocation study. The first model says that the mean log energy is different for the three types of flying animals, but that it does not involve the body weight: $\mu\{lenergy \mid lmass, TYPE\} = TYPE$. The second model—the parallel lines model—indicates that the regression has the same slope against log body weight in all three groups, but different intercepts: $\mu\{lenergy \mid lmass, TYPE\} = lmass + TYPE$. The third model—the separate lines regression model—allows the three groups to have completely different straight-line regressions of log energy on log body weight: $\mu\{lenergy \mid lmass, TYPE\} = lmass + TYPE + lmass \times TYPE$. Least squares fits to these three models yield the following results:

(1) $\hat{\mu}\{lenergy \mid lmass, TYPE\} = \quad 3.40 \quad - \quad 2.74ebat \quad - \quad 0.61bird$
$$\qquad\qquad\qquad\qquad\qquad\quad (0.42) \qquad (0.60) \qquad\qquad (0.49)$$

(2) $\hat{\mu}\{lenergy \mid lmass, TYPE\} = -1.58 \; + \; 0.08ebat \; + \; 0.10bird \; + \; 0.815lmass$
$$\qquad\qquad\qquad\qquad\qquad\quad (0.29) \qquad (0.20) \qquad\quad (0.11) \qquad\qquad (0.045)$$

(3) $\hat{\mu}\{lenergy \mid lmass, TYPE\} = -0.20 \; - \; 1.27ebat \; - \; 1.38bird \; + \; 0.59lmass$
$$\qquad\qquad\qquad\qquad\qquad\quad (1.26) \quad (1.29) \qquad\quad (1.30) \qquad\qquad (0.21)$$
$$\qquad\qquad\qquad\qquad + \; 0.21(ebat \times lmass) \; + \; 0.25(bird \times lmass)$$
$$\qquad\qquad\qquad\qquad\quad (0.22) \qquad\qquad\qquad\qquad (0.21)$$

where the parenthesized numbers beneath the coefficients are their standard errors.

The coefficient of the indicator variable *ebat* is -2.74 in (1), $+0.80$ in (2), and -1.27 in (3). These might appear to be contradictory findings, since the variable is highly significant in (1) but not significant in (2) and (3), and since the sign of the estimated coefficient differs in the three fits. However, a contradiction arises only if one takes the view that the variable *ebat* plays the same role in each equation, which it does not. In (1), the coefficient of *ebat* measures the difference between mean log energy among echolocating bats and mean log energy among non-echolocating bats, ignoring any explanation of differences based on body size (*lmass* is not in the equation to provide the control). It happened (as indicated in Display 10.3) that the echolocating bats were much smaller than the non-echolocating bats in this study. Without taking body size into account, the statistical model attributes all the difference to group differences.

In (2), however, the coefficient of *ebat* measures the difference between energy expenditure among echolocating bats after adjusting for body size. Interpreting the coefficient as the difference between energy expenditure of echolocating bats and non-echolocating bats of about the same size exceeds the scope of this study, because all the echolocating bats were small and all the non-echolocating bats were large. The echolocating versus non-echolocating difference is thus confounded with the differences based on body size. Because "after adjusting for body size" means that group differences are considered only after the best explanation for body size is taken into account, and because body size explains the differences well, it is easy to understand why the coefficient of *ebat* in (2) can be insignificant even though in (1) it is significant.

In (3) the coefficient of *ebat* measures the difference between the intercept parameter in the regression of echolocating bat log energy versus log body weight and the intercept parameter in the regression of non-echolocating bat log energy versus log body weight. This is a very different interpretation from the one in (2) because here the slopes of the regression equations are allowed to be different. Because the coefficient measures a different quantity in (3) than in (2), there is no reason to expect its statistical significance to be the same.

10.2.3 Tests and Confidence Intervals for Linear Combinations of Coefficients

In the parallel lines model for the echolocation study (represented in Display 10.5), the slope in the regression of log energy on log mass is β_1 for all three groups. The intercept for non-echolocating bats is β_0, the intercept for non-echolocating birds is $\beta_0 + \beta_2$, and the intercept for echolocating bats is $\beta_0 + \beta_3$.

Since the bird line coincides with the non-echolocating bat line if β_2 is 0, a test of the equality of energy distribution in birds and non-echolocating bats, after accounting for body size, is accomplished through a test of whether β_2 equals 0. Similarly, a test of whether the echolocating bat regression line coincides with the non-echolocating bat regression line is accomplished with a test of $\beta_3 = 0$. The bird and echolocating bat regression lines coincide when $\beta_2 = \beta_3$, however, so a test of the equality of these two groups involves the hypothesis that $\beta_3 - \beta_2 = 0$ (or equivalently, that $\beta_2 - \beta_3 = 0$). This is a hypothesis about a linear combination of regression coefficients.

One method of performing the test compares the estimate of $\beta_3 - \beta_2$ with its standard error. Calculating the standard error is a problem here, however, because the estimates are not statistically independent. The correct formula expands the formula for linear combinations encountered in Section 6.2.2 (page 147) to include the variances and *covariances* in the joint sampling distribution of the regression coefficient estimates. The correct formula appears in Section 10.4.3. For this problem—as for many similar problems—there is an easier solution.

Redefining the Reference Level

Recall that the choice of reference level for any categorical explanatory variable is arbitrary. Thus, any of the three types of flying vertebrates could be used as the reference level. It is a simple matter to refit the model with another choice for the reference level, and either the birds or the echolocating bats will do. The model will remain the same as the one fit in Display 10.6, although the names of the intercepts will change, and the comparison of the non-echolocating birds to the echolocating bats can be accomplished through a single coefficient—the test for which appears in the standard output.

Inference About the Mean at Some Combination of X's

One special linear combination is the mean of Y at some combination of the X's. For Galileo's data and the regression model

$$\mu\{distance \mid height\} = \beta_0 + \beta_1 height + \beta_2 height^2,$$

the mean distance at a height of 250 is

$$\mu\{distance \mid height = 250\} = \beta_0 + (\beta_1 \times 250) + (\beta_2 \times (250)^2),$$

which is estimated by the linear combination of the estimates

$$\hat{\mu}\{distance \mid height = 250\} = (\hat{\beta}_0 \times 1) + (\hat{\beta}_1 \times 250) + (\hat{\beta}_2 \times (250)^2).$$

A standard error for this estimate may be obtained by using the methods of Section 10.4.3, but it may also be extracted from a computer analysis, by redefining the reference level for height. The trick is to create a new explanatory variable by subtracting 250 from each height measurement. Let $ht250 = height - 250$, so $height = 250$ corresponds to $ht250 = 0$, and then fit the model

$$\mu\{distance \mid ht250\} = \beta_0^* + \beta_1^* ht250 + \beta_2^* (ht250)^2.$$

The mean distance at $height = 250$ is $\mu\{distance \mid ht250 = 0\} = \beta_0^*$. Redefining the model terms permits the question of interest to be worded in terms of a single parameter; the standard error for this parameter appears in the usual output. Use the estimated intercept as the estimated mean of interest, and use the reported standard error of the intercept as the standard error for this estimated mean. As with the indicator variable example, the two models are identical, but are parameterized differently. Display 10.7 shows the analysis for both ways of writing the quadratic regression model, emphasizing how to obtain both the estimate of the mean (at $height = 250$ in this example) and its standard error.

Display 10.7   Estimates of polynomial coefficients with two different references levels of height, in Galileo's study

---

Reference height = 0

Variable	Coefficient	Standard Error	*t*-Statistic	Two-Sided *p*-Value
CONSTANT	199.91	16.76	11.93	.0003
$height$	0.7083	0.0748	9.47	.0007
$height^2$	−0.0003437	0.0000668	5.15	.0068

R-squared = 99.0%   **Adj. R-squared** = 98.6%   **Estimate of SD** = 13.6

$\hat{\mu}\{distance \mid height = 250\}$ $SE(\hat{\mu}\{distance \mid height = 250\})$

Reference height = 250

Variable	Coefficient	Standard Error	*t*-Statistic	Two-Sided *p*-Value
CONSTANT	355.51	6.62	53.66	<.0001
$height - 250$	0.5365	0.0430	12.48	.0002
$(height - 250)^2$	−0.0003437	0.0000668	5.15	.0068

R-squared = 99.0%   **Adj. R-squared** = 98.6%   **Estimate of SD** = 13.6

Confidence Bands for Multiple Regression Surfaces

Display 10.8 shows the estimated regressions for the bat data (the parallel regression lines for the log of energy expenditure on the log of body mass) on the original scales of measurement.

Display 10.8   Estimated median energy expenditures for birds, echolocating bats, and non-echolocating bats as functions of body mass (parallel lines model on log–log scale, with 95% confidence bands)

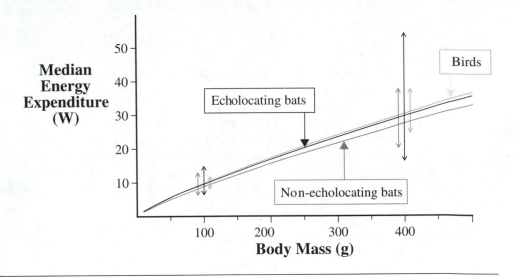

For non-echolocating bats, for example,

$$\text{Median}\{energy \mid mass\} = \exp(\beta_0)mass^{\beta_1}.$$

The curve for this group on the plot was drawn by computing the estimated median at many values of *mass* and connecting the resulting points.

The double arrows in Display 10.8 show the limits of a 95% confidence band for the median energy expenditure at 100 and 400 grams for each of the three groups. These intervals were constructed by finding confidence bands for the regression surface of log(*energy*) and then back-transforming the endpoints. The intervals were constructed by using the computer trick for finding the standard errors of estimated means. This required six different fits to the model, with different reference values, as shown in Display 10.9. The estimated mean for birds of 100-gram body mass, for example, was the estimated intercept in the regression of *lenergy* on *lmass*-100 and indicator variables for the two groups other than birds. This turned out to be 2.2789, with a standard error of 0.0604. From these, the confidence interval is calculated and then back-transformed for the light gray double-arrow line at Body mass = 100 in Display 10.8. The calculations appear in the bottom portion of Display 10.9.

Notice that the multiplier used in the confidence band is based on the 95th percentile from an *F*-distribution. (This example assumes that a confidence band over the entire region—all masses and types—is desired, rather than confidence intervals at a few specific

Display 10.9 Construction of the 95% confidence band, using repeated fits of the multiple regression model with different reference points

① Computer Work

Reference Point		Explanatory Variables			
TYPE	Body Mass	TYPE Indicators	Body Mass Variable	Intercept Estimate	Standard Error
birds	100	nbat, ebat	lmass −log(100)	2.2789	0.0604
	400	" "	lmass −log(400)	3.4087	0.0635
non-echo bats	100	ebat, bird	lmass −log(100)	2.1767	0.1144
	400	" "	lmass −log(400)	3.3064	0.0931
echo bats	100	nbat, bird	lmass −log(100)	2.2553	0.1277
	400	" "	lmass −log(400)	3.3851	0.1759

② Hand Calculations — an Example

$$\text{Multiplier} = \sqrt{4\, F_{4,16;\, 0.95}} = 3.468$$

$$\text{Lower limit} = \exp[2.2789 - (3.468)(0.0604)] = 7.9$$
$$\text{Upper limit} = \exp[2.2789 + (3.468)(0.0604)] = 12.0$$

values.) This type of multiplier was first introduced as the Scheffé multiple comparison procedure in Section 6.4.2 and again for confidence bands for simple regression in Section 7.4.2. Here, the numerator and denominator degrees of freedom in F are equal to the number of parameters in model (4) and to the residual degrees of freedom (16), respectively. The multiplier of F inside the square root sign is equal to the number of parameters in the model (4, in this case). Use of the F-based multiplier guarantees at least 95% confidence that the bands contain the regression surface throughout the experimental range. (*Note:* It is dangerous to draw conclusions about the regression outside of the scope of the available data. For example, no non-echolocating bats had body sizes of less than 258 grams. Inference about the distribution of energy expenditure for non-echolocating bats of 100-gram body size requires extrapolation of the model beyond the range within which it was estimated.)

10.2.4 Prediction

Prediction is an important objective in many regression applications. If it is the only objective, there is no need to interpret the coefficients. For example, a reseacher may be able to predict individuals' blood pressures from the number of bathrooms in their houses, without having to interpret or understand the relationship between those two variables.

Predicted values are the same as estimated means. They are calculated by evaluating the estimated regression at the desired explanatory variable values, as in the previous section. As in simple linear regression, the error in prediction comes from two sources: the error of estimating the population mean, and the error inherent in the fact that the new observation differs from its mean. Since $\hat{Y} = \text{pred}\{Y \mid X\} = \hat{\mu}\{Y \mid X\}$, the prediction error is

$$Y - \hat{Y} = Y - \hat{\mu}\{Y \mid X\} = [Y - \mu\{Y \mid X\}] - [\hat{\mu}\{Y \mid X\} - \mu\{Y \mid X\}].$$

The variance of a prediction error consists of two corresponding pieces:

$$\text{Prediction variance} = \sigma^2 + [\text{SD}(\hat{\mu}\{Y \mid X\})]^2$$

The variance of prediction at some explanatory variable value, therefore, may be estimated by adding the estimate of σ^2 to the square of the standard error of the estimated mean; see also Section 7.4.3. The standard error of prediction is the square root of the estimated variance of prediction. Prediction intervals are based on this standard error and on the appropriate percentile from the t-distribution with degrees of freedom equal to those associated with the estimate of σ. As discussed with regard to simple regression, predictions are valid only within the range of values of explanatory variables used in the study. The validity also depends fairly strongly on the assumption of normal distributions in the populations.

Galileo Example

Display 10.7 has all the information needed to predict the horizontal distance of a single ball released at a height of 250 punti. The predicted distance is 355.5 punti. The variance of prediction is $(13.6)^2 + (6.62)^2$, whose square root is the standard error of prediction $= 15.1$ punti. With $7 - 3 = 4$ degrees of freedom, the t-multiplier for a 95% prediction interval is 2.776, so the interval extends from $355.5 - 42.0 = 313.5$ punti to $355.5 + 42.0 = 397.5$ punti.

10.3 Extra-Sums-of-Squares F-Tests

In multiple regression, data analysts often need to test whether *several* coefficients are all zero. For example, the model

$$\mu\{lenergy \mid lmass, \; TYPE\} = \beta_0 + \beta_1 lmass + \beta_2 bird + \beta_3 ebat$$

presents the regression as three parallel straight lines. The lines are identical under the hypothesis

$$H\colon \beta_2 = 0 \text{ and } \beta_3 = 0.$$

The alternative hypothesis is that at least one of the coefficients—β_2 or β_3—is nonzero. *T*-tests, either individually or in combination, cannot be used to test such a hypothesis involving more than one parameter.

10.3.1 Comparing Sizes of Residuals in Hierarchical Models

The extra-sum-of-squares method, on the other hand, is ideally suited for this purpose. It directly compares a full model (*lmass* + *TYPE*) to a reduced model (*lmass*):

Full: $\mu\{lenergy \mid lmass,\ TYPE\} = \beta_0 + \beta_1 lmass + \beta_2 bird + \beta_3 ebat,$
Reduced: $\mu\{lenergy \mid lmass,\ TYPE\} = \beta_0 + \beta_1 lmass.$

If the coefficients in the last two terms of the full model are zero, their estimates should be close to zero, and fits to the two models should give about the same results. In particular, the residuals should be about the same size. If either β_2 or β_3 is not zero, however, the full model should do a better job of explaining the variation in the response variables, and the residuals should tend to be smaller in magnitude than those from the reduced model. Even if the reduced model is correct, however, the squared residuals in the full model must be somewhat smaller, since the full model has more flexibility to match chance variations in the data. The *F*-test is used to assess whether the difference between the sums of squared residuals from the full and reduced models is greater than can be explained by chance variation. The sum of squared residuals in the reduced model minus the sum of squared residuals in the full model (as in Section 5.3) is the *extra sum of squares:*

> *Extra sum of squares =*
> *Sum of squared residuals from reduced model –*
> *Sum of squared residuals from full model.*

The sum of squared residuals is a measure of the response variation that remains unexplained by a model. The extra sum of squares may be interpreted as being the amount by which the unexplained variation decreases when the extra terms are added to the reduced model. Or it may be interpreted as being the amount by which the unexplained variation increases when the extra terms are dropped from the full model.

10.3.2 *F*-Test for the Joint Significance of Several Terms

The extra-sum-of-squares *F*-test has been encountered previously as a one-way analysis-of-variance tool for comparing the separate-means model to the single-mean model (Section 5.3), and as a tool in simple regression for comparing the regression model to the single-mean model or to the separate-means model if there are replicates (Section 8.5). These are all special cases of a general form that is often called a *partial F-test*.

The F-statistic, based on the extra sum of squared residuals in a reduced model over a full model is defined as follows:

$$F\text{-statistic} \quad = \quad \cfrac{\left[\cfrac{Extra\ sum\ of\ squares}{Number\ of\ betas\ being\ tested} \right]}{Estimate\ of\ \sigma^2\ from\ full\ model}.$$

The part in brackets is the average size of the extra-sum-of-squares per coefficient being tested. If the reduced model is correct, then this per-coefficient variation should be roughly equal to the per observation variation, σ^2. The F-statistic, therefore, should be close to one.

When the reduced model is correct and the rest of the model assumptions (including normality) hold, the sampling distribution of this F-statistic is an F-distribution. (Although exact justification is based on normality, the F-test and the t-tests are robust against departures from normality.) The numerator degrees of freedom are the number of β's being tested. This is either the number of β's in the full model minus the number of β's in the reduced model or, equivalently, the residual degrees of freedom in the reduced model minus the residual degrees of freedom in the full model. The denominator degrees of freedom are those associated with the estimate of σ^2 in the full model. If the F-statistic is larger than expected from this F-distribution, this is interpreted as evidence that the reduced model is incorrect. The test's p-value—the chance that an F-variable exceeds the calculated value of the F-statistic—measures the strength of that evidence.

Example—Bat Echolocation Data

Computations in the extra-sum-of-squares F-test are detailed in Display 10.10, leading to a conclusion that there is no evidence of a group difference.

Special Case: Testing the Significance of a Single Coefficient

One special hypothesis is that a single coefficient is zero; for example, $H\colon \beta_3 = 0$. Since the t-test is available for this hypothesis, one might wonder whether the F-test leads to the same result. It does. The F-statistic is the square of the t-statistic, and the p-value from the F-test is the same as the two-sided p-value from the t-test. As a practical matter, the t-test results are more convenient to obtain from computer output.

Special Case: Testing the "Overall Significance" of the Regression

Another special hypothesis is that all regression coefficients except β_0 are zero. This hypothesis proposes that none of the considered explanatory variables are useful in explaining

Display 10.10 The extra-sum-of-squares *F*-test for testing equality of intercepts in the parallel regression lines model (bat echolocation data)

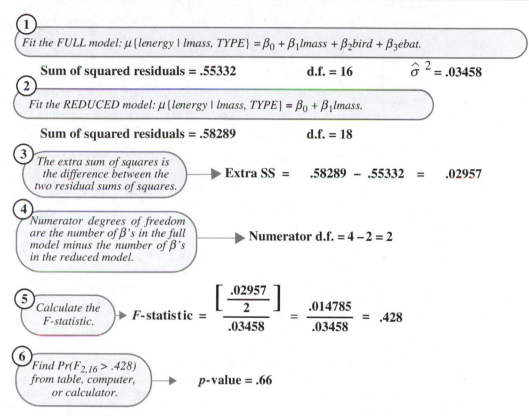

(1)

Fit the FULL model: $\mu\{lenergy \mid lmass, TYPE\} = \beta_0 + \beta_1 lmass + \beta_2 bird + \beta_3 ebat.$

Sum of squared residuals = .55332   **d.f. = 16** $\hat{\sigma}^2 = .03458$

(2)

Fit the REDUCED model: $\mu\{lenergy \mid lmass, TYPE\} = \beta_0 + \beta_1 lmass.$

Sum of squared residuals = .58289   **d.f. = 18**

(3) *The extra sum of squares is the difference between the two residual sums of squares.* → **Extra SS = .58289 − .55332 = .02957**

(4) *Numerator degrees of freedom are the number of β's in the full model minus the number of β's in the reduced model.* → **Numerator d.f. = 4 − 2 = 2**

(5) *Calculate the F-statistic.* → $\textbf{F-statistic} = \dfrac{\left[\dfrac{.02957}{2}\right]}{.03458} = \dfrac{.014785}{.03458} = .428$

(6) *Find Pr($F_{2,16} > .428$) from table, computer, or calculator.* → **p-value = .66**

Conclusion: There is no evidence that mean log energy differs for birds, echolocating bats, and non-echolocating bats, after accounting for body mass.

the mean response. Although this hypothesis only occasionally corresponds to a question of interest, the extra-sums-of-squares *F*-test for this hypothesis—often called the *F-test for overall significance of the regression*—is routine output in most statistical computer programs. As discussed in the following section, the component calculations of this *F*-test are routinely available in an analysis of variance table.

10.3.3 The Analysis of Variance Table

The reduced model for the mean contains only the constant term β_0. Hence the least squares estimate of β_0 is the average response ($\overline{Y}$), and the sum of squared residuals is $(n-1)s_Y^2$.

This is the "Total sum of squares" as it measures the total variation of the responses about their average. The sum of squared residuals from the model under investigation is labeled the "Residual sum of squares" (or sometimes the "Error sum of squares") in the analysis of variance table. The difference between these two—the extra-sum-of-squares—is the amount of total variation in the response variable that can be explained by the regression on the explanatory variables. It is called the "Regression sum of squares" or the "Model sum of squares." An analysis of variance table for the quadratic fit to Galileo's data appears in Display 10.11.

Display 10.11 Analysis of variance table for Galileo's data: fit of the data to the quadratic model: $\mu\{distance \mid height\} = \beta_0 + \beta_1\ height + \beta_2\ height^2$ (based on 7 observations)

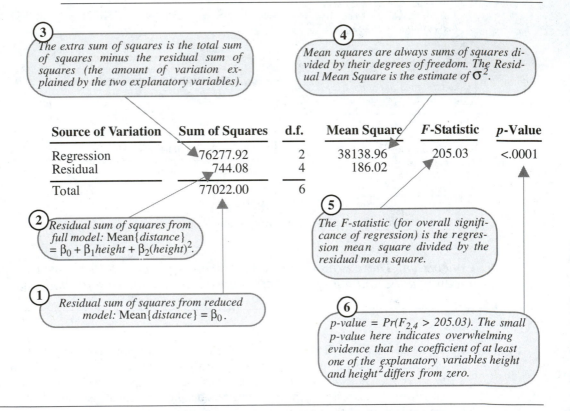

Using Analysis of Variance Tables for Extra-Sums-of-Squares Tests

Aside from the two special cases mentioned previously, the calculations for extra-sums-of-squares tests must be carried out manually, as was shown in Display 10.10. The sums of squared residuals in steps 1 and 2 are obtained by fitting the full and reduced models, obtaining the analysis of variance tables for each, and reading the sum of squared residuals from the tables.

Analysis of the Bat Echolocation Data

The analysis of the bat echolocation data involves several extra-sums-of-squares *F*-tests. The main question of interest is, "Is there a difference in the in-flight energy expenditures of echolocating and non-echolocating bats after body size is accounted for?" This question does not involve non-echolocating birds, but including them in the analysis is important for obtaining the best possible estimate of σ^2. A secondary question compares birds to the two bat groups.

Inspection of the data and initial graphical exploration indicated the need to analyze both energy and body mass on the log scale. The statistical analysis starts coming into focus with a graphical display like the coded scatterplot in Display 10.4. It is apparent that a straight-line regression of log energy on log body mass is probably appropriate for each of the three types, and there is no obvious evidence of a difference in the slopes of these lines for the three types of flying vertebrates. These results are fortunate, because the question of interest is most easily addressed if the group differences are the same for all body weights. Hence, the parallel lines model offers the most convenient inferential model.

To buttress the argument, the analyst must make sure that the regression lines are, indeed, parallel before testing whether the intercepts are equal. This involves first fitting a rich model that incorporates different intercepts and different slopes. The separate lines model is

$$\mu\{lenergy \mid lmass, TYPE\} =$$
$$\beta_0 + \beta_1 lmass + \beta_2 bird + \beta_3 ebat + \beta_4 (lmass \times bird) + \beta_5 (lmass \times ebat).$$

The next step is to fit the rich model and examine a residual plot (plot of residuals versus fitted values). This plot (not shown) indicates no problems with the model assumptions, so it is appropriate to perform the *F*-test for the hypothesis that the interaction terms can be dropped (namely, $H: \beta_4 = \beta_5 = 0$). This *F*-test, based on the analysis of variance tables, is shown in Display 10.12.

Since the *p*-value is .53, it is safe to drop the interaction terms and address the comparison of interest in terms of the parallel lines model. It is then appropriate to test whether the intercepts in the parallel regression lines model are the same for the three groups. This is accomplished by conducting another extra-sum-of-squares *F*-test, as previously illustrated in Display 10.10. The *p*-value of .66 implies that the data are consistent with the hypothesis that no difference in energy expenditure exists among the three groups, after the effect of body mass is accounted for. Nevertheless, confidence intervals for the group differences should still be presented, as in the summary of statistical findings in Section 10.1.2, to reveal other possibilities that are not excluded by the data.

Display 10.12   The extra sum of squares *F*-test comparing the separate regression lines
model to the parallel regression lines model—bat echolocation data

(1) FIT FULL MODEL: $\mu\{lenergy \mid lmass, TYPE\} =$
$\beta_0 + \beta_1 lmass + \beta_2 bird + \beta_3 ebat + \beta_4(lmass \times bird) + \beta_5 (lmass \times ebats).$

Source of Variation	Sum of Squares	d.f.	Mean Square	*F*-Statistic	*p*-Value
Regression	29.46993	5	5.89399	163.4	<.0001
Residual	.50487	14	.03606		
Total	29.97480	19			

Residual SS

Estimate of σ^2
d.f.

(2) FIT REDUCED MODEL: $\mu\{lenergy \mid lmass, TYPE\} =$
$\beta_0 + \beta_1 lmass + \beta_2 bird + \beta_3 ebat.$

Source of Variation	Sum of Squares	d.f.	Mean Square	*F*-Statistic	*p*-Value
Regression	29.42148	3	9.80716	283.6	<.0001
Residual	.55332	16	.03458		
Total	29.97480	19			

Residual SS

(3) The extra sum of squares
is the difference between   Extra SS = .55332 − .50487 = .04845
residual sums of squares.

*Numerator d.f. = Number of
β's in full model minus
Number of β's in reduced model.*

(5) Calculate the   *F*-statistic $=\dfrac{\left[\dfrac{.04845}{2}\right]}{.03606} = .672$
F-statistic.

(6) Look up $Pr(F_{2,14} > 0.672)$ ⟶ *p*-value = 0.53

Conclusion: *There is no evidence that the association between energy
expenditure and body size differs among the three types
of flying vertebrates (p-value = 0.53).*

10.4 Related Issues

10.4.1 Further Notes on the *R*-Squared Statistic

Example—Galileo's Data

From the analysis of variance table in Display 10.11, it is evident that the quadratic regression of *distance* on *height* explains 76277.92/77022.00, or 99.03% of the total variation in the observed distances. Only 0.97% of the variation in distances remains unexplained.

Display 10.13 shows the regression output for another model for these data: the regression of distance on *height*, *height-squared*, and *height-cubed*. The *p*-value for the coefficient of *height-cubed* (.0072) provides strong evidence that the cubic term is significant even when the linear and quadratic terms are in the model. On the other hand, the percentage of variation explained by the cubic model is 99.94%, only 0.91% more than the percentage of variation explained by the quadratic model. Of course that 0.91% accounts for 94% of the remaining variability from the quadratic model, which helps explain why the cubic term is statistically significant.

Display 10.13 Partial output from the regression of distance on *height*, *height-squared*, and *height-cubed*, for Galileo's data

Variable	Coefficient	Standard Error	*t*-Statistic	*p*-Value
Constant	155.78	8.33	18.71	0.0003
height	1.1153	0.0657	16.98	0.0004
height-squared	−0.001245	0.000138	−8.99	0.0029
height-cubed	5.477×10^{-7}	0.838×10^{-7}	6.58	0.0072

Estimate of standard deviation about the regression: 4.011 on 3 degrees of freedom
$R^2 = 99.94\%$.

Two very useful features of *R*-squared have been revealed with Galileo's data. First, the summary measure—for instance, 99.03% of the variation explained—provides a valuable image of the tightness of the fit. Second, the proportion of additional variation explained by one newly introduced variable after the others have been accounted for can be used in assessing its practical significance. Despite these useful summarizing applications, however, *R*-squared suffers considerable abuse. It is rarely an appropriate statistic to use for model checking, model comparison, or inference.

R-Squared Can Always Be Made 100% by Adding Explanatory Variables

Display 10.14 shows the scatterplot of distance versus height for Galileo's seven measurements. Drawn through the points is a sixth-order polynomial regression line:

$$\mu\{distance \mid height\} = \beta_0 + \beta_1 height + \beta_2 (height)^2 + \beta_3 (height)^3$$
$$+ \beta_4 (height)^4 + \beta_5 (height)^5 + \beta_6 (height)^6.$$

This fit produces residuals that are all zero and, consequently, an R^2 of 100%. In fact, for n data points, an $n - 1$ polynomial regression will always fit exactly. That does not imply that the equation has any usefulness, though. While the data at hand have been modeled exactly, the particular equation is unlikely to fit as well on future data.

Display 10.14   Scatterplot of Galileo's horizontal distances versus initial heights, with estimated sixth-order polynomial regression curve (R^2 = 100%)

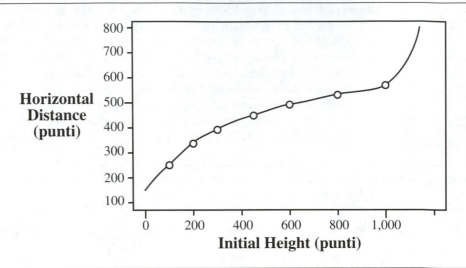

This same warning applies to multiple regression in general. R-squared can always be made 100% by adding enough explanatory variables. The question is whether the model is appropriate for the population or instead is only appropriate for the particular data on hand. If inference to some population (real or hypothetical) is desired, it is important not to craft a model that is overly customized to the data on hand. Tests of significance and residual plots are more appropriate tools for model building and checking.

The Adjusted R-Squared Statistic

The adjusted R-squared is a version of R-squared that includes a penalty for unnecessary explanatory variables. It measures the proportion of the observed spread in the responses that is explained by the regression model, where spread is measured by residual mean squares rather than by residual sums of squares. This approach provides a better basis for judging the improvement in a fit due to adding an explanatory variable, but it does not have the simple summarizing interpretation that R^2 has:

$$\text{Adjusted } R^2 = 100 \frac{(\text{Total mean square}) - (\text{Residual mean square})}{\text{Total mean square}}\%.$$

Thus adjusted R-squared is useful for casual assessment of improvement of fit, but R-squared is better for description, as illustrated in the summary of statistical findings in Section 10.1.1.

10.4.2   Improving Galileo's Design with Replication

Since Galileo's study contained no replication, it is impossible to separate the repeatable part of the relationship (if any) from the real variability in replicate measurements of horizontal distances of balls started at the same height. In other words, all estimates of σ are necessarily based on a particular form of the regression model and may be biased as a result of model inadequacies.

Suppose that Galileo instead had selected four evenly spaced heights—say 100, 400, 700, and 1,000 punti—and had replicated the experiment at each height. Using four heights would have allowed Galileo to fit a cubic polynomial, giving him a model on which to judge the proposed parabola (through a test of the cubic term). The revised design also yields 4 degrees of freedom for estimating residual variability free from possible model complications. The cubic model can be judged by comparing it to the separate-means model for the four groups.

A natural, *design-based* estimate of variance is obtained by pooling sample variances from groups of data at identical values of the explanatory variables. It can only be obtained if there are replicates. Without replication, the only available estimate of variance is the one obtained from the residuals about the fit to the presumed model for the mean. There is always a possibility of bias in this *model-based* estimate of variance, due to inadequacies of the regression model. With a design-based estimate no such bias can occur, because the size of the residuals does not hinge on the accuracy of any presumed regression model. This is a strong reason for incorporating some replication into an experimental design.

10.4.3   Variance Formulas for Linear Combinations of Regression Coefficients

On some occasions it may seem more convenient to estimate a linear combination of regression parameters directly than to instruct the computer to change the reference level (as in Section 10.2.3). In general, a linear combination can be written as

$$\gamma = C_0\beta_0 + C_1\beta_1 + C_2\beta_2 + \cdots + C_p\beta_p,$$

where the C's are known coefficients, some of which may be zeros. The estimate of this combination is

$$g = C_0\hat{\beta}_0 + C_1\hat{\beta}_1 + C_2\hat{\beta}_2 + \cdots + C_p\hat{\beta}_p.$$

The formula for the variance of this linear combination differs from the one for variance of a linear combination of independent averages in Section 6.2.2 (page 147). Since the estimated coefficients are not independent of one another, the formula here involves their covariances.

For a population of pairs U and V, the *covariance* of U and V is the mean of $(U - \mu_U)(V - \mu_V)$, and it describes how the two variables covary. The covariance of $\hat{\beta}_1$ and

$\hat{\beta}_2$ is the mean of $(\hat{\beta}_1 - \beta_1)(\hat{\beta}_2 - \beta_2)$ in their sampling distribution. A formula for the covariances of least squares estimates is available, and estimates are available from least squares computer routines. If Cov represents estimated covariance, the variance of the sampling distribution of the estimated linear combination is

$$\text{Var}\{g\} = C_0^2 \text{SE}(\hat{\beta}_0)^2 + C_1^2 \text{SE}(\hat{\beta}_1)^2 + \cdots + C_p^2 \text{SE}(\hat{\beta}_p)^2 +$$
$$2C_0C_1\text{Cov}(\hat{\beta}_0, \hat{\beta}_1) + 2C_0C_2\text{Cov}(\hat{\beta}_0, \hat{\beta}_2) + \cdots + 2C_{p-1}C_p\text{Cov}(\hat{\beta}_{p-1}, \hat{\beta}_p).$$

To see why this is so consider the linear combination $\gamma = C_1\beta_1 + C_2\beta_2$ and its estimate $g = C_1\hat{\beta}_1 + C_2\hat{\beta}_2$. Then

$$(g - \gamma) = C_1(\hat{\beta}_1 - \beta_1) + C_2(\hat{\beta}_2 - \beta_2)$$

and

$$(g - \gamma)^2 = C_1^2(\hat{\beta}_1 - \beta_1)^2 + C_2^2(\hat{\beta}_2 - \beta_2)^2 + 2C_1C_2(\hat{\beta}_1 - \beta_1)(\hat{\beta}_2 - \beta_2).$$

The variance of g is $\text{Mean}(g - \gamma)^2$ which can be expressed as

$$C_1^2\text{Mean}(\hat{\beta}_1 - \beta_1)^2 + C_2^2\text{Mean}(\hat{\beta}_2 - \beta_2)^2 + 2C_1C_2\text{Mean}(\hat{\beta}_1 - \beta_1)(\hat{\beta}_2 - \beta_2)$$
$$= C_1^2\text{Var}(\hat{\beta}_1) + C_2^2\text{Var}(\hat{\beta}) + 2C_1C_2\text{Cov}(\hat{\beta}_1, \hat{\beta}_2).$$

The estimate of this expression is obtained by substituting the estimates of the variances and covariance.

The Estimated Variance–Covariance Matrix for the Coefficients

The estimated covariances of the coefficients' sampling distributions are calculated by most statistical computer programs. They are stored as an array, or *matrix*, for use in formulas such as the one just discussed. See also Exercise 10.22 (page 287). The easy computer trick for finding the standard error of $\hat{\beta}_2 - \hat{\beta}_3$ for the parallel regression lines model fit to the echolocation data was shown in Section 10.2.3. Display 10.15 illustrates the alternative, direct approach, using the estimated coefficient covariances supplied by a statistical computer program.

10.4.4   Further Notes About Polynomial Regression

A Second-order Model in Two Explanatory Variables

A full, *second-order model* in explanatory variables X_1 and X_2 is

$$\mu\{Y \mid X_1, X_2\} = \beta_0 + \beta_1 X_1 + \beta_2 X_2 + \beta_3 X_1^2 + \beta_4 X_2^2 + \beta_5 X_1 X_2.$$

This is a useful model in *response surface studies*, where the goal is to model a nonlinear response surface in terms of a polynomial. It is rarely useful to attempt cubic and higher-order terms.

Display 10.15   Inference about $\beta_2 - \beta_3$, the coefficient of the indicator variable for birds minus the coefficient of the indicator variable for echolocating bats

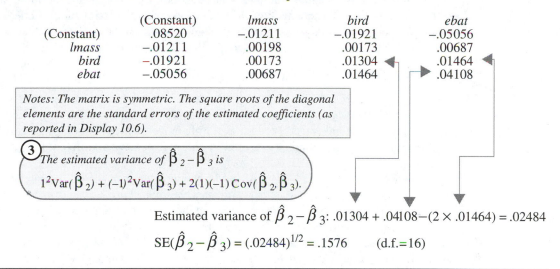

Estimate the linear combination of coefficients as the same linear combination of estimated coefficients.

Estimate of $\beta_2 - \beta_3$, from Display 10.6: $.1023 - .0787 = .0236$

Obtain the estimated variance–covariance matrix of the estimated regression coefficients.

Estimated variance–covariance matrix (from computer):

	(Constant)	lmass	bird	ebat
(Constant)	.08520	−.01211	−.01921	−.05056
lmass	−.01211	.00198	.00173	.00687
bird	−.01921	.00173	.01304	.01464
ebat	−.05056	.00687	.01464	.04108

Notes: The matrix is symmetric. The square roots of the diagonal elements are the standard errors of the estimated coefficients (as reported in Display 10.6).

The estimated variance of $\hat{\beta}_2 - \hat{\beta}_3$ is

$$1^2 \text{Var}(\hat{\beta}_2) + (-1)^2 \text{Var}(\hat{\beta}_3) + 2(1)(-1)\text{Cov}(\hat{\beta}_2, \hat{\beta}_3).$$

Estimated variance of $\hat{\beta}_2 - \hat{\beta}_3$: $.01304 + .04108 - (2 \times .01464) = .02484$

$$\text{SE}(\hat{\beta}_2 - \hat{\beta}_3) = (.02484)^{1/2} = .1576 \qquad (\text{d.f.} = 16)$$

When Should Squared Terms Be Included?

As with interaction terms, quadratic terms should not routinely be included. They are useful to consider in four situations: when the analyst has good reason to suspect that the response is nonlinear in some explanatory variable (through knowledge of the process or by graphical examination); when the question of interest calls for finding the values that maximize or minimize the mean response; when careful modeling of the regression is called for by the questions of interest (and presumably this is only the case if there are just a few explanatory variables); or when inclusion is used to produce a rich model for assessing the fit of an inferential model.

Finding the X that Maximizes the Mean Response with Quadratic Regression

A quadratic regression model is useful when the goal is to determine the explanatory variable value that maximizes or minimizes the mean response. Calculus permits calculating the value of X that maximizes (or minimizes) $\mu\{Y|X\} = \beta_0 + \beta_1 X + \beta_2 X^2$, which proves to be $X_{\max} = -\beta_1/(2\beta_2)$. This value maximizes if β_2 is negative and minimizes if β_2 is positive. The estimate, obtained by substituting the estimates in place of β_1 and β_2, is a *nonlinear* combination of the $\hat{\beta}$'s, so the method for obtaining standard errors for linear combinations will not help. Special tools are required.

10.4.5 The Principle of Occam's Razor

The principle of Occam's Razor is that simple models are to be preferred over complicated ones. Named after the fourteenth-century English philosopher, William of Occam, this principle has guided scientific research ever since its formulation. It has no underlying theoretical or logical basis; rather, it is founded in common sense and successful experience. It is often called the Principal of Parsimony.

In statistical applications, the idea translates into a preference for the more simple of two models that fit data equally well. One should seek a parsimonious model that is as simple as possible and yet adequately explains all that can be explained. Methods for paring down large sets of explanatory variables are discussed in Chapter 12.

10.4.6 Informal Tests in Model Fitting

Two tests for hypotheses about regression coefficients—t-tests and extra-sum-of-squares F-tests—are valuable, for two purposes: for formally providing evidence regarding questions of interest in a final model; and for exploring models by testing potential terms at the exploratory stage. The attitude toward the p-values is somewhat different for these two purposes.

In answering questions of interest, p-values are given their formal interpretation. A small p-value provides strong evidence against the null hypothesis and in favor of the alternative. The null hypothesis may be true and the particular sample may have happened by chance to be an unusual one; the p-value provides a measure of just how unusual it would be in such a case. A large p-value means either that the null hypothesis is correct or that the study was not powerful enough to detect a departure from it. Although it is common in the wording attached to such tests to include some indication of "statistical significance" or the degree to which a hypothesis can safely be "rejected," the p-value is reported, too, as opposed to simply communicating whether the hypothesis was or was not rejected at some level.

In the exploratory stage, however, the tests are performed so that some action may be taken. For example, a quadratic term may be added to a simple linear regression model to help determine whether the straight-line model is adequate. Based on a test of whether the coefficient of the quadratic term is zero, the straight-line model is either rejected or accepted for further use. The p-value in this case is interpreted much less formally, and the decision of whether to abandon a simple model depends to some extent on whether another simple model can be found and on the degree of adequacy of the simple model. Regarding the second of these, with very slight curvature, a small data set may not find convincing evidence that the x-squared coefficient differs from zero, but a large enough data set would (even if the quadratic term only explained an additional 0.1% of the response variation).

For answering questions of interest, tests should not be overemphasized. Even if the question of interest calls for a test, reporting a confidence interval to indicate the possible sizes of the effects of interest remains important. This is true whether the p-value for the test is small or large.

10.5 Summary

Galileo's Study

Galileo's data are used to find a polynomial describing the mean distance as a function of height. The scatterplot shows that the relationship is not a straight line. The coefficient of a height-squared term, when added to the simple linear regression model, significantly differs from zero. R^2 is a useful summary here: the quadratic regression model explains 99.03% of the variation in the horizontal distances. The large R^2 does not mean that all other terms are insignificant. A test of hypothesis is used to resolve that matter. In fact, when a height-cubed term is added to the model, the p-value for testing whether its coefficient is zero is .007, and the value of R^2 increases to 99.94%. This example demonstrates the benefit of R^2 for summarizing a fit and for indicating the degree of relevance of a significant term.

Echolocation by Bats

The goal is to see whether the in-flight energy expended by echolocating bats differs from the energy used by non-echolocating bats of similar body mass. There is no major question involving the birds, but their inclusion helps to clarify a common relationship between energy and body mass. A useful starting point in the analysis is a coded scatterplot of energy versus body mass. It is apparent by inspection that both energy and body mass should be transformed to their logarithms. A plot on this scale (Display 10.4) reveals a straight line relationship but very little additional difference in energy expenditure among the three types of flying vertebrates. Comparing the in-flight energy expenditures of echolocating bats and non-echolocating bats is difficult because the former are all small bats and the latter all big bats. Multiple linear regression permits a comparison after accounting for body mass, but the comparison must be made cautiously, since the ranges of body mass for the two types do not overlap. In light of this limitation, the comparison is made on the basis of an indicator variable in a parallel regression lines model. The data are consistent with the hypotheses that echolocating bats pay no extra energy price for their echolocating skills.

Further Reading

The inferential procedures given here are standard and are presented thoroughly in the references provided at the conclusion of Chapter 9. The extra-sum-of-squares F-test often goes by the name of partial F-test.

10.6 Exercises

Conceptual Exercises

1. **Galileo's Data.** Why is horizontal distance, rather than height, the response variable?

2. **Brain Weight.** Display 10.16 shows two possible *influence diagrams* relating the variables in the brain weight study of Section 9.1.2. If brain weight is directly associated with gestation period and litter size, as in part (a) of the figure, then animals that have approximately the same body size but different gestation period and different litter size should have different brain sizes, and scatterplots of brain size versus gestation and litter size individually should show some association. But the

scatterplots can also show indirect associations, as in part (b) of the figure. If brain weight, gestation period, and litter size are all driven by body size, they should show mutual associations, whether or not a direct association exists. Can a statistical analysis distinguish between direct and indirect association? Explain how or, if not, why not.

Display 10.16   Associations of brain weight with gestation period and litter size: direct or indirect?

(a) DIRECT ASSOCIATION

Animals with about the same body size will have different brain sizes when their litter sizes and gestation periods are different.

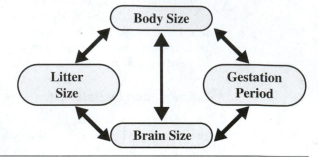

(b) INDIRECT ASSOCIATION

Animals with about the same body size will have about the same brain size even if their gestation periods and litter sizes are different. Any apparent association between brain size and litter size, say, is a result of their both being related to body size.

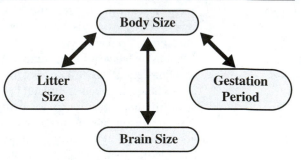

---

3.   **Brain Weight.** Consider the mammal brain weight data from Section 9.1.2, the model

$$\mu\{lbrain \mid lbody, lgest, llitter\} = \beta_0 + \beta_1 lbody + \beta_2 lgest + \beta_3 llitter,$$

and the hypothesis H: $\beta_2 = 0$ and $\beta_3 = 0$. (a) Why can this not be tested by the two t-tests reported in the standard output? (b) Why can this not be tested by the two t-tests along with an adjustment for multiple comparisons?

4.   **Stocks and Jocks.** Based on data from nine days in June 1994, a multiple regression equation was fit to the Dow Jones Index on the following seven explanatory variables: the high temperature in New York City on the previous day; the low temperature on the previous day; an indicator variable taking on the value 1 if the forecast for the day was sunny and 0 otherwise; an indicator variable taking on the value 1 if the New York Yankees won their baseball game of the previous day and 0 if not; the number of runs the Yankees scored; an indicator variable taking on the value of 1 if the New York Mets won their baseball game of the previous day and 0 if not; and the number of runs the Mets scored. As the chart in Display 10.17 shows, the predicted values of the stock market index were strikingly close to the actual values. R^2 was 89.6%. Why is this unremarkable?

Display 10.17 Actual values and predicted values of Dow Jones Index

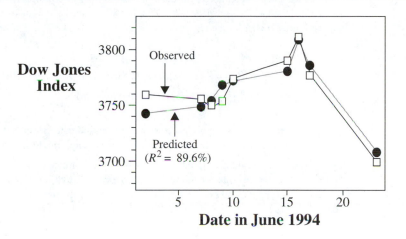

5. **Bat Echolocation.** Consider these three models:

$$\mu\{lenergy \mid lmass, TYPE\} = \beta_0 + \beta_1 lmass$$

$$\mu\{lenergy \mid lmass, TYPE\} = \beta_0 + \beta_1 lmass + \beta_2 bird + \beta_3 ebat$$

$$\mu\{lenergy \mid lmass, TYPE\} = \beta_0 + \beta_1 lmass + \beta_2 bird + \beta_3 ebat$$
$$+ \beta_4 (lmass) \times (bird) + \beta_5 (lmass \times (ebat).$$

(a) Explain why they can be described as representing the single line, parallel lines, and separate lines models, respectively. (b) Explain why the second model can be the "reduced model" in one F-test but the "full model" in another.

6. **Bat Echolocation.** A possible statement in conclusion of the analysis is this: "It is estimated that the median in-flight energy expenditure for echolocating bats is 1.08 times as large as the median in-flight energy expenditure for non- echolocating bats of similar body size." Referring to Display 10.4, explain why including the phrase, "for bats of similar body mass" in this statement is suspect. What alternative wording is available?

7. **Life is a Rocky Road.** A regression of the number of crimes committed in a day on volume of ice cream sales in the same day showed that the coefficient of ice cream sales was positive and significantly differed from zero. Which of the following is the most likely explanation? (a) The content of ice cream (probably the sugar) encourages people to commit crimes. (b) Successful criminals celebrate by eating ice cream. (c) A pathological desire for ice cream is triggered in a certain percentage of individuals by certain environmental conditions (such as warm days), and these individuals will stop at nothing to satisfy their craving. (d) Another variable, such as temperature, is associated with both crime and ice cream sales.

8. In the terminology of extra-sums-of-squares F-tests, does the reduced model correspond to the case where the null hypothesis is true? Is the full model the one that corresponds to the alternative hypothesis's being true?

Computational Exercises

9. Crab Claws. Reconsider the data on claw closing force and claw size for three species of crabs, shown in Display 7.15 on page 195. Display 10.18 shows output from the least squares fit to the separate-lines model for the regression of log force on log height. The regression model for log force was

$$\mu\{lforce \mid lheight, SPECIES\} = \beta_0 + \beta_1 lheight + \beta_2 lb + \beta_3 cp$$
$$+\beta_4(lheight \times lb) + \beta_5(lheight \times cp)$$

where *lheight* represents log height, *lb* represents an indicator variable for the second species, and *cp* represents an indicator variable for the third species. The sample size was 38.

(a) How many degrees of freedom are there in the estimate of σ?

(b) What is the *p*-value for the test of the hypothesis that the slope in the regression of log force on log height is the same for species 2 as it is for species 1?

(c) What is a 95% confidence interval for the amount by which the slope for species 3 exceeds the slope for species 1?

Display 10.18   Least Squares Output Data for Exercise 9

Variable	Estimate	SE	t-Stat	p-value
CONSTANT	0.5191	1.0000	0.5191	0.6073
lheight	0.4083	0.4868	0.8387	0.4079
lb	−4.2992	1.5283	2.8131	0.0083
cp	−2.2992	1.7606	1.4123	0.1675
lheight × lb	2.5653	0.7354	3.4885	0.0014
lheight × cp	1.6601	0.7889	2.1043	0.0433

---

10. Crab Claws. The sum of squared residuals from the fit described in Exercise 9 is 5.99713, based on 32 degrees of freedom. The sum of squared residuals from the fit without the last two terms is 8.38155, based on 34 degrees of freedom. Form an *F*-statistic and find the *p*-value for the test that the slopes are the same for the three species.

11. Butterfly Occurrences. Display 10.19 summarizes results from the regression of the log of the number of butterfly species observed on the log of the size of the reserve and the number of days of observations, from 16 reserves in the Amazon River Basin.

(a) What is the two-sided *p*-value for the test of whether size of reserve has any effect on number of species, after accounting for the days of observation? What is the one-sided *p*-value if the alternative is that size has a positive effect? Does this imply that there is no evidence that the median number of species is related to reserve size? The researchers tended to spend more days searching for butterflies in the larger reserves. How might this affect the interpretation of the results?

(b) What is a two-sided *p*-value for the test that the coefficient of *lsize* is 1? (This is simply a computational exercise; there is no obvious reason to conduct this test with these data.)

(c) What is a 95% confidence interval for the coefficient of *lsize*?

(d) What proportion of the variation in log number of species remains unexplained by log size and days of observations?

Display 10.19   Regression Output Data for Exercise 11

Variable	Estimate	SE	t-Stat	p-value
Constant	3.775	0.3881	9.7321	<0.0000
lsize	0.0809	0.1131	0.7139	0.2443
days	0.0774	0.1447	0.5346	0.5104

Estimated SD about the regression is 0.8234 on 13 degrees of freedom; $R^2 = 11.41\%$.

---

12. **Brain Weights.** With the data described in Section 9.1.2 (page 227), construct an extra sum of squares F-test for determining whether gestation period and litter size are associated with brain weight after body weight is accounted for.

13. **Bat Echolocation.** (a) Fit the parallel regression lines model to duplicate the results in Display 10.6. (b) From these results, what are the estimated intercept and estimated slope for the regression of log energy on log mass for (i) non-echolocating bats, (ii) non-echolocating birds, and (iii) echolocating bats? (c) Refit the model using, instead, the indicator variables *bird* and *nbat*, where *nbat* takes on the value 1 for species of non-echolocating bats and 0 for other species. (d) Based on the results in part (c), what are the estimated intercept and estimated slope for the regression of log energy on log mass for (i) non-echolocating bats, (ii) non-echolocating birds, and (iii) echolocating bats? How do these compare to the estimates obtained in part (b)? (e) With the results of (c), test whether the lines for the echolocating bats and the non-echolocating birds coincide.

14. **Toxic Effects of Copper and Zinc.** In a study of the joint toxicity of copper and zinc, researchers randomly allocated 25 beakers containing minnow larvae to receive one of 25 treatment combinations. The treatment levels were all combinations of 5 levels of zinc and 5 levels of copper added to a beaker. Following a four-day exposure, a sample of the minnow larvae were homogenized and analyzed for protein. The results are shown in Display 10.20. (Data from D. A. J. Ryan, J. J. Hubert, J. B. Sprague, and J. Parrott, "A Reduced-Rank Multivariate Regression Approach to Aquatic Joint Toxicity Experiments," *Biometrics* 48 (1992): 155–62.) Fit a full second-order model for the regression of protein on copper and zinc, and examine the plot of residuals versus fitted values. Repeat after taking the log of protein. Which model is preferable?

15. **Old Faithful.** Reconsider the Old Faithful eruption durations and intervals in Display 7.16. Fit the regression of Interval on Duration and Day treated as a factor (include seven indicator variables to distinguish the eight days). Obtain the analysis of variance table, and then fit the regression of interval on duration alone to obtain the analysis of variance table for this reduced model. Use the quantities listed in the tables to construct an F-statistic for the test of whether any difference in mean intervals is due to the particular day of recording. Find the p-value (or, at least, say whether the p-value is bigger than .05, between .05 and .01, or less than .01).

16. **Galileo's Data.** Use Galileo's data in Display 10.1 to perform the following operations.

(a) Fit the regression of distance on height and height-squared. Obtain the estimates, their standard errors, the estimate of σ^2, and the variance–covariance matrix of the estimated coefficients.

Display 10.20   Protein in minnow larvae exposed to copper and zinc

Copper (ppm)	Zinc (ppm)	Protein(μg/larva)	Copper (ppm)	Zinc (ppm)	Protein(μg/larva)
0	0	201	112.5	0	188
0	375	186	112.5	375	172
0	750	173	112.5	750	157
0	1125	110	112.5	1125	115
0	1500	115	112.5	1500	108
37.5	0	202	150	0	133
37.5	375	161	150	375	125
37.5	750	172	150	750	184
37.5	1125	138	150	1125	135
37.5	1500	133	150	1500	114
75	0	204			
75	375	165			
75	750	148			
75	1125	143			
75	1500	123			

(b)   Verify that the square roots of the diagonal elements are equal to the standard errors reported with the estimated coefficients.

(c)   Compute the estimated mean distance when the initial height is 500 punti.

(d)   Calculate the standard error for the estimated mean in part (c).

(e)   Use the answer to parts (a) and (d) and the relationship between the variance of the estimated mean and the variance of prediction to obtain the standard error of prediction at an initial height of 500 punti.

17.   Galileo's Data. Use Galileo's data in Display 10.1 to fit the regression of distance on (a) height; (b) height and height^2; (c) height, height^2, and height^3; (d) height, height^2, height^3, and height^4; (e) height, height^2, height^3, height^4, and height^5; (f) height, height^2, height^3, height^4, height^5, and height^6. For each part, find R^2 and R^2_{adj}.

18.   Corn Yield and Rainfall. Reconsider the corn yield and rainfall data (Display 9.17 on page 252). Fit the regression of yield on rainfall, rainfall-squared, and year. Use the approach of Section 10.4.4 to find the rainfall that maximizes mean yield.

19.   Meadowfoam. Carry out a *lack-of-fit* F-test for the regression of number of flowers on light intensity and an indicator variable for time, using the data in Display 9.2 (page 227): (a) Fit the regression of *flowers* on *light* and an indicator variable for *time* = 24, and obtain the analysis of variance table. (b) Fit the same regression except with *light* treated as a factor (using 5 indicator variables to distinguish the 6 groups), and with the interaction of these two factors, and obtain the analysis of variance table. (c) Perform an extra-sum-of-squares F-test comparing the full model in part (b) to the reduced model in part (a). (*Note:* The full model contains 12 parameters, which is equivalent to the model in which a separate mean exists for each of the 12 groups. No pattern is implied in this model. See Displays 9.7 and 9.8 for help.)

20.   Calculus Problem. The least squares problem in multiple linear regression is to find the parameter values that minimize the sum of squared differences between responses and fitted values,

$$\text{SS}(\beta_0, \beta_1, \ldots, \beta_p) - \sum_{i=1}^{n}(Y_i - \beta_0 - \beta_1 X_{1i} - \cdots - \beta_p X_{pi})^2.$$

Set the partial derivatives of SS with respect to each of the unknowns equal to zero. Show that the solutions must satisfy the set of *normal equations*, as follows:

$$\beta_0 n + \beta_1 \Sigma X_{1i} + \beta_2 \Sigma X_{2i} + \cdots + \beta_p \Sigma X_{pi} = \Sigma Y_i$$
$$\beta_0 \Sigma X_{1i} + \beta_1 \Sigma X_{1i}^2 + \beta_2 \Sigma X_{1i} X_{2i} + \cdots + \beta_p \Sigma X_{1i} X_{pi} = \Sigma X_{1i} Y_i$$
$$\beta_0 \Sigma X_{2i} + \beta_1 \Sigma X_{2i} X_{1i} + \beta_2 \Sigma X_{2i}^2 + \cdots + \beta_p \Sigma X_{2i} X_{pi} = \Sigma X_{2i} Y_i$$

$$\vdots \qquad\qquad \vdots \qquad\qquad \vdots \qquad\qquad \vdots \qquad\qquad \vdots$$

$$\beta_0 \Sigma X_{pi} + \beta_1 \Sigma X_{pi} X_{1i} + \beta_2 \Sigma X_{pi} X_{2i} + \cdots + \beta_p \Sigma X_{pi}^2 = \Sigma X_{pi} Y_i,$$

where each Σ indicates summation over all cases ($i = 1, 2, \ldots, n$). Show, too, that solutions to the normal equations minimize SS.

21.   Matrix Algebra Problem. Let $\mathbf{Y}$ be the $n \times 1$ column vector containing the responses, let $\mathbf{X}$ be the $n \times (p + 1)$ array whose first column consists entirely of ones and whose other columns are the explanatory variable values, and let $\mathbf{b}$ be the $(p + 1) \times 1$ column containing the resulting parameter estimates. Show that the normal equations in Exercise 20 can be written in the form

$$(\mathbf{X}^T \mathbf{X})\mathbf{b} = \mathbf{X}^T \mathbf{Y}.$$

Therefore, as long as the matrix inversion is possible, the least squares solution is

$$\mathbf{b} = (\mathbf{X}^T \mathbf{X})^{-1} \mathbf{X}^T \mathbf{Y}.$$

When is the inversion possible?

22.   Continuing Exercise 21, statistical theory says that the means of the estimates in the vector $\mathbf{AY}$, where $\mathbf{A}$ is a matrix, are the elements of the vector $\mathbf{A}\mu\{\mathbf{Y}\}$; and the matrix of covariances of these estimates is $\mathbf{A}\mathrm{Cov}(\mathbf{Y})\mathbf{A}^T$. Use the theory and the model $\mu\{\mathbf{Y}\} = \mathbf{X}\beta$, $\mathrm{Cov}(\mathbf{Y}) = \sigma^2 \mathbf{I}$ (where $\mathbf{I}$ is an $n \times n$ identity matrix) to show that the mean in the sampling distributions of the least squares estimate $\mathbf{b}$ is β. Then show that the matrix of covariances is $\mathrm{Cov}\{\mathbf{b}\} = \sigma^2 (\mathbf{X}^T \mathbf{X})^{-1}$.

Data Problems

23.   Thinning of Ozone Layer. Thinning of the protective layer of ozone surrounding the earth may have catastrophic consequences. A team of University of California scientists estimated that increased solar radiation through the hole in the ozone layer over Antarctica altered processes to such an extent that primary production of phytoplankton was reduced 6 to 12%.

   Depletion of the ozone layer allows the most damaging ultraviolet radiation—UVB (280–320 nm)—to reach the earth's surface. An important consequence is the degree to which oceanic phytoplankton production is inhibited by exposure to UVB, both near the ocean surface (where the effect should be slight) and below the surface (where the effect could be considerable).

   To measure this relationship, the researchers sampled from the ocean column at various depths at 17 locations around Antarctica during the austral spring of 1990. To account for shifting of the ozone hole's positioning, they constructed a measure of UVB exposure integrated over exposure time. The exposure measurements and the percentages of inhibition of normal phytoplankton production were extracted from their graph to produce Display 10.21. (Data from R. C. Smith, et al., "Ozone Depletion: Ultraviolet Radiation and Phytoplankton Biology in Antarctic Waters,"

Science 255 (1992): 952–57.) Does the effect of UVB exposure on the distribution of percentage inhibition differ at the surface and in the deep? How much difference is there? Analyze the data, and write a summary of statistical findings and a section of details documenting those findings.

Display 10.21 Exposure to ultraviolet B radiation and percentage inhibition of primary phytoplankton production in Antarctic water

Location	Percent Inhibition	UVB Exposure	Surface (S) or Deep (D)
1	0.0	.0000	D
2	1.0	.0000	D
3	6.0	.0100	D
4	7.0	.0150	S
5	7.0	.0185	S
6	7.0	.0335	S
7	9.0	.0435	S
8	9.5	.0090	D
9	10.0	.0025	D
10	11.0	.0255	S
11	12.5	.0280	S
12	14.0	.0055	D
13	20.0	.0285	D
14	21.0	.0435	S
15	25.0	.0180	D
16	39.0	.0325	D
17	59.0	.0300	D

---

24. Factors Affecting Extinction. The data in Display 10.22 are measurements on breeding pairs of land-bird species collected from 16 islands around Britain over the course of several decades. For each species, the data set contains an average time of extinction on those islands where it appeared (this is actually the reciprocal of the average of $1/T$, where T is the length of time the species remained on the island, and $1/T$ is taken to be zero if the species did not become extinct on the island); the average number of nesting pairs (the average, over all islands where the birds appeared, of the number of nesting pairs per year); the size of the species (categorized as large or small); and the migratory status of the species (migrant or resident). (Data from S. L. Pimm, H. L. Jones, and J. Diamond, "On the Risk of Extinction," *American Naturalist* 132 (1988): 757–85.) It is expected that species with larger numbers of nesting pairs will tend to remain longer before becoming extinct. Of interest is whether, after accounting for number of nesting pairs, size or migratory status has any effect. There is also some interest in whether the effect of size differs depending on the number of nesting pairs. If any species have unusually small or large extinction times compared to other species with similar values of the explanatory variables, it would be useful to point them out. Analyze the data. Write a summary of statistical findings and a section of details documenting the findings.

Display 10.22   Bird Extinction Data

Species	Ave. Extinction Time (years)	Ave. Number of Nesting Pairs	Size (Large or Small)	Migratory Status (Resident or Migrant)
Sparrowhawk	3.030	1.000	L	R
Buzzard	5.464	2.000	L	R
Kestrel	4.098	1.210	L	R
Peregrine	1.681	1.125	L	R
Grey partridge	8.850	5.167	L	R
Quail	1.493	1.000	L	M
Red-legged partridge	7.692	2.750	L	R
Pheasant	3.846	5.630	L	R
Water rail	16.667	3.000	L	R
Corncrake	4.219	4.670	L	M
Moorhen	8.130	4.056	L	R
Coot	5.000	1.000	L	R
Lapwing	7.299	6.960	L	M
Golden plover	1.000	1.670	L	M
Ringed plover	27.027	5.560	L	R
Curlew	3.106	2.830	L	M
Redshank	4.000	4.375	L	M
Snipe	16.129	4.125	L	M
Stock dove	3.484	3.670	L	R
Rock dove	37.037	8.330	L	R
Wood pigeon	7.299	2.750	L	R
Cuckoo	2.525	1.430	L	M
Short-eared owl	4.132	2.000	L	R
Little owl	2.000	2.750	L	R
Magpie	10.000	4.500	L	R
Jackdaw	2.667	7.120	L	R
Carrion crow	4.587	4.580	L	R
Raven	58.824	2.350	L	R
Skylark	32.258	6.870	S	R
Swallow	2.571	3.830	S	M
House martin	2.160	5.000	S	M
Yellow wagtail	1.000	1.250	S	M
Pied wagtail	2.967	2.270	S	R
Meadow pipit	9.524	5.350	S	R
Wren	11.111	8.700	S	R
Dunnock	7.299	6.100	S	R
Robin	4.000	3.330	S	R
Stonechat	2.381	3.640	S	R
Wheatear	2.611	4.830	S	M
Blackbird	3.257	4.670	S	R
Song thrush	1.701	1.700	S	R
Mistle thrush	1.795	1.330	S	R
Grasshopper warbler	1.198	1.000	S	M
Sedge warbler	3.185	1.900	S	M
Whitethroat	2.273	4.420	S	M
Willow warbler	1.111	1.250	S	M
Chiffchaff	1.000	1.000	S	M
Goldcrest	1.000	1.000	S	R
Spotted flycatcher	1.230	1.000	S	M
Great tit	6.061	2.500	S	R
Blue tit	3.175	1.500	S	R
Yellowhammer	2.000	2.500	S	R
Reed bunting	5.076	5.630	S	R
Chaffinch	1.934	2.370	S	R
Goldfinch	1.493	1.500	S	R
Redpoll	1.000	1.000	S	R
Linnet	5.102	6.500	S	R
House sparrow	3.003	4.500	S	R
Tree sparrow	1.898	2.170	S	R
Starling	41.667	11.620	S	R
Pied flycatcher	1.000	1.000	S	M
Siskin	1.000	1.000	S	R

Answers to Conceptual Exercises

1.   The initial heights were controlled by Galileo. Distance is the only random quantity.

2.   Yes. The extra-sum-of-squares F-test, which compares the model with all three explanatory variables to the model with *lbody* only, addresses precisely this issue.

3.   (a) The test where β_2 and β_3 both equal zero can be cast in terms of comparing the full model $(\beta_0 + \beta_1 lbody + \beta_2 lgest + \beta_3 llitter)$ to the reduced model $(\beta_0 + \beta_1 lbody)$. The t-test where β_2 is zero, on the other hand, implies a reduced model of $\beta_0 + \beta_1 lbody + \beta_3 llitter$; and the t-test where β_3 is

zero implies a reduced model of $\beta_0 + \beta_1 lbody + \beta_2 lgest$. Neither of these reduced models is the same as the one sought, nor can the results from them be combined in any way to give some answer. (b) Same reason. The t-tests consider models where only one parameter is zero, but the model with both β_1 and β_2 equal to zero does not enter the picture. Multiple comparison adjustment does nothing to resolve the fact that these tests are different.

4. The model has nearly as many free parameters (8) as it has observations (9). One should expect a good fit to the data at hand, even if the explanatory variables have little relationship to the response.

5. (a) Each of the models represents the relationship between *lenergy* and *lmass* as a straight line, within the groups. The first model says that the intercept and slope—and hence the full line—is the same in all groups. The second model says that the slope is the same in each model while the intercepts are different. The third model allows the slopes and the intercepts to differ among all groups. (b) The second model is a reduced model in a test for equal slopes (with possibly differing intercepts); it is the full model for a test of equal intercepts (given equal slopes).

6. No bats of comparable size could be found in both groups. A more correct wording here might be "after adjustment for body size," but the underlying difficulty is not avoided.

7. (d)

8. The reduced model takes the null hypothesis to be true. The full model, however, encompasses both the null hypothesis and the alternative hypothesis. Thus, the full model is also correct when the reduced model is correct. Put another way, the full model is thought to be adequate from the start; the reduced model is obtained by imposing the constraints of the null hypothesis on the full model.

Model Checking and Refinement

Multiple regression analysis takes time and care. Dead ends in the pursuit of models are expected and common, especially when many explanatory variables are involved. Frustration and wasted effort can be avoided, however, by going about the analysis in a proper order. In particular, transformations and outliers must be dealt with early on. Although each analysis will be guided by the peculiarities of the particular data and the questions of interest, initial assessment and graphical analysis are usually followed by the fitting of a rich, tentative model and by examination of residual plots. This part of the analysis should suggest whether more investigation into transformation or contemplation of outliers is needed. Some special tools for the latter are provided in this chapter. Once the data analyst has checked that the inferential tools are valid and not seriously influenced by one or two observations, the structure of the model itself can be refined by testing terms to see which should be included.

11.1 Case Studies

11.1.1 Alcohol Metabolism in Men and Women—An Observational Study

Women tend to exhibit a lower tolerance for alcohol and tend to develop alcohol-related liver disease more readily than men. When men and women of the same size and drinking history consume equal amounts of alcohol, the women on average carry a higher concentration of alcohol in their bloodstream. According to a team of Italian researchers, this occurs because alcohol-degrading enzymes in the stomach (where alcohol is partially metabolized before it enters the bloodstream and is eventually metabolized by the liver) are more active in men than in women. The researchers studied the extent to which the activity of the enzyme explained the first-pass alcohol metabolism and the extent to which it explained the differences in first-pass metabolism between women and men. Their data (read from a graph) are listed in Display 11.1. (Data from M. Frezza et al., "High Blood Alcohol Levels in Women," *New England Journal of Medicine* 322 (1990): 95–99.)

Display 11.1   First-pass metabolism of alcohol in the stomach (mmol/liter-hour) and gastric alcohol dehydrogenase activity in the stomach (μmol/min/g of tissue) for 18 women and 14 men

Subject	Metabolism	Gastric Activity	Female (1 = F, 0 = M)	Alcoholic (1 = A, 0 = N)	Subject	Metabolism	Gastric Activity	Female (1 = F, 0 = M)	Alcoholic (1 = A, 0 = N)
1	0.6	1.0	1	1	17	2.5	3.0	1	0
2	0.6	1.6	1	1	18	2.9	2.2	1	0
3	1.5	1.5	1	1	19	1.5	1.3	0	1
4	0.4	2.2	1	0	20	1.9	1.2	0	1
5	0.1	1.1	1	0	21	2.7	1.4	0	1
6	0.2	1.2	1	0	22	3.0	1.3	0	1
7	0.3	0.9	1	0	23	3.7	2.7	0	1
8	0.3	0.8	1	0	24	0.3	1.1	0	0
9	0.4	1.5	1	0	25	2.5	2.3	0	0
10	1.0	0.9	1	0	26	2.7	2.7	0	0
11	1.1	1.6	1	0	27	3.0	1.4	0	0
12	1.2	1.7	1	0	28	4.0	2.2	0	0
13	1.3	1.7	1	0	29	4.5	2.0	0	0
14	1.6	2.2	1	0	30	6.1	2.8	0	0
15	1.8	0.8	1	0	31	9.5	5.2	0	0
16	2.0	2.0	1	0	32	12.3	4.1	0	0

The subjects were 18 women and 14 men, all volunteers living in Trieste. Three of the women and five of the men were categorized as alcoholic. All subjects received ethanol, at a dose of 0.3 grams per kilogram of body weight, orally one day and intravenously another, in randomly determined order. Since the intravenous administration bypasses the stomach, the difference in blood alcohol concentration—the concentration after intravenous administration minus the concentration after oral administration—provides a measure of the "first-pass metabolism" in the stomach. In addition, gastric alcohol dehydrogenase (AD) activity (activity of the key enzyme) was measured in mucus samples taken from the stomach linings. The data are plotted in Display 11.2.

Display 11.2 First-pass metabolism and gastric alcohol dehydrogenase activity in alcoholic and nonalcoholic men and women

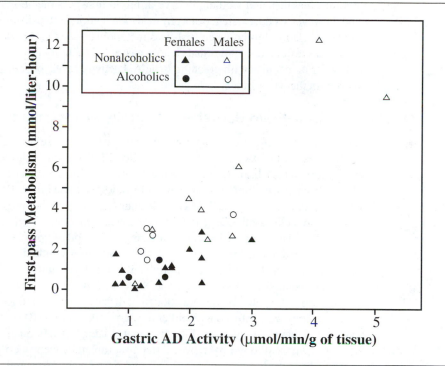

Several questions arise. Do levels of first-pass metabolism differ between men and women? Can the differences be explained by postulating that men have more dehydrogenase activity in their stomachs? Are the answers to these questions complicated by an alcoholism effect?

Summary of Statistical Findings

There was no evidence from these data that alcoholism was related to first-pass metabolism in any way (p-value $= .93$, from $F = 0.209$ with 4 and 22 d.f.). Convincing evidence exists that first-pass metabolism was larger for males than for females overall (two-sided p-value $= .0002$, from a rank-sum test) and that gastric AD activity was larger for males than for females (two-sided p-value $= .07$ from a rank-sum test). Although gastric AD activity (in the range of .8 to 3 μmol/min/g of tissue) explained 68% of the variation in first-pass metabolism, convincing evidence indicates that males had higher first-pass metabolism than females even after accounting for differences in gastric AD activity (two-sided *p*-value = .0003 from a *t*-test for equality of male and female slopes when both intercepts are zero). For a given level of gastric dehydrogenase activity, the mean first-pass alcohol metabolism for men is estimated to be 2.20 times as large as the mean first-pass alcohol metabolism for women.

Scope of Inference

Because the subjects were volunteers, no inference to a larger population is justified. The inference that men and women do have different first-pass metabolism is greatly strengthened, however, by the existence of a physical explanation for the difference. The conclusions about the relationship between first-pass metabolism, gastric AD dehydrogenase activity, and sex are restricted to individuals whose gastric AD activity is less than 3. The sparseness of data for individuals with greater gastric AD activity levels prevents any resolution of the answers in the wider range.

11.1.2 The Blood–Brain Barrier—A Controlled Experiment

The brain is protected from bacteria and toxins, which course through the bloodstream, by a system called the *blood–brain barrier*. Blood flowing through the brain's capillaries is sealed from outside brain tissue by a single layer of cells. This barrier normally allows only a few substances, including some medications, to reach the brain. Because chemicals used to treat brain cancer have such large molecular size, they cannot pass through the barrier to attack tumor cells. At the Oregon Health Sciences University, Dr. E. A. Neuwelt developed a method of disrupting the barrier by infusing a solution of concentrated sugars.

As a test of the disruption mechanism, researchers conducted a study on rats. (Data from P. Barnett et al., "Differential Permeability and Quantitative MR Imaging of a Human Lung Carcinoma Brain Xenograft in the Nude Rat," *American Journal of Pathology* 146(2) (1995): 436–49.) The rats were inoculated with cancer cells to induce brain tumors. After 9 to 11 days they were infused with either the barrier disruption (BD) solution or, as a control, a normal saline (NS) solution. Fifteen minutes later, the rats received a standard dose of the therapeutic antibody L6-F(ab′)$_2$. After a set time they were sacrificed, and the amounts of antibody in the brain tumor and in normal tissue were measured. The time line for the experiment is shown in Display 11.3. Measurements for the 34 rats are listed in Display 11.4.

Display 11.3 Time line for blood–brain barrier disruption experiment

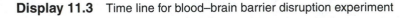
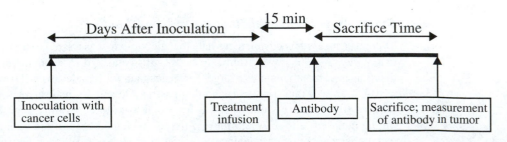

Display 11.4   Response variable, design variables, and several covariates for 34 rats in the blood–brain barrier disruption experiment

	Response Variable	*Design Variables*			*Covariates*			
Case	Brain Tumor Count (per gm) / Liver Count (per gm)	Sacrifice Time (hours)	Treatment	Days Post Inoculation	Sex	Tumor Weight (10^{-4} grams)	Weight Loss (grams)	Initial Weight (grams)
1	41081 / 1456164	0.5	BD	10	F	239	5.9	221
2	44286 / 1602171	0.5	BD	10	F	225	4.0	246
3	102926 / 1601936	0.5	BD	10	F	224	−4.9	61
4	25927 / 1776411	0.5	BD	10	F	184	9.8	168
5	42643 / 1351184	0.5	BD	10	F	250	6.0	164
6	31342 / 1790863	0.5	NS	10	F	196	7.7	260
7	22815 / 1633386	0.5	NS	10	F	200	0.5	27
8	16629 / 1618757	0.5	NS	10	F	273	4.0	308
9	22315 / 1567602	0.5	NS	10	F	216	2.8	93
10	77961 / 1060057	3	BD	10	F	267	2.6	73
11	73178 / 715581	3	BD	10	F	263	1.1	25
12	76167 / 620145	3	BD	10	F	228	0.0	133
13	123730 / 1068423	3	BD	9	F	261	3.4	203
14	25569 / 721436	3	NS	9	F	253	5.9	159
15	33803 / 1019352	3	NS	10	F	234	0.1	264
16	24512 / 667785	3	NS	10	F	238	0.8	34
17	50545 / 961097	3	NS	9	F	230	7.0	146
18	50690 / 1220677	3	NS	10	F	207	1.5	212
19	84616 / 48815	24	BD	10	F	254	3.9	155
20	55153 / 16885	24	BD	10	M	256	−4.7	190
21	48829 / 22395	24	BD	10	M	247	−2.8	101
22	89454 / 83504	24	BD	11	F	198	4.2	214
23	37928 / 20323	24	NS	10	F	237	2.5	224
24	12816 / 15985	24	NS	10	M	293	3.1	151
25	23734 / 25895	24	NS	10	M	288	9.7	285
26	31097 / 33224	24	NS	11	F	236	5.9	380
27	35395 / 4142	72	BD	11	F	251	4.1	39
28	18270 / 2364	72	BD	10	F	223	4.0	153
29	5625 / 1979	72	BD	10	M	298	12.8	164
30	7497 / 1659	72	BD	10	M	260	7.3	364
31	6250 / 928	72	NS	10	M	272	11.0	484
32	11519 / 2423	72	NS	11	F	226	2.2	168
33	3184 / 1608	72	NS	10	M	249	−4.4	191
34	1334 / 3242	72	NS	10	F	240	6.7	159

Since the amount of the antibody in normal tissue indicates how much of it the rat actually received, a key measure of the effectiveness of transmission across the blood–brain barrier is the ratio of the antibody concentration in the brain tumor to the antibody concentration in normal tissue outside of the brain. The brain tumor concentration divided by the liver concentration is a measure of the amount of the antibody that reached the brain relative to the amount of it that reached other parts of the body. This is the response variable; both the numerator and denominator of this ratio are listed in Display 11.4. The explanatory variables in the table comprise two categories: *design variables* are those that describe manipulation by the researcher; *covariates* are those measuring characteristics of the subjects that are not controllable by the researcher.

Was the antibody concentration in the tumor increased by the use of the blood–brain barrier disruption infusion? If so, by how much? Do the answers to these two questions depend on the length of time after the infusion (from 1/2 to 72 hours)? What is the effect of treatment on antibody concentration after weight loss, total tumor weight, and other covariates are accounted for? A coded scatterplot relating to the major questions is shown in Display 11.5.

Display 11.5 Log–log scatterplot of the ratio of antibody concentration in brain tumor to antibody concentration in liver versus sacrifice time, for 17 rats given the barrier disruption infusion and for 17 rats given a saline (control) infusion

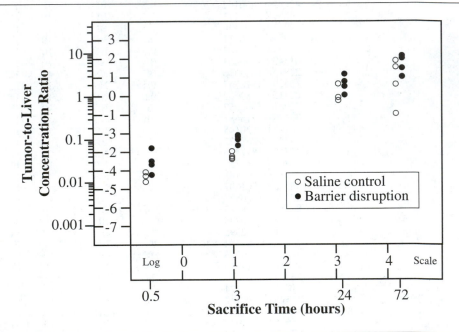

Summary of Statistical Findings

The median antibody concentration in the tumor (relative to that in the liver) was estimated to be 2.22 times as much for rats receiving the barrier disruption infusion than for those receiving the control infusion (95% confidence interval, from 1.56 to 3.15 times as much). This multiplicative effect appears to be constant between 1/2 and 72 hours after the infusion (the two-sided p-value for a test of interaction between treatment and sacrifice time is .92, from an F-test on 3 and 26 degrees of freedom).

Scope of Inference

One hitch in this study is that randomization was not used to assign rats to treatment groups. This oversight raises the possibility that the estimated relationships might be related to confounding variables over which the experimenter exercised no control. Including the measured covariates in the model helps alleviate some concern, and the results appear not to have been affected by these potential confounding variables. Nevertheless, causal implications can only be justified on the tenuous assumption that the assignment method used was as effect-neutral as a random assignment would have been.

11.2   Residual Plots

Faced with analyzing data sets like those involved in the blood–brain barrier and alcohol metabolism studies, a researcher must seek good-fitting models for answering the questions of interest, bearing in mind the model assumptions required for least squares tools, the robustness of the tools against violations of the assumptions, and the sensitivity of these tools to outliers. Since model-building efforts are wasted if the analyst fails to detect problems with nonconstant variance and outliers early on, it is wise to postpone detailed model fitting until after outliers and transformation have been thoroughly considered.

Much can be resolved from initial scatterplots and inspection of the data, but it is almost always worthwhile to obtain the finer picture provided by a residual plot. Creating this plot involves fitting some model in order to get residuals. On the basis of the scatterplots, the analyst can choose some tentative model or models and conduct residual analysis on these, recognizing that further modeling will follow.

Selecting a Tentative Model

A tentative model is selected with three general objectives in mind: the model should contain parameters whose values answer the questions of interest in a straightforward manner; it should include potentially confounding variables; and it should include features that capture important relationships found in the initial graphical analysis.

It is disadvantageous to start with either too many or too few explanatory variable terms in the tentative model. With too few, outliers may be induced simply because of omitted relationships, obscuring the information in a residual plot. With too many (lots of interactions and quadratic terms, for example), the analyst risks overfitting the data—causing real outliers to be explained away by complex, but meaningless, structural relationships. Overfitting becomes less of a problem when the sample sizes are substantially larger than the number of model parameters.

For large sample sizes, therefore, the initial tentative model for residual analysis can err on the side of being rich, including potential model terms that may not be retained in the end. For small sample sizes, several tentative models may be needed for residual analysis; and the data analyst must guard against including terms whose significance hinges on one or two observations. As evident in the strategy for data analysis laid out in Display 9.9, the process of trying a model and plotting residuals is often repeated until a suitable inferential model is determined.

Example—Preliminary Steps in the Analysis of the Blood–Brain Barrier Data

The coded scatterplot in Display 11.5 is a good starting point for the analysis. Apparently, the disruption solution does allow more antibody to reach the brain than the control solution does; this effect is about the same for all "sacrifice times" (actually time between antibody treatment and sacrifice); an increasing proportion of antibody reaches the brain with increasing time after infusion; and this increasing relationship appears to be slightly nonlinear. A matrix of scatterplots and a correlation matrix (an array showing the sample correlation coefficients for all possible pairs of variables), which are not shown here, indicate further that the covariates—days after inoculation, initial weight, and sex of the rat—are associated with the response. These covariates are also related to the treatment given. (Recall that randomization was not used.) In particular, rats treated at longer days after inoculation were also assigned to the longer sacrifice times. Furthermore, all male rats were assigned to the longer sacrifice times.

This initial investigation suggests the following tentative regression model (using the shorthand model specification of Section 9.3.5 on page 240):

$$\mu\{antibody \mid SAC, TREAT, DAYS, FEM, weight, loss, tumor\} =$$
$$SAC + TREAT + (SAC \times TREAT) + DAYS + FEM + weight + loss + tumor,$$

where *antibody* is the logarithm of the ratio of antibody in the brain tumor to that in the liver. *SAC* is the sacrifice time factor with four levels; *TREAT* is treatment, with two levels; *DAYS* is days after inoculation, with three levels; and *FEM* is sex, with two levels. *Weight*, *loss*, and *tumor* are the initial weight, weight loss, and tumor weight variables. Display 11.5 shows a strong linear effect of log sacrifice time on the response, but some additional curvature may be present as well. To avoid mismodeling the effect of sacrifice time at the start, it is treated as a factor with four levels. Similarly, the coded scatterplot suggests that the difference between the two treatments may be greater for the shorter sacrifice times than for the longer ones. Consequently, the sacrifice time by treatment interaction terms are included in the tentative model. Although more terms may be added to this model later, it captures the most prominent features of the scatterplot. Display 11.6 shows the plot of residuals versus the fitted values from the regression model (*Note:* Even if prior experience

---

Display 11.6 Scatterplot of residuals versus fitted values from the fit of the logged response on a rich model for explanatory variables—brain–barrier data

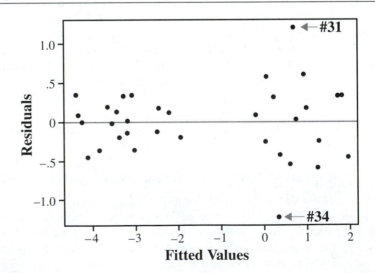

or initial inspection had not led the researchers to consider the logarithms of the response, the coded scatterplot and residual plot would have revealed that the variability increases with increasing response, leading them to the same consideration.)

The residual plot in Display 11.6 exemplifies the ambiguity that can arise with small data sets. Is there a funnel-shaped pattern, or is the apparent funnel only due to a few outliers? The usual course of action consists of three steps:

1. Examine the outliers for recording error or contamination.
2. Check whether a standard transformation resolves the problem.
3. If neither of these steps works, examine the outliers more carefully to see whether they influence the conclusions (following the strategy suggested in Section 11.3).

The residual plot is based on a response that has already been transformed into its logarithm. A reciprocal transformation corrects more pronounced funnel-shaped patterns than does the log. Here, however, it does not help, and there is no suggestion of a recording error. Consequently, the analyst must proceed with further model fitting, paying careful attention to the roles of observations 31 and 34.

Example—Preliminary Steps in the Analysis of the Alcohol Metabolism Data

Refer to the coded scatterplot in Display 11.2. The next step is to examine a residual plot for outliers and to assess the need for transformation. The plot in Display 11.7 is a residual

---

Display 11.7   Residual plot from the regression of first-pass metabolism on gastric activity, sex indicator, alcoholism indicator, and all second- and third-order interactions

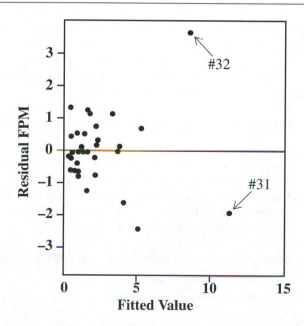

plot from the regression of first-pass metabolism on gastric AD activity (*gast*), an indicator variable for females (*fem*), an indicator for alcoholics (*alco*), and the interaction terms *gast* × *fem*, *fem* × *alco*, *gast* × *alco*, and *gast* × *fem* × *alco*. (The last term is a *three-factor interaction* term, formed as the product of three explanatory variables.)

The plot draws attention to two observations: one that has a considerably larger residual than the rest and one that has a fitted value quite a bit larger than the rest. These are cases 31 and 32, and they appear in the coded scatterplot of Display 11.2 in the upper right-hand corner, separated from the rest of the points. There appears to be a downward trend in the residual plot, excluding cases 31 and 32. This could reflect a model that is heavily influenced by one or two observations and consequently does not fit the bulk of the observations well.

11.3   A Strategy for Dealing with Influential Observations

Least squares regression analysis is not resistant to outliers. One or two observations can strongly influence the analysis, to the point where the answers to the questions of interest change when these isolated cases are excluded. Although any influential observation that comes from a population other than the one under investigation should be removed, removing an observation simply because it is influential is not justified. In any circumstance, it is unwise to state conclusions that hinge on one or two data points. Such a statistical study should be considered extremely fragile.

There are two approaches for dealing with excessively influential observations in regression analysis. One is to use a robust and resistant regression procedure. The other is to use least squares but to examine outliers and influence closely to see whether the suspect observations are indeed influential, why they are influential (this can help dictate the subsequent course of action), and whether they provide some interesting extra information about the process under study. Using a robust and resistant regression procedure is particularly useful if past experience indicates that the response distribution tends to have long tails and that outliers are an expected nuisance. When influential observations are discovered in the course of a regression analysis, however, *The Statistical Sleuth* recommends closer examination. This often clarifies the problems and sometimes reveals additional, unexpected information.

Assessment of Whether Observations Are Influential

The strategy for assessing influence involves temporarily removing suspected influential observations to see whether the answers to the questions of interest change. Does the evidence from a test change from slight evidence to convincing evidence? Does the decision to include a term in the model change? Does an important estimate change by a practically relevant amount? If not, the observation is not influential and the analysis can proceed as usual. If so, further action must be taken.

What to Do About Influential Observations

If an observation is influential and it substantially differs from the remaining data in its explanatory variable values, perhaps it is being accorded too much weight in fitting a model over a sparsely represented region of the explanatory variables. For example, given 30 observations for X between 0 and 5 and a single value of X at 15, it is unrealistic to try to model the regression of Y on X over the entire range of 0 to 15. The observations in the sparsely represented region can be removed, the model fit on the remaining data points, and conclusions stated only for the restricted range of explanatory variables. This restriction should be stated explicitly, with the added comment that more data would be needed to model the regression accurately over the broader range of explanatory variables.

If an influential observation is not particularly unusual in its explanatory variable values, and if no definitive explanation for its unique behavior can be found, omitting it cannot be justified. More data are needed to answer the questions of interest. As a last resort, the results can be reported with and without the influential observation.

A complete strategy is summarized in Display 11.8. Although this is not stated explicitly in the strategy, the analyst should first check any unusual observation for recording accuracy and for alternative explanations of its unusualness before proceeding to this statistical approach.

Display 11.8   A strategy for dealing with suspected influential cases

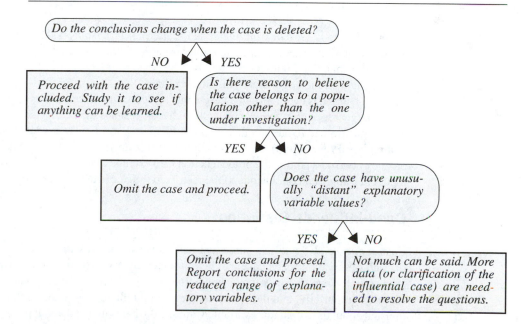

Example—Alcohol Metabolism Study

The tentative model is fit with and without the cases 31 and 32, to examine which aspects of the fit change and by how much. The resulting changes in estimates, standard errors, and p-values are shown in Display 11.9.

A striking consequence of the exclusion of cases 31 and 32 is the drop in significance of the interaction of gastric activity and sex from a p-value of .013 to one of .15. The reason

Display 11.9 Regression parameter estimates, standard errors, and *p*-values from the regression of first-pass metabolism on gastric activity, an indicator for female, an indicator for alcoholic, and all second- and third-order interactions: first with all observations and then with all observations except 31 and 32

Variable	All 32 Observations			Cases 31 and 32 Removed		
	Estimate	Standard Error	Two-sided *p*-value	Estimate	Standard Error	Two-sided *p*-value
Constant	−1.660	1.000	.11	−0.680	1.309	.61
Gastric activity (G)	2.514	0.343	<.0001	1.921	0.608	.0045
Female (F)	1.466	1.333	.28	0.486	1.467	.74
Alcoholic (A)	2.552	1.946	.20	1.572	1.812	.40
G×F	−1.673	0.620	.013	−1.081	0.721	.15
F×A	−2.252	4.394	.61	−1.272	3.467	.72
G×A	−1.459	1.053	.18	−0.866	0.963	.38
G×F×A	1.199	2.998	.69	0.606	2.316	.80

for this change is evident from the coded scatterplot in Display 11.2. Ignoring the effect of alcoholism, imagine the slope in the regression of first-pass metabolism on gastric AD activity for males and females separately. The slope for males is substantially greater with cases 31 and 32 included, than with them excluded.

What does this mean? Maybe the estimated slope with those cases is accurate and males have a larger slope than females; but alternatively, perhaps the relationship is not a straight line for gastric activity greater than 3. It is difficult to know how to model the relationship in that region. Proceeding from the guideline that it is unwise to state conclusions that hinge on one or two data points, the prudent action is to exclude cases 31 and 32, and to restrict the model building and conclusions to the restricted range of gastric AD activity less than 3.

11.4   Case-Influence Statistics

Case-influence statistics are numerical measures associated with the individual influence of each observation (each case). When provided by a statistical computer program, they are useful for two reasons: they can help identify influential observations that may not be revealed graphically; and they partition the overall influence of an observation into what is unusual about its explanatory variable values and what is unusual about its response relative to the fitted model. This partition may be useful in following the strategy for dealing with cases of suspect influence suggested in Display 11.8.

11.4.1   Leverages for Flagging Cases with Unusual Explanatory Variable Values

The *leverage* of a case is a measure of the distance between its explanatory variable values and the average of the explanatory variable values in the entire data set. As illustrated in the alcohol metabolism study, cases with high leverage may exert strong influence on the results of model fitting.

When only a single explanatory variable X is involved, the leverage of the ith case is

$$h_i = \frac{1}{(n-1)}\left[\frac{X_i - \overline{X}}{s_X}\right]^2 + \frac{1}{n} \quad \text{or} \quad \frac{(X_i - \overline{X})^2}{\sum(X - \overline{X})^2} + \frac{1}{n},$$

where s_X is the sample standard deviation of X. Aside from the parts involving n, the first formulation shows that the leverage is a distance of X_i from $\overline{X}$, in units of standard deviations; and the second formulation shows that the leverage is the proportion of the total sum of squares of the explanatory variable contributed by the ith case.

When two or more explanatory variables are involved, the leverage can only be expressed in matrix notation. The interpretation is a straightforward extension of the one-variable, however. The leverage for case i is a measure of the distance of case i from the average (in a multivariable sense). The one aspect of leverage in the multiple regression setting that is not an obvious extension of the preceding formulas is that the distance from the average is relative to the joint spread of the explanatory variables. Display 11.10 illustrates a problem in which one case is relatively far from the two-dimensional scatter of the other cases, although it is not particularly unusual in either dimension alone.

Display 11.10 An illustration of what is meant by "far from the average" of multiple explanatory variables when they are correlated

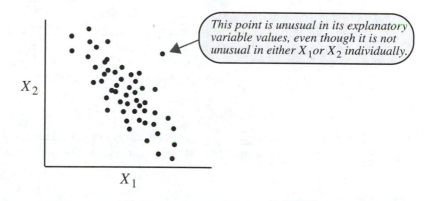

It is important to identify this case, since it stands alone. Because it would be largely responsible for dictating the location of the estimated regression surface in the surrounding region, the point has a high potential for influence. The multiple regression version of leverage would be large for such an observation.

Cases with Large Leverage

Leverages are greater than $1/n$ and less than 1 for each observation, and the average of all the h_i's in a data set is always p/n, where p is the number of regression coefficients. Leverages also depend on what model is being entertained. The standard deviation of the residual for the ith case is related to its leverage:

$$\text{SD}(\text{Residual}_i) = \sigma\sqrt{(1 - h_i)}.$$

A case with large leverage has a residual with low variability. Because its explanatory variable values are so unusual, it dictates the location of the estimated regression over the whole region in its vicinity; no other points in the region share the responsibility. Because its residual must be small, this case acts like a magnet on the estimated regression surface. If, however, its response falls close to the regression surface (as determined by the remaining observations alone), it is not necessarily influential. Therefore, while a large leverage does not necessarily indicate that the case is influential, it does imply that the case has a high *potential* for influence.

Cases with large leverage can easily be overlooked in scatterplots and other graphical displays. Display 11.10 shows how a case with high leverage can be visible in a two-dimensional scatterplot, even though it would not be exceptional in either of the component one-dimensional histograms. Similarly, a case with high leverage may be visible in a three-dimensional plot but not in any of the component two-dimensional scatterplots. Therefore, examining the numerical measures of leverage by themselves is important. The leverage measure h_i would identify the case in Display 11.10 because its distance takes into account the correlations among the variables.

It is difficult to say how large a value of h_i is sufficiently large to warrant further attention. Since the average of the h_i's is p/n, some statisticians (and statistical computer programs) use twice the average—$2p/n$—as a lower cutoff for flagging cases that have a high potential for excessive influence. This formulation is somewhat arbitrary. The main point is to use the leverage measure in conjunction with the other case influence statistics to get some overall assessment of influence. Display 11.11 shows one way to display the leverages in conjunction with other case influence statistics.

Leverage Computations

Most statistical computer programs will carry out the calculations for leverage as part of their regression routine. If not, the calculations may be made by using the formula

$$h_i = \left[\frac{\text{SE}(\text{fit}_i)}{\hat{\sigma}}\right]^2.$$

The values for $\text{SE}(\text{fit}_i)$ are often available. If not, they may be found with the method in Section 10.2.3.

11.4.2   Studentized Residuals for Flagging Outliers

A *studentized residual* is a residual divided by its estimated standard deviation. Some residuals naturally have less variation than others because of their leverages. Therefore, the usual residual plot may not direct attention to cases whose residuals lie much farther from zero than expected. The studentized residuals,

$$studres_i = \frac{res_i}{\hat{\sigma}\sqrt{1 - h_i}},$$

put all residuals on a common scale of number of standard deviations. Since roughly 95% of normally-distributed variables fall within two standard deviations of their mean, it is common to investigate, as possibly peculiar, observations whose studentized residuals are smaller than -2 or larger than 2. Of course, it is not unusual to find roughly 5% of observations outside this range, and 5% can be a sizeable number if the sample size is large. (Technically, this is called the *internally studentized residual*; see Note 1 at the end of Section 11.4.4.) As with leverages, studentized residual calculations are typically performed by the statistical computer program.

Example—Observation #31 from the First Pass Metabolism Study

The observed value for the 31st observation is 9.5 and its fitted value is 11.0024. This leaves the residual of $res_{31} = (9.5 - 11.0024) = -1.5024$. From computer calculations its leverage is $h_{31} = 0.5355$. Notice that the rough lower cutoff for flagging large influence, $2p/n$, is $2 \times 4/32 = 0.25$; according to that rule, case 31 has a high potential for influence. The estimated standard deviation of the residual is $1.20730(1 - 0.5355)^{1/2} = 0.8228$. The studentized residual is -1.8260. This is moderately far from zero, but not alarmingly so.

11.4.3   Cook's Distances for Flagging Influential Cases

Cook's Distance is a case statistic that measures *overall* influence—the effect that omitting a case has on the estimated regression coefficients. For case i, Cook's Distance can be represented as

$$D_i = \sum_{j=1}^{n} \frac{(\hat{Y}_{j(i)} - \hat{Y}_j)^2}{p\hat{\sigma}^2},$$

where $\hat{Y}_j$ is the jth fitted value in a fit using all the cases; $\hat{Y}_{j(i)}$ is the jth fitted value in a fit that excludes case i from the data set; p is the number of regression coefficients; and $\hat{\sigma}^2$ is the estimated variance from the fit, based on all observations. The numerator measures how much the fitted values (and, therefore, the parameter estimates) change when the ith case is deleted. The denominator scales this difference in a useful way. A case that is influential, in terms of the least squares estimated coefficients changing when it is deleted, will have a large value of Cook's Distance.

An equivalent expression for Cook's Distance is

$$D_i = \frac{1}{p}(studres_i)^2 \left(\frac{h_i}{1 - h_i} \right).$$

This alternate expression is useful for computing, since it does not require that the ith case actually be deleted. More importantly, it shows that an influential case—with a large

Cook's Distance—is influential because it has a large studentized residual, a large leverage, or both. Cook's Distance for case #31 is 0.9610. Some statisticians use a rough guideline that a value of D_i close to or larger than 1 indicates a large influence. Realize, however, that Cook's Distance measures the influence on all the regression coefficients. By identifying potentially problematic cases the data analyst can directly refit the model with and without the indicated observations to see whether the answers to the important questions of interest change. The effect of the case removal on a single coefficient of interest may be much more or much less dramatic than what is indicated by the general measure D_i. Nevertheless, Cook's Distance is very useful for calling attention to potentially problematic cases.

11.4.4 A Strategy for Using Case Influence Statistics

Inspecting graphical displays and scatterplots may alert the analyst to the need for case influence investigations. In addition, the presence of statistically significant, complex effects that make little sense may indicate problems with influence. But when the residual plot from fitting a good inferential model fails to suggest any problems, there is generally no need to examine case influence statistics at all.

The suggested strategy in Display 11.8 requires some assessment of influence, of the unusualness of a case's explanatory variable values, and of the degree to which a case is an outlier. Therefore, the trio of case influence measures—D_i, h_i, and studres$_i$ (or similar trios described later)—are examined jointly. Since values of these measures exist for every observation in the data set, the examination of a full list may be difficult. Some computer programs use rough guidelines (such as $D_i > 1$, $h_i > 2p/n$, or $|\text{studres}_i| > 2$) to flag the relevant cases. Ideally, a graphical display of the case statistics, like Display 11.11 for the alcohol metabolism data, makes case influence analysis convenient.

In the top plot, the values of Cook's Distance for cases 31 and 32 are obviously substantially larger than the rest, indicating what was previously observed: these two observations are influential. In addition, they both have high leverages, as one would expect from their gastric AD activity values' being so much larger than the rest. One other case—number 17—has a large leverage. This is the female with the largest gastric AD activity value. No evidence indicates that this observation is influential, however, so it need not be studied further.

Notes About Case Influence Statistics

1. *Externally studentized residuals.* A potential problem with the internally studentized residual is that the estimate of σ from the fit with all observations may be tainted by case i if i is indeed an outlier. The *externally studentized residual* is

$$\text{studres}_i^* = \frac{res_i}{\hat{\sigma}_{(i)}\sqrt{1 - h_i}},$$

Display 11.11 Case influence statistics for the fit of first-pass metabolism on gastric activity, sex indicator, and their interaction

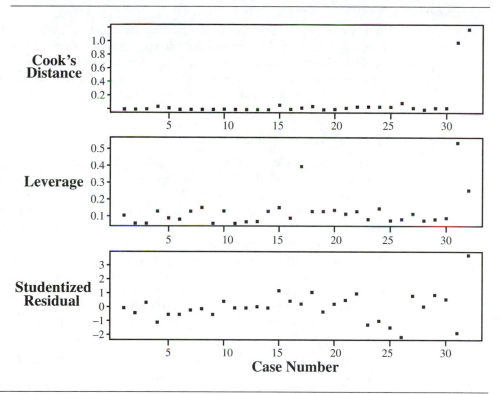

where $\hat{\sigma}_{(i)}$ is the estimate of the standard deviation about the regression line from the fit that excludes case i.

2. *External studentization in measures of influence.* If external studentization makes sense for residuals, it also makes sense for the measure of influence. Although it is not widely used, an externally studentized version of Cook's Distance, D_i^*, can be obtained by using $\hat{\sigma}_{(i)}$ in place of $\hat{\sigma}$ in the definition—or equivalently, by using $studres_i^*$ in place of $studres_i$ in the computing version. A measure that *is* widely used is DFFITS_i, which is $(pD_i^*)^{1/2}$. It measures influence in the same way as the externally studentized version of Cook's Distance, but on a slightly different scale. The decision about which of these measures to use is based largely on which is available in the statistical computer program.

3. *These case statistics address one-at-a-time influence.* If two or more cases are jointly influential and, in particular, if they are together in an unusual region of the explanatory variables, then Cook's Distance and the leverage measure may fail to detect their joint influence and joint leverage. The user must be on guard for this situation. It is always possible to remove a pair of observations to investigate their joint influence directly.

11.5 Refining the Model

11.5.1 Testing Terms

The key assumption for the correctness of the statistical conclusions is that the terms in the model sufficiently describe the regression of the response on the explanatory variables. It is misleading to leave out important explanatory variables. On the other hand it is also important to use Occam's razor (Section 10.4.5 on page 280) to trim away nonessential terms. Therefore, after transformations and outliers have been resolved, the analysis focuses on finding a simple, good-fitting model.

Example—Alcohol Metabolism Study

Once the decision is made to set cases 31 and 32 aside, the analyst must refit the model and examine the new residual plot. No additional problem is indicated, so model refinement may begin.

Since alcoholism is not of primary concern, and since so few alcoholics are included in the data set, it would be helpful if this variable could be ignored. One approach in the investigation is to test whether all the terms involving alcoholism in the tentative model can be dropped, using an extra-sum-of-squares F-test. There are four such terms: *alco*, *gast* × *alco*, *fem* × *alco*, and *gast* × *fem* × *alco*. The F-statistic is 0.21. By comparison to an F-distribution on 4 and 22 degrees of freedom, the p-value is .93, so the data are consistent with their being no alcoholism effect.

With alcoholism left out, the resulting model for consideration is

$$\mu\{metabolism \mid gast, fem\} = \beta_0 + \beta_1 gast + \beta_2 fem + \beta_3(gast \times fem),$$

the separate regression lines model. The estimates of regression parameters, their standard errors, and their p-values appear in Display 11.12.

Display 11.12   Least squares estimates for the regression of first-pass metabolism on gastric AD activity, sex, and their interaction (excluding cases 31 and 32)

Variable	Estimate	Standard Error	t-Statistic	Two-sided p-value
Constant	0.070	0.802	0.087	.93
gast	1.565	0.407	3.843	.0007
fem	−0.267	0.993	−0.269	.79
gast × *fem*	−0.728	0.539	0.114	.91

It may appear that sex plays no significant role, because the parameters β_2 (the amount by which the women's intercept exceeds the men's intercept) and β_3 (the amount by which the women's slope exceeds the men's slope), have nonsignificant p-values individually.

This is deceiving. A sex difference is distinctly noticeable on the scatterplot (Display 11.2). The *p*-values suggest, however, that there is no need for different slopes if different intercepts are included in the model, and no need for different intercepts if different slopes are included.

In Display 11.2, the intercepts for both men and women are near zero. Since, in addition, a zero intercept makes sense in this application (because there is no first-pass metabolism if there is no activity of the enzyme), it is useful to force both lines to go through the origin, by dropping the constant term and the female indicator variable:

$$\mu\{metabolism \mid gast, fem\} = \beta_1 gast + \beta_2(gast \times fem).$$

In this model, first-pass metabolism is directly proportional to gastric activity, but the constant of proportionality differs for men and for women. The *F*-statistic comparing this model to the one whose fit is summarized in Display 11.12 is 0.56 with 2 and 28 degrees of freedom, so the smaller model is adequate.

The fit to the new model appears in Display 11.13. In this model, both terms are essential, so this is accepted as the final version for inference. The conclusions stated in the summary of statistical findings are based on this fit. Notice in particular that, for any level of gastric AD activity (in the range of 0.8 to 3.0), the mean first-pass metabolism for males divided by the mean first-pass metabolism for females is $\beta_1/(\beta_1 + \beta_2)$, which is estimated as 2.20. Thus, the mean for males is estimated to be 2.20 times the mean for females, even after accounting for gastric dehydrogenase activity.

Display 11.13   Results for mean metabolism being proportional to gastric activity

Variable	Estimate	Standard Error	*t*-Statistic	Two-sided *p*-value
gast	1.599	0.125	12.800	<.0001
gast × *fem*	−0.873	0.174	−5.019	<.0001

11.5.2 Partial Residual Plots

A scatterplot exhibits only the marginal association of two variables, which may depend heavily on their mutual association with a third variable. The question of interest would be better addressed by a plot showing the association of the two variables, after getting the effect of the third variable out of the way.

The top scatterplot in Display 11.14 shows log brain weight versus log gestation length for the 96 species of mammals examined in the study discussed in Section 9.1.2. An important question of interest is whether an association exists between brain weight and gestation, after body weight is accounted for. The multiple regression coefficient addresses this directly, but a graphical display in conjunction with it would be helpful.

Display 11.14 Scatterplot of log brain weight versus log gestation length, and scatterplot of the partial residuals of log brain weight (adjusted for log body weight) versus log gestation length—mammal brain weight data from Section 9.1.2

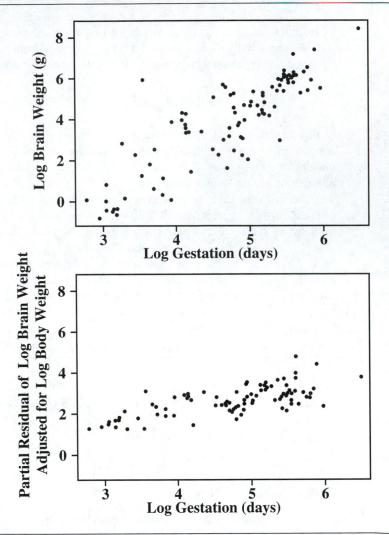

Suppose that

$$\mu\{lbrain \mid lbody, lgest\} = \beta_0 + \beta_1 lbody + f(lgest),$$

where *lbrain*, *lbody*, and *lgest* represent the logarithms of brain weight, body weight, and gestation length, and where the last term is some unspecified function of *lgest*. The purpose of a *partial residual plot* is to explore the nature of the function $f(lgest)$, including whether it can be adequately approximated by a single linear term $\beta_3 lgest$ and whether the estimated relationship might be affected by one or several influential cases.

In this setup,

$$f(lgest) = \mu\{lbrain \mid lbody, lgest\} - (\beta_0 + \beta_1 lbody),$$

so it would be useful to plot $lbrain - (\beta_0 + \beta_1 lbody)$ versus $lgest$ to visually explore the function $f(lgest)$. Since the β's are unknown, however, this is impossible. They can be replaced by estimates, but not just any estimates will work. It would not be appropriate to estimate them by the regression of $lbrain$ on $lbody$ alone, since the necessary coefficients β_0 and β_1 are those in the regression of $lbrain$ on $lbody$ and $lgest$. On the other hand, since $f(lgest)$ is unspecified, it is unclear how to include it in the model.

Partial Residuals

The idea behind a partial residual plot is to approximate $f(lgest)$ with the linear function $\beta_3 lgest$. This may be a crude approximation, but it is often good enough to estimate the correct β_0 and β_1 for plotting purposes. The following steps are used to draw a *partial residual plot* of $lbrain$ versus $lgest$, adjusting for $lbody$:

1.   Obtain the estimated coefficients in the linear regression of $lbrain$ on $lbody$ and $lgest$: $\mu\{lbrain \mid lbody, lgest\} = \beta_0 + \beta_1 lbody + \beta_2 lgest$.
2.   Compute the *partial residuals* as $pres = lbrain - \hat{\beta}_0 - \hat{\beta}_1 lbody$.
3.   Plot the partial residuals versus $lgest$.

A partial residual plot is shown in the lower scatterplot of Display 11.14. After getting the effect of log body weight out of the way, less of an association remains between log brain weight and log gestation, but some association does persist and a linear term should be adequate to model it.

When the effects of many other explanatory variables need to be "subtracted," an easier calculation formula is available. Steps 1 and 2 of the method just described can be replaced with these:

1.   Obtain the residuals, res, from the fit to the linear regression of $lbrain$ on $lbody$ and $lgest$ (include all explanatory variables—those whose effects are to be subtracted, and the one that is to be plotted).
2.   Compute the partial residuals as $pres = res + \hat{\beta}_2 lgest$ (the residuals from the fit with all explanatory variables plus the estimated component associated with the effect of the explanatory variable under question).

The meaning of the partial residuals is clearer in the first version, but the calculation is often more straightforward with the second. Because of this calculating formula, partial residual plots are sometimes referred to as *component plus residuals plots*.

Notes About Partial Residuals

When Should Partial Residual Plots Be Used?   Partial residuals are primarily useful when analytical interest centers on one explanatory variable whose effect is expected to be small relative to the effects of others. They are also useful when uncertainty exists about a particular explanatory variable that needs to be modeled carefully or when the underlying explanation for why an observation is influential on the estimate of a single coefficient needs to be understood.

Augmented Partial Residuals.   Rather than using $\beta_2 lgest$ as an approximation to $f(lgest)$, some statisticians prefer to use $\beta_2 lgest + \beta_3 lgest^2$. Here, partial residuals are obtained just as in the preceding algorithms, except that $lgest^2$ is also included as an explanatory variable in step 1. (In step 2 of the component-plus-residual version, $pres = res + \hat{\beta}_2 lgest + \hat{\beta}_3 lgest^2$.) If they are equally convenient to use, the augmented partial residuals are preferred. In many cases, however, the difference between the partial residual and the augmented partial residual is slight.

Example—Blood–Brain Barrier

The residual plot in Display 11.7 indicated some potential outliers, but further investigation does not show that these points are influential in determining the structure of the model or in answering the questions of interest (see Exercise 11.17).

The key explanatory variables are the indicator variable for whether the rat received the disruption infusion or the control infusion; the length of time after infusion that the rat was sacrificed, and the interaction of these. The additional covariates should be given a chance to be included in the model, for two reasons. First (and most importantly), since randomization was not used, it behooves the researchers to demonstrate that the differences in treatment effects cannot be explained by differences in the types of rats that received the various treatments. Second, even if randomization had been used, including important covariates can yield higher resolution. If the covariates have some additional association with the response, smaller standard errors and more powerful tests should result from their inclusion.

Among the covariates, sex and days after inoculation are associated with both the response and the design variables. To some extent the effects of these variables are confounded, since their effects on the response cannot be separated. On the other hand, the effects of the design variables can be examined after the covariates are accounted for, and the effects of the covariates can be examined after the design variables are accounted for. This is shown graphically in the partial residual plots of Display 11.15.

The top scatterplot indicates that the relationship between the response and the design variables (sacrifice time and treatment) is much the same when the effects of the covariates are included as when they are ignored (Display 11.5). The lower two plots show that, after the effects of the design variables are accounted for, little evidence exists of a sex effect, although slight visual evidence exists of a days after inoculation effect.

This conclusion is further investigated through model fitting. A search through possible models that contain covariates shows that sex and days after inoculation (treated as a factor) are the only ones associated with the response. When the design variables are included as

Display 11.15 Some partial residual plots for the blood–brain barrier data, with log of antibody concentration ratio (brain tumor-to-liver) as response

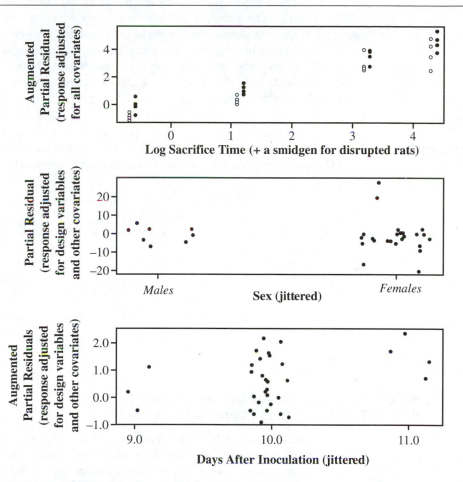

well, three conclusions are supported:

1. The covariates are not significant when the design variables are also included in the model.
2. The design variables *are* significant when the covariates are also included in the model.
3. The conclusions regarding the design variables depend very little on whether the covariates are in the model.

These results suggest that the conclusions can be based satisfactorily on the model without the covariates.

Since the effect of log sacrifice time is not linear (and since the addition of a quadratic term does not remedy the lack-of-fit), sacrifice time is treated as a factor with four levels. Therefore, the final model used to estimate the treatment effect has the following terms: *TIME + TREAT*. The estimates and standard errors are shown in Display 11.16. The coefficient of the indicator variable for the blood–brain barrier disruption treatment is .797. So, expressed in accordance with the interpretation for log-transformed responses, the median ratio of antibody concentration in the brain tumor to antibody concentration in the liver is estimated to be $\exp(.797) = 2.22$ times greater for the blood–brain barrier diffusion treatment than for the control.

Display 11.16   Results from the regression of log ratio of antibody concentration (brain tumor-to-liver) on sacrifice time (treated as a factor) and treatment

Variable	Estimate	Standard Error	t-Statistic	Two-sided p-value
Constant	−3.505	0.195	−17.94	<.0001
Indicator for time = 3	1.341	0.252	4.50	.0001
Indicator for time = 24	4.257	0.259	16.43	<.0001
Indicator for time = 72	5.154	0.259	19.89	<.0001
Indicator for treatment = BD	0.797	0.183	4.35	.0002

11.6   Related Issues

11.6.1   Weighted Regression for Certain Types of Nonconstant Variance

Although nonconstant variance can sometimes be corrected by a transformation of the response, in many situations it cannot. If enough information is known about the form of the nonconstant variance, the method of *weighted least squares* may be used.

The *weighted regression* model, written here with two explanatory variables, is

$$\mu\{Y_i \mid X_{1i}, X_{2i}\} = \beta_0 + \beta_1 X_{1i} + \beta_2 X_{2i}$$
$$\text{Var}\{Y_i \mid X_{1i}, X_{2i}\} = \sigma^2/w_i,$$

where the w_i's are known constants called *weights* (because cases with larger w_i's have smaller variances and should be weighted more in the analysis).

This model arises in at least three practical situations:

1. *Responses are estimates; SEs are available.* Sometimes the response variable values are measurements whose estimated standard deviations, $\text{SE}(Y_i)$, are available. In the

preceding model, the w_i's are taken to be $1/[SE(Y_i)]^2$; that is, the responses with smaller standard errors should receive more weight.

2. *Responses are averages; only the sample sizes are known.* If the responses are averages from samples of different sizes and if the ordinary regression model applies for the individual observations (the ones going into the average), then the weighted regression model applies to the averages, with weights equal to the sample sizes. The averages based on larger samples are given more weight.

3. *Variance is proportional to X.* Sometimes, while the regression of a response on an explanatory variable is a straight line, the variance increases with increases in the explanatory variable. Although a log transformation of the response might correct the nonconstant variance, it would induce a nonlinear relationship. A weighted regression model, with $w_i = 1/X_i$ (or possibly $w_i = 1/X_i^2$) may be preferable.

The weighted regression model can be estimated by *weighted least squares* within the standard regression procedure in most statistical computing programs. The estimated regression coefficients are chosen to minimize the weighted sum of squared residuals (see Exercise 21 for the calculus). It is necessary for the user to specify the response, the explanatory variables, and the weights.

11.6.2   Measurement Errors in Explanatory Variables

Sometimes a theoretical model specifies that the mean response depends on certain explanatory variables that cannot be measured directly. This is called the *errors-in-variables problem.* If, for example, the dietary intake of polyunsaturated fat for an individual is estimated by a survey of what the individual eats during one week, the regression on the latter (surrogate) variable may not correspond to the regression of interest on what it is measuring.

If the purpose of the regression is prediction, and if the prediction of future responses is to be based on measured explanatory variables that have the same kind of measurement errors as the ones used to estimate the regression, then measurement errors in explanatory variables do not present a problem. The model of interest in this case is the one for the regression of the response on the measured explanatory variables. The regression on the exact explanatory variables is not relevant.

For other purposes, however, measurement errors in explanatory variables present a problem. The least squares estimates based on the imprecisely measured explanatory variables are biased estimates of the coefficients in the regression of the response on the precisely measured explanatory variables. If there is a single explanatory variable, the estimate tends to be closer to zero than it should be. If there are many explanatory variables, some of which are measured with error, the individual coefficients (including the coefficients of variables that are free of measurement error) may be over- or underestimated, depending on the correlation structure of the explanatory variables.

Alternatives to least squares that account for measurement errors in explanatory variables do exist, but they are sophisticated and require extra information about the measurement errors (for example, that replicate measurements are available on a subset of subjects). The advice of a professional statistician may be required for this problem.

11.7 **Summary**

This chapter focuses on the first three steps—exploring the data, fitting a model, and check-ing the model—of the strategy for data analysis using statistical models (Display 9.9 on page 241). After initial exploration, the analyst fits a tentative model and uses the residuals to check on the need for transformation or outlier action. The exploration continues by entertaining various models and testing terms within the model. There is not necessarily any "best" or "final" model. Often different questions of interest are addressed through different models. In the blood–brain barrier data set, for example, one of the questions of interest is investigated via interaction terms. When these are found to be insignificant, they are dropped in favor of a different model for making inferences about the treatment effect.

Alcohol Metabolism Study

Several questions are asked of these data. For two-sample questions like "Does the first-pass metabolism of alcohol differ between males and females?" the rank-sum test is used to avoid considering the outliers. To investigate whether metabolism differs between males and females once the effect of gastric AD activity has been accounted for, a regression model for metabolism as a function of gastric AD activity and an indicator variable for sex should be used. Finding a model is made difficult by the need to consider quite a few terms on the basis of a small data set, as well as by the related influence of two observations with atypical values of the explanatory variable. Since the data are not capable of resolving a regression relationship for gastric AD activities larger than 3, the two extreme cases are set aside and the subsequent analysis and conclusions are restricted to the reduced range. Within that reduced range (and ignoring the effect of alcoholism), either of two models can be used: the parallel regression lines model or the regression model forced through the origin with different slopes for males and females. There is no objective basis for preferring one of these models over the other. The latter is selected for inference primarily because it seems somewhat more likely to apply in the broader range.

Blood–Brain Barrier Study

The lack of random assignment of the rats to the treatment groups is a major concern. In partial compensation for this design flaw, tentative regression models include potentially confounding covariates—measured but uncontrolled characteristics of the experimental units. Although some covariate differences are detected among the treatment groups, the regression analysis is able to clarify the treatment effects after accounting for the covariates. Without randomization, however, the causal conclusions remain tied to the adequacy of the models and to the speculation that no additional unmeasured covariates that might explain the treatment effects differ between the groups.

Further Reading

Discussions of case influence statistics are provided in all modern regression textbooks, including Neter, Wasserman, and Kutner (1989). One method of robust regression—the method of least absolute deviations—is also discussed in that text, as is weighted regression.

Partial residual plots and augmented partial residual plots are the subject of a paper by C. L. Mallows (1986). Few nontechnical accounts of measurement errors in explanatory variables exist. The book *Measurement Error Models* by W. Fuller (1987), however, is an authoritative reference for statisticians.

11.8 Exercises

Conceptual Exercises

1. Alcohol Metabolism. The subjects in the study were given alcohol on two consecutive days. On one of the days it was administered orally and on the other it was administered intravenously. The type of administration given on the first day was decided by a random mechanism. Why was this precaution taken?

2. Alcohol Metabolism. Here are two models for explaining the mean first-pass metabolism:

Model 1: $\beta_0 + \beta_1 gast + \beta_2 fem$
Model 2: $\beta_0 + \beta_1 gast + \beta_2 gast \times fem$.

(a) Why are there no formal tools for comparing these two models? (b) For a given value of gastric activity, what is the mean first-pass metabolism for men minus the mean first-pass metabolism for women (i) from Model 1? (ii) from Model 2?

3. Alcohol Metabolism. What would be the meaning of a third-order interactive effect of gastric activity, sex, and alcoholism on the mean first-pass metabolism?

4. Blood–Brain Barrier. (a) How should rats have been randomly assigned to treatment groups? How many treatment groups were there? What is the name of this type of experimental design and this type of treatment structure? (b) How should rats have been randomly assigned to treatments if the researchers suspected that the sex of the rat might be associated with the response? What is the name of this type of experimental design?

5. Blood–Brain Barrier. The residual plot in Display 11.6 contains two distinct groups of points: a cluster on the left and a cluster on the right. (a) Why is this? (b) Does it imply any problems with the model?

6. Robustness and resistance are different properties. (a) What is the difference? (b) Why are they both relevant when the response distribution is "long-tailed"?

7. Display 11.17 shows a hypothetical scatterplot of salary versus experience, with different codes for males and females. The male and female slopes differ significantly if the male with the most experience is included, but not if he is excluded. What course of action should be taken, and why?

8. (a) Why does a case with large leverage have the *potential* to be influential? Why is it not necessarily influential? (b) Draw a hypothetical scatterplot of Y versus a single X, where one observation has a high leverage but is not influential. (c) Draw a hypothetical scatterplot of Y versus a single X, where one observation has a high leverage and is influential.

9. Suppose it is desired to obtain partial residuals in order to plot Y versus X_2 after getting the effect of X_1 out of the way. The first task is to fit the regression of Y on X_1 and X_2. Let the estimated coefficients be $\hat{\beta}_0, \hat{\beta}_1$, and $\hat{\beta}_2$, and let the residuals from this fit be represented by res_i. The definition of the ith partial residual is $pres_i = Y_i - \hat{\beta}_0 - \hat{\beta}_1 X_{1i}$. The alternative computational formula is $res_i + \hat{\beta}_2 X_{2i}$. Why are these formulas equivalent?

Display 11.17   Hypothetical scatterplot of salary versus experience for males and females

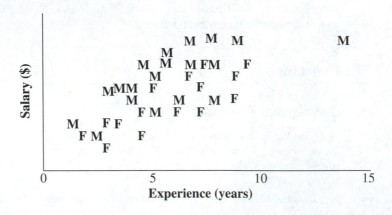

Computational Exercises

10.   Pollen removal. Reconsider the pollen removal data in Display 3.11 on page 77. Draw a coded scatterplot of the log of the proportion of pollen removed relative to the proportion unremoved (if p is the proportion removed, take $Y = \log[p/(1-p)]$) versus the log of the duration of visit, with a code to distinguish queens from workers. (b) Fit the regression of Y on the two explanatory variables and their interaction, and obtain a residual plot. Does the residual plot indicate any problems? (c) Obtain a set of case influence statistics. Are there any problem observations? What is the most advisable course of action? (d) Does a significant interaction appear to exist, or can the simpler parallel regression lines model be used?

11.   Chernobyl Fallout. Fit the simple regression of cesium concentration in mushrooms versus cesium concentration in the soil, from the data listed in Exercise 8.18 on page 221. (a) Obtain a set of case influence statistics. What conclusion can be drawn about case number 17? (b) Repeat part (a) after taking the logarithms of both variables.

12.   Brain Weights. Reconsider the brain weight data of Display 9.4 on page 229. (a) Fit the regression of brain weight on body weight, gestation, and log litter size, using no transformations. Obtain a set of case-influence statistics. Is any mammal influential in this fit? (b) Refit the regression without the influential observation, and obtain the new set of case influence statistics. Are there any influential observations from this fit? (c) What lessons about the connection between the need for a log transformation and influence can be discerned?

13.   Brain Weights. Identifying which mammals have larger brain weights than were predicted by the regression model might point the way to further variables that can be examined. Fit the regression of log brain weight on log body weight, log gestation, and log litter size, and compute the studentized residuals. Which mammals have substantially larger brain weights than were predicted by the model? Do any mammals have substantially smaller brain weights than were predicted by the model?

14.   Corn Yield and Rainfall. Reconsider the data in Display 9.17 on page 252. Fit the regression of corn yield on rainfall, rainfall-squared, and year. (a) Obtain the partial residuals of corn yield, adjusted for rainfall, and plot them versus year. (b) Obtain the augmented partial residual of corn yield, adjusted for year, and plot these values versus rainfall. (c) In your opinion, do these plots provide any clarification over the ordinary scatterplots?

15. Election Fraud. The data in Display 11.18 are Democratic minus Republican votes in Philadelphia counted by absentee ballot and by voting machine, previously described in Exercise 8.20 on page 222. This table duplicates Display 8.19, except that the counts from the year in controversy (1993, which is the 22nd observation) are included here. (a) Plot the absentee count versus the machine count. (b) Fit the regression of absentee count on machine count and obtain internally studentized residuals. (c) Obtain the externally studentized residual for case 22. (d) For case 22, are the internally and externally studentized residuals much different? Why might they differ? (e) Is there any evidence of lack-of-fit or of a potentially long-tailed response distribution? (f) How much (if at all) greater than expected is the Democratic minus Republican difference in election 22, according to the pattern between absentee and machine counts established by the previous 21 elections?

Display 11.18 Democratic minus Republican vote counts by absentee ballot and by voting machine, for 22 elections in Philadelphia's senatorial districts over the last 10 years

Election	Absentees	Machine	Election	Absentees	Machine
1	−841	−1375	12	364	2620
2	−436	−599	13	384	3596
3	−224	−129	14	320	3619
4	160	47	15	240	4195
5	597	7132	16	1365	4395
6	−356	1116	17	392	4512
7	300	940	18	404	5088
8	288	987	19	408	5570
9	609	1927	20	693	6286
10	412	2150	21	400	6533
11	167	2573	22	1025	−564

16. First-pass Metabolism. Calculate the leverage, the studentized residual, and Cook's Distance for the 32nd case. Use the model with gastric activity, a sex indicator variable, and the interaction of these two.

17. Blood–Brain Barrier. (a) Using the data in Display 11.4, compute "jittered" versions of treatment, days after inoculation, and an indicator variable for females by adding small random numbers to each (uniform random numbers between −.15 and .15 work well). (b) Obtain a matrix of scatterplots for the following variables: log sacrifice time, treatment (jittered), days after inoculation (jittered), sex (jittered), and the log of the brain tumor-to-liver antibody ratio. (b) Obtain a matrix of the correlation coefficients among the same five variables (not jittered!). (c) In pencil, write the relevant correlation (two digits is enough) in a corner of each of the scatterplots in the matrix of scatterplots. (d) On the basis of this, what can be said about the relationship between the covariates (sex and days after inoculation), the response, and the design variables (treatment and sacrifice time)?

18. Blood–Brain Barrier. Using the data in Display 11.4, fit the regression of the log response (brain tumor-to-liver antibody ratio) on all covariates, the treatment indicator, and sacrifice time, treated as a factor with four levels (include three indicator variables, for sacrifice time = 3, 24, and

72 hours). (a) Obtain a set of case influence statistics, including a measure of influence, the leverage, and the studentized residual. (b) Discuss whether any influential observations or outliers occur with respect to this fit.

19. Blood–Brain Barrier. (a) Using the data in Display 11.4, fit the regression of the log response (brain tumor-to-liver antibody ratio) on an indicator variable for treatment and on sacrifice time treated as a factor with four levels (include three indicator variables, for sacrifice time = 3, 24, and 72 hours). Use the model to find the estimated mean of the log response at each of the eight treatment combinations (all combinations of the two infusions and the four sacrifice times). (b) Let X represent log of sacrifice time. Fit the regression of the log response on an indicator variable for treatment, X, X^2, and X^3. Use the estimated model to find the estimated mean of the log response at each of the eight treatment combinations. (c) Why are the answers to parts (a) and (b) the same?

20. Warm-blooded T. Rex? The data in Display 11.19 are the isotopic composition of structural bone carbonate (X) and the isotopic composition of the coexisting calcite cements (Y) in 18 bone samples from a specimen of the dinosaur *Tyrannosaurus rex*. Evidence that the mean of Y is positively associated with X was used in an argument that the metabolic rate of this dinosaur resembled warm-blooded more than cold-blooded animals. (Data from R. E. Barrick and W. J. Showers, "Thermophysiology of *Tyrannosaurus rex*: Evidence from Oxygen Isotopes," *Science* 265 (1994): 222–24.) (a) Examine the effects on the p-value for significance of regression and on R-squared of deleting (i) the case with the smallest value of X, and (ii) the two cases with the smallest values of X. (b) Why does R-squared change so much? (c) Compute the case influence statistics, and discuss interesting cases. (d) Recompute the case statistics when the case with the smallest X is deleted. (e) Comment on the differences in the two sets of case statistics. Why may pairs of influential observations not be found with the usual case influence statistics? (f) What might one conclude about the influence of the two unusual observations in this data set?

Display 11.19 Isotopic composition of carbonate and of calcite cements in 18 samples of bone from a *Tyrannosaurus rex* specimen

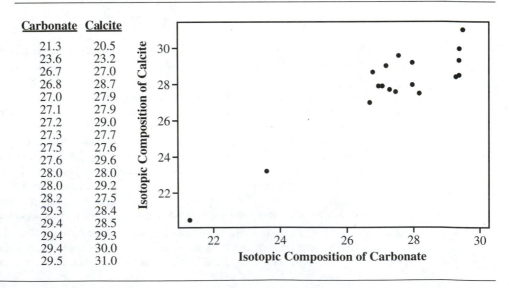

Carbonate	Calcite
21.3	20.5
23.6	23.2
26.7	27.0
26.8	28.7
27.0	27.9
27.1	27.9
27.2	29.0
27.3	27.7
27.5	27.6
27.6	29.6
28.0	28.0
28.0	29.2
28.2	27.5
29.3	28.4
29.4	28.5
29.4	29.3
29.4	30.0
29.5	31.0

21. Calculus Problem. The weighted least squares problem in multiple linear regression is to find the parameter values that minimize the weighted sum of squares,

$$SS_w(\beta_0, \beta_1, \ldots, \beta_p) = \sum_{i=1}^{n} w_i(Y_i - \beta_0 - \beta_1 X_{1i} - \cdots - \beta_p X_{pi})^2,$$

with all $w_i > 0$. (a) Setting the partial derivatives of SS_w with respect to each of the parameters equal to zero, show that the solutions must satisfy this set of normal equations:

$$
\begin{aligned}
\beta_0 \sum w_i + \beta_1 \sum w_i X_{1i} + \beta_2 \sum w_i X_{2i} + \cdots + \beta_p \sum w_i X_{pi} &= \sum w_i Y_i \\
\beta_0 \sum w_i X_{1i} + \beta_1 \sum w_i X_{1i}^2 + \beta_2 \sum w_i X_{1i} X_{2i} + \cdots + \beta_p \sum w_i X_{1i} X_{pi} &= \sum w_i X_{1i} Y_i \\
\beta_0 \sum w_i X_{2i} + \beta_1 \sum w_i X_{2i} X_{1i} + \beta_2 \sum w_i X_{2i}^2 + \cdots + \beta_p \sum w_i X_{2i} X_{pi} &= \sum w_i X_{2i} Y_i \\
\vdots \qquad\qquad \vdots \qquad\qquad \vdots \qquad\qquad \vdots \qquad\qquad \vdots \\
\beta_0 \sum w_i X_{pi} + \beta_1 \sum w_i X_{pi} X_{1i} + \beta_2 \sum w_i X_{pi} X_{2i} + \cdots + \beta_p \sum w_i X_{pi}^2 &= \sum w_i X_{pi} Y_i.
\end{aligned}
$$

(b) Show that solutions to the normal equations minimize SS.

Data Problems

22. Deforestation and Debt. It has been theorized that developing countries cut down their forests to pay off foreign debt. Two researchers examined this belief using data from 11 Latin American nations. (Data from R. T. Gullison and E. C. Losos, "The Role of Foreign Debt in Deforestation in Latin America," *Conservation Biology* 7(1) (1992): 140–47.) The data on debt, deforestation, and population appear in Display 11.20. Does the evidence significantly support the theory that debt causes deforestation? Does debt exert any effect after the effect of population on deforestation is accounted for? Describe the effect of debt, after accounting for population.

Display 11.20 Foreign debt, annual deforestation area, and population for 11 Latin American countries

Country	Debt (millions of dollars)	Deforestation (thousands of hectares)	Population (thousands of people)
Brazil	86,396	12,150	128,425.0
Mexico	79,613	2680	74,194.5
Ecuador	6990	1557	8750.5
Colombia	10,101	1500	27,254.0
Venezuela	24,870	1430	16,170.5
Peru	10,707	1250	18,496.5
Nicaragua	3985	550	3021.5
Argentina	36,664	400	29,400.5
Bolivia	3810	300	5970.5
Paraguay	1479	250	3424.5
Costa Rica	3413	90	2439.5

23. Air Pollution and Mortality. Does pollution kill people? Data in one early study designed to explore this issue came from 60 Standard Metropolitan Statistical Areas (SMSA) in the United States, obtained for the years 1959–1961. (Data from G. C. McDonald and J. A. Ayers, "Some Applications of the 'Chernoff Faces': A Technique for Graphically Representing Multivariate Data," in *Graphical Representation of Multivariate Data* (New York: Academic Press, 1978).) Total age-adjusted mortality from all causes, in deaths per 100,000 population, is the response variable. The explanatory variables listed in Display 11.21 include mean annual precipitation (in inches); median number of school years completed, for persons of age 25 years or older; percentage of 1960 population that is nonwhite; relative pollution potential of oxides of nitrogen, NO_X; and relative pollution potential of sulfur dioxide, SO_2. "Relative pollution potential" is the product of the tons emitted per day per square kilometer and a factor correcting for SMSA dimension and exposure. The first three explanatory variables are a subset of climate and socioeconomic variables in the original data set. (*Note:* Two cities—Lancaster and York—are heavily populated by members of the Amish religion, who prefer to teach their children at home. The lower years of education for these two cities do not indicate a social climate similar to other cities with similar years of education.) Is there evidence that mortality is associated with either of the pollution variables, after the effects of the climate and socioeconomic variables are accounted for? Analyze the data and write a report of the findings, including any important limitations of this study. (*Hint:* Consider looking at case-influence statistics.)

Display 11.21 Air pollution and mortality data for 60 U.S. cities, 1959–1961

City	Mortality	Precipitation	Education	Nonwhite	NO_X	SO_2
San Jose, CA	790.733	13	12.2	3.0	32	3
Wichita, KS	823.764	28	12.1	7.5	2	1
San Diego, CA	839.709	10	12.1	5.9	66	20
Lancaster, PA	844.053	43	9.5	2.9	7	32
Minneapolis, MN	857.622	25	12.1	2.0	11	26
Dallas, TX	860.101	35	11.8	14.8	1	1
Miami, FL	861.439	60	11.5	13.5	1	1
Los Angeles, CA	861.833	11	12.1	7.8	319	130
Grand Rapids, MI	871.338	31	10.9	5.1	3	10
Denver, CO	871.766	15	12.2	4.7	8	28
Rochester, NY	874.281	32	11.1	5.0	4	18
Hartford, CT	887.466	43	11.5	7.2	3	10
Fort Worth, TX	891.708	31	11.4	11.5	1	1
Portland, OR	893.991	37	12.0	3.6	21	44
Worcester, MA	895.696	45	11.1	1.0	3	8
Seattle, WA	899.264	35	12.2	5.7	7	20
Bridgeport, CT	899.529	45	10.6	5.3	4	4
Springfield, MA	904.155	45	11.1	3.4	4	20
San Francisco, CA	911.701	18	12.2	13.7	171	86
York, PA	911.817	42	9.0	4.8	8	49
Utica, NY	912.202	40	10.3	2.5	2	11
Canton, OH	912.347	36	10.7	6.7	7	20
Kansas City, MO	919.729	35	12.0	12.6	4	4
Akron, OH	921.870	36	11.4	8.8	15	59
New Haven, CT	923.234	46	11.3	8.8	3	8
Milwaukee, WI	929.150	30	11.1	5.8	23	125
Boston, MA	934.700	43	12.1	3.5	32	62
Dayton, OH	936.234	36	11.4	12.4	4	16
Providence, RI	938.500	42	10.1	2.2	4	18
Flint, MI	941.181	30	10.8	13.1	4	11

Display 11.21 Air pollution and mortality data for 60 U.S. cities, 1959–1961—Continued

City	Mortality	Precipitation	Education	Nonwhite	NO$_X$	SO$_2$
Reading, PA	946.185	41	9.6	2.7	11	89
Syracuse, NY	950.672	38	11.4	3.8	5	25
Houston, TX	952.529	46	11.4	21.0	5	1
Saint Louis, MO	953.560	34	9.7	17.2	15	68
Youngstown, OH	954.442	38	10.7	11.7	13	39
Columbus, OH	958.839	37	11.9	13.1	9	15
Detroit, MI	959.221	31	10.8	15.8	35	124
Nashville, TN	961.009	45	10.1	21.0	14	78
Allentown, PA	962.354	44	9.8	0.8	6	33
Washington, DC	967.803	41	12.3	25.9	28	102
Indianapolis, IN	968.665	39	11.4	15.6	7	33
Cincinnati, OH	970.467	40	10.2	13.0	26	146
Greensboro, NC	971.122	42	10.4	22.7	3	5
Toledo, OH	972.464	31	10.7	9.5	7	25
Atlanta, GA	982.291	47	11.1	27.1	8	24
Cleveland, OH	985.950	35	11.1	14.7	21	64
Louisville, KY	989.265	30	9.9	13.1	37	193
Pittsburgh, PA	991.290	36	10.6	8.1	59	263
New York, NY	994.648	42	10.7	11.3	26	108
Albany, NY	997.875	35	11.0	3.5	10	39
Buffalo, NY	1001.902	36	10.5	8.1	12	37
Wilmington, DE	1003.502	45	11.3	12.1	11	42
Memphis, TN	1006.490	50	10.4	36.7	18	34
Philadelphia, PA	1015.032	42	10.5	17.5	32	161
Chattanooga, TN	1017.613	52	9.6	22.2	8	27
Chicago, IL	1024.885	33	10.9	16.3	63	278
Richmond, VA	1025.502	44	11.0	28.6	9	48
Birmingham, AL	1030.380	53	10.2	38.5	32	72
Baltimore, MD	1071.289	43	9.6	24.4	38	206
New Orleans, LA	1113.056	54	9.7	31.4	17	1

Answers to Conceptual Exercises

1. By randomly determining order, the researchers avoid bias in determining first-pass metabolism that would occur if an order effect existed.

2. (a) Neither of the models is a subset of the others, so it is impossible to test a term in a "full model" to see whether the "reduced model" does just as well. (Incidentally, the models explain the data about equally well.) (b) (i) β_2 (ii) $\beta_2 gast$.

3. It would mean that the effect of gastric AD activity on first-pass metabolism differed between males and females, and that the amount of the sex difference differed between alcoholics and non-alcoholics. (Two-factor interactions are hard enough to describe in words. Three- and higher-factor interactions typically involve very long and confusing sentences. A theory without interactions is obviously simpler than one with interactions.)

4. (a) Rats should have been randomly assigned to one of eight groups, corresponding to the eight combinations of treatment and sacrifice time. This is a completely randomized design with factorial (2×4) treatment structure. (b) If the response is suspected to be related to the sex of the rat, a randomized block experiment should be performed. The procedure in part (a) can be followed separately for male and female rats.

5. (a) The fitted values are noticeably larger for the rats in the groups with longer sacrifice times. (b) No.

6. (a) *Robustness* describes the extent to which inferential statements are correct when specific assumptions are violated. *Resistance* describes the extent to which the results remain unchanged when a small portion of the data is changed, perhaps drastically (and therefore describes the extent to which individual observations can be influential). (b) When the distribution is long-tailed, the robustness against departures from normality cannot be guaranteed. Put another way, there are likely to be outliers, which can have undue influence on the results, since the least squares method is not resistant.

7. In the absence of further knowledge, it is safest to exclude the very experienced male from the data set and restrict conclusions about the differences between male and female salaries to individuals with 10 years of experience or fewer. It may be that males and females have different slopes over the wider range of experience, or it may be that the straight line is not an adequate model over the wider range of experiences. There is certainly insufficient data in the more-than-10-years range to resolve this issue.

8. (a) A large leverage indicates that a case occupies a position in the "X-space" that is not densely populated. It therefore plays a large rule in shaping the estimated regression model in that region. Since it does not share its role (much) with other nearby points, it must draw the regression surface close to it. For this reason it has a high potential for influence. If, however, the fit of the model without that point is about the same as the fit of the model with it, it is not influential.

9. Since $res_i = Y_i - \hat{\beta}_0 - \hat{\beta}_1 X_{1i} - \hat{\beta}_2 X_{2i}$, the alternative calculating formula is $Y_i - \hat{\beta}_0 - \hat{\beta}_1 X_{1i} - \hat{\beta}_2 X_{2i} + \hat{\beta}_2 X_{2i}$. The last two terms cancel, leaving the original definition.

Strategies for Variable Selection

There are two good reasons for paring down a large number of explanatory variables to a smaller set. The first reason is somewhat philosophical: simplicity is preferable to complexity. Thus, redundant and unnecessary explanatory variables should be excluded on principle. The second reason is more concrete: unnecessary terms in the model yield less precise inferences.

Various statistical tools are available for choosing a good subset from a large pool of explanatory variables. Discussed in this chapter are sequential variable selection techniques and comparisons among all possible subsets through examination of Cp and the Bayesian Information Criterion. The most important practical lessons are that the variable selection process should be sensitive to the objectives of the study, and that the particular subset chosen is relatively unimportant.

12.1   Case Studies

12.1.1   State Average SAT Scores—An Observational Study

When, in 1982, average Scholastic Aptitude Test (SAT) scores were first published on a state-by-state basis in the United States, the huge variation in the scores was a source of great pride for some states and of consternation for others. Average scores ranged from a low of 790 (out of a possible 1,600) in South Carolina to a high of 1,088 in Iowa. This 298-point spread dwarfed the 20-year national decline of 80 points. Two researchers set out to "assess the extent to which the compositional/demographic and school-structural characteristics are implicated in SAT differences." (Data from B. Powell and L. C. Steelman, "Variations in State SAT Performance: Meaningful or Misleading?" *Harvard Educational Review* 54(4) (1984): 389–412.)

Display 12.1 contains state averages of the total SAT (verbal + quantitative) scores, along with six variables that may be associated with the SAT differences among states. Some explanatory variables come from the Powell and Steelman article, while others were obtained from the College Entrance Examination Board (by Robert Powers). The variables are: *takers* is the percentage of the total eligible students (high school seniors) in the state who took the exam; *income* is the median income of families of test-takers, in hundreds of dollars; *years* is the average number of years that the test-takers had formal studies in social sciences, natural sciences, and humanities; *public* is the percentage of the test-takers who attended public secondary schools; *expend* is the total state expenditure on secondary schools, expressed in hundreds of dollars per student; *rank* is the median percentile ranking of the test-takers within their secondary school classes.

Notice that the states with high average SATs had low percentage takers. One reason is that these are mostly midwestern states that administer other tests to students bound for college in-state. Only their best students planning to attend college out of state take the SAT exams. As the percentage of takers increases for other states, so does the likelihood that the takers include lower-qualified students. After accounting for the percentage of students who took the test and the median class rank of the test-takers (to adjust, somewhat, for the selection bias in the samples from each state), which variables are associated with state SAT scores? After accounting for the percentage of takers and the median class rank of the takers, how do the states rank? Which states perform best for the amount of money they spend?

Summary of Statistical Findings

The percentage of eligible students taking the test and the median class rank of the students taking the test explain 81.5% of the variation in state average test scores. This confirms the expectation that much of the between-states variation in SAT scores can be explained by the relative quality of the students within the state who decide to take the test. After the percentage of students in a state who take the test and the median class rank of these students are accounted for, convincing evidence exists that both state expenditures (one-sided p-value $< .0001$) and years of formal study in social sciences, natural sciences, and humanities (one-sided p-value $= .0005$) are associated with SAT averages. Alaska had a substantially higher expenditure than other states and was excluded from the analysis.

Display 12.1   Average SAT scores by U.S. state in 1982, and possible associated factors

	State	SAT	Takers	Income	Years	Public	Expend	Rank
1	Iowa	1088	3	326	16.79	87.8	25.60	89.7
2	South Dakota	1075	2	264	16.07	86.2	19.95	90.6
3	North Dakota	1068	3	317	16.57	88.3	20.62	89.8
4	Kansas	1045	5	338	16.30	83.9	27.14	86.3
5	Nebraska	1045	5	293	17.25	83.6	21.05	88.5
6	Montana	1033	8	263	15.91	93.7	29.48	86.4
7	Minnesota	1028	7	343	17.41	78.3	24.84	83.4
8	Utah	1022	4	333	16.57	75.2	17.42	85.9
9	Wyoming	1017	5	328	16.01	97.0	25.96	87.5
10	Wisconsin	1011	10	304	16.85	77.3	27.69	84.2
11	Oklahoma	1001	5	358	15.95	74.2	20.07	85.6
12	Arkansas	999	4	295	15.49	86.4	15.71	89.2
13	Tennessee	999	9	330	15.72	61.2	14.58	83.4
14	New Mexico	997	8	316	15.92	79.5	22.19	83.7
15	Idaho	995	7	285	16.18	92.1	17.80	85.9
16	Mississippi	988	3	315	16.76	67.9	15.36	90.1
17	Kentucky	985	6	330	16.61	71.4	15.69	86.4
18	Colorado	983	16	333	16.83	88.3	26.56	81.8
19	Washington	982	19	309	16.23	87.5	26.53	83.2
20	Arizona	981	11	314	15.98	80.9	19.14	84.3
21	Illinois	977	14	347	15.80	74.6	24.41	78.7
22	Louisiana	975	5	394	16.85	44.8	19.72	82.9
23	Missouri	975	10	322	16.42	67.7	20.79	80.6
24	Michigan	973	10	335	16.50	80.7	24.61	81.8
25	West Virginia	968	7	292	17.08	90.6	18.16	86.2
26	Alabama	964	6	313	16.37	69.6	13.84	83.9
27	Ohio	958	16	306	16.52	71.5	21.43	79.5
28	New Hampshire	925	56	248	16.35	78.1	20.33	73.6
29	Alaska	923	31	401	15.32	96.5	50.10	79.6
30	Nevada	917	18	288	14.73	89.1	21.79	81.1
31	Oregon	908	40	261	14.48	92.1	30.49	79.3
32	Vermont	904	54	225	16.50	84.2	20.17	75.8
33	California	899	36	293	15.52	83.0	25.94	77.5
34	Delaware	897	42	277	16.95	67.9	27.81	71.4
35	Connecticut	896	69	287	16.75	76.8	26.97	69.8
36	New York	896	59	236	16.86	80.4	33.58	70.5
37	Maine	890	46	208	16.05	85.7	20.55	74.6
38	Florida	889	39	255	15.91	80.5	22.62	74.6
39	Maryland	889	50	312	16.90	80.4	25.41	71.5
40	Virginia	888	52	295	16.08	88.8	22.23	72.4
41	Massachusetts	888	65	246	16.79	80.7	31.74	69.9
42	Pennsylvania	885	50	241	17.27	78.6	27.98	73.4
43	Rhode Island	877	59	228	16.67	79.7	25.59	71.4
44	New Jersey	869	64	269	16.37	80.6	27.91	69.8
45	Texas	868	32	303	14.95	91.7	19.55	76.4
46	Indiana	860	48	258	14.39	90.2	17.93	74.1
47	Hawaii	857	47	277	16.40	67.6	21.21	69.9
48	North Carolina	827	47	224	15.31	92.8	19.92	75.3
49	Georgia	823	51	250	15.55	86.5	16.52	74.0
50	South Carolina	790	48	214	15.42	88.1	15.60	74.0

The ranking of states after accounting for the relative quality of students in the state who take the exam (as represented by the percentage of takers and the median class rank of the students taking the exam) is shown on the left side of Display 12.2. Here, the states are

Display 12.2   State SAT scores after adjustment (in points above or below average)

SATs Adjusted for % Taking Exam and Their Median Class Rank			SATs Adjusted for % Taking Exam, Rank, and Expenditure		
Rank	**State**	**Adjusted SAT Average**	**Rank**	**State**	**Adjusted SAT Average**
1	New Hampshire	49	1	New Hampshire	67
2	Iowa	41	2	Tennessee	49
3	Montana	38	3	Vermont	39
4	Connecticut	38	4	Connecticut	28
5	Washington	34	5	Nebraska	22
6	Minnesota	34	6	Virginia	20
7	Wisconsin	31	7	Maine	18
8	Colorado	31	8	Arizona	18
9	New York	29	9	Minnesota	17
10	Kansas	29	10	North Dakota	17
11	Massachusetts	27	11	Illinois	16
12	Illinois	25	12	Ohio	15
13	Nebraska	24	13	Washington	15
14	Vermont	21	14	Iowa	14
15	North Dakota	20	15	Colorado	11
16	Tennessee	16	16	Idaho	10
17	Delaware	13	17	Utah	10
18	Maryland	12	18	Missouri	8
19	Virginia	11	19	Maryland	7
20	Ohio	11	20	New Mexico	5
21	New Mexico	8	21	South Dakota	4
22	Rhode Island	8	22	Indiana	4
23	New Jersey	8	23	Wisconsin	3
24	South Dakota	7	24	Rhode Island	3
25	Arizona	6	25	Kentucky	1
26	Pennsylvania	4	26	Montana	0
27	Missouri	4	27	Florida	−1
28	Oregon	3	28	Kansas	−1
29	Maine	2	29	Hawaii	−5
30	Michigan	−1	30	Delaware	−5
31	Wyoming	−2	31	Alabama	−5
32	Utah	−3	32	Massachusetts	−6
33	Idaho	−5	33	New Jersey	−8
34	Florida	−6	34	Oklahoma	−11
35	California	−6	35	New York	−13
36	Oklahoma	−14	36	Arkansas	−13
37	Hawaii	−19	37	Pennsylvania	−14
38	Kentucky	−23	38	Michigan	−14
39	Indiana	−24	39	California	−18
40	Nevada	−28	40	West Virginia	−19
41	West Virginia	−32	41	Texas	−22
42	Louisiana	−33	42	Georgia	−23
43	Arkansas	−34	43	Nevada	−25
44	Alabama	−38	44	Wyoming	−27
45	Texas	−40	45	Louisiana	−28
46	Georgia	−58	46	Oregon	−29
47	Mississippi	−60	47	Mississippi	−40
48	North Carolina	−61	48	North Carolina	−42
49	South Carolina	−94	49	South Carolina	−55

ordered according to the size of their residuals in the regression of SAT scores on percentage of takers and median class rank. For example, the SAT average for New Hampshire is 49 points higher than the estimated mean SAT for states with the same percentage of students taking the exam and median class rank. This ranking and the ranking based on raw averages show some dramatic differences. South Dakota ranks second in its raw SAT scores, for example (see Display 12.1), but only 2% of its eligible students took the exam. After percentage of takers and median class rank are accounted for, South Dakota ranks 24th.

The ranking of states after additionally accounting for state expenditure is shown on the right side of Display 12.2. This ranking provides a means of gauging how states perform for the amount of money they spend (but see the cautionary notes in the next subsection regarding scope of inference). For example, Oregon ranks 31st in its unadjusted average SAT. But after accounting for the percentage of takers, their median class rankings, and the state's expenditure per student, it ranks 46th, indicating that its students performed below what might have been expected given the amount of money the state spends on education.

Scope of Inference

The participating students were *self-selected*; that is, each student included in a state's sample decided to take the SAT, as opposed to being a randomly selected student. Therefore, a state's average does not represent all its eligible students. Including the percentage of takers and the median class rank of takers in the regression models adjusts for this difference, but only to the extent that these variables adequately account for the relative quality of students taking the test. No check on this assumption is possible with the existing data. In addition, these are observational data, for which many possible confounding factors exist, so the statistical statements of significance (as for the effect of expenditure, for example) do not imply causation.

12.1.2 Sex Discrimination in Employment— An Observational Study

Display 12.3 lists data on employees from one job category (skilled, entry-level clerical) of a bank that was sued for sex discrimination. (Data from a file made public by the defense, as documented in Harry V. Roberts, "Harris Trust and Savings Bank: An Analysis of Employee Compensation," Report 7946, Center for Mathematical Studies in Business and Economics, University of Chicago Graduate School of Business, 1979.) These are the same 32 male and 61 female employees, hired between 1965 and 1975, who were considered in Section 1.1.2. The measurements are of annual salary at time of hire, salary as of March 1977, sex (1 for females and 0 for males), seniority (months since first hired), age (months), education (years), and experience prior to employment with the bank (months).

Did the females tend to receive lower starting salaries than similarly qualified and similarly experienced males? After accounting for measures of performance, did females tend to receive smaller pay increases than males?

Summary of Statistical Findings

There is convincing evidence that the median starting salary for females was lower than the median starting salary for males, even after the effects of age, education, previous

Display 12.3 Sex Discrimination Data

Beginning Salary	1977 Salary	FSex(1=F)	Seniority	Age	Education	Experience
5,040	12,420	0	96	329	15	14
6,300	12,060	0	82	357	15	72
6,000	15,120	0	67	315	15	35.5
6,000	16,320	0	97	354	12	24
6,000	12,300	0	66	351	12	56
6,840	10,380	0	92	374	15	41.5
8,100	13,980	0	66	369	16	54.5
6,000	10,140	0	82	363	12	32
6,000	12,360	0	88	555	12	252
6,900	10,920	0	75	416	15	132
6,900	10,920	0	89	481	12	175
5,400	12,660	0	91	331	15	17.5
6,000	12,960	0	66	355	15	64
6,000	12,360	0	86	348	15	25
5,100	8,940	1	95	640	15	165
4,800	8,580	1	98	774	12	381
5,280	8,760	1	98	557	8	190
5,280	8,040	1	88	745	8	90
4,800	9,000	1	77	505	12	63
4,800	8,820	1	76	482	12	6
5,400	13,320	1	86	329	15	24
5,520	9,600	1	82	558	12	97
5,400	8,940	1	88	338	12	26
5,700	9,000	1	76	667	12	90
3,900	8,760	1	98	327	12	0
4,800	9,780	1	75	619	12	144
6,120	9,360	1	78	624	12	208.5
5,220	7,860	1	70	671	8	102
5,100	9,660	1	66	554	8	96
4,380	9,600	1	92	305	8	6.25
4,290	9,180	1	69	280	12	5
5,400	9,540	1	66	534	15	122
4,380	10,380	1	92	305	12	0
5,400	8,640	1	65	603	8	173
5,400	11,880	1	66	302	12	26
4,500	12,540	1	96	366	8	52
5,400	8,400	1	70	628	12	82
5,520	8,880	1	67	694	12	196
5,640	10,080	1	90	368	12	55
4,800	9,240	1	73	590	12	228
5,400	8,640	1	66	771	8	228
4,500	7,980	1	80	298	12	8
5,400	11,940	1	77	325	12	38
5,400	9,420	1	72	589	15	49
6,300	9,780	1	66	394	12	86.5
5,160	10,680	1	87	320	12	18
5,100	11,160	1	98	571	15	115
4,800	8,340	1	79	602	8	70
5,400	9,600	1	98	568	12	244
4,020	9,840	1	92	528	10	44
4,980	8,700	1	74	718	8	318
5,280	9,780	1	88	653	12	107
5,700	8,280	1	65	714	15	241
4,800	8,340	1	87	647	12	163
4,800	13,560	1	82	338	12	11
5,700	10,260	1	82	362	15	51

Display 12.3 Sex Discrimination Data—Continued

Beginning Salary	1977 Salary	FSex (1=F)	Seniority	Age	Education	Experience
4,380	9,720	1	93	303	12	4.5
4,380	10,500	1	89	310	12	0
5,400	10,680	0	88	359	12	38
5,400	11,640	0	96	474	12	113
5,100	7,860	0	84	535	12	180
6,600	11,220	0	66	369	15	84
5,100	8,700	0	97	637	12	315
6,600	12,240	0	83	536	15	215.5
5,700	11,220	0	94	392	15	36
6,000	12,180	0	91	364	12	49
6,000	11,580	0	83	521	15	108
6,000	8,940	0	80	686	12	272
6,000	10,680	0	87	364	15	56
4,620	11,100	0	77	293	12	11.5
5,220	10,080	0	85	344	12	29
6,600	15,360	0	83	340	15	64
5,400	12,600	0	78	305	12	7
6,000	8,940	0	78	659	8	320
5,400	9,480	0	88	690	15	359
6,000	14,400	0	96	402	16	45.5
5,700	10,620	1	88	410	15	61
5,400	10,320	1	78	584	15	51
4,440	9,600	1	97	341	15	75
6,300	10,860	1	84	662	15	231
6,000	9,720	1	69	488	12	121
5,100	9,600	1	85	406	12	59
4,800	11,100	1	87	349	12	11
5,100	10,020	1	87	508	16	123
5,700	9,780	1	74	542	12	116.5
5,400	10,440	1	72	604	12	169
5,100	10,560	1	84	458	12	36
4,800	9,240	1	84	571	16	214
6,000	11,940	1	86	486	15	78.5
4,380	10,020	1	93	313	8	7.5
5,580	7,860	1	69	600	12	132.5
4,620	9,420	1	96	385	12	52
5,220	8,340	1	70	468	12	127

experience, and time at which the job began are taken into account (one-sided p-value $< .0001$). The median beginning salary for females was estimated to be only 89% of the median salary for males, after accounting for the variables mentioned above (a 95% confidence interval for the ratio of adjusted medians is 85% to 93%). There is little evidence that the pay increases for females differed from those for males (one-sided p-value $= .27$ from a t-test for the difference in mean log of average annual raise; one-sided p-value $= .72$ after further adjustment for other variables except beginning salary; one-sided p-value $= .033$ after further adjustment for other variables including beginning salary).

Inferences

Since these are observational data, the usual dictum should apply: no cause-and-effect inference of discrimination can be drawn. Of course that argument disallows nearly all claims of discrimination, because supporting data are always observational. To evaluate such evidence, U.S. courts have adopted a method of burden-shifting. Plaintiffs present statistical analyses as prima facie evidence of discrimination. If the evidence is substantial, the burden shifts to the defendants to show either that these analyses are flawed or that alternative, nondiscriminatory factors explain the differences. If the defendants' arguments are accepted, the burden shifts back to the plaintiffs to show that the defendants' analyses are wrong or that their explanations were pretextual. If plaintiffs succeed in this, the inference of discrimination stands.

12.2   Specific Issues Relating to Many Explanatory Variables

12.2.1   Objectives

Several convenient tools are available for paring down large sets of explanatory variables. But without understanding what they offer, one many incorrectly suppose that the computer is uncovering some law of nature. Like any other statistical tool, these paring tools are most helpful when used to address well-defined questions of interest. Although it may be tempting to think that finding "the important set" of explanatory variables is a question of interest, the actual set chosen is usually one of several (or many) equally good sets.

Objective I: Adjusting for a Large Set of Explanatory Variables

In the sex discrimination example, the objective is to examine the effect of sex after accounting for other legitimate determinants of salary. A game plan for the analysis is to begin by finding a subset of the other explanatory variables with a variable selection technique, and then to add the indicator variable for sex into the model. The variable selection techniques are entirely appropriate for this purpose. The set of explanatory variables chosen is evidently one of several (or perhaps many) equally useful sets. Although a particular explanatory variable may not be included in the final model, it has still been adjusted for, since it was given a chance to be in the model (prior to inclusion of the sex indicator). No interpretation is made of the particular set chosen, nor is any interpretation made of the coefficients. The coefficient of the sex indicator variable is interpreted, but this is straightforward: it represents the association between sex and salary *after* accounting for the effects of the other explanatory variables.

Objective II: Fishing for Explanation

In many studies, unfortunately, no such well-defined question has been posed. One question likely to be asked of the SAT data is vague: "Which variables are important in explaining state average SAT scores, after accounting for percentage of takers and median class rank?" Variable selection techniques can be used to identify a set of explanatory variables, but the analyst may be tempted to attach meaning to the particular variables selected and to interpret coefficients—temptations not encountered in Objective I.

There are several reasons to exercise great caution when using regression for this purpose:

1. The explanatory variables chosen are not necessarily special. Inclusion or exclusion of individual explanatory variables is affected strongly by the correlations between them.

2. Interpreting the coefficients in a multiple regression with correlated explanatory variables is extremely difficult. For example, an estimated coefficient may have the opposite sign of the one suggested by a scatterplot, since a coefficient shows the effect of an explanatory variable on a response after accounting for the other variables in the model. It is often difficult to recognize what "after accounting for the other variables in the model" means when many variables are included.

3. A regression coefficient continues to be interpreted (technically) as a measure of the effect of one variable while holding all other variables fixed. Unfortunately, increases in one variable are almost always accompanied by changes in many others, so the situation described by the interpretation lies outside the experience provided by the data.

4. Finally, of course, causal interpretations from observational studies are always suspect. The selected variables may be related to confounding variables that are more directly responsible for the response.

Objective III: Prediction

If the purpose of the model is prediction, no interpretation of the particular set of explanatory variables chosen or their coefficients is needed, so there is little room for abuse. The variable selection techniques are useful for showing a few models that work well. From these, the researcher selects one with explanatory variables that can be obtained conveniently for the future predictions.

Different Questions Require Different Attitudes Toward the Explanatory Variables

With regard to the SAT score example, a business firm looking for a place to build a new facility may have as its objective in analyzing the data a ranking of states in terms of their ability to train students. The only relevant question to this firm is whether the raw SAT averages accurately reflect the educational training. The problem of selection bias is critical, so inclusion of the percentage of SAT takers or of the median rank is essential. All other variables reflect factors that will not affect the firm's decision, so they are not relevant. This firm could fit a model using the percentage of takers as an explanatory variable and use the residuals as a way of ranking the states.

A legislative watchdog committee, on the other hand, may have the objective of determining the impact of state expenditures on state SAT scores. It might include all variables that could affect SAT scores before including expenditures. Under this arrangement, the committee would claim an effect for expenditures only when the effect could not be attributed to some other related factor.

If prediction is the purpose, all explanatory variables that can be used in making the prediction should be considered eligible for inclusion. (Prediction is not reasonable in the SAT score example, because one would not know the explanatory variables until the SAT scores themselves were available.)

12.2.2 Loss of Precision

A loss in precision—in prediction or in answering questions based on coefficients—occurs when too many explanatory variables are included in the model. The variance of the sampling distribution of the least squares estimate of a single coefficient β_j in a multiple regression model can be expressed as

$$\text{Var}(\hat{\beta}_j) = \frac{\sigma^2}{(n-1)s_j^2(1 - R_j^2)},$$

where s_j^2 is the sample variance of X_j, and R_j^2 is the R-squared term from a regression of X_j on the other explanatory variables included in the model (that is, treating X_j as if it were a response variable). This formula emphasizes the difficulties that correlations among the explanatory variables create. If one of the explanatory variables can be predicted fairly well from some combination of the others (so that its R_j^2 is large), the variance of its coefficient in the regression will be substantially larger than it would have been if it were the only explanatory variable ($\sigma^2/[(n-1)s_j^2]$). As additional explanatory variables are included in the model, the R_j^2 cannot get smaller, and it typically increases.

Multicollinearity

The term *multicollinearity* describes the situation in which $s_j^2(1 - R_j^2)$ in the preceding formula is small for one or more of the coefficients. It is unnecessary to calculate these quantities in practice, but, it is important to realize that highly correlated explanatory variables pose a problem (in decreasing the precision of individual estimated coefficients), and that the problem gets worse as more explanatory variables are included in the model.

Inflated variances of estimated coefficients mean wider confidence intervals and a diminished ability of tests to find significant results. Besides leading to inflated variances multicollinearity has several related consequences:

1. The variances of predicted values will also be inflated.
2. The chance of having influential observations will be greater.
3. The effects of measurement errors in any of the explanatory variables will be more severe.

12.2.3 A Strategy for Dealing with Many Explanatory Variables

A strategy for dealing with many explanatory variables might include the following elements:

1. Identify the key objectives.
2. Screen the available variables, deciding on a list that is sensitive to the objectives and excludes obvious redundancies.
3. Perform exploratory analysis, examining graphical displays and correlation coefficients.
4. Perform transformations as necessary.

5. Examine a residual plot after fitting a rich model, performing further transformations and considering outliers.
6. Use a computer-assisted technique for finding a suitable subset of explanatory variables, exerting enough control over the process to be sensitive to the questions of interest.
7. Proceed with the analysis, using the selected explanatory variables.

Example—Preliminary Analysis of the State SAT Data

The objectives of the SAT study were set out earlier. The variables in Display 12.1 were screened from the original source variables. (Several irrelevant variables were not retained.)

Display 12.4 shows the matrix of scatterplots for the SAT data. Examination of the bottom row indicates some nonlinearity in the relationship between SAT and percentage

Display 12.4   Matrix of scatterplots for SAT scores and six explanatory variables

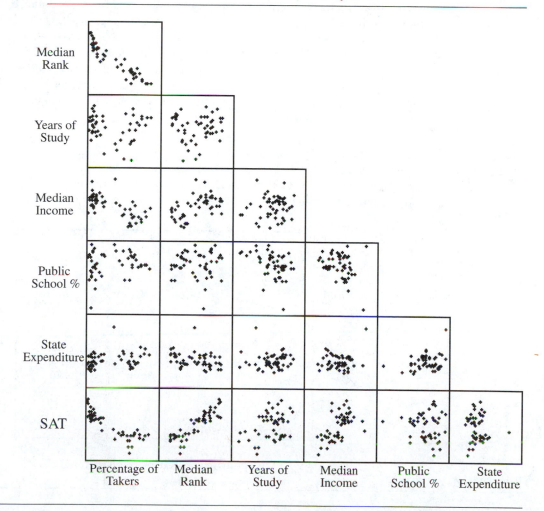

of takers and shows some potential outliers with respect to the variables public school percentage and state expenditure. Using the logarithm of the percentage of takers responds to the first problem, yielding a relationship that looks like a straight line. The outliers were identified: Louisiana is the state with the unusually low percentage of students attending public schools, and Alaska is the state with the very high state expenditure. These require further examination after a model is fitted.

The questions of interest call for adjustment with respect to percentage of SAT takers and median class rank of those takers in each state. When a preliminary regression of SAT was fitted on both of these variables (a fit unaffected by Louisiana and Alaska), the variables were found to explain 81.5% of the variation between state SAT averages.

Partial residual plots at this stage help the researcher evaluate the effects of some of the other explanatory variables, after the first two are accounted for. Display 12.5 shows a partial residual plot of the relationship between SAT and expenditure, after the effects of percentage of takers and class rank have been cleared out of the way. This plot has two striking features: It reveals the noticeable effect of expenditure, after the two adjustment variables are accounted for; and it demonstrates that Alaska is very influential in any fit involving expenditure. To clarify this, the diagram shows two possible straight line regression fits—one with and one without Alaska. The slopes of these two lines differ so markedly that neither value would lie within a confidence interval based on the other. Case influence statistics confirm that Alaska is influential. It has a large leverage ($h_i = 0.458$, from a statistical computer program) and does not fit well to the model that fits the other states (studentized residual $= -3.28$).

Display 12.5 Partial residual plot of state average SAT scores (adjusted for percentage of students in the state who took the test and for median class rank of the students who took the test) versus state expenditure on secondary education

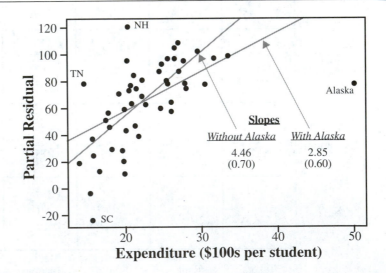

The prudent step is to set Alaska aside for the remainder of the analysis. It can be argued that the role of expenditure is somewhat different in that state, since the costs of heating school buildings and transporting teachers and students long distances are far greater in Alaska, and these expenses do not contribute directly to educational quality. Furthermore, too few states have similar expenditures to permit an accurate determination of the proper form of the model for expenditures greater than $3,500 per student. A similar plot of partial residuals versus percentage of students in public schools provides no reason to be concerned about Louisiana. The effect of the percentage of students who attend public schools in a state appears to be insignificant whether Louisiana is included or not.

12.3 Sequential Variable Selection Techniques

The search for a suitable subset of explanatory variables may encompass a large array of possible models. Computers facilitate the process immensely, since hundreds or even thousands of models can be analyzed in a short period of time. Yet in some instances the search may be so wide as to tie up even the best computer for an intolerable amount of time. Sequential variable selection procedures offer the option of exploring some (but not all) of the possible models.

A step in any sequential (or *stepwise*) procedure begins with a tentative *current model*. Several other models in the neighborhood of the current model (but differing from the current model in having one of its variables excluded or having one additional variable included) are examined to determine whether they are in some sense superior. If so, the best is chosen as the tentative model for the next step. Procedures may differ in their definitions of neighborhood, in their criteria for superiority, and in their initial tentative model.

12.3.1 Forward Selection

The *forward selection* procedure starts with a constant mean as its current model and adds explanatory variables one at a time until no further addition significantly improves the fit. Each forward selection step consists of two tasks:

1. Consider all models obtained by adding one more explanatory variable to the current model. For each variable not already included, calculate its "F-to-enter" (the extra-sum-of-squares F-statistic for testing its significance). Identify the variable with the largest F-to-enter.
2. If the largest F-to-enter is greater than 4 (or some other user-specified number), add that explanatory variable to form a new current model.

Tasks 1 and 2 are repeated until no additional explanatory variables can be added.

12.3.2 Backward Elimination

In the *backward elimination* method, the initial current model contains *all* possible explanatory variables. Each step in this procedure consists of two tasks:

1. For each variable in the current model, calculate the F-to-remove (the extra-sum-of-squares F-statistic for testing its significance). Identify the variable with the smallest F-to-remove.
2. If the smallest F-to-remove is 4 (or some other user-specified number) or less, then remove that explanatory variable to arrive at a new current model.

Tasks 1 and 2 are repeated until none of the remaining explanatory variables can be removed.

12.3.3 Stepwise Regression

The constant mean model with no explanatory variables is the starting current model. Each step consists of the following tasks.

1. Do one step of forward selection.
2. Do one step of backward elimination.

Tasks 1 and 2 are repeated until no explanatory variables can be added or removed.

Inclusion of Factors

Each categorical factor is represented in a regression model by a set of indicator variables. In the absence of a good reason for doing otherwise, this set should be added or removed as a single unit. Some computer packages allow this automatically.

No Universal Best Model

Forward selection, backward elimination, and stepwise regression can lead to different final models. This fact should not be alarming, since no single best model can be expected. It does, however, emphasize a disadvantage of these methods: only one model is presented as an answer, which may give the false impression that this model alone explains the data well. As a result, unwarrantedly narrow attention may be accorded to the particular explanatory variables in that single model.

12.3.4 Sequential Variable Selection with the SAT Data

Display 12.6 is a skeleton of all possible models in the SAT example (with Alaska excluded). Such a display would not be used in practice, but it illustrates how sequential techniques find a good-fitting model even though they search through only a handful of all possible models.

The models on the display are labeled with one-character symbols associated with each explanatory variable. The model labeled 1 has only the intercept. The model labeled T has the intercept and t (the log of the percentage of takers) only. The model labeled ET has the intercept, the log of percentage of takers (t), and state expenditure (e). And so on. All models in the same row contain the same number of explanatory variables. Models are coded so that the variables whose F-statistics are greater than 4 are listed in uppercase letters, and those with F-statistics less than or equal to 4 are listed in lowercase letters. So, for example, in the fit to the four-variable model with income, years, public, and

Display 12.6 Anatomy of sequential variable selection—SAT example

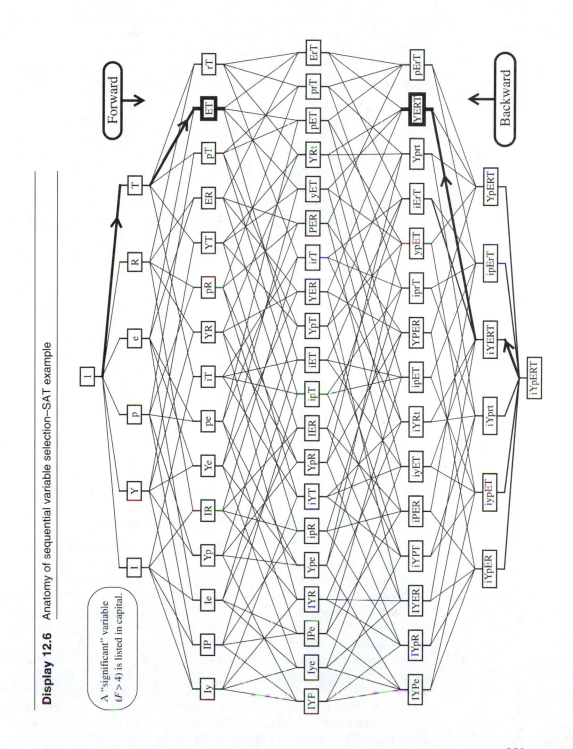

A "significant" variable ($F > 4$) is listed in capital.

expenditure, listed as *IYPe*, expenditure is not significant, but the other variables are. The lines connecting the boxes in the display represent the paths for getting from one model to another that has one fewer or one more explanatory variable.

Forward Selection

Model 1 is taken as the initial current model, and all single-variable models are considered. In the second row of Display 12.6, models I, Y, R, and T all have coefficients whose F-statistics are larger than 4. The one of these with the largest F-to-enter is T, so it is taken as the next current model. Next, all models that include T and one other variable are examined. Only Y and E add significantly when added to the model with T. Of these, E has the larger F-to-enter, so the next current model is ET. Next, all three-variable models that include E and T are considered, but in no case is the additional variable significant. Therefore, the forward selection process stops, and the final model is ET. The entire procedure considered only 16 of the 64 possible models.

Backward Elimination

The model with all variables, *iYpERT*, is taken as the starting point. Of the variables in the model, p (public) has the smallest F-statistic, and p is dropped because its F is less than 4. The resulting model is *iYERT*. The least significant coefficient in the new current model is the one for i (income), and its F-statistic is less than 4, so it is dropped. The resulting model is *YERT*. All of the variables in this model have F-statistics greater than 4, so this is the final model. This procedure considered only 16 of the 64 possible models, and none of the 16 was in the set considered by forward selection.

Stepwise Regression

Stepwise regression starts with one step of forward selection to arrive at the model T. No variables can be dropped, so it tries to add another. It arrives at ET. It then checks whether any of the variables in this model can be dropped. None can, so it seeks to add another variable, but no others add significantly, so the procedure stops at ET.

12.3.5 Compounded Uncertainty in Stepwise Procedures

The cutoff value of 4 for the F-statistic (or 2 for the magnitude of the t-statistic) corresponds roughly to a two-sided p-value of less than .05. The notion of "significance" cannot be taken seriously, however, because sequential variable selection is a form of data snooping.

At step 1 of a forward selection, the cutoff of $F = 4$ corresponds to a hypothesis test for a single coefficient. But the actual statistic considered is the largest of several F-statistics, whose sampling distribution under the null hypothesis differs sharply from an F-distribution.

To demonstrate this, suppose that a model contained ten explanatory variables and a single response, with a sample size of $n = 100$. The F-statistic for a single variable at step 1 would be compared to an F-distribution with 1 and 98 degrees of freedom, where only 4.8% of the F-ratios exceed 4. But suppose further that all eleven variables were generated completely at random (and independently of each other), from a standard normal distribution. What should be expected of the largest F-to-enter?

This random generation process was simulated 500 times on a computer. Display 12.7 shows a histogram of the largest among ten F-to-enter values, along with the theoretical F-distribution. The two distributions are very different. At least one F-to-enter was larger than 4 in 38% of the simulated trials, even though none of the explanatory variables was associated with the response.

Display 12.7   Simulated distribution of the largest of ten F-statistics

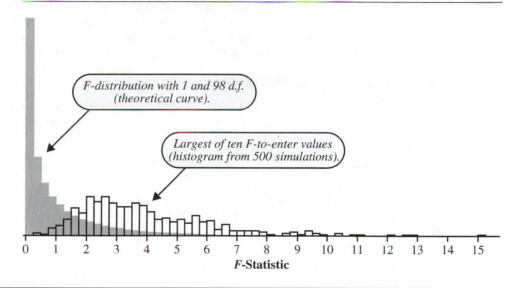

This and related arguments imply that sequential procedures tend to select models that have too many variables if the set contains unimportant explanatory variables. This may not be a serious problem for some data problems, but better techniques are available. In part, variable selection routines are important for historical reasons, since they were the only tool available when computing time was substantially slower. They are also important because they extend in an obvious way to other types of modeling. Now that computers can look at every possible regression model, however, criteria not based on sequential significance address more pertinent features in desired models.

12.4   Model Selection Among All Subsets

A second approach to model selection involves fitting all possible subset models and identifying the ones that best satisfy some model-fitting criterion. Since sequential selection is avoided, the criteria can be based directly on the key issues: including enough explanatory variables to model the response accurately without the loss of precision that occurs when essentially redundant and unnecessary terms are included.

The two criteria presented in this section are the Cp statistic and Schwarz's Bayesian Information Criterion. These are calculated for the fit to a model, so each of the possible regression models receives a numerical score. Before these are examined, some simpler choices should be mentioned.

R-Squared and Adjusted R-Squared

Comparing two models with the same number of parameters is relatively easy: the one with the smaller residual mean square $\hat{\sigma}^2$ is preferred. When applied to comparisons of models that have different numbers of parameters, R-squared (not adjusted) always leads to selecting the model with all variables. Because R^2 is incapable of making a sensible selection, some researchers use the adjusted R^2. This is equivalent to selecting the model with the smallest $\hat{\sigma}^2$—a criterion that generally favors models with too many variables.

12.4.1   The Cp Statistic and Cp Plot

The Cp criterion focuses directly on the trade-off between bias due to excluding important explanatory variables and extra variance due to including too many. The bias in the ith fitted value is

$$\text{Bias}\{\hat{Y}_i\} = \mu\{\hat{Y}_i\} - \mu\{Y_i\}.$$

This is the amount by which the mean in the sampling distribution of the ith fitted value differs from the mean it is attempting to estimate. The mean squared error is the squared bias plus the variance:

$$\text{MSE}\{\hat{Y}_i\} = [\text{Bias}\{\hat{Y}_i\}]^2 + \text{Var}\{\hat{Y}_i\}.$$

The *total mean squared error* (*TMSE*) for a given model is the sum of these over all observations:

$$\text{TMSE} = \sum_{i=1}^{n} \text{MSE}\{\hat{Y}_i\}.$$

It is desired to find a subset model with a small value of TMSE. The squared bias terms are small if no important explanatory variables are left out, and the variance terms are small if no unnecessary explanatory variables are included.

The *Cp statistic* is an estimate of TMSE/σ^2, based on assuming that the model with all available explanatory variables has no bias. For a model that has p regression coefficients, the Cp statistic is computed as

$$\text{Cp} \;=\; p + (n-p)\frac{(\hat{\sigma}^2 - \hat{\sigma}^2_{\text{full}})}{\hat{\sigma}^2_{\text{full}}},$$

where $\hat{\sigma}^2$ is the estimate of σ^2 from the tentative model, and $\hat{\sigma}^2_{full}$ is the estimate of σ^2 from the fit with all possible explanatory variables. One Cp statistic can be calculated for each possible model.

Models with small Cp statistics are looked on more favorably. If a model lacks important explanatory variables, it will show greater residual variability than the full model with all explanatory variables will show. Thus, $\hat{\sigma}^2 - \hat{\sigma}^2_{full}$ will be large. On the other hand, if the difference in estimates of σ^2 is close to zero, including p in the formula will add a penalty in the Cp statistic for having more explanatory variables than necessary.

The Cp Plot

A Cp plot is a scatterplot with one point for each subset model. The y-coordinate is the Cp statistic, and the x-coordinate is p—the number of coefficients in the model. Each point in the plot is labeled to show the model to which it corresponds, often by including one-letter codes corresponding to each explanatory variable that appears in the model.

The Cp plot, which is available in many statistical computer packages, can be scanned to identify several models with low Cp statistics. The line at Cp $= p$ is often included on the plot, since a model without bias should have Cp approximately equal to p. Some users treat this line as a reference guide for visual assessment of the breakdown of the Cp statistic into bias and variance components; but the Cp statistic itself is the criterion. Thus, the models with smallest Cp statistics are considered. There is nothing magical about the model with the smallest Cp, since the actual ordering is likely to be affected by sampling variability; thus, a different set of data may result in a different ordering. Picking a single model from among all those with small Cp statistics is usually a matter of selecting the most convenient one whose coefficients all differ significantly from zero.

SAT Example

Display 12.8 shows a Cp plot for the SAT data (without Alaska), including only models with Cp less than 10. The models are identified on the plot by the variables included, with $t = $ log percentage of takers, $i = $ income, and so on. Observe that the Cp statistic for the general model that includes all variables—*tiyper*—equals the number of parameters, 7. The model *tyer*, which was identified by backward elimination, has the smallest Cp statistic.

12.4.2   Schwarz's Bayesian Information Criterion

G. Schwarz (in "Estimating the Dimension of a Model," *Annals of Statistics* 6 (1978): 461–64) derived a criterion that imposes a heavy penalty on models that contain large numbers of parameters. Schwartz's solution is called the Bayesian Information Criterion (BIC) and is calculated as

$$\mathrm{BIC} = n \log(\hat{\sigma}^2) + p \log(n).$$

Display 12.8   Cp plot for state SAT averages (showing only those models with Cp < 10); t = log percentage of takers, i = income, y = years, p = public, e = expend, and r = rank

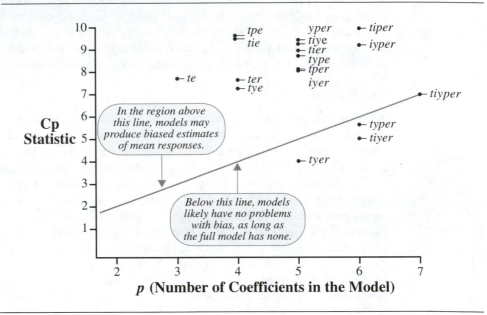

According to this criterion, the best models are the ones with the smallest BIC, which effectively means those with small $\hat{\sigma}^2$ and small p. The rationale for this particular function of $\hat{\sigma}^2$ and p as a criterion is discussed later. Theoretical reasoning and computer simulation suggest that it identifies good subset models.

SAT Example

The model *te* has the smallest BIC, at 321.9. It is followed by *tyer*, with BIC = 323.8; *tye*, with BIC = 324.5; *ter*, with BIC = 324.8; *tie*, with BIC = 326.8; *tpe*, with BIC = 326.9; and *tiyer*, with BIC = 327.7. (Compare these results with the Cp plot, Display 12.8.)

Beliefs About Different Models

One branch of statistical reasoning, called *Bayesian statistics*, assumes that beliefs in various hypotheses can be expressed on a probability scale. The elegant theorem of English mathematician Thomas Bayes (1702–1761), known as Bayes' Theorem, specifies precisely how beliefs held prior to seeing the data should be altered by statistical evidence to arrive at a new set of posterior beliefs. In the present setting, Bayesian statistics can be used to update the probability that each subset model is correct, given the available data.

Suppose that consideration is confined to models $M_1, M_2, \ldots, M_K$, and suppose that prior probabilities (probabilities of belief prior to seeing the data) on these models are

$pr\{M_1\}$, $pr\{M_2\}$, ..., $pr\{M_K\}$. Applying a Bayesian formulation, Schwarz showed that the updated probabilities for the models, given the data (D), were approximately

$$pr\{M_j \,|\, D\} \;=\; pr\{M_j\} \times \exp\{-BIC_j\} \,/\, SUM$$

where

$$SUM \;=\; \sum_{i=1}^{K} pr\{M_i\} \times \exp\{-BIC_i\},$$

where $pr\{M_j|D\}$, called a *posterior probability*, is the probability of M_j after seeing the data. When no prior reasons exist for believing that one model is better than any other, $pr\{M_j\}$ is the same for all j, and the posterior probabilities are simple functions of the BIC values for each model.

For the SAT example, the values of BIC have been calculated for each model and then used in this formula to identify the posterior probabilities. The posterior probability that the model is *te* is 0.764, followed by *tyer* with 0.115, *tie* with 0.060, and *tpe* with 0.041. The posterior probability that the model is any of the other 60 models is 0.020, and no other single model has a posterior probability greater than 0.006.

12.5   Analysis of the Sex Discrimination Data

To investigate whether males tended to receive larger beginning salaries than females requires a strategy that convincingly accounts for variables other than the sex indicator before adding it to the equation.

Initial investigation indicates that beginning salary should be examined on the log scale. Scatterplots of log beginning salary versus other variables show that the effect of experience is not linear. Beginning salaries increase with increasing experience up to a certain point; then they level off and even drop down for individuals with more experience. A quadratic term could reproduce this behavior well. A similar effect is seen with age: a correlation between age and experience may be responsible for that effect, but it is wise initially to consider a squared term for both variables. It may seem that seniority (number of months working with the company) is an inappropriate term to include for modeling beginning salary, but its inclusion accounts for increasing beginning salaries over time. Rather than as seniority, it might be thought of as time prior to March 1977 that the individual was hired.

Such considerations point to the necessity of including in the investigation a model rich enough to be extremely unlikely to miss an important relationship between log beginning salary and all explanatory variables other than the sex indicator. A saturated second-order model that includes all quadratic terms and all interaction terms satisfies that criterion. (See Section 12.6.3 for further discussion of second-order models.)

Notation

Fourteen variables appear in the saturated second-order model, so naming variables with single letters utilizes space well, even though it sacrifices clarity. The full complement of explanatory variables is given in Display 12.9. A word thus represents a model whose explanatory variables consist of the letters that make up the word. For example, the word *saebc* represents the model with seniority, age, education, the square of age, and the product of age with education—a model with six parameters in all (including the constant).

Display 12.9   Explanatory variables for the sex discrimination data

---

Main Effect Variables	Quadratic Variables	Interaction Variables	
s = Seniority	$t = s^2$	$m = s \times a$	$c = a \times e$
a = Age	$b = a^2$	$n = s \times e$	$k = a \times x$
e = Education	$f = e^2$	$v = s \times x$	$q = e \times x$
x = Experience	$y = x^2$		

A total of $16{,}384 (= 2^{14})$ models are formed by all possible combinations of the 14 terms in the preceding box. The goal is to find a good subset of these terms for explaining log salary, to which the sex indicator variable may be added. Although quadratic and interaction terms should be considered, it is not good practice to include quadratic terms without the corresponding linear term; neither is it good practice to include interaction terms without the two corresponding main effects. The preferred approach, therefore, is to identify the better models (of the 16,384) and then to identify the best subset of these that meets the narrower criteria restrictions of "good practice."

Identifying Good Subset Models

Display 12.10 shows a Cp plot. The Cp statistics were based on the full model containing all 14 terms. To reduce clutter, the plot displays Cp statistics only for models that meet the criteria discussed in the previous paragraph and, in addition, have small Cp values. Only a few good-fitting models with more than seven parameters are shown on this plot.

Models *saexck* and *saexnck* have lower Cp statistics than any others, and they appear to have no bias relative to the saturated model (since they fall near or below the line at $Cp = p$). The BIC clarifies the issue further: among 382 models examined, the BIC assigns a posterior probability of .7709 to *saexck*, with *saexyc* receiving the second highest posterior probability of .0625. Display 12.11 shows the 20 models that recorded the highest posterior probabilities.

Evaluating the Sex Effect

When the sex indicator variable is added into model *saexck*, its estimated coefficient is -0.1196, with a standard error of 0.0229. The t-statistic is 5.22, and the test that the

Display 12.10   Cp plot for the sex discrimination study

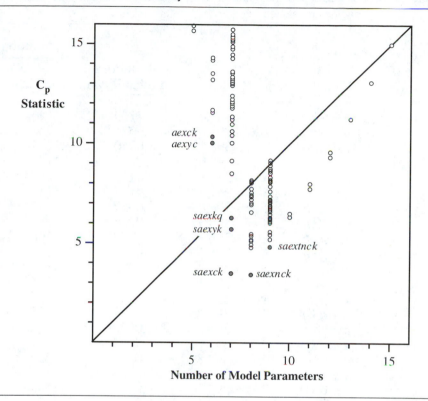

coefficient is zero against the alternative that it is negative has a one-sided p-value $= 6.3 \times 10^{-7}$. Thus, conclusive evidence exists that sex is associated with beginning salary, even after seniority, age, education, and experience are accounted for in this model. The median female salary is estimated to be $\exp(-0.1196) = 0.887$ times as large as (that is, about 11% less than) the median male salary, adjusted for the other variables.

Accounting for Model Selection Uncertainty Using the BIC

The preceding inferential statements pertain specifically to the single model *saexck*, selected to represent the influence of the potentially confounding variables. One criticism of this result is that it fails to account for the uncertainty involved in the model selection process. A second criticism is that the addition of the sex indicator variable to other good-fitting models might produce less convincing evidence of a sex effect.

The posterior probabilities for various models allow the analyst to address both criticisms. Suppose that δ represents the coefficient of the sex indicator variable and that $f(\delta|M_j)$ represents some function $f(\delta)$ based on the assumption that model M_j is correct. The Bayesian analysis permits an average estimate of $f(\delta)$ over different models.

Display 12.11   Bayesian posterior analysis of the difference between male and female log beginning salaries

---

Model	p	BIC	Posterior Probability	Addition of Sex Indicator coeff	SE	One-sided p-value
saexck	7	−401.40	.7709	−.1196	.0229	6.27E-7
saexyc	7	−398.89	.0625	−.1287	.0226	8.42E-8
saexkq	7	−398.28	.0340	−.1244	.0221	1.18E-7
saexnck	8	−398.08	.0279	−.1173	.0229	9.48E-7
aexyc	6	−397.81	.0213	−.1247	.0238	5.59E-7
aexck	6	−397.51	.0157	−.1135	.0246	6.94E-6
saexckb	8	−396.49	.0057	−.1195	.0229	6.70E-7
saexckt	8	−396.37	.0051	−.1189	.0232	9.10E-7
saexkqb	8	−396.36	.0050	−.1206	.0221	2.41E-7
saexycn	8	−396.33	.0048	−.1258	.0225	1.37E-7
saexk	6	−396.26	.0045	−.1331	.0221	1.96E-8
sexyq	6	−396.15	.0040	−.1345	.0201	1.02E-9
saexckf	8	−396.12	.0039	−.1196	.0230	6.93E-7
saexckq	8	−396.05	.0037	−.1208	.0230	5.54E-7
exyq	5	−395.93	.0032	−.1302	.0211	1.11E-8
saexcky	8	−395.91	.0032	−.1257	.0232	2.81E-7
saexyq	7	−398.89	.0031	−.1328	.0218	1.51E-8
saexckm	8	−395.84	.0030	−.1195	.0231	7.46E-7
saexckv	8	−395.80	.0028	−.1196	.0231	7.31E-7
saexbc	7	−395.20	.0016	−.1230	.0237	6.95E-7

s = Seniority	$t = s^2$	$m = s \times a$	$c = a \times e$
a = Age	$b = a^2$	$n = s \times e$	$k = a \times x$
e = Education	$f = e^2$	$v = s \times x$	$q = e \times x$
x = Experience	$y = x^2$		

Furthermore, the average is weighted to represent the strength of belief in each of the various models. The estimate of $f(\delta)$ in this way incorporates the uncertainty in model selection. The weighted average is

$$f\{\delta \mid D\} = \sum_{i=1}^{K} f(\delta \mid M_j) \times pr\{M_j \mid D\}.$$

Two choices for $f(\delta)$ are δ itself (the estimate of the sex effect) and the p-value for testing its significance. The sex indicator variable was added to each of the 382 models and combined as shown in the preceding equation to produce the single (posterior) estimate of -0.1206 and a single (posterior) p-value of 6.7×10^{-7} (this requires some programming). A listing of the top 20 candidate models, together with their contributions to the posterior information, is shown in Display 12.11.

The use of Bayesian statistics requires that somewhat different interpretations be given to the estimate and to the p-value. The average estimate is termed a *Bayesian posterior-mean estimate*, while the average p-value is called the *posterior probability* that δ is greater than zero. The probability refers to a measure of belief about the hypothesis that the sex coefficient is zero. This is a different philosophical attitude than any previously considered, but two issues argue in its favor for this application. First the formula itself gives the p-value for every model a chance to contribute in the final estimate, with a weight appropriate to the support given the model by the data. Second, no known statistical procedures based on other philosophies incorporate model selection uncertainty in this way.

12.6   Related Issues

12.6.1   The Trouble with Interpreting Significance When Explanatory Variables Are Correlated

When explanatory variables are correlated, interpretation of regression coefficients and their significance is quite difficult and perhaps too convoluted to be of use. The central difficulty is that interpreting a single coefficient as the amount by which the mean response changes as the single variable changes by one unit, holding all other variables fixed, does not apply; this is because the data generally lack experience with situations where one variable can be changed in isolation.

Consider the variable *expend*—the per student state expenditure—in the SAT example. This variable can be added to any model that does not already contain it. This was done for all models not containing *expend*, with the results shown in Display 12.12.

The first three columns of Display 12.12 are self-explanatory, the key point being the fact that the p-values vary widely. The last column is the addition to R^2 resulting from the addition of *expend* to the model. The variation in these percentages can be traced to the correlation between *expend* and the other variables used in the model. The last column can be interpreted as the extra sum of squares due to *expend*, expressed as a percentage of the total sum of squares. The penultimate column is the extra sum of squares due to *expend*, expressed as a percentage of the sum of squared residuals from the model without it. The huge variations in that column arise because of the differing amounts of variability explained by the different models. This is most directly related to the variation in p-values for *expend*, but the correlational aspect also plays a role.

Ultimately, the answer to the often-asked question, "Which variables are significant?" has to be "It depends." Similarly, there is usually no definitive answer to the question, "Which variable is most significant?" It is best to try to focus attention on questions

Display 12.12   Statistical measures for the contribution of *expend* to different models (Alaska removed)

			% Variation Explained by *Expend*	
Model	*t*-Statistic	Two-sided *p*-value	Previous Residual	Total Variation
E	0.27	.79	0.2	0.2
I+E	0.40	.69	0.3	0.2
Y+E	0.86	.39	1.6	1.4
P+E	0.18	.86	0.1	0.1
R+E	4.78	.00002	49.6	7.4
T+E	6.04	.0000002	79.4	8.4
IY+ E	0.07	.95	0.0	0.0
IP+E	0.11	.92	0.0	0.0
IR+E	4.88	.000001	52.9	6.8
IT+E	5.88	.0000005	76.7	8.1
YP+E	1.05	.30	2.5	2.1
YR +E	4.20	.0001	39.2	4.3
YT+E	5.31	.000003	62.6	6.3
PR+E	6.02	.0000003	80.5	9.4
PT+E	5.91	.0000004	77.7	8.2
RT+E	6.13	.0000002	83.5	8.4
IYP+E	0.92	.36	1.92	0.9
IY R+E	4.35	.00008	43.1	4.2
IYT+ E	5.20	.000005	61.4	6.2
IPR+E	5.57	.000001	70.6	8.0
IPT+E	5.59	.000001	71.1	7.5
IRT+E	5.92	.0000004	79.7	7.9
YPR +E	4.80	.00002	52.3	5.3
YPT+E	4.78	.00002	51.9	5.1
Y RT+E	5.23	.000005	62.1	5.6
PRT+E	6.27	.0000001	89.5	8.8
IY PR+E	4.27	.0001	42.4	4.1
IYPT+E	4.33	.00009	43.7	4.3
IYR T+E	5.00	.00001	58.2	5.1
IPRT+E	5.94	.0000004	82.0	8.0
Y PRT+E	5.13	.000007	61.3	5.4
IYPR T+E	4.64	.00003	51.2	4.5

that ask specifically for the assessment of significance after accounting for other specified variables or groups of variables.

12.6.2 Regression for Adjustment and Ranking

One valuable use of regression is for adjustment. Consider the regression of state SAT averages on percentage of eligible students in the state who decide to take the test and on the median class rank of the students who take the exam. A residual plot from the fit (excluding Alaska) is shown in Display 12.13. The residuals are the state SATs with the estimated linear effects of log percentage of takers and median class rank removed. Thus, they serve as the SATs adjusted for these two variables (and are scaled so that the average is

Display 12.13 Scatterplot of residuals versus fitted values from the regression of state SAT average on percentage of takers and median class rank of takers

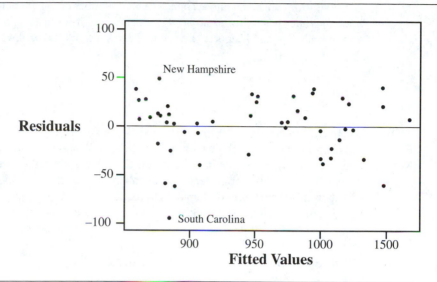

zero). New Hampshire, for example, whose "raw" SAT average is 925 (which is 23 points less than the overall average of all states) has a very high percentage of eligible students who take the exam and a fairly low median class rank among the takers. The estimated mean SAT score for a state with these values is 876. Since New Hampshire's actual SAT score was 925, its value exceeded the prediction (according to the model with percentage of takers and median class rank) by 49 points. It is therefore appropriate to say that New Hampshire's SAT adjusted for percentage of takers and median class rank of takers is 49 points above average.

The ranking on the left side of Display 12.2 is based directly on the residuals shown in Display 12.13. New Hampshire has the largest residual, and therefore the largest SAT adjusted for percentage of takers and median class rank. South Carolina has the smallest.

The ranking on the right side of Display 12.2 is based on the residuals from a model that also includes expenditure. In this case, New Hampshire's SAT, adjusted for percentage of takers, median class rank, and expenditure, is 67 points above the average predicted for a state with the same values of percentage of takers, median rank, and expenditure.

12.6.3 Saturated Second-order Models

A saturated second-order model (SSOM) includes the squares of all explanatory variables and all cross-products of pairs of explanatory variables. It describes $\mu(Y)$ as a completely arbitrary parabolic surface. The SSOM should contain a subset model that describes the regression surface well.

Numbers of Subset Models

Given K original variables, the SSOM itself contains $[K(K+3)/2]+1$ parameters. When all possibilities are taken into account, the total number of p-parameter hierarchical models (meaning ones where, for example, X_1^2 appears only if X_1 also appears, and X_1X_2 appears only if both X_1 and X_2 appear) available from K original variables is

$$\sum_{j=0}^{K} C_{K,j} \times C_{(C_{j+1,2}),(p-1-j)},$$

where $C_{n,m}$, recall, is $n!/[m!(n-m)!]$ and $C_{n,m}=0$ whenever $n<m$. Holding out the sex indicator in the sex discrimination problem leaves $K=4$ explanatory variables: *seniority*, *age*, *education*, and *experience*. The full SSOM has 15 parameters, and subset models have 1, 2, 3, ..., 14 parameters. The number of distinct hierarchical models that can be considered is the number of hierarchical modes with one parameter plus the number with two parameters and so on up to the number of hierarchical models with fourteen parameters. According to the formula, a total of 1,337 distinct models can be considered.

Strategies for Exploring Subsets of the SSOM

The SAT study featured $K=6$ explanatory variables. The SSOM for that problem contains 28 parameters. The number of hierarchical models with 17 parameters, for example, is 352,716, and the total of all hierarchical models to consider is 2,104,489. Things rapidly get out of hand, and it becomes necessary to plan a strategy for sorting through some—but not all—of the models for promising candidates.

The strategy employed in the sex discrimination study was to examine all subset models up to some level ($p=7$) first. Forward selection from the best models with $p \leq 7$ came next, followed by backward elimination from the saturated second-order model. Finally, models in the neighborhood of promising models were included. A *neighborhood* of one model consists of all models that can be obtained by adding or dropping one variable.

Another strategy involves grouping variables into sets that have a common theme and exploring each set separately to determine its best subset. When these are combined into an overall model, subsets consisting of products of retained variables from one set with those from another may also be examined.

A third strategy is to identify which of the original variables are important, by examining all subsets with main effects only (i.e., excluding squared and interaction terms), and then to build the SSOM from those variables alone and explore subsets.

12.6.4 Cross Validation

Inference after the use of a variable selection technique is tainted by the data snooping involved in the process. The selected model is likely to fit much better to the data that gave it birth than to fresh data. Thus, p-values, confidence intervals, and prediction intervals should be used cautiously.

If the data set is very large, the analyst may benefit from dividing it at random into separate model construction and validation sets. A variable selection technique can be used on the construction set to determine a set of explanatory variables. The selected model can then be refit on the validation set, without any further exploration into suitable explanatory variables, and inferential questions can be investigated on this fit, ignoring the construction data. When the purpose of the regression analysis is prediciton, it is recommended that the validation data set be about 25% of the entire set. Although the saving of the 25% of the data for validation seems reasonable for other purposes, the actual benefits are not very well understood.

12.6.5 Bias and Variance of Fitted Values

The bias and variance of fitted values associated with the fit of a model are important measures of how well a model fits the data. These concepts arise from the behavior of statistical estimates when a procedure is repeated over and over to form a sampling distribution. Suppose that the proposed model to be fit to the mean response is a straight line in the explanatory variable—$\mu\{response \mid explan\} = \beta_0 + \beta_1 explan$—whereas the correct model is quadratic: $\mu\{response \mid explan\} = \beta_0 + \beta_1 explan + \beta_2 explan^2$. Repeatedly fitting straight lines tends to produce fitted values that consistently overshoot the true mean at some point (Display 12.14). The variation of the fitted values around their average is termed the variance of the fitted values. The difference between the average of the fitted values and the true mean is the bias. The TMSE measures the typical distance from all fitted values to the

Display 12.14 Bias and variance of fitted values with a poorly specified model (hypothetical example)

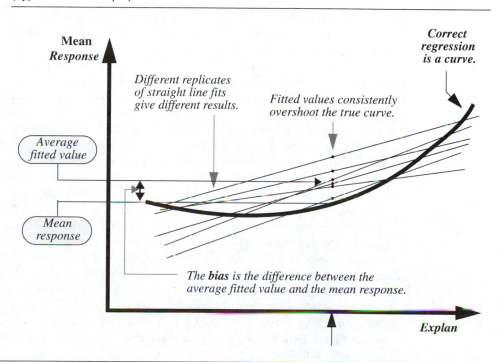

true means; thus, it combines both the variances and the biases of fitted values at different points along the curve.

12.7 Summary

Model selection requires an overall strategy for analyzing the data with regression tools (see Display 9.9 on page 241). After giving initial thought to a game plan for investigating the questions of interest, the preliminary analysis consists of a combination of exploration, model fitting, and model checking. Once some useful models have been identified, the answers to the questions of interest can be addressed through inferences about their parameters.

Tools for initial exploration include graphical methods (such as a matrix of scatterplots, coded scatterplots, jittered scatterplots, and interactive labeling) and correlation coefficients between various variables. Certain tricks for modeling are used extensively in the case studies—indicator variables, quadratic terms, and interaction terms. For model checking and model building, a number of other tools are suggested: residual plots, partial residual plots, informal tests of coefficients, case influence statistics, the Cp plot, the BIC, and sequential variable selection techniques. Finally, some inferential tools are presented: t-tests and confidence intervals for individual coefficients and linear combinations of coefficients, extra sums-of-squares F-tests, prediction intervals, and calibration intervals.

SAT Study

Although this example is used to demonstrate variable selection techniques, the actual analysis is guided by the objectives, and variable selection played only a minor role. These data have been used to rank the states on their success in secondary education, but in this regard selection bias poses a serious problem. In some states, for example, a high SAT average reflects the fact that only a small proportion of students—the very best ones— took the test, rendering the self-selected sample far from representative of high school students in the state overall. One goal of the regression analysis is to establish a ranking that accounts for this selection bias. Although overcoming the limitations of a self-selected sample is impossible, it is possible to rank the states after subtracting out the effects of the different proportions of students taking the test and their different median class rankings. Accomplishing this involves fitting the regression of SAT scores on these two explanatory variables, and ranking the states according to the sizes of their residuals.

A further question is exploratory in nature: are any other variables associated with SAT score? For example, is the amount of money spent on secondary education related to it? Variable selection techniques may be useful for sorting through various models to identify promising predictors. Whatever models are suggested, though, only the ones that include percentage of takers and median class rank should be chosen, since the question does not make sense unless it addresses the effect on SAT after these two variables associated with the selection bias are accounted for. After performing a transformation of the percentage takers and setting Alaska aside, the Cp plot selects the model *tyer* with log takers, years (studying natural science, social science, and humanities), expenditure, and median rank as explanatory variables.

Sex Discrimination Study

Model selection in this example is motivated by the need to account fully for the nondiscriminatory explanatory variables before the sex indicator variable is introduced. Initial scatterplots indicate the important explanatory variables, which permit fitting a tentative model on which residual analysis can be conducted. After transformation, the analysis consists of using a variable selection technique to identify a good set of variables for explaining beginning salary. An inferential model is formed by this set plus the sex indicator variable, making precise the objective of seeing whether sex constitutes an important explanatory factor *after* everything else is accounted for. It may be disconcerting to realize that the different variable selection routines indicate different subset models; but it is reassuring that the estimated sex effect is about the same in all cases.

Further Reading

Discussions of multicollinearity, the Cp plot, sequential variable selection tools, and other methods and issues associated with variable selection are found in most modern books on regression; see, for example, Draper and Smith (1981) and Neter, Wasserman, and Kutner (1990). An example of cross-validation is provided in the latter text. For a more in-depth discussion of cross-validation, see Mosteller and Tukey (1977) and Picard and Berk (1990).

12.8 Exercises

Conceptual Exercises

1. **State SATs.** True or false? If the coefficient of the public school percentage is not significant (p-value $> .05$) in one model, it cannot be significant in any model found by adding new variables to it.

2. **State SATs.** True or false? If the coefficient of income is significant (p-value $< .05$) in one model, it will also be significant in any model that is found by adding new variables to it.

3. **State SATs.** Why are partial residual plots useful for this particular data problem?

4. **Sex Discrimination.** Suppose that another explanatory variable available on each employee is length of hair at time of hire. (a) Will the estimated sex difference in beginning salaries be greater, less, or unchanged when length of hair is included in the set of explanatory variables for adjustment? (b) Why should this variable not be used?

5. The following formula defines the variance of the sampling distribution of a least squares estimate of a regression coefficient in multiple regression:

$$\text{Var}(\hat{\beta}_j) = \frac{\sigma^2}{(n-1)s_j^2(1 - R_j^2)}.$$

(a) Is this equal to the formula for the estimated slope in simple regression if there is one X?

(b) Suppose that the variance of the estimated slope in the simple regression of Y on X_1 is 10. Suppose that X_2 is added to the model, and that X_2 is uncorrelated with X_1. Will the variance of the coefficient of X_1 still be 10?

6. In a study of mortality rates in major cities, researchers collected four weather-related variables, eight socioeconomic variables, and three air-pollution variables. Their primary question concerned

the effects, if any, of air pollution on mortality, after accounting for weather and socioeconomic differences among the cities. (a) What strategy for approaching this problem should be adapted? (b) What potential difficulties are involved in applying a model-selection strategy similar to the one used in the sex-discrimination study?

7. In the Cp plots shown in Display 12.8 and Display 12.10, the model with all available explanatory variables falls on the line (with intercept 0 and slope 1), meaning that the value of Cp for this model is exactly p. Will this always be the case? Why?

8. What is the usual interpretation of the probability of an event E? How does a Bayesian interpretation differ?

9. How does the posterior function, discussed in Section 12.5, account for model-selection uncertainty?

Computational Exercises

10. A, B, and C are three explanatory variables in a multiple linear regression with $n = 28$ cases. Display 12.15 shows the residual sums of squares and degrees of freedom for all models.

Display 12.15   Data for Exercise 10

Model Variables	Residual Sum of Squares	Degrees of Freedom
none	8,100	27
A	6,240	26
B	5,980	26
C	6,760	26
AB	5,500	25
AC	5,250	25
BC	5,750	25
ABC	5,160	24

(a) Calculate the estimate of σ^2 for each model. (b) Calculate the adjusted R^2 for each model. (c) Calculate the Cp statistic for each model. (d) Calculate the BIC for each model. (e) Which model has (i) the smallest estimate of σ^2? (ii) the largest adjusted R^2? (iii) the smallest Cp statistic? (iv) the smallest BIC?

11. Using the residual sums of squares from Exercise 10, find the model indicated by forward selection. (Start with the model "none," and identify the single-variable model that has the smallest residual sum of squares. Then perform an extra-sum-of-squares F-test to see whether that variable is significant. If it is, find the two-variable model that includes the first term and has the smallest residual sum of squares. Then perform an extra-sum-of-squares F-test to see whether the additional variable is significant. Continue until no F-statistics greater than 4 remain for inclusion of another variable.)

12. Again referring to Exercise 10, calculate $\exp\{-BIC + BIC_{min}\}$ for each model. Add these up and divide each by the sum. What is the resulting posterior distribution on the models?

13. Use the computer to simulate 100 data points from a normal distribution with mean 0 and variance 1. Store the results in a column called Y. Repeat this process 10 more times, storing results in $X_1, X_2, \ldots, X_{10}$. Notice that the Y should be totally unrelated to the explanatory variables. (a) Fit the regression of Y on all 10 explanatory variables. What is R^2? (b) What model is suggested by forward selection? (c) Which model has the smallest Cp statistic? (d) Which model has the smallest BIC? (e) What danger (if any) is there in using a variable selection technique when the number of explanatory variables is a substantial proportion of the sample size?

14. Blood–Brain Barrier. Using the data in Display 11.3, perform the following variable selection techniques to find a subset of the covariates days after inoculation, tumor weight, weight loss, initial weight, and sex for explaining log of the ratio of brain tumor antibody count to liver antibody count. (a) Cp plot (b) forward selection (c) backward elimination (d) stepwise regression

15. Blood–Brain Barrier. Repeat Exercise 14, but include sacrifice time (treated as a factor with three levels), treatment, and the interaction of absorption and treatment with the other explanatory variables.

16. Sex Discrimination. The analysis in this chapter focused on beginning salaries. Another issue is whether annual salary increases tended to be higher for males than for females. If an annual raise of $100r\%$ is received in each of N successive years of employment, the salary in 1977 is: Sal77 = SalBeg $\times (1+r)^N$. Seniority measures the number of months of employment, so the number of years of employment is $N =$ seniority/12; and

$$\log(1+r) = (12/\text{seniority}) \times (\text{Sal77}/\text{SalBeg}) = z(\text{say})$$

for each individual. Calculate z from beginning salary, 1977 salary, and seniority; then calculate r as $\exp(z) - 1$. Now consider r as a response variable (the average annual raise). (a) Use a two-sample t-test to see whether the distribution of raises is different for males than for females. (Is a transformation necessary?) (b) What evidence is there of a sex effect after the effect of age on average raise has been accounted for? (c) What evidence is there of a sex effect after the effects of age and beginning salary have been accounted for?

17. Pollution and Mortality. Display 12.16 shows the complete set of variables for the problem introduced in Exercise 11.23 on page 322. The 15 variables for each of 60 cities are (1) mean annual precipitation (in inches); (2) percent relative humidity (annual average at 1 P.M.); (3) mean January temperature (in degrees Fahrenheit); (4) mean July temperature (in degrees Fahrenheit); (5) percentage of the population aged 65 years or over; (6) population per household; (7) median number of school years completed by persons of age 25 years or more; (8) percentage of the housing that is sound with all facilities; (9) population density (in persons per square mile of urbanized area); (10) percentage of 1960 population that is nonwhite; (11) percentage of employment in white-collar occupations; (12) percentage of households with annual income under $3,000 in 1960; (13) relative pollution potential of hydrocarbons (HC); (14) relative pollution potential of oxides of nitrogen (NO_X); and (15) relative pollution potential of sulphur dioxide (SO_2). (See Display 11.21 on page 322 for the city names.) It is desired to determine whether the pollution variables (13, 14, and 15) are associated with mortality, after the other climate and socioeconomic variables are accounted for. [*Note:* These data have problems with influential observations and with lack of independence due to spatial correlation; these problems are ignored for purposes of this exercise.]

(a) With mortality as the response, use a Cp plot and the BIC to select a good-fitting regression model involving weather and socioeconomic variables as explanatory. To the model with the lowest Cp, add the three pollution variables (transformed to their logarithms) and obtain the p-value from the extra-sum-of-squares F-test due to their addition.

(b) Repeat part (a) but use a sequential variable selection technique (forward selection, backward elimination, or stepwise regression). How does the p-value compare?

18. Suppose that a problem involves four explanatory variables (like the weather variables in Exercise 17). How many variables would be included in the corresponding saturated second-order model (Section 12.6.3)?

Display 12.16   Pollution and mortality data for 60 cities

Column headers (left to right): SMSA; Mortality (d/100,000); *Weather* — Precipitation (in/yr), Humidity (annual av %), January Temperature (F), July Temperature (F); *Demographic* — Over 65 (%), Population per Household, Education (years), Housing (% sound), Density (popn/mi^2), Nonwhite (%), White Collar (%), Poor (% < \$3,000); *Pollution Potentials* — Hydrocarbons, Nitrogen Oxides, Sulfur Dioxide

SMSA	Mortality (d/100,000)	Precip	Humid	Jan Temp	Jul Temp	Over 65 (%)	Pop/Household	Education	Housing	Density	Nonwhite (%)	White Collar (%)	Poor (%)	Hydrocarbons	Nitrogen Oxides	Sulfur Dioxide
SJCA	790.733	13	71	49	68	7.0	3.36	12.2	90.7	2702	3.0	51.9	9.7	105	32	3
WIKS	823.764	28	54	32	81	7.0	3.27	12.1	81.0	3665	7.5	51.6	13.2	4	2	1
SDCA	839.709	10	61	55	70	7.3	3.11	12.1	88.9	3033	5.9	51.0	14.0	144	66	20
LAPA	844.053	43	54	32	74	10.1	3.38	9.5	79.2	3214	2.9	43.7	12.0	11	7	32
MIMN	857.622	25	58	12	73	9.2	3.28	12.1	83.1	2095	2.0	51.9	9.8	20	11	26
DATX	860.101	35	54	46	85	7.1	3.22	11.8	79.9	1441	14.8	51.2	16.1	1	1	1
MIFL	861.439	60	60	67	82	10.0	2.98	11.5	88.6	4657	13.5	47.3	22.4	3	1	1
LACA	861.833	11	47	53	68	9.2	2.99	12.1	90.6	4700	7.8	48.9	12.3	648	319	130
GRMI	871.338	31	61	24	72	9.0	3.37	10.9	82.8	3226	5.1	45.2	12.3	5	3	10
DECO	871.766	15	38	30	73	8.2	3.15	12.2	84.2	4824	4.7	53.1	12.7	17	8	28
RONY	874.281	32	60	25	72	10.9	3.21	11.1	82.5	4355	5.0	46.4	10.8	7	4	18
HACT	887.466	43	56	27	72	9.0	3.25	11.5	87.1	2909	7.2	51.6	9.5	7	3	10
FWTX	891.708	31	53	45	85	7.3	3.22	11.4	80.7	1844	11.5	48.1	18.5	1	1	1
POOR	893.991	37	73	38	67	11.3	2.99	12.0	81.5	3387	3.6	50.3	13.5	56	21	44
WOMA	895.696	45	56	24	70	11.8	3.25	11.1	79.8	3678	1.0	44.8	14.0	7	3	8
SEWA	899.264	35	72	40	64	9.6	3.02	12.2	82.5	3626	5.7	54.3	10.1	20	7	20
BRCT	899.529	45	56	30	73	9.3	3.29	10.6	86.0	2140	5.3	40.4	10.5	6	4	4
SPMA	904.155	45	56	28	74	10.6	3.21	11.1	82.6	1883	3.4	41.9	12.3	5	4	20
SFCA	911.701	18	71	48	63	9.2	2.92	12.2	87.7	4253	13.7	51.2	12.0	311	171	86
YOPA	911.817	42	54	33	76	9.7	3.22	9.0	76.2	9699	4.8	42.2	14.5	8	8	49
UTNY	912.202	40	60	23	71	11.3	3.28	10.3	73.8	1671	2.5	47.4	13.5	5	2	11
CAOH	912.347	36	59	27	72	9.5	3.36	10.7	79.3	4213	6.7	41.0	13.2	12	7	20
KCMO	919.729	35	55	31	81	9.2	3.10	12.0	78.3	3262	12.6	48.6	13.9	7	4	4
AKOH	921.870	36	59	27	71	8.1	3.34	11.4	81.5	3243	8.8	42.6	11.7	21	15	59
NHCT	923.234	46	58	30	72	10.2	3.16	11.3	83.2	3327	8.8	45.3	12.2	4	3	8
MIWI	929.150	30	64	20	69	8.8	3.26	11.1	85.4	2934	5.8	44.0	9.4	33	23	125
BOMA	934.700	43	56	30	74	10.9	3.23	12.1	83.9	4679	3.5	49.2	11.3	21	32	62
DAOH	936.234	36	58	30	75	7.5	3.35	11.4	81.9	4029	12.4	44.0	12.0	6	4	16
PRRI	938.500	42	56	29	72	10.7	3.19	10.1	79.5	3508	2.2	38.8	15.7	6	4	18
FLMI	941.181	30	61	24	72	6.5	3.53	10.8	79.5	3694	13.1	33.8	12.4	11	4	11
REPA	946.185	41	54	33	77	11.2	3.08	9.6	79.9	4843	2.7	38.6	14.1	11	11	89
SYNY	950.672	38	61	24	72	9.8	3.34	11.4	78.0	4923	3.8	50.5	11.1	8	5	25
HOTX	952.529	46	59	55	84	5.6	3.35	11.4	79.7	2647	21.0	46.9	17.9	6	5	1
SLMO	953.560	34	57	32	79	9.3	3.23	9.7	76.8	5160	17.2	45.1	15.3	31	15	68
YOOH	954.442	38	58	28	72	8.9	3.48	10.7	79.8	3451	11.7	37.5	13.0	14	13	39
COOH	958.839	37	58	31	75	8.0	3.26	11.9	78.4	4259	13.1	49.6	13.9	23	9	15
DEMI	959.221	31	59	27	74	7.2	3.44	10.8	87.0	4834	15.8	43.5	13.6	52	35	124
NATN	961.009	45	56	40	80	8.3	3.32	10.1	70.3	2682	21.0	46.1	24.1	17	14	78
ALPA	962.354	44	54	29	74	10.4	3.21	9.8	81.6	4260	0.8	39.4	12.4	6	6	33
WADC	967.803	41	52	37	78	6.2	3.25	12.3	89.5	5308	25.9	59.7	10.3	65	28	102
ININ	968.665	39	60	29	75	8.7	3.23	11.4	78.6	4412	15.6	46.6	13.2	13	7	33
CIOH	970.467	40	57	34	77	9.2	3.21	10.2	77.0	4101	13.0	45.7	15.1	26	26	146
GRNC	971.122	42	53	40	77	6.1	3.45	10.4	71.8	2269	22.7	41.4	19.5	8	3	5
TOOH	972.464	31	59	26	73	9.3	3.22	10.7	81.3	3249	9.5	43.9	13.6	11	7	25
ATGA	982.291	47	56	45	79	6.5	3.41	11.1	77.5	3125	27.1	50.2	20.6	18	8	24
CLOH	985.950	35	60	28	71	8.8	3.29	11.1	86.3	3042	14.7	44.6	11.4	31	21	64
LOKY	989.265	30	57	35	71	8.3	3.37	9.9	77.4	4474	13.1	42.6	17.7	38	37	193
PIPA	991.290	36	56	29	72	9.5	3.32	10.6	77.6	3437	8.1	45.5	13.8	45	59	263
NYNY	994.648	42	58	33	77	9.7	3.03	10.7	83.5	7462	11.3	48.7	12.4	41	26	108
ALBY	997.875	35	57	23	72	11.1	3.14	11.0	78.8	4281	3.5	50.7	14.4	8	10	39
BUNY	1001.902	36	61	24	70	9.0	3.31	10.5	83.2	6582	8.1	42.5	12.6	18	12	37
WIDE	1003.502	45	56	33	76	7.7	3.39	11.3	82.2	3152	12.1	47.3	10.9	14	11	42
METN	1006.490	50	59	42	82	7.3	3.49	10.4	72.5	3497	36.7	43.3	26.4	15	18	34
PHPA	1015.032	42	54	32	76	9.1	3.32	10.5	87.5	6092	17.5	45.3	13.2	29	32	161
CHTN	1017.613	52	56	42	79	7.7	3.39	9.6	69.2	2302	22.2	41.3	24.2	18	8	27
CHIL	1024.885	33	58	26	76	8.6	3.20	10.9	83.4	6122	16.3	44.9	10.7	88	63	278
RIVA	1025.502	44	53	39	78	8.2	3.32	11.0	79.9	3768	28.6	49.5	17.5	12	9	48
BIAL	1030.380	53	54	45	80	7.7	3.45	10.2	66.8	3235	38.5	43.1	25.5	30	32	72
BAMD	1071.289	43	55	35	77	7.6	3.44	9.6	84.6	6441	24.4	43.7	14.3	43	38	206
NOLA	1113.056	54	62	54	81	7.4	3.36	9.7	72.8	3172	31.4	45.5	24.2	20	17	1

19. In the expression (Section 12.6.3) for the number of subset models with p parameters, the index j of summation refers to the number of original variables in a particular model. If there are a total of five original explanatory variables, the total number of subset models with seven parameters that have exactly three original variables is the jth term in the sum, $C_{5,3} \times C_{6,3} = 10 \times 20 = 200$. The reasoning here is that there are ten ways to select three of the original five variables; and that, with three variables, there are three quadratic plus three product terms, or a total of six second-order terms; so, to make up a model with seven parameters, you have the constant [1] and the original variables [3], so you need $7 - 1 - 3 = 3$ second-order terms from the six available. Problem: Let the original variables be a, b, c, d, and e. Select any three of them. Then write down all 20 of the seven-parameter models involving only those three original variables.

Data Problems

20. Galapagos Islands. The data in Display 12.17 come from a 1973 study. (Data from M. P. Johnson and P. H. Raven, "Species Number and Endemism: The Galapagos Archipelago Revisited," *Science* 179 (1973): 893–95.) The number of species on an island is known to be related to the island's area. Of interest is what other variables are also related to the number of species, after island area is accounted for, and whether the answer differs for native species and nonnative species. (*Note:* Elevations for five of the islands were missing and have been replaced by estimates for purposes of this exercise.)

21. River Nitrogen. The rise in abundance of algae in coastal waters is thought to be due to increases in nutrients such as nitrate and other forms of nitrogen. It is theorized that the excessive amounts of nitrate are due to human influences. Researchers gathered the data shown in Display 12.18 to gauge the evidence that nitrates in the discharges of rivers around the world are associated with human population density. (Data from J. L. Cole et al., "Nitrogen Loading of Rivers as a Human-driven Process," in M. J. McDonnell and S. T. A. Pickett, eds., *Humans as Components of Ecosystems: The Ecology of Subtle Human Effects and Populated Areas* (New York: Springer-Verlag, 1993), p. 141–57.) Human populations can affect nitrogen inputs to rivers through industrial and automobile emissions to the atmosphere (causing the nitrogen to enter the river through rainfall), through fertilizer runoff, through sewage discharge, and through watershed disturbance. The nine variables listed are (1) discharge, the estimated annual average discharge of the river into an ocean (in m^3/sec); (2) runoff, the estimated annual average runoff from the watershed (in liters/(sec $\times$ k $\times$ m^2)); (3) precipitation (in cm/yr); (4) area of watershed (in km^2); (5) density (people/km^2); (6) nitrate concentration (in $\mu M/l$); (7) nitrate export, which is the product of runoff times nitrate concentration; (8) deposition, which is the product of precipitation times nitrate concentration; and (9) nitrate precipitation, the concentration of nitrate in wet precipitation at sites located near the watersheds (in μmol NO_3/(sec $\times$ km^2). The response variables are nitrate concentration and nitrate export. It is desired to determine whether these (separately) are associated with deposition or nitrate precipitation, after accounting for discharge, runoff, precipitation, and area of watershed. It is also of interest to estimate the effect that human population density has, over and above its influence on the previous variables. Thus the analysis attempts to determine the extent to which the human effect is from pollutants discharged into the river directly as opposed to those discharged indirectly through atmospheric pollution. [*Note:* Three variables were not available for five rivers and have been replaced by estimates for the purposes of this data problem.]

Answers to Conceptual Exercises

1. False. In the model with P and E, neither variable is significant (Display 12.6). But both are significant in the model *PER*.

2. False again. Both income and rank are significant in the model *IR*. When T is included, however, neither is significant. (Again, see Display 12.6.)

Display 12.17 Plant species and geography of the Galapagos Islands

Island	Observed Species Total	Native	Area (km²)	Elevation (m)	Distance (km) From Nearest Island	From Santa Cruz	Area of Nearest Island (km²)
Baltra	58	23	25.09	332	0.6	0.6	1.84
Bartolome	31	21	1.24	109	0.6	26.3	572.33
Caldwell	3	3	0.21	114	2.8	58.7	0.78
Champion	25	9	0.10	46	1.9	47.4	0.18
Coamano	2	1	1.05	130	1.9	1.9	903.82
Daphne Major	18	11	0.34	119	8.0	8.0	1.84
Daphne Minor	24	12	0.08	93	6.0	12.0	0.34
Darwin	10	7	2.33	168	34.1	290.2	2.85
Eden	8	4	0.03	46	0.4	0.4	17.95
Enderby	2	2	0.18	112	2.6	50.2	0.10
Espanola	97	26	58.27	198	1.1	88.3	0.57
Fernandina	93	35	634.49	1,494	4.3	95.3	4,669.32
Gardner (Esp.)	58	17	0.57	49	1.1	93.1	58.27
Gardner (S.M.)	5	4	0.78	227	4.6	62.2	0.21
Genovesa	40	19	17.35	76	47.4	92.2	129.49
Isabela	347	89	4,669.32	1,707	0.7	28.1	634.49
Marchena	51	23	129.49	343	29.1	85.9	59.56
Onslow	2	2	0.01	25	3.3	45.9	0.10
Pinta	104	37	59.56	777	29.1	119.6	129.49
Pinzon	108	33	17.95	458	10.7	10.7	0.03
Las Plazas	12	9	0.23	84	0.5	0.6	25.09
Rabida	70	30	4.89	367	4.4	24.4	572.33
San Cristobal	280	65	551.62	716	45.2	66.6	0.57
San Salvador	237	81	572.33	906	0.2	19.8	4.89
Santa Cruz	444	95	903.82	864	0.6	0.0	0.52
Santa Fe	62	28	24.08	259	16.5	16.5	0.52
Santa Maria	285	73	170.92	640	2.6	49.2	0.10
Seymour	44	16	1.84	154	0.6	9.6	25.09
Tortuga	16	8	1.24	186	6.8	50.9	17.95
Wolf	21	12	2.85	253	34.1	254.7	2.33

There are two Gardners—one near Espanola and one near Santa Maria.

3. There are two reasons. (i) Percentage of takers and median rank explain so much of the variation that any additional effect of the other variables is hidden in the ordinary scatterplots. (ii) The questions of interest call for the examination of some of the variables after getting percentage of takers and median rank out of the way. The partial residual plots allow visual investigation into these questions.

4. (a) The estimated sex difference will probably be less after length of hair is accounted for. If there are differences between male and female hair lengths, then that variable is picking up the sex differences and the sex indicator variable will be less meaningful when it is included. (b) The males and females should be compared after adjustment for nondiscriminatory determinants of salary.

5. (a) Yes. R_j^2 is 0 if there are no other explanatory variables. (b) The denominator will not change; but if X_2 explains some variation in Y in addition to what is explained by X_1, σ will be smaller, so the variance of the coefficient of X_1 might decrease.

6. (a) Use model selection tools to find a good-fitting model involving the weather and socioeconomic variables. Then include the pollution variables, using t- and F-statistics to judge importance. As

Display 12.18   River nitrogen data

River, Location	Discharge	Runoff	Prec	Area	Density	NO$_3$	Export	Dep	NPrec
Adige, Italy	223	18.3	84.8	1220	102	67	1224.7	1237.5	46
Amazon, S. America	175000	24.8	181.1	7050000	1	3	74.5	120.6	2.1
Caragh, Ireland	7.29	5.6	104.9	160	7.15	3.6	164	86.5	2.6
Columbia, USA	7900	11.8	99.1	670000	10	26.6	313.6	62.8	2
Danube, Rumania	6500	8.1	57.9	805000	90	46	371.4	826.4	45
Delaware, USA	336	19.1	107.4	17600	100	61	1167.2	851.7	25
Fraser, Canada	3550	16.1	145.8	220000	2	6.4	103.3	739.7	16
Ganges, India	16000	14.9	160	1070000	300	91.3	1361.4	294.3	5.8
Glaama, Norway	706	16.9	68.3	41770	12	24	405.7	975	45
Huanghe, China	1470	2	32.3	750000	200	139	272.6	286.4	28
Hudson, USA	560	16.1	107.4	34700	150	47.8	771.4	851.7	25
Kazan and Back, Canada	1900	6.1	27.4	312000	.4	1.1	6.7	60.9	7
Mackenzie, Canada	10600	5.9	33.3	1787000	.15	5.7	33.8	73.9	7
Magdalenawe, Colombia	7500	31.3	106.2	240000	30	17	531.3	87.5	2.6
Mekong, SE Asia	15000	19.2	139.2	783000	43	17	325.7	334.1	7.6
Mersey, England	21	17.5	100.3	1200	200	156	2730	919.4	28.9
Meuse, Netherlands	317	9.1	65	34900	250	230	2089.1	742.3	36
Mississippi, USA	16100	5	114.8	3220000	30	63	315	691.7	19
Murray–Darling, Australia	318.2	.3	53.6	1073000	1.5	15	4.4	74.8	4.4
Nelson, Canada	2370	2.2	37.3	1070000	2	5	11.1	248.6	21
Niger, W Africa	7000	6.2	181.6	1125000	20	7	43.6	555.2	9.6
Nile, NE Africa	950	.3	15.7	2960000	50	20	6.4	50.9	10.2
Orange, S Africa	170	.2	18.1	1020000	20	50	8.3	154.9	23
Orinoco, Venezuela	33900	33.9	97.3	1000000	2	6	203.4	92.5	3
Parana, Argentina	15900	5.7	75.8	2800000	10	14.2	80.6	216.2	9.9
Po, Italy	1470	22	84.8	66700	232	102	2247.3	1237.5	46
Rhine, Europe	2200	11.9	86.6	185300	300	286	3395.6	1647.9	60
Rhone, France	1700	17.7	73.2	96000	100	57.2	1012.9	695.9	30
Shannon, Ireland	190	13.5	92.7	14000	35	54	727.7	252.8	8.6
Stikine, Canada/USA	1100	22	242.1	50000	1	6.1	134.2	76.8	1
St. Lawrence, Canada/USA	10700	10.4	101.1	1025000	15	16	167	673.2	21
Susquehanna, USA	1100	15.1	103.6	73000	100	66	994.5	821.5	25
Tees, England	50	27.7	58.2	1806	100	75	2076.5	608.7	33
Thames, England	78	7.8	58.2	9950	400	520	4076.4	1125.1	61
Tiber, Italy	230	13.5	84.8	17000	262	100	1352.9	1237.5	46
Uruguay, S America	3850	10.5	86.1	365000	10	29	305.9	355.6	13.7
Vistula, Poland	1100	5.5	55.9	200000	120	70.5	387.8	832.8	47
Volga, Russia	8200	6.1	36.8	1350000	50	30	182.2	151.8	13
Yangtze, China	29000	15.4	116.8	1900000	200	58.2	897	370.5	10
Yukon, Canada	6180	7.4	78.5	831000	.4	9.3	69.2	185.4	7.8
Zaire, Zaire	39730	10.4	147.3	3820000	11.7	6	62.4	467.2	10
Zambezi, SE Africa	3200	2.5	51.8	1300000	15	9.3	22.9	138.5	8.4

an alternative, employ the Bayesian strategy of averaging the pollution effects over a wide range of models. (b) A key weather variable (like humidity) may not be directly related to mortality, but it may interact with air pollution variables to affect mortality. A model selection method that chooses one "best" model may leave the key variable out. (See Exercise 17.)

7.   Yes. Notice what happens to the formula for Cp when the model under investigation is also the "full" model: the second term is zero, so Cp = p.

8.   The probability of E is taken to be the proportion of times when E occurs in a long run of trials. Bayesian statistics also interprets probability as a measure of belief. If M is a particular model among many, a Bayesian probability for M would be the proportion of one's total belief that is assignable to the belief that M is the correct model.

9.   Estimates of the treatment effect can be made with all models involving confounding variables. The posterior function draws its conclusions based on all these estimates, with weights assigned according to the strength of the evidence supporting each model.

The Analysis of Variance for Two-way Classifications

Two-way analysis of variance is a special case of multiple regression analysis with two categorical explanatory factors. This data structure arises in both observational studies and in randomized experiments. In cases where the data are balanced (i.e., the same number of observations in each cell of the table), the regression analysis can be simplified in two respects: (1) a single analysis of variance table conveniently shows F-tests for row, column, and interactive effects, and (2) the least squares estimated means are straightforward combinations of row and column averages. Although there is some special terminology associated with analysis of variance tools and some special considerations due to the particular structure of the data, the steps in the analysis are very similar to those used for multiple regression analysis.

13.1   Case Studies

13.1.1   Intertidal Seaweed Grazers—A Randomized Experiment

To study the influence of ocean grazers on regeneration rates of seaweed in the intertidal zone, a researcher scraped rock plots free of seaweed and observed the degree of regeneration when certain types of seaweed-grazing animals were denied access. The grazers were limpets (L), small fishes (f), and large fishes (F).

A plot was taken to be a square rock surface, 100 cm on each side. Each plot received one of six treatments, named here by which grazers were allowed access:

LfF: All three grazers were allowed access.
fF: Limpets were excluded by surrounding the plot with caustic paint.
Lf: Large fish were excluded by covering the plot with a course net.
f: Limpets and large fish were excluded.
L: Small and large fish were excluded by covering the plot with a fine net.
C: Control: limpets, small fish, and large fish were all excluded.

The fish enclosures were made of wire netting mounted on cages set in holes drilled in the rock. Cages were also mounted over *LfF* and *fF* plots to eliminate confusion of the results with a possible cage effect where fish grazers might avoid a plot simply because of the presence of a cage. These treatments nearly constitute a *factorial* arrangement of three factors (limpets, large fish, small fish) where each factor can be either present or absent and all combinations of presence/absence are used. However, two combinations are missing because it was impossible to exclude small fish without also excluding large fish. Display 13.1 presents a diagram of the six treatments.

Because the intertidal zone is a highly variable environment, the researcher applied the treatments in eight blocks of twelve plots each. Within each block she randomly assigned treatments to plots so that each treatment was applied to two plots. The blocks covered a wide range of tidal conditions.

Block #1: Just below high tide level, exposed to heavy surf
Block #2: Just below high tide level, protected from the surf
Block #3: Midtide, exposed
Block #4: Midtide, protected
Block #5: Just above low tide level, exposed
Block #6: Just above low tide level, protected
Block #7: On near-vertical rock wall, midtide level, protected
Block #8: On near-vertical rock wall, above low tide level, protected

This was a *randomized block* experiment: separate randomizations were used to assign the six treatment levels to plots within each of the blocks. Under control conditions, the seaweed regeneration should be less variable among the plots within a block than it is across plots in different blocks, so blocking makes it possible to look at treatment differences on

Display 13.1   Six treatments excluding three kinds of intertidal grazers from regenerating seaweed on the Oregon coast

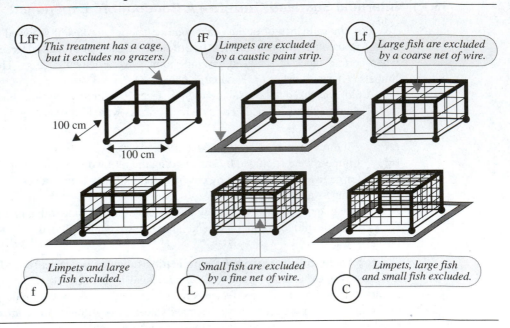

similar plots. Selecting a wide range of exposure conditions also broadens the scope of inference. If the effects of the treatments are similar in all blocks, one can make a more universal claim about them.

Four weeks after treating the plots, the researcher returned to determine rates of regeneration. She did this by covering each plot with a square piece of metal sheeting with 100 small holes drilled in it. She used the number of holes directly over regenerating seaweed as an estimate of the percentage of seaweed cover on the plot. Display 13.2 shows percentages of seaweed cover on all of the plots. (Data based on the study of Annette Olson, Evolutionary and Ecological Interactions Affecting Seaweeds, Ph.D. Thesis. Oregon State University, 1993).

The overall goal of the study was to determine the impacts of the three different grazers on regeneration rates of seaweed. Which one consumes the most seaweed? Do the different grazers influence the impacts of each other? Are the grazing effects similar in all microhabitats?

Summary of Statistical Findings

There is little evidence that treatment differences changed from one block to another (p-value for interaction $= 0.12$ from an F-statistic of 1.437 with 35 and 48 degrees of freedom). Effects of treatments are best described as multiplicative factors changing the regeneration ratio $R =$ percent covered/percent uncovered. Allowing limpets access to plots caused

Display 13.2   Percentage of regenerated seaweed cover on plots with different grazers excluded, in eight blocks of differing tidal situation and exposure

		Treatment: Grazers with Access									
Block #	*Control*		*L*		*f*		*Lf*		*fF*		*LfF*	
1	14	23	4	4	11	24	3	5	10	13	1	2
2	22	35	7	8	14	31	3	6	10	15	3	5
3	67	82	28	58	52	59	9	31	44	50	6	9
4	94	95	27	35	83	89	21	57	57	73	7	22
5	34	53	11	33	33	34	5	9	26	42	5	6
6	58	75	16	31	39	52	26	43	38	42	10	17
7	19	47	6	8	43	53	4	12	29	36	5	14
8	53	61	15	17	30	37	12	18	11	40	5	7

a significant reduction in the regeneration of seaweed (two-sided p-value < 0.0001 from a t-statistic of 14.97 with 83 d.f.). The median regeneration ratio when limpets were present is estimated to be only 0.161 times as large as the median regeneration ratio when they are excluded (95% confidence interval: 0.126 to 0.205). Although there was conclusive evidence that the presence of small and large fish also reduced the regeneration ratio (two-sided p-values were 0.0001 and 0.010, respectively), the effects were smaller than for limpets. The median regeneration ratio when large fish were present is estimated to be .541 times as large as the regeneration ratio when they are excluded (95% CI: 0.402 to 0.729). The median regeneration ratio when small fish were present is estimated to be 0.685 times as large as the median regeneration ratio when they were excluded (95% CI: 0.501 to 0.909). There is no evidence that the effect of limpets depended on whether small fish or large fish were present (two-sided p-values = 0.50 and 0.71, respectively.)

13.1.2   The Pygmalion Effect—A Randomized Experiment

Pygmalion was a mythical king of Cyprus who, according to one account, sculpted a figure of the ideal woman and then fell in love with his own creation. The Pygmalion effect in psychology refers to a situation where the high expectations of a supervisor or teacher translate into improved performance by subordinates or students. It is a special case of the self-fulfilling prophecy. Many experiments have demonstrated the validity of the effect. In a typical experiment, a supervisor is placed in charge of a new unit with, say, 20 workers. In a private conversation, the supervisor's manager explains that 5 of these workers are exceptional, whereas in fact the 5 were selected at random. Later, an independent test to evaluate work performance compares the 5 exceptional workers with the other 15. If the Pygmalion effect is present, the five will significantly outperform the 15.

A researcher observed that such experiments involve interpersonal contrasts: the other 15 may perceive the supervisor's increased expectations of the 5 and the reduced expectations of them and may therefore oblige with a reduced performance. The researcher proceeded to conduct a different experiment involving no interpersonal contrasts.

Ten companies of soldiers in an army training camp were selected for the study. Each company had three platoons. Each platoon had its own platoon leader. Using a random mechanism, one of the three platoons was selected to be the Pygmalion platoon. The randomization was conducted separately for each company to create a randomized block experiment with companies as blocks.

Prior to assuming command of a platoon, each leader met with an army psychologist. To each of the Pygmalion treatment platoon leaders, the psychologist described a nonexistent battery of tests that had predicted superior performance from his or her platoon.

At the conclusion of basic training, soldiers took a battery of real tests. The Practical Specialty Test evaluated soldiers on their ability to operate weapons and answer questions about their use. This test was chosen for the study because it was considered particularly vulnerable to the Pygmalion effect. Display 13.3 shows average scores on the Practical Specialty Test for the soldiers in the different platoons. (Data based on the study of Dov Eden, "Pygmalion Without Interpersonal Contrast Effects: Whole Groups Gain from Raising Manager Expectations," *Journal of Applied Psychology*) 75(4) (1990): 395–98.)

Display 13.3 Average scores of soldiers on the Practical Specialty Test, for platoons given the Pygmalion treatment and for control platoons

	Treatments		
Company	*Pygmalion*	*Control*	
1	80.0	63.2	69.2
2	83.9	63.1	81.5
3	68.2	76.2	
4	76.5	59.5	73.5
5	87.8	73.9	78.5
6	89.8	78.9	84.7
7	76.1	60.6	69.6
8	71.5	67.8	73.2
9	69.5	72.3	73.9

Observe that the entire platoon's average is taken as one observation. This may appear strange because individual scores were measured for all the soldiers in each of the platoons. But the treatments were randomized and assigned to platoons as a whole—not to individual soldiers—so platoons are the experimental units. Notice also that one company had only two platoons, so it had only one control.

Summary of Statistical Findings

The Pygmalion treatment adds an estimated 7.22 points to a platoon's score (95% interval: 1.80 to 12.64 points). The evidence strongly suggests the effect is real (one-sided p-value = .0060), and the experimental design allows for a causal inference.

13.2   Additive and Nonadditive Models for Two-way Tables

13.2.1   The Additive Model

Both data sets may be viewed as two-way tables, meaning that each observation is identified with two categorical explanatory factors. In the seaweed grazers study, the factors are *block* (with eight levels) and *treatment* (with six levels). The design is *balanced*, with two observations in each of the $6 \times 8 = 48$ treatment combinations (cells). In the Pygmalion study the categorical explanatory factors are *company* (with ten levels), and *treatment* (with two levels). It has an unbalanced design, because unequal numbers of observations occur in the $2 \times 10 = 20$ cells.

Research questions pertaining to two-way tables usually involve an additive model with no interactive effects of the two categorical explanatory variables. In an additive model, the effects of one factor are the same at all levels of the other factor. For the 20 populations defined by all combinations in the Pygmalion study, a set of hypothetical additive means would appear as in the table in Display 13.4. The mean score in Pygmalion-treated platoons is 8 units more than the mean score of the control platoons, within each company. It is also true that the difference between the mean scores of control platoons in two different companies is the same as the difference between the mean scores of the Pygmalion platoons in the two; in other words, block effects are the same within all treatments.

Display 13.4   Hypothetical mean scores on the Practical Specialty Test, illustrating additivity of treatment and company effects

	Treatments		Treatment Effects
	Pygmalion	*Control*	*(Pygmalion - Control)*
Company			
1	78	70	8
2	85	77	8
3	79	71	8
4	72	64	8
5	84	76	8
6	89	81	8
7	73	65	8
8	76	68	8
9	75	67	8
10	82	74	8

The additive model says that the Pygmalion effect adds the same value to the score of a treated platoon regardless of the company. Because this result gives the Pygmalion effect a degree of universality, the additive model is an important inferential setting to consider. The model allows the researcher to make statements about the effect without any qualifications regarding the individual companies involved. If the Pygmalion data, however, do not fit an additive model, then the discussion of treatment effects will necessarily be cluttered by repeated references to the individual companies.

With *score* representing a platoon's average score and *TREAT* and *COMP* representing the factors treatment and company (with 2 and 10 levels, respectively), a representation of the additive model is

$$\mu\{ score \mid TREAT, COMP\} = TREAT + COMP.$$

A Regression Parameterization for the Additive Two-way Model

A *parameterization* is a particular way to assign parameters to represent the means in a model. One convenient parameterization results from using indicator variables to model the categorical explanatory variables in a multiple regression model. If *pyg* is an indicator for platoons that received the Pygmalion treatment, and if *cmp2, cmp3, ..., cmp10* are indicator variables for companies 2 through 10, then a multiple linear regression expression for the additive model is

$$\mu\{score \mid TREAT, COMP\} = \beta_0 + \beta_1 pyg + \beta_2 cmp2 + \beta_3 cmp3 + \cdots + \beta_{10} cmp10.$$

The means for each cell in the table (see Display 13.5) are functions of the regression parameters. For example, the mean score for Pygmalion platoons in the third company is obtained by setting $pyg = 1$, $cmp3 = 1$, and all the other indicator variables to 0; $\mu\{score \mid pyg = 1, cmp3 = 1\} = \beta_0 + \beta_1 + \beta_3$. The entire table of means, expressed in terms of the regression coefficients, appears in Display 13.5. In terms of the regression coefficients, the Pygmalion treatment effect is β_1 units, in any company. Similarly, the

---

Display 13.5   Mean scores on the Practical Specialty Test according to the additive model, in terms of coefficients in a multiple regression model with indicators

| Company | Treatments | | Treatment Effects |
	Pygmalion	Control	(Pygmalion - Control)
1	$\beta_0 + \beta_1$	β_0	β_1
2	$\beta_0 + \beta_2 + \beta_1$	$\beta_0 + \beta_2$	β_1
3	$\beta_0 + \beta_3 + \beta_1$	$\beta_0 + \beta_3$	β_1
4	$\beta_0 + \beta_4 + \beta_1$	$\beta_0 + \beta_4$	β_1
5	$\beta_0 + \beta_5 + \beta_1$	$\beta_0 + \beta_5$	β_1
6	$\beta_0 + \beta_6 + \beta_1$	$\beta_0 + \beta_6$	β_1
7	$\beta_0 + \beta_7 + \beta_1$	$\beta_0 + \beta_7$	β_1
8	$\beta_0 + \beta_8 + \beta_1$	$\beta_0 + \beta_8$	β_1
9	$\beta_0 + \beta_9 + \beta_1$	$\beta_0 + \beta_9$	β_1
10	$\beta_0 + \beta_{10} + \beta_1$	$\beta_0 + \beta_{10}$	β_1

mean score in company 3 is $\beta_3 - \beta_2$ units more than the mean score in company 2, for treated as well as for control platoons.

Notice that 11 regression coefficients are needed. For I rows and J columns, there are $(I - 1)$ terms for row effects (for example, the coefficients of the $I - 1$ indicator variables needed to distinguish I levels); there are $(J - 1)$ terms for column effects; and there is the constant term β_0, so there are $(I + J - 1)$ distinct parameters used to describe the means in an additive model.

13.2.2   The Saturated, Nonadditive Model

A regression model that includes interactions between treatment and companies is

$$\mu\{score \mid TREAT, COMP\} = TREAT + COMP + TREAT \times COMP,$$

which in the regression parameterization is

$$\beta_0 + (\beta_1 pyg) + (\beta_2 cmp2 + \beta_3 cmp3 + \cdots + \beta_{10} cmp10)$$

$$+ (\beta_{11}[pyg \times cmp2] + \beta_{12}[pyg \times cmp3] + \cdots + \beta_{19}[pyg \times cmp10]),$$

where parentheses have been included to distinguish the terms for treatment main effects, for company main effects, and for treatment-by-company interactive effects. The explanatory variables for describing interactions are formed as all possible products of the treatment and company indicator variables.

The table of means for this saturated model looks like Display 13.6. The mean score for treated platoons is β_1 units more than the mean score of control platoons in company

Display 13.6   Mean scores on the Practical Specialty Test, in terms of the parameters in a saturated multiple linear regression model with interaction

Company	Treatments		Treatment Effects
	Pygmalion	Control	(Pygmalion − Control)
1	$\beta_0 + \beta_1$	β_0	β_1
2	$\beta_0 + \beta_2 + \beta_1 + \beta_{11}$	$\beta_0 + \beta_2$	$\beta_1 + \beta_{11}$
3	$\beta_0 + \beta_3 + \beta_1 + \beta_{12}$	$\beta_0 + \beta_3$	$\beta_1 + \beta_{12}$
4	$\beta_0 + \beta_4 + \beta_1 + \beta_{13}$	$\beta_0 + \beta_4$	$\beta_1 + \beta_{13}$
5	$\beta_0 + \beta_5 + \beta_1 + \beta_{14}$	$\beta_0 + \beta_5$	$\beta_1 + \beta_{14}$
6	$\beta_0 + \beta_6 + \beta_1 + \beta_{15}$	$\beta_0 + \beta_6$	$\beta_1 + \beta_{15}$
7	$\beta_0 + \beta_7 + \beta_1 + \beta_{16}$	$\beta_0 + \beta_7$	$\beta_1 + \beta_{16}$
8	$\beta_0 + \beta_8 + \beta_1 + \beta_{17}$	$\beta_0 + \beta_8$	$\beta_1 + \beta_{17}$
9	$\beta_0 + \beta_9 + \beta_1 + \beta_{18}$	$\beta_0 + \beta_9$	$\beta_1 + \beta_{18}$
10	$\beta_0 + \beta_{10} + \beta_1 + \beta_{19}$	$\beta_0 + \beta_{10}$	$\beta_1 + \beta_{19}$

1, but it is $\beta_1 + \beta_{11}$ units more in company 2. Treatment effects depend on company. The effects are not necessarily additive, and no interpretation for a main effect of treatment can be formulated.

This model has 20 regression coefficients. In general, for I rows and J columns, there is a β_0 term, there are $(I - 1)$ row main effect terms, $(J - 1)$ column main effect terms, and $(I - 1) \times (J - 1)$ interaction terms. The total number is $I \times J$. Since one coefficient appears in each cell in the table, this is called a *saturated* model. It is the most general model possible for these data, and it is equivalent to modeling them as a one-way table with $I \times J$ groups.

In the saturated model, the cell means are completely unrelated. Therefore, the least-squares estimate of the mean in any cell is the sample average of responses in that cell. Since the cell averages are the fitted values, the residuals are the differences between responses and cell averages. The estimate of σ^2 from the saturated model is therefore the pooled estimate of variance from all cells, with pooled degrees of freedom: $n - I \times J$.

13.2.3 A Strategy for Analyzing Two-way Tables with Several Observations per Cell

The fixed-effects analysis of two-way tables is a regression analysis based on the constant variance model of Section 9.2.1 (page 230). The analysis proceeds in much the same way as other regression analyses. Typically, a graphical procedure is used for initial exploration, including consideration of outliers and the need for transformation. Then a tentative rich model, including interactions, is fit and a residual plot is examined for further exploration of outliers and transformation.

The presence of interactions complicates the task of answering questions of interest about the main factor effects. When there are significant interactions, the best forms of communication are a table of estimated means and a graph of the means against one factor showing separate curves for the levels of the other factor.

If there is no evidence that interaction terms are needed, the questions of interest may be addressed through the parameters of the additive model. For example, a further F-test can be used to test whether the additive effects of the row categories are zero. Special questions of interest, such as "which seaweed grazers reduce regeneration of seaweed the most and by how much?" can be investigated through particular terms in the additive model or by examining linear combinations of group means.

13.2.4 The Analysis of Variance F-Test for Additivity

If the coefficients of all the interaction terms are zero then the nonadditive model reduces to the additive model. Thus, the check on the additive model is accomplished by comparing the (full) nonadditive model,

$$\mu\{score \mid TREAT, COMP\} = TREAT + COMP + TREAT \times COMP,$$

to the (reduced) additive model,

$$\mu\{score \mid TREAT, COMP\} = TREAT + COMP,$$

with an extra-sum-of-squares F-test. There are $I \times J$ parameters in the nonadditive model, so $(n - I \times J)$ degrees of freedom go into the estimate of σ^2. Since there are $(I-1) \times (J-1)$ interaction terms, the F-test for additivity has $(I-1) \times (J-1)$ numerator degrees of freedom and $n - (I \times J)$ denominator degrees of freedom. [*Note*: If there is only one observation per cell then $n = I \times J$ with, consequently, no degrees of freedom to estimate σ^2 in the full model. Special considerations are needed for the analysis of this type of two-way table; see Chapter 14].

13.3 Analysis of the Seaweed Grazer Data

13.3.1 Initial Assessment of Additivity, Outliers, and the Need for Transformation

A coded scatterplot of the cell averages versus block number appears in Display 13.7. Blocks appear along the horizontal axis according to the average size of the responses, because the blocks themselves are of little interest and because they have no natural order. Connecting the treatment averages with lines makes it easy to track treatment differences in the different blocks.

Treatments with limpets excluded (represented by darkened symbols) had a higher rate of seaweed regeneration than those with limpets present (represented by open symbols), for all blocks. Limpets are apparently the heaviest grazers. It appears that the other treatments consistently differ across blocks, as well.

Display 13.7 shows evidence of nonadditivity. For example, the difference between average responses to the *fF* and *f* treatments appears to be much larger in block 4 than in block 1. A consistent pattern to these effects can be observed—treatment differences appear to be larger in blocks with larger response means. This type of nonadditivity can often be removed by transformation.

Before a formal assessment of interaction, the residual plot from the saturated model is examined (Display 13.8). It appears to have a funnel-shape at both the high and the low ends of the percentage scale. The funnel-shape is more noticeable at the low end of the scale where the majority of the observations lie. This pattern, typical of residual plots when the responses are between 0 and 1 (or between 0 and 100%), suggests the need for transformation. (*Note*: the symmetry of the residual plot about the zero line is simply an artifact of having two observations in each cell and is not a pattern that requires special attention.)

Transformation

Two transformations often work well for responses that are proportions: the *arcsine square root* transformation, arcsine $(Y^{1/2})$; and the *logit* transformation, $\log[Y/(1-Y)]$. The first of these, although sometimes quite useful, does not provide a particularly appealing measure for interpretation, nor can it be back-transformed to any convenient statement. The logit, on the other hand, provides for a useful interpretation as a *regeneration ratio*—here, the proportion of the plot covered by seaweed divided by the proportion remaining uncovered. The logit is the log of the regeneration ratio.

Display 13.7 Average percentages of seaweed regeneration with different grazers allowed

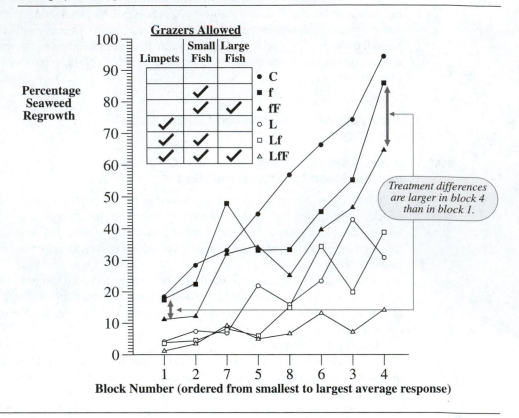

Display 13.8 Residual plot from the saturated model fit to the seaweed grazer data

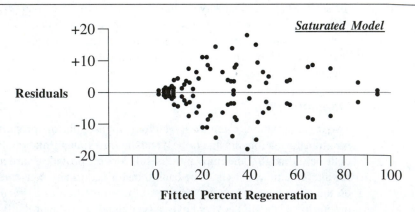

A coded scatterplot of group averages of the log of the regeneration ratio versus block number is shown in Display 13.9. If the averages exhibited perfect additivity then the treatment averages would be parallel. Although this is not exactly true in the plot, the differences between treatments appear to be roughly constant across blocks, and the lack of parallelism may be due to sampling variation. Although not shown, the residual plot from the fit to the nonadditive model after transformation indicates no evidence of nonconstant variance or outliers.

Display 13.9   Averages of the log of the seaweed regeneration ratio versus block number, with code for treatment

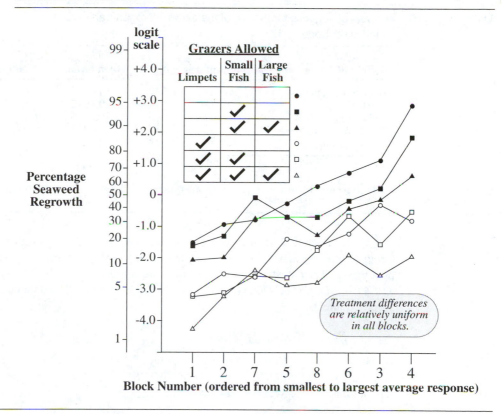

13.3.2   The Analysis of Variance Table from the Fit to the Saturated Model

The analysis of variance table, Display 13.10, simultaneously displays the F-test for the interaction terms, an F-test for the block effects, and an F-test for the treatment effects. The F-statistic in the Between Groups row is the *one-way* analysis of variance F-statistic for testing equality of means in the 48 groups (all combinations of blocks and treatments).

It compares the full model,

$$\mu\{Y \mid BLOCK, TREAT\} = BLOCK + TREAT + (BLOCK \times TREAT),$$

to the reduced model,

$$\mu\{Y \mid BLOCK, TREAT\} = \mu$$

(the constant term only).

Display 13.10 Analysis of variance for the log of the seaweed regeneration ratio: nonadditive model

Source of Variation	Sum of Squares	d.f.	Mean Square	F-Statistic	p-Value
Between Groups	188.4622	47	4.0098	13.2407	<0.0001
Blocks	76.2386	7	10.8912	35.9634	<0.0001
Treatments	96.9932	5	19.3986	64.0554	<0.0001
Interactions	15.2304	35	0.4352	1.4369	0.1209
Within Groups	14.5364	48	0.3028		
Total	202.9986	95			

R-squared = 92.84%   **adj. R-squared** = 85.83%   **Estimated SD** = 0.5503

What is new in this analysis of variance table is the decomposition of the Between Groups sum of squares into three separate components designed to test three different hypotheses. The F-tests in the indented rows are comparisons of the full, nonadditive model $TREAT + BLOCK + (TREAT \times BLOCK)$ to the reduced models obtained by omitting the specified effects.

Tested Effects	Reduced Model
Blocks	$TREAT + (TREAT \times BLOCK)$
Treatments	$BLOCK + (TREAT \times BLOCK)$
Interactions	$TREAT + BLOCK.$

Row 4 of the table in Display 13.10 shows the extra-sum-of-squares F-test comparing the nonadditive to the additive model. The residual sum of squares from the nonadditive model is the Within Groups sum of squares. The residual sum of squares from the additive (reduced) model is not shown, but the difference, the sum of squares explained by the interaction terms, is listed as the Interaction sum of squares, with $(I - 1) \times (J - 1) = 7 \times 5 = 35$ degrees of freedom. The F-statistic is the Interaction mean square divided by the estimate of σ^2 from the saturated model. The large p-value, .1209, suggests that the data are consistent with an additive model.

13.3.3 The Analysis of Variance Table for the Fit to the Additive Model

The analysis of variance table in Display 13.11 comes from the fit to the additive model. This table is the same as the one in Display 13.10, except that the sum of squares for Interaction has been pooled with (added to) the Within Groups sum of squares to form a single source called Residual. The estimate of σ^2 in the additive model is the Residual mean square.

Display 13.11 Analysis of variance for the log of the seaweed regeneration ratio: additive model

Source of Variation	Sum of Squares	d.f.	Mean Square	F-Statistic	p-Value
Model	173.2318	12	14.4360	40.2520	<0.0001
Blocks	76.2386	7	10.8912	30.3684	<0.0001
Treatments	96.9932	5	19.3986	54.0900	<0.0001
Residual	29.7668	83	0.35864		
Total	202.9986	95			

R-squared = 85.34%   **adj. R-squared** = 83.22%   **Estimated SD** = 0.5989

Which Test for Treatment Effects?

Displays 13.10 and 13.11 offer two different tests for the hypothesis of no treatment effects. The first test for no treatment effects from the fit to the nonadditive model is the following comparison:

$$\text{Full model:} \quad TREAT + BLOCK + (TREAT \times BLOCK)$$
$$\text{Reduced model:} \quad BLOCK + (TREAT \times BLOCK),$$

The second test for no treatment effects from the additive model is the following comparison:

$$\text{Full model:} \quad TREAT + BLOCK$$
$$\text{Reduced model:} \quad BLOCK.$$

Opinions differ about which is appropriate. The second is more transparent: given an additive model with treatment and block effects, is there any need to include treatment effects in a model that also has block effects? Such a question would be asked once it was determined that there were no significant interactive effects. The argument against this test is that the nonsignificance of the interaction terms does not prove that they are zero, but only that there is insufficient evidence to justify concluding otherwise. Therefore, the estimate of σ^2 used in the denominator of the second test may be biased due to the omission of relevant terms. The estimate of σ^2 from the saturated model, on the other hand, does not depend on any model's being correct. The trouble with the first test, however, is that the reduced model rarely makes much sense. It is contradictory to state that there are no treatment effects while simultaneously saying that treatment effects change from block to block.

When the questions of interest involve main effects, the authors of this book consider that the proper inferential model is the additive one. The nonadditive model is used tentatively for model checking (just as other regression models were used tentatively in previous multiple regression examples), and a test for the significance of interaction follows in the model refinement stage to see if the additive model is appropriate for answering the questions of interest.

For the seaweed data, the nonsignificance of the interaction terms provides evidence in support of using the additive model for estimating and testing treatment effects. The model may not be entirely correct, but its structure allows further questions to be answered directly.

The Test for Blocks

Since the researcher selected the eight different blocks to span a wide range of intertidal environments and wanted to account for them in the treatment comparisons, there is little use for a formal test of block effects here. The researcher already knows that there are differences between blocks; formal significance tests do not answer a question of interest. (In other two-way tables both row and column effects may be of interest.)

13.3.4 Answers to Specific Questions of Interest Using Contrasts

The analysis of variance table in Display 13.11 demonstrates significant treatment differences (p-value $< .0001$), but it would be a mistake to stop the analysis without further investigation and clarification. Specific questions concerning the roles of limpets, small fish, and large fish as grazers can now be addressed in the context of the additive model for the log of the seaweed regeneration ratio. This section shows how the questions can be investigated through contrasts of treatment effects. Identical results can also be obtained using indicator variables in multiple regression, as discussed in the next section.

The table in Display 13.12 shows the cell averages for the 48 cells of the table, along with row and column averages. From this, the estimates of the additive block and treatment effects (in the classical parameterization described in Section 13.5.6) are calculated as the differences of row or column averages from the overall average (which would not be the case if the data were unbalanced.) The fitted value for each cell is the sum of the overall average, a block effect ($\overline{Y}_i.. - \overline{Y}...$), and a treatment effect ($\overline{Y}_{.j}. - \overline{Y}...$). For example, the fitted value corresponding to block 1 and the control treatment is $-1.23 - 1.40 + 1.41 = -1.22$. (The fitted value from the saturated model, on the other hand, is -1.51.) The logs of regeneration ratios in block 1 tend to be smaller than the overall average by about 1.40, but those in the control treatment tend to be about 1.41 larger than the overall average.

Recall that a contrast is a linear combination of means such that the coefficients add up to zero. Since the additive model applies, it is convenient to think of treatment means, such as μ_{fF} for the mean log regeneration ratio when small fish and large fish are allowed

Display 13.12 Table of averages of log percentage of seaweed regeneration ratio with different grazer combinations in eight blocks

Block	Treatment: Grazers with Access						Block Average	Block Effect
	Control	L	f	Lf	fF	LfF		
1	−1.51	−3.18	−1.62	−3.21	−2.05	−4.24	−2.64	−1.40
2	−0.94	−2.51	−1.31	−3.11	−1.97	−3.21	−2.18	−0.94
3	1.11	−0.31	0.22	−1.56	−0.12	−2.53	−0.53	0.70
4	2.85	−0.81	1.84	−0.52	0.64	−1.93	0.34	1.58
5	−0.27	−1.40	−0.69	−2.63	−0.68	−2.83	−1.42	−0.19
6	0.71	−1.23	−0.18	−0.66	−0.41	−1.89	−0.61	0.62
7	−0.79	−2.60	−0.08	−2.59	−0.74	−2.38	−1.53	−0.29
8	0.28	−1.66	−0.64	−1.75	−1.25	−2.77	−1.31	−0.07
Treatment Average	0.18	−1.71	−0.31	−2.00	−0.82	−2.72	−1.23	
Treatment Effect	1.41	−0.48	0.92	−0.77	0.41	−1.49		

access. One measure of the effect of large fish on the response is $\mu_{fF} - \mu_f$, the effect of large fish in the presence of small fish. Another is $\mu_{LfF} - \mu_{Lf}$, the effect of large fish in the presence of small fish and limpets. If there is no difference between these two large fish effects, a contrast measuring the common large fish effect is the average of the two: $(1/2)(\mu_{fF} - \mu_f) + (1/2)(\mu_{LfF} - \mu_{Lf})$. A similar strategy can be used for determining appropriate contrasts for four additional questions about treatment effects. The meanings for all five are described next. Their estimates are summarized in Display 13.13, along with standard errors and t-statistics (for the hypotheses that the contrasts are zero).

Display 13.13 Separate effects of grazers using linear combinations of treatment means

Treatment:	LfF	fF	Lf	f	L	C	Contrast Summary		
Sample size:	16	16	16	16	16	16	Estimate	Standard Error	t-Stat
Average:	−2.7247	−0.8214	−2.0044	−0.3137	−1.7120	+0.1805			
Large Fish:	$+\dfrac{1}{2}$	$+\dfrac{1}{2}$	$-\dfrac{1}{2}$	$-\dfrac{1}{2}$	0	0	−0.6140	0.1497	4.10
Small Fish:	0	0	$+\dfrac{1}{2}$	$+\dfrac{1}{2}$	$-\dfrac{1}{2}$	$-\dfrac{1}{2}$	−0.3933	0.1497	2.63
Limpets:	$+\dfrac{1}{3}$	$-\dfrac{1}{3}$	$+\dfrac{1}{3}$	$-\dfrac{1}{3}$	$+\dfrac{1}{3}$	$-\dfrac{1}{3}$	−1.8288	0.1222	14.97
Limpets × Small:	$+\dfrac{1}{2}$	$-\dfrac{1}{2}$	$+\dfrac{1}{2}$	$-\dfrac{1}{2}$	−1	+1	+0.0955	0.2593	0.37
Limpets × Large:	+1	−1	−1	+1	0	0	−0.2126	0.2994	0.71

1. *Do large fish have an effect on the regeneration ratio? If so, how much?* The difference between means from *fF* and *f* treatments measures this effect in the presence of small fish only; the difference between means from the *LfF* and *Lf* treatments measures the effect in the presence of both small fish and limpets. The large fish effect is taken to be the average of those two separate effects: $\gamma_1 = (1/2)(\mu_{fF} - \mu_f) + (1/2)(\mu_{LfF} - \mu_{Lf})$. This effect averages over different limpet conditions, so it measures a meaningful effect only if there is no limpet-by-big-fish interaction. If there is an interaction, it is better to estimate separate big fish effects for the cases where limpets are present and absent (see question 5).

2. *Do small fish have an effect on the regeneration ratio? If so, how much?* This is investigated through the average of the difference between the *f* and *C* treatment means (the effect of small fish when neither of the other grazers is present) and the *Lf* and *L* treatment means (the effect of small fish in the presence of limpets): $\gamma_2 = (1/2)(\mu_f - \mu_C) + (1/2)(\mu_{Lf} - \mu_L)$. This also averages over different limpet conditions, so it measures a meaningful effect only if there is no small fish-by-limpet interaction (see question 4).

3. *Do limpets have an effect on the regeneration ratio? If so, how much?* This is investigated with: $\gamma_3 = (1/3)(\mu_L - \mu_C) + (1/3)(\mu_{Lf} - \mu_f) + (1/3)(\mu_{LfF} - \mu_{fF})$. Again, this averaging is appropriate only when there are no limpet-by-big-fish or limpet-by-small-fish interactions.

4. *Do limpets have a different effect when small fish are present than when small fish are not present?* The average of the two limpet effects where small fish are present, $[(\mu_{LfF} - \mu_{fF}) + (\mu_{Lf} - \mu_f)]/2$ is contrasted with the limpet effect where small fish are excluded, $(\mu_L - \mu_C)$, giving a measure of how the effect of limpet exclusion changes with small fish exclusion: $\gamma_4 = (1/2)(\mu_{LfF} - \mu_{fF}) + (1/2)(\mu_{Lf} - \mu_f) - (\mu_L - \mu_C)$. The averaging here means that this measures a meaningful effect only when there is no limpet-by-big-fish interaction.

5. *Do limpets have a different effect when large fish are present than when large fish are not present?* The difference $\mu_{LfF} - \mu_{fF}$ measures the limpet effect where all fish are allowed, and the difference $\mu_{Lf} - \mu_f$ measures the limpet effect where only small fish are allowed. The difference between the two effects estimates how much the limpet effect changes when large fish are allowed versus when they are excluded: $\gamma_5 = (\mu_{LfF} - \mu_{fF}) - (\mu_{Lf} - \mu_f)$.

The contrasts are estimated by substituting the appropriate treatment averages (over all 16 observations) for the corresponding means. For example, the Large Fish estimate of -0.6140 in Display 13.13 is $(1/2)(\overline{Y}_{fF} - \overline{Y}_f) + (1/2)(\overline{Y}_{LfF} - \overline{Y}_{Lf})$. *As long as the data are balanced*, the effects of the blocks need not be considered. Using the estimate of σ that comes from the model with blocks included, the standard error of a linear combination (Section 6.2.2 on page 147) has systematic block effects removed.

Drawing Conclusions about Effects

For the seaweed grazers data the estimate of σ is 0.5989 with 83 degrees of freedom. The standard errors of the estimated contrasts in Display 13.13 come from the formula in Section 6.2.2 using this estimate of σ. The *t*-tests indicate the evidence regarding the various contrasts.

Since the questions dealing with grazer main effects (questions 1–3) involve assumptions about treatment interactions, it is important to examine the interaction questions first. The limpet-by-large-fish interaction comes first, because the other depends on it. The estimate provides no evidence that the limpet effects depend on whether large fish are excluded or allowed (two-sided p-value $= 0.48$). Next, the limpet-by-small-fish contrast does not indicate that the limpet effects depend on whether small fish are excluded (two-sided p-value $= .71$). Therefore, the contrasts that involve averaging of effects—like the average of the limpet effect in the absence of large fish and the limpet effect in the presence of large fish—can be easily interpreted.

All grazer main effects are statistically significant. Allowing limpets (two-sided p-value $< .0001$) produces the largest and most statistically significant effect, followed by that of allowing large fish (two-sided p-value $= .0001$) and small fish (two-sided p-value $= .01$). Since the limpet contrast is the mean log regeneration ratio when limpets are present minus the mean log regeneration ratio when limpets are absent, the antilog of the estimate, $e^{-1.8288}$, estimates the ratio of median regeneration ratios. The median regeneration ratio when limpets are allowed is estimated to be only .161 of the median regeneration ratio when they are excluded. A 95% confidence interval for this ratio of medians is obtained, as usual, by taking the antilogarithms of the endpoints of a 95% confidence interval for the limpet contrast (.126 to .205). The numerical conclusions reported in the summary of statistical findings about small fish and large fish were found in a similar fashion.

13.3.5   Answers to Specific Questions of Interest Using Multiple Regression with Indicator Variables

The inferences about the seaweed grazers above can also be derived by fitting a multiple linear regression analysis using indicators as explanatory variables. Since the computer does more of the work, this can be an easier approach to answering the questions of interest. The trick is to set up the indicator variables properly and to recognize the interpretations of the regression coefficients.

Question-oriented Indicator Variables

The following variables are indicators of presence/absence, rather than of treatment combination:

lmp: an indicator that is 1 for plots in which limpets were allowed to graze;
sml: an indicator that is 1 for plots in which small fish were allowed to graze;
big: an indicator that is 1 for plots in which large fish were allowed to graze.

Each variable sets up a contrast measuring the difference in mean responses between where the grazer is present (Indicator $= 1$) and where it is absent (Indicator $= 0$). Two additional variables measuring interactions between grazers are the products, *lmp* $\times$ *sml* and *lmp* $\times$ *big*. Other interactions are not estimable because of the constraints of the experimental design. Little fish could not be excluded without also excluding large fish, and a computer regression procedure would show an error message if the *sml* $\times$ *big* interaction is included.

A multiple linear regression analysis that parallels the previous analysis of contrasts begins by fitting the model:

$$\beta_0 + BLOCK + \beta_8 lmp + \beta_9 sml + \beta_{10} big + \beta_{11}(lmp \times sml) + \beta_{12}(lmp \times big).$$

This model does not directly reproduce the contrast estimates in Display 13.13. The estimate of the limpet-by-large-fish interaction, β_{12}, is the same as the contrast estimate of the limpet-by-large-fish interaction effect, but estimates of the other coefficients do not match the corresponding contrast effects. The reason for this is that the other contrast effects involve some averaging of simple effects, and the regression analysis does not average the simple effects if the model specifies that the simple effects are different because of interaction. However, the model

$$\beta_0 + BLOCK + \beta_8 lmp + \beta_9 sml + \beta_{10} big + \beta_{11}(lmp \times sml)$$

specifies that no limpet-by-large-fish interaction exists, so the regression analysis does the appropriate averaging and comes up with an estimate of the limpet-by-small-fish interaction parameter, β_{11}, that agrees with the contrast estimate of the limpet-by-small-fish interaction effect. By fitting the model

$$\beta_0 + BLOCK + \beta_8 lmp + \beta_9 sml + \beta_{10} big$$

with no grazer interactions whatsoever, multiple linear regression averages all the appropriate simple effects to produce estimates of the main effect parameters—β_8, β_9, and β_{10}—that agree with the contrast estimates of main effects. The regression analyses, however, produce slightly different standard errors for the coefficient estimates.

13.4   Analysis of the Pygmalion Data

The Pygmalion data are also from a randomized block experiment. Although there is no particular interest in the different response means for the different companies, it is important to compare the treatment difference after possible company differences are accounted for. One difference from the seaweed grazers data, however, is that the data here are *unbalanced*, since some of the treatment-block combinations had two experimental units and some only had one.

13.4.1   Initial Exploration and Check on Additive Model

The starting point in the analysis is a visual display of the data, as in Display 13.14. From this several features emerge: the Pygmalion platoons do seem to score higher than the control platoons in most companies; some companies seem to score higher than others; there are no obvious outliers and no obvious need for a transformation.

Next it is appropriate to fit the saturated model, check residual plots, and examine the significance of interactions. Since the data are unbalanced and since the Pygmalion effect can be conveniently examined through the coefficient of an indicator variable, a regression approach is chosen.

Display 13.14 Average scores for platoons on the Practical Specialty Test

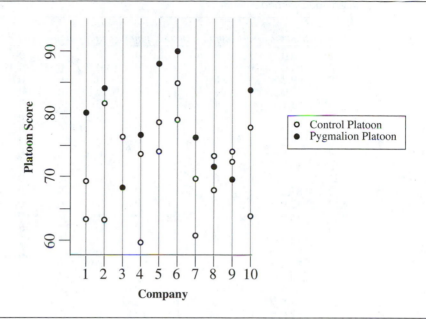

Multiple Linear Regression Model

Let *pyg* be the variable indicating the Pygmalion treatment ($pyg = 1$ for those platoons that were reported by the psychologist as likely to perform well and $pyg = 0$ for the other platoons), and let *cmp2*, *cmp3*, and so on be indicator variables for companies 2, 3, . . . , 10 (with the first company as the reference level). The full data layout based on these indicator variables appears in Display 13.15.

The full, nonadditive model for the mean platoon score given company and treatment is

$$\mu\{score \mid PYG, COMP\} = PYG + COMP + (PYG \times COMP).$$

A residual plot from the fit to this model (not shown) does not indicate any problems, but it is only partially useful since the residuals for those treatment-company combinations with only a single platoon are necessarily zero in the full model. Next, examine the F-test for the interaction terms (shown in Display 13.16) and, if the additive model seems appropriate given this F-test, reexamine a residual plot after fitting the additive model.

The large p-value of .72 suggests no problem with using the additive model for the analysis. Although a slight funnel shape appears in the residual plot from the additive model (Display 13.17), the funnel is headed in the opposite direction from what would be corrected by the log transformation. No suitable transformation applies here and, since the degree of nonconstant variance is small, the analysis will be pursued with the additive model on the original scale.

Display 13.15   The Pygmalion data with indicator variables defining treatment and companies in an additive model

---

Case	Score	pyg	cmp2	cmp3	cmp4	cmp5	cmp6	cmp7	cmp8	cmp9	cmp10
1	80.0	1	0	0	0	0	0	0	0	0	0
2	63.2	0	0	0	0	0	0	0	0	0	0
3	69.2	0	0	0	0	0	0	0	0	0	0
4	83.9	1	1	0	0	0	0	0	0	0	0
5	63.1	0	1	0	0	0	0	0	0	0	0
6	81.5	0	1	0	0	0	0	0	0	0	0
7	68.2	1	0	1	0	0	0	0	0	0	0
8	76.2	0	0	1	0	0	0	0	0	0	0
9	76.5	1	0	0	1	0	0	0	0	0	0
10	59.5	0	0	0	1	0	0	0	0	0	0
11	73.5	0	0	0	1	0	0	0	0	0	0
12	87.8	1	0	0	0	1	0	0	0	0	0
13	73.9	0	0	0	0	1	0	0	0	0	0
14	78.5	0	0	0	0	1	0	0	0	0	0
15	89.8	1	0	0	0	0	1	0	0	0	0
16	78.9	0	0	0	0	0	1	0	0	0	0
17	84.7	0	0	0	0	0	1	0	0	0	0
18	76.1	1	0	0	0	0	0	1	0	0	0
19	60.6	0	0	0	0	0	0	1	0	0	0
20	69.6	0	0	0	0	0	0	1	0	0	0
21	71.5	1	0	0	0	0	0	0	1	0	0
22	67.8	0	0	0	0	0	0	0	1	0	0
23	73.2	0	0	0	0	0	0	0	1	0	0
24	69.5	1	0	0	0	0	0	0	0	1	0
25	72.3	0	0	0	0	0	0	0	0	1	0
26	73.9	0	0	0	0	0	0	0	0	1	0
27	83.7	1	0	0	0	0	0	0	0	0	1
28	63.7	0	0	0	0	0	0	0	0	0	1
29	77.7	0	0	0	0	0	0	0	0	0	1

13.4.2   Answering the Question of Interest with Regression

The results of fitting the additive multiple linear regression model appear in Display 13.18. The estimate of the treatment effect and its standard error, t-statistic, and two-sided p-value appear in boldface.

A 95% confidence interval on the Pygmalion effect completes the standard analysis. With 18 degrees of freedom, the appropriate multiplier is $t_{18}(.975) = 2.101$, the interval half-width is $(2.101)(2.5795) = 5.4195$, and the interval goes from a low of $(7.2205 - 5.4195) = 1.8$ points to a high of $(7.2205 + 5.4195) = 12.6$ points. The mean score for the Pygmalion platoons was estimated to be 7.22 points higher than the mean score for the control platoons, after accounting for the effects of company (95% confidence interval: 1.8 to 12.6 points).

13.4.3   A Closer Look at the Regression Estimate of Treatment Effect

To see how the regression analysis controls for differences between companies, consider Display 13.19, which shows separate estimates of treatment effect from each company.

Display 13.16   *F-test for interactions between companies and treatment: Pygmalion data*

Analysis of variance table from regression fit to the full, nonadditive model,
     $PYG + COMP + (PYG \times COMP)$:

Source of Variation	Sum of Squares	d.f.	Mean Square	F-Statistic	p-Value
Regression	1321.3221	19	69.5433	1.3401	.1747
Residual	467.0400	9	51.8933		
Total	1,788.3621	28			

Analysis of variance table from regression fit to the additive model,
     $PYG + COMP$:

Source of Variation	Sum of Squares	d.f.	Mean Square	F-Statistic	p-Value
Regression	1,009.8581	10	100.9858	2.3349	.0564
Residual	778.5039	18	43.2502		
Total	1,788.3621	28			

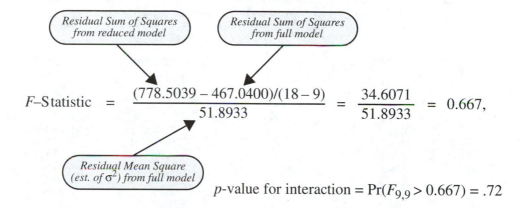

Residual Sum of Squares from reduced model

Residual Sum of Squares from full model

$$F\text{–Statistic} \; = \; \frac{(778.5039 - 467.0400)/(18-9)}{51.8933} = \frac{34.6071}{51.8933} = 0.667,$$

Residual Mean Square (est. of σ^2) from full model

p-value for interaction = $\Pr(F_{9,9} > 0.667) = .72$

The estimate for any particular company equals the difference between the score in the one treated platoon and the average of the two scores from the control platoons (except for company three, where the average control score is from one platoon). Because each of these estimates is constructed *within* a company, differences between companies cannot affect them. That is, these within company estimates of treatment effect *control* for company differences.

The regression analysis produces a composite (as in a paired *t*-analysis), but it gives slightly different weight to company 3 than to the others because of the different numbers of control observations. The least squares estimate of the treatment effect in the additive

Display 13.17   Residual plot from the fit of the additive model to the Pygmalion data

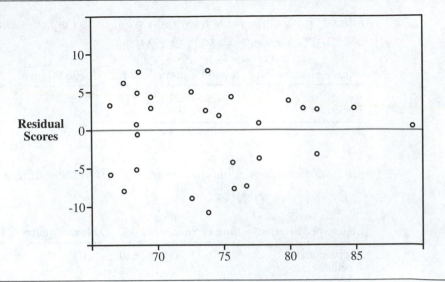

Display 13.18   Multiple linear regression output from the fit of the additive model to the Pygmalion data: $\mu \{score \mid PYG, COMP\} = PYG + COMPANY$

Variable	Coefficient	Standard Error	t-Statistic	Two-Sided p-Value
CONSTANT	75.6137	4.1682	18.1405	<.0001
pyg	**7.2205**	**2.5795**	**2.7992**	**.0119**
cmp2	5.3667	5.3697	0.9994	.3308
cmp3	0.1966	6.0189	0.0327	.9743
cmp4	−0.9667	5.3697	−0.1800	.8591
cmp5	9.2667	5.3697	1.7257	.1015
cmp6	13.6667	5.3697	2.5452	.0203
cmp7	−2.0333	5.3697	−0.3787	.7094
cmp8	0.0333	5.3697	0.0062	.9951
cmp9	1.1000	5.3697	0.2049	.8400
cmp10	4.2333	5.3697	0.7884	.4407

Estimated SD = 6.576 on 18 d.f.

model is equivalent to the weighted average

$$\hat{\delta}_W = \frac{(w_1 \hat{\delta}_1 + w_2 \hat{\delta}_2 + \cdots + w_{10} \hat{\delta}_{10})}{(w_1 + w_2 + \cdots + w_{10})},$$

where the weights are proportional to $1/\mathrm{Var}(\overline{Y}_{i1.} - \overline{Y}_{i2.})$, which, with the constant $1/\sigma^2$ dropped, is $= (1/n_{i1} + 1/n_{i2})^{-1}$. Equivalently:

Display 13.19   Different estimates of the treatment effect, from each company and from the combined data ignoring company differences

Company	Averages		Difference
i	**Pygmalion**	**Control**	$\hat{\delta}_i$
1	80.0	66.2	13.8
2	83.9	72.3	11.6
3	68.2	76.2	−8.0
4	76.5	66.5	10.0
5	87.8	76.2	11.6
6	89.8	81.8	8.0
7	76.1	65.1	11.0
8	71.5	70.5	1.0
9	69.5	73.1	−3.6
10	83.7	70.7	13.0
All	78.7000	71.6316	7.0684

$$w_i = \frac{n_{i1} n_{i2}}{(n_{i1} + n_{i2})} = \begin{cases} \dfrac{(1)(1)}{(1+1)} = \dfrac{1}{2}, \text{ for company } i = 3 \\[2ex] \dfrac{(1)(2)}{(1+2)} = \dfrac{2}{3}, \text{ for the other 9 companies} \end{cases}$$

(and the weighted average turns out to be 7.2205 points, as given in the regression output).

Because the regression estimate is a linear combination of averages in the 20 different groups, the standard deviation can be determined in its sampling distribution directly with the tools in Chapter 6:

$$SD(\hat{\delta}_W) = \sigma \sqrt{\frac{2}{13}} = \sigma \times 0.39223.$$

Here, using the different weights gives a slight improvement over the standard deviation in the sampling distribution of the *unweighted* average of estimates; the latter would be

$$SD(\hat{\delta}_U) = \sigma \sqrt{\frac{31}{200}} = \sigma \times 0.39370.$$

In fact, the multiple linear regression estimate will always give the most efficient weighting to estimates from different levels of a confounding variable in unbalanced situations. Although the difference in this case between using the regression estimate and using the unweighted average of effects is small, that is because this study was nearly balanced. In less balanced examples, the difference can be substantial.

13.4.4   The *p*-Value in the Randomization Distribution

Randomization tests were introduced in Chapter 4, where they helped make precise the notion of inferences for randomized experiments. With any data based on randomization, it is possible to construct a randomization distribution by considering the distribution of a test statistic over all possible ways the randomization could have turned out. Since the actual calculation of the randomization *p*-value is usually quite difficult, however, the usual analysis based on a population model is often accepted as an approximation to the exact measure of uncertainty based on a randomization distribution.

For the Pygmalion randomized block design an exact randomization test requires considering the numerical results under all possible rearrangements of the platoon scores within a given block (since the randomization was carried out separately for each block). The randomization distribution of the *t*-statistic for treatment effect refers to the distribution of the *t*-statistic under all possible groupings of the platoons within a company into a control group (of size 1 for company 3 and size 2 for other companies) and a pygmalion group (of size 1). There are $2 \times 3^9 = 39{,}366$ such arrangements. This is the product of 3 arrangements within company 1, times 3 arrangements within company 2, times 2 arrangements within company 3, times 3 arrangements within company 4, and so on. Calculating the *t*-statistic for each arrangement of the observed scores (with a computer program) yields the histogram in Display 13.20.

Display 13.20   Randomization distribution for the Pygmalion effect *t*-statistic in a multiple linear regression analysis with no interactions

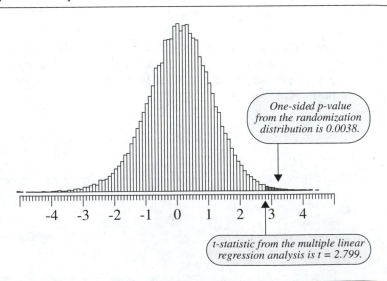

The actual observed t-statistic (from the output in Display 13.18) was 2.799. The proportion of t-statistics in the randomization distribution that exceed or are equal to this is 0.0038. The chance that the evidence of a Pygmalion effect is as large as it is, simply as a consequence of randomization, is .0038. Although this p-value is the most meaningful for this problem, of course it is difficult to compute. The approximation based on a comparison to the t-distribution on 18 degrees of freedom is .0060. Thus the conclusions using the standard regression analysis can be viewed as approximate results for the randomization test, if that interpretation is desired.

13.5 Related Issues

13.5.1 Additivity and Nonadditivities

As the term suggests, *nonadditivity* means anything other than additivity. There are $(I - 1) \times (J - 1)$ parameters more in the saturated model than in the additive model, and all describe nonadditivity. When nonadditivities are present, although computation of row and column averages and row and column effects is still possible, these summaries do not measure meaningful parameters. Consequently, making global statements about treatment differences can be highly misleading.

Nonadditivity may speak in many languages, not all of which are unintelligible. Display 13.21 shows several instances of a simple structure in a nonadditive situation. The upper left panel shows what the mean response to a control treatment might be at seven different levels of a categorical explanatory variable (plotted along the horizontal scale). The upper right panel shows how the means plot would appear in an additive situation with two different treatments applied at the same levels of the factor. The situation is additive because the vertical change from control to each treatment (the treatment effect) is the same at all levels of the factor. Notice that *additive* in the language of algebra translates to *parallel* in the language of geometry. If the panel is turned on its side to show seven different factor curves, each tracing the mean response from the three treatments, that picture would also show parallel lines.

The remaining four panels of Display 13.21 depict nonadditive situations. In each, the treatment effects change with the different factor levels. (But in all, the control treatment curve is the same.)

The nonadditive (I) panel shows a case where the treatment effects change systematically; they increase steadily with increasing levels of the explanatory factor. Assuming that the factor levels as listed are ordered according to the values of an explanatory variable, then including treatment-by-explanatory variable product terms in a multiple linear regression is likely to account for the nonadditivity.

The nonadditive (II) panel shows a situation in which treatment differences increase systematically as the mean response increases. At factor levels where the control mean is small, treatment effects are small. Where the control mean is larger, so are the treatment effects. On a transformed scale, mean responses are likely to exhibit additivity. Compare this panel with Display 13.7.

Display 13.21   Hypothetical treatment curves plotted against another factor, illustrating additive and some nonadditive conditions

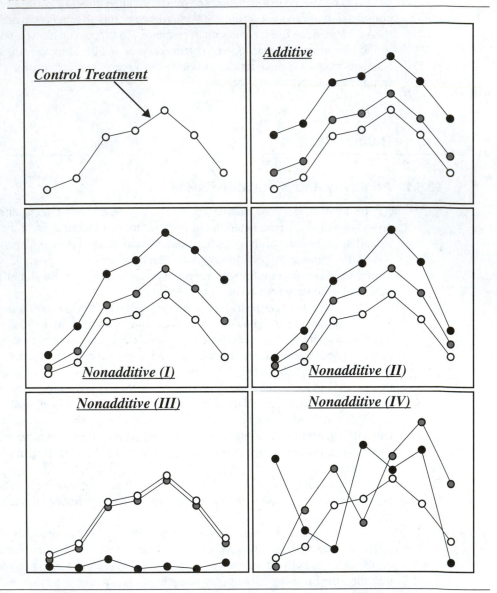

The nonadditive (III) panel illustrates a case where one of the treatments has a drastically different effect from the others. It may be that this treatment killed most of the subjects. Clearly it has destroyed the association between the mean response and the explanatory factor. Here the best strategy is to present separate accounts of what happened for each of the different treatments.

Only the nonadditive (IV) panel appears highly complex. No simple pattern emerges for the nonadditivities. Treatment effects do not vary systematically with either the explanatory factor or the mean response. These situations do arise, sometimes showing statistically significant nonadditivities. Several explanations are possible. First, there may be *intraclass correlation*, where observations within groups are not independent and tend to be more closely concentrated than a truly random sample would be. Second, the residual estimate of variability may underestimate the population variability, just by chance. The estimate is, of course, an estimate sometimes above its target and sometimes below. If it is below, all comparisons between cell means tend to look more significant. (The reason for pooling sources of variability to estimate σ is to get as much information about it as possible to avoid this problem.) Third, other important explanatory factors neither controlled in the design nor represented in the analysis may be confounding the issue.

As these examples illustrate, nonadditivity can arise in meaningful ways. Methods for dealing with it depend on what kind of nonadditivity is involved. The common thread in all these situations is that simply reporting treatment main effects fails to tell the whole story.

13.5.2   Orthogonal Contrasts

Two contrasts are *orthogonal* if the sum of the products of their coefficients equals zero. For example, the large fish and limpets contrasts in Display 13.13 are orthogonal because $(+1/2)(+1/3) + (+1/2)(-1/3) + (-1/2)(+1/3) + (-1/2)(-1/3) + (0)(+1/3) + (0)(-1/3) = 0$. A set of contrasts is considered mutually orthogonal if the contrasts are orthogonal pairwise to each other.

For balanced studies, estimates of orthogonal contrasts are statistically independent of each other, which is of course desirable. Indeed whenever there are I different treatments, it is possible to construct a set of $(I - 1)$ mutually orthogonal contrasts, and whenever the questions of interest are easily phrased in terms of orthogonal contrasts, using them is a good plan. On the other hand, when the questions of interest are most succinctly described in terms of contrasts that are not orthogonal, these should be used. In the seaweed study, the large fish and small fish contrasts were not orthogonal. Yet they are clearly the contrasts of interest, so they were estimated rather than any orthogonal contrasts (even though these might be straightforwardly constructed).

13.5.3   Randomized Blocks and Paired-*t* Analyses

Randomized block experiments extend the paired-*t* analysis of Chapter 3 to the case where there are more than two units per block. An ordinary paired-*t* analysis is a test of treatment effect in a two-way analysis of variance in which the pair (block) is one of the categorical factors and the treatment (with two levels) is the other. In a randomized block study, more generally, several comparisons are made within each block.

13.5.4   Should Insignificant Block Effects Be Eliminated from the Model?

The general advice for regression analysis has been to discard insignificant terms. For data from a randomized block experiment, however, block effects should be retained in the model, even though they may not be significant. This may seem at first to be an inconsistent policy.

The reason for retention is to ensure that the control exercised over block differences in the design is maintained in the analysis. By including the block *main effects*, the inferences are the ones that most closely approximate the randomization tests (Section 13.4.4).

13.5.5 Multiple Comparisons

The seaweed grazers study included specific planned comparisons among the six treatments, so no adjustments for multiple comparisons were needed. For unplanned comparisons the multiple comparison adjustment procedures of Section 6.3 (page 153) can be applied but the estimate of σ^2 and the associated degrees of freedom are from the two-way analysis of variance.

If the study merely asks which treatments differ from which other treatments, or if attention is drawn to a specific comparison by looking at the data (data snooping), then the Tukey–Kramer approach is recommended. If there are I levels of a factor, then confidence intervals will use a multiplier from the Studentized Range distribution corresponding to I groups and whatever degrees of freedom are associated with the estimate of σ^2.

13.5.6 An Alternate Parameterization for the Additive Model

It is convenient to represent the additive model generally as

$$\mu\{Y \mid ROW, COL\} = ROW + COL.$$

The parameterization in Section 13.2.1 used $I - 1$ indicator variables for *ROW* and $J - 1$ indicator variables for *COL*, with a convenient choice for the reference level. A traditional parameterization for the mean of the row i and column j cell is

$$\mu_{ij} = \mu + \alpha_i + \beta_j \quad (\text{for } i = 1, \dots, I \text{ and } j = 1, \dots, J)$$

subject to

$$\alpha_1 + \alpha_2 + \cdots + \alpha_I = 0 \text{ and } \beta_1 + \beta_2 + \cdots + \beta_J = 0.$$

The parameter μ is thought of as an overall mean; α_i is the ith row effect; and β_j is the jth column effect. Although there are $1 + I + J$ parameters in this version of the model, the constraints imply two restrictions, so there are actually only $1 + (I - 1) + (J - 1)$ independent parameters, just as discussed before.

The benefit of this parameterization is that, with balanced data, the least squares estimates of the parameters are simple functions of the row, column, and overall averages. Let Y_{ijk} denote the kth listed response in the ith row and jth column. *Dot notation* signifies summation over all levels of a subscript that is replaced by a dot. So $Y_{5..}$ represents the sum over all responses in row 5, and $Y_{.2.}$ represents the sum of all responses in column 2. This notation combines with the overbar notation for averages, so that $\overline{Y}_{5.}$ represents the average of all responses in row 5. The fitted value based on the fit of the additive model to a *balanced* two-way table is

$$\textit{fit}_{ijk} = \hat{\mu}\{Y_{ijk}\} = \overline{Y}\dots + (\overline{Y}_{i..} - \overline{Y}\dots) + (\overline{Y}_{.j.} - \overline{Y}\dots),$$

where $\overline{Y}\ldots$ is the overall average, and $\overline{Y}_i..$ and $\overline{Y}._j.$ are the average of the ith row the jth column, respectively. The fitted value is the sum of the overall average, the ith estimated *row effect* (the deviation of the ith block average from the overall average), and the jth estimated *column effect* (the deviation of the jth treatment average from the overall average). These are the estimates of the parameters above and satisfy the given constraints. An example of this breakdown is shown in Display 13.12.

Fitting Models with ANOVA or Regression Routines

Statistical computer programs have separate routines for analysis of variance (ANOVA) and for regression. Since the additive and nonadditive models can both be fit with multiple regression using indicator variables, it is possible to analyze two-way tables with either procedure. ANOVA routines use the parameterization discussed in this section and regression routines use a parameterization corresponding to whatever indicator variables are specified.

An important consideration is whether the data are unbalanced. Several choices are available for the analysis of variance decomposition of Between Group sums of squares into row, column, and interactive effects. The user should clearly understand what analysis is being performed. With regression, on the other hand, since the user is forced to specify the full and reduced models in order to carry out the calculations of F-tests, the user has less chance of misunderstanding the hypothesis being tested.

13.6 Summary

Two-way analysis of variance is regression analysis with two categorical explanatory variables. The term *analysis of variance* stems from the central role of the analysis of variance F-tests associated with the row, column, and interaction effects. Two-way tables of data arise from randomized block designs in which the levels of a single treatment are randomly assigned to experimental units within each block separately, completely randomized designs in which all combinations of levels of two treatments are randomly assigned to the experimental units, and observational studies in which responses are identified with each of two categorical explanatory variables. As usual, causal statements can be made from randomized experiments but not from observational studies.

Although the structure of the data may be quite similar for different data problems, the focus of inference (as in regression analysis more generally) depends on the questions of interest being asked. For some problems, a test of the significance of the interactive effects is of primary interest. For others, the interactions may be tested in a first stage of the analysis to check that the additive model, which is considerably more interpretable, can be pursued. In these cases there may be interest in the column effects after accounting for the row effects with no particular interest in the row effects themselves (or vice versa), or there may be interest in both row and column main effects.

F-tests are used for initial screening. Subsequent attention to contrasts, or equivalently to coefficients of indicator variables in the multiple regression analysis of the data, is required to answer particular questions of interest. If the data are unbalanced, then regression analysis with indicator variables is particularly helpful in avoiding confusion and the computer can do the work. If there are no particular questions of interest other than which treatment levels are different from which others, a multiple comparison adjustment, such as the Tukey–Kramer method is appropriate.

Seaweed Grazers

An important first step in the analysis is assessment of the need for transformation, based on graphical displays of the data and residuals. When the log of the regeneration ratio is used as a response, the variance of the residuals is roughly constant and there is no evidence of an interaction between blocks and treatments. The interaction is investigated via the extra sum of squares F-test comparing the nonadditive to the additive model. This is most conveniently obtained with an analysis of variance procedure in a statistical computer program. Analysis of variance is then used on a fit to the additive model, to show that the treatment effects are significant. Although the subsequent analysis can be based on tests and confidence intervals for appropriate linear combinations of treatment means (Section 13.3.4), it is more easily couched in terms of a regression model with indicator variables (Section 13.3.5). In particular, indicator variables for the presence of large fish, small fish, and limpets permit the analyst to obtain confidence intervals and tests that can be directly reported in the summary of statistical findings.

Pygmalion Effect

Since the F-test for interaction between companies (blocks) and treatment is insignificant, the additive model is used for inferences. A single term describes the additional average score of a pygmalion-treated platoon over a control platoon. Section 13.4.4 demonstrates, for this example, how a p-value is tied directly to the chance involved in the randomization process. Although this is a very good approach, it is typically easier and quite adequate to use the usual regression approach as an approximation.

Further Reading

Nonparametric techniques are generally not as useful for two-way classifications as they are for simpler data structures, but Friedman's rank test is occasionally employed. See Miller (1986). A thorough discussion of analysis of variance may be found in Kuehl (1994). More on randomization tests for this and other settings is provided by Good (1994).

13.7   Exercises

Conceptual Exercises

1.   **Seaweed.** If, instead of the five treatment variables in the regression formulation of Section 13.3.5, the control treatment is used as a reference and five indicators for the other five treatments (grazer combinations) are defined, how would the results be different?

2.   **Seaweed.** Display 13.11 shows that R^2 from the fit to the nonadditive model is 92.84%. R^2 from the additive model is 85.34%. Does this indicate that the nonadditive model is better?

3.   **Pygmalion.** Why are the average scores of the platoon used as the response variable, rather than the scores of the individual soldiers?

4.   Why is there so little interest in how the mean response is associated with blocks in a randomized block experiment?

5. Why is balance important?

6. What does it mean when there are significant interactions but no significant main effects?

7. If the F-test for treatments is not significant but the t-test for one of the contrasts is significant, is it proper to report the contrast?

8. Is it possible to examine whether an additive model is appropriate in cases where there is only one observation in each cell?

9. **Seaweed.** Why is the residual SD from the additive model a better choice than the residual SD from the full interactive model for calculating standard errors for the seaweed study linear combinations (Section 13.3.4)?

10. What is the difference between a randomized block design and a completely randomized design with factorial treatment structure? How might the analyses differ?

Computational Exercises

11. **Pygmalion.** Analyze the Pygmalion data using the regression approach described in Section 13.4. (a) Obtain the p-value for the test that interactions are zero. (b) Obtain a p-value, from the additive model, that the treatment effect is zero. (c) Analyze the data with an analysis of variance routine to see if similar results emerge.

12. **Seaweed.** (a) Calculate the sample variance of the six treatment averages in Display 13.13. Multiply this by 16, and verify that the result equals the Treatments Mean Square in Display 13.11. (b) Verify that the Blocks Mean Square in Display 13.11 equals 12 times the sample variance of the block averages from Display 13.12. (c) Verify that the Between Groups Mean Square (i.e., the Model Mean Square) in Display 13.10 is 2 times the sample variance of the 48 cell averages from Display 13.12. *Note: The operative rule here is that the sample variance of a set of averages gets multiplied by the number of observations going into each average.* (d) You can easily work backwards from the mean squares to the sums of squares. Then verify that the Interactions Sum of Squares is the difference between the Between Groups Sum of Squares and the sum of the Blocks and Treatments Sum of Squares.

13. **Seaweed.** Carry out the analysis of the seaweed data with multiple linear regression, as outlined in Section 13.3.5. Using the regression fit, obtain a set of fitted values and residuals. Construct a residual plot. Take the antilogarithms of the fitted values to get estimated medians of percent regeneration scale and construct a plot of these versus block number, ordered as in Display 13.7.

14. **Seaweed.** Using the fit to the regression model described in Section 13.3.5, estimate the mean log regeneration ratio on plots in the reference block with both small and large fish allowed but limpets excluded. Back-transform the estimate and word a statement about the median regeneration ratio.

15. **Blood–brain Barrier.** Analyze the effect of the design variables—sacrifice time and treatment—on the log of the ratio of brain count to liver count in the data set described in Section 11.1.2 (page 294). (a) Ignore the covariates and use an analysis of variance procedure to fit the data. Fit a model that includes interaction terms; plot the residuals versus the fitted values (estimated means). (b) Test whether there is an interactive effect of treatment and sacrifice time. What are the F-statistic, the degrees of freedom, and the p-value? (c) If there are no interactive effects, test whether there are main effects of treatment and sacrifice time. (d) Complete the analysis by describing the effects of treatment and sacrifice time, either by estimating the appropriate contrasts or by using a regression procedure with indicator variables to model treatment (one indicator) and sacrifice time (three indicators).

16. **Toxic Effects of Copper and Zinc.** Reconsider the data in Display 10.18 (page 284). (a) Is the experiment a randomized block design or a completely randomized design with factorial treatment structure? (b) Analyze the data using two-way analysis of variance. (c) How does the analysis of

variance compare to the regression analysis, i.e., what are the issues in deciding which analysis is more appropriate?

Data Problems

17.   Dinosaur Extinctions—An Observational Study. About 65 million years ago, the dinosaurs suffered a mass extinction virtually overnight (in geologic time). What happened in this period, the Cretaceous–Tertiary (KT) boundary, that could have produced such calamity? Among many clues, one that all scientists regard as crucial is a layer of iridium-rich dust that was deposited over much of the earth at that time. Data from one of the first discoveries of this layer are shown in Display 13.22. The diagram traces iridium concentrations in soil samples taken from extensive shale and limestone deposits at Gubbio, Italy. Iridium (parts per trillium) is graphed against the depth at which samples were taken, with depth giving a crude measure of historic time.

Display 13.22   Iridium trace at Gubbio, Italy

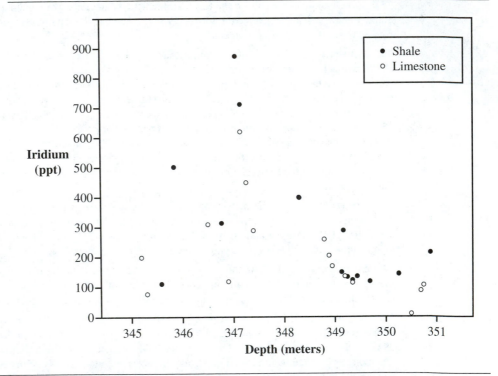

Iridium is a trace element present in most soils everywhere. But concentrations as high as those at the peak near 347 meters, at the KT boundary, can only occur in association with highly unusual events, such as volcanic eruptions or meteor impacts. So the theory is that such an event caused a massive dust cloud that blanketed the earth for years, killing off animals and their food sources. But was the cause a meteor (coming down on the Yucatan peninsula in central America) or volcanic eruptions (centered in southern China)? Two articles debating the issue appeared in the October 1990 issue of *Scientific American*—W. Alvarez and F. Asaro, "What Caused the Mass Extinction?

An Extraterrestrial Impact," *Scientific American* 263(4): 76–84, and E. Courtillot, "What Caused the Mass Extinction? A Volcanic Eruption," *Scientific American* 263(4): 85–93. A crucial issue in the debate is the shape of the iridium trace because the timing and extent of the source give clues to its nature.

(Several iridium traces of better quality have been developed since the Gubbio series was obtained. Since the purpose here is to illustrate certain statistical techniques, rather than discover why the dinosaurs crashed, the Gubbio series will be adequate.)

The objective of the exercise is to estimate the mean (or median) iridium concentration at different depths. To this end, construct a plot showing the estimates and their 95% confidence limits for both the limestone and the shale traces. This will amount to smoothing out the traces in Display 13.22. Are the traces in limestone and shale parallel? Are they the same? (*Hint:* Do the data require transformation?)

The data are organized in Display 13.23 as a two-way table. Notice that the table is not balanced. You should use both the analysis of variance computer tools and the multiple linear regression tools to analyze these data. Compare results to see how they differ in an unbalanced situation.

Display 13.23   Iridium (ppt) in samples from two strata at Gubbio, Italy, by depth category

	Depth Category (meters)					
	345–346	346–347	347–348	348–349	349–350	350–351
Limestone	75 200	120 310	290 450 620	170 205 260	120 135	5 90 105
Shale	110 501	315	710 875	400	120 130 135 150 290	145 215

18.   **Nature–Nurture.** A 1989 study investigated the effect of heredity and environment on intelligence. From adoption registers in France, researchers selected samples of adopted children whose biological parents and adoptive parents came from either the very highest or the very lowest socio-economic status (SES) categories (based on years of education and occupation). They attempted to obtain samples of size 10 from each combination: (1) high adoptive SES and high biological SES, (2) high adoptive SES and low biological SES, (3) low adoptive SES and high biological SES, and (4) low SES for both parents. It turned out, however, only eight children belonged to combination 2. The 38 selected children were given intelligence quotient (IQ) tests. The scores are reported in Display 13.24. (Data from C. Capron and M. Duyme, "Children's IQs and SES of Biological and Adoptive Parents in a Balanced Cross-fostering Study," *European Bulletin of Cognitive Psychology* 11(3) (1991): 323–48.) Does the difference in mean scores for those with high and low SES biological parents depend on whether the adoptive parents were high or low SES? If not, how much is the mean IQ score affected by the SES of adoptive parents, and how much is it affected by the SES of the biological parents? Is one of these effects larger than the other? Analyze the data and write a report of the findings.

Display 13.24   IQ scores for adopted children whose biological and adoptive parents were categorized either in the highest or the lowest socioeconomic status category

SES of *Adoptive* Parents	SES of *Biological* Parents	IQ Scores of Adopted Children									
High	**High**	136	99	121	133	125	131	103	115	116	117
Low	**Low**	94	103	99	125	111	93	101	94	125	91
High	**High**	98	99	91	124	100	116	113	119		
Low	**Low**	92	91	98	83	99	68	76	115	86	116

Answers to Conceptual Exercises

1.   This would be almost exactly like the analysis of Section 13.3.5. Instead of measuring interesting contrasts, the coefficients would measure effects of applying treatment combinations. To tease out the separate effects of each grazer, it would be necessary to consider linear combinations of the regression coefficients. (Notice how the judicious choice of indicator variables in Section 13.3.5 avoided the messy contrast approach.)

2.   No. As usual, R^2 is always larger for a model with more terms, whether or not those terms are significantly different from zero.

3.   The experimental unit—that which receives the treatment—is the platoon, not the individual soldier.

4.   The association is not an issue. Blocks are formed to provide similar units on which to compare treatments. Also, the randomization is not conducive to drawing inferences about block differences.

5.   The simple row and column means in a balanced table produce simple estimates of model parameters. To see that the treatment effects are not influenced by block differences, try adding a certain fixed amount to all responses in one block. It will not change any of the treatment differences in a balanced table. But it will in an unbalanced table.

6.   It simply means that there are significant interactions. When there is nonadditivity, the best strategy is to avoid talking about main effects.

7.   If the contrast was planned in advance, report it. If it is a comparison that the data suggested, you should account for the data snooping when attaching any significance to it.

8.   The analysis of variance table will have no within-group entry if there is only one observation in each cell. So there will be no formal analysis of variance test for nonadditivity. But the answer to the question is yes, and the next chapter will discuss how.

9.   If there are interactions, there is no sense in estimating the linear combinations.

10.   In the randomized block design the random allocation of treatment levels to experimental units is conducted separately for each block. In a completely randomized design the treatment levels are randomly allocated to all experimental units at once, but the levels correspond to all possible combinations of the levels of two distinct treatments. In the randomized block data there is generally no interaction expected and there is little interest in the block effects. In the design with factorial treatment arrangement there may be interest in an interactive effect and there is typically interest in the effects of both treatments.

CHAPTER 14

Multifactor Studies Without Replication

Chapter 13 emphasized that the analysis of variance for two-way tables is regression with categorical indicator variables. As long as factor effects are treated as fixed (not random; see Section 14.5.1), any number of factors and numerical explanatory variables—and their interactions—can be accommodated in a regression model. There is little need to distinguish between analyses that focus on numerical explanatory variables and those that focus on categorical factors, except to recognize that analysis of variance F-tests play a more central role in the latter.

One important strategic issue, however, is associated with the presence or absence of replicates in a multifactor study. Given more than one observation at some explanatory variable combinations, sample variances may be computed and pooled together as an estimate of σ^2 that does not hinge on the accuracy of any model for the mean. In regression this is called the *pure error* estimate of σ^2, which is used to perform a lack-of-fit test. In the absence of replicates, the residuals from a saturated model are necessarily zero, with no degrees of freedom for estimating σ^2 from the residuals.

Following a common tactic in regression, the analysis may proceed if a simpler model with fewer parameters can be assumed to fit, thus leaving degrees of freedom for estimating residual variation. For two-way tables, this means proceeding as if the interaction terms can be ignored. If one of the factors has three or more numerical levels, this means modeling the effect of the factor as a linear function of the numerical variable. The estimate of σ^2 is based on the sum of squared residuals. Even though some questions cannot be answered under these circumstances, much can be learned from such analyses.

14.1 Case Studies

14.1.1 Chimpanzees Learning Sign Language— A Controlled Experiment

An irony of the space age is that, while man beams coded messages toward distant galaxies in search of responsive intelligent life forms, communication with highly intelligent animals on this planet remains a mystery. In one study, however, a researcher taught 10 signs of American Sign Language (ASL) to four chimpanzees. (Data from R. S. Fouts, "Acquisition and Testing of Gestural Signs in Four Young Chimpanzees," *Science* 180 (1973): 978–80.) The study's goals were to determine whether some signs were more easily acquired than others and whether some chimps tended to learn signs more quickly than other chimps.

The subjects were four young chimpanzees—males Bruno and Booee, and females Cindy and Thelma—at the Institute for Primate Studies at the University of Oklahoma. The ASL signs—*hat, shoe, fruit, drink, more, look, key, listen, string,* and *food*— covered a wide range of objects, actions, and concepts. Chimpanzees were taught individually, using a system of rewards (and in the case of a reluctant Bruno, threats), until they could successfully produce unprompted responses on five consecutive occasions. Display 14.1 shows the times, in minutes, required to teach each sign to each subject, according to that criterion. Both factors are arranged according to increasing average times.

Display 14.1 Minutes to acquisition of American Sign Language signs by four chimps

	Listen	*Drink*	*Shoe*	*Key*	*More*	*Food*	*Fruit*	*Hat*	*Look*	*String*
Booee	12	15	14	10	10	80	80	78	115	129
Cindy	10	25	18	25	15	55	20	99	54	476
Bruno	2	36	60	40	225	14	177	178	345	287
Thelma	15	18	20	40	24	190	195	297	420	372

Large differences between times required to acquire different signs appear. *Listen, shoe,* and *drink* were acquired quickly by all four chimps, but *look* and *string* were seemingly quite difficult. There are some less apparent but consistent differences among the chimps. Booee, who was "willing to sell his soul for a raisin," seemed to acquire the signs quickly; while Thelma took considerably longer, being easily distracted by such intrusions as a fly in the cage. What can be said about whether such differences are systematic?

Summary of Statistical Findings

Inconclusive evidence suggested that some chimpanzees were faster than others at learning words (p-value = .064 from an F-test on 3 and 27 degrees of freedom). Convincing evidence indicated that some signs were learned more quickly than others (p-value = .00002 from an F-test on 9 and 27 degrees of freedom). Display 14.2 shows the mean completion times for each word. The double-arrow line is the half-width of a 95% confidence interval (based on the Tukey–Kramer procedure for multiple comparisons of means). A test of equality of any two sign means will have a p-value of less than .05 if the difference between the means is greater than the length of the HSD ("honest significant difference") line (Section 6.4.1).

Display 14.2   Multiple comparisons of sign means on the log scale

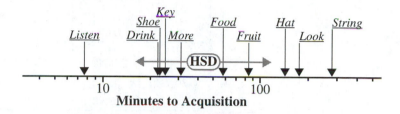

14.1.2   Effects of Ozone in Conjunction with Sulfur Dioxide and Water Stress on Soybean Yield—A Randomized Experiment

The National Crop Loss Assessment Network (NCLAN) was a series of studies designed to estimate the benefits to agriculture in the United States that followed from reducing levels of ambient ozone. These studies, conducted from 1980 to 1987, involved many crops in many locations. One of the most damaged crops was found to be soybeans, but serious questions were raised about the effects and possible interactions of ozone with sulfur dioxide and with water stress.

   This 1982 study at the Beltsville, Maryland, Agricultural Research Center was a randomized experiment involving two different soybean cultivars. The 30 possible combinations of ozone (at five levels), sulfur dioxide (at three levels), and water stress (at two levels) were randomly assigned to 30 open-topped chambers specially constructed to maintain stable conditions. Two soybean cultivars—*Forrest* and *Williams*—were randomly assigned to two rows within each chamber, with separate randomizations within each chamber. The soybeans were planted in July; sulfur dioxide and ozone were piped in and monitored daily; soil moisture was maintained by watering and was measured at a depth of 0.25 to 0.45 meters. In September, the crops were harvested, each row producing a quantity of soybean seeds that was converted into kilograms per hectare. These data appear in Display 14.3. (Data from H. E. Heggestad and V. M. Lesser, "Effects of Chronic Doses of Sulfur Dioxide, Ozone, and Drought on Yields and Growth of Soybeans Under Field Conditions," *Journal of Environmental Quality* 19 (1990): 488–95.)

   Many previous studies had found that ozone and sulfur dioxide reduce yield. Evidence about interactions between ozone and sulfur dioxide had not been consistent, nor was it clear how moisture stress influenced the results. Three questions were of interest in this study: (1) is there an additional influence of soil moisture stress? (2) are there interactive effects among the three factors? and (3) do the stress variables have different effects on the two cultivars?

Display 14.3   Seed yields for soybean cultivars *Forrest* and *Williams* from chambers kept
under varying conditions of ozone, sulfur dioxide, and water stress

Water Stress	SO_2 (μL/L)	O_3 (μL/L)	Yields (kg/ha) *Forrest*	*Williams*
Well-watered (WW) (−0.05 MPa)	0.0045	0.017	4376	5561
		0.049	4544	5947
		0.067	2806	4273
		0.084	3339	3470
		0.099	3320	3080
	0.0170	0.017	3747	5092
		0.049	4570	4752
		0.067	4635	4232
		0.084	3613	2867
		0.099	3259	3106
	0.0590	0.017	4179	4736
		0.049	5077	3672
		0.067	3401	3386
		0.084	3371	2854
		0.099	2158	2557
Soil moisture stress (SMS) (−0.40 MPa)	0.0045	0.017	4977	4520
		0.049	3780	3047
		0.067	3804	3526
		0.084	3941	3357
		0.099	2863	2663
	0.0170	0.017	5573	4869
		0.049	3555	3774
		0.067	3340	2955
		0.084	3243	3513
		0.099	2802	2838
	0.0590	0.017	4589	4056
		0.049	3250	2758
		0.067	3045	3094
		0.084	2827	2398
		0.099	2979	2101

Summary of Statistical Findings

Main Effects for Forrest *Cultivar*   There is strong evidence of an ozone effect on yield
(two-sided p-value $< .0001$). A 0.01 μL/L increase in ozone was estimated to decrease
median yield by 5.3% (the 95% confidence interval for median reduction per 0.01 μL/L
increase in ozone is 3.4% to 7.0%). These data are consistent with there being no effect of
sulfur dioxide on seed yield (two-sided p-value $= .13$) but the best estimate of the effect
of a 0.01 μL/L increase in sulfur dioxide is a 1.6% reduction in median yield (the 95%
confidence interval for median reduction is -0.5% to 3.5%). The data are also consistent
with there being no effect of water stress (two-sided p-value $= .55$), but the best estimate
is that the median yield was 3.3% greater for the stressed condition than for the well-

watered condition (the 95% confidence interval for increase due to water stress is -7.3% to 15.3%).

Interactions Between Ozone and the Other Factors for Forrest *Cultivar*   Although the data are consistent with there being no interactive effects of ozone, sulfur dioxide, and water stress on yield from the *Forrest* cultivar, these data have very little power for detecting interactions. The following 95% confidence intervals exhibit the range of nonzero interactive effects that are consistent with the data.

Ozone-by-sulfur dioxide: The ozone effect (the reduction in median seed yield per 0.01 μL/L increase in ozone) when the sulfur dioxide concentration is 0.0690 μL/L is estimated to be only 14.9% of the ozone effect when sulfur dioxide is 0.0045 μL/L (95% confidence interval: 0.2% to 1,450.0%).

Ozone-by-water stress: The ozone effect when the crop is exposed to water stress $(-.40\text{MPa})$ is estimated to be 430.3% of the ozone effect under well-watered conditions (95% confidence interval: 9.2% to 20,140.0%). Even though both intervals include 100% (which corresponds to no interaction), they also include an extremely wide range of other values.

Main Effects for Williams *Cultivar*   There is strong evidence of an ozone effect on yield (two-sided p-value $< .0001$). A 0.01 μL/L increase in ozone was estimated to decrease median yield by 6.6% (95% confidence interval for median reduction per 0.01 μL/L increase in ozone is 5.3% to 7.9%). There is also strong evidence of a sulfur dioxide effect on yield (two-sided p-value $< .0001$), with a 0.01 μL/L increase in sulfur dioxide estimated to cause a 3.5% reduction in median yield (95% confidence interval for median reduction is 2.0% to 4.9%). There is also strong evidence of an effect of water stress on yield (two-sided p-value $= .0001$), with stress at -0.40MPa estimated to reduce seed yield by 19.4% (the 95% confidence interval for median reduction is 10.2% to 29.4%).

Interactions for Williams *Cultivar*   The following 95% confidence intervals exhibit the range of nonzero interactive effects that are consistent with the data.

Ozone-by-sulfur dioxide: The ozone effect (the reduction in median seed yield per 0.01 μL/L increase in ozone) when the sulfur dioxide concentration is 0.0690 μL/L is estimated to be only 40.8% of the ozone effect when sulfur dioxide is 0.0045 μL/L (95% confidence interval: 1.4% to 1,197.5%).

Ozone-by-water stress: The ozone effect when the crop is exposed to water stress (-0.40 MPa) is estimated to be 23.8% of the ozone effect under well-watered conditions (95% confidence interval: 1.5% to 385.6%).

Note

The NCLAN soybean study had a split-plot design, which will be discussed further in Chapter 16. Chambers were *whole plots*, receiving treatment combinations of ozone, sulfur dioxide, and water. Within each chamber were two *split plots*, one receiving each cultivar. A split plot analysis-of-variance could be conducted and would lead to direct comparisons of the cultivars and their interactions with the whole-plot treatments. The analysis in this chapter illustrates the alternative approach of conducting separate univariate analyses for each cultivar, which is adequate for answering most of the questions asked.

14.2 Strategies for Analyzing Tables with One Observation per Cell

14.2.1 Rationale for Designs with One Observation per Cell

Both case studies in this chapter are multifactor studies with a single observation at each combination of the factors. The chimp data arose from a *repeated measures* experiment, meaning that the experimental units (the chimps) were observed under each of several conditions (the signs). A more general discussion of repeated measures is provided in Chapter 16. For now, the important factor of the experiment was the effort to teach each sign to each chimp: treating chimp as a block permits comparing signs without interference from chimp-to-chimp learning variability. One consequence of this design is the impossibility of obtaining replicates: a given sign cannot be taught independently to the same chimp twice.

In the soybean study the experimental units were the growing chambers. In this case more than one chamber could have been assigned to receive each combination of the experimental factors; but given the limited number of available chambers, the question here is whether replication constitutes the wisest allocation of the resources. The use of replicates with the same number of chambers would have precluded simultaneous examination of the three factors of interest and their interactions. The researchers elected to examine the questions of interest without the benefits of replication, realizing that much could be learned anyway.

The major benefit of replication is that it permits model-free estimation of variability of the responses; that is, the adequacy of the estimate of variance does not hinge on the correctness of any presumed model. Replication allows formal testing of the lack of fit for any model (as with the lack-of-fit test for regression discussed in Section 8.5.3 on page 210). Sometimes, however, it is more important to expend resources on aspects of the design that pertain to the questions of interest than on aspects that enable the analyst to test it for lack of fit. This is especially so when either of the following conditions is true: prior information strongly supports a simple model or structure, which need not be checked prior to being used; or exploratory and diagnostic procedures can be introduced to sort out the model without reliance on a formal lack-of-fit test.

Without replication, estimates of σ^2 (and hence, tools of inference) are only possible by fitting nonsaturated models. For the chimp data, this is accomplished by assuming that no *chimp*-by-*sign* interactions exist. For the soybean data, it is accomplished by treating ozone as a numerical explanatory variable, rather than as a categorical one, and by using a linear model to describe the effect of ozone.

14.2.2 Strategy for Data Analysis in the Absence of Replicates

The suggested analysis here is much the same as for regression analysis in general (see Display 9.9 on page 241). Graphical displays of the data, such as coded scatterplots of the response versus a code for one of the factors, are usually helpful. A plot of residuals versus fitted values from a fairly rich model should be used to investigate the need for transformation and the possible presence of outliers. After transformation and outliers have been considered and resolved, model refinement may take place with informal testing of model terms.

One aspect of this strategy requires more skill than do corresponding strategies for problems with replication: the choice of a rich model on which residual examination is conducted. Including too few terms might mask relevant patterns in the residual plot, if some of the omitted terms are actually important. Including too many terms will produce a fit tied too closely to the particular data. It is important to identify the major sources of variation in the exploratory analysis, and then to fit a model that explains most of the variability with as few terms as possible.

In a forward selection approach, main effects are considered and retained if significant. Then the *second-order interactions* (interaction between two factors) are investigated, but only between factors whose main effects have already been found to be significant. If appropriate, *third-order interactions* (interactions among three factors) are considered next, but only among factors whose main effects and second-order interactions have been determined to be significant.

In a backward elimination approach, a rich model is fit. Then testable highest-order interactions are considered and dropped if not significant. Then next-highest-order interactions are considered; and so on. Under either approach, including an interaction term is generally inappropriate unless the main effects of all the variables in the interaction are also included in the model.

If any of the explanatory variables can be treated as numerical, as opposed to categorical, it is usually appropriate to attempt the simpler fit involved in treating them as such. If this works, it will result in a model with fewer parameters and a simpler interpretation. Furthermore, some problems (like the soybean data problem) simply do not have enough degrees of freedom to permit estimation of the interactions of interest, so using a linear effect of the numerical variables (like ozone) at an early stage in the analysis helps to "save degrees of freedom" for subsequent investigation of interactions.

14.3   Analysis of the Chimpanzee Learning Times Study

Exploratory Analysis

There is no obvious ordering to either the chimpanzees or the different signs. Hence, it is convenient to order the levels of both factors according to increasing average learning times, as in Display 14.1. This helps the analyst visualize how steadily and regularly the times increase.

A *coded two-way table* can enhance such visualization by replacing numbers in the table with symbols that show relative position in the overall distribution. Display 14.4 shows a box plot of chimp learning times treated as a single sample. To the right of the box plot, positions expressed in terms of quartiles and extreme values are assigned suggestive labels that can be substituted for the values in the data table of Display 14.1. The resulting coded table appears as Display 14.5. Notice, for example, that Booee required 12 minutes to learn the sign for *listen*. Since 12 resides in the whisker extending below the lower quartile, it is replaced by the symbol =.

In an additive table, one should expect a steady progression from minus codes to plus codes as one goes from the upper left of the table to the lower right. This pattern holds reasonably well here; but the sign for *more* seems to be an anomaly, and Bruno appears to have had mixed success with the signs in comparison to the other chimps.

Display 14.4   Box plot of the chimpanzee sign acquisition times, and suggestive codes

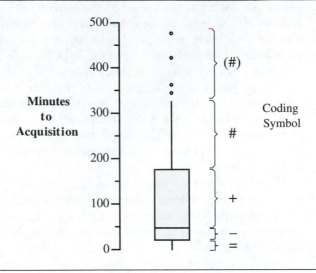

Display 14.5   Coded two-way table for the chimpanzee data

	Listen	*Drink*	*Shoe*	*Key*	*More*	*Food*	*Fruit*	*Hat*	*Look*	*String*
Booee	=	=	=	=	=	+	+	+	+	+
Cindy	=	–	–	–	=	+	–	+	+	(#)
Bruno	=	–	+	–	#	=	+	#	(#)	#
Thelma	=	–	–	–	–	#	#	#	(#)	(#)

Fitting the Additive Model

The numerical analysis begins with a fit to the additive model. Because the data are balanced (with one observation in each cell), fitted values can be calculated easily, without the use of a computer, as shown in Section 13.3.3 (page 375). The first step in obtaining fitted values is to calculate average times for each sign (row average), for each chimp (column average), and for all observations in the table (overall average). The estimated sign effects are the sign averages minus the overall average, as shown in the last column of Display 14.6. The estimated chimp effects are the chimp averages minus the overall average, as shown in the last row of that figure. The fitted values from the additive model are then

$$\text{Fitted value} = \text{Overall average} + \text{Chimp effect} + \text{Sign effect,}$$

and the residuals are found by subtracting fitted values from observed values.

Display 14.6   Observed values, fitted values, and residuals for the additive fit to the chimpanzee sign acquisition times (all in minutes)

Y Fitted Value Residual	*Booee*	*Cindy*	*Bruno*	*Thelma*	Average	*Sign Effect*
Listen	12 −43.325 +55.325	10 −17.925 +27.925	2 38.775 −36.775	15 61.475 −46.475	9.75	−97.625
Drink	15 −29.575 +44.575	25 −4.175 +29.175	36 52.525 −16.525	18 75.225 −57.225	23.50	−83.875
Shoe	14 −25.075 +39.075	18 0.325 +17.675	60 57.025 +2.975	20 79.725 −59.725	28.00	−79.375
Key	10 −24.325 +34.325	25 1.075 +23.925	40 57.775 −17.775	40 80.475 −40.475	28.75	−78.625
More	10 15.425 −5.425	15 40.825 −25.825	225 97.525 +127.475	24 120.225 −96.225	68.50	−38.875
Food	80 31.675 +48.325	55 57.075 −2.075	14 113.775 −99.775	190 136.475 +53.525	84.75	−22.625
Fruit	80 64.925 +15.075	20 90.325 −70.325	177 147.025 +29.975	195 169.725 +25.275	118.00	+10.625
Hat	78 109.925 −31.925	99 135.325 −36.325	178 192.025 −14.025	297 214.725 +82.275	163.00	+55.625
Look	115 180.425 −65.425	54 205.825 −151.825	345 262.525 +82.475	420 285.225 +134.775	233.50	+126.125
String	129 262.925 −133.925	476 288.325 +187.675	287 345.025 −58.025	372 367.725 +4.275	316.00	+208.625
Average	54.3	79.7	136.4	159.1	107.375	
Chimp effect	−53.075	−27.675	+29.025	+51.725		

The fitted value for Bruno's learning the sign for *hat*, for example, is the overall average time (107.375 minutes) adjusted for the fact that Bruno on average takes 29.025 minutes longer than all chimps do and for the fact that the sign for *hat* takes 55.625 minutes longer than the average sign. Fitted value = 107.375 + 29.025 + 55.625 = 192.025 minutes. The observed time was 14.025 minutes below the fitted time.

This table casts serious doubt on the additive model, because negative means are estimated for several cells, when the actual data must be positive. Occasionally this happens even with good data from a truly additive model. Here, however, many such cells are in evidence, and they occur in a definite pattern, as revealed by the residual plot in Display 14.7.

Display 14.7   Residual plot for the additive model fit to chimpanzee acquisition times

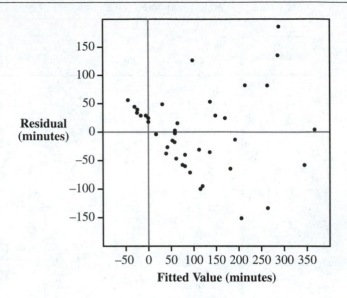

The pattern is horn-shaped—showing a curve trend and increasing variability with increasing estimated means. The curved pattern in the plot is due to the points in the upper left quadrant, which correspond to observations with negative fitted values. The additive model is evidently not correct since it leads to estimates for these chimp–sign combinations that are much lower than expected. This pattern, however, is typical of residual plots from additive fits to positive data where nonadditivity may be eliminated by a suitable transformation.

Transformation

The logarithm is usually the transformation to try first, because it provides the most desirable interpretation. After making the transformation, redraw the residual plot, and then try another transformation if there is still some pattern in this residual plot. Display 14.8 shows the residual plot from the additive fit after a log transformation. This plot indicates that no further transformations are needed; the convenient log transformation is satisfactory.

Analysis of Variance

With log of acquisition times as responses, the analysis of variance table appears in Display 14.9. Since there is no within-group ("pure error") mean square, the mean square for residuals from the additive model is used as the estimate of σ^2 and, consequently, as the denominator in F-statistics for tests about the main effects. The hope is that an additive model is correct on this scale. Is it? There is no clear way to tell, but plots of the raw

Display 14.8   Residual plot for the additive model fit to log(acquisition times)

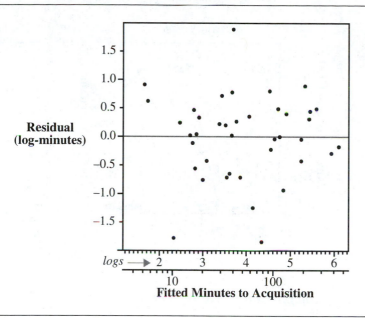

data on the minutes scale and on the log scales (shown in Display 14.10) suggest that the transformation has yielded in some degree of parallelism.

Display 14.9   Analysis of variance for the additive model fit to log(acquisition times)

Source of Variation	Sum of Squares	d.f.	Mean Square	F-Statistic	p-Value
Signs	45.6900	9	5.0767	7.7649	0.00001
Chimpanzees	5.3329	3	1.7776	2.7190	0.0642
Residual	17.6526	27	0.6538		
Total	68.6755	39			

R-squared = 74.3%   **Estimated SD** = 0.8086

The initial screening provided by the F-tests, assuming the additive model is correct, indicates convincing evidence of a sign effect (p-value $= .00001$) and suggestive but inconclusive evidence of a chimp effect (p-value $= .0642$). In fact, 66.5% of the variability in log acquisition times is explained by differences between signs (45.69/68.6755), whereas 7.8% is explained by chimp differences.

Display 14.10   Chimpanzee data plots: natural scale and logarithmic scale

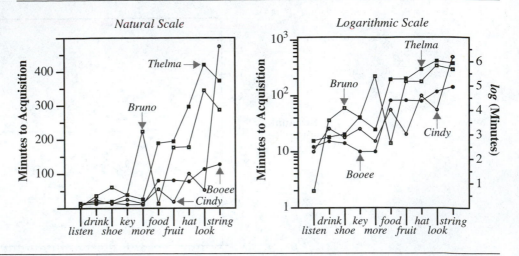

Multiple Comparisons of Sign Effects

There are 45 differences between averages in pairs of signs. For a test asking "Which signs are different from which other signs?" the Tukey–Kramer multiple comparison adjustment is appropriate (see Section 6.4.1 on page 154). The confidence interval for comparing sign j to sign k is

$$(\overline{Y}_{.j} - \overline{Y}_{.k}) \pm M \times \mathrm{SE}(\overline{Y}_{.j} - \overline{Y}_{.k}),$$

where $M = q(.05; I, \mathrm{d.f.})/\sqrt{2}$ and $q(.05; I, \mathrm{d.f.})$ is the 95th percentile of the studentized range distribution for I groups and d.f. degrees of freedom in the estimate of σ. Since each sign average is based on four observations,

$$\mathrm{SE}(\overline{Y}_{.j} - \overline{Y}_{.k}) = \hat{\sigma}\,(1/4 + 1/4)^{1/2}.$$

With $\hat{\sigma} = 0.8086$, $I = 10$, and d.f. $= 27$, the half-width of the confidence interval for all 45 comparisons—also called the HSD, for "honest significant difference"—is found to be HSD $= 1.95$.

It is not very effective to list all 45 confidence intervals. Instead, a chart of the sign averages, on the log scale, with an indication of the size of the HSD can be displayed, as in Display 14.2. The reader can visually compare the different signs to one another, realizing that differences between averages exceeding the HSD correspond to two-sided p-values of less than .05. Although this plot was constructed on the log scale, the labels for the sign effects are shown in minutes.

To illustrate the relative effects of chimps and signs further, and to highlight the multiplicative nature of the model on the original scale, Display 14.11 shows the antilogarithms of the fitted values from the additive fit to the logarithms of acquisition times. This can be compared to the plot in Display 14.10.

Display 14.11   Cell median estimates from additive model on the log scale

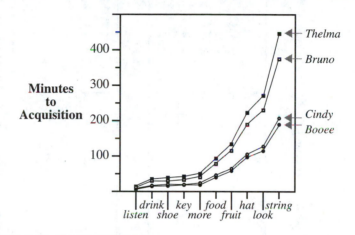

14.4   Analysis of the Soybean Data

The data in Display 14.3 come from a completely randomized experiment with a $2 \times 3 \times 5$ factorial treatment structure. Each of the 30 treatment combinations corresponds to a single response for each of the two cultivars, which are analyzed here as separate response variables. An estimate of σ based on replication is impossible, but linear effects of the explanatory variables sulfur dioxide and ozone may be exploited to obtain an estimate of σ in a reasonably rich regression model. One should check the linearity assumption with scatterplots, residual plots, and tests of quadratic terms so the inferential model can confidently be substantiated.

14.4.1   Exploratory Analysis

Previous studies have suggested that the variance of soybean seed yield is roughly constant on the log scale. The analysis begins, therefore, with scatterplots of log soybean seed yield versus each of the three design variables. Display 14.12 suggests that yield decreases linearly with ozone increase for both cultivars. The effect of sulfur dioxide might be linear, but the plots are inconclusive.

The next step is to examine a residual plot based on a fairly rich model, but one that exploits the patterns in the scatterplots. The scatterplot in Display 14.13 shows the residuals from the regression of yield on ozone (treated as a numerical explanatory variable), on sulfur

Display 14.12   Scatterplots of log soybean seed yield versus ozone, sulfur dioxide, and water stress

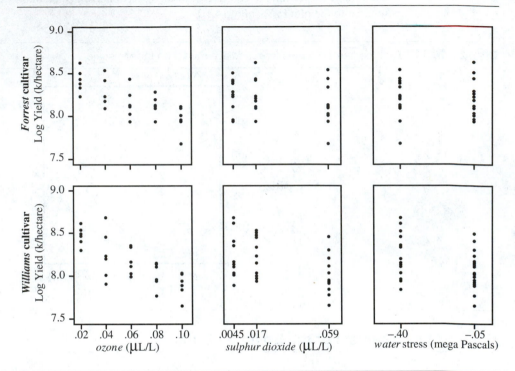

dioxide and water (both treated as categorical), and on all second-order interactions of these. The plots show no signs of problems.

Assessment of Linearity in Ozone

The linear patterns in the scatterplots of log yield versus ozone are convincing. It is possible to test informally for linearity by including a quadratic term in ozone in the models used to obtain the residual plots of Display 14.12. The p-values for the tests that the coefficients of $ozone^2$ are 0 are .49 for the *Forrest* cultivar and .56 for the *Williams* cultivar. As a final check, to ensure that the nonsignificance of the quadratic effect is not due to one or two outliers, a plot of the residuals (from the model with only the linear effect) versus *ozone* are examined (not shown here). This does not indicate any pattern or potentially influential observation. It seems appropriate to proceed with modeling the effect of *ozone* as linear.

Assessment of Interactions Between Ozone and the Other Treatments

If an interaction between ozone and sulfur dioxide exists, the effect of ozone—the change in mean response per unit increase in ozone—depends on the level of sulfur dioxide. These

Display 14.13   Residual plots from the regression of log soybean seed yield on ozone (numerical) sulfur dioxide (categorical), water stress (categorical), and second-order interactions

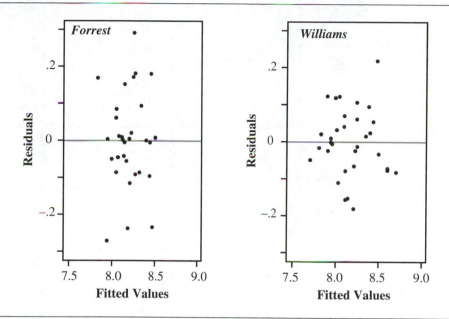

interactions, and similarly the ozone-by-stress interactions, can be investigated graphically by plotting the log yield versus ozone separately for each combination of sulfur dioxide and water stress, as in Display 14.14. Also shown on these plots are the least squares fits to the linear regression of log yield on ozone. An interaction between ozone and the other variables would be indicated if the slopes on these plots manifestly differed.

Notice that the slopes for the *Williams* cultivar are remarkably similar. For the *Forrest* cultivar, the slopes show some variability, but they do not depend in any systematic fashion on sulfur dioxide or water stress. Whether the observed differences in slopes represent a true interaction or simply sampling variability can be checked with tests for the appropriate interaction terms.

If all three factors—ozone, sulfur dioxide, and water stress—were treated as categorical explanatory variables, it would be impossible to test for a third-order interaction between them. Since the effect of ozone can be safely modeled as linear, all interactions can be tested with extra-sums-of-squares F-tests that compare full models with the specific interactions (and treat ozone as a numerical explanatory variable) to the reduced models without them. Shown in Display 14.15 are analysis of variance tables that summarize F-tests for all effects. Each sum of squares is the sum of squared residuals from the full model minus the sum of squared residuals from the reduced model that excludes the effect designated in that row.

Display 14.14 Scatterplots of log yield against ozone at all combinations of sulfur and water for both cultivars, with simple linear regression estimates

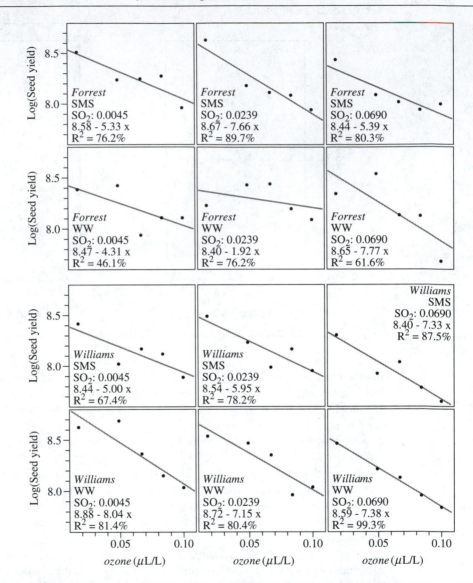

The full model for all these tests is the one that includes ozone as a numerical explanatory variable, sulfur and stress as factors, and all second- and third-order interactions between these three.

Display 14.15 Analysis of variance tables for screening effects on log soybean seed yield

Forrest Cultivar

Source	d.f.	Sum of Squares	Mean Square	F-stat	p-Value
ozone	1	.7208	.7208	29.4	.00003
SULFUR	2	.0635	.0317	1.30	.30
water	1	.0080	.0080	.328	.57
ozone × *SULFUR*	2	.0173	.0087	.353	.71
ozone × *water*	1	.0136	.0136	.556	.46
SULFUR × *water*	2	.0285	.0143	.583	.57
ozone × *SULFUR* × *water*	2	.0683	.0342	1.46	.26
residuals	18	.4211	.0234		

Williams Cultivar

Source	d.f.	Sum of Squares	Mean Square	F-stat	p-Value
ozone	1	1.150	1.150	86.8	<.00001
SULFUR	2	.2780	.1390	10.5	.001
water	1	.2376	.2376	17.9	.0005
ozone × *SULFUR*	2	.0037	.0019	.140	.87
ozone × *water*	1	.0128	.0128	.964	.34
SULFUR × *water*	2	.0263	.0131	.004	.39
ozone × *SULFUR* × *water*	2	.0093	.0047	.352	.71
residuals	18	.2384	.0132		

The *p*-values indicate that ignoring the interaction terms has little consequence. While there is no evidence of a *sulfur* or *water* effect on the *Forrest* cultivar yields, there is strong evidence that these are important factors on yield of the *Williams* cultivar.

Assessment of Linearity of Sulfur Dioxide

To this point sulfur dioxide has been treated as a factor because of the evidence in the original scatterplots of a possible deviation from linearity and because the linearity of ozone sufficiently freed up degrees of freedom for testing other effects. It is appropriate now to see whether the sulfur dioxide effect can be represented by a linear term. When sulfur dioxide is treated as a numerical explanatory variable and when linear and quadratic terms are included, the *p*-value for a test that the coefficient of *sulfur*^{2} is zero is .6. Thus the model with just a single term to represent the linear effect of sulfur dioxide will be used. (*Note:* Since water stress has only two levels, the fit is the same whether it is treated as a numerical explanatory variable or as a categorical one, and it is impossible to fit a quadratic effect.)

14.4.2 Answering Questions of Interest with the Fitted Models

Conclusions About Main Effects

Display 14.16 shows the least squares estimates of the coefficients of *ozone*, *sulfur*, and *water* in the multiple regression models in which all three explanatory variables are numerical.

The conclusions and confidence intervals reported in the summary of Section 14.1.2 follow from these statistics via back-transforming.

Display 14.16 Coefficient estimates and standard errors for the linear soybean models, with Y = log(soybean seed yield)

	Forrest			Williams		
Variable	**Coefficient**	**St. Error**	**2-Sided p-Value**	**Coefficient**	**St. Error**	**2-Sided p-Value**
CONSTANT	8.608	0.080		8.825	0.058	
ozone	−5.397	0.929	<.0001	−6.806	0.679	<.0001
sulphur	−1.566	0.989	.1252	−3.512	0.723	<.0001
water	0.094	0.153	.5453	0.507	0.112	.0001

A graph that shows the fitted values converted to the original scale of measurement (kilograms per hectare) provides a convenient summary of the final model and offers an indication of the relative importance of the various effects. Display 14.17 was constructed by choosing a set of values that span the range of ozone and calculating estimated median yields at these values of ozone and at each of two values of sulfur dioxide and water stress. For each of the four combinations of these last two variables, the estimated means were plotted versus the ozone values and connected by lines. As an advanced addition to this plot, confidence intervals for the median yield were computed at each of several values and drawn on the plot. The standard errors for the estimated means were found with the centering procedure, and confidence intervals were widened by the Scheffé method so that the intervals indicate the uncertainty in estimation over the entire region (see Section 10.2.3 on page 264).

Descriptive Summaries of Interactions

To gain some insight into the precision of this study's estimates of interaction, suppose that there actually was an interaction between ozone and sulfur dioxide, as in the model

$$\mu\{log(yield)\} = \beta_0 + \beta_1 ozone + \beta_2 sulfur + \beta_3 water + \beta_4 (ozone \times sulfur).$$

If so, the slope against *ozone* is $(\beta_1 + \beta_4 sulfur)$. The difference between the slopes against ozone for sulfur at its highest level (0.069 μL/L) and at its lowest level (0.0045 μL/L) is $\beta_4(0.069 - 0.0045) = \beta_4(0.0645)$.

The estimate of β_4 for the *Forrest* cultivar is −29.517 with a standard error of 34.544. The difference between slopes over the sulfur dioxide range is estimated to be $(-29.517)(0.0645) = -1.904$, with a 95% confidence interval from −6.484 to +2.676. After exponentiation, the rate of decline with increasing ozone for sulfur dioxide = 0.069μL/L is estimated to be 0.149 times (14.9% of) the rate of decline with increasing ozone for sulfur dioxide = 0.0045μL/L. This ratio may be as low as 0.002 (0.2%) or as high as

Display 14.17 Estimated median seed yields of *Forrest* and *Williams* cultivars under different ozone, sulfur dioxide, and water deprivation regimes

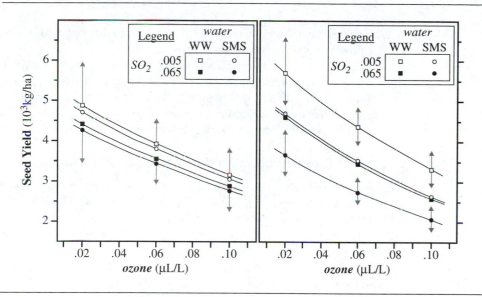

14.525 (1,450.0%). Similar calculations reveal the other confidence intervals noted in Section 14.1.2.

14.5 Related Issues

14.5.1 Random Effects Models

In an investigation like the chimp study, one potentially important issue involves the inferences (if any) that can be drawn to a wider population of chimps. If such inferences are desired, the most appealing way to structure the problem is to assume that chimp effects are random, rather than fixed. Each chimp is assumed to have an effect randomly selected from a population of such effects. The population has mean zero, and the overall contribution of chimp differences is represented by the variance σ_C^2 in the population of chimp effects. If there is little difference among chimps, after accounting for sign differences, σ_C^2 will be small.

The model for the responses, *ltime* = *log*(learning time), is

$$\mu\{ltime \mid SIGNS\} = SIGNS$$

with

$$\text{Var}\{ltime \mid SIGNS\} = \sigma^2 + \sigma_C^2,$$

where the responses from the same chimp are correlated as a result of the chimp effect:

$$\text{Corr}\{ltime_{ij}, ltime_{ik}\} = \sigma_C^2 / \{\sigma^2 + \sigma_C^2\}.$$

The parameter σ^2 represents replication variation (although it is not possible to replicate learning the same sign on the same chimp).

Analysis of the random effects model still centers on the analysis of variance table, (Display 14.9). Inferences about the sign effects are unchanged by assuming random rather than fixed chimp effects. Estimation of the chimp-effect variance proceeds as in Section 5.5.3. Inferences are subject to the reservation that one must justify the assumption that these chimps truly represent the population to which inference is desired.

14.5.2 Nested Classifications in the Analysis of Variance

Do learning times of male and female chimps consistently differ? Answering that question requires introducing a *SEX* factor into the model along with the *CHIMP* and *SIGN* factors. In this case, however, the levels of *CHIMP* are *nested* within the levels of *SEX*.

When every level of one factor appears in combination with every level of another factor, the two factors are said to be *crossed*. This is impossible with the factors *SEX* and *CHIMP*, since two of the *CHIMP* levels, Booee and Bruno, can only appear in combination with the *male* level of *SEX*; and similarly, the levels Thelma and Cindy can only appear in combination with the *female* level of SEX. Chimp individuals are nested within the levels of the *SEX* factor, rather than being crossed with them.

The nested model may be represented symbolically by

$$\mu\{ltime \mid SIGNS, SEX, CHIMPS\} = SIGNS + SEX + SEX/CHIMPS.$$

The last two terms are read as *SEX* plus *CHIMP* nested within levels of *SEX*. An indicator variable representation for these last two parts is

$$SEX + SEX/CHIMPS = \beta_1 female + \beta_2 Bruno + \beta_3 Thelma.$$

where *female* takes on the value 1 for females and 0 for males, and *Bruno* and *Thelma* are indicator variables for Bruno and Thelma. The rule for this representation is that a reference level of the nested factor (*CHIMPS*) will exist for every level of the factor it is nested within. Thus Booee is the reference level for male chimps, and Cindy is the reference level for female chimps. Neither of their indicator variables is included in the model.

This reparameterization produces the following mean responses for the reference sign:

For Booee: β_0
For Bruno: $\beta_0 + \beta_2$
For Cindy: $\beta_0 + \beta_1$
For Thelma: $\beta_0 + \beta_1 + \beta_3$

Here β_1 describes the sex effect—the amount by which the mean log learning time for females exceeds that for males; β_2 is the difference between Bruno's mean and Booee's mean (the reference male); and β_3 is the difference between Thelma's mean and Cindy's mean (the reference female).

An extra-sum-of-squares F-test for the two parameters, β_2 and β_3, examines the evidence for differences between individuals of the same sex. If there is none, the *CHIMP* effects may be dropped and a subsequent test of the parameter β_1 may be used to assess the female–male difference. If there is evidence of within-sex differences, the subsequent test should not be performed, for lack of clear interpretation.

The general preparatory procedure for nested classifications is to create a set of primary class indicators, then to select a reference within each primary class, and finally to define indicators for secondary levels within each primary level. At this point, one can test for secondary level variation within primary level and (if there is none) examine variation across primary levels.

14.5.3   Further Rationale for One Observation per Cell

Replication is good; but sometimes, when presented with limited experimental units, the researcher may be wiser to forgo replication in favor of increased blocking. The objective of an experimental design is to learn as much as possible with as little outlay as possible. This translates into making the standard deviations in sampling distributions small. The SD of a cell average—$\text{SD}(\overline{Y}) = \sigma/\sqrt{n}$—can be made smaller by increasing the number of observations (n) in the cell. But $1/\sqrt{n}$ does not decrease very fast as n increases (as can be seen by plotting). Experience has shown that a more effective way to reduce the SD is by reducing the σ—the standard deviation of the response, given the explanatory variables. This can be accomplished by improving experimental measurement, but also by including major sources of variation as design factors in the experiment.

Hypothetical Example #1

Suppose that a researcher wants to test a method for reducing aggressive behavior in preschool children. She plans to allocate four highly aggressive children randomly to the new method and to a control method and then to measure their posttreatment aggressiveness. Prior to randomization, she measures their aggressiveness (0–8 scale), with the results listed in Display 14.18. If the four are randomized, two to each treatment, the standard deviation of the difference between sample means is estimated to be $1.19 \times \sqrt{1/2 + 1/2} = 1.19$. Part of the variation, however, is a sex variation. If separate randomizations are conducted for girls and for boys, treating sex as blocks, this leaves one child per cell; yet the estimated SD of a cell entry is the pooled SD across sexes (0.47). The treatment main effect will still be estimated by the difference between the average of the two treated children and the average of the two control children, so the SE of the difference will be about $0.47 \times \sqrt{1/2 \times 1/2} = 0.47$. That is a 60% reduction. By identifying an important source of variation, one gains substantial advantage from performing the experiment in randomized blocks, even though doing so may mean forgoing replication.

Display 14.18   Aggressive children data with standard deviations

Child	Aggressiveness	Standard Deviations
Dawn	4.6	0.35 (girls) — 1.19 (all four children)
Astrid	5.1	
Ibrahim	7.2	0.57 (boys)
Michael	6.4	

Hypothetical Example #2

A researcher wants to compare two treatments for high blood pressure. Individuals whose ages differ considerably are readily available, but it is known that blood pressure generally changes with age. How should the researcher choose six subjects? Suppose that the researcher expects something like the situation illustrated in Display 14.19 in the absence of treatment. If six members of the age 50–75 group are selected for the study, and if age is ignored, the kind of variability to expect is the variability around the average blood pressure of 122.2 (A); that is, SD = 6.1 units. It appears, however, that selecting six subjects of identical age may result in a variability in blood pressure more like SD = 3.8 units (the typical deviation of blood pressure from the mean for that age group). This is a substantial reduction, but it comes at a price: the inferential base is restricted to the one age chosen, and nothing is learned about whether the treatment effects are age-dependent.

Display 14.19   Diastolic blood pressure example

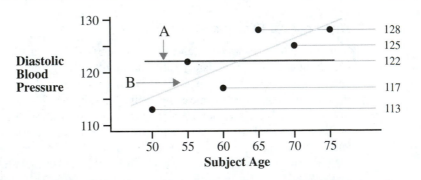

Suppose instead that the researcher retains subjects of different ages but pairs them by similar ages: the two youngest form one block, the two next youngest form the second block, and so on; with one member of each block randomly selected to receive the treatment. An analysis of the resulting data that includes age as a covariate (which has a linear effect in both groups) enables one to anticipate a residual SD closer to 3.8 units than to 6.1, while still having the generality of using individuals of different ages.

More Reasons for Keeping Replications Low

Cost is always a factor in designing studies. Whenever subjects are difficult or costly to obtain, designs that utilize few subjects are desirable. This is also true when treatments are expensive to apply or when sampling is destructive. (Destructive sampling usually means that the subject must be killed or rendered nonfunctional to measure variables of interest.)

In addition to these factors, larger groups of subjects generally tend to exhibit more heterogeneity. When subjects are collected over longer time intervals or over greater spatial distances, they tend to be more variable. Keeping studies on a small scale can be an efficient means to avoid these kinds of variation.

14.5.4   Uniformity Trials

In the hypothetical situations of Section 14.5.3, the researchers had a good idea about the kinds of variation they were likely to encounter, because they could envision what the responses would be when no treatments were applied. In practice, mock studies are sometimes run with no treatments applied just so the researchers can learn what kinds of variability to anticipate. Such studies are called *uniformity trials*. Uniformity trials are particularly valuable when conducted with the subjects or plots that will be used in the study. The set of responses obtained in the uniformity trial can then be used as a covariate in the actual study.

14.6   Summary

Does the adequacy of an estimate of σ^2, and hence the adequacy of the inferential tools based on that estimate, depend on the correctness of some specified model for the mean? If yes, the estimate and resulting inference are said to be model-based. When only one observation is made per cell, the only possible way of proceeding is with model-based inference. In some two-way tables with one observation per cell, like the chimp data, inferences can be made about row and column effects if the analyst can assume that there is no interaction. For some data problems, like the soybean data, inferences regarding questions of interest can be made if the effect of one of the factors of the table can be modeled as a linear effect, using a single parameter. In general, the statistical analyst constructs an estimate of variance from pieces of information that would otherwise estimate interaction or nonlinear effects.

No new tools are needed for analyzing single-replicate studies. Multiple linear regression is the tool of choice, even though the studies considered employ multifactor analysis of variance structures.

Lack of replication imposes two costs: an inability to rely on the p-value from a formal statistical test in assessing model adequacy; and severely reduced precision in estimating some interactions or nonlinearities in which one may have some interest. To deal with model adequacy, the analyst can rely on graphical procedures, searching for clues that suggest transformation and unsuspected effects. By starting with uncomplicated models and building up, one can maintain enough information in the residuals to provide tests for some of the effects. The lack of precision on interactive and nonlinear effects cannot be counteracted with clever analysis, however. If interactions are indeed of interest, additional studies will be required.

Further Reading

J. W. Tukey devised a one-degree-of-freedom test for transformable nonadditivity in a randomized block design. Kotz, Johnson, and Read (1988, vol. 9). Further discussion is offered by Box, Hunter, and Hunter (1978).

14.7 Exercises

Conceptual Exercises

1. **Chimp Learning Times.** Why is the learning time data not analyzed as a 10-group analysis of variance (Chapters 5 and 6), with four observations on each sign?

2. **Chimp Learning Times.** What would have to be true to support the inference that these learning time patterns are representative of learning times for these signs in a wider population of chimpanzees?

3. **Chimp Learning Times.** (a) Can you suggest simplifications to the additive model discussed in Section 14.3? (b) Can you suggest a model that is not additive but still has degrees of freedom available for residuals?

4. **Chimp Learning Times.** Why were Tukey–Kramer intervals used in the comparison of sign means?

5. **Soybeans.** If the seed yield had been recorded for each soybean plant, several observations would have been made in each of the groups. Would this have permitted analysts to assess interactions with more precision?

6. **Soybeans.** Why was cultivar not used as an explanatory factor along with the other factors?

7. In a nested classification model (see Section 14.5.2), is it possible to consider interactions between within-female variables and within-male variables?

Computational Exercises

8. **Chimp Learning Times.** Compute 95% confidence interval halfwidths for differences between signs, using (a) the Tukey–Kramer method and (b) the LSD method. Do the conclusions differ?

9. **Soybeans.** (a) Fit a model for the mean of log(yield) for the *Williams* cultivar that includes all linear main effects and the ozone-by-water interaction. (b) Estimate the difference between ozone slope parameters for stressed versus well-watered plots. Convert the estimate and its 95% confidence interval to the scale of seed yield. (c) Write a summary of what the interval represents. (d) Compare your answer with the summary in Section 14.1.2.

10. **Soybeans.** Fit the full model shown in Display 14.15 for the *Forrest* cultivar. Verify the results in the table.

11. Consider the data from hypothetical example #1 (Section 14.5.3). (a) Roll a die (or use a random number table to select a number from 1 to 6), and select the treatment group according to the outcome, as follows: 1 = Dawn & Astrid; 2 = Dawn & Ibrahim; 3 = Dawn & Michael; 4 = Astrid & Ibrahim; 5 = Astrid & Michael; and 6 = Ibrahim & Michael. Subtract 1.5 units from the aggressiveness scores

of the treated group. Then perform a two-sample t-test for a treatment difference. Note the t-statistic value and the standard error of the estimated treatment effect. (b) Flip a coin to decide whether to treat Dawn (heads) or Astrid (tails). Flip the coin again to select Ibrahim (heads) or Michael (tails). Again, subtract 1.5 units from the scores of the two in the treatment group. Then calculate the t-statistic for a treatment difference in the additive model with sex and treatment. Note the standard error. (c) Comment on the difference between the results.

12. Consider hypothetical example #2 of Section 14.5.3. (a) There are 20 ways to choose three subjects from the six. Record and order the combinations by listing the possible combinations for a treatment group of three subjects. Pick a random number between 1 and 20 to determine what treatment group to use. Subtract 8.0 units from the blood pressure of the treated group. Analyze the results with two-sample t-tools. (b) Flip a coin. If it is heads, use ages 50, 65, and 70 as the treatment group; if it is tails, treat ages 55, 60, and 75. Subtract 8.0 units from the blood pressure of the treated group. Analyze the results using a model that assumes equal slopes against age in the treated and control groups but has different intercepts. Estimate the treatment effect as the difference between intercepts. (c) Compare the results from (a) and (b).

13. Soybeans. In comparing the two soybean cultivars, using cultivar as an explanatory factor is not a valid approach. Here is how the comparison might be made. (a) To get an overall comparison of median yields of the two cultivars, perform a paired t-test (or signed rank test), using the chambers for pairing. (b) To see how the relative yield of the cultivars is influenced by the experimental factors, construct the response $Y = \log(Forrest$ yield/$Williams$ yield) and perform a multiple linear regression analysis similar to the one done for each cultivar separately. Summarize your conclusions in a paragraph.

Data Problems

14. Following the study described in Section 11.1.2, researchers at the Oregon Health Sciences University designed an experiment to investigate how delivery of brain cancer antibody is influenced by tumor size, antibody molecular weight, blood–brain barrier disruption, and delivery route. Subjects for the study were female rats of approximately the same age and weight. Tumor size was varied by starting treatments 8, 12, or 16 days after inoculation with tumor cells; three antibodies—AIB, MTX, and DEX7—having very different molecular weights were used; disruption (BD) was accomplished with an intra-arterial injection of mannitol to one side of the brain and was compared to a similar injection of a saline solution (NS); and the antibody was delivered either by intra-arterial (IA) or intravenous (IV) injection. One key measure of the delivery of antibody to the affected half of the brain is the ratio of the concentration of antibody in the brain around the tumor (BAT) to the concentration in the other, lateral half (LH) of the brain. Display 14.20 shows the data from a single replication of all 36 treatment combinations. Treatments were randomized to rats in this study. (Data from P. A. Barnett et al., "Differential Permeability and Quantitative MR Imaging of a Human Lung Carcinoma Brain Xenograft in the Nude Rat," *American Journal of Pathology* 146(2)(1995):436–449.)

Analyze these data and write a summary report. Consider the BAT/LH ratio and first determine whether a transformation is appropriate. Consider whether the response could be linear in days. Summarize the evidence for main effects. Explore possible interactions.

15. The researchers in Exercise 14 were unsure about the ability of a single replicate to provide sufficient evidence for effects, so they ran a second replicate (Display 14.21). Follow the same instructions as in Exercise 14.

16. Analyze both replicates of the barrier disruption as a single experiment, including an indicator variable to signify replicate. Assess the significance of effects by using a pooled within-group SD to estimate σ. Compare the results of this analysis with the two separate analyses in Exercises 14 and 15. What has been gained by the second replicate?

Display 14.20   One replicate of a study of the effects of four factors on the delivery of brain cancer antibody: *Route* at levels IV and IA; barrier disruption *Treatment* at levels NS and BD; antibody *Agent* (AIB, MTX, or DEX7); and *Days* after inoculation (8, 12, or 16). Y = Counts of radio-tagged molecules

Route:		IV				IA			
Treatment:		NS		BD		NS		BD	
Agent	*Days*	BAT	LH	BAT	LH	BAT	LH	BAT	LH
AIB	8	12025	10510	70034	10611	20870	8944	129747	9992
	12	10327	10345	129134	12311	27708	9242	68775	6316
	16	15852	13255	156655	178303	22006	7237	53808	4333
MTX	8	3995	1285	10049	1388	5598	1724	9683	1424
	12	1786	1077	10540	1781	2393	1070	23173	1520
	16	1828	1731	4808	1680	5076	1228	22293	1777
DEX7	8	6097	7997	109594	11191	18172	8148	159458	13460
	12	4684	5721	23254	7325	20636	6041	45550	9113
	16	7927	7120	17224	7996	15959	7016	96754	8070

Display 14.21   Second replicate of the barrier disruption study

Route:		IV				IA			
Treatment:		NS		BD		NS		BD	
Agent	*Days*	BAT	LH	BAT	LH	BAT	LH	BAT	LH
AIB	8	23866	10758	123721	10294	18722	7214	303377	6646
	12	10220	8771	88594	6665	24372	9514	144556	5942
	16	9582	7538	142743	8713	15983	6753	177892	6629
MTX	8	2335	1055	5542	1128	4698	1306	17594	695
	12	1206	1009	33214	2808	4191	1335	51134	2431
	16	2433	1094	7791	1460	7028	1206	6554	803
DEX7	8	4627	5704	127876	10498	8871	7283	33875	10237
	12	5533	4780	92892	13104	16864	7377	238691	16669
	16	6873	6161	7027	6008	23455	5938	25069	7948

Answers to Conceptual Problems

1.   The same chimpanzees were trained on all 10 signs, so the observation times should be correlated; that is, a chimp whose learning time on one sign is longer than average is more likely to have a longer-than-average learning time on other signs.

2.   The chimpanzees used in the study would have to have been randomly selected from the population of interest.

3.   (a) Several are possible. Chimps of the same sex could have the same median learning times. There may also be sign groupings in which the median times for signs within groups are the same.

(b) One form of nonadditivity might occur if, for example, the male chimps had more difficulty with conceptual signs than did the female chimps.

4.   The researchers had no preplanned comparisons. They were looking for any differences that might show up. The Tukey–Kramer intervals protect against compounded uncertainty.

5.   Any gain in precision is irrelevant, since such an analysis would not be valid. To be appropriate, the individual soybean plants would have to have been randomized to treatment. This was not done. Treatments were applied to rows of plants, so the row measure is the proper response to use.

6.   Considering responses of the two cultivars separately, with a cultivar indicator variable distinguishing between them, would yield two "cases" from each experimental unit. The cases would be correlated, violating the independence assumption.

7.   Conceptually, such an interaction makes no sense. Technically, the problem is that the products of the indicator variables are identically zero (because no chimps are both male and female).

C H A P T E R 15

Adjustment for Serial Correlation

The methods discussed in previous chapters have been based on an assumption of independence. Response variables in regression analysis, for example, were assumed to be independent of each other, after accounting for the effects of the explanatory variables. Sometimes this assumption is not appropriate if data are collected over time (or space) and the observations that are taken close together are related. Time series analysis is a specialized field of statistics that deals with this type of problem.

This chapter presents two extensions of regression methods to situations where the responses form a time series of the *first-order autoregressive* type. One method makes a multiplicative adjustment to the usual standard errors and the other performs the regression in the usual way after a special transformation of response and explanatory variables. In addition, diagnostic tools are presented for determining whether serial correlation may be a problem in ordinary regression and whether the first-order autoregressive model is adequate.

15.1 Case Studies

15.1.1 Logging Practices and Water Quality—An Observational Study

Douglas Fir forests in the Pacific Northwest are logged by clear-cutting entire sections of the forest, a practice that involves stripping the land of all vegetation and burning what is left before replanting. The rationale for clear-cutting is that regenerating Douglas Fir seedlings thrive in full sunlight and low competition, and in any case the alternative logging methods are equally devastating. One consequence of the practice is its effect on water quality in the streams draining the logged areas. Higher peak runoff, increased silt, and higher water temperatures are factors that reduce the quality of the habitat for anadromous (river ascending) fish.

One of the alternative methods is patch-cutting, where a watershed is cut in a patchwork pattern of clear-cut and unlogged sections. Display 15.1 shows the results of a study comparing the effects of patch-cutting a 68.4 hectare watershed to leaving a similar 48.6 hectare watershed undisturbed. (Data from R. D. Harr, R. L. Fredriksen, and J. Rothacher,

Display 15.1 Nitrates (NO_3-N) in runoff from patch-cut and undisturbed watersheds, for five years after logging

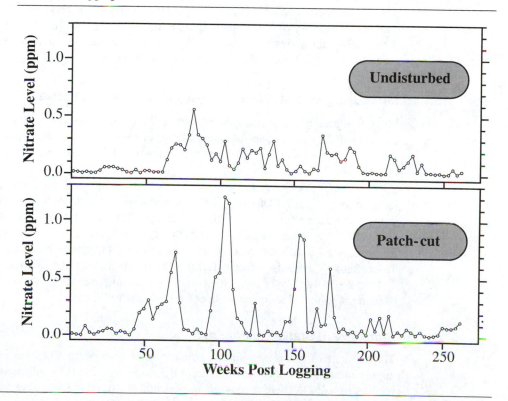

"Changes in Streamflow Following Timber Harvest in Southwestern Oregon," USDA/USFS Research Paper PNW-249 (1979), Pacific NW Forest and Range Experiment Station, Portland, Oregon.)

Nitrate (NO_3-N) levels are considered to be key indicators of the damage to water quality. The nitrate levels, in parts per million (ppm), of Display 15.1 were measured at approximately 3-week intervals at stream gauges in each watershed for a period of 5 years after logging occurred. In Display 15.1 the peaks in the patch-cut series appear to be somewhat higher than those of the undisturbed control. Is there sufficient evidence here that the mean (or median) nitrate levels differ in the two series?

Summary of Statistical Findings

There is not sufficient evidence to conclude that the median levels in the two series differ (one-sided p-value = .39, adjusted for an observed serial correlation of .64). The median nitrate in the patch-cut watershed is estimated to be 53% to 236% as large as the median nitrate concentration in the undisturbed watershed (approximate 95% confidence interval).

Inferences

Since the data are observational, no causal interpretation of the statistical results can be made. Furthermore, without random sampling from populations of patch-cut and undisturbed watersheds, the results only pertain to these two particular watersheds. The reader should note the desirability of having prelogging nitrate measurements taken on the two streams. If the prelogging series were comparable and the treatment stream selected randomly, the case for logging-induced effects could be much stronger.

15.1.2 Measuring Global Warming—An Observational Study

Most countries of the world now recognize the dangers associated with the greenhouse effect, and they are taking measures to curtail those activities that might contribute to further global warming. What is the true extent of global warming? Is it a demonstrable effect?

There have been several attempts at reconstructing mean temperature profiles over long periods of time. One of these is reproduced as Display 15.2. Each measurement is the temperature (in degrees Celsius) averaged for the northern hemisphere over a full year. The series begins in 1880 and runs through 1987. All measurements are expressed as differences from their 108-year mean. (Data from P. D. Jones, "Hemispheric Surface Air Temperature Variations—Recent Trends Plus an Update to 1987," *Journal of Climatology* 1 (1988): 654–60.) Is the mean temperature increasing over the 88 years? What is the rate of increase in global temperature over the past century?

Summary of Statistical Findings

The estimate of global warming is an increase of 0.46 degrees Celsius per century. (An approximate 95% confidence interval for the slope is 0.22 to 0.70, adjusted for a serial correlation of 0.452). There is convincing evidence of a trend of increasing mean temperature between 1880 and 1987 (two-sided p-value < 0.0001).

Display 15.2   Annual temperature anomalies in the Northern Hemisphere—1880 to 1987

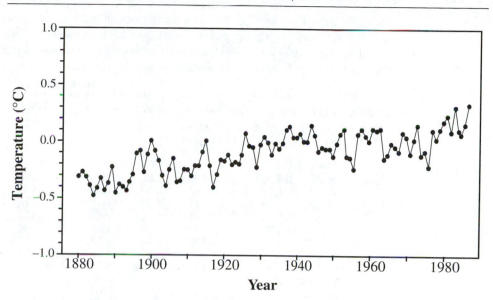

Scope of Inference

The inference one would like to draw here concerns the projection of this particular series into the future. To draw such an inference, it would be necessary to assume the trend observed over 1980 to 1987 will continue into the future.

15.2   Comparing the Means of Two Time Series

The Independence Assumption

The methods of Chapters 2–6 for comparing several samples only apply if the observations within a group are independent of one another. Similarly, the regression methods in Chapters 7–14 require that, after accounting for the effects of the explanatory variables, the responses be independent. The two major sources of nonindependence encountered in practice are block effects and serial effects. Serial effects require time series tools.

In many time series situations the comparison of the means of two (or more) time series can be accomplished using a straightforward correction of the standard error that accounts for serial correlation. This section explains how to measure serial correlation, why a correction is necessary, and how to make the correction.

15.2.1   Serial Correlation and Its Effect on the Average of a Time Series

Serial correlation occurs when the course of a time series is influenced by its recent past. The typical behavior of a time series with serial correlation is that its values will go on extended excursions away from the long-run mean. This pattern is present in the logging and stream quality data. Let N_t be the concentration of nitrates in parts per million at the time t. The transformation $Y_t = \log(1 + 100 \times N_t)$ was made in order to adjust for skewness in the nitrate concentrations. (*Note:* this is essentially a log transformation, with a small amount added to handle the zeros.) Next, each of the time series was mean-corrected by subtracting the series average from each observation, producing *residuals* listed in Display 15.3.

Display 15.3   Residual nitrate readings (ppm) from two watersheds—one logged by patch-cutting and the other undisturbed

Week	Patch-Cut	Undis-turbed	Week	Patch-Cut	Undis-turbed	Week	Patch-Cut	Undis-turbed
1	−1.32	−1.21	88	−0.92	1.56	175	2.08	0.98
4	−2.02	−1.91	91	−1.32	1.35	178	0.93	1.09
7	−2.02	−1.91	94	1.12	0.66	181	−0.41	0.66
10	0.18	−1.91	97	1.94	1.04	184	0.18	0.80
13	−0.92	−1.21	100	2.01	0.58	187	−0.63	1.31
16	−2.02	−1.91	103	2.78	1.46	190	−0.41	1.19
19	−0.92	−0.81	106	2.74	0.29	193	−2.02	0.29
22	−0.63	−0.11	109	1.70	−0.30	196	−0.07	−0.81
25	−0.22	−0.11	112	0.76	0.49	199	−1.32	−1.21
28	−0.22	−0.11	115	0.55	1.23	202	0.76	−0.81
31	−0.92	−0.30	118	−0.92	0.80	205	−0.07	−1.21
34	−0.63	−0.52	121	−1.32	1.19	208	0.82	−1.21
37	−0.92	−1.21	124	1.35	1.09	211	−0.63	−1.21
40	−2.02	−1.91	127	−2.02	1.27	214	0.98	1.04
43	−0.07	−0.52	130	−2.02	0.04	217	−2.02	0.80
46	1.03	−1.91	133	−0.41	0.98	220	−0.41	0.04
49	1.16	−0.81	136	−0.92	1.46	223	−0.92	0.29
52	1.45	−0.81	139	−0.63	0.29	226	0.06	0.66
55	0.69	−0.81	142	−1.32	0.73	229	−0.41	0.98
58	1.20	−1.21	145	0.62	−0.30	232	−2.02	−0.52
61	1.32	−1.21	148	0.62	−0.81	235	−0.41	0.49
64	1.42	0.66	151	1.75	−0.52	238	−1.32	−0.81
67	2.01	1.23	154	2.46	0.29	241	−2.02	−0.81
70	2.27	1.35	157	2.43	−0.52	244	−1.32	−0.81
73	1.35	1.35	160	−0.41	−1.21	247	−0.92	−1.21
76	−0.07	1.19	163	−0.41	0.04	250	0.29	−1.21
79	−0.41	1.65	166	1.20	−0.30	253	0.18	−1.91
82	−0.92	2.14	169	0.38	1.65	256	0.18	−0.11
85	−0.22	1.65	172	0.47	1.14	259	0.29	−0.81
						262	0.62	−0.52

Each residual series exhibits *runs*, where the values tend to be consistently positive or consistently negative for long periods. This is more clearly evident in Display 15.4, a time plot of the residuals in the undisturbed watershed.

To demonstrate the trouble with using an average and its usual standard error with time series data, suppose that the average of the full five-year series is the long-run mean. In Display 15.4, this is represented by the zero line. The shaded region in the display

Display 15.4 Mean-corrected nitrate concentrations after transformation, and a demonstration that the average of a segment of a time series may grossly misrepresent the full series mean

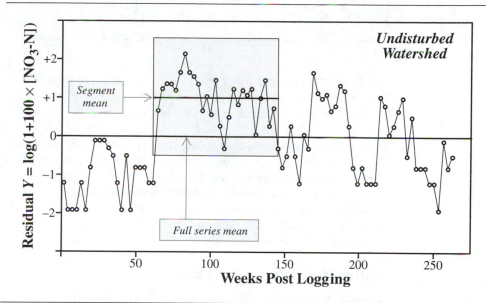

highlights a segment of 27 values where the series has gone on an excursion away from its long-run mean. *What would happen if this segment were the only sample available and if it was treated as a series of independent measurements?* The sample average is 1.017, and the sample standard deviation is 0.551, giving a standard error of 0.106 for the average. The sample average is therefore 9.59 standard errors away from the long-run mean. If this sample were used to estimate the long-run mean, zero would be (incorrectly) considered an impossible value.

Serial correlation in the series created two problems. First, a sample average in a serially correlated time series tends to be further from the long-run mean than expected, because the series is inclined to drift away from its long-run mean. Secondly, the serial correlation makes the values in the sample tend to be closer to each other than would be the case with an equal number of independent measurements, which makes them appear much less variable than they are. Since the estimate of how close the sample average is to the long-run mean is based upon sample variability, serial correlation engenders an overly optimistic view of the true variation in the sample average.

15.2.2 The Standard Error of an Average in a Serially Correlated Time Series

Fortunately, there is a simple way to adjust the standard error formula to account for serial correlation.

$$\text{SE}(\overline{Y}) \;=\; \sqrt{\frac{1 + r_1}{1 - r_1}}\;\frac{s}{\sqrt{n}}\;.$$

In this formula, the factor $s/\sqrt{n}$ is the standard error calculated as if the data were independent (s is the usual sample standard deviation for one sample, or the pooled estimate for two or more independent samples). The quantity r_1 is the sample *first serial correlation coefficient*. Notice that if r_1 is zero (no serial correlation), then the adjustment factor in front of the usual standard error is one. If r_1 is positive, then the adjustment factor is larger than 1; and if r_1 is negative the adjustment factor is less than 1. Like other correlation coefficients, r_1 must fall between -1 and 1.

The nitrate series from the undisturbed watershed, shown in Display 15.4, has $r_1 = 0.744$ (calculation of r_1 shown in next section), so the adjustment factor is $2.61 = \{(1 + 0.744)/(1 - 0.744)\}^{1/2}$. The standard error of the serially correlated series is two and a half times what it would be if the series had independent measurements.

First-order Autoregression Model

This adjustment to the standard error is appropriate under the following ideal model, called the *autoregressive model of lag 1*, or AR(1):

1. The series, $\{Y_t\}$, is measured at equally-spaced points in time.
2. Let $(Y_t - v)$ be the deviation of an observation at time t from the long-run series mean v. Let $\mu\{(Y_t - v) \mid past\ history\}$ be the mean of the tth deviation as a function of all previous deviations. Then,

$$\mu\{(Y_t - v) \mid past\ history\} = \alpha(Y_{t-1} - v),$$

where the parameter α is the *autoregression coefficient*. In other words, the regression of a deviation on all previous deviations depends only on the most recent one. Hence the term *lag 1*.

All values in an AR(1) series are correlated with each other, yet the entire structure of the correlation is contained in the single parameter α. The autoregression coefficient, also called the population first serial correlation, is estimated by the sample first serial correlation coefficient, r_1.

15.2.3   The First Serial Correlation Coefficient

In a time series that embarks on excursions from the mean, the value of a deviation at one point in time tends to be similar to the value of the deviation at the most recent point in time. This behavior is evident in the scatterplot of residuals adjacent in time. Display 15.5 shows the scatterplot for the series from the undisturbed watershed. The plotted data are the 87 pairs of consecutive residuals in Display 15.3: $(-1.21, -1.91)$, $(-1.91, -1.91)$, and so on.

Display 15.5 The relationship between adjacent residuals in the undisturbed watershed time series

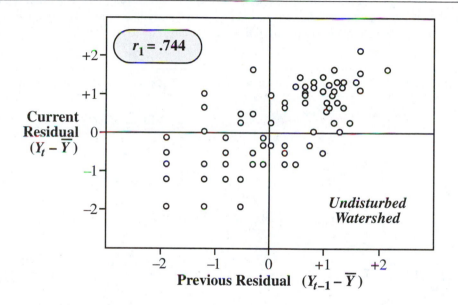

The coefficient r_1 provides a numerical summary measure of the correlation between adjacent residuals, apparent in Display 15.5. From a single time series, r_1 is the sample correlation of the consecutive residuals, like those plotted in Display 15.5. The formula provided below is general enough to be extended easily to get a pooled estimate from several time series. The calculation of r_1 in a single time series is the ratio:

$$r_1 = \frac{c_1}{c_0},$$

which involves the two *autocovariance* estimates,

$$c_1 = \frac{1}{n-1} \sum_{t=2}^{n} (Y_t - \overline{Y})(Y_{t-1} - \overline{Y})$$

$$c_0 = \frac{1}{n-1} \sum_{t=2}^{n} (Y_t - \overline{Y})^2.$$

Note that c_0 is just the sample variance of the residuals, and that r_1 is similar to the estimated slope in a regression of Y_t on Y_{t-1}.

15.2.4 Pooling Estimates and Comparing Means of Two Independent Time Series with the Same First Serial Correlation

To arrive at a standard error for the difference in averages of two time series, a pooled estimate of the first serial correlation is required. This is carried out by obtaining pooled estimates of c_1 and c_0 and then taking r_1 as their ratio. The calculations for the logging case study are summarized in Display 15.6.

Display 15.6   Pooling procedure for estimating the first lag serial coefficient—logging data

$$\text{Pooled Estimate} \quad = \quad \frac{(\text{d.f.}_1 \times \text{est}_1) \; + \; (\text{d.f.}_2 \times \text{est}_2)}{\text{d.f.}_1 + \text{d.f.}_2}$$

Patch-cut: $\overline{Y}_C = 2.016, \quad c_0 = 1.628, \quad c_1 = 0.932, \quad \text{d.f.} = 87$

Undisturbed: $\overline{Y}_U = 1.905, \quad c_0 = 1.172, \quad c_1 = 0.872, \quad \text{d.f.} = 87$

Pooled: $c_1 = \dfrac{(87 \times 0.932) + (87 \times 0.872)}{87 + 87}$ $c_0 = \dfrac{(87 \times 1.628) + (87 \times 1.172)}{87 + 87}$

$$r_1 = 0.902/1.400 = 0.644$$

A standard error of the difference in averages uses the same adjustment factor that is applied to the standard error of a single average. The adjusted standard error is the usual standard error (Chapter 2) times the adjustment factor:

$$\text{SE}(\overline{Y}_C - \overline{Y}_U) \quad = \quad \sqrt{\frac{1 + r_1}{1 - r_1}} \; s_p \sqrt{\frac{1}{n_C} + \frac{1}{n_U}} .$$

For the logging data this standard error is

$$\text{SE}(\overline{Y}_C - \overline{Y}_U) = \sqrt{\frac{1 + 0.644}{1 - 0.644}} 1.183 \sqrt{\frac{1}{88} + \frac{1}{88}} = 0.383,$$

or 2.145 times larger than the SE of the difference that would be used if the data were treated as independent. An approximate 95% confidence interval for the difference may be formed by using the normal multiplier, 1.96. The interval half-width is $(1.96)(0.383) = 0.751$ and the estimated difference is 0.111, so the interval is $(-0.64, +0.86)$. Similarly, a z-statistic for testing equality of means is $z = 0.111/0.383$.

15.3 Regression After Transformation in the AR(1) Model

The global temperature series exhibits a steady increase over the past century. A simple linear regression of the temperatures on the year number can be used to estimate the average increase per year, but the residuals from this regression display serial correlation—a feature that could have been guessed from the excursions the series takes away from a straight-line in Display 15.2.

It is possible to calculate the adjusted standard error for the estimated slope as the correction factor of the previous section times the standard error from the regression. This type of adjustment, however, is only appropriate if the explanatory variable does not fluctuate cyclically with time. Even cyclical fluctuation would not be a problem if time itself were the explanatory variable, but it could be a problem for other explanatory variables (such as daily maximum temperature). Fortunately, a more general approach that does not require much extra work is available. This section describes the approach.

15.3.1 The Serial Correlation Coefficient Based on Regression Residuals

Let Y_t be the temperature for year t ($t = 1, 2, \ldots, 108$). The average increase in temperature per year may be represented by a regression coefficient, β_1, in the simple linear regression model that states the average temperature for year t is $\mu\{Y_t \mid t\} = \beta_0 + \beta_1 t$. This is a special case of the more general model considered here: $\mu\{Y_t \mid X_t\} = \beta_0 + \beta_1 X_t$. The explanatory variable X_t may be t, but it may also be another explanatory variable measured at the same times as Y_t.

Analysis begins with the estimation of the parameters in this model by ordinary least squares regression. Display 15.7 summarizes the fit and the analysis of variance. Display 15.8 contains a time series plot of the residuals from the regression, in which a pattern of drifting from the mean is apparent. The plot suggests positive serial correlation, perhaps represented by a first-order autoregressive model for the series of deviations. The estimated first serial correlation coefficient is calculated by applying the method of section 15.2.3 to the series of residuals from the regression fit of Y_t on X_t. The estimate is $+0.452$.

15.3.2 Regression with Filtered Variables

In addition to correcting the standard error directly, serial correlation can be handled by a special transformation called *filtering*. The idea is to create a new response variable and a

---

Display 15.7   Fit of a linear trend to the global temperature data

Parameter	Estimate	Standard Error			
Constant	−0.10011	0.01096			
time	0.00449	0.00035			
Residual SD	0.11386				

Source	Sum of Squares	d.f.	Mean Square	F-Statistic	P-Value
Regression	2.1195	1	2.1195	163.50	<.0001
Residual	1.3742	106	0.0130		
Total	3.4937	107			

Display 15.8   Residual time series plot for global temperature data

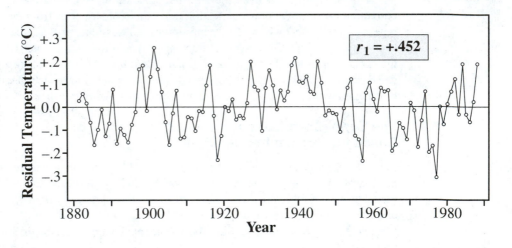

new explanatory variable such that the regression coefficient is the same but the deviations are independent. If the AR(1) model is adequate, then by taking

$$V_t = Y_t - \alpha Y_{t-1}, \text{ and } U_t = X_t - \alpha X_{t-1},$$

it happens that $\mu\{V_t \mid U_t\} = \gamma_0 + \beta_1 U_t$ (a simple linear regression with the same slope as in the regression of Y on X), and there is no serial correlation in the deviations about the regression. The new intercept is related to the original intercept by $\gamma_0 = \beta_0(1 - \alpha)$.

Since the independence assumption is met for the filtered variables, ordinary least squares regression can be used to arrive at an efficient estimate of β_1. The only hitch is that one must know α to construct the filtered variables V and U.

The simplest approach is to use r_1 (the sample serial correlation coefficient, estimated from the residuals in the regression of Y on X) as an estimate of α. Then the filtered response variable $V_t = Y_t - r_1 Y_{t-1}$ and the filtered explanatory variable $U_t = X_t - r_1 X_{t-1}$ may be used to estimate β_1 and its standard error with an ordinary least squares program. For the Jones data, this procedure gives $\hat{\beta}_1 = 0.00460$, with a standard error of 0.00058. The residuals from this regression fit, plotted in Display 15.9, show less of a pattern of serial correlation.

Display 15.9   Time plot of residuals after regression with filtered variables: global temperature data

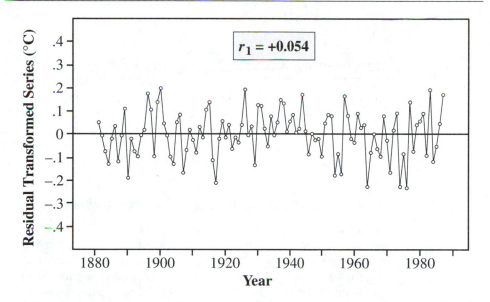

Comparison of Filtering Method with Standard Error Adjustment

With the untransformed residuals, the serial correlation of 0.452 would have given an adjustment factor (Section 15.2) of 1.628. Multiplying this by the standard error of the slope estimate, 0.00035, gives the same result as the standard error based on the filtered variables. The two methods constitute two approaches to the same result. However, the transformation approach is preferred because, as mentioned above, it extends to time series models that include any explanatory variables that fluctuate cyclically.

Filtering with More Than One Explanatory Variable

The extension to multiple regression, when the deviations follow the AR(1) model, is straightforward. For each explanatory variable X_t, a filtered version, $U_t = X_t - r_1 X_{t-1}$ is created. The filtered response is regressed on the filtered explanatory variables using a regular regression routine.

To summarize the procedure for filtering with more than one explanatory variable: if it can be determined that the time series of residuals about the regression is of the AR(1) type (checking for AR(1) is discussed in Section 15.5), then:

(1)   Fit the regression of the response on the explanatory variables and obtain residuals.

(2)   Calculate the autocovariance estimates c_0 and c_1 from the residuals. From these calculate the first serial correlation coefficient $r_1 = c_1/c_0$.

(3)   Compute the filtered versions of the response and explanatory variables.

(4)   Fit the regression of the filtered response on the filtered explanatory variables, and use the usual tools to make inferences about the coefficients (but not the intercept). The intercept for the model of interest, if desired, is estimated by the reported intercept estimate divided by $(1 - r_1)$.

15.4   Determining if Serial Correlation is Present

Even though observations come in the form of a time series, it may not be necessary to adjust for serial correlation. In this section, three tests are introduced for detecting serial correlation. Each of these tests should be performed using residuals from proposed models, rather than using the response variable directly. The primary function of the tests is diagnostic, to see if serial correlation is present.

15.4.1   An Easy, Large-sample Test for Serial Correlation

If one estimates a serial correlation coefficient from a series of n *independent* variables with constant variance, the estimate has an approximate normal sampling distribution centered at zero with standard deviation $1/\sqrt{n}$. This assumes that n is large—say 100 or so. The test statistic, $Z = r_1\sqrt{n}$, has an approximate standard normal distribution when no serial correlation is present. This simple test for serial correlation consists of determining whether $|Z|$ is too large.

The serial correlation coefficient, $r_1 = .452$, in the Jones global temperature series comes from $n = 108$ regression residuals. The resulting $Z = 4.70$ is much larger than one would expect if Z came from a standard normal distribution. (A p-value is $\Pr(|Z| > 4.70)$, which is less than .0001.) One may safely conclude that serial correlation is present.

15.4.2   The Nonparametric Runs Test

A *run* is a string of values that are all above (or all below) the estimated mean. That is, a run is another term for an excursion. In serially correlated time series, runs tend to be long, so the number of different runs tends to be small. Consider a small example involving five residuals: $-2.75, -2.31, +0.27, +1.66$ and $+3.13$. The magnitudes do not affect the number of runs, so these residuals can be described by the sequence $(-, -, +, +, +)$, containing $p = 3$ plus values and $m = 2$ minus values. Here are all possible sequences having 3 plus and 2 minus signs.

Sequence	Number of Runs in This Sequence
(+,+,+,-,-)	2
(+,+,-,+,-)	4
(+,+,-,-,+)	3
(+,-,+,+,-)	4
(+,-,+,-,+)	5
(+,-,-,+,+)	3
(-,+,+,+,-)	3
(-,+,+,-,+)	4
(-,+,-,+,+)	4
(-,-,+,+,+)	2

Notice that there are 10 different sequences. Ten is $C_{5,2}$, the number of ways to select two positions for the minus values from the five available positions. (Recall that $C_{a,b} = a!/[b!(a - b)!]$.) If the residuals have no serial correlation—if they are independent—then each of these sequences will have the same chance (1 in 10) of occurring. This independence condition is summarized in the null hypothesis distribution of the number of runs, which is:

Number of Runs	Chance of Occurrence
2	0.2
3	0.3
4	0.4
5	0.1

Low numbers of runs suggest positive serial correlation, so a p-value for the test comes from the lower tail of a distribution such as this. The example sequence had two runs, which is the most extreme case. But there is a good chance (0.2) that this might happen with independence, so the strength of the statistical evidence for serial correlation is weak.

Tables of critical values for the number of runs are available for sample sizes up to about 20. (See Swed and Eisenhart (1943).) The concern here, however, is with longer series, where the sampling distribution of the number of runs may be approximated by a normal distribution. With a series of independent residuals, p of which are plus and m of which are minus, the mean number of runs is

$$\mu = \frac{2mp}{m + p} + 1,$$

and the standard deviation of the number of runs is

$$\sigma = \sqrt{\frac{2mp(2mp - m - p)}{(m + p)^2(m + p - 1)}}.$$

A test statistic is

$$Z = \frac{(\text{number of runs}) - \mu + C}{\sigma},$$

where C is a continuity correction: $C = +0.5$ if the number of runs is less than μ, and $C = -0.5$ if the number of runs is greater than μ. In a series of independent residuals, Z will have an approximate standard normal distribution.

To illustrate the runs test, consider the residual nitrate levels from the undisturbed (control) watershed, in Display 15.3. The residuals begin with a run of 21 minuses, followed by another run of 15 pluses. Next come a run of a single minus, another long run of 11 pluses, and so on. In all, the pluses and minuses are evenly divided as $p = m = 44$, and the series has 15 distinct runs. For series of 88 independent observations having 44 pluses and minuses, the mean number of runs is 45 and the SD is 4.66. So the observed number of runs is very low. Formally, $Z = -6.33$, which represents convincing evidence of serial correlation.

15.4.3 The Durbin–Watson Test Statistic

In a time series with positive serial correlation, today's observation tends to be like yesterday's. J. Durbin and G. S. Watson used this property as the basis of a test for comparing how close current residuals are to their past values with how close they are to their average (zero). The Durbin–Watson test statistic is:

$$\text{DW} = \frac{\sum_{t=2}^{n}(e_t - e_{t-1})^2}{\sum_{t=1}^{n}e_t^2}.$$

The calculated value of this statistic is compared with critical values in the null distribution of the statistic (See Durbin and Watson (1951).) Durbin–Watson tables are specifically designed for residuals from regression. For a given level of significance, for a series of length n, and for a regression equation with k explanatory variables, the tables contain *two* critical values, d_L and d_U, with the prescription that $\text{DW} < d_L$ leads to rejecting independence, $\text{DW} > d_U$ leads to accepting independence, and for $d_L \leq \text{DW} \leq d_U$ the test is inconclusive (L and U refer to lower and upper.). This test does not help if the serial correlation is negative.

For example, consider the Jones global temperature series, which has $k = 1$ explanatory variable (time) with a series of length $n = 108$. The 1% critical values for the Durbin–Watson test statistic are, respectively, $d_L = 1.52$ and $d_U = 1.56$. Calculation of the statistic, which is accomplished in most statistical software packages, gives $\text{DW} = 1.069$. Since $\text{DW} < d_L$, the DW test confirms the strength of the evidence for serial correlation.

15.5   Diagnostic Procedures for Judging the Adequacy of the AR(1) Model

The main issue in this section is how to determine whether the first-order autoregressive model, AR(1), fully accounts for all serial correlation. If it does not, then more sophisticated methods than the ones so far presented will be needed. Other issues of model checking also arise in regression analysis when serial correlation is present: normality of errors, equal variances, and so on. These can be tackled using tools already developed, such as plots of residuals versus fitted values. Before launching into the situations that may call for higher-order models for correlation, however, one piece of diagnostic advice about transformations is mentioned.

15.5.1   When Is a Transformation of a Time Series Indicated?

Why were the nitrate measurements transformed to the log scale? One clue comes from an examination of a time plot of the raw data, Display 15.1. The tools presented in this chapter apply well to stationary, normally distributed time series. (Stationary means that, aside from the effects of the explanatory variables, the behavior of the time series is the same at different segments of time.) The plots characteristic of such series show no obvious long-term trends in either mean or variability, and *the shapes of the peaks and the valleys are roughly the same*. Imagine turning the time plot upside down. Does it look about the same? If not, a transformation may be advisable. In Display 15.1, both series show very sharp peaks, but the valleys are broad and shallow. Because the logarithm squeezes large numbers together and spreads small numbers apart, it is a good candidate for toning down the peaks and deepening the valleys. Since there are zeros, a small number may be added to all measurements before taking the logarithms (1 has been added here). After transformation (Display 15.4) the peaks and valleys have about the same shape to them, which is an indication that the chosen transformation has been successful.

15.5.2   The Partial Autocorrelation Function (PACF)

Using the first serial correlation coefficient usually accounts for most of the serial correlation present in time series residuals. However, situations arise where the structure of the time relationship is more complex. The partial autocorrelation function is a tool for examining two aspects of this problem: Is the situation more complex than a first-order autoregressive model; and, if so, will a modest extension of the first-order model suffice?

In the first-order autoregressive model,

$$\mu\{(Y_t - \nu) \mid past\ history\} = \alpha(Y_{t-1} - \nu).$$

In a second-order autoregressive model,

$$\mu\{(Y_t - \nu) \mid past\ history\} = \alpha_1(Y_{t-1} - \nu) + \alpha_2(Y_{t-2} - \nu).$$

The last term is a lag 2 deviation. Similarly, it is possible to have higher-order autoregressive models by including higher-order lagged deviations. The first *partial autocorrelation* is the estimate of α in the first-order autoregressive model, and this is already known to be r_1. The *second partial autocorrelation* is the estimate of α_2 in the second-order autoregressive model, AR(2). Other, higher-order partial autocorrelations are similarly defined.

The *partial autocorrelation function* (PACF) is a plot of partial autocorrelations versus lags. It is used to see which partial autocorrelations are close enough to zero to be ignored. If they can all be ignored, then no time series procedures are needed. If all but the first can be ignored, then the methods of this chapter will suffice.

Display 15.10 is the partial autocorrelation function (PACF) for the global temperature residuals. The thin lines in the display indicate two-SD limits around zero (based on the method of Section 15.4.1). Partial correlations falling within the limits do not significantly differ from zero—at least their *p*-values are greater than 0.05—while those outside the limits do.

Display 15.10   Partial autocorrelations in the Jones temperature residual series, to lag 10

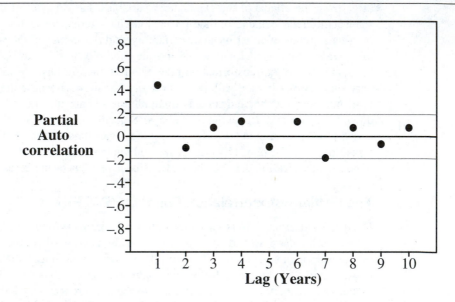

Knowing only that much, it is apparent from the display that only one partial auto-correlation is significantly different from zero. It is the first partial autocorrelation. Thus it is apparent from Display 15.10 that the first-order autoregressive model of Section 15.3 adequately accounts for all serial correlation in the residuals from regression.

It is worthwhile to notice that the lag 2 partial autocorrelation is not simply the correlation of the residuals with the lag 2 residuals. It is the correlation of the residuals with the lag 2 residuals after accounting for the effects of the lag 1 residuals (by regression methods).

This is why the term *partial* is present. Each partial autocorrelation measures the association between residuals spaced a certain number of lags apart in time, after accounting for the effects of the residuals between them. Typically partial autocorrelations are plotted for lags up to the square root of the series length.

For a series without serial correlation, the sampling distribution of the PACF at any lag is approximately normal with mean zero and SD $= 1/\sqrt{n}$. So the error limits in Display 15.10 are placed at $\pm 1.96/\sqrt{n}$. Display 15.11 shows some common patterns for sample PACFs.

Display 15.11   Partial autocorrelation functions for four different types of time series

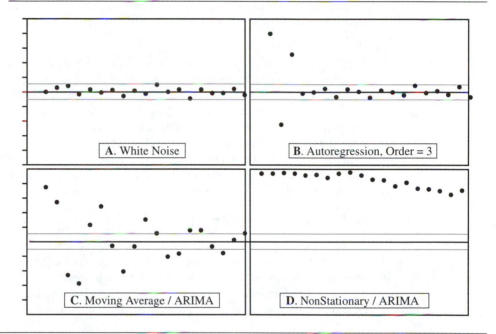

PACF A   With no serial correlation at all, the residuals are all independent with zero means and the same variance. In time series terminology, this is known as *white noise*. If this is the pattern, there is no need to consider time series adjustments to standard analyses.

PACF B   *Autoregressive* models are characterized by a PACF that is nonzero out to a certain lag and then abruptly becomes zero beyond that lag. The maximum lag where partial autocorrelation exists ($p = 3$, in the example shown) is the *order* of the autoregression.

This is the modest extension of the AR(1) model mentioned earlier. If one fits a regression model for the mean and finds that the residuals have a PACF indicating a pth order autoregression, then a two-step filtering procedure similar to that in Section 15.3 will give more efficient estimates of the regression parameters along with approximately correct standard errors. The first step is to fit the parameters in the autoregressive model. Most computer software packages can do this, once the number of lags in the model is specified. The computation of the partial autocorrelations is similar to fitting a regression of residuals on lagged residuals. (Details are not given here.)

The next step is to perform an autoregressive transformation of the response variable and all explanatory variables. If the second-order autoregressive model is appropriate then the filtered response variable is $V_t = Y_t - \hat{\alpha}_1 Y_{t-1} - \hat{\alpha}_2 Y_{t-2}$. A similar filtering is used on the explanatory variables. The final step is to recompute regression coefficients and perform the necessary tests and confidence intervals using the regression of the filtered variables.

PACF C   Another class of stationary time series is called *moving averages*. This PACF shows the typical behavior of a series with a moving average component. After perhaps a few values having no particular pattern to them, the partial autocorrelations either decay exponentially toward zero, or they follow some decaying, cyclical pattern (as does the example given).

PACF D   When the PACF fails to disappear or decay after the first several lags, the most likely explanation is large trends in the series that have not been accounted for in the initial regression model. In this case another time-related explanatory variable may be available for use in the regression equation. Alternatively, it may be that taking successive time differences will result in a series that fits the assumptions of a simple time series.

Such time series that require differencing belong to a general class of time series called Autoregressive Integrated Moving Averages, or ARIMA models. These (such as PACF C and PACF D) require very specialized statistical tools. Researchers with such problems should seek technical assistance from a professional statistician.

15.5.3   Bayesian Information Criterion

The aim of time series analysis is to develop a model in which residuals, after filtering and regression, have the same variability and have no serial correlation remaining. The distribution of these residuals is summarized by an estimate, $\hat{\sigma}^2$, of their variance.

Schwarz introduced the Bayesian Information Criterion (Section 12.4.2) as a criterion for deciding on the order of an autoregressive model in time series analysis. The criterion,

$$\text{BIC} = n \log(\hat{\sigma}^2) + k \log(n),$$

can, however, be used to compare the fits of any models in the ARIMA class. Here k is the number of parameters in the model, including both the parameters describing serial correlation and the regression parameters.

In autoregressive models, the BIC can be calculated directly from the partial autocorrelation function. If $\hat{\phi}_j$ is the estimate of the jth partial autocorrelation, then the BIC for an autoregressive model of order p for the residuals from a regression with q parameters is

$$\text{BIC} = (p + q) \log(n) + \sum_{j=1}^{p} \log(1 - \hat{\phi}_j^2).$$

The model with the smallest BIC gives the best fit to the data.

15.6   **Related Issues**

15.6.1   **Time Series Analysis Is a Large-sample Game**

When the length of the time series is less than around $n = 100$, one must exercise care in verifying that assumptions are met. With n less than 50, the tools described in this chapter are unlikely to yield reliable results. As Display 15.4 shows, the mean value can be surprisingly misleading in short series.

In the absence of a long series the analysis of serial correlation requires advanced techniques.

15.6.2   **The Autocorrelation Function**

Time series analysis usually employs the *autocorrelation function* (ACF) more than the partial autocorrelation function. The ACF describes the correlation between outcomes separated by various time lags. The information in the ACF is the same as that in the PACF, but is presented in a form that allows the user to spot moving averages.

15.6.3   **Time Series Without a Time Series Package**

Only the simplest adjustment for serial correlation should be attempted without computer software specifically designed for the task. Although it might seem possible to organize time series calculations—by creating lagged variables as new variables and computing correlations—in computer software packages that do not have a formal time series capability, this is not advised. The correlation formulas are slightly different from those used in the time series packages and the PACF is especially difficult to calculate outside a package.

15.6.4   **Negative Serial Correlation**

The examples in this chapter exhibit drift, with fewer crossings of the mean level than one would expect in a purely random series. Although this is the most commonly encountered situation, there are some cases with more crossings of the mean level than one would expect in a random series. Observations that are unusually high tend to be followed by compensating values that are low and vice-versa. These series possess *negative* serial correlation.

A short-run average in a series with negative serial correlation tends to be closer to its long-run mean than expected, because the values on both sides of the mean average out well. But the sample SD in a short-run series tends to look larger than expected because of the swings back and forth across the mean. This produces an effect that is similar to, but opposite from, the effect of positive serial correlation. The mean is closer than expected, while the SE is too large. Detecting negative serial correlation and adjusting the standard errors for it may be done exactly as has been outlined above for positive serial correlation.

15.7   **Summary**

This chapter is concerned with regression analysis when the data are collected over time or space. In some situations no special adjustment for serial correlation is necessary. In others a simple adjustment along the lines of those described in Sections 15.2 and 15.3 is satisfactory. For others, more advanced techniques are required.

After the usual exploration, transformation, and model fitting, the diagnostic tests of Section 15.4 can be applied to the residuals to see whether serial correlation is present. If it is, then the tools of Section 15.5 can be used to explore possible models for the serial correlation and, in particular, to see if the AR(1) model is adequate. The partial autocorrelation function plot indicates whether the AR(1) model is appropriate. If it is, then the regression coefficients can be estimated by regressing a filtered response variable on filtered explanatory variables, as shown in Section 15.3.2.

Logging Practices

The scatterplot in Display 15.1 (or any other initial graphical display) indicates the need for a log transformation of nitrate level. Since the data were collected over time, next it is important to check for serial correlation. A runs test (Section 15.4.2) indicates conclusively that there is. Examination of a partial autocorrelation function (PACF) confirms that the autocorrelation can be represented by an AR(1) model. This permits the simple adjustment to the two-sample t-test, obtained by modifying the standard error of the difference in averages (Section 15.2.2) after obtaining a pooled estimate of the first serial correlation coefficient. The one-sided p-value for comparing the mean log nitrate level in the undisturbed watershed to the patch-cut watershed is .39.

Global Warming

The initial scatterplot in Display 15.2 suggests that the mean annual temperatures are increasing linearly with year. A linear regression of temperature on year can be used to examine this, but it is essential first to consider possible serial effects, and adjust for them if necessary. The partial autocorrelation function (Display 15.10) indicates the presence of a first-order lag, but no second- or higher-order lags in the residuals (from an initial linear regression fit without attention to serial correlation). Since the AR(1) model is indeed appropriate, the simple linear regression model is refit with the filtering method (Section 15.3.2). It is estimated that the mean annual temperature is increasing by 0.0046 degrees Celsius per year.

Further Reading

Further discussion of diagnostic procedures for assessing the evidence of serial correlation from regression residuals is provided in most texts on regression. More discussions of time series problems are available in Chatfield (1984) and Bowerman and O'Connell (1993).

15.8 Exercises

Conceptual Exercises

1. Logging Practices. (a) Discuss the benefit of replication for the problem in Section 15.1.1. That is, how would the question of interest be better answered if there were data on several undisturbed watersheds and several patch-cut watersheds? (b) Could the methods of this chapter be used to analyze data from such a study? (c) Would causal conclusions be possible?

2. Global Warming. What specifically is the problem with fitting the simple regression of temperature on time and using the slope as an estimate of the change in mean temperature per year?

3. To contrast the economic strength of England under each of the two political parties, a political scientist determined the monthly unemployment rates over the past 45 years. Is it reasonable to make a comparison by looking at the difference between the average unemployment rates under Labour and those under Conservatives in a two-sample t-test (Chapter 2)?

4. To test a new pollution control device at a factory, dioxin concentrations at a downstream location were recorded daily for two full years before installation and for another two full years after installation. The pre- and postinstallation series showed similar strong serial correlation. So the t-test comparing the two groups (of 730 observations each) was adjusted, and it gave convincing evidence of a decrease in dioxin concentrations (one-sided p-value $= .00003$). What is the scope of the inference that can be drawn from this?

5. To compare two diet preparations for before-school breakfasts, researchers plan to provide menu A on 75 school mornings and menu B on 75 other school mornings. At the end of each day, the students' teachers will evaluate the students' general level of accomplishment as the response variable. The researchers fully expect strong positive serial correlation in these responses from day to day. Knowing that, which of the following allocations is the best? and why?

    (a) Give the students menu A for the first 75 days and menu B for the last 75 days.
    (b) Alternate menu A on one day with menu B on the next throughout the 150 days.
    (c) Select 75 of the 150 days *at random* for providing menu A.

6. Suppose that Y_t is the gross national product of the United States in year t and X_t is the interest rate set by the Federal Reserve Board at the beginning of the year. If a time series adjustment is used to estimate the regression of Y_t on X_t for the years 1950 to 1995, is there anything wrong with using the results to predict (forecast) Y for 1996 once X_{1996} is known?

Computational Exercises

7. The data on successive intervals and durations of eruptions by Old Faithful geyser in Yellowstone National Park (Chapter 7, Exercise 24) constitute a time series. (a) Calculate the residuals from the regression of the interval on the previous eruption duration. (b) Count the numbers of positive and negative residuals and the number of runs. Use the runs test (Section 15.4.2) to see if there is any serial correlation. (c) Calculate the first serial correlation coefficient, and perform the large-sample Z-test for serial correlation. (d) Fit the regression of interval on duration, duration(-1), and interval(-1), where duration(-1) and interval(-1) are the lagged versions of these variables (the most recent duration and interval). Is there serial correlation among the residuals from this regression? Are the lagged explanatory variables significant predictors of interval? By how much is R^2 increased when the lagged explanatory variables are included?

8. The zero-lag and first-lag covariances from four time series appear in Display 15.12. Assuming all four series have the same first-order serial correlation, determine the pooled estimate of the first serial correlation coefficient.

9. Perhaps the most intensively studied time series consists of annual counts of sunspots, begun in 1610 by Wolfer and continued to the present time; 150 years of the sunspot series appear in Display 15.13. (Data from M. Waldmeier, *The Sunspot Activity in the Years 1610–1960*, Swiss Federal Observatory, Zurich (1961).) (a) Construct a time plot of this series. If this were a response variable, would it require a transformation? (b) After transformation (if necessary), use a computer to construct

Display 15.12   Data for Exercise 8

Series	Length	c_0	c_1
1	35	201.45	49.83
2	38	183.08	106.37
3	33	162.96	85.49
4	41	190.53	78.83

Display 15.13   Sunspot counts for 1749–1948

1749–1773	1774–1798	1799–1823	1824–1848	1849–1873	1874–1898	1899–1923	1924–1948
89	11	6	10	78	33	11	24
70	11	24	23	67	13	5	61
47	41	41	47	60	13	3	72
41	139	43	60	45	7	11	69
17	142	45	61	30	2	33	68
11	107	48	71	16	16	51	57
11	80	35	62	3	44	62	28
17	51	20	41	9	62	60	15
45	31	7	14	36	55	51	9
46	14	7	7	76	74	51	5
62	12	0	24	95	55	33	15
72	46	0	93	91	41	13	57
72	108	3	139	68	14	3	101
53	138	7	126	55	11	3	110
49	127	15	82	43	5	4	103
26	109	20	82	41	6	24	76
18	76	48	50	24	17	59	58
19	63	43	29	7	56	69	48
50	55	36	19	17	77	98	21
78	41	24	12	57	87	78	9
112	33	23	28	106	71	51	19
94	20	11	47	135	53	31	56
80	10	6	65	98	35	23	126
53	5	1	123	92	26	7	145
39	7	6	121	52	20	8	139

the PACF up to a lag of 20 years for the residuals from the average level. Interpret the result in terms of a time series model. (c) Use the Bayesian Information Criterion to select the best-fitting order for an autoregressive model.

10.   An observer timed the dives of feeding ducks. There were four ducks each of two species, Lesser Scaup and Greater Scaup. The observer recorded the times (in seconds) of 10 consecutive dives by each bird. When the data were analyzed as a several-sample problem (the eight different birds being the groups), the analyst isolated different group comparisons in the following analysis of variance. This analysis suggests that all the ducks were doing their own thing. However, the analyst noted that the dives were serially correlated, with a pooled estimate of the first serial correlation being $r_1 = 0.4483$.

Source of variation	Sum of squares	d.f.	Mean square	F-Statistic	p-Value
Between Birds	1,469.950	7	209.993	3.923	0.0011
Between Species	305.200	1	305.200	5.702	0.0196
Between Lessers	675.833	3	255.278	4.209	0.0084
Between Greaters	488.862	3	162.954	3.044	0.0342
Within Birds	778.5039	72	53.525		
Total	1,788.3621	28			

(a) Determine the correction factor.

(b) Thinking of this as a correction factor for s_p, the corresponding correction factor for $s_p{}^2$ would be its square. Calculate this, and apply it to the Within Birds mean square to get a mean square adjusted for serial correlation. Put an asterisk by that mean square to indicate that it has been adjusted.

(c) Recompute the F-statistics and the p-values to give an adjusted table (using the modified s_p^2, rather than the unmodified version, as the divisor in F-statistics).

(d) Interpret the results.

Data Problems

11. Melanoma and Sunspot Activity—An Observational Study. Several factors suggest that the incidence of melanoma (skin cancer) is related to solar radiation. Melanoma is more common in fair-skinned individuals, and among them it is more common nearer the equator. Display 15.14 gives data on age-adjusted melanoma incidence among males from the Connecticut Tumor Registry from 1936 to 1972, along with the annual sunspot series. (Data from Houghton, Munster and Viola, "Increased Incidence of Malignant Melanoma After Peaks of Sunspot Activity," *Lancet* (April 8, 1978): 759–60.) Is there any evidence that melanoma incidence is related to sunspot activity in the same year, or to sunspot activity in the previous one or two years?

12. Many animal populations experience approximately 10-year cycles of boom and bust. An example is the Canadian lynx. Annual numbers of lynx trapped in the Mackenzie River district of northwest Canada from 1821 to 1934 appear in Display 15.15. (Data from C. Elston and M. Nicholson, "The Ten Year Cycle in Numbers of the Lynx in Canada," *Journal of Animal Ecology* 11 (1942): 215–44.) One conceivable explanation is that these fluctuations are related to solar activity, as described by the sunspot series. Is there evidence that lynx numbers are related to sunspot activity in the same year or to sunspot activity in the previous one or two years?

Answers to Conceptual Exercises

1. (a) With only one watershed of each type, it is impossible to separate the effect of patch-cutting from natural variation between any two watersheds. There may be a confounding factor. Using many watersheds provides information about the natural between-watershed variation and gives a comparison for the variation from logging practice differences. (b) Yes, as long as each watershed has the same autocorrelation structure. (c) No.

2. There is no problem with estimating the slope, but the standard error from the regression output is not an accurate measure of the uncertainty. Tests and confidence intervals could be misleading if serial correlation is present.

Display 15.14   Age-adjusted melanoma incidence among males from Connecticut Tumor
Registry, 1936–1972

Year	Male Melanoma Incidence*	Sunspot Relative Number	Year	Male Melanoma Incidence*	Sunspot Relative Number
1936	1.0	40	1956	2.4	60
1937	0.9	115	1957	2.6	190
1938	0.8	100	1958	2.6	180
1939	1.4	80	1959	4.4	175
1940	1.2	60	1960	4.2	120
1941	1.0	40	1961	3.8	50
1942	1.5	23	1962	3.4	35
1943	1.9	10	1963	3.6	20
1944	1.5	10	1964	4.1	10
1945	1.5	25	1965	3.7	15
1946	1.5	75	1966	4.2	30
1947	1.6	145	1967	4.1	60
1948	1.8	130	1968	4.1	105
1949	2.8	130	1969	4.0	105
1950	2.5	80	1970	5.2	105
1951	2.5	65	1971	5.3	80
1952	2.4	20	1972	5.3	65
1953	2.1	10			
1954	1.9	5			
1955	2.4	10			

*Incidence is the number of cases per 100,000 population.

---

3. The monthly unemployment rate is a time series. There are strong a priori reasons for expecting positive serial correlation. So the two-sample t-test should not be used without correction.

4. Two questions to consider are: (a) Can one attribute the difference to the device? and (b) Can one expect the device to give the same reduction at another similar factory? A positive response to question (a) amounts to a cause-and-effect inference. This would require random assignment of treatment conditions, which is lacking here. Indeed, there may be many other reasons why emission measurements can drop from one period to the next (increased streamflow, for example). To infer yes to (b) does not require cause-and-effect necessarily. But it would require that this factory be representative of the group of factories under consideration, which would require that it be randomly selected from that group. Again, this is lacking. So the inference is strictly internal. Chance allocation does not seem to have caused the difference, but that is about the only agent that can be eliminated.

5. Option (a) is clearly wrong. If the accomplishment drifts, they might end up with a large imbalance in the numbers of higher-than-average accomplishment days being in the first half of the school year. Option (c) for random selection may seem the best because the random assignment controls for a variety of factors that are impossible to account for in the analysis. But if the researchers are certain about the strong positive serial correlation, they should expect today's accomplishment level to be very close to yesterday's (without treatment), so any differences that appear will be more easily attributed

Display 15.15   Numbers of lynx trapped annually in the Mackenzie River district, Canada, and the corresponding sunspot numbers (from Display 15.13)

1821–1840		1841–1860		1861–1880		1881–1900		1901–1920		1921–1934	
lynx	spots	lynx	spots	lynx	spots	lynx	spots	lynx	spots	lynx	spots
269	6	151	29	236	68	469	62	758	3	229	23
321	1	45	19	245	55	736	55	1307	11	399	7
585	6	68	12	552	43	2042	74	3465	33	1132	8
871	10	213	28	1623	41	2811	55	6991	51	2432	24
1475	23	546	47	3311	24	4431	41	6313	62	3574	61
2921	47	1033	65	6721	7	2511	14	3794	60	2935	72
3928	60	2129	123	4254	17	389	11	1836	51	1537	69
5943	61	2536	121	687	57	73	5	345	51	529	68
4950	71	957	78	255	106	39	6	382	33	485	57
2577	62	361	67	473	135	49	17	808	13	662	28
523	41	377	60	358	98	59	56	1388	3	1000	15
98	14	225	45	784	92	188	77	2713	3	1590	9
184	7	360	30	1594	52	377	87	3800	4	2657	5
279	24	731	16	1676	33	1292	71	3091	24	3396	15
409	93	1638	3	2251	13	4031	53	2985	59		
2285	139	2725	9	1426	13	3495	35	3790	69		
2685	126	2871	36	756	7	587	26	674	98		
3409	82	2119	76	299	2	105	20	81	78		
1824	82	684	95	201	16	153	11	80	51		
409	50	299	91	229	44	387	5	108	31		

to differences in the diet. In effect, adjacent days are paired in (b) in a manner similar to a paired-t analysis. In engineering, this design strategy is known as putting the signal where the noise is lowest.

6.   No, except that the prediction is outside the scope of the data on which the regression was based. Structural changes occurring in 1996 cannot be anticipated, and the inference based on the assumption of no structural changes is speculative and uncheckable.

Repeated Measures

Responses with several components are called *multivariate responses*. A *repeated measure* is a special kind of multivariate response obtained by measuring the same variable on each subject several times, possibly under different conditions.

Many kinds of studies lead to repeated measures. Some approaches to analyzing such measures are discussed in this chapter, but there is no single way to handle them all. The best policy is to let the question of interest and the structure of the repeated measure guide the analysis.

One highlighted method involves reducing the several response variables to one or two summaries that address the questions of interest, followed by performing analyses that use those summaries as responses. The summaries might consist of, for example, the average, the maximum, or the change in the repeated measure.

If there are several distinct research questions then ordinary univariate analyses on distinct summaries are usually appropriate. If, however, there is a single research question that involves several facets—so the question of interest is truly a multivariate one—then the answer obtained by separate univariate analyses can be misleading. Some multivariate tools for dealing with these situations are illustrated in this chapter.

16.1   Case Studies

16.1.1   Sites of Short- and Long-term Memory— A Controlled Experiment

Studies on memory-impaired human patients and primates suggest that short-term and long-term memory reside in separate locations in the brain. Patients suffering retrograde amnesia often lose access to the recent past more readily than to the remote past. Patients with damage to the hippocampal formation and with loss of short-term memory have demonstrated unimpaired memory of the remote past. These observations led S. M. Zola-Morgan and L. R. Squire (12 Oct. 1990, "The Primate Hippocampal Formation: Evidence for a Time-limited Role in Memory Storage," *Science* 250: 288–290) to hypothesize that the hippocampal formation is not a repository of permanent memory.

To test the hypothesis, they trained 18 monkeys to discriminate 100 pairs of objects, 20 pairs each at 16, 12, 8, 4, and 2 weeks prior to a treatment. The objects were toys, cans, odd pieces of wood, etc., randomly selected from a population of such objects used for behavioral studies. After learning to distinguish each of the 100 pairs of objects, 11 monkeys were treated by blocking access to their hippocampal formations. The remaining seven monkeys were untreated controls. After treatment, all monkeys were retested on distinguishing the object pairs. The response measurements were the percentages of correctly discriminated pairs among the 20 pairs learned at each of the five pretreatment times. Display 16.1 shows the average percentages for the treated and control monkeys, with bars indicating the

Display 16.1   Percentage correct identifications by treated and control monkeys of objects learned at five different pretreatment times

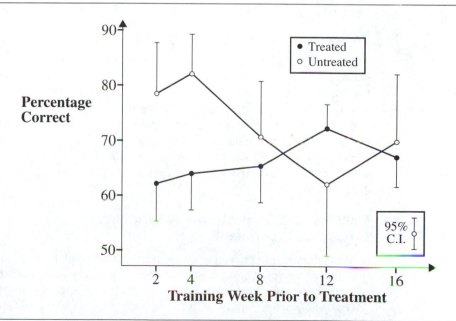

half-widths of 95% confidence intervals constructed for each mean using only the group average and standard deviation. The raw data appear in Display 16.4.

The two central questions are: (1) is there evidence of a difference between groups in their abilities to discriminate pairs learned in the recent past—the short-term? and (2) is there evidence of such a difference in abilities to discriminate pairs learned in the remote past—the long-term?

Summary of Statistical Findings

For each monkey, the percentage of correctly identified pairs learned 2 and 4 weeks prior to treatment was designated as a short-term memory response, while the percentage correct of those learned at 8, 12, and 16 weeks prior to treatment was designated as a long-term memory response. Hotelling's T^2 statistic showed convincing evidence that the means of the responses differed in the control and the treated groups (p-value $= .0007$). The treatment was associated with a decrease of 18.6% correct in short-term memory score (95% confidence interval: 14.2 to 23.0%) and a decrease of 2.3% correct in the long-term memory score (95% confidence interval: -2.6 to 7.2%). Two of the monkeys were moderate outliers from their groups, but their presence did not affect the conclusions drawn.

Scope of Inference

The statistical analysis suggests that blocking the hippocampal formation affected short-term but not long-term memory for the monkeys in this experiment. The random selection of object pairs and the presence of controls make it very unlikely that these effects might have been caused by more difficult object pairs being presented in the recent past. It is difficult to draw inferences beyond the specific monkeys used in the experiment, however, because the authors do not describe how the monkeys were selected. Furthermore, randomization was not used for treatment assignment, so any causal inference is speculative.

Added Note Concerning Confidence Intervals

The research hypothesis states that the group difference on one response (short-term) is nonzero while the difference on a second response (long-term) is zero. The exclusion of 0 from the 95% confidence interval for the difference in short-term score means is interpreted as strong evidence that the first hypothesis is true. That 0 is included in the 95% confidence interval for the difference in long-term score means implies that the data are consistent with the second research hypothesis, but does not prove that the difference is zero. The range of the interval plays a key role here by giving bounds for the difference. Readers of the study's findings can assess whether differences in the range of -2.6% and 7.2% are practically meaningful.

16.1.2 Oat Bran and Cholesterol—A Randomized Crossover Experiment

Studies attempting to prove that supplementing a diet with fiber reduces serum cholesterol have had mixed results. Water-soluble fibers, such as the kind available in oat bran, seem to be effective while water-insoluble fibers are much less so. An article in the *New England Journal of Medicine* (Jan. 18, 1990, "Comparison of the Effects of Oat Bran and Low-fiber Wheat on Serum Lipoprotein Levels and Blood Pressure," by J. F. Swain, I. L. Rouse,

C. B. Curley and F. M. Sacks) suggested a mechanism: "Oat bran could lower serum cholesterol levels by replacing foods in the diet that contain saturated fat and cholesterol, rather than by direct action of the soluble fiber."

Twenty volunteer hospital employees were selected as subjects, after screening for excess weight, medications, and hypertension. They entered a randomized, double-blind, crossover trial using diet manipulation. Two dietary supplements were prepared as entrees or muffins. A high-fiber preparation used oat bran, and a low-fiber preparation used refined, low-fiber wheat. All participants consumed their normal diet during a one-week baseline period, at the end of which they had blood chemistries taken. Then subjects were randomly assigned to either the high- or low-fiber dietary supplement, to be included in their diets for 6 weeks. At the end of 6 weeks, a second blood chemistry was obtained. Subjects returned to their normal diets for another 2 weeks, then crossed over to the opposite supplement from the one that they had received first. After 6 weeks on that diet a third chemistry was obtained.

Considerable numbers of measurements were taken during the course of the experiment. Display 16.2 shows only the three readings on total serum cholesterol (in milli-

Display 16.2 Total serum cholesterol before treatment (baseline) and after dietary supplements of high- and low-fiber content: subjects ordered roughly from smallest to largest cholesterol levels

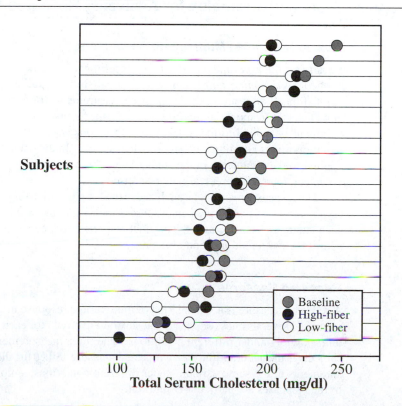

grams per deciliter) prior to treatment, after the high-fiber diet supplement period, and after the low-fiber supplement period. The subjects in Display 16.2 have been ordered roughly by their general cholesterol levels, which vary considerably. The data appear in Display 16.9.

Two important questions are: is there evidence that the high-fiber supplements lowered serum cholesterol? If they did, but if the reduction is due to a reduction in fat consumption, then there might also be lower cholesterol after the low-fiber diet supplements. So, is there any difference between cholesterol measurements taken after the low-fiber and those taken after the high-fiber supplements?

Summary of Statistical Findings

It is estimated that the high-fiber supplement caused a 13.85 mg/dl reduction in cholesterol (95% confidence interval for reduction is 4.17 to 23.53 mg/dl) and that the low-fiber supplement caused a 13.00 mg/dl reduction (95% confidence interval is 3.48 to 22.52 mg/dl). The data are consistent with there being no difference in high-fiber and low-fiber effects (two-sided p-value $= .81$ and the 95% confidence interval for high-fiber effect minus low-fiber effect is -10.5 to 8.8 mg/dl).

16.2   Tools and Strategies for Analyzing Repeated Measures

How to Recognize Multivariate Responses

Recall that the *experimental unit* in an experiment is the object to which the treatment is applied. A *sampling unit* in an observational study is the object that is selected from a population. If there are several response measurements for each experimental or sampling unit, then the response is *multivariate*. In the monkey memory data, for example, the treatment is applied to the monkeys, so the monkeys are the experimental units and the five proportions constitute a five-variable response. In the oat bran study the treatments are applied to the volunteers in the study. Since there are three cholesterol readings for each volunteer, there is a three-variable response.

The responses need not be analyzed in the form in which they come. A few summarizing combinations of them may better address the questions asked. The objective in seeking such summaries is to reduce the original responses to a new set, where each new response addresses a different question of interest.

Repeated Measures

A repeated measure is a special kind of multivariate response in which the same variable is measured several times (or possibly at several locations) for each sampling or experimental unit, possibly under different conditions. The three measurements of cholesterols for the subjects in the oat bran study are a repeated measure. Since the different responses are measurements of the same variable under different conditions, simple summarizing functions of the responses are particularly relevant.

16.2.1   Types of Repeated Measures Studies

Longitudinal Studies

In a *longitudinal study* the repeated measurements are of the same variable on the same subjects at different points in time. The sampling units in an *observational longitudinal study* typically belong to different groups and have different explanatory variable values. The experimental units in a *randomized longitudinal experiment* are given the treatments at the beginning of the study.

In one observational longitudinal study, IQ scores were obtained at four different ages from a sample of adopted children. It was desired to see how their IQ scores were related to the IQ scores of their natural and adoptive parents. In one randomized longitudinal experiment, hypertensive patients were assigned to receive either a new treatment or a standard treatment for reducing blood pressure. Their blood pressure was then measured every 3 months for 5 years.

Crossover Experiments

In a *crossover experiment* each subject receives more than one treatment level. Their first measure is the response after one level of the treatment. Then they cross over to another level, and their second measure is the response after that level of the treatment, and so on. The subjects are randomly assigned to one of several orders for receiving the treatment levels.

The oat bran study is an example of a crossover experiment with a two-level treatment. Half of the subjects were given the high-fiber diet first and the low-fiber diet second and the other half were given the diets in the opposite order. Cholesterol level was measured at the end of each diet period.

Split-plot Experiments with Repeated Measures over Time

A *split-plot experiment*, for examining two separate treatments, has features of both the longitudinal study and the crossover experiment. A first randomization is used to allocate experimental units to one of several levels of the first treatment. Then, in a second randomization, the order of presentation of levels of a second treatment is determined and *all* levels of the second treatment are applied as in a crossover experiment. Notice that all experimental units receive all levels of the second treatment (the split-plot treatment) but only one level of the first treatment (the whole-plot treatment).

For example, suppose that patients with high blood pressure are randomly assigned to receive either a new drug or a standard drug for reducing blood pressure. Furthermore, the blood pressure of each patient is measured after 6 weeks on a standard diet and after 6 weeks on a special diet—with random assignment to the order in which the diets were given. Each subject receives only one of the drugs, but both of the diets. A key feature of split-plot experiments is that they involve two stages of randomization.

Split-plot Experiments with Repeated Measures at Several Locations

Split-plot experiments are often used for agricultural trials when the repeated measure is over space rather than time. Large fields are randomly assigned to receive one of several levels of the first treatment. Each field is composed of a number of plots, which are randomly assigned to receive levels of the second treatment.

To explore the combined effects of type of fertilizer and crop variety on crop yield, for example, the fields are randomly assigned to receive one of the fertilizers. Then the plots within the fields are randomly assigned to receive one of the crop varieties. Each field is assigned to receive only one of the fertilizers, but each field receives every crop variety.

16.2.2   Profile Plots for Graphical Exploration

A profile plot is a scatterplot showing the response versus time or versus condition—whichever is appropriate—drawn separately for each subject. This plot should be customized to best address the questions of interest without undue clutter. For example, if there are only a few subjects, draw the profile plots on the same graph with lines connecting the points belonging to each subject and different line types (dashed or dotted, for example) showing the different treatment groups. With many subjects, however, this type of display is messy. A reasonable display strategy is to separate the profiles into different panels.

Display 16.3 is a profile plot showing the percentage of objects correctly identified by the 18 monkeys in the memory-loss experiment, versus the weeks prior to treatment when the objects were learned. The broken lines show the short-term and long-term averages. The control monkeys, typified by Spank, tended to identify about 80% of those objects taught within 2 to 4 weeks prior to treatment, but only about 70% of objects taught between 8 and 16 weeks prior to treatment. For the treated monkeys, typified by Irv, the corresponding percentages were roughly 60% and 70%. The preliminary indication, therefore, is that the treatment may have had an adverse effect on short-term memory but no effect on long-term memory.

Profile plots are valuable for preliminary analysis of repeated measures because they often exhibit a transparent relationship that is not evident from a perusal of the raw, multivariate data; and they often help focus the analysis on what is important. A plot of average responses versus time or condition, as in Display 16.1, is useful for gleaning an overall effect and structuring a model. The profile plots complement this by showing the subject-to-subject variability and clarifying to what extent the general pattern is evident in different subjects.

16.2.3   Strategies for Analyzing Repeated Measures

Univariate Analysis on a Summary of the Multivariate Response

The most practical strategy is to perform a univariate analysis on a single summary of the multivariate repeated measure, if a summary can be found that directly addresses a question of interest. Typical summaries are the profile average, the final response, the minimum, the maximum, a treatment difference, the time at which the maximum occurs or at which another condition is attained, and the estimated slope in a regression of the response on time.

The univariate analysis will have one of two forms: (1) In longitudinal studies, the focus is on comparing subjects that receive different treatments, so the appropriate tools are

Display 16.3 Profile plots: percentage correct identification of 20 pairs of objects, versus week prior to treatment when learned (memory data)

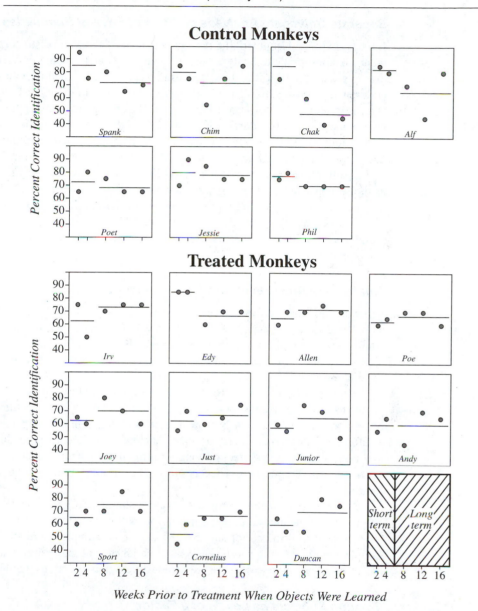

those that compare several groups, possibly after adjustment for related covariates. (2) For crossover or split-plot studies, the focus is on comparing treatments on a single unit and summarizing response variables as differences—such as score after treatment minus score after control—and analyses focus on whether the mean differences are zero, possibly after accounting for covariates.

Separate Univariate Analyses on Each of Several Summaries

If there are several summarizing responses that address several distinct research questions, then it is appropriate to analyze each response separately with a univariate analysis. For example, suppose IQ test scores are measured at ages 2, 4, 8, and 13 for a sample of adopted children. It is desired to relate these repeated measures to the IQ test scores of the natural and foster mothers. Two separate questions concerning the relative effects of heredity and environment might be addressed: (1) Is the child's overall IQ score associated with the foster mother's IQ score, after accounting for the effect of natural mother's IQ? (2) Is the degree of change in IQ score from age 2 to age 13 associated with the foster mother's IQ? The first question may be addressed by taking the average of the four IQ measures as the response, and regressing that response on the two mothers' IQ scores. The second may be addressed by taking as a response either the difference (IQ at age 13 minus IQ at age 2) or perhaps the slope in the regression of IQ on year, computed separately for each child. The usual regression analyses will provide measures of uncertainty pertaining to each question individually.

Multivariate Analysis on Several Summaries

Separate univariate analyses on each response are appropriate if they address separate research questions. If, on the other hand, there is a single research question with several components, then univariate analyses can provide misleading answers. Multivariate techniques are needed to supply measures of uncertainty pertaining to the joint analysis of several correlated responses.

In the memory loss study, for example, the researchers hypothesized that the hippocampal formation is not a repository of long-term memory. They believed their hypothesis could be supported by simultaneously showing that blocking the hippocampal formation *does* have an effect on short-term memory (to show that their experimental procedure is working) and that it *does not* have an effect on long-term memory. There is a single research issue involving a two-part question. The measures of uncertainty should pertain to the two answers jointly.

A multivariate analysis provides a way of summarizing measures of uncertainty from several univariate analyses and adjusting for their correlations. These techniques parallel those already considered for single responses. For example, there are extensions of one-sample and two-sample *t*-tests, analysis of variance, and regression to handle multivariate responses. The first two of these are demonstrated in the sections that follow.

Treating Subjects as Levels of a Factor

The chimp learning time study in Section 14.1.1 (page 398) was a repeated measures study with chimps as subjects. For each chimp, learning times were measured on each of 10 different signs of American Sign Language, producing a 10-variable response for each subject. Two-way analysis of variance was used to estimate sign effects on learning times, treating chimps and signs as factors. This approach is justifiable when, after accounting for treatment (sign) effects, the different observations on the subjects (chimps) are independent. This would not be the case if there were carryover effects depending on the order in which the treatments are given.

There is often no direct interest in the particular subjects used in a repeated measures study. If the independence assumption is satisfied subjects may be treated as blocks. If, however, the subjects are randomly selected from some broader population of subjects to which inference is desired, then one should treat subject effects as random effects (see Section 14.5.1 on page 415).

16.3 Comparing the Means of Bivariate Responses in Two Groups

Why Separate Univariate Analyses May Fail to Address a Multivariate Question

It is supposed in this section that the question of interest is truly a multivariate one. For example, if researchers are investigating the effect of two treatments on the size of animals, and size is thought to be measured by both weight and height, then there is a single question about size, which happens to have two components. The question is a multivariate one.

The two-response problem differs from two separate one-response problems in that the analysis must allow for possible correlation between the response variables. A naive solution is to obtain separate confidence intervals for each response individually, using the two-sample t-tools. This conveys the impression that the region of plausible differences in means is a rectangle bounded by the confidence intervals. However, it ignores the correlation between the two estimated differences. The correct confidence region for both differences in means, simultaneously, has the shape of an ellipse.

Unfortunately, an ellipse is both difficult to compute and to present (and the difficulty is even more severe with more than two responses). An adjustment, however, can be used that modifies the naive one-response confidence intervals to conservatively approximate the correct confidence ellipse. Thus, the recommended practical procedure consists of calculating univariate confidence intervals for each response component individually, but with Hotelling's T^2 adjustment to account for the number of components in the multivariate response and their correlations.

16.3.1 Summary Statistics for Bivariate Responses

It will help to refer to Display 16.4 showing the data on the memory example. There are two groups of subjects: the control group with seven monkeys and the treated group with 11. A short-term memory score is defined as the percentage of object pairs learned 2 and 4 weeks prior to treatment that were correctly distinguished. So Spank, the first control monkey, had short-term memory score $= (95 + 75)/2 = 85.00$ percent. The second response variable, the long-term memory score, is the percentage of object pairs taught 8, 12, and 16 weeks prior to treatment that were correctly distinguished. Spank's long-term memory score was $(80 + 65 + 70)/3 = 71.67$ percent.

Display 16.4 Calculation of summary statistics for the bivariate response, short-term memory (weeks 2 and 4 average) and long-term memory (weeks 8, 12, and 16 average): memory data

		In Pairs from Week #					Memory Scores		Deviations from Group Average		Product
		2	4	8	12	16	SHORT	LONG	SHORT	LONG	
	Spank	95	75	80	65	70	85.00	71.67	4.64	4.05	18.79
	Chim	85	75	55	75	85	80.00	71.67	−0.36	4.05	−1.45
Control	*Chak	75	95	60	40	45	85.00	48.33	4.64	−19.29	−89.54
Group	Alf	85	80	70	45	80	82.50	65.00	2.14	−2.62	−5.61
	Poet	65	80	75	65	65	72.50	68.33	−7.86	0.71	−5.61
	Jessie	70	90	85	75	75	80.00	78.33	−0.36	10.71	−3.83
	Phil	75	80	70	70	70	77.50	70.00	−2.86	2.38	−6.80

Averages: 80.36   67.62

Sample variances: 19.64   88.76 (6 d.f.)

Sample covariance: −15.67 $= \dfrac{-94.05}{6}$ −94.05

		In Pairs from Week #					Memory Scores		Deviations from Group Average		Product
		2	4	8	12	16	SHORT	LONG	SHORT	LONG	
	Irv	75	50	70	75	75	62.50	73.33	−0.68	5.00	−3.41
	*Edy	85	85	60	70	70	85.00	66.67	21.82	−1.67	−36.36
	Allen	60	70	70	75	70	65.00	71.67	1.82	3.33	6.06
	Poe	60	65	70	70	60	62.50	66.67	−0.68	−1.67	1.14
Treated	Joey	65	60	80	70	60	62.50	70.00	−0.68	1.67	−1.14
Group	Just	55	70	60	65	75	62.50	66.67	−0.68	−1.67	1.14
	Junior	60	55	75	70	50	57.50	65.00	−5.68	−3.33	18.94
	Andy	55	65	45	70	65	60.00	60.00	−3.18	−8.33	26.52
	Sport	60	70	70	85	70	65.00	75.00	1.82	6.67	12.12
	Cornelius	45	60	65	65	70	52.50	66.67	−10.68	−1.67	17.80
	Duncan	65	55	55	80	75	60.00	70.00	−3.18	1.67	−5.30

Averages: 63.18   68.33

Sample variances: 65.11   17.78 (10 d.f.)

Sample covariance: 3.750 $= \dfrac{37.50}{10}$ 37.50

*Possible outliers

Let Y_{ijk} denote the measurement of the ith response variable on the kth subject within group j. Here i is 1 for short-term or 2 for long-term memory score; j is 1 for the control or 2 for the treated group; and k is 1, 2, ..., 7 (in the control group) or 1, 2, ..., 11 (in the treated group). Thus $Y_{214}(= 65.00)$ is the long-term memory score for the fourth control monkey, Alf. In the dot notation, $\overline{Y}_{ij}.$ represents the average of the ith response variable for all subjects in group j.

The remainder of Display 16.4 exhibits the summary statistics required for the multivariate analysis. The averages and sample variances for each response are shown for each group. $\overline{Y}_{ij}.$ represents the average of the ith response in group j and s_{ij}^2 represents the sample variance for the ith response in group j, based on $n_j - 1$ degrees of freedom; for response variable $i = 1$ or 2 and group $j = 1$ or 2.

The final column of Display 16.4 contains the products of residuals (deviations from group average) from the two response variables. The *sample covariance* between the two response variables is the sum of these products divided by the degrees of freedom:

$$c_j = \frac{1}{(n_j - 1)} \sum_{k=1}^{n_j} (Y_{1jk} - \overline{Y}_{1j}.)(Y_{2jk} - \overline{Y}_{2j}.)$$

calculated separately for groups $j = 1$ and 2.

16.3.2 Pooled Variability Estimates

Sample variances and covariances are pooled across groups in the same way that variances were pooled (see Section 5.2.2 on page 114). The pooled estimate of variance for response i is obtained by combining s_{i1}^2 from group 1 and s_{i2}^2 from group 2 by

$$S_i^2 = \frac{(n_1 - 1)s_{i1}^2 + (n_2 - 1)s_{i2}^2}{(n_1 + n_2 - 2)}$$

for responses $i = 1$ and 2 separately. Covariances are pooled in the same way:

$$\text{Pooled estimate of covariance} = C = \frac{(n_1 - 1)c_1 + (n_2 - 1)c_2}{(n_1 + n_2 - 2)}.$$

These calculations for the memory loss example appear in Display 16.5.

Display 16.5   Pooled estimates of variance and covariance for the memory example

	Control	Treated	Pooled	
Sample Size (n)	7	11	18	= 7 + 11
Degrees of Freedom	6	10	16	= 6 + 10
Sample variances				
Short-term:	19.64	65.11	48.06	$= \dfrac{\{6(19.64)+10(65.11)\}}{16}$
Long-term:	88.76	17.78	44.40	= ↗
Sample covariance	−15.67	3.75	−3.53	= ⸺ ↗ *similar pooling*

Pooled Estimate of Correlation

The sample correlation coefficient of Section 7.5.4 (page 189) between two variables is a scaled version of the sample covariance. For each group a sample correlation between the two responses may be calculated as

$$r_j = \frac{c_j}{s_{1j}s_{2j}}, \qquad (j = 1, 2).$$

If it is assumed that the correlation between the two responses is the same in both groups, then a pooled estimate is appropriate. This is not obtained by pooling individual sample correlation coefficients, but by using the formula based on the pooled estimates of covariance and standard deviations:

$$R = \frac{C}{S_1 S_2}.$$

This estimate is useful for assessing the degree of correlation between the two responses, and it also plays an important role in the Hotelling's T^2 test for comparing the two bivariate means.

16.3.3 Hotelling's T^2 Statistic

The summary statistics in Display 16.5 give estimates of population quantities in a specific model. One ideal random sampling model has two populations of subjects, with two response variables measured on each, with the following assumptions:

1. The mean of response i in population j is $\mu_{ij} (i = 1, 2; j = 1, 2)$.
2. The standard deviation, σ_i, of response variable i is the same in both populations, for both responses $i = 1$ and 2.
3. The population correlation between the two response variables is the same in both populations.
4. Subjects' responses are distributed around their means according to a bivariate normal distribution.
5. Subjects are randomly selected from each population.
6. The two groups of sampled subjects are selected independently.

A *bivariate distribution* describes relative frequencies of occurrence in the population of pairs of values for the bivariate response. In a bivariate normal distribution each component, individually, has a normal distribution and every possible linear combination of the components also has a normal distribution.

Hotelling's T^2 statistic is a tool for drawing joint inferences about differences in group means for both variables:

$$\delta_1 = \mu_{11} - \mu_{12} \text{ and } \delta_2 = \mu_{21} - \mu_{22}.$$

The Student's t-ratio may be used to draw inferences about each difference, δ_i, individually:

$$t_i = \frac{(\overline{Y}_{i1.} - \overline{Y}_{i2.}) - \delta_i}{\text{SE}(\overline{Y}_{i1.} - \overline{Y}_{i2.})}.$$

The standard error of the difference between group averages is, as before,

$$\text{SE}(\overline{Y}_{i1.} - \overline{Y}_{i2.}) = S_i \sqrt{\frac{1}{n_1} + \frac{1}{n_2}}$$

for responses $i = 1$ and 2.

Hotelling's T^2 combines the two individual t-ratios into a single quantity, adjusting for the correlation of the two response components:

$$\text{Hotelling's } T^2 = \frac{t_1^2 + t_2^2 - 2R\, t_1 t_2}{1 - R^2}.$$

When T^2 is multiplied by a degrees of freedom factor, its sampling distribution is the F-distribution with numerator degrees of freedom 2 and denominator degrees of freedom $n_1 + n_2 - 3$ (see Section 16.5.1 for an extension to more than two responses):

$$F = \frac{(n_1 + n_2 - 3)}{2\,(n_1 + n_2 - 2)}\, T^2 \quad \text{has an } F_{2,(n_1 + n_2 - 3)} \text{ distribution.}$$

One use of T^2 is for testing to see if the data show group differences on *either* response variable. This is a screening test, with the null hypothesis that

$$H_0\text{: } \delta_1 = 0 \quad \text{and} \quad \delta_2 = 0,$$

and the alternative hypothesis that at least one of the differences is nonzero. To perform the test, first calculate the usual Student's t-statistics using the hypothesized (zero) differences, then calculate the T^2-statistic based on these and the pooled estimate of correlation, R. Convert it to an F-statistic, and see if the result looks like a typical F-value. Failure of the hypothesis in any way should result in a larger than anticipated F-statistic, so the appropriate p-value is the tail area in the F-distribution above the observed statistic. Display 16.6 develops this test for the memory-loss example.

Display 16.6   Hotelling's T^2 calculations for the memory example

	Short-term	**Long-term**
① (Control − Treatment) Average Differences	$17.18 = (80.36 - 63.18)$	$-0.71 = (67.62 - 68.33)$
② Standard Errors	$3.35 = 6.93\sqrt{\dfrac{1}{7} + \dfrac{1}{11}}$	$3.22 = 6.66\sqrt{\dfrac{1}{7} + \dfrac{1}{11}}$
③ t-statistics for individual hypotheses of no difference	$5.12 = \dfrac{17.18}{3.35}$	$-0.22 = \dfrac{-0.71}{3.22}$

④ Hotelling's T^2

$$26.28 = \frac{(5.12)^2 + (-0.22)^2 - 2(5.12)(-0.22)(-.0765)}{1 - (-.0765)^2}$$

⑤ F-statistic

$$12.32 = \frac{(15)(26.28)}{(2)(16)}, \qquad \text{d.f.} = 2, 15$$

The F-statistic with 2 and 15 degrees of freedom is 12.32 and has associated p-value = .0007, showing strong evidence against the hypothesis of no treatment effects. The individual t-statistics suggest that there is a treatment effect on short-term memory but contain no evidence of an effect on long-term memory. Before making these conclusions more precise with confidence intervals, it is wise to examine the assumptions (see page 462).

16.3.4 Checking on Assumptions

The distributional assumptions for the ideal model include the assumptions underlying the individual t-tools for each response. Consequently, all diagnostic tools introduced earlier, such as the residual plots, apply here. Assumption 3—that the correlation between response components is the same in both groups—is new and important, so a new check is needed. Since the multivariate comparison is not very robust to departures from normality, careful attention to the normality of each component is also required. Normal probability plots are useful for this purpose (see Section 8.6.3 on page 215).

Scatterplots of the response variables or their residuals show the nature of the joint distributions. Display 16.7, for example, is a coded scatterplot of the two responses in the memory problem, with distinct symbols for each group.

When data conform to a bivariate normal distribution (assumption 4), each group forms an elliptical cluster of points. The highest concentrations of points are at the centers and the concentrations taper off from the centers. When the groups have the same standard deviations for each variable (assumption 2), the spread along each of the axes of the ellipse

Display 16.7   Scatterplot of short- and long-term memory scores, by treatment group

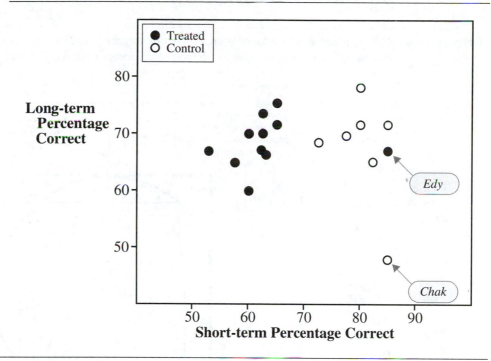

is roughly the same for both groups. When both groups have the same standard deviations and the same correlation between responses (assumption 3), the axes of orientation of the group ellipses are the same.

Looking at Display 16.7, all these features would appear plausible were it not for Edy and, perhaps, Chak. Edy's responses appear atypical of the treated group (and more typical of the controls). In comparison to the variability in the rest of the diagram, Chak's responses may also be anomalous. Examination of the resulting multivariate analysis with and without these outliers is considered in Exercise 10 of this chapter.

16.3.5   Confidence Ellipses and Individual Confidence Intervals for Differences in Bivariate Means

Hotelling's T^2-statistic also forms the basis for establishing confidence regions on the differences in means. The details of the calculations are difficult and not discussed here, but the rationale is fairly straightforward. A method for obtaining a 95% confidence region jointly for the difference in means of the first component and the second component of the response, δ_1 and δ_2, requires testing all possible hypothesized values of δ_1 and δ_2, and including those in the confidence region for which Hotelling's T^2-test produces a p-value greater than .05. Methods for calculation do not require actually testing all possible hypothesized values. When there are two components to the response this procedure produces an elliptical confidence region, as shown by the bold ellipse in Display 16.8.

Display 16.8 95% confidence ellipse for differences in long-term and short-term memory means, and approximate confidence rectangles, constructed with T^2 and t multipliers

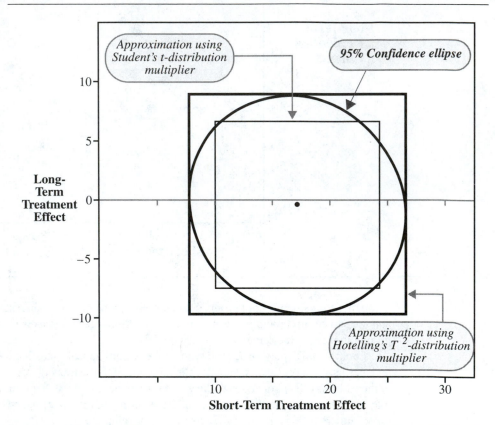

Elliptical confidence regions are the best technical description of what are and are not plausible values for the mean differences. Any values inside the ellipse are more consistent with the data than any values outside it. The confidence coefficient correctly assesses the frequency with which the procedure captures the correct population mean differences.

If there are more than two questions of interest, there will be more than two responses, and the confidence regions will be *ellipsoids* in higher dimensions. Ellipsoids are difficult to summarize in lower dimensional form for publication and are also difficult to interpret. Therefore, it is important to devise a sensible way of conveying approximately the same information using separate intervals for each parameter, even though ellipsoids are technically superior.

Separate Confidence Intervals for Each Component

Separate confidence intervals for differences in means δ_1 and δ_2 have the form:

$$(\overline{Y}_{i1.} - \overline{Y}_{i2.}) \pm \textit{Multiplier} \times \text{SE}(\overline{Y}_{i1.} - \overline{Y}_{i2.})$$

for components $i = 1$ and 2. *Multiplier* is a multiplier chosen to give the desired coverage. Display 16.8 exhibits separate 95% confidence intervals (all subjects included) corresponding to two different choices for the multiplier.

One choice of multiplier is based on Hotelling's T^2-distribution:

$$\textit{Multiplier} = \sqrt{\frac{2(n_1 + n_2 - 2)}{n_1 + n_2 - 3} F_{2,n_1+n_2-3}(1 - \alpha)}.$$

The sample sizes were 7 and 11, so the degrees of freedom for F are 2 and 15. The 95th percentile of $F_{2,15}$ is 3.682 so *Multiplier* = 2.803. The bold outer rectangle in Display 16.8 is the region where both the mean differences are within their separate intervals, with at least 95% confidence, as constructed from this T^2 multiplier.

The naive choice of multiplier—the Student's t-distribution percentile—ignores the multivariate nature of the response. Its pooled degrees of freedom are 16, giving the *multiplier* 2.120. The inner rectangle in Display 16.8 shows the region where both mean differences are within their individual t-based intervals.

Comparing the two rectangles with the elliptical 95% confidence region, one first notices that the t-based intervals have short ranges that exclude values that are in agreement with the data, according to the ellipse. This naive confidence region roughly approximates the ellipse when the correlation between responses is close to zero, but can seriously err when the correlation is substantially non-zero.

The T^2-based rectangle exactly matches the highest and lowest possible values admitted for each difference by the ellipse. The rectangle, therefore, includes all pairs of values within the ellipse, but it also includes many other pairs that are not in the ellipse. Consequently, the T^2-based intervals have considerably more confidence attached and should be reported as having "at least 95% confidence." They are conservative but are safer than the naive intervals when the correlation between responses is substantially nonzero.

Other multipliers, intermediate between the two displayed above, are possible. They arise from considerations similar to those in the multiple comparisons problem of Section 6.4. Bonferroni and other multipliers all produce confidence rectangles, and each excludes some values that are plausible by the ellipse method while including some that are not.

16.4 One-sample Analysis with Bivariate Responses

16.4.1 Treatment Differences in the Oat Bran Study

The question "Does the high-fiber diet reduce cholesterol?" can be investigated by testing whether the mean of the difference, cholesterol after high-fiber diet minus cholesterol after control diet, is zero. The question "Is the reduction in cholesterol the same after the low-fiber diet as after the high-fiber diet?" can be investigated by testing whether the mean of the difference, cholesterol after high-fiber diet minus cholesterol after low-fiber diet, is zero.

Whether each paired *t*-test should be conducted separately or whether a multivariate adjustment for the correlation between the differences is necessary depends on whether these are considered as separate research questions or as two components of a single research question. The position taken here is that it is the latter. That is, to establish that a reduction in cholesterol after a high-fiber diet is due to the elimination of other foods from the diet, rather than due to direct effects of high-fiber, the experiment must show that there is a high-fiber effect and that the effect is not different from the low-fiber effect. Both together support the research hypothesis.

There are two responses for each subject, and the inferential goal is to test whether the means of each are zero. The problem, therefore, is one of testing whether the means in a bivariate response are zero, from a single sample. Display 16.9 shows the raw cholesterol measurements and the computed differences of interest. The analysis could use separate

---

Display 16.9   Cholesterol measurements for subjects after three diet regimes, with two comparison measures

Base Line	High Fiber	Low Fiber	Comparisons Hi - Base ◀▶ Hi - Lo		Fiber Order
205	187	193	−18	−6	HL
161	145	138	−16	7	HL
166	168	169	2	−1	HL
195	167	176	−28	−9	HL
206	174	203	−32	−29	HL
135	102	130	−33	−28	HL
172	175	156	3	19	HL
172	157	161	−15	−4	HL
234	202	198	−32	4	HL
175	155	169	−20	−14	HL
200	185	193	−15	−8	LH
151	160	127	9	33	LH
188	168	163	−20	5	LH
204	182	163	−22	19	LH
128	132	149	4	−17	LH
202	218	197	16	21	LH
165	163	171	−2	−8	LH
190	180	182	−10	−2	LH
225	220	216	−5	4	LH
246	203	206	−43	−3	LH

Average:	−13.85	−0.85
Sample SD:	15.80	15.78
Correlation:	+.516	

one-sample t-tests on the two columns of differences (i.e., two separate paired t-tests). To account for correlation of the separate responses, a Hotelling's T^2 adjustment will be used.

16.4.2   Summary Statistics for a Single Sample of Bivariate Responses

Let Y_{ij} be the jth measurement on the ith response. For the oat bran example, Y_{1j} refers to the cholesterol after the high-fiber diet minus the baseline cholesterol for the jth subject; and Y_{2j} is the cholesterol after the high-fiber diet minus the cholesterol after the low-fiber diet, for the jth subject. The sample averages along with the sample standard deviations and correlation between these two variables appear in Display 16.9. Since there is only a single sample, no pooling is needed.

16.4.3   Hotelling's T^2 Test That the Means of a Bivariate Response Are Both Zero

Based on a model in which subjects are randomly selected from a population and where their responses have a bivariate normal distribution with means δ_1 and δ_2, the Hotelling's T^2 test statistic for the hypothesis

$$H: \delta_1 = 0 \qquad \text{and} \qquad \delta_2 = 0,$$

may be constructed from the individual, univariate t-statistics and the sample correlation coefficient for the two response components. If t_1 and t_2 are the one-sample t-statistics for $H: \delta_1 = 0$ and $H: \delta_2 = 0$, respectively, and if r is the sample correlation coefficient, then

$$T^2 = \frac{t_1^2 + t_2^2 - 2rt_1t_2}{1 - r^2}.$$

This formula is the same as for the two-sample combination of t-statistics, but the degrees of freedom differ here. The statistic converts to an F-statistic as follows:

$$F\text{-statistic} = \frac{(n-2)}{2(n-1)} \; T^2; \; \text{d.f.} = 2, n-2.$$

The p-value is the proportion of values from the F-distribution that are greater than this test statistic. Small p-values are taken as evidence that at least one of δ_1 and δ_2 is nonzero. The procedure is demonstrated on the oat bran data in steps 1 through 6 of Display 16.10.

Approximate Confidence Intervals

More specific conclusions can be made from constructing separate confidence intervals, using multipliers based on Hotelling's T^2. The multiplier is

$$Multiplier = \sqrt{\frac{2(n-1)}{(n-2)} F_{2,n-2}(.95)},$$

Display 16.10   Inferential tools applied to the oat bran data

① Summary statistics:

	Hi-Base	Hi-Lo
Average:	−13.85	−0.85
Sample SD:	15.80	15.78
Sample Size:	20	
Correlation:	+.516	

② Calculate standard errors for the average effects.

$$SE_1 = \frac{15.80}{\sqrt{20}} = 3.533 \quad ; \quad SE_2 = \frac{15.78}{\sqrt{20}} = 3.528$$

③ Calculate individual one-sample t-statistics.

$$t_1 = \frac{-13.85 - 0.0}{3.533} = -3.920 \quad ; \quad t_2 = \frac{-0.85 - 0.0}{3.528} = -0.241$$

④ Combine into Hotelling's T^2.

$$T^2 = \frac{(-3.920)^2 + (-0.241)^2 - 2(+.516)(-3.920)(-0.241)}{1 - (.516)^2} = 19.698$$

⑤ Convert to an F-statistic.

$$F\text{-statistic} = \frac{(18)}{(19)(2)} \ (19.698) = 9.331$$

⑥ Look up the p-value.

numerator d.f. = 2   ;   denominator d.f. = $n - 2 = 18$

p-value = .0 017 ◄── *provides convincing evidence that both mean effects are not zero*

⑦ Construct 95% separate confidence intervals.

$F_{2,18}(.95) = 3.555$ ◄── *from table*

$$T^2 = \frac{(19)(2)}{(18)} (3.555) = 7.505 \quad ; \quad Multiplier = \sqrt{7.504} = 2.740$$

Hi – Base: −13.85 ± (2.740)(3.533) = ↗ −23.5 mg/dl
                                       ↘ −4.2   mg/dl

Hi – Lo: −0.85 ± (2.740)(3.529) = ↗ +8.8   mg/dl
                                       ↘ −10.5 mg/dl

where $F_{2,n-2}(.95)$ is the 95th percentile in the F-distribution. As shown in step 7 of Display 16.10, the high-fiber diet apparently caused a very significant reduction in cholesterol, but so too did the low-fiber diet. The difference in cholesterol reduction from the two diets appears to be small.

16.4.4 Checking on Assumptions

With crossover designs, there is a danger that the responses of a subject to treatments may depend on the order in which the treatments are presented. This should be anticipated at the design stage, where different orders of presentation can be arranged and then randomly assigned to subjects. The oat bran study included three different diets. These were presented in two orders: (1) control followed by high-fiber followed by low-fiber (HL) and (2) control followed by low-fiber followed by high-fiber (LH). Random assignment of the two orders ensures that an order effect will not be confounded with the treatment effect. Nevertheless, it is important to check for an order effect. Its presence could invalidate the Hotelling's T^2 comparison.

Display 16.11 is a scatterplot of the first response component versus the second, with codes indicating order of treatment presentation. The group that received the high-fiber diet first has slightly smaller averages on both components (suggesting that this group may exhibit somewhat lower mean cholesterol measurements after their high-fiber supplement period). A formal test has p-value = .41 indicating that any apparent carryover effect could just be due to sampling error. This test is the two-sample Hotelling's T^2 test of Section 16.3.

Display 16.11   High-baseline and high-low cholesterol levels, by the order of assignment

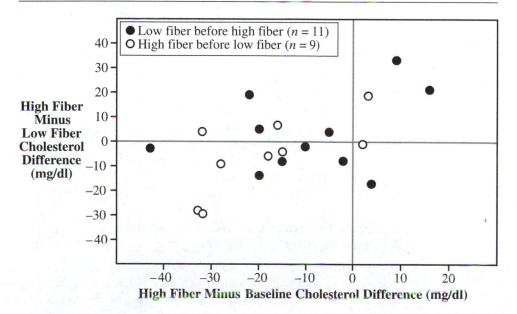

16.5 Related Issues

16.5.1 Two-sample Analysis with More Than Two Responses

Extending the Hotelling's T^2 tools to situations involving more than two response variables is far from trivial. It is nearly impossible to tackle the calculations for T^2 without an understanding of linear algebra, including the use of matrix algebra. Nevertheless the Hotelling T^2 test statistic for comparing equality of two q-dimensional means from populations with the same variance and correlation structure, can be calculated with statistical computer packages. Suppose that there are q responses for each subject and that there are n_1 subjects in group 1 and n_2 subjects in group 2. A p-value is obtained by comparing

$$F\text{-statistic} = (n_1 + n_2 - q - 1)T^2/[q(n_1 + n_2 - q)]$$

to an F-distribution with q and $(n_1 + n_2 - q - 1)$ degrees of freedom. Drawbacks of this tool are that it is not robust if the variances and correlations differ in the two populations and it is not robust if the population distributions are not normal.

16.5.2 One-sample Analysis with More Than Two Responses

The Hotelling's T^2 test that the means of each response are zero also extends to the case of q response variables per subject. The T^2, once calculated by the computer, converts to an F-statistic by:

$$F\text{-statistic} = (n - q)T^2/[q(n - 1)],$$

which is compared to an F-distribution with q and $n - q$ degrees of freedom.

More on T^2 Multipliers

In the summary of the statistical findings for the oat bran study, 95% confidence intervals were stated for *three* different response features. One desirable feature of using the T^2 multipliers is that confidence intervals may be constructed for any and *all* linear combinations of the basic response variables, which maintains at least 95% confidence that all population combinations are captured by their sample intervals. To do this requires that q be one less than the number of repeated measures (as was the case in the oat bran example).

16.5.3 Multivariate Regression and Multivariate Analysis of Variance (MANOVA)

The conceptual difficulties of dealing with regression problems that have multivariate responses are generally similar to those that arise with one-sample and two-sample analysis. From multiple responses several interesting comparison measures are constructed and each is analyzed separately by multiple linear regression. The role of multivariate analysis is to provide an overall assessment of the strength of the evidence about all comparisons, while taking into account that they are correlated.

Extra Sums-of-Squares (and Cross-Products) Tests

The extra sum-of-squares F-test for comparing a full to a reduced model generalizes directly to multivariate testing. In order to compare a model for all the responses with a more restrictive model, residuals of each response are calculated from the fits of both models. If the residual sum-of-squares and cross-products *matrices* are $\mathbf{T}$ in the reduced model and $\mathbf{R}$ in the larger version, the extra sum-of-squares and cross-products matrix is $(\mathbf{T} - \mathbf{R})$. Tests of hypothesis are based on the matrix $\mathbf{R}^{-1}(\mathbf{T} - \mathbf{R})$. Several test statistics, such as Wilk's lambda, can be constructed from the characteristic roots (eigenvalues) of the matrix.

16.5.4   Planned and Unplanned Summaries of Multivariate Measurements as Response Variables

In the two examples discussed in this chapter, more than two response variables were originally measured for each subject, but these were condensed to just two. These two responses were specifically related to the twofold nature of the research hypotheses. These are not instances of data snooping, however, since the decisions on how to summarize were based on the research question or on the inherent structure of the data, rather than on which summaries seemed to work best.

In contrast, some studies intentionally employ a strategy where a multitude of responses are measured to see what happens when experimental factors are varied. A few summary response variables may synthesize what the researcher observes. Using this data snooping strategy greatly diminishes the inferential strength of the study. It is best to treat the conclusions from such a study as suggestive and in need of objective verification by more focused research.

16.5.5   Planning an Experiment: Benefits of Repeated Measurements

It is sometimes possible to give every treatment to every subject. In this case the subject essentially serves as a block, so for this case the benefits of the repeated measures design parallel those of a randomized block design. If substantial variation occurs between subjects, a more precise assessment of treatment effects can be achieved by within-subject comparisons. Thus, the between-subject variability is eliminated from the estimated treatment effects.

Consider the problem of designing the oat bran study (Section 16.1.2). The variability of baseline cholesterol levels (SD = 31 mg/dl) might be available in advance. Previous studies probably suggest the amount by which an oat bran based high-fiber diet supplement can lower cholesterol. Suppose this is 14 mg/dl. If it is desired to randomly assign three different diets—control, high-fiber, and low-fiber—to different subjects, how many subjects should be assigned to each treatment group? With n subjects per treatment, the standard error of a treatment difference will be about $(31.0)\sqrt{(2/n)}$. Getting a treatment difference of 14 mg/dl declared statistically significant requires the t-ratio $14/\{31.0\sqrt{(2/n)}\}$ to be around 2.0 or larger. With $n = 40$ subjects in each group—a total of 120 subjects overall—the anticipated t-statistic will be 2.02, which is barely significant. Taking $n = 40$ is risky, because each of the two treatment comparisons will have to be modified to account for the multiple comparisons problem. A larger n is also advisable to protect against getting a somewhat unlucky sample. All in all, it would seem reasonable to involve 150 or more subjects in the study.

The actual study, however, involved only 20 subjects. The high-fiber average minus the baseline average was 13.85 mg/dl; and the baseline sample standard deviation was indeed 31.0 mg/dl. How is it that the observed difference gave a highly significant t-statistic of 3.92? The design assigned all three treatments to each subject, allowing each of the two treatment differences to be compared to a standard error that did not involve the subject-to-subject variability in cholesterol levels. Thus the standard deviation of the baseline measurements was irrelevant. The sample standard deviations for the *differences* were considerably smaller.

Assigning all treatments to each subject has other advantages. First, the key question in most people's minds is, "What can I expect to happen to my cholesterol if I change my diet in this way?" The design focuses precisely on that kind of change. Second, the fact that subject effects are eliminated means that one is less concerned about how representative the subjects are of a target population.

16.6 Summary

A valuable approach in the analysis of multivariate responses is the reduction of the responses to one or several summaries that address specific questions of interest. If there is one such summary then it may be used as the response in an ordinary univariate analysis.

Hotelling's T^2 tests are discussed in this chapter for multivariate one-sample and two-sample analyses. Extensions to multivariate regression and analysis of variance are straightforward with statistical computer packages. The main adjustment provided by the multivariate analyses is for the correlations among the responses. Consequently, separate univariate analyses may be used instead of a multivariate analysis if it can be demonstrated that the correlations among the responses, after accounting for the effects of the explanatory variables, are all small. (This is checked by computing the correlation between *residuals* from separate univariate analyses.) If they are not, then univariate analyses give unreliable answers to the multivariate question.

Memory Data

Although the monkeys are tested on their memory of items taught at five different times prior to the treatment, the researchers reduce the scores to two summaries, representing long-term and short-term memory. The analysis focuses on a comparison of the bivariate responses in the treated and control groups. Most of the exploratory analysis for this multivariate data problem involves standard techniques for univariate analysis, applied to each response component individually. An exception is the scatterplot addressing the bivariate normality of the responses and the common standard deviations and correlations in the two groups (Display 16.7). A Hotelling's T^2 test and corresponding confidence intervals provide the primary basis of inference for the results stated in the summary of statistical findings.

Oat Bran Data

The main step in the analysis is the recognition of the appropriateness of converting the original three-dimensional response into a two-dimensional response of differences. Initial scatterplots show no need for transformation and do not indicate any crossover effect. Consequently, the formal analysis proceeds with a test that the mean differences are both zero and construction of confidence intervals for the mean differences. Hotelling's T^2 procedure is used.

Further Reading

References on multivariate analysis include Mardia, Kent, and Bibby (1979); Morrison (1990); and Johnson and Wichern (1990). Some specific analysis of variance tables for several types of repeated measures are provided in Neter, Wasserman, and Kutner (1990). See also Hand and Taylor (1987).

16.7 Exercises

Conceptual Exercises

1. What is the distinction between multivariate analysis and multiple linear regression analysis?

2. What is the conceptual difference between the two examples given in this chapter?

3. How would the robustness (to lack of normality, say) of Hotelling's T^2 analysis compare to the robustness of Student's t analysis? More robust? About the same? Less robust?

4. **Nature–Nurture.** The data in Display 16.12 are a subset from an observational, longitudinal study on adopted children (M. Skodak and H. M. Skeels, 1949, "A Final Follow-up Study of One

Display 16.12 Skeels and Skodak data on 63 adopted children

AMED	BMIQ	Age 2 IQ	Age 4 IQ	Age 8 IQ	Age 13 IQ
10	100	120	115	109	106
10	71	131	109	113	95
14	89	126	115	113	90
7	73	120	102	111	121
14	64	126	125	114	96
8	64	125	109	96	87

Display 16.12   Skeels and Skodak data on 63 adopted children—Continued

AMED	BMIQ	Age 2 IQ	Age 4 IQ	Age 8 IQ	Age 13 IQ
13	104	105	107	106	104
16	76	130	112	124	125
10	81	107	120	109	115
8	78	104	108	125	124
14	79	120	117	114	109
12	128	120	128	148	127
9	65	114	102	112	122
12	71	122	100	128	119
17	75	119	101	102	97
13	109	102	107	113	108
14	88	133	121	115	97
14	90	95	89	115	97
15	96	82	106	105	105
9	95	136	115	118	104
15	80	104	107	107	96
13	102	119	103	101	86
6	92	116	121	119	109
12	88	90	94	95	77
12	100	104	114	104	100
14	91	99	102	128	126
14	70	135	112	118	118
15	84	125	108	116	101
13	78	102	90	99	73
13	78	108	90	86	80
16	87	113	97	101	109
15	63	127	121	119	101
10	67	116	113	113	91
15	83	101	93	99	88
16	109	99	126	139	132
13	54	117	114	119	98
9	66	105	109	90	105
13	109	112	113	125	128
12	88	114	138	124	122
8	95	140	130	126	118
13	92	120	113	114	127
15	65	110	111	114	95
8	63	110	113	107	101
19	113	128	112	114	114
4	110	116	92	105	103
10	105	125	111	129	110
16	96	128	139	118	115
12	78	138	125	139	116
14	67	109	92	87	74
16	80	109	112	127	131
12	53	81	87	80	66
13	74	121	132	132	113
12	91	120	105	131	123
12	98	142	135	147	123
13	109	128	145	125	119
14	92	115	113	113	112
11	91	105	130	115	111
11	88	112	107	110	103
13	99	117	112	109	101
10	90	122	127	129	126
15	104	108	124	116	113
8	88	122	112	119	97

Hundred Adopted Children," *Journal of Genetic Psychology*, 75: 85–125). Is child's intelligence, as measured by IQ test scores at four different ages, related to intelligence of the biological mother (measured by *BMIQ*, IQ test score at the time the child was placed for adoption) and the intelligence of the adoptive mother (measured by the surrogate variable *AMED*, years of education)?

(a) Is the multivariate response IQ at age 2, IQ at age 4, IQ at age 8, and IQ at age 13 a repeated measure?

(b) What reductions of this multivariate response might be helpful for addressing questions about the relative importance of the biological mother's and adoptive mother's intelligence?

(c) Suppose the following two reductions are used: the average of the four IQ scores and the change (IQ at age 13 minus IQ at age 2). One strategy for analysis is to conduct separate (univariate) regression analyses on these two responses (regressing each response on *AMED* and *TMIQ*). Under what conditions would this be a safe alternative to multivariate regression?

(d) How could a check be made on the conditions of part (c)?

Computational Exercises

5. **Nature–Nurture.** Reconsider the adopted children IQ data in Display 16.12 and described in Exercise 4. Compute the average IQ score for each child. Compute the difference, IQ at age 13 minus IQ at age 2, for each child. Regress each of these, separately, on adoptive mother's years of education (*AMED*) and biological mother's IQ (*BMIQ*). (a) Find the correlation coefficient between the residuals from these two regressions. (b) If the correlation is fairly small, between $-.2$ and $.2$ say, then use the separate regression analyses to answer these questions: (i) What proportion of variation in average IQ can be explained by the biological mother's IQ? (ii) After accounting for the effect of the biological mother's IQ, is there any additional effect of the adoptive mother's years of education? (Provide a *p*-value.) (iii) What proportion of variation in IQ change can be explained by the adoptive mother's years of education? (iv) After accounting for the effect of the adoptive mother's years of education, is there any additional effect of the biological mother's IQ? (Provide a *p*-value.)

6. Suppose that two treatments (*A* and *B*) plus a control (*C*) applied to each of 18 subjects produced the following summary statistics.

	A minus C	*B minus C*
Average:	2.4	3.7
Sample SD:	9.8	11.6
Sample size:	18	
Correlation:	$-.696$	

(a) For each of the indicated comparisons, (*A* minus *C*) and (*B* minus *C*), test a hypothesis that the mean effect is zero using one-sample *t*-tools.

(b) Compute the Hotelling's T^2 statistic for testing the composite hypothesis that both mean effects are zero.

(c) Sketch a picture of a 95% confidence ellipse with a computer program or with the following steps: (i) For each of several trial values of δ_1 (the mean of *A* minus *C*)—say, $-5, -3, -1,$ $1, 3, 5, 7, 9$—compute the *t*-statistic, t_1, for the hypothesis that δ_1 is the trial value, using the summary statistics for the *A* minus *C* differences. (ii) Using $F_{2,16}(.95) = 3.6337$, find, for each value of t_1 from part (i), the two values of t_2 that satisfy

$$F\text{-statistic} = 16/(2 \times 17)T^2 = F_{2,16}(.95),$$

where $T^2 = (t_1^2 + t_2^2 - 2rt_1t_2)/(1 - r^2)$ (solve the quadratic equation). (iii) From the values of t_2 and the summary statistics for the B minus C differences, calculate the hypothesized value of δ_2 (the B minus C mean) to which each value of t_2 corresponds. (iv) Sketch the ellipse by plotting the two values of δ_2 versus the corresponding trial value of δ_1. (*Note:* All pairs for δ_1 and δ_2 within this ellipse correspond to hypothesized values that jointly produce a p-value greater than .05 by Hotelling's T^2 test.)

(d) Does the sketch reveal any explanation about the apparently contradictory results in parts (a) and (b) above?

7. Two species of flycatchers in the genus *Empidonax* are so similar that certain identification in the field is next to impossible. In hand, slight but consistent differences in the lengths of the first and the third primary flight feathers appear. The following summary statistics depict measurements (in millimeters) of feather lengths taken from samples of males of the two species. (a) Construct t-statistics for comparing the mean lengths of the feathers separately. Does a 95% confidence interval on the difference between mean first primary lengths include zero? Does the 95% confidence interval for the difference between mean third primary lengths include zero? (b) Compute Hotelling's T^2 statistic for comparing mean lengths of both feathers simultaneously. (*Note:* This will require the calculation of sample variances and covariances from the tabled values before pooling.) (c) Does the 95% confidence ellipse for mean differences include both being zero? (This can be answered from part (b), without computing the actual ellipse.) (d) Interpret the results.

	E. hammondii (n = 14)		E. oberholseri (n = 17)	
	1st primary	*3rd primary*	*1st primary*	*3rd primary*
Average:	53.41	40.51	59.83	43.62
Sample SD:	8.98	3.50	7.73	4.25
Correlation:	$r = +.4655$		$r = +.5303$	

8. Multivariate outliers are harder to spot than outliers on single variables. Consider the following $(A - \text{Control})$ and $(B - \text{Control})$ responses on 20 subjects.

```
A − Control:  5  7  3  7  4  4  2  6  2  5  8  4  3  2  6  6  8  7  3  5
B − Control:  3  6  3  5  4  2  2  5  7  4  6  3  1  1  4  6  7  7  2  5
```

(a) Construct stem-and-leaf diagrams for each variable separately. Can you see any outliers? (b) Construct a scatterplot for the two variables. Now can you see any outliers?

9. **Memory Data.** Pick several hypothetical short-term and long-term effects (pick some hypothetical values for δ_1 and δ_2) in the memory example. (See Display 16.8.) For each, confirm whether the hypothetical values are within or outside the 95% confidence ellipse.

10. **Memory Data.** (a) Find the Hotelling's T^2 adjusted confidence intervals for the difference in mean long-term percentage correct and the difference in mean short-term correct, between the treated and control groups, without the outlier Edy. (b) Repeat (a), but excluding Chak instead of Edy. (c) Exclude both. (d) By comparing these results to the ones reported in the summary of statistical findings, state whether either of these cases seems to be influential.

11. **Religious Competition.** Adam Smith, in *Wealth of Nations*, observed that even religious monopolies become weak when they are not challenged by competition. To illustrate the case, A. J. Gill gathered data from 21 countries in which the percentages of Catholics in the populations varied from a low 1.2% to a high 97.6%. (Data from A. J. Gill, "Rendering unto Caesar? Religious Competition and Catholic Political Strategy in Latin America, 1962–79," *American Journal of Political Science* 38(2) (1994): 403–25.) In addition to the percentage Catholic, he determined the number of clergy per 10,000 Catholic parishioners (the priest-to-parishioner ratio) and the percentage of the Catholic clergy that was indigenous. His data, in Display 16.13, come from the early 1970s. The last two columns are thought of as responses (that measure the strength of the church in that country) and percentage Catholics is an explanatory variable.

Display 16.13   Catholic clergy in Latin America and selected countries

Country	% Catholics in Population	Priest-to-Parishioner Ratio	% Clergy Indigenous
Argentina	95.8	2.0	59
Bolivia	93.2	1.8	30
Brazil	90.2	1.5	58
Chile	84.3	3.0	45
Ecuador	96.7	2.7	76
El Salvador	96.8	1.1	61
Guatemala	95.0	1.2	13
Honduras	96.3	0.9	21
Nicaragua	95.5	1.4	NA
Uruguay	61.0	3.8	64
Australia	28.6	10.8	90
France	80.3	2.0	94
Great Britain	13.0	10.5	90
India	1.2	11.9	88
Poland	82.1	6.7	100
South Korea	3.6	8.0	71
Spain	97.6	4.4	99
Sweden	1.2	10.4	NA
Switzerland	49.6	8.3	90
United States	28.1	9.8	95
(West) Germany	48.5	4.9	96

(a) Construct a scatterplot of the priest-to-parishioner ratio versus the percentage Catholic and fit by simple linear regression, saving the residuals. (Call these res$_{priest|catholic}$.)

(b) Construct a scatterplot of the percent nonindigenous versus the log of the priest-to-parishioner ratio. Fit a regression equation, again saving the residuals. (Call these res$_{clergy|priest}$.)

(c) Construct a scatterplot of res$_{priest|catholic}$ versus res$_{clergy|priest}$ and estimate the correlation between them.

(d) The purpose of this exercise is to emphasize the conditions under which two separate regression analyses are appropriate. (i) Given that the researcher wishes to know whether the strength of the Catholic church in the country can be explained by the percentage of Catholics in the country, and given that strength is measured by two components, priest-to-parishioner ratio and percentage indigenous clergy, what is called for—a multivariate regression or two separate univariate regressions? (ii) The inference from multivariate regression can be approximated by the separate inferences from univariate regression only if the correlation between residuals from the two univariate regressions is close to zero. Does it appear that this is the case for this data problem? (iii) What kinds of conclusions regarding the relationship of church strength to percentage Catholics are suggested?

12. **Wastewater.** Municipal wastewater treatment plants are required by law to monitor their discharges into rivers and streams on a regular basis. Concern about the reliability of data from one of these self-monitoring programs led to a study in which samples of effluent were divided and sent to two laboratories for testing. Display 16.14 contains measurements of biochemical oxygen demand (BOD) and suspended solids (SS) obtained for $n = 11$ sample splits, from the two laboratories. (Data from R. A. Johnson and D. W. Wichern, *Applied Multivariate Statistical Analysis*, Englewood Cliffs, N.J.: Prentice-Hall, 1988.) Do the two laboratories' analyses agree? If they are different, how? Answer with the following steps:

(a) Create two laboratory difference measurements, one for BOD and one for SS.
(b) Determine a paired t-statistic for each difference measurement.
(c) Determine the correlation between differences.
(d) Calculate Hotelling's T^2 statistic for the joint hypothesis that the means of both difference measures are zero.
(e) Demonstrate that the joint 95% confidence ellipse does not contain the possibility that both means are zero. [*Hint:* Exploit the connection between the 95% confidence ellipse and the set of parameter values that are rejected by a 5% level hypothesis test.]
(f) Demonstrate that both t-ratio 95% confidence intervals for each mean separately do include zero.
(g) What should be concluded from the apparent disagreement of the confidence ellipse and the confidence rectangle? Do the laboratories agree or disagree?

Display 16.14   Biochemical oxygen demand and suspended solid measurements from two laboratories

---

	Commercial Laboratory		State Laboratory	
Sample	**BOD**	**SS**	**BOD**	**SS**
1	6	27	25	15
2	6	23	28	13
3	18	64	36	22
4	8	44	35	29
5	11	30	15	31
6	34	75	44	64
7	28	26	42	30
8	71	124	54	64
9	43	54	34	56
10	33	30	29	20
11	20	14	39	21

Data Problems

13. Flea Beetle Distinction. Two flea beetle species—*Ch. concinna* and *Ch. heikertingeri*—of the genus *Chaetocnema* are so similar that the most reliable method of distinguishing between them is to measure the sum of the widths (in micrometers) of the first joints of the first two feet and the sum of the widths of the second joints of the first two feet. A. A. Lubischew reported measurements on 18 beetles from each species (Display 16.15). (Data from A. A. Lubischew, "On the Use of Discriminant Functions in Taxonomy," *Biometrics* 18 (1962): 455–77.) Estimate the differences in the means. In doing this, assess whether the methodology is appropriate.

Display 16.15   Measurements from two species of flea beetles

Ch. concinna		Ch. heikertingeri	
First Joints	**Second Joints**	**First Joints**	**Second Joints**
191	131	186	107
185	134	211	122
200	137	201	114
173	127	242	131
171	118	184	108
160	118	211	118
188	134	217	122
186	129	223	127
174	131	208	125
163	115	199	124
190	143	211	129
174	131	218	126
201	130	203	122
190	133	192	116
182	130	195	123
184	131	211	122
177	127	187	123
178	126	192	109

14. Psychoimmunology. Can you will yourself to fend off a cold? Can your mental state contribute to your getting sick? Recent studies in the field of psychoimmunology suggest a link exists between behavioral events and the functioning of one's immune system. Some have even suggested that the immune function might be conditioned to produce antibodies when challenged by antigens. In a study by Dr. William Keppel, 12 subjects, aged 20–50, were monitored during three distinct activities. The first activity, a baseline period, Phase A, consisted of neutral activity such as reporting tasks and answering questions. During the second activity, Phase B, subjects listened through headphones to a cassette tape of exercises including images of heaviness and warmth in the body, relaxation-deepening images, suggestions to remember happy events, and suggestions to remember recovering from an illness. The third activity, Phase C, was a nonaudiotape follow-up stimulus consisting of continued relaxation as in Phase B and a verbal discussion of the positive aspects of the audiotape.

During each phase, the investigator measured Interleukin-1 (IL-1) activity from blood samples. IL-1 plays a central role in immune system regulation, and it has a short half-life. Display 16.16

shows the median IL-1 measurements, assayed using D-10.S thymocyte and reported in counts per minute. (Data from W. Keppel, "Effects of Behavioral Stimuli on Plasma Interleukin-1 Activity in Humans at Rest," *Journal of Clinical Psychology* 49(6) (1993): 777–85.)

Display 16.16   IL-1 levels (counts per minute) from twelve subjects during three activities

Subject	Phase A	Phase B	Phase C
1	6,850	29,100	34,300
2	27,400	41,100	47,000
3	20,700	30,700	33,600
4	3,000	4,900	5,900
5	22,100	25,700	27,100
6	14,300	17,200	25,700
7	24,300	33,600	31,400
8	2,400	4,700	9,400
9	6,500	7,900	6,400
10	4,800	4,700	5,700
11	2,000	3,000	3,700
12	2,200	5,100	6,800

The principal issues are (1) whether and how much difference there is between Phase B and the baseline Phase A levels of IL-1, and (2) whether and how much difference there is between the Phase C and Phase B responses. After answering the questions, comment on what inferences can be drawn from this study—in which the subjects were volunteers and the treatments were always given in the order A, B, C.

Solutions to Conceptual Exercises

1. The distinction is the number of *response* variables. In multiple linear regression, there may be many explanatory variables but only one response. Multivariate analysis means there are two or more different responses.

2. The difference concerns the relationships of treatments and subjects. In the oat bran study, all subjects received each of the treatments (but in different orders). In the memory study, subjects were divided into two groups that received different treatments.

3. Because T^2 involves the separate t-ratios, it can hardly be more robust. The additional bit of statistical information in T^2 is the correlation coefficient estimate, which is a second-order estimate. Like a variance estimate, it should be expected to be somewhat less robust than estimates of the mean. So T^2 should be expected to be a bit less robust.

4. (a) Yes. (b) If IQ is inherited to any extent, some measure of overall IQ, such as the measurement at age 2 or the average at all ages, is most likely to be related to the biological mother's IQ. The adoptive mother's education level is most likely related to some measure of the change in IQ brought about by learning, such as the final measurement at age 13 minus the initial measurement at age 2. (c) If the residuals from these regressions have little correlation, conclusions drawn from separate regressions are close approximations to the multivariate analysis conclusions. (d) Save the residuals from each regression, and calculate their correlation.

Exploratory Tools for Summarizing Multivariate Responses

In the case studies of Chapter 16, the researchers had little question about how to summarize the repeated measures into variables that addressed their questions of interest. In other studies, a useful reduction of the multivariate response is less obvious but still desirable. The decision about what summaries to construct in such cases may be postponed until after the researcher has had the opportunity to look at the data for clues. This chapter illustrates two methods that rely on the data to suggest summarizing linear combinations.

Principal components analysis (PCA) is a way to select several linear combinations that capture most of the variation of the multivariate responses. This is most useful if relatively few linear combinations explain most of the variability and if the linear combinations lend themselves to some useful interpretation. In the first case study of this chapter, principal components are used to characterize an eleven-dimensional response in terms of three principal linear combinations, which are seen to describe meaningful features. The linear combinations suggested by the principal components are then used as the responses in a multivariate comparison, as in Chapter 16.

Canonical correlations analysis (CCA) is a method for describing and testing the linear relationships between two sets of variables. As a starting point, CCA finds a pair of linear combinations—one from each of the two sets—such that the correlation between the two combinations is as large as possible; then it finds another pair of linear combinations that have the highest possible correlation, but under the constraint that they be independent of the first pair; and so on.

17.1   Case Studies

17.1.1   Magnetic Force on Rods in Printers—A Controlled Experiment

Engineers manipulated three factors to enhance the magnetic force surrounding a metal rod in an electronic printer: the electric current passing through the rod (0, 250, or 500 milliamperes); the configuration of components (0 or 1); and the type of metal from which the rod was made (1, 2, 3, or 4). Arranged factorially the current, configuration, and metal options yield $3 \times 2 \times 4 = 24$ possible treatment combinations.

With a given treatment combination, a printer operated for 2 minutes; then the engineers measured the magnetic force at 11 equally spaced points along the rod. Two replicates were attempted for each treatment combination. Four trials failed, so the full data set in Display 17.1 contains only 44 trials. These unpublished data were provided by Dr. V. J. Wildman.

The main issue was this: how are the variations in magnetic forces along the rods related to the design factors? The phrase "variations in magnetic forces" needed clarification, however, because 11 different and highly correlated responses were involved. This raised further issues. Can the number of responses be reduced from 11 to just a few summary variables without eliminating important sources of variation? Can summary variables be chosen that are less highly correlated?

Summary of Statistical Findings

Principal components analysis identified three uncorrelated linear combinations of responses that account for 99.73% of the total variation in the original 11 measurements. The first principal component, accounting for 96.46%, measured the average magnetic force over all 11 positions on a rod. The second principal component, accounting for 2.57%, measured the difference between magnetic forces at the ends of a rod. The third principal component, accounting for 0.70%, measured the difference between the magnetic force at position #6 (in the middle of the rod) and the magnetic force at the first several positions.

No evidence indicated that the average magnetic force was affected by any of the design factors (p-value $= .97$, from the analysis of variance F-test for initial screening). Suggestive but inconclusive evidence supported the hypothesis that the difference between magnetic forces at the ends was affected by the level of current (p-value $= .051$, from the analysis of variance F-test). Finally, no evidence indicated that the variable measuring the difference between the magnetic force at position #6 and the first positions was affected by any of the treatment factors (p-value $= .82$, from the analysis of variance F-test for screening). A plot of the rods in the space of the principal components (Display 17.17) showed that the responses fell into four distinct groups. As the analyses of variance indicate, however, the groupings were not related to current, configuration, or metal. (When this analysis was presented to the researchers, a technician remembered that four different machines had been used in the study. No record of which rods were used in the different machines was kept, because the machines were thought to be identical. This appears to have been an incorrect assumption.)

Display 17.1 Magnetic force measurements at 11 positions on each of 44 rods

Rod #	L1	L2	L3	L4	L5	L6	L7	L8	L9	L10	L11	Metal	Configuration	Current
1	136	142	139	131	122	118	134	138	148	149	171	0	0	1
2	639	723	782	756	804	804	909	962	1042	1058	1022	250	0	1
3	673	709	709	719	682	681	759	912	1122	1121	900	500	0	1
4	471	501	521	519	528	521	540	523	548	525	513	0	1	2
5	578	617	650	625	632	634	677	695	733	735	747	250	1	2
6	120	124	110	110	101	104	126	140	158	164	182	500	1	2
7	542	578	583	598	587	618	654	696	737	726	752	0	0	3
8	597	632	631	655	659	699	792	890	972	960	953	250	0	3
9	615	641	634	649	648	699	809	892	975	974	968	500	0	3
10	487	486	485	489	488	487	491	490	500	483	471	0	1	4
11	113	127	124	116	107	110	131	143	161	163	171	250	1	4
12	563	579	586	613	635	689	806	907	1013	1025	973	500	1	4
13	436	456	455	465	469	468	472	471	486	464	457	0	1	1
14	492	527	546	565	583	602	658	692	748	756	744	250	1	1
15	109	119	113	101	94	100	122	136	155	160	178	500	1	1
16	589	589	568	584	589	626	652	663	699	710	726	0	0	2
17	623	624	608	630	657	710	795	865	960	971	961	250	0	2
18	492	497	482	487	487	497	492	492	492	482	471	500	0	2
19	452	462	466	481	485	484	493	487	496	479	473	0	1	3
20	665	670	670	681	681	702	791	864	954	980	969	250	1	3
21	147	151	143	130	120	123	145	160	179	187	209	500	1	3
22	466	471	466	477	487	482	482	477	487	482	471	0	0	4
23	568	600	601	612	617	618	661	687	735	752	773	250	0	4
24	581	613	607	628	639	628	692	723	792	803	803	0	1	1
25	634	701	706	721	742	736	846	924	1038	1048	1001	500	1	1
26	684	694	694	710	821	1236	963	857	936	952	921	0	0	2
27	112	120	115	106	96	99	121	138	156	162	179	250	0	2
28	608	623	634	655	665	712	802	933	1048	1027	995	500	0	2
29	477	466	471	482	487	482	482	487	492	477	461	0	1	3
30	545	561	572	599	604	615	658	721	774	764	770	250	1	3
31	584	588	582	612	611	621	667	697	733	732	752	500	1	3
32	100	118	115	109	101	106	130	145	162	164	172	0	0	4
33	573	584	605	626	652	642	684	694	721	705	721	500	0	4
34	589	603	627	656	681	689	782	869	967	970	936	0	0	1
35	521	551	571	596	616	647	772	861	981	964	963	250	0	1
36	447	446	445	465	469	457	472	476	481	464	463	500	0	1
37	494	478	472	493	492	481	491	491	490	463	473	0	1	2
38	99	112	109	102	94	99	122	137	153	157	168	250	1	2
39	587	617	616	631	640	676	769	878	987	986	980	500	1	2
40	471	471	465	480	480	484	489	489	488	472	471	0	0	3
41	542	562	561	581	585	610	651	692	728	722	747	250	0	3
42	545	545	545	566	587	592	618	649	697	686	702	0	1	4
43	692	687	703	720	747	768	843	891	992	1014	982	250	1	4
44	655	696	679	694	734	864	837	847	924	923	964	500	1	4
AVE:	479.8	496.8	499.2	509.7	518.3	541.4	579.1	613.2	664.5	662.8	656.3			
SD:	185.2	191.5	196.5	204.0	215.4	245.0	249.8	271.6	307.4	309.9	294.1			

17.1.2 Love and Marriage—An Observational Study

Thirty couples participated in a study of love and marriage, conducted by E. Hatfield. (Data reported in R. A. Johnson and D. W. Wichern, *Applied Multivariate Statistical Analysis*, 2d ed. (Englewood Cliffs, N.J.: Prentice-Hall, 1988), Chapter 6.) Wives and husbands responded separately to four questions:

1. What is the level of passionate love you feel for your spouse?
2. What is the level of passionate love your spouse feels for you?
3. What is the level of compassionate love you feel for your spouse?
4. What is the level of compassionate love your spouse feels for you?

Each response was recorded on the following five-point scale: 1= None, 2 = Very Little, 3 = Some, 4 = A Great Deal, and 5 = A Tremendous Amount. The responses appear in Display 17.2.

The average responses of subjects in the study suggest that they are generally happy in their marriages. Can anything be said about the relationships between categories of responses? Is there a relationship between husbands' responses and wives' responses? (A "relationship" in this context would mean that the husbands of wives who scored above average on some combination of their responses would themselves score consistently above average on a combination of their own responses.) Is there a relationship between responses concerning passionate love and responses concerning compassionate love? Is there a relationship between love toward spouse and perception of spouse's love in return? What is the nature of these relationships?

Summary of Statistical Findings

There is no evidence of a correlation between the husbands' responses and the wives' responses (p-value = .45, from a canonical correlation analysis). There is also no evidence of a correlation between responses to questions about passionate love and responses to questions about compassionate love (p-value = .43, from a canonical correlation analysis). There is, however, convincing evidence of a correlation between husbands' and wives' feelings of love and their *perceptions* of the love their spouses feel for them (p-value < .0001 from a canonical correlation analysis). Specifically, three points are noteworthy:

1. The degree of a husband's compassionate love for his wife is highly correlated with the degree of compassionate love he believes his wife has for him (estimated correlation = .93, p-value = .0001).
2. The degree of a wife's compassionate love for her husband is correlated with the degree of compassionate love she believes her husband has for her (estimated correlation = .81, p-value = .0001).
3. The degree of a husband's passionate love for his wife is correlated with the degree of perceived passionate love he receives from her (estimated correlation = .65, p-value = .0001).

Display 17.2 Responses to questions about love with partner

	Level of love you feel for your spouse				Level of love your spouse feels for you			
Type of Love:	Passionate		Compassionate		Passionate		Compassionate	
Respondent:	Husband	Wife	Husband	Wife	Husband	Wife	Husband	Wife
Couple # 1	2	4	5	5	3	4	5	5
2	5	4	4	5	5	5	4	5
3	4	4	5	5	5	4	5	5
4	4	4	4	5	3	5	4	5
5	3	4	5	5	3	4	5	5
6	3	3	4	4	3	3	5	4
7	3	4	4	5	4	3	4	4
8	4	3	5	5	4	4	5	5
9	4	4	5	5	5	4	5	4
10	4	3	3	4	4	4	3	4
11	4	4	5	5	4	5	5	5
12	5	5	4	5	5	5	4	5
13	4	4	4	5	4	4	4	5
14	4	4	5	4	3	4	5	4
15	4	4	5	5	4	4	5	5
16	3	3	4	4	3	4	5	4
17	4	5	4	5	5	5	4	5
18	5	4	5	4	5	5	5	4
19	5	3	4	4	5	4	4	4
20	4	5	4	4	4	3	4	4
21	4	5	4	4	4	3	4	4
22	4	4	4	4	4	5	4	4
23	3	2	5	5	4	5	5	5
24	5	3	5	5	3	4	5	5
25	5	4	3	5	5	3	3	5
26	3	4	4	4	3	4	4	4
27	4	4	4	5	4	4	4	5
28	3	3	5	4	3	4	5	4
29	4	4	3	5	4	4	3	4
30	4	4	5	5	4	4	5	5
Average:	3.90	3.83	4.33	4.63	3.97	4.10	4.40	4.53
Sample SD:	0.76	0.70	0.66	0.49	0.76	0.66	0.67	0.51
Variable Abbr:	HP	WP	HC	WC	PW	PH	CW	CH

On the other hand, there is no evidence that the degree of a wife's passionate love for her husband is correlated with the perceived passionate love she receives from him (estimated correlation = −.04, p-value = .85).

Scope of Inference

The interesting aspect of this is that both husbands and wives seem to think their spouses have compassionate and passionate feelings similar to their own, when in fact there is no evidence of actual similarity. (Differences in the levels of love for spouse were related only to differences in the *perceived* levels of love returned—not to differences in the *actual* levels returned.) The apparent implication that love is a creation of one's own mind should be tempered by two considerations: the usual reservations about causal interpretations from observational studies; and the roughness of the p-values, whose validity is tied to a questionable assumption of normality. Since these subjects were not a random sample, it is difficult to justify the inference that a similar relationship holds in a more general population.

17.2 Linear Combinations of Variables

It is desirable to replace the multivariate responses—the eleven-dimensional measurement on the printer rods or the eight-dimensional response for each couple—with some summarizing quantities of smaller dimension. Here, however, unlike in the last chapter, no obvious summary suggests itself. This chapter presents two methods for finding linear combinations suggested by the data. These linear combinations may in turn suggest meaningful reductions of the multivariate responses, which can then be analyzed in accordance with Chapter 16 techniques.

A linear combination of several variables is a new variable. It is constructed from other variables to represent a specific feature. Each subject is assigned a specific numerical value for the feature, and different subjects generally have different values. For example, the percentage correct of objects learned at 8, 12, and 16 weeks in the memory study of Section 16.1.1 were averaged to form a single long-term memory measurement.

Formula

A linear combination L of variables $Y_1, Y_2, \ldots, Y_q$ is

$$L = C_1 Y_1 + C_2 Y_2 + \cdots C_q Y_q.$$

This formula is to be used for each subject. The values of the Y-variables for a specific subject go into the formula; each variable value is multiplied by its corresponding coefficient (C), and the products are added. The result is one number, which is the value of the linear combination for that subject. The process is repeated, using the Y's unique to each subject but using the same coefficients throughout.

The coefficients—$C_1, C_2, \ldots, C_q$—determine what important feature the linear combination measures. Here are some examples:

1. $C_1 = C_2 = \cdots = C_q = 1$ gives the sum of all responses.
2. $C_1 = C_2 = \cdots = C_q = 1/q$ gives the average of all responses.
3. $C_1 = 1, C_2 = -1, C_3 = C_4 = \cdots = C_q = 0$ gives the difference between the first two responses.
4. $C_1 = C_2 = 1/2, C_3 = C_4 = C_5 = -1/3$ gives the difference between the average of the first two responses and the average of the next three responses.

See Chapter 6 for other examples.

Graphical Representation

When thinking about a linear combination, try to visualize a scatterplot of the original variables. That may entail visualizing a seven-dimensional point cloud—a difficult feat. To get the idea, consider first a case involving just two responses, Y_1 and Y_2.

Determining the new variable $L = C_1Y_1 + C_2Y_2$ is like taking a train ride on a straight-line track in the (Y_1, Y_2) plane that goes through the origin $(0, 0)$ and through the point (C_1, C_2) whose coordinates are the linear combination coefficients (see Display 17.3). As you travel, you can only see out the side windows, glimpsing objects as they pass your view (which is perpendicular to your line of travel). When a subject (or point) is passed, you record your location along the track; that location is the value of the linear combination for that subject. In this way, the point cloud is reduced to a set of locations along the track that constitute the linear combination variable.

Display 17.3   The train ride through the coefficient point to determine the value of a linear combination

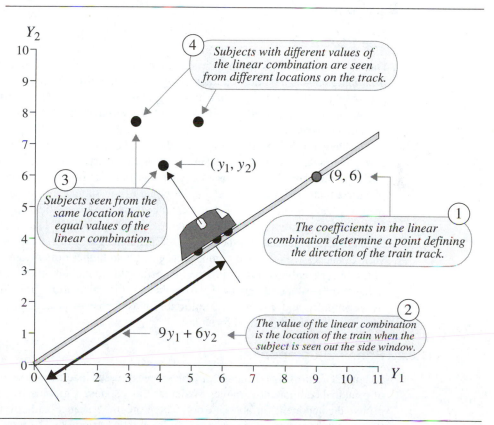

This view of a linear combination helps clarify what the resulting variable is like and what role it plays. Often visualizing a desirable direction for the track is easy. Any point on the line can be used to determine the coefficients of the linear combination. The first coefficient is the first coordinate of the point on the line and the second coefficient is the second coordinate. The set of coefficients is not unique, since all points on the line will define the same direction.

17.3   Principal Components Analysis

When researchers change experimental conditions, they expect to see changes in the system responses. The engineers who investigated the printer rods of Section 17.1.1 changed the current, configuration, and metal to see how the magnetic forces along the rod would change. Naturally, the forces changed at all 11 positions, so visualizing what was happening would have involved thinking about an eleven-dimensional point cloud. If one direction through the cloud could be chosen to view the changes, a natural choice would be the direction along which variation was most dramatic. This is the direction of the first principal component of variation.

17.3.1   The PCA Train

Display 17.4 illustrates the principal components solution in two dimensions. The *first principal component* follows a track along the main axis of the point cloud, giving values shown by the black dots along the axis. The axis perpendicular to the principal component axis is the direction of least variation, and the variable that follows it is called the *last principal component*.

In three dimensions, the passenger views a full two-dimensional plane from any location along the track. As the train goes along the first PCA track, imagine that the points adhere to the plane when the train passes. After the train has passed the cloud completely, the plane contains a two-dimensional point cloud of the subject's positions, much like a flat map of a geographic surface. Another train track can be established along the direction of maximum variation on the plane, giving the *second principal component* variable. The third principal component is the last one, perpendicular to both the first and the second. This view of successive principal components extends to higher dimensions. The computation of the linear combinations giving the first principle component, the second one, and so on is beyond the scope of this book. Nonetheless, the calculations are easily accomplished by a computer, and a user can comfortably make sense of principle components without understanding the numerical formulas.

17.3.2   Principal Components

In a full PCA, the set of response variables is reexpressed in terms of a set of an equal number of principal component variables. Whereas the response variables are intercorrelated, the principal component variables are not. Therefore, the variance-and-covariance structure of the principal components is described fully by their variances. The total variance—that is, the sum of their variances—is the same as the total variance of the original variables. For

Display 17.4 The PCA train, illustrating the principal component directions through a point cloud in two dimensions

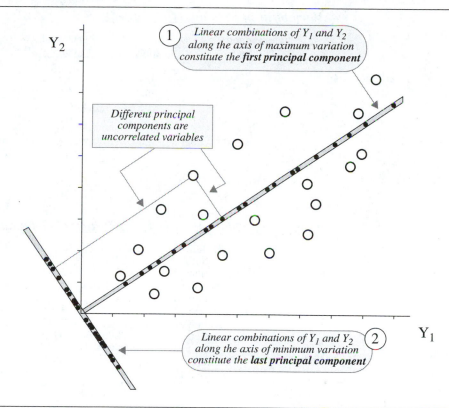

this reason, the variances of the principal components are usually expressed as percentages of the total variation.

Each principal component variable is a linear combination of the original variables. Typically, a computer analysis provides directional information by the sets of coefficients and variational information as either individual variations or percentages of the total. All principal component variances and coefficients in the linear combinations leading to the first three principal components for the magnetic force example appear in Display 17.5.

The first principal component accounts for 96.46% of the variation in all 11 responses. The first three principal components together account for 99.73%, making it safe to conclude that virtually all the variation in the 11 dimensions is explained by a three-dimensional summary.

Interpreting Coefficients

An essential part of the analysis consists of making sense out of the coefficients for each of the principal components that are examined. Two general facts about these coefficients

Display 17.5   Summary for principal components analysis of magnetic force responses

a. Variability of Principal Components

Component Number	Variance	Percent of Variance	Cumulative Percentage
1	646,948	96.46	96.46
2	17,247	2.57	99.03
3	4,684	0.70	99.73
4	901	0.13	99.86
5	448	0.07	99.93
6	240	0.04	99.97
7	99	0.01	99.98
8	63	0.01	99.99
9	28	0.00	100.00
10	17	0.00	100.00
11	13	0.00	100.00

b. Coefficient Table for First Three Principal Components

Variable	PC #1	PC #2	PC #3
Location 1	.2228	−.3041	−.2588
Location 2	.2327	−.2653	−.2699
Location 3	.2390	−.2596	−.2909
Location 4	.2486	−.2639	−.2926
Location 5	.2627	−.3068	−.0670
Location 6	.2901	−.3875	.7913
Location 7	.3095	−.0832	.2005
Location 8	.3360	.1785	−.0787
Location 9	.3769	.3697	−.0546
Location 10	.3792	.4044	−.0002
Location 11	.3601	.3419	.0995

identified by PCA are as follows:

1. *The coefficients are normalized so that the sum of their squares equals one.* The coefficients describe a direction, and any coefficients that specify the same directions would work. This normalization, however, guarantees that the total variance of the principal components equals that of the original variables.
2. *The coefficients for successive principal components are constrained by the requirement that the variables be uncorrelated with previous components.* More will be said about this in Section 17.3.3.

First Principal Component in the Magnetic Force Example

Examining the coefficients facilitates understanding what each principal component represents. The first principal component (PC#1) assigns nearly equal weight to the magnetic forces developed at each of the 11 locations. The principal component variable is roughly the average magnetic force over all 11 locations. Any rod that has higher than average magnetic forces at all locations will have a PC#1 score that is above average; a rod that has lower than average magnetic forces at all locations will have a PC#1 score that is below average. Rods that are about average at all locations or have a mixture of above-average and below-average locations will have PC#1 scores near average.

Second Principal Component in the Magnetic Force Example

Coefficients for the second principal component (PC#2) are positive for locations 9–11 and negative for locations 1–3. This suggests that the component represents a contrast between the average magnetic forces developed at opposite ends of the rod. (The coefficients are also negative for positions 4–6. The principle component seems to be contrasting the two ends of the rod, and making a substitute combination contrasting the first three positions with the last three positions is particularly convenient.) A rod will have a higher-than-average PC#2 score if the magnetic force at locations 9–11 exceeds the force at locations 1–3 by a greater-than-average amount.

Third Principal Component in the Magnetic Force Example

The third principal component (PC#3) assigns a large positive weight to the magnetic force developed at location 6 and small negative weights to most other locations. This component measures a pattern observed in a single rod (#26; see Display 17.17) that had a large peak in the magnetic force at location 6.

17.3.3 Variables Suggested by PCA

Principal components should seldom be used directly as responses in further analysis. This has to do with repeatability: if the study is repeated, the same coefficients would never appear in the analysis. For this reason, it is usually better to use easily comprehended linear combinations that capture the features identified by the principal components. This requires thinking about what the principal components really measure.

For further analysis, replace PC#1 by the average of all responses; replace PC#2 by the difference between the average forces at positions 9–11 and at positions 1–3; replace PC#3 by the difference between the force at position 6 and the average force at the first two positions. These replacements are suggested by the coefficients, but they should be checked to verify that together they measure the same variational features as the first three principal components. To this end, Display 17.6 shows a scatterplot of the first principal component against its proposed replacement. The principal component appears to be a nearly exact linear function of the average.

Constraint #2 in Section 17.3.2 requires the second principal component to be uncorrelated with the first and the third to be uncorrelated with the first two. The surrogates are not so constrained, and the correlation that exists will cloud the issue if a scatterplot is constructed for PC#2 versus its proposed surrogate. The method around this is to determine the part of the surrogate (the "end-contrast" summary) that is constrained to be uncorrelated with the first surrogate. Accomplishing this involves regressing the end-contrast variable against the rod average variable and retaining the residuals. Similarly, taking the third surrogate (the "peak 6" summary) and retaining its residuals from a regression on the first two surrogates produces the result shown in Display 17.7, where close agreement is evident between the surrogates and the principal components they replace.

To complete the analysis, the three surrogates are used as response variables, with the design factors as explanatory variables. The analysis of the three-dimensional multivariate response proceeds in accordance with the methods of Chapter 16. Since the three components are nearly uncorrelated, however, there is little need to adjust confidence intervals

---

Display 17.6   Scatterplot of PC#1 against the average magnetic force across all positions

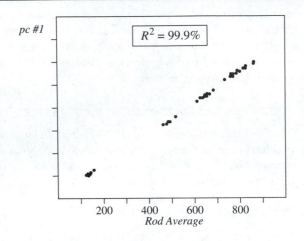

Display 17.7   Scatterplots of PC#2 and PC#3 against residuals of their surrogates on lower-order surrogates

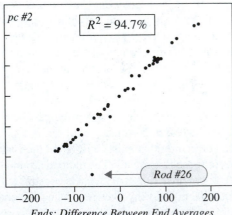
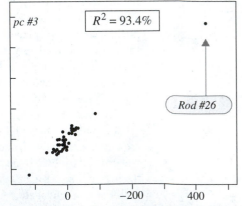

Ends: Difference Between End Averages
(residual from rod average)

Peak: Difference Between L6 and Initial Positions
(residual from rod average and end comparison)

to account for correlations. Therefore, the analysis can be accomplished by three separate univariate analyses that examine the effects of the design variables on each of the three linear combinations of rod measurements.

17.3.4   Scatterplots in Principal Component Space

Although principal components are uncorrelated, their scatterplots sometimes reveal important structures in the data other than linear correlation. Display 17.8 shows a matrix of scatterplots for the three principal components in the magnetic forces example. The first two principal components reveal that there are four distinct groups of rods and that the patterns of within-group variation are somewhat different. The latter revelation suggests that some transformation of the data may be appropriate prior to PCA (a suggestion not pursued here). The plots show that variation in the third principal component consists primarily of the difference between rod #26 and all the rest.

17.3.5   The Factor Analysis Model and Principal Components Analysis

Factor analysis provides a useful conceptual model for interpreting PCA. Rather than thinking of the principal components as being extracted or distilled from the original set of variables, imagine them as real underlying factors that are not directly measurable but nonetheless contribute to all measurements. For example, scores on tests and homework are all related to a person's general subject mastery; subject mastery, therefore, is the factor. This cannot be measured exactly, so the teacher instead takes the exam scores, which are strongly influenced by the factor. By thinking about what the underlying factors in a problem might be, one often gains insight into what the principal components are measuring.

17.3.6   PCA Usage

Scale Sensitivity

Principal component analysis is scale-sensitive. If Y_j is measured in centimeters, then $10Y_j$ is the same variable measured in millimeters. The variance of $10Y_j$ is 100 times the variance in Y_j. Thus, changing Y_j from centimeters to millimeters while leaving other variables unchanged causes PCA to give more attention to it.

Some researchers use PCA to summarize sets of variables that have very different scales—liters, millimeters, kg/sec, megaparsecs—but they *standardize* all the variables beforehand by dividing them by their sample standard deviations. This does not evade the necessity of determining the meaning of a linear combination of, for example, a length measurement and a momentum measurement.

PCA on Explanatory Variables Prior to Regression

PCA is sometimes recommended as a way of avoiding problems of multicollinearity in multiple linear regression (Section 12.2 on page 332). Many interrelated explanatory variables may be grouped by general theme—socio-economic, weather, geographic, or physiologic, for example. PCA is then used to select a small number of uncorrelated variables to

Display 17.8 Matrix of scatterplots for the first three principal components of the magnetic force measurements

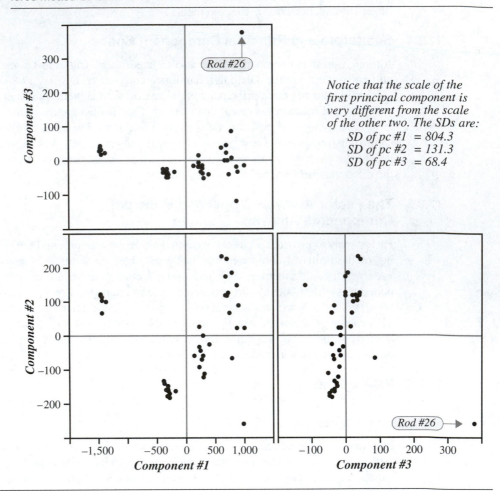

Notice that the scale of the first principal component is very different from the scale of the other two. The SDs are:
SD of pc #1 = 804.3
SD of pc #2 = 131.3
SD of pc #3 = 68.4

represent each group in the overall regression. Unfortunately, by regressing a response on the first or first few principle components, a researcher may overlook a linear combination of explanatory variables that is highly correlated with the response, if that linear combination lacks high variability.

Recommended Usage

Using PCA to find low-dimensional representations of very diverse variables can run into many problems. PCA works best in situations where all variables are measurements of the same quantity, but at different times or locations, as in the magnetic force study. It may also be used in reducing the dimension of a multivariate response, in reducing the dimension of a set of explanatory variables, and in summarizing a single multivariate sample.

17.4 **Canonical Correlations Analysis**

Canonical correlations analysis (CCA) is used when multivariate measurements are partitioned into two sets. CCA finds linear combinations in the two sets—L_1 and L_2, say—that have the largest possible correlation. In this sense, CCA is a correlation analysis not between two variables but between two sets of variables. In some problems, one set consists of response variables $Y_1, Y_2, \ldots, Y_q$, and the other consists of explanatory variables $X_1, X_2, \ldots, X_p$; but as with ordinary correlation, one need not make explicit "response" and "explanatory" designations. The form of the linear combinations will, hopefully, reveal some meaningful relationship between the two sets of variables.

17.4.1 **Canonical Variables**

One question asked about the love and marriage data was whether they revealed a relationship between love toward one's spouse and perceived love returned by the spouse. For each couple in the data set, there are four variables pertaining to love toward spouse and four pertaining to perceived love from spouse; therefore, the question can be investigated through the canonical correlations between these two sets. CCA selects one linear combination from each set to form a pair of *canonical variables.* The coefficients determining the pair are chosen to give maximum correlation between the two variables in the pair.

Next, CCA selects a second pair of linear combinations, one from each set. The variables in this pair have the largest correlation possible, given that they are not correlated to the variables already chosen to form the first pair. CCA continues to construct variable pairs in this way until the number of pairs reaches the smaller of the number of variables in the two sets.

The sets of variables regarding love toward spouse and perceived love from spouse were labeled as "To" and "From" respectively. Canonical correlations analysis provides four linear combinations from each set. The CCA computer output appears in Display 17.9.

According to the table of coefficients in Display 17.9, the first pair of linear combinations (referred to as the *first pair of canonical variables*) is

$$\text{To}_1 = 0.118(HP) - 0.951(HC) + 0.113(WP) + 0.130(WC)$$
$$\text{From}_1 = 0.040(PW) - 0.982(CW) - 0.068(PH) + 0.039(CH)$$

(see Display 17.9 for the meanings of the variable abbreviations).

The second pair of canonical variables is

$$\text{To}_2 = 0.375(HP) + 0.136(HC) + 0.042(WP) + 0.866(WC)$$
$$\text{From}_2 = 0.404(PW) + 0.021(CW) - 0.038(PH) + 0.876(CH).$$

Four such pairs of linear combinations are provided in the output of Display 17.9.

Interpreting the Statistical Tests

The upper table in Display 17.9 shows that the correlation between To_1 and From_1 is 0.9507—the *first canonical correlation.* The correlation between To_2 and From_2 is 0.8666; and so on.

Display 17.9   Canonical variables and their correlations for the relationships between responses concerning love given and responses concerning perceived love returned

Estimates of Canonical Correlations and their Statistical Significance

Pair	Canonical Correlation	Chi-square	d.f.	p-Value
1	0.9507	100.83	16	.0000
2	0.8666	43.46	9	.0000
3	0.5572	9.41	4	.0516
4	0.1107	0.30	1	.5827

Coefficients for Canonical Variables

Love Toward Spouse ("To")	Pair 1	Pair 2	Pair 3	Pair 4
HP: *Husband's passionate*	0.118	0.375	−0.898	−0.369
HC: *Husband's compassionate*	−0.951	0.136	−0.387	0.207
WP: *Wife's passionate*	0.113	0.042	−0.165	1.025
WC: *Wife's compassionate*	0.130	0.866	0.538	−0.102

Perceived Love in Return ("From")	Pair 1	Pair 2	Pair 3	Pair 4
PW: *Passionate love from wife*	0.040	0.404	−0.904	0.475
CW: *Compassionate love from wife*	−0.982	0.021	−0.335	0.333
PH: *Passionate love from husband*	−0.068	−0.038	−0.282	−1.059
CH: *Compassionate love from husband*	0.039	0.876	0.621	0.135

The chi-square statistic for pair j is a test statistic for the hypothesis that the correlation between the jth pair is zero and that the correlations between *all subsequent pairs* are also zero. The top row, therefore, contains a chi-square statistic for the test that *all four* canonical correlations are zero. It is not a test of correlation for the first pair alone, although strong evidence that some correlation exists points directly to at least the first pair. The second row's chi-square is for the hypothesis that the *last three* canonical correlations are zero. The third chi-square tests for the last two canonical correlations, and the final chi-square tests the fourth canonical correlation. The p-value for each hypothesis is found as the proportion of values from a chi-squared distribution with the appropriate degrees of freedom (see Appendix A.3) that exceed the test statistic. The degrees of freedom are presented in the Display 17.9 output; their derivations are not discussed here.

To determine how many correlations the data support as being nonzero, see where the break occurs between significantly large chi-square statistics to nonsignificant ones. In this example, overwhelming evidence supports two canonical correlations and suggestive but not conclusive evidence supports a third.

If the first test finds no significant evidence of any canonical correlation (as is the case when the variables are grouped as Husband's and Wife's or as Passionate and Compassionate), CCA permits the strong conclusion that no evidence exists of any correlations between linear combinations of variables in the sets.

17.4.2 Variables Suggested by CCA

Once again, it is desirable to select *meaningful* linear combinations that capture the essence of the canonical variables. Since CCA is not scale-sensitive, one has the option of inspecting coefficients of the original variables or of their standardized versions. Thus, a CCA combination may be written in either of the following forms:

$$\text{Canonical variable } = \sum (C_j \times Y_j) = \sum [(C_j s_j) \times (Y_j / s_j)].$$

Here Y_j is an original scaled variable, and s_j is its sample standard deviation. Then (Y_j / s_j) is the standardized variable, with standard deviation $= 1$. The unstandardized coefficient is C_j, and the standardized coefficient is $C_j s_j$. The magnitude of the standardized coefficient is the amount of change in the canonical variable that is associated with a one-SD change in the variable (other variables held constant), so coefficients of the standardized variables are best for identifying which variables contribute most to a linear combination.

Coefficients are constrained in two ways:

1. *Canonical variables have standard deviations equal to one* (1.0).
2. *Canonical variables in one pair are uncorrelated with the canonical variables in pairs previously selected.*

It becomes increasingly difficult to interpret succeeding pairs, because the second constraint places heavy conditions on the coefficients. To judge whether a particular surrogate linear combination adequately approximates a canonical variable, look at the relationship between the canonical variable and the *residuals* from the regression of that surrogate on the previously determined surrogates (for the lower-order canonical variables) in the same set.

17.4.3 Love and Marriage Example

Consider the coefficients for the first canonical variable pair for direct versus perceived feelings. The coefficients in Display 17.9 are for the nonstandardized variables, but this does not cause a problem because all variables are measured on the same scales. In both canonical variables, the magnitude of the coefficient of the husband's compassionate love responses is very large relative to the magnitudes of the other coefficients; the coefficient of HC dominates in To_1, and the coefficient of CW dominates in $From_1$. To verify that these canonical variables primarily measure the husbands' compassionate love responses, Display 17.10 shows scatterplots of the canonical variables along with the original variables HC and CW.

Display 17.10(A) shows the pair of computer-selected canonical variables having the highest correlation. Part (B) illustrates that the canonical variable $From_1$ is indeed measuring the same thing as the variable CW (although the sign of the coefficient of CW in $From_1$ produces an inverse relationship). Similarly, part (C) shows that To_1 closely corresponds to HC. Finally, part (D) illustrates the scatter of HC against CW, showing that CCA has identified this as the strongest relationship between responses on direct and perceived love questions.

Display 17.10 Scatterplots of the first canonical variables for the From and To sets of responses, and their proposed surrogates HC (husband's compassionate love) and CW (husband's perceived compassionate love from wife)

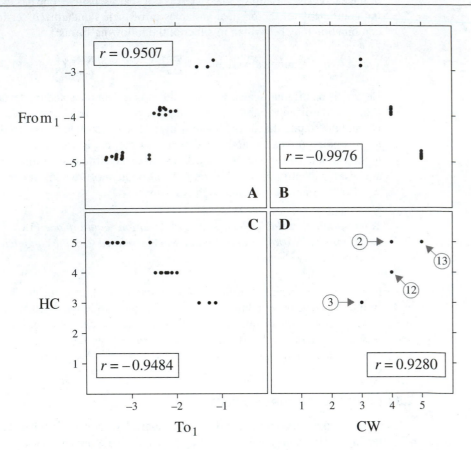

Because CCA identified canonical variables as being so closely related to single original variables, it is sensible to report the original variables, HC and CW, along with the correlation, +0.93. To explain the results of the CCA, reporting the natural pair (HC, CW) and its internal correlation is preferable to reporting the canonical variables. Although CCA was used as an exploratory tool, the summary of statistical findings reports the results for the ordinary correlation analysis between HC and CW: their sample correlation coefficient is .93. Overwhelming evidence indicates that the population correlation is nonzero (p-value < .001; see Section 7.5.4 on page 189 for tests of correlation).

The CCA also produced convincing evidence of a correlation between a second pair of canonical variables. Inspection of coefficients for pair 2 in Display 17.9 suggests that these linear combinations might simply be picking up WC from the To set and CH from the

From set; that is, the wife's compassionate love is correlated with the compassionate love she perceives her husband to return. A scatterplot of the residuals (from the regression of WC on HC) versus the second canonical variable from the To set, together with a scatterplot of the residuals (from the regression of CH on CW) versus the second canonical variable in the From set, indicates that these components capture most of what is explained by the second canonical variables (plots not shown here). Their correlation analysis is reported in the summary of statistical findings, as is the correlation analysis between HP and PW, which is suggested by the third canonical variables.

17.5   Introduction to Other Multivariate Tools

The following subsections sketch several other popular multivariate tools. The brief discussions merely convey the purposes and interpretations of the techniques. References to more detailed explanations are provided at the end of the chapter.

17.5.1   Discriminant Function Analysis (DFA)

The objective of *discriminant function analysis* (DFA) is to find a rule involving multivariate responses that best discriminates the subjects in different groups. This problem can be approached in many ways, and vast literature surrounds it. This section describes one solution—Fisher's linear discriminant function—featured by most statistical computer packages. Logistic regression, presented in Chapter 20, provides a more flexible approach to meet the same purpose.

The DFA Train

The hypothetical scatterplot of Display 17.11 contains two distinct groups of subjects. In such a situation, a variable that best enables one to distinguish group membership is called a *discriminant function*. Taking the train along the track indicated in the display, one passes by all the subjects in group #1 before coming to any subject from group #2. Histograms of the locations along that track for the two groups do not overlap. Therefore a new point, whose group membership is not known, could be classified with considerable confidence based on its location along that axis.

There would be overlap of the two groups in each of the separate variables in Display 17.11. Applying *t*-tools to test for group differences on each variable separately could easily fail to detect the distinction between the groups.

Fisher's Linear Discriminant Function

The approach taken by R. A. Fisher was to find the linear combination that maximizes the one-way analysis of variance F-statistic. (When two groups are involved, the t-statistic for comparing the values of the linear combination in the two groups is maximized.) Most computer software packages provide this solution. Relevant output consists of the coefficients of the discriminant functions, the discriminant function scores for each subject, measures of the percentage of cases correctly classified by the analysis, and an overall measure of

Display 17.11   The discriminant function analysis (DFA) train travels a route that best separates two clusters of subjects

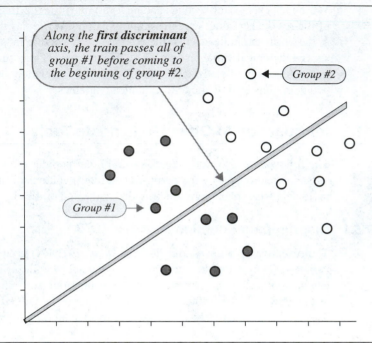

group differences. The maximum number of discriminant functions available in a general problem is the smaller of the number of variables and one less than the number of groups.

A *confusion matrix* is an array showing the proportions of subjects in the actual group that are classified in the predicted groups by the rule. A confusion matrix applies the rule to the same data used in devising the rule, so the proportions tend to be overly optimistic as estimates of the reliability of future predictions. Consequently, with large data sets, it is advisable to employ *cross-validation*, where subjects are divided into a *training set* used to devise the rule and a cross-validation set used to estimate the rule's reliability.

17.5.2 Cluster Analysis

The objective of *cluster analysis* is to classify sampled subjects into homogeneous groups, or *clusters*. The nature of the clusters is not known in advance. The data are used to construct distances (d) separating subjects, much like the city-to-city distance tables that are found on most maps. Clustering consists of grouping subjects that are close to each other and separating these from subjects that are more distant. Numerous clustering algorithms are available. Cluster analysis does not yield linear combinations of variables to use in classification, nor does it ordinate the subjects in a coordinate system.

Example

Diet analysis examines who eats what. Suppose that researchers identify N consumers and M food categories. A field study measures the proportion p_{ij} of the diet for consumer i that falls in the jth food category. Two consumers have identical diets if their proportions match; thus, a measure of the squared distance between diets might be the sum over the different food categories of the squared differences in proportions, $d_{ik}^2 = 100 * \Sigma (p_{ij} - p_{kj})^2$. Display 17.12 shows such an analysis of the diets of nine shorebirds foraging in the intertidal zone of the U.S. Pacific coast. (Data from F. L. Ramsey and C. P. Marsh, "Diet Dissimilarity," *Biometrics* 40 (1984): 707–15). The left half of the display shows the distances between pairs of diets, and the right half shows a *dendogram* linking the birds into groups. In this diagram, the two species of birds represented—Surfbird (SB) and Black Turnstone (BT)—were separated into distinct clusters at a distance of approximately 120 units.

Display 17.12   Dietary distances of nine consumers and a resulting cluster analysis

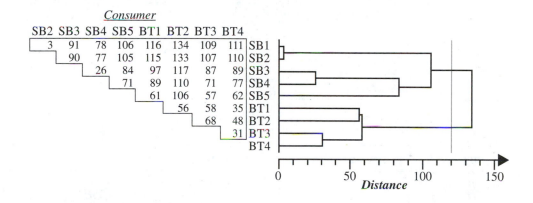

17.5.3 Multidimensional Scaling

Multidimensional scaling (MDS) is a tool for arranging all subjects on a low-dimension map. As in cluster analysis, a measure of the pairwise distance between subjects is available or is constructed from the multivariate responses. The objective is to assign each subject a location in, say, two dimensions, so that the ordinary straight-line distances between their locations match the distances in the data as closely as possible. MDS provides a map in which clustering often appears and in which axes distinguish subjects at extreme distances apart.

Example

Cox and Brandwood examined sentence endings in seven works of Plato—*Republic, Laws, Critias, Philebus, Politicus, Sophist,* and *Timaeus*—with the objective of determining the chronological order of their writing. A sentence ending consists of the last five syllables, and classifying each syllable as long or short yields a set of 32 possible ending types. The researchers counted the proportions of endings of each type based on a large selection from each work. With p_{ij} representing the proportion of type j endings in work i, one distance between works i and k is defined by $d_{ik} = \sum |p_{ij} - p_{kj}|$ as shown in Display 17.13. (Data from D. R. Cox and L. Brandwood, "On a Discriminatory Problem Connected with the Works of Plato," *Journal of the Royal Statistical Society* B21 (1959): 195–200.) A one-dimensional MDS consists of assigning a time t_i to each work in such a way that the time differences closely match the distances $\{d_{ik}\}$. A two-dimensional MDS expands on that to describe another dimension of the differences between works. The MDS solution appears in Display 17.14.

Display 17.13   Percentage discrepancies between sentence endings in seven works of Plato

	Laws	Critias	Philebus	Politicus	Sophist	Timaeus
Republic	55.2	44.7	60.0	52.8	37.8	39.1
Laws		48.7	25.6	33.6	41.6	54.3
Critias			47.7	52.3	43.9	44.2
Philebus				31.2	44.0	56.1
Politicus					32.8	48.9
Sophist						27.1

Display 17.14   Two-dimensional MDS map of Plato's works

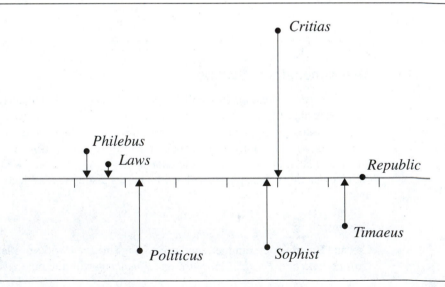

17.5.4   Correspondence Analysis

Correspondence analysis is a method of ordinating categorical attributes. Each attribute has several levels that are not necessarily ordered. The objective is to assign a numerical value to each level of each factor in such a way that the correlations between the numerical scales are maximized. The method emerged in the 1930s and is a popular tool in botany and plant ecology.

Example

R. A. Fisher introduced the method in connection with measurements of hair and eye color in a sample of school children from Caithness, Scotland. Display 17.15 shows the array of counts. (Data from R. A. Fisher, "The Precision of Discriminant Functions," *Annals of Eugenics* 10 (1940): 422–29.) The correspondence analysis results are the eye-color scores in the right margin and the hair-color scores across the bottom. Assigning children the scores indicated by their eye and hair color results in two variables that have correlation 0.4464, the largest possible correlation achievable by such a scoring procedure.

Display 17.15   Hair color and eye color of school children in Caithness, Scotland

		Hair Color					**Total**	**Score**
Eye Color	*Fair*	*Red*	*Medium*	*Dark*	*Black*			
Blue	326	38	241	110	3		718	−0.897
Light	688	116	584	188	4		1,580	−0.987
Medium	343	84	909	412	26		1,774	+0.075
Dark	98	48	403	681	85		1,315	+1.574
Total:	1,455	286	2,137	1,391	118		5,387	
Score:	−1.219	−0.523	−0.094	+1.319	+2.452			$r = 0.4464$

17.5.5   PCA and Empirical Orthogonal Functions (EOFs)

Display 17.16 shows several magnetic force measurement profiles over the 11 positions, along with the profile of rod averages. Measurements at different locations are highly correlated. They all tend to be higher than average or lower than average at all positions on a given rod.

A model for the profiles in Display 17.16 might be, approximately,

$$\text{Force on rod } i \text{ at location } j \; = \; \text{Mean at location } j + \beta_i f(j).$$

Here, $f(j)$ is a function that describes a pattern of variation common in varying degrees to all rods, and β_i specifies the strength of the pattern in the ith rod.

If, for example, $f(j) = 1$ at all positions, the model would say that all rods have profiles that approximately parallel the average profile. The value of β_i, then, would measure the

Display 17.16 Magnetic force profiles for a sample of rods, and the average profile for all rods

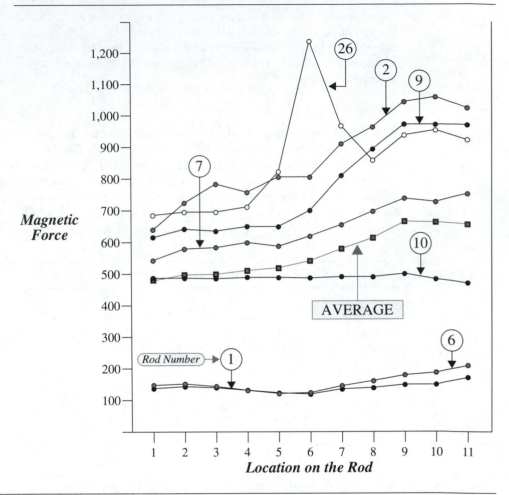

distance between the profile of rod i and the average profile. As another example, if $f(j) = (j - 6)$ for each j, departures from the average would be minimal in the center and would increase or decrease linearly away from the center. Magnetic force on a rod with a large positive β_i would rise much more sharply than on the average rod; conversely, magnetic force on a rod with a negative β_i would either rise less sharply or decrease.

If the choice of a pattern for $f(j)$ is known, the problem is one of simple linear regression analysis (separately for each rod) to determine the rod constants. If the choice is not obvious, the data can help the examiner choose the pattern. If the problem is posed as one of finding the pattern that most closely matches the variation in rod profiles—in the sense that the variability in the β_i's is largest—the solution is as follows: $f(j)$ is the jth coefficient in the linear combination forming the first principal component; furthermore, β_i is just the value of the first principal component for rod i.

An extension of the model for the profiles may now be considered. If the difference between the profile for rod i and the prediction from the average and the first principal component has a similar form—namely, $\beta_{2i} f_2(j)$—involving a constant multiple of another pattern, the best choice is to have β_{2i} represent the scores and $f_2(j)$ the coefficients from the second principal component; and so on.

Plotting the coefficients of the principal components, as in Display 17.17, shows the patterns of variation in the rods. These functions of position are called *empirical orthogonal functions* (EOF).

Display 17.17   Empirical orthogonal functions depicting modes of variation in magnetic forces along printer rods

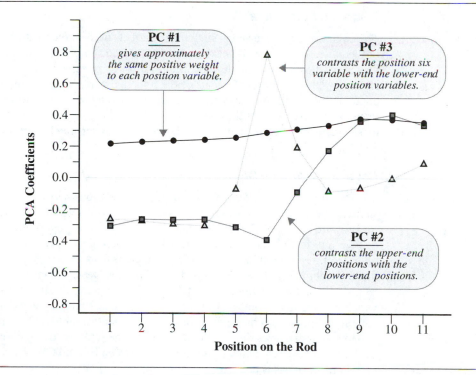

An expression that accounts for 99.73% of all the variation is then

$$\text{Force on rod } i \text{ at location } j = \mu_j + \beta_{1i} f_1(j) + \beta_{2i} f_2(j) + \beta_{3i} f_3(j).$$

17.6   Summary

When researchers do not know the best way to summarize responses before looking at the data, they specify some desirable objective for the summary measure to have. Several special statistical tools create summaries to meet special objectives. Principal components analysis reduces the dimensionality of a set of responses, without significantly reducing their overall variation. Canonical correlations analysis reduces the dimensionality in two sets of variables, while preserving the relationships between them.

PCA, CCA, and similar multivariate tools are designed to point in general directions. Although they provide detailed sets of coefficients and scores, the linear combinations they identify are rarely used directly. Thus, the user must determine what sensible variables the tools are attempting to identify, by examining the coefficients and correlating proposed meaningful linear combinations with the indicated linear combinations.

The tools in this chapter are exploratory: They are used to suggest meaningful linear combinations, which may then be analyzed more formally. Since this is a form of data snooping, the user should report that the tools were used to suggest reductions in multivariate responses prior to the onset of formal data analysis.

Magnetic Force Data

The engineers are interested in studying the effects of the three treatments on the variability of the eleven-dimensional measurement on magnetic force. A principle components analysis is used to reduce the eleven-variable response to three variables that capture nearly all the variability. The coefficients of the principal components suggest meaningful linear combinations of the response components: the average magnetic force; the difference between magnetic forces at the ends of the rod; and the magnetic force in the middle. Since these are nearly uncorrelated with each other, each is separately regressed on the treatment factors.

Love and Marriage Study

The questions of interest call for identifying correlations between sets of multivariate responses. Section 17.4 focuses on the correlation between variables representing feelings of love toward spouse (HP, WP, HC, and WC) and variables representing perceived feelings of love returned by the spouse (PW, PH, CW, CH). Canonical correlations analysis provides strong evidence that two linear combinations from the first set are correlated with two linear combinations from the second set. Suggestive but inconclusive evidence is found of a significant correlation of a third pair. Investigation of coefficients of the canonical variables suggests meaningful linear combinations. In this case the linear combinations corresponding to each of the three pairs of canonical variables are simply individual components. For example, the correlation between the first pair of canonical variables can be explained almost entirely by the correlation between HC and CW. Thus, a formal, ordinary correlation analysis is performed on this pair.

Further questions asked of the data require a different partitioning of the eight responses. For the question of whether a husband's responses correlate with a wife's responses, the variables are categorized into husband's (HP, HC, PW, CW) and wife's (WP, WC, PH, CH). Although not shown in this chapter, the canonical correlation indicates no significant correlations between these two groups, nor between groups corresponding to passionate and compassionate love.

Further Reading

Any text book on multivariate analysis should discuss principle components, canonical correlations analysis, and discriminant analysis. A reference for clustering is VanRyzin,

(1977). Multidimensional scaling is discussed and additional references are provided in the passage on multidimensional scaling in Kotz, Johnson, and Read (1985, vol. 5). A discussion of correspondence analysis may be found in Hill (1974).

17.7 Exercises

Conceptual Exercises

1. In a study of the atmospheric concentrations of air pollutants in the Los Angeles area, daily measurements of seven pollution-related variables were recorded over an extended period of time. In addition, five weather related variables and three variables measuring the city's activity level were taken. The main question: was there a relationship between the pollution levels and the weather/activity variables? What analysis tool from this chapter seems the best to start with?

2. Track-and-field events have frequently been examined with the hope of identifying primary skills related to success in various events. Data from eight different Olympic decathlon championships were gathered to analyze the performances of entrants in the 10 decathlon events. The objective was to determine whether a few common factors account for the variation in all 10 events. What analysis procedure do you recommend?

3. A researcher obtained five variables measuring job characteristics (*feedback; task significance; task variety; task identity; autonomy*) and seven variables measuring job satisfaction (*supervisor satisfaction, career-future satisfaction, financial satisfaction, workload satisfaction, company identification, kind-of-work satisfaction, general satisfaction*) from 784 executives in the corporate branch of a large retail merchandizing corporation. (Data from R. B. Dunham, "Reaction to Job Characteristics: Moderating Effects of the Organization," *Academy of Management Journal* 20(1) (1977): 42–65.) The researcher had two questions: (a) Are measures of job satisfaction associated with measures of job characteristics? (b) If so, what features of job satisfaction are related to what features of job characteristics? What analysis procedure(s) do you recommend?

4. A survey collected annual financial data for 21 businesses 2 years away from bankruptcy and for 25 financially sound businesses. The data included measurements of *cash flow, total debt, net income, total assets, net sales, current assets,* and *current liabilities*. Two questions were to be investigated: (a) Are the financial characteristics of the businesses headed for bankruptcy different from the financial characteristics of sound businesses? (b) If so, is there a way to predict from its financial characteristics whether a business is headed for bankruptcy? What analysis procedures do you recommend?

5. Biologists gathered habitat association information from 250 plots in Madagascar. They measured the population densities of 14 key vertebrates, and they measured 23 variables describing the plots' terrain and vegetational structure. Which of the following analyses is preferable, and why? (a) PCA for the terrain and vegetation variables; PCA for the densities; then regression analysis to relate the two. (b) PCA for the terrain and vegetation variables; then regression to relate the densities to the principal components. (c) CCA.

6. Researchers constructed a questionnaire containing 200 questions about job satisfaction in the nursing profession. To reduce its length, they administered it to a sample of 500 current and retired nurses and used PCA to identify 15 main factors accounting for 98.3% of the variation in all 200 questions. Subsequently, they reduced the questionnaire to a manageable 30 questions specifically designed to measure those 15 factors. Now the researchers want to use DFA to identify factors that distinguish between nurses who will stay in the profession and nurses who will leave the profession for work elsewhere. Should the researchers be concerned that the reduction from 200 to 30 questions might have overlooked key factors that could distinguish these two groups?

7.   One hundred students each took five exams, each of which was scored on a percentage scale (0–100). The exams were: Physics (C), Biology (C), Geometry (O), Algebra (O), and Statistics (O). The letter in parentheses indicates whether the exam format was closed book or open book. PCA gave the coefficients and principal component variances listed in Display 17.18.

Display 17.18   Data for Exercise 7

| | Coefficients for Principal Component Number | | | | |
Test (Open/Closed)	1	2	3	4	5
Physics (C)	0.51	0.75	−0.30	0.30	0.08
Biology (C)	0.37	0.21	0.42	−0.78	0.19
Geometry (O)	0.35	−0.08	0.15	−0.00	−0.92
Algebra (O)	0.45	−0.30	0.60	0.52	0.29
Statistics (O)	0.53	−0.55	−0.60	−0.18	0.15
Variances:	679.2	199.8	102.6	83.7	31.8

(a) Which, if any, of the principal components measures a difference in performance on open-book versus closed-book exams? (b) Which, if any, of the principal components measures overall performance on all exams? (c) What percentage of the variability of all five exam scores is explained by the first two principal components?

8.   **Pig Fat.** Magnetic resonance imaging (MRI) can be used to obtain cross-sectional images through living tissue. One application of MRI is in estimating pig fatness. MRI images were collected at 13 equal spacings (approximately 8 cm apart) along the full-body lengths of 12 pigs. The pigs were subsequently killed and dissected to determine their actual fat percentages. The data are shown in Display 17.19. (Data from C. A. Glasbey and P. A. Fowler, "Regression Models Fitted Using Conditional Independence to Estimate Pig Fatness from Magnetic Resonance Images," *Statistician* 41 (1992): 179–84.) It is desired to use MRI for predicting pig fat prior to killing a pig. Some redundancies exist in the MRI measurements, and not all 13 measurements are needed. Two general approaches are possible for developing a prediction equation: use a variable selection procedure, such as forward selection, to identify a subset of measurements and a regression model for predicting pig fat; or use principle components analysis on the MRI measurements to identify a few linear combinations that explain most of the variability, and then regress actual pig fat percentage on the meaningful combinations suggested by these. (a) What are the relative advantages and disadvantages of these two approaches? (b) As a practical matter, it is time-consuming to interpret all 13 images. If the goal is to develop a prediction model that requires the fewest MRI measurements, which of the two strategies is preferable?

Computational Exercises

9.   **Pig Fat.** For the data in Display 17.19 (and described in Exercise 8), use principle components analysis to find two or three linear combinations of the MRI measurements of pig fat that explain most of the variability in the measurements (M1 through M13). Try to find meaningful linear combinations (for example, the average of all 13) that these suggest.

Display 17.19 Actual pig fat (in percent) and measurements of pig fat (also in percent) from magnetic resonance images at 13 locations, for 12 pigs

Fat	M1	M2	M3	M4	M5	M6	M7	M8	M9	M10	M11	M12	M13
17.13	29.24	29.09	10.01	15.11	14.15	18.13	14.16	10.77	17.47	19.34	12.98	12.18	10.67
17.29	18.81	9.65	11.90	14.47	13.89	13.72	14.31	12.37	17.28	19.62	20.25	18.70	16.90
23.06	35.53	14.16	12.47	19.22	21.26	18.04	14.41	18.35	18.63	28.52	25.96	14.05	18.91
23.30	26.01	17.22	15.21	19.68	17.74	20.94	21.17	18.03	17.49	20.76	13.85	16.88	17.94
27.21	39.49	26.16	19.90	24.97	27.99	19.92	17.99	21.79	25.23	31.52	32.84	17.68	22.41
28.15	37.74	15.60	20.93	26.69	22.98	19.21	22.64	20.44	23.59	21.35	29.77	18.50	15.94
36.00	52.07	39.14	40.94	43.26	34.65	36.59	35.22	31.90	40.10	34.07	34.65	23.62	21.95
38.45	44.53	35.33	37.65	38.69	39.28	35.56	31.03	35.08	37.89	39.00	39.20	29.40	29.76
39.29	41.74	32.63	39.31	35.19	44.36	33.14	29.75	38.48	33.87	41.01	38.72	25.13	29.00
39.69	39.64	20.61	38.03	39.09	31.07	34.38	37.69	32.30	41.36	37.24	37.94	24.55	23.65
40.55	37.95	32.30	31.78	41.26	35.69	32.49	30.57	32.60	42.20	39.25	44.14	28.06	27.02
43.72	53.11	29.29	36.57	40.43	37.52	45.89	45.15	43.76	45.01	43.46	40.03	29.25	26.42

10. **Magnetic Force.** Carry out the principal components analysis, as in this chapter, but use the square roots of magnetic force as the response. Replace the first two principal components with variables that have similar, but more comprehensible coefficients. Construct a scatterplot of the subjects in the coordinates of these two variables. Use the first variable (average magnetic force, on the square-root scale) as a response variable, with the design variables (current, configuration, and material) as explanatory variables. Try combinations such as interactions—or even indicator variables—for the different levels of current. What do you conclude from this? Repeat the analysis with the second constructed variable as the response. What do you conclude?

11. **Love and Marriage.** Rework the CCA analysis for the data on love between spouses. (a) Group variables on the basis of passionate love versus compassionate love. Is there any relationship between the two sets? If so, characterize the related variables. (b) Group variables on the basis of questions asked of husband versus questions asked of wife. Is there any relationship between the two sets? If so, characterize the related variables.

12. **Nature–Nurture.** Reconsider the IQ scores for adopted children from Display 16.12 on pages 475–476. The goal here is to consider the associations between the response variables (each child's IQ at ages 2, 4, 8, and 13) and the two explanatory variables relating to foster mother and natural mother. In Chapter 16, several linear combinations of the responses were considered for addressing different questions of interest. Do the data themselves suggest useful linear combinations?

(a) Perform a principal components analysis on the four responses to see whether one or two useful linear combinations that explain most of the variation in the four measurements present themselves. Try to express the linear combinations in simpler meaningful variables.

(b) Perform a canonical correlations analysis with the response variables in one set and the explanatory variables in another. Does this suggest any useful univariate regression analyses to try?

Data Problems

13. **Church Distinctiveness.** Why do some religious denominations require their members to adopt distinctive, stigmatizing behavior—having shaved heads, wearing pink robes, shunning telephones and modern medical practice? Why is membership in these extreme religious denominations growing

while membership in traditional religious denominations declines? L. R. Iannaccone argues that differences between mainstream churches and extreme sects represent neither historical coincidences nor expressions of irrational fanaticism. He proposes that individuals' rational choice to participate in a collective activity results in a need for successful groups to eliminate "free riders"—uncommitted people who join to enjoy group benefits without contributing to its maintenance. Free riding causes committed members to choose low levels of involvement, thereby weakening the group. Requiring nonconformist behavior imposes a cost on membership that discourages uncommitteds from joining. Iannaccone argues that this explanation for differences between denominations has predictive power and is "unidimensional."

In support of the theory, he supplied data on 17 Protestant denominations plus Catholics. (Data from L. R. Iannaccone, "Why Strict Churches Are Strong," *American Journal of Sociology* 99(5) (1994): 1180–1211.) To get a measure of "distinctiveness" (strictness of discipline), he averaged scores from 16 experts on a seven-point scale for the criteria: "Does the denomination emphasize maintaining a separate and distinctive life style or morality in personal and family life, in such areas as dress, diet, drinking, entertainment, uses of time, marriage, sex, child rearing, and the like? Or does it affirm the current American mainline life style in these respects?" These scores and four survey-based measures appear in Display 17.20. *Attendance* is the average percentage of weeks that individuals attended a church meeting. *Nonchurch* is the average number of secular organizations to which members belong. *Strong* is the average percentage of members that describe themselves as being strong church members. *Income* is the average annual income of members. The last four variables come from an extensive survey of church members.

Treat the final four variables as responses characterizing different aspects of the churches' memberships. To what extent does a single dimension describe differences on all four scales? (Use PCA.) Draw a scatterplot of the first versus the second principal components, and label the points with the church denomination. Interpret the first principal component by comparing the denominations at the

Display 17.20   Measures that differ among denominations of American Protestant and Catholic churches

Church Denomination	Distinctiveness	Church Attendance (% weekly)	Number of Nonchurch Memberships	Strong Member (%)	Annual Income ($US)
American Baptist	2.5	25.6	1.01	50.6	24,000
Assemblies of God	4.8	35.4	0.68	58.6	27,100
Catholic	3.0	26.4	1.43	40.0	32,900
Disciples of Christ	2.1	24.3	2.58	47.0	28,600
Episcopal	1.1	17.3	1.93	32.0	39,000
Evangelical Lutheran	2.7	23.0	1.71	41.5	33,700
Jehovah's Witness	6.0	33.6	0.38	60.6	26,300
Methodist	1.8	19.1	1.56	30.6	32,800
Missouri Synod Lutheran	3.6	27.5	1.76	47.7	35,100
Mormon	5.4	37.8	1.73	70.2	31,600
Nazarene	4.5	33.1	0.86	48.1	31,600
Presbyterian	1.6	21.2	1.88	32.4	37,100
Quaker	4.1	29.6	1.89	58.3	32,500
Reformed Church	2.8	36.7	1.12	61.4	30,400
Seventh Day Adventist	5.8	28.5	0.61	58.7	29,700
Southern Baptist	4.0	25.0	1.13	44.8	30,400
Unitarian	1.6	13.2	2.79	40.8	42,700
United Church of Christ	1.3	19.2	1.56	33.6	40,200

ends of this scale. How does that single dimension relate to the author's distinctiveness scale? To what extent do these data support his theory?

14.   Insurance. In the 1970s, the U.S. Commission on Civil Rights investigated charges that insurance companies were attempting to redefine Chicago "neighborhoods" in order to cancel existing homeowner insurance policies or refuse to issue new ones. Data on homeowner and residential fire insurance policy issuances from 47 zip codes in the Chicago area appear in Display 17.21. Six variables describe general zip code features: *fire* = fires per 1,000 housing units; *theft* = thefts per

Display 17.21   Voluntary and involuntary homeowner policy issuances

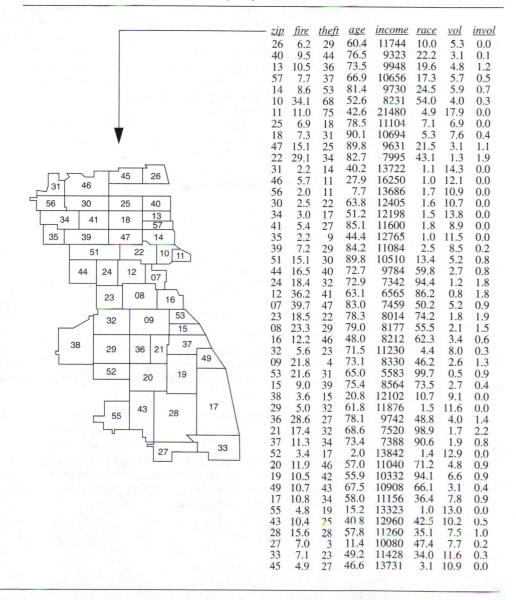

zip	fire	theft	age	income	race	vol	invol
26	6.2	29	60.4	11744	10.0	5.3	0.0
40	9.5	44	76.5	9323	22.2	3.1	0.1
13	10.5	36	73.5	9948	19.6	4.8	1.2
57	7.7	37	66.9	10656	17.3	5.7	0.5
14	8.6	53	81.4	9730	24.5	5.9	0.7
10	34.1	68	52.6	8231	54.0	4.0	0.3
11	11.0	75	42.6	21480	4.9	17.9	0.0
25	6.9	18	78.5	11104	7.1	6.9	0.0
18	7.3	31	90.1	10694	5.3	7.6	0.4
47	15.1	25	89.8	9631	21.5	3.1	1.1
22	29.1	34	82.7	7995	43.1	1.3	1.9
31	2.2	14	40.2	13722	1.1	14.3	0.0
46	5.7	11	27.9	16250	1.0	12.1	0.0
56	2.0	11	7.7	13686	1.7	10.9	0.0
30	2.5	22	63.8	12405	1.6	10.7	0.0
34	3.0	17	51.2	12198	1.5	13.8	0.0
41	5.4	27	85.1	11600	1.8	8.9	0.0
35	2.2	9	44.4	12765	1.0	11.5	0.0
39	7.2	29	84.2	11084	2.5	8.5	0.2
51	15.1	30	89.8	10510	13.4	5.2	0.8
44	16.5	40	72.7	9784	59.8	2.7	0.8
24	18.4	32	72.9	7342	94.4	1.2	1.8
12	36.2	41	63.1	6565	86.2	0.8	1.8
07	39.7	47	83.0	7459	50.2	5.2	0.9
23	18.5	22	78.3	8014	74.2	1.8	1.9
08	23.3	29	79.0	8177	55.5	2.1	1.5
16	12.2	46	48.0	8212	62.3	3.4	0.6
32	5.6	23	71.5	11230	4.4	8.0	0.3
09	21.8	4	73.1	8330	46.2	2.6	1.3
53	21.6	31	65.0	5583	99.7	0.5	0.9
15	9.0	39	75.4	8564	73.5	2.7	0.4
38	3.6	15	20.8	12102	10.7	9.1	0.0
29	5.0	32	61.8	11876	1.5	11.6	0.0
36	28.6	27	78.1	9742	48.8	4.0	1.4
21	17.4	32	68.6	7520	98.9	1.7	2.2
37	11.3	34	73.4	7388	90.6	1.9	0.8
52	3.4	17	2.0	13842	1.4	12.9	0.0
20	11.9	46	57.0	11040	71.2	4.8	0.9
19	10.5	42	55.9	10332	94.1	6.6	0.9
49	10.7	43	67.5	10908	66.1	3.1	0.4
17	10.8	34	58.0	11156	36.4	7.8	0.9
55	4.8	19	15.2	13323	1.0	13.0	0.0
43	10.4	25	40.8	12960	42.5	10.2	0.5
28	15.6	28	57.8	11260	35.1	7.5	1.0
27	7.0	3	11.4	10080	47.4	7.7	0.2
33	7.1	23	49.2	11428	34.0	11.6	0.3
45	4.9	27	46.6	13731	3.1	10.9	0.0

1,000 population; *age* = percentage of housing units built prior to 1940; *income* = median family income; *race* = percentage minority; and *zip* = the last two digits of the zip code (the first three being 606). Chicago's FAIR plan provided a way for households that were rejected by the voluntary insurance market to obtain insurance coverage. These policies were issued involuntarily by the insurance industry. Voluntary (*vol*) is the number of new policies (plus renewals, minus cancellations and nonrenewals) per 100 housing units, while involuntary (*invol*) is the number of FAIR plan policies and renewals per 100 housing units. (Data from D. F. Andrews and A. M. Herzberg, *Data* (New York: Springer-Verlag, 1985).)

(a) Perform a canonical correlations analysis with Set₁ = {*fire, theft, age, income, race*} and with Set₂ = {*vol, invol*}. (b) Draw a matrix of scatterplots for Set₁ variables, including the derived canonical variables. Characterize the first canonical variable from Set₁ by contrasting the *fire, theft, age, income*, and *race* characteristics of zip codes on the high end of its scale with the characteristics of zip codes on the low end of its scale. (c) Draw a matrix of scatterplots for Set₂ variables, also including its canonical variables. Once again, describe the canonical variables in terms of *vol* and *invol* characteristics associated with extreme ends of the scale. (d) These exercises provide small insight into the second canonical variable from Set₁. To see that this variable has some spatial consistency, categorize zip codes into quartiles based on this variable and color the zip codes in a reproduction of the map in Display 17.21. Repeat the color coding for the quartiles of the first canonical variable from Set₁ to show its spatial pattern. (d) Write a short summary of the analysis, providing a descriptive interpretation of the canonical variables and assessing the strength of the relationship between canonical variable pairs.

Answers to Conceptual Exercises

1. Canonical correlations analysis.

2. Principal components analysis.

3. CCA again. Question (b) can also be answered by using multivariate regression.

4. Discriminant function analysis.

5. CCA. It preserves and describes whatever relationships exist between the densities and the habitat variables. A habitat feature crucial to determining vertebrate densities may be discarded by PCA if its variability is not large in relation to other features less important to the animals.

6. It should not be a major concern. If nurses fall into two distinct groups of nurses, depending on their job satisfaction, the between-group variability will have been picked up as a major source of variability by PCA. (This is what happened in the magnetic force study.) The advantages of using a standard instrument for the survey far outweigh the disadvantages of its possible oversights.

7. (a) PC #2. (b) PC #1. (c) 80.12%.

8. (a) The PCA approach reduces the set of explanatory variables to a manageable set prior to the use of regression, which frees the regression analysis from having too many explanatory variables for the number of observations. PCA, on the other hand, does not necessarily find the combinations of explanatory variables that are best for predicting actual fat; it only ensures finding the combinations that explain most of the variability in the MRI measurements. (b) The variable selection procedure would be better, since it would identify one or two of the measurements, as opposed to one or two linear combinations of quite a few measurements.

Comparisons of Proportions or Odds

This chapter returns to the elementary setting of comparing two groups, but for the special case where the response measurement on each subject is binary (0 or 1). A binary variable is a way of coding a two-group categorical response—like dead or alive, or diseased or not diseased—into a number. Statistical analysis of binary responses leads to conclusions about two population proportions or probabilities. Alternatively, conclusions may be stated about odds—like the odds of death or the odds of disease.

The discussion here involves large-sample tests and confidence intervals for comparing two proportions or two odds. The next three chapters provide extensions that are analogous to the analysis of variance and regression extensions of the two-sample t-tools. In particular, logistic regression in Chapters 20 and 21 permits regression modeling with binary response variables.

18.1   Case Studies

18.1.1   Obesity and Heart Disease—An Observational Study

There are two schools of thought about the relationship of obesity and heart disease. Proponents of a physiological connection cite several studies in North America and Europe that estimate higher risks of heart problems for obese persons. Opponents argue that the strain of social stigma brought on by obesity is to blame. One study sheds some light on this controversy, because it was conducted in Samoa, where obesity is not only common but socially desirable. As part of that study, the researcher categorized subjects as obese or not, according to whether their 1976 weight divided by the square of their height exceeded 30 kg/m². Shown in Display 18.1 are the numbers of women in each of these categories who died and did not die of cardiovascular disease (CVD) between 1976 and 1981. (Data from D. E. Crews, "Cardiovascular Mortality in American Samoa," *Human Biology* 60 (1988): 417–33.) Is CVD death in the population of Samoan women related to obesity?

Display 18.1   Cardiovascular deaths and obesity among women in American Samoa

	CVD Death		
	Yes	No	Totals
Obese	16	2,045	2,061
Not Obese	7	1,044	1,051
Totals	23	3,089	3,112

Summary of Statistical Findings

The proportion of CVD deaths among the obese women (0.00776 or 7.76 deaths per 1,000 women) was slightly higher than the corresponding proportion among nonobese women (0.00666). However, the estimated difference in population proportions, 0.00110, is small relative to its standard error (0.00325). The data are consistent with the hypothesis of equal proportions of CVD deaths in the populations of obese and nonobese Samoan women (one-sided p-value $= .37$).

Scope of Inference

The sample was large, comprising approximately 60% of the adult population of Tutuila, the main island. The sample's age distribution was similar to the population's, so the 3,112 women may be assumed to constitute a representative sample from the population of all American Samoan women in 1976. Although the results cannot be extrapolated to populations of women outside of American Samoa, the small difference between the obese and nonobese groups here lends credibility to the theory that results in other countries may be linked to cultural attitudes toward obesity.

18.1.2 Vitamin C and the Common Cold— A Randomized Experiment

Linus Pauling, recipient of Nobel Prizes in Chemistry and in Peace, advocated the use of vitamin C for preventing the common cold. A Canadian experiment examined this claim, using 818 volunteers. At the beginning of the winter, subjects were randomly divided into two groups. The *vitamin C* group received a supply of vitamin C pills adequate to last through the entire cold season at 1,000 mg per day. The *placebo* group received an equivalent amount of inert pills. At the end of the cold season, each subject was interviewed by a physician who did not know the group to which the subject had been assigned. On the basis of the interview, the physician determined whether the subject had or had not suffered a cold during the period. The results are shown in Display 18.2. (Data from T. W. Anderson, D. B. W. Reid, and G. H. Beaton, "Vitamin C and the Common Cold," *Canadian Medical Association Journal* 107 (1972): 503–8.) Can the risk of a cold be reduced by using vitamin C?

Display 18.2 Vitamin C and the common cold

	Outcome		
	Cold	No Cold	Totals
Placebo	335	76	411
Vitamin C	302	105	407
Totals	637	181	818

Summary of Statistical Findings

Of the 411 subjects who took the placebo pill, 82% caught colds at some time during the winter. Among the 407 who took vitamin C pills, however, only 74% caught colds. The difference of 7.3 percentage points was judged to be statistically significant (one-sided p-value = .0059), lending strong support to Dr. Pauling's convictions. The odds of catching a cold while on the placebo regimen are estimated to be 1.10 times to 2.14 times as large as the odds for catching one while on the vitamin C regimen (approximate 95% confidence interval).

Scope of Inference

Because the treatments were randomly assigned to the available volunteers, the difference in cold rates in these groups can safely be attributed to differences in treatments, but it is important to understand exactly what the treatments entail. One essential point is that the randomized experiment was *double-blind*, meaning that neither the subject nor the doctor evaluating the subject knew the treatment that the subject received. Without this precautionary step, knowledge of the treatment received would become another difference between the two groups and might be responsible for statistical differences. In other vitamin C experiments that were constructed to be double-blind, subjects correctly guessed—by taste—which treatment they were receiving. Surprisingly, the proportions with colds in that study depended significantly on which treatment the subjects *thought* they were receiving. Care was taken in the Canadian study to put citric acid and orange flavoring in the placebo pills, so the subjects were less likely to "break the blind."

18.1.3   Smoking and Lung Cancer—A Retrospective Observational Study

In an investigation of the association of smoking and lung cancer, 86 lung cancer patients (thought of as a random sample of all lung cancer patients at the hospitals studied) and 86 controls (thought of as a random sample of subjects without lung cancer from the same communities as the lung cancer patients) were interviewed about their smoking habits. Display 18.3 shows the numbers in each of these groups who did and did not smoke. (Data from H. F. Dorn, "The Relationship of Cancer of the Lung and the Use of Tobacco," *American Statistician* 8 (1954): 7–13.) Are the odds of lung cancer different for smokers than for nonsmokers? By how much?

Display 18.3   Smoking and lung cancer

	Outcome		
	Cancer	Control	Totals
Smokers	83	72	155
Nonsmokers	3	14	17
Totals	86	86	172

Summary of Statistical Findings

There is convincing evidence that the odds of lung cancer are greater for smokers than for nonsmokers (approximate one-sided p-value = .0005). The odds of lung cancer for smokers are estimated to be 5.4 times the odds of lung cancer for nonsmokers (approximate 95% confidence interval: 1.5 times to 19.5 times).

Scope of Inference

This is an observational study, so whether smoking causes lung cancer cannot be addressed by the statistical analysis. Furthermore, the inference to a wider population is only appropriate to the extent that the samples of cancer patients and controls constitute random samples from populations with and without cancer. These data are *retrospective*, meaning that samples were taken from each response group rather than from each explanatory group. The inferences allowed from this kind of sampling are discussed in Section 18.4.1.

18.2   Inferences for the Difference of Two Proportions

18.2.1   The Sampling Distribution of a Sample Proportion

Binary Response Variables

Suppose that a population of units is classified as either yes or no. For example, each member of the population of Samoan women described in Section 18.1.1 may be classified as yes if she died of cardiovascular disease (CVD) in the six years of the study or no if she did not. Although this type of categorical response does not provide a numerical outcome, it is useful to manufacture one. For each member of the population a response variable Y is defined to be 1 for a yes and 0 for a no. Of course, yes and no may be replaced by any other label that distinguishes two levels of a categorical response.

Indicator variables with values of 0 or 1 have been used in previous chapters as *explanatory* variables in regression. They may also be used as *response* variables, in which case they are referred to as *binary response variables*. The average value of a binary response variable in a population is the proportion of members of the population that are classified as yes. The proportion of yes responses in the population is represented by the symbol π and is called the *population proportion*. If the population is hypothetical or if proportions are being compared in a randomized experiment, it is more common to refer to π as the *probability* of a yes response.

The Variance of a Binary Response Variable

Because a binary response variable can have only two numerical values, its variance is an exact function of its mean. Imagine what a listing of the responses for a population of $N = 1,000,000$ subjects might look like. If $\pi = 0.72$ then 720,000 responses would be 1's and the remaining 280,000 would be 0's. All the Y's would sum to 720,000, so the mean of the Y's would be $0.72 = \pi$. The population variance of a variable is the average of the squared difference of the Y's from their mean. In this population, the value of $(Y - \pi)^2$ is either $(1 - \pi)^2$ or $(0 - \pi)^2$, depending on whether Y is 1 or 0; so the set of all $(Y - \pi)^2$ values adds to $720,000(1 - \pi)^2 + 280,000(0 - \pi)^2$. Dividing by 1,000,000 produces the variance. Since 720,000 divided by 1,000,000 is π and 280,000 divided by 1,000,000 is $(1 - \pi)$, the operation yields

$$\text{Variance}(Y) = \text{Mean}\{(Y - \pi)^2\} = \pi(1 - \pi)^2 + (1 - \pi)\pi^2$$
$$= \pi(1 - \pi)[(1 - \pi) + \pi] = \pi(1 - \pi).$$

> *If Y is a binary response variable with population mean π, then:*
>
> $$\text{Variance}\{Y\} = \pi(1-\pi).$$

Two features make this unlike the models for response variables considered in earlier chapters: there is no additional parameter, like σ^2; and the variance is a function of the mean.

The Sample Total and the Binomial Distribution

Suppose that $\{Y_1, Y_2, \ldots, Y_n\}$ is a random sample from a population of binary response variables. The sum of these binary responses, $S = Y_1 + Y_2 + \cdots + Y_n$, is a count of the number of subjects, for which $Y = 1$. The variable S has one of the oldest known and most intensively studied statistical distributions—the *binomial distribution*. S will have one of the values: $0, 1, \ldots, n$. The exact probability for S to be equal to the integer k is given by the formula

$$\Pr\{S = k\} = \frac{n!}{k!(n-k)!}\pi^k(1-\pi)^{n-k}.$$

The mean of S is $n\pi$; the variance of S is $n\pi(1-\pi)$; and the distribution of S is skewed except when $\pi = 1/2$. When the products $n\pi$ and $n(1-\pi)$ are both of moderate size (> 5), the normal distribution approximates binomial probabilities very closely—a fact that motivates many of the succeeding developments in this chapter.

The Sample Proportion

Let $\hat{\pi}$ represent the average of the binary responses in a random sample. Although it is an average, the symbol $\hat{\pi}$ is used rather than $\overline{Y}$ to emphasize that it is also the sample proportion. In the sample of 2,061 obese Samoan women, 16 died of CVD during the six-year study period. The sample proportion $\hat{\pi}$ is therefore 0.00776.

The Sampling Distribution of the Sample Proportion

Since $\hat{\pi}$ is an average, standard statistical theory about the sampling distribution of an average applies:

> If $\hat{\pi}$ is a sample proportion based on a sample of size n from a population with population proportion π, then
>
> 1. Mean$\{\hat{\pi}\} = \pi$.
> 2. Var$\{\hat{\pi}\} = \pi(1-\pi)/n$.
> 3. If n is large enough, the sampling distribution of $\hat{\pi}$ is approximately normal.

Procedures exist for drawing inferences based on the exact distribution of the sample proportion, but methods that rely on the normal approximation are adequate for many applications. *How large must the sample size be?* This is not an easy question to answer, because the answer depends on π. If π is near one-half, the sampling distribution is nearly normal for sample sizes as low as 5 to 10. If π is extremely close to one or zero, a sample size of 100 may not be adequate. For testing a hypothesis that π is some number π_0, it is generally safe to rely on the normal approximation if $n\pi_0 > 5$ and $n(1 - \pi_0) > 5$. A normal approximation confidence interval for π is generally safe if $n\pi_0 > 5$ and $n(1 - \pi_0) > 5$ are satisfied by both the lower and upper endpoints of the confidence interval. A confidence interval, with these informal checks included, is demonstrated in Display 18.4. It is important to realize that the t-distribution is not involved here. The t-tools discussed earlier are based on normally-distributed responses. In particular, the degrees of freedom attached to the t-tools are those associated with an estimate of σ. For binomial responses the variance is a known function of the mean, so there are no analogous degrees of freedom.

Display 18.4   Confidence interval for the proportion of cardiovascular disease deaths in the population of obese Samoan women

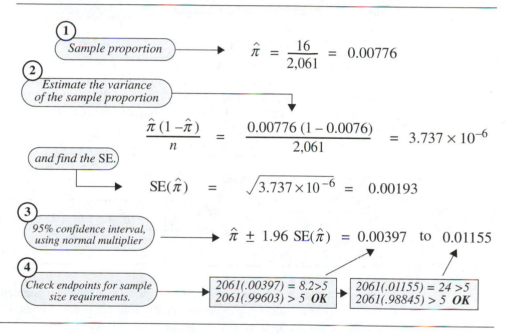

18.2.2   Sampling Distribution for the Difference Between Two Sample Proportions

The question of interest in the Samoan women study is whether the proportion of CVD deaths in the population of obese Samoan women differs from the proportion of CVD deaths in the population of nonobese Samoan women. What can be inferred about the difference in population proportions from the sample statistics? If $\hat{\pi}_1$ and $\hat{\pi}_2$ are computed from independent random samples, the sampling distribution of their difference has these properties:

1. $\text{Mean}\{\hat{\pi}_2 - \hat{\pi}_1\} = \pi_2 - \pi_1$
2. $\text{Variance}\{\hat{\pi}_2 - \hat{\pi}_1\} = \dfrac{\pi_1(1 - \pi_1)}{n_1} + \dfrac{\pi_2(1 - \pi_2)}{n_2}$
3. If n_1 and n_2 are large, the sampling distribution of $\hat{\pi}_2 - \hat{\pi}_1$ is approximately normal.

These properties lead to a test of the hypothesis that $\pi_2 - \pi_1 = 0$ and to construction of a confidence interval for $\pi_2 - \pi_1$, based on a normal approximation. For the test, check whether $n_s \hat{\pi}_c$ and $n_s(1 - \hat{\pi}_c)$ are both greater than 5, where n_s is the smaller of the two sample sizes and π_c is the sample proportion when the two samples are combined into one. A simple rule for judging the adequacy of the normal approximation for confidence interval construction is not available. Usually, the procedure of checking whether $n\hat{\pi}$ and $n(1 - \hat{\pi})$ are larger than 5 in both samples is adequate.

Two Standard Errors for $\hat{\pi}_2 - \hat{\pi}_1$

The standard error of $\hat{\pi}_2 - \hat{\pi}_1$ is the square root of the variance of the sampling distribution, where the unknown parameters are replaced by their best estimates. Values of the best estimates, however, depend on what model is being fit. If the purpose is to get a confidence interval for $\pi_2 - \pi_1$, then the model being fit is the one with separate population proportions; π_1 is then estimated by $\hat{\pi}_1$, and π_2 is estimated by $\hat{\pi}_2$.

For confidence interval:

$$\text{SE}(\hat{\pi}_2 - \hat{\pi}_1) = \sqrt{\frac{\hat{\pi}_1(1 - \hat{\pi}_1)}{n_1} + \frac{\hat{\pi}_2(1 - \hat{\pi}_2)}{n_2}}.$$

If, on the other hand, the purpose is to test the equality of the two population proportions, the null hypothesis model is that the two proportions are equal. The best estimate of both π_1 and π_2 in this model is $\hat{\pi}_c$, the sample proportion from the combined sample.

For testing equality:

$$\text{SE}_0(\hat{\pi}_2 - \hat{\pi}_1) = \sqrt{\frac{\hat{\pi}_c(1 - \hat{\pi}_c)}{n_1} + \frac{\hat{\pi}_c(1 - \hat{\pi}_c)}{n_2}}.$$

The use of this different standard error for testing underscores the need to describe the sampling distribution of a test statistic supposing the null hypothesis is true.

18.2.3 Inferences About the Difference Between Two Population Proportions

Approximate Test for Equal Proportions

To the extent that the normal distribution adequately describes the sampling distribution of $\hat{\pi}_2 - \hat{\pi}_1$, the ratio

$$z\text{-ratio} = \frac{(\hat{\pi}_2 - \hat{\pi}_1) - (\pi_2 - \pi_1)}{\text{SE}(\hat{\pi}_2 - \hat{\pi}_1)}$$

has a standard normal distribution (a normal distribution with Mean $= 0$ and Standard deviation $= 1$). To test the null hypothesis $H: \pi_2 - \pi_1 = 0$, substitute the hypothesized value into the z-ratio, along with the sample information and the test version of the standard error. The hypothesis is rejected if the numerical result—the z-statistic—looks as though it did not come from a standard normal distribution:

$$z\text{-statistic} = \frac{(\hat{\pi}_2 - \hat{\pi}_1) - 0}{\text{SE}_0(\hat{\pi}_2 - \hat{\pi}_1)}.$$

The p-value is the proportion of a standard normal distribution that is more extreme than the observed z-statistic. The computations for the heart disease and obesity study are shown in Display 18.5.

Display 18.5   Test for equality of two population proportions: obesity and CVD deaths

---

① **Sample proportions of CVD deaths**

group 2 (obese): $\hat{\pi}_2 = \dfrac{16}{2,061} = 0.00776$

group 1 (nonobese) $\hat{\pi}_1 = \dfrac{7}{1,051} = 0.00666$

② **Difference** $\hat{\pi}_2 - \hat{\pi}_1 = 0.00110$

③ **Proportion from combined sample** $\hat{\pi}_c = \dfrac{16 + 7}{2,061 + 1,051} = 0.00739$

④ **Check adequacy of normal approximation.**

$1051(.00739) = 7.8 > 5$
$1051(.99261) > 5 \ \textbf{OK}$

⑤ **Standard error (test version)** $\sqrt{\dfrac{(0.00739)(0.99261)}{2,061} + \dfrac{(0.00739)(0.99261)}{1,051}} = 0.00325$

⑥ **z-statistic** $z = \dfrac{0.00110}{0.00325} = 0.340$

⑦ **p-value** $\Pr(Z > 0.340) = .37$

Conclusion: There is no evidence that the two proportions differ.

Approximate Confidence Interval for the Difference Between Two Proportions

When the behavior of the z-ratio is adequately described by a standard normal distribution, the following procedure constructs a confidence interval with approximate coverage of $100(1 - \alpha)\%$.

> *An approximate $100(1-\alpha)\%$ confidence interval for $\pi_2 - \pi_1$ is*
>
> $$(\hat{\pi}_2 - \hat{\pi}_1) \quad \pm \quad [\, z(1-\alpha/2) \times \text{SE}(\hat{\pi}_2 - \hat{\pi}_1) \,].$$

Here, $z(1 - \alpha/2)$ is the $100(1 - \alpha/2)$th percentile in the standard normal distribution. For a 95% confidence interval, for example, α is 0.05 and $z(.975)$ is 1.96. A confidence interval for the vitamin C study is demonstrated in Display 18.6.

Display 18.6   95% confidence interval for the difference in two population proportions: vitamin C and the common cold

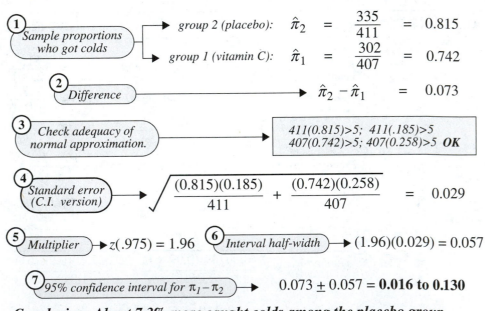

① *Sample proportions who got colds*

group 2 (placebo): $\hat{\pi}_2 \quad = \quad \dfrac{335}{411} \quad = \quad 0.815$

group 1 (vitamin C): $\hat{\pi}_1 \quad = \quad \dfrac{302}{407} \quad = \quad 0.742$

② *Difference* $\hat{\pi}_2 - \hat{\pi}_1 \quad = \quad 0.073$

③ *Check adequacy of normal approximation.*

$411(0.815)>5;\ 411(.185)>5$
$407(0.742)>5;\ 407(0.258)>5$ **OK**

④ *Standard error (C.I. version)* $\sqrt{\dfrac{(0.815)(0.185)}{411} + \dfrac{(0.742)(0.258)}{407}} \quad = \quad 0.029$

⑤ *Multiplier* $z(.975) = 1.96$ **⑥** *Interval half-width* $(1.96)(0.029) = 0.057$

⑦ *95% confidence interval for $\pi_1 - \pi_2$* $0.073 \pm 0.057 = \textbf{0.016 to 0.130}$

Conclusion: About 7.3% more caught colds among the placebo group than among the vitamin C group (95% CI: 1.6% to 13.0%).

18.3   Inference About the Ratio of Two Odds

18.3.1   A Problem with the Difference Between Proportions

There are difficulties with interpreting the difference between two population proportions or two probabilities. If π_1 is the probability of a disease in the absence of treatment, and if π_2 is the probability of a disease under treatment with a preventive drug, then a difference of 0.05 may be rather small if the true probabilities are 0.50 and 0.45, yet the same difference can have considerable practical significance if the true probabilities are 0.10 and 0.05. In both cases the disease rate can be reduced by 5% if the drug is used; but in the first case the treatment will prevent the disease in 1 out of 10 people who would have gotten the disease otherwise, whereas in the second case the treatment will prevent the disease in 1 out of 2 people who would have gotten it.

The point is that the *difference* between two population proportions may not be the best way to compare them. Differences tend to differ in meaning when the proportions are near 0.5 from when the proportions are near 0 or 1. An alternative to comparing proportions is to compare their corresponding odds. This makes sense over a wider range of possibilities and forms the basis for interpreting regression models for binary data.

18.3.2   Odds

If π is a population proportion (or probability) of yes outcomes, the corresponding *odds* of a yes outcome in the population are $\pi/(1-\pi)$, represented by the symbol ω (the Greek letter *omega*).

The *sample odds* are $\hat{\omega} = \hat{\pi}/(1-\hat{\pi})$. Instead of saying, for example, that the proportion of cold cases among the vitamin C group was 0.742, one can convey the same information by saying that the odds of getting a cold were 2.876 to 1. The number 2.876 is the ratio of the proportion of cold cases to the proportion of noncold cases, $2.876 = 0.742/0.258$. These odds make the 0.742 easier to visualize: there are 2.876 cold cases for every 1 noncold case (or about 23 cold cases for every 8 noncold cases). Odds are particularly useful in visualizing proportions that are close to 0 or 1.

It is customary to cite the larger number first when stating odds, so an event with chances of 0.95 has odds of 19 to 1 *in favor* of its occurrence while an event with chances of 0.05 has the same odds, 19 to 1, *against* it.

Here are some numerical facts about odds:

1. Whereas a proportion must be between 0 and 1, odds must be greater than or equal to 0 but have no upper limit.
2. A proportion of 1/2 corresponds to odds of 1. Some common nonscientific descriptions of this situation are "equal odds," "even odds," and "the odds are fifty-fifty."
3. Odds are not defined for proportions that are exactly 0 or 1.
4. If the odds of a yes outcome are ω, the odds of a no outcome are $1/\omega$.
5. If the odds of a yes outcome are ω, the probability of yes (or the population proportion) is $\pi = \omega/(1+\omega)$.

18.3.3 The Ratio of Two Odds

When two populations have proportions π_1 and π_2, with corresponding odds ω_1 and ω_2, a useful alternative to the difference in proportions is the *odds ratio*, $\phi = \omega_2/\omega_1$. (ϕ is the Greek letter *phi*.) If π_2 is larger than π_1, then ω_2 is larger than ω_1 and the odds ratio is greater than 1. If the odds ratio is some number—say, $\phi = 3$—then $\omega_2/\omega_1 = 3$ is equivalent to $\omega_2 = 3\omega_1$. In words, the odds of a yes outcome in the second group are three times the odds of a yes outcome in the first group. If, for example, $\omega_1 = 2.5$, then group 1 has 5 yes for every 2 no outcomes. With $\phi = 3$, group 2 would have 15 yes for every 2 no outcomes.

The sample odds ratio, or *estimated odds ratio* is $\hat{\phi} = \hat{\omega}_2/\hat{\omega}_1$. In the vitamin C and cold study, the odds of a cold in the placebo group are estimated as $\hat{\omega}_2 = 335/76 = 4.408$. (There were 4.408 people with colds for every 1 without a cold.) The odds of a cold in the vitamin C group are estimated as $\hat{\omega}_1 = 302/105 = 2.876$. The estimated odds ratio is therefore $4.408/2.876 = 1.53$. *The odds of getting a cold on the placebo regimen are estimated to be 1.53 times as large as the odds of getting a cold on the vitamin C regimen.* Equivalently, the odds of a cold on placebo are 53% greater than the odds of a cold on vitamin C.

When the counts are arranged in a 2×2 table, the odds ratio may quickly be estimated by dividing the product of the upper left and lower right counts by the product of the upper right and lower left counts, as shown in Display 18.7. One caution: the calculation in Display 18.7 provides the odds ratio estimate for getting a cold on the placebo regimen relative to the vitamin C regimen. To estimate the odds ratio for not getting a cold on the placebo regimen relative to the vitamin C regimen, one must invert the ratio in Display 18.7.

Display 18.7 Shortcut method for estimating the odds ratio in a 2×2 table

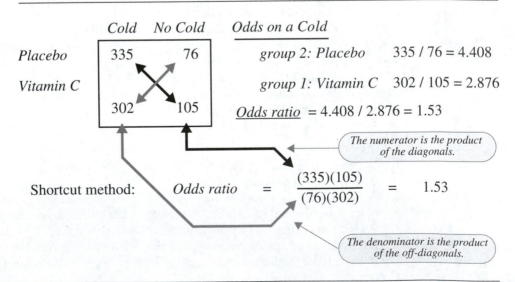

The odds ratio is a more desirable measure than the difference in population proportions, for three reasons:

1. In practice, the odds ratio tends to remain more nearly constant over levels of confounding variables.
2. The odds ratio is the *only* parameter that can be used to compare two groups of binary responses from a retrospective study (see Section 18.4).
3. The comparison of odds extends nicely to regression analysis (see Chapters 20 and 21).

18.3.4   Sampling Distribution of the Log of the Estimated Odds Ratio

If $\hat{\omega}_1$ and $\hat{\omega}_2$ are computed from independent samples, the sampling distribution of the *natural log of their ratio* has these properties:

1. Mean$\{\log(\hat{\omega}_2/\hat{\omega}_1)\} = \log(\omega_2/\omega_1)$ (approximately).
2. Var$\{\log(\hat{\omega}_2/\hat{\omega}_1)\} = [n_1\pi_1(1 - \pi_1)]^{-1} + [n_2\pi_2(1 - \pi_2)]^{-1}$ (approximately).
3. If n_1 and n_2 are large, the sampling distribution of $\log(\hat{\omega}_2/\hat{\omega}_1)$ is approximately normal.

Of course, inference is desired for the odds ratio, not for its logarithm (which has no useful interpretation). The sampling distribution of the log of the estimated odds ratio, however, is more closely approximated by the normal distribution than is the sampling distribution of the estimated odds ratio. Consequently, statistical inferences will be carried out for the log odds ratio and subsequently reexpressed in terms of odds ratios.

Statisticians do not have a good general guideline for checking whether the sample sizes are large enough in this problem. Most authorities agree that one can get away with smaller sample sizes here than for the difference of two proportions. Therefore, if the sample sizes pass the rough checks discussed in Section 18.2.2, they should be large enough to support inferences based on the approximate normality of the log of the estimated odds ratio, too.

Two Standard Errors for the Log of the Odds Ratio

The estimated variance is obtained by substituting sample quantities for unknowns in the variance formula in the preceding box. As before, the sample quantities used to replace the unknowns depend on the usage. For a confidence interval, π_1 and π_2 are replaced by their sample estimates.

Confidence interval:

$$SE[\log(\hat{\omega}_1/\hat{\omega}_2)] = \sqrt{\frac{1}{n_1\hat{\pi}_1(1-\hat{\pi}_1)} + \frac{1}{n_2\hat{\pi}_2(1-\hat{\pi}_2)}}.$$

To test equality of odds in two populations, one must estimate the common proportion from the combined sample and compute the standard error based on it.

Testing equality:

$$SE_0[\log(\hat{\omega}_1/\hat{\omega}_2)] = \sqrt{\frac{1}{n_1\hat{\pi}_c(1-\hat{\pi}_c)} + \frac{1}{n_2\hat{\pi}_c(1-\hat{\pi}_c)}}.$$

Display 18.8   Test for equality of two population odds: heart disease and obesity data

(1) *Estimate the odds on CVD death.*

group 1 (obese): $\hat{\omega}_1 = \dfrac{16}{2{,}045} = 0.00782$

group 2 (nonobese): $\hat{\omega}_2 = \dfrac{7}{1{,}044} = 0.00670$

(2) *Odds ratio and its log*

$\hat{\phi} = \dfrac{0.00782}{0.00670} = 1.1669 \longrightarrow \log(\hat{\phi}) = 0.154$

(3) *Proportion from the combined sample*

$\hat{\pi}_c = \dfrac{16+7}{2{,}061+1{,}051} = 0.00739$

(4) *SE for the log odds ratio estimate (test version)*

$\sqrt{\dfrac{1}{2{,}061(0.00739)(1-0.00739)} + \dfrac{1}{1{,}051(0.00739)(1-0.00739)}} = 0.443$

(5) *z-statistic*

$z = \dfrac{0.154}{0.443} = 0.349$

(6) *one-side p-value*

$\Pr(Z > 0.349) = \mathbf{0.36}$

Conclusion: There is no evidence that the odds ratio differs from 1.

If the odds are equal, then the odds ratio is 1 and the log of the odds ratio is 0. Therefore, a test of equal odds is carried out by testing whether the log of the odds ratio is 0. The test calculations for the heart disease and obesity data are shown in Display 18.8. The resulting p-value is nearly identical to that obtained with the z-test for equal proportions, as is the case when the sample sizes are large.

To get a confidence interval for the odds ratio, construct a confidence interval for the log of the odds ratio and take the antilogarithm of the endpoints. A shortcut formula for the standard error is the square root of the sum of the reciprocals of the four cell counts in the 2×2 table, as illustrated in step 2 of Display 18.9.

Display 18.9   Confidence interval for an odds ratio: vitamin C and cold data

---

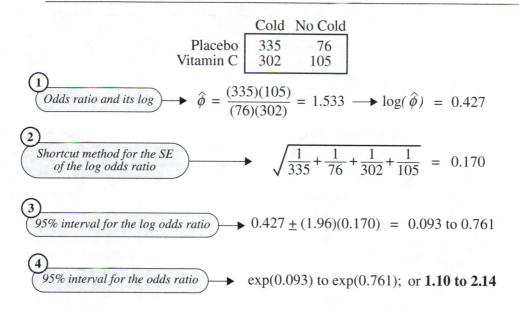

	Cold	No Cold
Placebo	335	76
Vitamin C	302	105

1 Odds ratio and its log $\longrightarrow$ $\hat{\phi} = \dfrac{(335)(105)}{(76)(302)} = 1.533 \longrightarrow \log(\hat{\phi}) = 0.427$

2 Shortcut method for the SE of the log odds ratio $\longrightarrow$ $\sqrt{\dfrac{1}{335} + \dfrac{1}{76} + \dfrac{1}{302} + \dfrac{1}{105}} = 0.170$

3 95% interval for the log odds ratio $\longrightarrow$ $0.427 \pm (1.96)(0.170) = 0.093$ to 0.761

4 95% interval for the odds ratio $\longrightarrow$ $\exp(0.093)$ to $\exp(0.761)$; or **1.10 to 2.14**

Conclusion: The odds of a cold for the placebo group are estimated to be 1.53 times the odds of a cold for the vitamin C group (approximate 95% CI: 1.10 to 2.14).

18.4   Inference from Retrospective Studies

18.4.1   Retrospective Studies

Each of the case studies of Section 18.1 involves an explanatory factor with two levels and a response factor with two levels. In the obesity study, random samples were taken from both explanatory variable categories, obese and nonobese. In the vitamin C study, subjects were randomly assigned to an explanatory category, either vitamin C or placebo. Either plan is called *prospective sampling* because it sets the explanatory situation and then awaits the response.

In the smoking and lung cancer data of Section 18.1.3, the question of interest requires that the smoking behavior is explanatory and lung cancer is the response. However, the sampling was carried out for each level of the *response* factor—a sample of 86 lung cancer patients and a sample of 86 control subjects were obtained, and all subjects were asked whether they smoked. This is an example of *retrospective sampling*. Samples were taken for each level of the response variable and the levels of the explanatory factor were determined for subjects in these samples. A comparison of prospective and retrospective sampling is provided in Display 18.10.

Display 18.10   Prospective and retrospective sampling

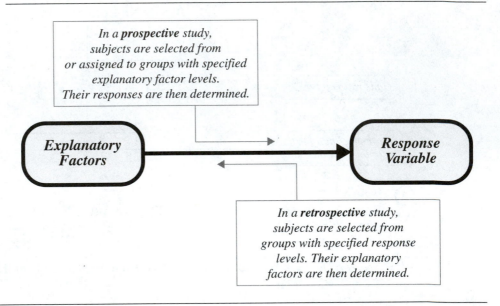

Labeling of one of the factors a "response" and the other "explanatory" is a matter of choice. One could simply call lung cancer the explanatory variable and smoking the response, but comparing proportions of smokers in the populations of lung cancer patients and non-lung-cancer patients is less pertinent than comparing proportions of lung cancer victims in the populations of smokers and nonsmokers. The latter is of direct interest with regard to the health effects of smoking.

Reasons for Retrospective Sampling

In many areas of study, both prospective and retrospective studies play important roles. A prospective study that follows random samples of smokers and nonsmokers would be very useful, but a retrospective study can be accomplished without having to follow the subjects through their entire lifetimes. Furthermore, if the response proportions are small then huge samples are needed in a prospective study to get enough yes outcomes to permit inferences to be drawn. This problem is bypassed in a retrospective study.

18.4.2 Why the Odds Ratio is the Only Appropriate Parameter if the Sampling Is Retrospective

The odds ratio is the only parameter that describes binary response outcomes for the explanatory categories that can be estimated from retrospective data. The smoking and lung cancer data, for example, cannot be used to estimate the individual proportions of smokers and nonsmokers who get lung cancer or the difference between the proportions.

   The odds ratio is the same regardless of which factor is considered the response. As a practical demonstration of this (without a formal proof), a hypothetical population of smokers and nonsmokers is shown in Display 18.11. Notice that the proportions of smokers among lung cancer patients and among controls tell nothing about the proportions of lung cancer patients among smokers and among nonsmokers. Yet the odds ratio is the same regardless of which factor is thought of as the response: the odds of smoking among lung cancer patients is two times the odds of smoking among controls; and the odds of lung cancer among smokers is two times the odds of lung cancer among nonsmokers. Consequently, the odds ratio can be estimated either by observing cancer incidence in independent samples of smokers and nonsmokers or by observing proportions of smokers in independent samples of cancer victims and controls. The odds ratio is the only quantity pertaining to the prospective populations that can be estimated from retrospective studies.

Display 18.11   Illustration that the odds ratio is the same regardless of which factor is thought of as the response, with a hypothetical population

	Lung Cancer	*No Cancer*
Smokers	1,000	2,000,000
Nonsmokers	4,000	16,000,000

$$\frac{\text{Odds of cancer among smokers}}{\text{Odds of cancer among nonsmokers}} = \frac{1{,}000/2{,}000{,}000}{4{,}000/16{,}000{,}000} = 2$$

$$\frac{\text{Odds that a cancer victim was a smoker}}{\text{Odds that a person without cancer was a smoker}} = \frac{1{,}000/4{,}000}{2{,}000{,}000/16{,}000{,}000} = 2$$

Example—Smoking and Lung Cancer

From the data described in Example 18.1.3 the estimated odds ratio is 5.38, the natural log of the estimated odds ratio is 1.68, and the standard error of the estimated log odds ratio is

0.656 (confidence interval version). An approximate 95% confidence interval for the log odds ratio is 0.396 to 2.969. Therefore, the odds of lung cancer for smokers are estimated to be 5.38 times the odds of lung cancer for nonsmokers. An approximate 95% confidence interval for the odds ratio is 1.5 to 19.4. As usual, no causal interpretation can be drawn from the statistical analysis of observational data, although the data are consistent with the theory that smoking causes lung cancer.

18.5 Summary

Populations of binary responses can be compared by estimating the difference in population proportions or the ratio of population odds. Tests and confidence intervals are based on the approximate normality of the difference in sample proportions or on the approximate normality of the logarithm of the estimated odds ratio. These techniques also apply to randomized experiments. The choice of whether to compare proportions or odds is subjective. The odds are often more appropriate if the proportions are close to 0 or 1. The proportions are more familiar to some audiences. If the data are sampled retrospectively, only the odds ratio can be estimated.

Obesity and Death from Cardiovascular Disease

Crews's study of obesity and heart disease deaths in American Samoa includes both women and men (see Exercise 9). For both women and men, the odds of CVD death are 15 to 20% higher among obese persons, but neither result is statistically significant. Confidence intervals are more explicit about whether any practically significant difference is likely. The similarity of results for men and women suggests asking whether a common odds ratio exists and is greater than 1.0. Tools for handling these questions are introduced in the next chapter.

Vitamin C and the Common Cold

The odds of getting a cold while using the placebo are estimated to be 4.4 to 1 in this Canadian study; use of vitamin C reduces those odds by approximately 35%. Increasingly, scientific journals and the popular press describe studies of risk factors in terms of multiplicative effects on odds ratios. This natural terminology comes from an increasing reliance on statistical analysis at the log-odds scale.

Smoking and Lung Cancer

A randomized experiment to determine a causal effect of smoking on lung cancer will never be performed. Nor does the time scale involved lend itself to prospective sampling. Retrospective sampling, however, is a powerful tool for studying associations in such situations. Here, the odds ratio of 5.4 to 1 offers strong evidence of some association between smoking and lung cancer.

Further Reading

The properties of the binomial probability distribution are detailed in books on mathematical statistics, such as the one by Rice (1995). Futher discussion of odds and odds ratios, and of retrospective sampling, are provided in textbooks on biostatistics, such as the one by Fisher and Van Belle (1993).

18.6 Exercises

Conceptual Exercises

1. **Obesity.** For the heart disease and obesity data (Section 18.1.1), does the large p-value for the test of equal population proportions prove that six-year cardiovascular disease death and obesity were unrelated over the six-year study period in the population of American Samoan women? Why not? How does a confidence interval clarify the picture?

2. **Vitamin C.** The vitamin C and cold experiment (Section 18.1.2) was double-blind, meaning that neither the volunteers nor the doctors who interviewed them were told whether they were taking vitamin C or placebo. If many of the volunteers correctly guessed which group they were in, and if incidence of colds is highly associated with *perceived* treatment, do these considerations invalidate the assertion that cause-and-effect statements can be made from randomized experiments?

3. During investigation of the U.S. space shuttle *Challenger* disaster, it was learned that project managers had judged the probability of mission failure to be 0.00001, whereas engineers working on the project had estimated failure probability at 0.005. The difference between these two probabilities, 0.00499 was discounted as being too small to worry about. Is a different picture provided by considering odds? How is that interpreted?

4. **Perry Preschool Project.** In a 1962 social experiment, 123 three- and four-year-old children from poverty-level families in Ypsilanti, Michigan, were randomly assigned either to a treatment group receiving 2 years of preschool instruction or to a control group receiving no preschool. The participants were followed into their adult years. The following table shows how many in each group were arrested for some crime by the time they were 19 years old. (Data reported in *Time*, July 29, 1991).

Arrested for Some Crime?

	Yes	No
Preschool	19	42
Control	32	30

(a) Is it possible to use these data to see whether preschool instruction can *cause* a lower rate of criminal arrest for some populations? (b) Is this a retrospective or a prospective study?

5. **Violence Begets Violence.** It is often argued that victims of violence exhibit more violent behavior toward others. To study this hypothesis a researcher searched court records to find 908 individuals who had been victims of abuse as children (11 years or younger). She then found 667 individuals, with similar demographic characteristics, who had not been abused as children. Based on a search through subsequent years of court records, she was able to determine how many in each of these groups became involved in violent crimes, as shown in the following table. (Data from C. S. Widom, "The Cycle of Violence," *Science* 244 (1989): 160). (a) Is this a randomized experiment or an observational study? (b) Is this a prospective or a retrospective study? (c) Consider the populations of abused victims and controls from which these samples were selected. Let Y be a binary response variable taking the value 1 if an individual became involved in violent crime and 0 if not, in each of these populations. If π_1 and π_2 are the means of Y in the two populations, what is the variance of Y in each population? (d) What

Involved in a Violent Crime?

	Yes	No
Abuse victim	102	806
Control	53	614

is the variance of the sampling distribution of $\hat{\pi}_1 - \hat{\pi}_2$? (e) Let Z be a binary response variable that takes on the value 1 if an individual did not become involved in violent crime and 0 if the individual did (so Z is $1 - Y$). What are the means of Z in the two populations, in terms of π_1 and π_2 (the means of Y)? (f) What are the variances of Z in the two populations, in terms of π_1 and π_2? (g) How does the analysis of the data differ if Z is used as a response variable instead of Y (in terms of estimating a difference in population proportions and estimating a ratio of odds)?

6. In Exercise 5, there are many ways of wording an odds ratio statement. For example:
   (i) The odds of Yes among victims relative to controls.
   (ii) The odds of Yes among controls relative to victims.
   (iii) The odds of No among victims relative to controls.
   (iv) The odds of No among controls relative to victims.
   (v) The odds of victims among Yes relative to No.
   (vi) The odds of victims among No relative to Yes.
   (vii) The odds of controls among Yes relative to No.
   (viii) The odds of controls among No relative to Yes.
   (ix) The odds of victims among Yes'es relative to controls.
(a) Which, if any, of these are the same? (b) Do all of them make sense?

7. **Alcohol and Breast Cancer.** The following are partial results from a *case–control* study involving a sample of *cases* (women with breast cancer) and a sample of *controls* (demographically similar women without breast cancer). (Data from L. Rosenberg, J. R. Palmer, D. R. Miller, E. A. Clarke, and S. Shapiro, "A Case–Control Study of Alcoholic Beverage Consumption and Breast Cancer," *American Journal of Epidemiology* 131 (1990): 6–14). The women were asked about their drinking habits, from which the following table was constructed. (a) For assessing the

Breast Cancer

	Cases	Controls
Fewer than 4 drinks per week	330	658
4 or more drinks per week	204	386

risk of drinking on breast cancer, which factor is the response? (b) Is this a retrospective or a prospective study? (c) Is this a randomized experiment or an observational study?

8. **Salk Polio Vaccine.** The Salk polio vaccine trials of 1954 included a double-blind experiment in which elementary school children of consenting parents were assigned at random to injection with the Salk vaccine or with a placebo. Both treatment and control groups were set at 200,000 because the target disease, infantile paralysis, was uncommon (but greatly feared). (Data from

Infantile Paralysis Victim?

	Yes	No
Placebo	142	199,858
Salk polio vaccine	56	199,944

J. M. Tanur et al., *Statistics: A Guide to the Unknown* (San Francisco: Holden-Day, 1972). (a) Is this a randomized experiment or an observational study? (b) Is this a retrospective or a prospective study? (c) Describe the population to which inferences can be made. (d) Would it make more sense to look at the difference in disease probabilities or the ratio of disease odds here?

Computational Exercises

9. **Heart Disease and Obesity in American Samoan Men.** The study described in Section 18.1.1 also included men. The data for men are shown in the accompanying table. (a) Compute (i) the sample

	CVD Death	
	Yes	*No*
Obese	22	1,179
Not obese	22	1,409

proportions of CVD deaths for the obese and nonobese groups; (ii) the standard error for the difference in sample proportions; (iii) a 95% confidence interval for the difference in population proportions. (b) Find a one-sided p-value for the test of equal population proportions (using the standard error already computed). (c) Compute (i) the sample odds of CVD death for the obese and nonobese groups; (ii) the estimated odds ratio; (iii) the standard error of the estimated log odds ratio; (iv) a 95% confidence interval for the odds ratio. (d) Write a concluding sentence that incorporates the confidence interval from part (c).

10. **Salk Polio Vaccine.** (a) For the data in Exercise 7 find a one-sided p-value for testing the hypothesis that the odds of infantile paralysis are the same for the placebo and for the Salk vaccine treatments. (b) An alternative method for testing equal proportions is to look only at the 198 children who contracted polio. Each of them had a 50% chance of being assigned to the Salk group or to the placebo group, unless group membership altered their chances. Under the null hypothesis of no difference in disease probabilities, the proportion of the 198 who are in either treatment group has a mean of 1/2. Is the observed proportion in the control group, 142/198, significantly larger than would be expected under the null hypothesis? Use the results about the sampling distribution of a single population proportion (in Section 18.2.1) to test this hypothesis. (This is a conditional test—conditional on the number getting the disease. It is useful here because the disease probabilities are so small.)

11. In the hypothetical example of Display 18.11, calculate each of the following quantities in the population, in the expected results from prospective sampling, and in the expected results from retrospective sampling. (a) The proportion of lung cancer patients among smokers. (b) The proportion of lung cancer patients among nonsmokers. (c) The difference between parts (a) and (b). (d) Verify that retrospective sampling results cannot be used to estimate these population properties.

12. Suppose that the probability of a disease is .00369 in a population of unvaccinated subjects and that the probability of the disease is .001 in a population of vaccinated subjects.

    (a) What are the odds of disease without vaccine relative to the odds of disease with vaccine?

    (b) How many people out of 100,000 would get the disease if they were not treated?

    (c) How many people out of 100,000 would get the disease if they were vaccinated?

    (d) What proportion of people out of 100,000 who would have gotten the disease would be spared from it if all 100,000 were vaccinated? (This is called the *protection rate*.)

    (e) Follow the steps in parts (a)–(d) to derive the odds ratio and the protection rate if the unvaccinated probability of disease is .48052 and the vaccinated probability is .2. (The point is that the odds ratio is the same in the two situations, but the total benefit of vaccination also depends on the probabilities.)

13.   An alternative to the odds ratio and the difference in probabilities is the relative risk, which is a common statistic for comparing disease rates between risk groups. If π_u and π_v are the probabilities of disease for unvaccinated and vaccinated subjects, the relative risk (due to not vaccinating) is $\rho = \pi_u/\pi_v$. (So if the relative risk is 2, the probability of disease is twice as large for unvaccinated as for vaccinated individuals.)

> **(a)**   Show that the odds ratio is very close to the relative risk when very small proportions are involved. (You may do this by showing that the two quantities are quite similar for a few illustrative choices of small values of π_v and π_u.)
>
> **(b)**   Calculate the relative risk of not vaccinating for the situation introduced at the beginning of Exercise 12.
>
> **(c)**   Calculate the relative risk of not vaccinating for the situation described in Exercise 12(e).
>
> **(d)**   Here is one situation with a relative risk of 50: the unvaccinated probability of disease is .0050 and the vaccinated probability of disease is .0001 (the probability of disease is 50 times greater for unvaccinated subjects than for vaccinated ones). Suppose in another situation the vaccinated probability of disease is 0.05. What unvaccinated probability of disease would imply a relative risk of 50? (Answer: None. This shows that the *range* of relative risk possibilities depends on the baseline disease rate. This is an undesirable feature of relative risk.)

14.   The Pique Technique was developed because a target for solicitation is more likely to comply if mindless refusal is disrupted by a strange or unusual request. Researchers had young people ask 144 targets for money. The targets were asked either for a standard amount—a quarter—or an unusual amount—seventeen cents. 43.1% of those asked for seventeen cents responded, compared to 30.6% of those asked for a quarter. Each group contained 72 targets. What do you make of that? (Data from M. D. Santos, C. Leve, and A. R. Pratkanis, "Hey Buddy, Can You Spare Seventeen Cents?," *Journal of Applied Social Psychology* 24(9) (1994): 755–64.)

Data Problems

15.   Perry Preschool Project.   Analyze the data shown in Exercise 4 to determine whether a preschool program like the one used can lead to lower levels of criminal activity; and, if so, by how much.

16.   Violence Begets Violence.   Reconsider the data described in Exercise 5. The researcher concluded: "Early childhood victimization has demonstrable long-term consequences for ... violent criminal behavior." Conduct your own analysis of the data and comment on this conclusion. In particular, is there a statistically significant difference in population proportions, and is the strength of the causal implication of this statement justified by the data from this study?

17.   Alcohol and Breast Cancer.   In Exercise 7 the researcher is interested in the question of whether drinking can cause breast cancer. Analyze the data to see whether the data support the theory that it does, and quantify the association with an appropriate measure.

Answers to Conceptual Exercises

1.   No. There could be a difference, but the sample was too small to detect it. A confidence interval would reveal a set of possible values for the difference in population proportions that remain plausible in light of these data.

2.   No, but in this case a confounding factor (whether the subject correctly guessed the treatment) arose after the randomization. The effect of treatment may be confounded with this factor, so causal conclusions must be made cautiously.

3. The odds of failure estimated by the project managers were 99,999 to 1 against. Engineers estimated them to be 199 to 1 against. The ratio is about 500. Thus, for every failure envisioned by the project managers, the engineers saw the possibility of 500 failures.

4. (a) Yes. (b) Prospective.

5. (a) Observational study. (b) Prospective. (c) $\pi_1(1 - \pi_1)$ and $\pi_2(1 - \pi_2)$. (d) $\pi_1(1 - \pi_1)/908 + \pi_2(1 - \pi_2)/667$. (e) $(1 - \pi_1)$ and $(1 - \pi_2)$. (f) $\pi_1(1 - \pi_1)$ and $\pi_2(1 - \pi_2)$ (the same as for Y) (g) The analysis will be the same, except that the sign of the difference in population proportions will be reversed and the odds ratio will be the reciprocal of the odds ratio based on Y.

6. (a) i = iv = v = viii; ii = iii = vi = vii (= the reciprocal of the value in the first group). (b) ix makes no sense.

7. (a) Breast cancer. (b) Retrospective. (c) Observational.

8. (a) Randomized experiment. (b) Prospective. (c) Elementary school children of consenting parents (in 1954). (d) There is nothing technically wrong with either measure in this prospective study, but the odds ratio offers a more suitable interpretation, since the probabilities are so small.

CHAPTER 19

More Tools for Tables of Counts

The data sets in the previous chapter can be displayed as 2×2 tables of counts, which list the numbers of subjects falling in each cross-classification of a row factor (such as obese or not obese) and a column factor (such as heart disease or no heart disease). This chapter elaborates on sampling schemes leading to tables of counts, the appropriate wording of hypotheses based on the sampling scheme and the question of interest, an exact test for 2×2 tables, and two extensions to larger tables.

Fisher's Exact Test is the gold standard of testing tools for 2×2 tables, appropriate under any of the sampling schemes that lead to 2×2 tables. It is exact in the sense that the p-value is based on a permutation distribution and requires no approximations. An approximate test, the *chi-squared test*, is also presented for 2×2 tables. It extends in a straightforward way to larger tables of counts. Sometimes several 2×2 tables of counts are used, with each separate table corresponding to a different level of a blocking or potentially confounding variable. If the odds ratio is the same in each 2×2 table—even though the individual odds may vary—the *Mantel–Haenszel Test* is appropriate for testing whether the common odds ratio is one.

19.1 Case Studies

19.1.1 Sex Role Stereotypes and Personnel Decisions—A Randomized Experiment

Psychologists performed an experiment on male bank supervisors attending a management institute, to investigate biases against women in personnel decisions. Among other tasks in their training course, the supervisors were asked to make a decision on whether to promote a hypothetical applicant based on a personnel file. For half of them, the application file described a female candidate; for the others it described a male. The files were identical in all other respects. Results on the promotion decisions for the two groups are shown in Display 19.1. (Data from B. Rosen and T. Jerdee, "Influence of Sex Role Stereotypes on Personnel Decisions," *Journal of Applied Psychology* 59 (1974): 9–14.) Was there a bias against female applicants, or can the observed difference in promotion rates be explained just as reasonably by the chance involved in random allocation of subjects to treatment groups?

Display 19.1 Decisions of 48 male supervisors to promote a fictitious candidate, listed separately for those receiving "male" and "female" personnel folders

	Promoted	
	Yes	No
Male	21	3
Female	14	10

Summary of Statistical Findings

Substantial evidence supports the conclusion that the promotion rate is higher for male applicants than for female applicants. The one-sided *p*-value from Fisher's Exact Test is .0245.

Scope of Inference

Since randomization was used, the inference is that the sex listed on the applicant file was responsible for the difference observed with these 48 supervisors. The inference derives from a probability model based on the randomization in the following way. If there had been no sex effect, each supervisor would have reached the same decision regardless of the sex listed on the application. Fisher's Exact Test examines all possible assignments of applications to supervisors and finds that only 2.45% of them result in a difference as great as that observed. Since the 48 supervisors were not a random sample from any population, extension of the inference to any broader group is speculative.

19.1.2 Death Penalty and Race of Murder Victim—An Observational Study

In a 1988 law case, *McCleskey v. Zant*, lawyers for the defendant showed that, among black convicted murderers in Georgia, 35% of those who killed whites and 6% of those who killed blacks received the death penalty. Their claim, that racial discrimination was a factor in sentencing, was contested by lawyers for the state of Georgia. Murders with black victims, they argued, were more likely to be unaggravated "barroom brawls, liquor-induced arguments, or lovers' quarrels"—crimes which rarely result in the death sentence anywhere. Among white-victim murders, on the other hand, there was a higher proportion of killings committed in the course of an armed robbery or involving torture—which more often result in death sentences.

To examine the validity of this argument, researchers collected data on death penalty sentencing in Georgia, including the race of victim, categorized separately for six progressively serious types of murder. Category 1 comprises barroom brawls, liquor-induced arguments, lovers' quarrels, and similar crimes. Category 6 includes the most vicious, cruel, cold-blooded, unprovoked crimes. (Data from G. C. Woodworth, "Statistics and the Death Penalty," *Stats* 2 (1989): 9–12). Display 19.2 shows, for each category, a 2×2 table categorizing convicted black murderers according to the race of their victim and whether they received the death penalty. Is there evidence that black murderers of white victims are more likely to receive the death penalty than black murderers of black victims, even after accounting for the level of aggravation?

Display 19.2 Death sentences for black convicted murderers of white and black victims in Georgia, by aggravation category

Aggravation Level	Race of Victim	Death Penalty Yes	Death Penalty No
1	white	2	60
	black	1	181
2	white	2	15
	black	1	21
3	white	6	7
	black	2	9
4	white	9	3
	black	2	4
5	white	9	0
	black	4	3
6	white	17	0
	black	4	0

Summary of Statistical Findings

There is convincing evidence that convicted black murderers were more likely to receive the death sentence if their victim was white than if their victim was black, even after accounting for the aggravation level of the crime (p-value $= .0004$ from a Mantel–Haenszel Test). The odds of a white-victim murderer receiving the death penalty are estimated to be 5.5 times the odds for a black-victim murderer in the same aggravation category.

Scope of Inference

Since the data are observational, one cannot discount the possibility that other confounding variables may be associated with both the victim of the crime and the sentence. However, it is safe to exclude the level of aggravation and chance variation as determinants of the disparity in sentencing.

19.2   Population Models for 2 × 2 Tables of Counts

19.2.1   Hypotheses of Homogeneity and of Independence

The last chapter presented tests for the following identical hypotheses:

$$H: \pi_2 - \pi_1 = 0 \qquad \text{and} \qquad H: \omega_2/\omega_1 = 1,$$

where π_2 and π_1 are population proportions or probabilities and ω_2 and ω_1 are the corresponding odds. These are often called hypotheses of *homogeneity* because interest is focused on whether the distribution of the binary response is homogeneous—the same— across populations.

The hypothesis of *independence*, appropriate for some sampling schemes, is used to investigate an association between row and column factors without specifying one of them as a *response*. Although the hypothesis can be expressed in terms of parameters, the actual parameters used differ for different sampling schemes, and it is more convenient to use words:

H: The row categorization is independent of the column categorization.

For example, individuals in a population may be categorized according to whether they favor legalized abortion or not and according to whether they favor the death penalty or not. It is natural to ask whether an association exists between the two factors—whether people who answer yes on one of the questions are more (or less) likely to answer yes on the other. If there is no association, the factors are independent. With the test of independence, the two factors are treated equally, and neither needs to be specified as the response.

One difference between hypotheses of homogeneity and of independence lies in whether one thinks of two populations, each with two categories of response, or of a single population with four categories of response (formed by two binary factors in a 2 × 2 table). The homogeneity hypothesis is appropriate for the first, and independence is appropriate for the second. Whether the two-population or single-population model (or both) is appropriate depends on the research question and on how the data were collected. In the next section, several sampling schemes are described, and the appropriateness of the two hypotheses is discussed for each.

19.2.2   Sampling Schemes Leading to 2 × 2 Tables

Poisson Sampling

In this scheme (named after the French mathematician Siméon Denis Poisson, 1781–1840), a fixed amount of time (or space, volume, money, etc.) is devoted to collecting a random sample from a single population, and each member of the population falls into one of the four cells of the 2 × 2 table. An example is the heart disease and obesity data of Section 18.1.1 (page 516). In that case, the researchers spent a certain amount of time sampling health records in American Samoa and ended up with 3,112 women. The women were categorized as having been obese or not, and as having died of cardiovascular disease (in six years) or not. One easy way to recognize this sampling scheme is to observe that none of the marginal totals (see Display 19.3) is known in advance of data collection.

Multinomial Sampling

This scheme is very similar to Poisson sampling except that the total sample size is determined in advance. If a list of all women in American Samoa existed, the researcher could have selected a random sample of 5,000 from this list. The term *multinomial* means that each subject selected in the sample is categorized into one of several cells—in this case, one of four cells in the 2 × 2 table. The only marginal total known in advance of data collection is the grand total.

Prospective Product Binomial Sampling

This sampling scheme formed the basis for deriving the tools in the previous chapter. The two populations defined by the levels of the explanatory factor are identified, and random samples are selected from each. If separate lists of the populations of obese American Samoan women and nonobese American Samoan women were available, a random sample of 2,500 could be selected from each. The term *binomial* means that each woman (in each population) falls into one of *two* categories (see Section 18.2.1 on page 519). *Product binomial* is a term whose statistical meaning reflects there being more than one binomial population from which independent samples have been taken. The sizes of the samples from each of the populations are chosen by the researcher.

Retrospective Product Binomial Sampling

Retrospective studies were discussed in Section 18.4.1 (page 529). Random samples are selected from the subpopulations defined by each level of the response factor. The sampling scheme is technically the same as in the prospective product binomial, except that the roles of response and explanatory factors (according to the questions of interest) are reversed. The researcher specifies the sample totals to be obtained at each level of the response factor.

Display 19.3   Recognizing the sampling schemes and appropriate test hypotheses by noticing which marginal totals are fixed

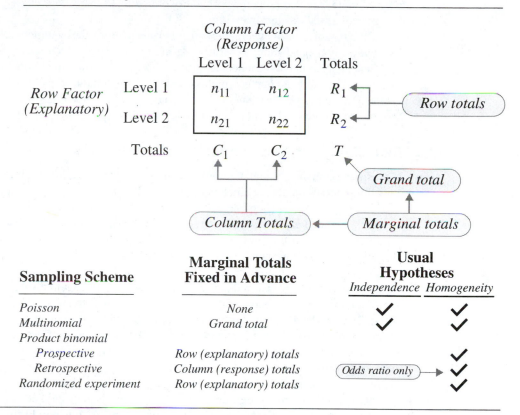

Sampling Scheme	Marginal Totals Fixed in Advance	Usual Hypotheses	
		Independence	Homogeneity
Poisson	None	✓	✓
Multinomial	Grand total	✓	✓
Product binomial			
Prospective	Row (explanatory) totals		✓
Retrospective	Column (response) totals	Odds ratio only →	✓
Randomized experiment	Row (explanatory) totals		✓

Randomized Binomial Experiment

In a randomized experiment, the subjects—regardless of how they were obtained—are randomly allocated to the two levels of the explanatory factor. Aside from the fact that randomization rather than random sampling is used, the setup here is essentially the same as in prospective product binomial sampling. The explanatory factor totals are fixed by the researcher.

The Hypergeometric Probability Distribution

Sometimes an experiment can be conducted in such a way that the row and column totals are *both* fixed by the researcher. In a classic example, a woman is asked to taste 14 cups of tea with milk and categorize them according to whether the tea or the milk was poured first. She is told that there were seven of each, so she tries to separate them into two groups of size seven. If a 2 × 2 table is formed, with the row factor being actual status for each of the 14 cups (milk first or tea first) and the column factor being guessed status (milk first or tea first), then each row and each column total will be seven. If the subject cannot correctly guess the order of milk and tea, then the seven cups she guesses to have had milk first are like a random sample from the 14 eligible cups. The hypergeometric probability distribution is appropriate for calculating the probability of getting as many correct assignments as she did, given that she was just guessing.

Although this type of *hypergeometric experiment* is rare, the hypergeometric probability distribution is applied more generally. If interest is strictly focused on the odds ratio, statistical analysis may be conducted conditionally on the row and column totals that were observed. Therefore, no matter which sampling scheme produced the 2×2 table, one may use the hypergeometric distribution to calculate a *p*-value. This is known as *Fisher's Exact Test*, which is described in more detail in Section 19.4.

19.2.3   Testable Hypotheses and Estimable Parameters

In prospective product binomial sampling and randomized experiments, it is very natural to draw inferences about the parameters π_1 and π_2 (or ω_1 and ω_2) describing the two binomial distributions. A test of homogeneity is used to determine whether the populations are identical. A confidence interval for $\pi_1 - \pi_2$ or for the odds ratio, ω_1/ω_2, is used to describe the difference. As discussed in Chapter 18, ω_1/ω_2 is the only comparison parameter that may be estimated if the sampling is retrospective.

With Poisson and multinomial sampling, information is available about the single population with four categories, and independence may be tested without specifying either row or column factor as a response. Some further options are also available. Either the row or the column may be designated as a response, thus dividing the population into two subpopulations for each level of the explanatory factor. Although the sample sizes may differ, the samples are indeed random samples from each subpopulation. Consequently, all inferential tools for prospective product binomial sampling may be used with Poisson and multinomial sampling as well. The choice of testing for independence or homogeneity should depend on whether treating one of the categories as a response is appropriate. A summary of the hypotheses that may be tested for each sampling scheme is shown in Display 19.3.

The heart disease and obesity study employed Poisson sampling. Since CVD death is the response factor, one would phrase inferences in terms of proportions of CVD deaths in the two subpopulations, thus testing homogeneity and providing a confidence interval for the difference in proportions or the ratio of odds.

Estimable Parameters

Although testing independence or estimating an odds ratio is often a primary goal, additional interest may be focused on other population features. The question of interest may involve the overall proportion of yes responses in the entire population or the proportions of units having different levels of the explanatory factor. For example, one might have interest in estimating the overall proportion of American Samoan women who were obese or the overall proportion of CVD deaths. These are marginal proportions, in the sense that estimates are constructed from ratios of marginal totals in the table of counts. Display 19.4 indicates whether the proportion of the population in row 1 (call this population quantity ρ) and in column 1 (call this γ) can be estimated from the different sampling procedures.

Display 19.4 Estimable parameters for sampling schemes leading to 2 × 2 tables of counts (the check mark indicates the feature is estimable)

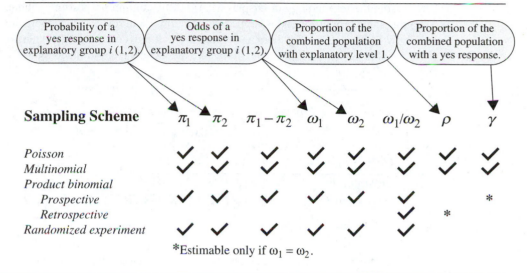

*Estimable only if $\omega_1 = \omega_2$.

19.3 The Chi-Squared Test

19.3.1 The Pearson Chi-Squared Test for Goodness of Fit

Karl Pearson (1857–1936) devised a general method for comparing fact with theory. His approach assumes that sampled units fall randomly into cells and that the chance of a unit's falling into a particular cell can be estimated from the theory under test. The number of units that fall in a cell is the cell's *observed* count, and the number predicted by the theory to do so is the cell's *expected* count. Pearson combined these into the single statistic

$$\chi^2 = \sum \frac{(Observed - Expected)^2}{Expected}.$$

The sum is over all cells of the table. The name *Pearson chi-square statistic* attached to χ^2 (the square of the Greek letter *chi*) reflects its approximate sampling distribution, which is chi-squared if the theory (null hypothesis) is correct. The degrees of freedom are equal to (Number of cells) minus (Number of parameters estimated) minus 1.

19.3.2   Chi-Squared Test of Independence in a 2 × 2 Table

A 2 × 2 table contains four cells. The proportion of counts that fall in column 1 is C_1/T (using the terminology in Display 19.3). If the column proportion is independent of row, this same proportion should appear in both row 1 and row 2. Since there are R_1 subjects in row 1, $R_1 \times (C_1/T)$ of them are expected to fall into column 1. By extension, the expected count in cell (i, j), if row and column are independent, is estimated by $R_i C_j/T$, as follows.

Observed and Estimated Expected Cell Counts
for Testing Independence

Cell (i,j):	$(1,1)$	$(1,2)$	$(2,1)$	$(2,2)$
Observed:	n_{11}	n_{12}	n_{21}	n_{22}
Expected (est.):	$R_1 C_1/T$	$R_1 C_2/T$	$R_2 C_1/T$	$R_2 C_2/T$

The pattern for determining an estimate for an expected cell count is to multiply the marginal totals in the same row and the same column together, then divide by the grand total.

Testing the hypothesis of independence involves the following steps:

1. Estimate the expected cell counts from the preceding formulas.
2. Compute the Pearson X^2 statistic from the formula in the previous box.
3. Find the p-value as the proportion of values from a chi-squared distribution on one degree of freedom that are greater than χ^2, using calculator, computer, or the table in Appendix A.3.

A small p-value provides evidence that the row and column categorizations are not independent. The example in Display 19.5 tests independence of the death penalty sentence and the race of the victim, using all six aggravation levels combined.

19.3.3   Equivalence of Several Tests for 2 × 2 Tables

The Chi-Squared Test for Homogeneity

Since the expected cell counts one obtains under the independence hypothesis are also the expected cell counts under the homogeneity hypothesis, the chi-squared test of independence in Section 19.3.2 is also a test of homogeneity. The chi-squared statistic is identical to the square of the Z-statistic for testing equal population proportions (see Section 18.2.3 on page 523). Furthermore, the χ^2 distribution on one degree of freedom is identical to the distribution of the square of a standard normal variable. Consequently, the p-values from the chi-squared test of independence and the chi-squared test of homogeneity are identical to the two-sided p-value from the Z-test of equal population proportions.

Display 19.5 Chi-squared test of independence between death penalty and race of victim (ignoring aggravation level of the crime)

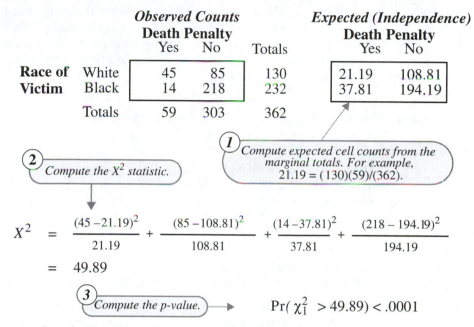

		Observed Counts **Death Penalty**		**Totals**	**Expected (Independence)** **Death Penalty**	
		Yes	No		Yes	No
Race of	White	45	85	130	21.19	108.81
Victim	Black	14	218	232	37.81	194.19
	Totals	59	303	362		

2 *Compute the X^2 statistic.*

1 *Compute expected cell counts from the marginal totals. For example,* $21.19 = (130)(59)/(362)$.

$$X^2 = \frac{(45-21.19)^2}{21.19} + \frac{(85-108.81)^2}{108.81} + \frac{(14-37.81)^2}{37.81} + \frac{(218-194.19)^2}{194.19}$$

$$= 49.89$$

3 *Compute the p-value.* → $\Pr(\chi_1^2 > 49.89) < .0001$

Conclusion: There is convincing evidence that death penalty sentence and race of victim are not independent.

Calculation and Continuity Correction

The most convenient formula for the chi-squared statistic from a 2×2 table is the following:

$$\chi^2 = \frac{T(n_{11}n_{22} - n_{12}n_{21})^2}{R_1 R_2 C_1 C_2}.$$

The statistic has approximately a chi-squared distribution with one degree of freedom, under the independence or homogeneity hypothesis. The approximation is premised on there being large cell counts; all expected counts larger than 5 is the customary check for validity.

An adjustment—the *continuity correction*—improves the approximation when sample sizes are not large:

$$\chi^2 = \frac{T(|n_{11}n_{22} - n_{12}n_{21}| - T/2)^2}{R_1 R_2 C_1 C_2}.$$

Limitations

The chi-squared tests for independence and homogeneity extend to $r \times c$ tables of counts (see Section 19.6.1). Although the chi-squared test is one of the most widely used of statistical tools, it is also one of the least informative. First, the only product is a p-value; there is no associated parameter to describe the degree of dependence. Second, the alternative hypothesis—that row and column are not independent—is very general. When more than two rows and columns are involved, there may be a more specific form of dependence to explore (see Chapters 21 and 22).

19.4   Fisher's Exact Test: The Randomization (Permutation) Test for 2×2 Tables

Although there are several statistical motivations for using Fisher's Exact Test, it is explained here as a randomization test for the difference in sample proportions. It is identical to the randomization test presented in Chapter 4, except that the responses are binary and the statistic of interest is the difference in sample proportions. Fisher's Exact Test is appropriate for any sample size, for tests of homogeneity or of independence.

19.4.1   The Randomization Distribution of the Difference in Sample Proportions

This section illustrates the randomization distribution of the difference of sample proportions on a small, hypothetical example. While it is more convenient to use the formulas in the next section or a computer program to obtain the same p-value, the approach here is helpful for visualizing the randomization distribution on which the calculations are based. The p-value has a concise meaning related to the probabilities involved in the random allocation of subjects to treatment groups.

Consider a small, hypothetical version of the sex role stereotype study of Section 19.1.1. Suppose each of seven men is randomly assigned to one of two groups such that four are in group 1 and three are in group 2. Each man decides whether to promote the applicant, based on the personnel file. Suppose that the men in group 1 are A, B, C, and D, while E, F, and G are in group 2. Suppose further that A, B, C, and E decide to promote while the others decide not to promote. Then the proportions of promotions in groups 1 and 2 are, respectively, 3/4 and 1/3. This evidence suggests that group 1 has a larger promotion rate than group 2; but the observed difference may be the result of the chance allocation of subjects to groups. The p-value measures how likely that explanation is.

The randomization distribution of the difference in sample proportions is a listing of the differences in proportions, given the observed outcomes, for every possible allocation of the men into groups of size four and three. This randomization distribution is shown in Display 19.6. The groupings in Display 19.6 are ordered according to the difference between promotion proportions.

The promotion decision for each man remains the same, but his group status is allowed to vary. Therefore, the differences in observed proportions in the final column are due entirely to the different group assignments. The one-sided p-value from the randomization test is

Display 19.6   The randomization distribution of the difference in two sample proportions

			Hypothetical Grouping		Number of 1's		Difference in sample
			Group 1	Group 2	in 1	in 2	Proportions
(1) List the subjects with their observed responses.			ADFG	BCE	1	3	−0.750
			BDFG	ACE	1	3	−0.750
			CDFG	ABE	1	3	−0.750
			DEFG	ABC	1	3	−0.750
Subject Code	Actual Group	Response	ABDF	CEG	2	2	−0.167
			ABDG	CEF	2	2	−0.167
A	1	1	ABFG	CDE	2	2	−0.167
B	1	1	ACDF	BEG	2	2	−0.167
C	1	1	ACDG	BEF	2	2	−0.167
D	1	0	ACFG	BDE	2	2	−0.167
E	0	1	ADEF	BCG	2	2	−0.167
F	0	0	ADEG	BCF	2	2	−0.167
G	0	0	AEFG	BCD	2	2	−0.167
(2) List all possible ways that the subjects can be arranged into two groups.			BCDF	AEG	2	2	−0.167
			BCDG	AEF	2	2	−0.167
			BCFG	ADE	2	2	−0.167
			BDEF	ACG	2	2	−0.167
			BDEG	ACF	2	2	−0.167
(3) For each arrangement, calculate the difference between the sample proportions; e.g. (2/4 − 2/3) = −0.167			BEFG	ACD	2	2	−0.167
			CDEF	ABG	2	2	−0.167
			CDEG	ABF	2	2	−0.167
			CEFG	ABD	2	2	−0.167
			ABCD	**EFG**	**3**	**1**	**0.417**
			ABCF	DEG	3	1	0.417
			ABCG	DEF	3	1	0.417
(4) The P-value is the proportion of groupings with differences in proportions ≥ 0.417			ABDE	CFG	3	1	0.417
			ABEF	CDG	3	1	0.417
			ABEG	CDF	3	1	0.417
			ACDE	BFG	3	1	0.417
			ACEF	BDG	3	1	0.417
One-sided p-value = 13/35 = 0.37			ACEG	BDF	3	1	0.417
			BCDE	AFG	3	1	0.417
			BCEF	ADG	3	1	0.417
			BCEG	ADF	3	1	0.417
			ABCE	DFG	4	0	1.000

the proportion of groupings that lead to values of $\hat{\pi}_1 - \hat{\pi}_2$ greater than or equal to the one actually observed ($\hat{\pi}_1 - \hat{\pi}_2 = 0.417$). Since 13 groupings out of the 35 lead to differences as great as or greater than the one observed, the p-value is .37. Interestingly, if the alternative hypothesis had been two-sided in this example, the p-value might have been defined as the proportion of groupings that have $\hat{\pi}_1 - \hat{\pi}_2$ greater than or equal to 0.417 in magnitude. The p-value in this case is .49, which is quite different from twice the one-sided p-value.

This motivation suggests an appealing interpretation of p-values from randomized, binomial experiments—as the probability that randomization alone is responsible for the observed disparity between proportions. If an applicant's sex had no effect in the sex role stereotype study, the promotion decisions would be unaffected by whether a male or female folder was received. However, the probability that chance allocations led to a disparity in proportions as great as or greater than the observed difference—assuming decisions are unaffected by sex of the applicant—is only .0245.

19.4.2 The Hypergeometric Formula for One-sided *P*-Values

Although the randomization distribution can always be listed as in Display 19.6, the number of possible outcomes can be quite large even for small sample sizes. There is, however, a short-cut formula for the p-value, based on the *hypergeometric probability function*:

1. Focus on the upper left-hand cell of the table in Display 19.1. (It makes no difference which cell is selected, but the customary choice is the upper left-hand cell, and the instructions that follow are based on that choice.)

2. Determine the *expected* number of counts in the upper left-hand cell of the table, as in the computation of the chi-squared statistic: $Expected = R_1 C_1 / T$.

3. Identify values (k) for the upper left-hand cell of the table that are as extreme as or more extreme than the observed value n_{11} in evidence against the null hypothesis. If n_{11} exceeds *Expected*, the values are $k = n_{11}, n_{11} + 1, \ldots, \min(R_1, C_1)$ (i.e., the smaller of the totals in the first row and first column). If n_{11} is less than *Expected*, the values are $k = 0, 1, \ldots, n_{11}$.

4. For each value of k, determine the proportion of possible randomizations leading to that value, according to the hypergeometric probability formula:

$$\Pr\{k\} = \frac{R_1! R_2! C_1! C_2!}{T! k! (R_1 - k)! (C_1 - k)! (R_2 - C_1 + k)!}.$$

(Recall that $5! = 5 \times 4 \times 3 \times 2 \times 1$ and that $0!$ is taken to be 1.)

5. Calculate the p-value by adding up these probabilities. Display 19.7 illustrates the calculations for the sex role stereotype data.

19.4.3 Fisher's Exact Test for Observational Studies

Fisher's Exact Test has been presented here as a randomization test based on the statistic $\hat{\pi}_1 - \hat{\pi}_2$. If the data are observational, however, it is thought of as a permutation test, which is a useful interpretation when the entire population has been sampled or when the sample is not random. The permutation test here is also equivalent to a sampling distribution, thus allowing inference about population parameters with random samples from the Poisson, multinomial, or product binomial sampling schemes. As demonstrated, the calculations can be used to obtain exact p-values for tests of equal population proportions, of equal population odds, or for independence.

19.4.4 Fisher's Exact Test Versus Other Tests

If Fisher's test gives exact p-values for all hypotheses, why should one ever consider using the Z-test for equal population proportions, the Z-test for equal population odds, or the

Display 19.7   Fisher's exact test: calculations for testing equal promotion probabilities for bank supervisors given "male" and "female" application folders

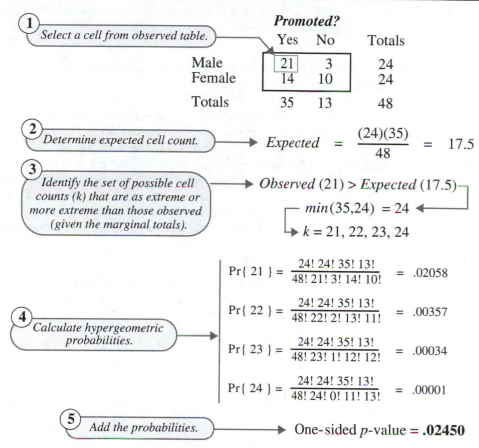

1. Select a cell from observed table.

Promoted?

	Yes	No	Totals
Male	21	3	24
Female	14	10	24
Totals	35	13	48

2. Determine expected cell count.

$$Expected = \frac{(24)(35)}{48} = 17.5$$

3. Identify the set of possible cell counts (k) that are as extreme or more extreme than those observed (given the marginal totals).

Observed (21) > Expected (17.5)

min(35,24) = 24

k = 21, 22, 23, 24

4. Calculate hypergeometric probabilities.

$$Pr\{ 21 \} = \frac{24!\ 24!\ 35!\ 13!}{48!\ 21!\ 3!\ 14!\ 10!} = .02058$$

$$Pr\{ 22 \} = \frac{24!\ 24!\ 35!\ 13!}{48!\ 22!\ 2!\ 13!\ 11!} = .00357$$

$$Pr\{ 23 \} = \frac{24!\ 24!\ 35!\ 13!}{48!\ 23!\ 1!\ 12!\ 12!} = .00034$$

$$Pr\{ 24 \} = \frac{24!\ 24!\ 35!\ 13!}{48!\ 24!\ 0!\ 11!\ 13!} = .00001$$

5. Add the probabilities.

One-sided *p*-value = **.02450**

Conclusion: There is substantial evidence that the observed division of favorable promotion decisions was not the result of random allocation.

chi-squared test for independence? If a test is all that is desired and if a computer program is available, there is indeed no reason to use a test other than Fisher's. But confidence intervals are commonly desired. Furthermore, if the sample sizes are large, the various tests differ very little, so choosing one over the others tends not to be a matter of major concern.

Report p-Values

It is important to provide a *p*-value rather than simply to state whether the *p*-value exceeds or fails to exceed some prespecified significance level (such as .05). It is especially important in categorical data analysis. For example, consider the test of whether a flipped coin has probability 1/2 of turning up heads. Based on 10 tosses, the probability of getting eight or more heads is .055 and the probability of getting nine or more heads is 0.011 (from binomial probability calculations—see Section 18.2.1 on page 520). Thus, the null hypothesis would be rejected at the 5% level of significance if nine or more heads were obtained, but only 1.1% of all possible outcomes lead to such a rejection.

19.5 Combining Results from Several Tables with Equal Odds Ratios

Cause-and-effect conclusions cannot be drawn from observational studies, because the effect of confounding variables associated with both the indicator variable (defining the populations) and the binary response cannot be ruled out. As with regression, however, tools are available for taking the next step: conducting comparisons that account for the effect of a confounding variable. The Mantel–Haenszel procedure may be used for this purpose when several 2×2 tables are involved, one for each level of a confounding variable, if the odds ratio is the same for all levels of the confounding variable. It is also appropriate for randomized block experiments in which a separate 2×2 table exists for each block.

19.5.1 The Mantel–Haenszel Excess

The *excess* is a name given to the observed count minus the expected count (according to the hypothesis) in one of the cells of the 2×2 table. It is only necessary to consider the excess in one of the cells, and the convention used here will be to focus on the upper left cell. If

$$Observed = n_{11}$$

then

$$Expected = \frac{R_1 C_1}{T}$$

and

$$Excess = Observed - Expected.$$

The relevant properties of the excess are displayed in the following box.

When the null hypothesis is correct,

$$\text{Mean}\{Excess\} = 0 \quad \text{and} \quad \text{Variance}\{Excess\} = \frac{R_1 R_2 C_1 C_2}{T T (T - 1)}.$$

And when, in addition, the sample sizes are large, *the sampling distribution of the excess is approximately normal.*

An approximate p-value is obtained in the usual way, based on a Z-test. This is illustrated in Display 19.8 in connection with the sex role stereotype data. The one-sided p-value is .012, which is somewhat smaller than the exact one-sided p-value of .024.

The essential features of this statistic are as follows:

1. For 2×2 tables of counts, the excess is an approximation to Fisher's Exact Test.
2. The p-value is close but not identical to the one from the Z-test for equal population proportions and the chi-squared test for independence.
3. The excess statistics from several 2×2 tables can be added up to get combined evidence in 2×2 tables for each of several levels of a confounding or blocking factor.

Display 19.8 Excess test as an approximation to Fisher's Exact Test: calculations for testing equal promotion probabilities for bank supervisors given "male" and "female" application folders

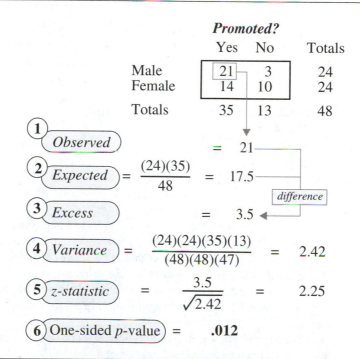

The Mantel–Haenszel excess test provides nothing new for analyzing 2×2 tables, but it is the only test that can be combined over several tables.

19.5.2 The Mantel–Haenszel Test for Equal Odds in Several 2 × 2 Tables

The Mantel–Haenszel testing procedure applies when several 2×2 tables are involved—one for each level of a confounding or blocking factor. Such is the case in the death penalty and race study of Section 19.1.2, where the confounding variable is the aggravation level of the crime. The proportion of death penalty sentences is expected to be larger for higher aggravation categories, but the relative odds—the odds of death penalty for white-victim murderers relative to the odds of death penalty for black-victim murderers—may nonetheless remain the same for all aggravation levels. If so, the Mantel–Haenszel test is appropriate for evaluating the hypothesis that the common odds ratio is 1.

The test statistic, a single *Excess* statistic, is the sum of the excesses from the individual tables. The procedure for running the test is as follows:

1. Select one cell in the 2×2 table, where the excess will be computed.
2. Compute *Excess* and Var{*Excess*} for that cell in each table separately.
3. The combined *Excess* is the sum of the individual ones.
4. The variance of the combined *Excess* is the sum of the variances of the individual ones. The standard deviation is the square root of this.
5. A Z-statistic is the combined *Excess* divided by its standard deviation.

This procedure is demonstrated in Display 19.9. Even after the aggravation level of the crime is accounted for, convincing evidence remains that the odds of a black murderer's getting the death penalty if the victim was white are greater than the odds of death penalty if the victim was black (one-sided p-value = .0004).

Notes About the Mantel–Haenszel Test

1. One assumption is that the odds ratio is the same in each 2×2 table. For one thing, the hypothesis that the odds ratio is 1 does not make sense if the odds ratio differs in the different 2×2 tables. Fortunately, even though the odds and the proportions may differ considerably in the different tables, the relative odds often remain roughly constant. A way to check the assumption is given in Chapter 21.
2. The p-value is approximate, based on large sample size. Fortunately, the total sample size of all tables is the important measure. A safe guideline is that the sum of the expected cell counts (added over all tables) should be at least 5, for each of the four cell positions.
3. The Mantel–Haenszel test treats the different levels of the confounding variable as nominal, ignoring any quantitative structure. It is possible to use *logistic regression* (see Chapters 20 and 21) to include aggravation level, for example, as a numerical explanatory variable.
4. The Mantel–Haenszel test is appropriate for prospective and retrospective observational data, as well as for randomized experiments.
5. Often a chi-squared version of the Mantel–Haenszel test is used. It is equal to the square of the Z-statistic.

19.5.3 Estimate of the Common Odds Ratio

The Mantel–Haenszel theory also provides an estimate of the common odds ratio from all tables combined. The estimated odds ratio in a single 2×2 table is $(n_{11}n_{22})/(n_{12}n_{21})$. In aggravation level 1 of Display 19.9, for example, this value is $[(2)(181)]/[(60)(1)] = 6.03$.

Display 19.9 Mantel–Haenszel calculations for death penalty and race data
from Section 19.1.2

① *Select a cell.* ⟶ White-victim murderers given the death penalty

② *Find the excess and its variance for each table.*

Aggravation Level	Race of Victim	Death Penalty Yes	No	Expected	Excess	Variance
1	white	2	60	0.762	1.238	0.564
	black	1	181			
2	white	2	15	1.308	0.692	0.699
	black	1	21			
3	white	6	7	4.333	1.667	1.382
	black	2	9			
4	white	9	3	7.333	1.667	1.007
	black	2	4			
5	white	9	0	7.313	1.688	0.640
	black	4	3			
6	white	17	0	17.000	0.000	0.000
	black	4	0			

③ *Add the excesses and their variances.* ⟶ 6.951   4.291

④ *Calculate the z-statistic.* ⟶ $z\text{-statistic} = \dfrac{6.951}{\sqrt{4.291}} = 3.356$

⑤ *One-sided p-value from the normal distribution* ⟶ $\Pr\{Z > 3.356\} = $ **.0004**

Conclusion: Nearly 7 more white-victim murderers received the death penalty than would have been expected if the odds of receiving the death penalty were the same for white-victim and black-victim murders. This cannot reasonably be attributed to chance.

To get an estimated odds ratio from all tables together, one sums the numerators in these expressions, weighted by the reciprocal of their table totals. Then one sums the denominators, similarly weighted. The estimate of the common odds ratio from several tables is

$$\hat{\omega} = \frac{\text{Sum of } \{n_{11}n_{22}/T\} \text{ over all tables}}{\text{Sum of } \{n_{12}n_{21}/T\} \text{ over all tables}}.$$

Use of this estimate assumes that the odds ratio is the same in all tables.

Example—Death Penalty and Race

The Mantel–Haenszel estimated odds ratio from the death penalty data is

$$\hat{\omega} = \frac{\dfrac{(2)(181)}{244} + \dfrac{(2)(21)}{39} + \dfrac{(6)(9)}{24} + \dfrac{(9)(4)}{18} + \dfrac{(9)(3)}{16} + \dfrac{(17)(0)}{21}}{\dfrac{(60)(1)}{244} + \dfrac{(15)(1)}{39} + \dfrac{(7)(2)}{24} + \dfrac{(3)(2)}{18} + \dfrac{(0)(4)}{16} + \dfrac{(0)(4)}{21}} = 5.49.$$

After the aggravation level of the crime has been accounted for, the odds of a black killer's receiving the death penalty are estimated to be 5.49 times greater if the victim was white than if the victim was black. Software that computes this estimate also provides a standard error or a confidence interval. The method is complex and will not be described here.

19.6   Related Issues

19.6.1 $r \times c$ Tables of Counts

More general tables arise from counts of subjects falling into cross-classifications of several factors, each with many levels. Display 19.10 shows a table listing the number of homicides of children by their parents, over 10 years in Canada, categorized according to the parent–offspring sex combination and according to age category. (Data from M. Daly and M. Wilson, "Evolutionary Social Psychology and Family Homocide," *Science* 242 (1988): 519–24.)

Display 19.10   Parent–offspring homicides in Canada, 1974 to 1983, by sexes of the parent killer and the child killed, and by age class of the child

Sex of Parent/Child	Age Classification of Child				
	Infantile (0–1)	Oedipal (2–5)	Latency (6–10)	Circumpubertal (11–16)	Adult (≥ 17)
Male/male	24	21	21	29	104
Male/female	17	27	10	14	47
Female/male	53	21	19	9	8
Female/female	50	27	5	4	15

The sampling scheme for this example is Poisson, so one should either test for independence of the row and column factors or declare one of these as a response factor and test for homogeniety of the (multinomial) responses across levels of the explanatory factor. For this problem it is unnecessary to specify either factor as response.

The chi-squared test of independence is conducted just as for 2×2 tables; that is, the *observed* cell count in row i and column j is n_{ij}, and the *expected* cell count (supposing that the null hypothesis of independence or homogeneity is true) is the row total times the column total, divided by the grand total: $R_i C_j / T$. The chi-squared statistic is the sum, for all cells in the table, of

$$(Observed - Expected)^2 / Expected.$$

The p-value is found by comparing this test statistic to a chi-squared distribution on $(r - 1) \times (c - 1)$ degrees of freedom, where r is the number of rows and c is the number of columns in the table.

19.6.2 Higher-dimensional Tables of Counts

Some additional structure is present in the table because the four levels of the row factor correspond to the cross-classifications of parent (male or female) and child (male or female). The table should therefore more properly be considered a $2 \times 2 \times 5$ table of counts. The chi-squared analysis has been historically used for this type of table. An initial screening test assesses independence between all three factors. If the independence hypothesis is rejected, the chi-square statistic is decomposed into a set of component chi-squared statistics with smaller degrees of freedom, each testing different aspects of the dependency structure. This is accomplished by means of a series of slices and lumpings of categories. Carried out fully, the original chi-squared statistic is decomposed into chi-squared statistics (with 1 degree of freedom), each of which tests a hypothesis of interest.

Because the analysis sequence resembles an analysis of variance so closely, it suggests that there may be a comparable regression approach. There is. Chapter 22 introduces Poisson log-linear regression, which offers a more flexible approach to analyzing tables of counts, using regression techniques. For example, age of the offspring may be modeled either as a factor or as a numerical explanatory variable.

19.7 Summary

Fisher's Exact Test can be used for testing in 2×2 tables, no matter how the data arose, no matter which of the two hypotheses is being tested, and no matter what the sample sizes are. The interpretation of the p-value, however, should be customized to the particular hypothesis, which is chosen according to the question of interest and the sampling scheme. For large sample sizes, the easier-to-compute approximations are adequate.

Pearson's chi-squared statistic is a convenient approximation to the exact test. A principal advantage of the Pearson method is that it permits testing independence in larger tables. The principal disadvantage is that it is a test procedure only.

The Mantel–Haenszel Test compares the odds in two groups after accounting for the effect of a blocking or confounding variable. If the blocking variable is numerical, however, there is some merit in attempting a more specific model, using logistic regression as in Chapter 21. Logistic regression can also be used to test for the condition that the odds ratio is the same in all the 2 × 2 tables: if it isn't, the Mantel–Haenszel test is inappropriate.

Sex Role Stereotypes in Promotions

Fisher's Exact Test is appropriate for testing whether the probability of promotion is the same regardless of the sex of the applicant. The *p*-value in this case, .0245, is exact and provides an inferential measure tied directly to the probabilities of assignment in the randomized experiment.

Death Penalty Sentencing Study

The Mantel–Haenszel procedure applied to these data computes an overall excess of death penalty sentences given to murderers of white victims. *Excess* means the additional number of murderers receiving the death penalty over what would be expected if death penalty and race of victim are independent. The excess is a direct measure of the practical significance of the disparity in sentencing and provides the basis for the measure of statistical significance.

Further Reading

For more on the mathematics of the hypergeometric probability distribution, see Rice (1995). For a discussion of Fisher's Exact Test and the chi-squared test, see Yates (1984). Accounts of chi-squared tests are provided in most introductory statistics text books. Most of the topics in this chapter are also covered in books on categorical data analysis, such as the one by Agresti (1990).

19.8 **Exercises**

Conceptual Exercises

1.   Sex Role Stereotypes. (a) From this study, what can be inferred about the attitudes of bank managers toward promotion of females? (b) How would you respond to someone who asserted that the higher proportion of the "male" folders promoted occurred simply by chance, because the tougher managers happened to be placed in the "female" group?

2.   Death Penalty and Race. (a) Lawyers for the defense showed that only 6% of black-victim murderers received the death penalty, while 35% of white-victim murderers received the death penalty. They claimed that the pattern of sentencing was racially biased. How should the (statistically knowledgeable) lawyers for the state respond? (b) What can the lawyers for the defense say to rebut the state lawyers' response? (c) How should the lawyers for the state respond to this further analysis?

3.   (a) Why is the hypothesis $H: \pi_1 - \pi_2 = 0$ identical to the hypothesis $H: \omega_1/\omega_2 = 1$. (b) Does this mean that retrospective data can be used to estimate $\pi_1 - \pi_2$?

4.   Coffee Drinking and Sexual Habits. From a sample of 13,912 households in Washtenaw County, Michigan, 2,993 persons were identified as being 60 years old or older. Of these, 1,956 agreed to participate in a study. The following table is based on data from the married women in this sample who answered the questions about whether they were sexually active and whether they drank coffee. (Data from A. C. Diokno et al., "Sexual Function in the Elderly," *Archives of Internal*

Medicine 150 (1990): 197–200.) (a) What type of sampling scheme was this? To what population can inferences be made? (b) Are these data more useful for investigating whether coffee drinking causes decreased sexual activity or whether sexual activity causes decreased coffee drinking? (c) If it is desired to treat sexual activity as a categorical response, does the inequality of the row totals create a problem?

| | Sexually Active | |
	Yes	No
Coffee drinker	15	25
Not coffee drinker	115	70

5.   Crab Parasites. Crabs were collected from Yaquina Bay, Oregon, during June 1982 and categorized by their species (Red or Dungeness) and by whether a particular type of parasite was found. (a) What type of sampling would you call this? (b) What tool would you use to determine whether the parasite is more likely to be present on Red Crab than on Dungeness Crab?

| | Parasite Present | |
	Yes	No
Red Crab	5	312
Dungeness Crab	0	503

6.   Hunting Strategies. A popular way to synthesize results in a research area is to assemble a bibliography of all similar studies and to use a random sample of them for statistical analysis. The following table categorizes a random sample of 50 (out of 1,000) published references relating to the hunting success of predators. Each article was categorized according to whether the animal under discussion hunted alone or in groups and whether it hunted single or multiple prey. (Data from R. E. Morris, Oregon State University Department of Fisheries and Wildlife.) (a) What type of sampling would you call this? (b) What is the population to which inferences can be made?

| | Prey Sought | |
	Single	Multiple
Hunt alone	17	12
Hunt in groups	14	7

7.   (a) In what way is the relationship between the normal and chi-squared distributions similar to the relationship between the t- and F-distributions? (b) How could one obtain a p-value for the chi-squared test of independence in a 2×2 table, using the standard normal (and not the chi-squared) distribution?

8.   (a) In investigating the odds ratio from a randomized experiment, why must one account for the effects of an additional variable (with the Mantel–Haenszel procedure, for example), if it is important? (b) What further reason is there if the data are observational?

9.   The mathematics and physics departments of Crane and Eagle colleges (fictitious institutions) had a friendly competition on a standard science exam. Of the two math departments, the Crane students had a higher percentage of passes; of the two physics departments, the Crane students again had a higher percentage of passes. But for the two departments combined, the students at Eagle had a higher percentage of passes. (No joke: this is an example of *Simpson's paradox*.) The results are shown in the following table. Which college "won"? Why? Can you explain the apparent paradox in nonstatistical language?

	Mathematics		*Physics*	
	Pass	*Fail*	*Pass*	*Fail*
Crane College	12	6	8	14
Eagle College	20	16	1	3

Computational Exercises

10. Coffee Drinking and Sexual Habits. Find the *p*-value for the chi-squared test of independence in the coffee drinking and sexual habits data (Exercise 4).

11. Crab Parasites. Find the one-sided *p*-value for testing whether the odds of a parasite's being present are the same for Dungeness Crab as for Red Crab (Exercise 5), using Fisher's Exact Test. (Notice that only one hypergeometric probability needs to be computed.)

12. Hunting Strategy. For the hunting strategy data of Exercise 6, find the *p*-value for testing independence by (a) the chi-squared test of independence, and (b) Fisher's Exact Test.

13. Smoking and Lung Cancer. For the data in Section 18.1.3 on page 518, test whether the odds of lung cancer for smokers are equal to the odds of lung cancer for nonsmokers, using (a) the excess statistic, and (b) Fisher's Exact Test.

14. Mantel–Haenszel Test for Censored Survival Times: Lymphoma and Radiation Data. Central Nervous System (CNS) lymphoma is a rare but deadly brain cancer; its victims usually die within one year. Patients at Oregon Health Sciences University all received chemotherapy administered after a new treatment designed to allow the drugs to penetrate the brain's barrier to large molecules carried in the blood. Of 30 patients as of December 1990, 17 had received only this chemotherapy; the other 13 had received radiation therapy prior to their arrival at OHSU. The table in Display 19.11 shows survival times, in months, for all patients. Asterisks refer to censored times; in these cases, the survival time is known to be at least as long as reported, but how much longer is not known, because the patient was still living when the study was reported. (Data from E. A. Neuwelt et al., "Primary CNS Lymphoma Treated with Osmotic Blood-brain Barrier Disruption: Prolonged Survival and Preservation of Cognitive Function," *Journal of Clinical Oncology* 9(9) (1991): 1580–90.)

The Mantel–Haenszel procedure can be used to construct a test for equality of survival patterns in the two groups. It involves constructing separate 2 × 2 tables for each month in which a death occurred, as shown in Display 19.12. The idea is to compare the numbers surviving and the numbers dying in the two groups, accounting for the length of time until death. The data are arranged in several 2 × 2 tables, as in Display 19.12. The censored observations are included in all tables for which the patients are known to be alive and are excluded from the others. Compute the Mantel–Haenszel test statistic (equivalently called the *log-rank statistic*), and find the one-sided *p*-value. What can be inferred from the result?

15. Bats. Bat Conservation International (BCI) is an organization that supports research on bats and encourages people to provide bat roost boxes (much like ordinary birdhouses) in backyards and gardens. BCI phoned people across Canada and the United States who were known to have put out boxes, to determine the sizes, shapes, and locations of the boxes and to identify whether the boxes were being used by bats. Among BCI's findings was the conclusion, "Bats were significantly more likely to move into houses during the first season if they were made of old wood. Of comparable

Display 19.11   Survival times (months) for two groups of lymphoma patients

Status (Dead or Alive) and Survival Times as of December 1990
for CNS lymphoma patients with and without prior radiation

Radiation		**No Radiation**	
Months	*Status*	*Months*	*Status*
3	D	3	D
3	D	7	D
10	D	9	D
11	D	9	D
12	D	16	D
18	D	*17	A
21	D	*18	A
22	D	*20	A
23	D	24	D
33	D	*25	A
36	D	*30	A
48	D	32	D
*62	A	*38	A
		41	D
		*54	A
		*71	A
		*99	A

*Censored survival time: patient was alive at end of study; actual survival time is unknown,
but is at least as large as what is listed.

Display 19.12   Several 2 × 2 tables, formed from the censored and uncensored survival times

Month *After Diagnosis*	**Radiation Group**		**No Radiation Group**	
	Known to **Survive** *Beyond This Month*	*Known to* **Die** *After This Many Months*	*Known to* **Survive** *Beyond This Month*	*Known to* **Die** *After This Many Months*
3	11	2	16	1
7	11	0	15	1
9	11	0	13	2
10	10	1	13	0
11	9	1	13	0
12	8	1	13	0
16	8	0	12	1
18	7	1	10	0
21	6	1	9	0
22	5	1	9	0
23	4	1	9	0
24	4	0	8	1
32	4	0	5	1
33	3	1	5	0
36	2	1	5	0
41	2	0	3	1
48	1	1	3	0

houses first made available in spring and eventually occupied, eight houses made of old wood were all used in the first season, compared with only 32 of 65 houses made of new wood." (Data from "The Bat House Study," *Bats* 11(1) (Spring 1993): 4–11.) The researchers required that, to be significant, a chi-square test meet the 5% level of significance. (a) Is the chi-squared test appropriate for these data? (b) Evaluate the data's significance, using Fisher's Exact Test.

16. **Vitamin C and Colds.** Pursuing an experiment similar to the one in Section 18.1.2 on page 517, skeptics interviewed the same 800 subjects to determine who knew and who did not know to which group they had been assigned. Vitamin C, they argued, has a unique bitter taste, and those familiar with it could easily recognize whether their pills contained it. If so, the blind was broken, and the possibility arose that subjects' responses may have been influenced by the knowledge of their group status. The following table presents a fictitious view of how the resulting study might have turned out. Use the Mantel–Haenszel procedure (a) to test whether the odds of a cold are the same for the placebo and Vitamin C users, after accounting for group awareness, and (b) to estimate the common odds ratio.

Vitamin C and Colds Data

	Knew their group		Did not know	
	Cold	No cold	Cold	No cold
Placebo	266	46	62	26
Vitamin C	139	29	157	75

Data Problems

17. **Alcohol Consumption and Breast Cancer—A Retrospective Study.** A 1982–1986 study of women in Toronto, Ontario, assessed the added risk of breast cancer due to alcohol consumption. (Data from Rosenberg et al., "A Case–Control Study of Alcoholic Beverage Consumption and Breast Cancer," *American Journal of Epidemiology* 131 (1990): 6–14.) A sample of confirmed breast cancer patients at Princess Margaret Hospital who consented to participate in the study was obtained, as was a sample of cancer-free women from the same neighborhoods who were close in age to the cases. The following tables show data only for women in two categories at the ends of the alcohol consumption scale: those who drank less than one alcoholic beverage per month (3 years before the interview), and those that drank at least one alcoholic beverage per day. The women are further categorized by their body mass, a possible confounding factor. What evidence do these data offer that the risk of breast cancer is greater for heavy drinkers (≥ 1 drink/day) than for light drinkers (< 1 drink/month)?

Alcohol Consumption and Breast Cancer Study

	$<21 \ kg/m^2$		$21–25 \ kg/m^2$		$>25 \ kg/m^2$	
	Cases	Controls	Cases	Controls	Cases	Controls
At least 1 drink per day	38	52	65	147	30	42
Less than one drink per month	26	61	94	153	56	102

18. **The Donner Party.** In 1846 the Donner party became stranded while crossing the Sierra Nevada mountains near Lake Tahoe. (Data from D. J. Grayson, "Donner Party Deaths: A Demographic Assessment," *Journal of Anthropological Research* 46 (1990): 223–42. See also Section 20.1.1 on page 565.) Shown in the following tables are counts for male and female survivors for six age groups.

Is there evidence that the odds of survival are greater for females than for males, after the possible effects of age have been accounted for?

							Age Category						
	15–19			*20–29*			*30–39*			*40–49*			*50–59*	
	Lived	Died		Lived	Died		Lived	Died		Lived	Died		Lived	Died
Male	1	1		5	10		2	4		2	1		0	1
Female	1	0		6	1		2	0		1	3		0	1

	60–69	
	Lived	Died
Male	0	3
Female	0	0

Answers to Conceptual Exercises

1. (a) Any inference to a population other than the 48 males in the training course is speculative because these subjects were not randomly selected from any population. (b) They are possibly correct. But the chance of such an extreme randomization is about 2 out of 100 (the p-value).

2. (a) They argue that no inference that the victims' races cause different death penalty rates can be drawn from this observational data. Potentially confounding variables, such as aggravation category, might be associated with both the race of the victim and the likelihood of a death sentence. (b) The defense lawyers compare the odds of white-victim and black-victim death penalty sentences, after accounting for a conspicuous confounding variable—aggravation level (using the Mantel–Haenszel procedure). (c) The state lawyers respond that further confounding variables may exist, whether they can be identified or not, that might explain the difference.

3. (a) If $\pi_1 = \pi_2$, then $\pi_1/(1 - \pi_1) = \pi_2/(1 - \pi_2)$ (so their ratio is 1). (b) No. Unless $\omega_1/\omega_2 = 1$, the value of ω_1/ω_2 does not identify the value of $\pi_1 - \pi_2$.

4. (a) It is not clear whether the sampling scheme was Poisson or multinomial. The 225 women are a sample from a population of married women of age 60 or older in the particular county, who agreed to participate in the study and who answered the questions about sexual activity and coffee drinking. Extrapolating to any other population introduces the potential for bias. (b) Although the observational data can be examined for consistency with the causal theories, the statistical analysis alone cannot be used to imply causation, one way or another. (c) No.

5. (a) Poisson. (b) Fisher's Exact Test, since there is a problem with small cell counts.

6. (a) Multinomial. (b) The population of 1,000 published articles about hunting success.

7. (a) The F-distribution on 1 and k degrees of freedom describes the distribution of the square of a variable with the t-distribution on k degrees of freedom. Similarly, the chi-square distribution on 1 degree of freedom describes the square of a variable with the standard normal distribution. (b) Take the square root of χ^2, and double the area under a standard normal curve to the right of this.

8. (a) Accounting for the confounding variable in the analysis will make the inference about the association under study more precise. (b) The study's lack of control over the confounding variable at the design stage provides an added incentive.

9. Among math students, 67% passed at Crane, compared to 56% at Eagle. Among physics students, 36% passed at Crane, compared to 25% at Eagle. Among all students (ignoring department), 50% passed at Crane compared to 53% at Eagle. Crane won. The paradox arises because the math students had an easier time with their exam than the physics students had with theirs, coupled with the fact that Eagle had a very high proportion of participants from its math department (90%, compared to 45% at Crane). Crane won, because it won both match-ups involving students with supposedly comparable training.

Logistic Regression for Binary Response Variables

Logistic regression analysis describes how a binary (0 or 1) response variable is associated with a set of explanatory variables. The mean of a binary response is a probability, so the logistic regression model specifies that a probability—such as the probability of lung cancer—is related to a regression-like function of explanatory variables, such as age or smoking habits.

Interpretations of logistic regression coefficients are made in terms of statements about odds and odds ratios. Although the model, estimation, and inference procedure differ from those of ordinary regression, the practical use of logistic regression analysis with a statistical computing program parallels the usual regression analysis closely.

20.1   Case Studies

20.1.1   Survival in the Donner Party—An Observational Study

In 1846 the Donner and Reed families left Springfield, Illinois, for California by covered wagon. In July, the Donner Party, as it became known, reached Fort Bridger, Wyoming. There its leaders decided to attempt a new and untested route to the Sacramento Valley. Having reached its full size of 87 people and 20 wagons, the party was delayed by a difficult crossing of the Wasatch Range and again in the crossing of the desert west of the Great Salt Lake. The group became stranded in the eastern Sierra Nevada mountains when the region was hit by heavy snows in late October. By the time the last survivor was rescued on April 21, 1847, 40 of the 87 members had died from famine and exposure to extreme cold.

Display 20.1 shows the ages and sexes of the adult (over 15 years) survivors and non-survivors of the party. These data were used by an anthropologist to study the theory that females are better able to withstand harsh conditions than are males (Data from D. K. Grayson, 1990, "Donner Party Deaths: A Demographic Assessment," *Journal of Anthropological Research* 46: (1990): 223–42.) For any given age, were the odds of survival greater for women than for men?

Summary of Statistical Findings

The odds of survival for females were estimated to be 4.9 times the odds of survival for males of the same age. An approximate 95% confidence interval for this odds ratio is 1.1 to 21.6.

Scope of Inference

Since the data are observational, the result cannot be used as proof that women were more apt to survive than men; the possibility of confounding variables cannot be excluded. Furthermore, since the 45 individuals were not drawn at random from any population, inference to a broader population is not justified.

20.1.2   Birdkeeping and Lung Cancer— A Retrospective Observational Study

A 1972–1981 health survey in The Hague, Netherlands, discovered an association between keeping pet birds and increased risk of lung cancer. To investigate birdkeeping as a risk factor, researchers conducted a *case–control* study of patients in 1985 at four hospitals in The Hague (population 450,000). They identified 49 cases of lung cancer among patients who were registered with a general practice, who were age 65 or younger, and who had resided in the city since 1965. They also selected 98 controls from a population of residents having the same general age structure. (Data based on P. A. Holst, D. Kromhout, and R. Brand, 1988, "For Debate: Pet Birds as an Independent Risk Factor for Lung Cancer," *British Medical Journal* 297 (1988): 13–21.) Display 20.2 shows the data gathered on the following variables:

Display 20.1   Sex, age, and survival status of Donner Party members, 15 years or over

Name	Sex	Age	Survival
Antoine	male	23	no
Breen, Mary	female	40	yes
Breen, Patrick	male	40	yes
Burger, Charles	male	30	no
Denton, John	male	28	no
Dolan, Patrick	male	40	no
Donner, Elizabeth	female	45	no
Donner, George	male	62	no
Donner, Jacob	male	65	no
Donner, Tamsen	female	45	no
Eddy, Eleanor	female	25	no
Eddy, William	male	28	yes
Elliot, Milton	male	28	no
Fosdick, Jay	male	23	no
Fosdick, Sarah	female	22	yes
Foster, Sarah	female	23	yes
Foster, William	male	28	yes
Graves, Eleanor	female	15	yes
Graves, Elizabeth	female	47	no
Graves, Franklin	male	57	no
Graves, Mary	female	20	yes
Graves, William	male	18	yes
Halloran, Luke	male	25	no
Hardkoop, Mr.	male	60	no
Herron, William	male	25	yes
Noah, James	male	20	yes
Keseberg, Lewis	male	32	yes
Keseberg, Phillipine	female	32	yes
McCutcheon, Amanda	female	24	yes
McCutcheon, William	male	30	yes
Murphy, John	male	15	no
Murphy, Lavina	female	50	no
Pike, Harriet	female	21	yes
Pike, William	male	25	no
Reed, James	male	46	yes
Reed, Margaret	female	32	yes
Reinhardt, Joseph	male	30	no
Shoemaker, Samuel	male	25	no
Smith, James	male	25	no
Snyder, John	male	25	no
Spitzer, Augustus	male	30	no
Stanton, Charles	male	35	no
Trubode, J.B.	male	23	yes
Williams, Baylis	male	24	no
Williams, Eliza	female	25	yes

FM = Sex (1 = F, 0 = M)

AG = Age, in years

SS = Socioeconomic status (1 = High, 0 = Low), determined by occupation of the household's principal wage earner

YR = Years of smoking prior to diagnosis or examination

CD = Average rate of smoking, in cigarettes per day

BK = Indicator of birdkeeping (caged birds in the home for more than 6 consecutive months from 5 to 14 years before diagnosis (cases) or examination (controls)

Display 20.2   Birdkeeping and lung cancer data: FM = female indicator; AG = Age, in years; SS = High socioeconomic status indicator; YR = Years of smoking; CD = Average number of cigarettes per day; BK = Birdkeeper indicator

FM	LC	BK	SS	AG	YR	CD		FM	LC	BK	SS	AG	YR	CD		FM	LC	BK	SS	AG	YR	CD
0	1	1	0	59	39	20		1	1	0	0	63	29	20		0	1	1	0	64	42	20
0	0	0	0	61	42	12		0	0	1	1	62	0	0		0	0	1	0	63	40	10
1	0	1	0	58	36	15		0	0	0	1	63	0	0		0	0	0	1	64	39	25
1	1	1	0	61	38	15		0	1	1	0	37	19	12		0	1	1	0	60	38	15
0	0	1	0	60	36	25		1	0	0	0	40	16	2		0	0	1	1	60	25	15
1	0	0	0	59	0	0		0	0	1	0	38	20	20		1	0	0	1	59	41	12
0	1	1	0	63	41	10		0	1	0	0	61	28	15		0	1	1	1	63	41	40
1	0	0	0	64	45	20		0	0	0	1	60	0	0		0	0	0	0	64	44	20
0	0	0	0	63	22	20		0	0	0	1	59	34	1		1	0	1	0	66	38	25
1	1	1	0	64	42	20		0	1	0	0	62	43	20		1	1	1	0	44	22	15
0	0	1	0	63	0	0		0	0	1	0	62	38	20		0	0	0	0	43	21	20
1	0	1	0	64	40	20		0	0	0	0	61	44	30		0	0	1	0	45	0	0
1	1	1	0	60	0	0		1	1	1	0	50	28	20		1	1	1	0	49	27	20
0	0	0	0	61	0	0		0	0	0	0	51	23	12		1	0	0	0	48	25	15
0	0	1	0	62	20	10		0	0	1	0	50	0	0		0	0	1	0	49	25	15
1	1	1	0	47	25	25		0	1	0	0	64	32	3		0	1	1	0	41	22	15
1	0	0	1	47	0	0		1	0	1	1	63	0	0		1	0	1	1	42	21	8
0	0	1	0	46	0	0		0	0	0	1	63	41	20		0	0	0	1	40	13	25
0	1	0	0	56	36	25		0	1	0	0	53	33	20		0	1	0	0	58	35	25
1	0	0	0	55	41	30		1	0	0	1	52	29	10		0	0	0	0	58	22	10
0	0	0	0	57	18	10		1	0	0	0	54	19	15		0	0	0	0	59	37	15
0	1	0	0	63	45	20		0	1	1	1	62	39	20		0	1	1	0	66	47	10
0	0	0	0	62	22	15		0	0	0	0	61	43	20		1	0	0	0	66	0	0
0	0	0	0	63	20	15		0	0	0	0	62	28	18		0	0	0	0	65	44	6
1	1	0	0	64	40	25		0	1	1	0	64	44	15		0	1	0	0	58	38	20
1	0	0	1	64	0	0		1	0	0	0	65	34	10		1	0	0	0	58	38	20
0	0	0	0	65	7	2		0	0	0	1	63	42	10		0	0	0	0	59	34	20
0	1	1	1	64	13	30		1	1	1	0	58	37	20		0	1	1	1	59	40	15
0	0	0	0	64	41	6		0	0	0	0	64	39	20		0	0	1	1	60	39	12
0	0	0	1	65	39	20		1	0	1	0	58	34	15		0	0	0	1	59	7	1
0	1	0	1	51	24	15		0	1	1	1	65	43	30		1	1	1	0	46	24	15
1	0	0	1	52	5	4		0	0	0	1	66	30	20		1	0	1	0	45	24	4
0	0	0	1	50	17	10		1	0	0	1	63	47	45		0	0	1	1	47	16	5
0	1	0	1	58	42	30		1	1	1	0	49	23	20		0	1	1	1	64	47	16
0	0	0	0	58	15	40		0	0	1	0	50	0	0		0	0	0	0	64	36	15
0	0	1	0	59	35	20		1	0	0	0	48	27	20		0	0	0	1	63	43	20
0	1	1	0	56	36	20		0	1	0	1	56	26	25		0	1	1	0	49	31	20
1	0	0	0	58	39	20		1	0	1	0	56	22	25		0	0	1	0	49	15	10
1	0	1	0	55	39	15		1	0	1	0	55	36	20		1	0	1	0	48	28	10
0	1	1	0	66	50	25		0	1	1	0	62	40	15		0	1	1	1	52	31	20
1	0	0	0	65	46	20		1	0	0	1	63	14	30		0	0	1	0	53	28	25
1	0	0	1	67	0	0		1	0	0	1	61	0	0		1	0	1	1	52	30	37
0	1	1	0	46	24	15		0	1	1	0	57	39	25		0	1	0	1	43	19	15
1	0	0	1	46	24	20		1	0	0	0	55	35	15		1	0	0	0	43	25	25
0	0	1	0	47	28	40		0	0	1	0	58	24	8		0	0	0	1	42	17	15
1	1	1	0	54	33	6		0	1	1	0	56	33	10								
1	0	0	1	55	24	15		1	0	0	0	56	32	30								
1	0	1	0	53	9	10		0	0	0	0	57	24	15								
0	1	0	1	65	45	10		0	1	0	0	56	35	40								
0	0	0	1	66	42	18		0	0	0	1	57	19	5								
0	0	1	0	64	46	20		1	0	0	0	54	39	25								

Age and smoking history are both known to be associated with lung cancer incidence. After age, socioeconomic status, and smoking have been controlled for, is an additional risk associated with bird keeping?

Summary of Statistical Findings

The odds of lung cancer among birdkeepers are estimated to be 4.1 times as large as the odds of lung cancer among nonbirdkeepers, after accounting for the effects of smoking, sex, age, and socioeconomic status. An approximate 95% confidence interval for this odds ratio is 1.8 to 9.1. The data provide convincing evidence that increased odds of lung cancer were associated with keeping pet birds, even after accounting for the effect of smoking (one-sided p-value = .0004).

Scope of Inference

Inference extends to the population of lung cancer patients and unaffected individuals in The Hague in 1985. Statistical analysis of these observational data cannot be used as evidence that birdkeeping causes the excess cancer incidence among birdkeepers, although the data are consistent with that theory. [As further evidence in support of the causal theory, the researchers investigated other potential confounding variables (beta-carotene intake, vitamin C intake, and alcohol consumption). They also cited medical rationale supporting the statistical associations: people who keep birds inhale and expectorate excess allergens and dust particles, increasing the likelihood of dysfunction of lung macrophages, which in turn can lead to diminished immune system response.]

20.2 The Logistic Regression Model

The response variables in both case studies—*survival* in the Donner Party study and *lung cancer* in the birdkeeping study—are *binary*, meaning that they take on values either 0 or 1. Both studies involve several explanatory variables, some categorical and some numerical. Logistic regression is the appropriate tool in such situations.

20.2.1 Logistic Regression as a Generalized Linear Model

A generalized linear model (GLM) is a probability model in which the mean of a response variable is related to explanatory variables through a regression equation. Let $\mu = \mu\{Y|X_1, \ldots, X_p\}$ represent the mean response. The regression structure is linear in unknown parameters $(\beta_0, \beta_1, \ldots, \beta_p)$:

$$\beta_0 + \beta_1 X_1 + \cdots + \beta_p X_p;$$

and some specified function of μ—called the *link function*—is equal to the regression structure

$$g(\mu) = \beta_0 + \beta_1 X_1 + \cdots + \beta_p X_p.$$

The strength of a generalized linear model and analysis comes from the regression equation. Virtually all of the regression technology developed in Chapters 7 through 11 carries over with minor modification, allowing users to relate explanatory information to a much wider class of response variables. The special variables—indicator variables for categorical factors, polynomial terms for curvature, products for interactions—provide flexible and sensible ways to incorporate and evaluate explanatory information.

Link Functions

The appropriate link function depends on the distribution of the response variable. The natural choice of link for a normally distributed response, for example, is the identity link, $g(\mu) = \mu$, leading to ordinary multiple linear regression. Other response variables such as binary responses or counts have different natural links to the regression structure.

The Logit Link for Binary Responses

The natural link for a binary response variable is the *logit*, or log-odds, function. The symbol π (rather than μ) is used for the population mean, to emphasize that it is a proportion or probability. The logit link is $g(\pi) = \text{logit}(\pi)$, which is defined as $\log[\pi/(1 - \pi)]$, and logistic regression is

$$\text{logit}(\pi) = \beta_0 + \beta_1 X_1 + \cdots + \beta_p X_p.$$

The inverse of the logit function is called the *logistic function*. If $\text{logit}(\pi) = \eta$, then

$$\pi = \exp(\eta)/[1 + \exp(\eta)].$$

Display 20.3 illustrates how the logit and logistic functions connect the regression structure (represented by η) with the population proportion.

Nonconstant Variance

In normal regression, $\text{Var}\{Y|X_1, \ldots, X_p\} = \sigma^2$, where σ^2 is a constant that does not depend on the explanatory variable values. As was shown in Section 18.2.1 on page 519, however, the variance of a population of binary response variables with mean π is $\pi(1-\pi)$. Therefore, the mean and variance specifications of the logistic regression model are as given in the following box.

Logistic Regression Model

$$\mu\{Y|X_1, \ldots, X_p\} = \pi; \qquad \text{Var}\{Y|X_1, \ldots, X_p\} = \pi(1 - \pi);$$

$$\text{logit}(\pi) = \beta_0 + \beta_1 X_1 + \cdots + \beta_p X_p.$$

Logistic regression is a kind of nonlinear regression, since the equation for $\mu\{Y|X_1, \ldots, X_p\}$ is not linear in β's. Its nonlinearity, however, is solely contained in the link function—hence the term *generalized linear*. The implication of this is that the regression structure (the part with the β's) can be used in much the same way as ordinary linear regression. The model

Display 20.3   The standard logistic function

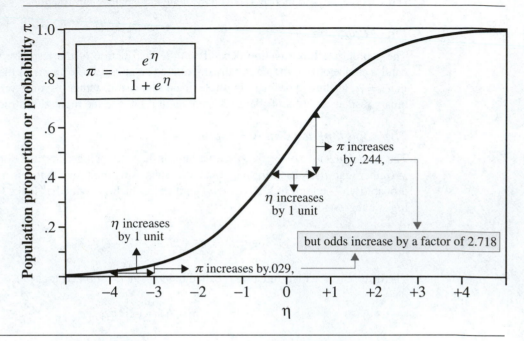

also differs from ordinary regression in variance structure: $\text{Var}\{Y|X_1, \ldots, X_p\}$ is a function of the mean and contains no additional parameter (like σ^2).

20.2.2   Interpretation of Coefficients

Although the preceding formulation is useful from a technical viewpoint because it casts logistic regression into the framework of a generalized linear model, it is important as a matter of practical interpretation to remember that the logit is the log odds function. Exponentiating the logit yields the odds. So the odds of a *yes* response (the odds that $Y = 1$) at the levels $X_1, \ldots, X_p$ are

$$\text{Odds that } Y = 1: \omega = \exp(\beta_0 + \beta_1 X_1 + \cdots + \beta_p X_p).$$

For example, the odds that the response is 1 at $X_1 = 0$, $X_2 = 0, \ldots, X_p = 0$ are $\exp(\beta_0)$, provided that the regression model is appropriate for these values of X's.

The ratio of the odds at $X_1 = A$ relative to the odds at $X_1 = B$, for fixed values of the other X's, is

$$\omega_A / \omega_B = \exp[\beta_1(A - B)].$$

In particular, if X_1 increases by 1 unit, the odds that $Y = 1$ will change by a multiplicative factor of $\exp(\beta_1)$, other variables being the same.

Example—Donner Party

The fit of a logistic regression model to the Donner Party data, where π represents survival probability, gives

$$\text{logit}(\hat{\pi}) = 1.63 - 0.078\, age + 1.60\, fem.$$

For comparing women 50 years old (A) with women 20 years old (B), the odds ratio is estimated to be $\exp[-0.078(50 - 20)] = 0.096$, or about 1/10. So the odds of a 20-year-old woman's surviving were about 10 times the odds of a 50-year-old woman's surviving. Comparing a woman ($fem = 1 = A$) with a man ($fem = 0 = B$) of the same age, the estimated odds ratio is $\exp(1.60[1 - 0]) = 4.95$—the woman's odds were about 5 times the odds of survival of a man of the same age.

Retrospective Studies

In logistic regression problems, probabilities for binary responses are modeled prospectively as functions of explanatory variables, as if responses were to be measured at set levels of the explanatory factors. In some studies—particularly those in which the probabilities of yes responses are very small—independent samples are drawn retrospectively from the populations with different response outcomes. As discussed in connection with the case of a single explanatory factor with two levels (Section 18.4), prospective probabilities cannot be estimated from these retrospective samples. Yet retrospective odds ratios are the same as prospective odds ratios, so they can be estimated from retrospective samples. In a logistic regression model for a retrospective study, the estimated intercept is not an estimate of the prospective intercept. However, estimates of coefficients of explanatory variables from a retrospective sample estimate the corresponding coefficients in the prospective model and may be interpreted in the same way. For example, it is possible with retrospective samples of lung cancer patients and of patients with no lung cancer to make a statement such as: the odds of lung cancer are estimated to increase by a factor of 1.1 for each year that a person has smoked. This result has tremendous bearing on medical case-control studies and the field of epidemiology.

20.3 Estimation of Logistic Regression Coefficients

In generalized linear models, the method of least squares is replaced by the *method of maximum likelihood*, which is outlined here for the logistic regression problem.

20.3.1 Maximum Likelihood Parameter Estimation

When values of the parameters are given, the logistic regression model specifies how to calculate the probability that any outcome (such as $y_1 = 1$, $y_2 = 0$, $y_3 = 1, \ldots$) will occur. A convenient way to express the model begins with the convention that raising any number to the power of zero results in the number one ($x^0 = 1$, for any x.) Then the probability model for a single binary response Y can be written as

$$\Pr\{Y = y\} = \pi^y (1 - \pi)^{(1-y)},$$

where the substitution of $y = 1$ results in π and the substitution of $y = 0$ results in $(1 - \pi)$. Now suppose that there are n such responses, with π_i being the parameter for $Y_i (i = 1, \ldots, n)$. If the responses are independent, their probabilities multiply to give the probability

$$\Pr\{Y_1 = y_1, \ldots, Y_n = y_n\} = \prod_{i=1}^{n} \pi_i^{y_i} (1 - \pi_i)^{1-y_i}$$

for the outcomes $y_1, \ldots, y_n$.

Examples—Donner Party

Suppose that this model is plausible for how the Donner Party data came about. It is as though a coin flip of fate had occurred for each party member. If the ith flip came up heads—which it would do with probability π_i—the ith member survived. Suppose, too, that an additive logistic model is correct, with the following hypothetical parameter values:

$$\text{logit}(\pi_i) = 1.50 - 0.080\, age_i + 1.25\, fem_i.$$

Then, referring to Display 20.1, Antoine (23-year-old male: $age_1 = 23$, $fem_1 = 0$) would have

$$\text{logit}(\pi_1) = 1.50 - (0.080 \times 23) + (1.25 \times 0) = -0.340,$$

so

$$\pi_1 = \exp(-0.340)/[1 + \exp(-0.340)] = 0.416.$$

Antoine had a 0.416 chance of surviving. Mary Breen (40-year-old female) would have a survival probability of 0.389; Patrick Breen (40-year-old male) would have a survival probability of 0.154; and so on down the list to Eliza Williams, with her survival probability of 0.679.

Now the probability for any combination of independent outcomes can be calculated by multiplying individual probabilities. For example, the probability that all persons survive (all the y_i are $= 1$) is $(0.416)(0.389)(0.154)\cdots(0.679) = \exp(-53.3631)$. Conversely, the probability that all persons die (all $y_i = 0$) is $(1-0.416)(1-0.389)(1-0.154)\cdots(1-0.679) = \exp(-26.1331)$. The probability that all the women survive but none of the men do is $(1-0.416)(0.389)(1-0.154)\cdots(0.679) = \exp(-21.1631)$; while the probability that the individuals survive who actually did survive is $(1-0.416)(0.389)(0.154)\cdots(0.679) = \exp(-26.1531)$. The best way to make these calculations is to add the logarithms of the individual probabilities.

The outcome in which everyone survives may be rosier, but it holds little interest because it did not actually occur. What is of particular interest is the single outcome that did occur. When attention is focused on that one outcome, the probability formula assumes a different role.

Likelihood

The same formula can be used to calculate the probability of the observed outcome under various possible choices for the β's. The probability of the known outcome, calculated earlier as $\exp(-26.1531)$ when $\beta_0 = 1.50$, $\beta_1 = -0.080$, and $\beta_2 = 1.25$, is called the *likelihood* of the values $(1.50, -0.080, 1.25)$ for the three parameters.

> The **likelihood** of any specific values for the parameters is the probability of the actual outcome, calculated with those parameter values.

According to this terminology, one set of parameter values is more likely than another set if it gives the observed outcome a higher probability of occurrence.

Maximum Likelihood Estimation

Examples of likelihood values (on the logarithmic scale) for several possible values of β_0, β_1, and β_2 in the logistic regression model for the Donner Party data are shown in Display 20.4. Notice that the largest likelihood listed is for the values $\beta_0 = 1.63$, $\beta_1 = -0.078$, and $\beta_2 = 1.60$. These values are in fact the most likely of all possible; they are called the *maximum likelihood estimates* of the parameters. As a guide for how to select parameter estimates, the *maximum likelihood method* says to choose as estimates of parameters the values that assign the highest probability to the observed outcome.

Display 20.4 Some possible parameter values and their log-likelihoods, for the Donner Party data (logit of probability of survival $= \beta_0 + \beta_1 age + \beta_2 fem$)

β_0	β_1	β_2	log (Likelihood)
1.50	−.050	1.25	−27.7083
		1.80	−28.7931
	−.80	1.25	−26.1531
		1.80	−25.7272
1.70	−.050	1.25	−29.0467
		1.80	−30.3692
	−.80	1.25	−25.7972
		1.80	−25.6904
1.63	−.078	1.60	−25.6282

Notes on Calculation

Although a computer routine could be used to search through many possible parameter values to find the combination that produces the largest likelihood (or equivalently, the largest log-likelihood), this is unnecessary. As with the method of least squares, calculus provides a way to find the maximizing parameter values. Unlike least squares for linear regression, however, calculus does not provide closed-form expressions for the answers. Some iterative computational procedures are used. See the references at the end of the chapter for more details on these methods.

Properties of Maximum Likelihood Estimators

Statistical theory reveals that the maximum likelihood method leads to estimators that generally have good properties.

Properties of Maximum Likelihood Estimates

If a model is correct and the sample size is large enough, then

1. The maximum likelihood estimators are essentially unbiased.
2. Formulas exist for estimating the standard deviations of the sampling distributions of the estimators.
3. The estimators are about as precise as any nearly unbiased estimators that can be obtained.
4. The shapes of the sampling distributions are approximately normal.

The upshot of properties (1) and (3) is that the maximum likelihood principle leads to pretty good estimates, at least when the sample size is large. Properties (1), (2), and (4) provide a way to carry out tests and construct confidence intervals. Both of these aspects are dampened somewhat by the requirement that the sample size be large enough. For small samples, it is best to label confidence intervals and test results as approximate.

20.3.2   Tests and Confidence Intervals for Single Coefficients

The properties imply that each estimated coefficient β_j in logistic regression has a normal sampling distribution, approximately, and therefore that

$$\textbf{Z-ratio} = (\hat{\beta}_j - \beta_j)/[\text{SE}(\hat{\beta}_j)].$$
$$\textbf{has a standard normal distribution (approximately).}$$

The standard error is the estimated standard deviation of the sampling distribution of the estimator. Its computation is carried out by the statistical computer package. The known likely values of the standard normal distribution may be used to construct confidence intervals or to obtain p-values for tests about the individual coefficients.

A confidence interval is the estimate plus and minus the half-width, which is a z-multiplier times the standard error of the estimate; and the z-multiplier is the appropriate percentile of the standard normal distribution. For a 95% confidence interval, it is the 97.5th percentile. Similarly, a test statistic is the ratio of the estimate minus the hypothesized value to the standard error, and the p-value is obtained by comparing this to a standard normal distribution. A test based on this approximate normality of maximum likelihood estimates is referred to as *Wald's test*. Since the t-theory only applies to normally distributed response variables, t-distributions are not involved here.

Example of Wald's Test for a Single Coefficient (Donner Party Data)

Display 20.5 demonstrates how Wald's test can be used to determine whether the log odds of survival are associated with age differently for men than for women. The display shows typical output from a logistic regression fit to a model that includes age, sex, and the interaction between sex and age. Notice that the summaries of the estimated coefficients, their standard errors, and the z-statistics (for the tests that each coefficient is zero) parallel the standard summary for multiple regression analysis. The quantity at the bottom labeled *deviance* is new, but it is analogous to the sum of squared residuals in regression (and is discussed in Section 20.4).

Example of a Confidence Interval for a Coefficient (Donner Party Data)

Display 20.6 shows the results of fitting the additive logistic regression of survival on age and the indicator variable for sex. Although some suggestive evidence was found in support

Display 20.5   Wald's test for the hypothesis that the coefficient of the interaction term is zero in the logistic regression of survival (1 or 0) on *age*, *fem* (= 1 for females), and *age* x *fem*: Donner Party data

Variable	Coefficient	Standard Error	z-Statistic
Constant	0.318	1.131	0.28
age	−0.032	0.035	−0.92
fem	6.927	3.354	2.06
age × *fem*	−0.162	0.093	−1.73

Deviance = 47.346   **Degrees of freedom** = 41

From the normal distribution ⟶ Two-sided p-value = $2 \times Pr(Z > 1.73) = .085$

Conclusion: There is suggestive but inconclusive evidence of an interaction.

Display 20.6   Confidence intervals for the odds of survival for females divided by the odds of survival for males, accounting for age, from the model without interaction (Donner Party data)

Variable	Coefficient	Standard Error	z-Statistic
Constant	1.633	1.105	1.48
age	−0.078	0.037	−2.11
fem	1.597	0.753	2.10

Deviance = 51.256   **Degrees of freedom** = 42

95% Confidence Interval for the coefficient of *fem* ⟶ $1.597 \pm (1.96 \times 0.753) = 0.121$ to 3.073 z(.975)

Take anti-logarithms of endpoints to get interval for the odds ratio. ⟶ $exp(0.121)$ to $exp(3.073) = 1.13$ to 21.60

Conclusion: The odds of survival for females are estimated to have been 4.9 times the odds of survival for males of similar age (95% CI: 1.1 times to 21.6 times).

of an interactive effect of sex and age, the coefficient of *fem* in this parallel lines model is a convenient approximation to an average difference between the sexes. The anti-logarithm of the coefficient of the indicator variable should be used in a summarizing statement. A confidence interval should be obtained for the coefficient of *fem*, and the anti-logarithm of the endpoints should be taken as a confidence interval for the odds ratio.

Example of Confidence Interval for a Multiple of a Coefficient (Donner Party Data)

From the output in Display 20.6, the estimated coefficient of age is -0.078. A 95% confidence interval is $-0.078 \pm (1.96 \times 0.037)$, which is -0.1505 to -0.0055. By taking anti-logarithms of the estimate and the interval endpoints, one obtains the following summary statement results: it is estimated that the odds of survival change by a factor of 0.92 for each one year increase in age (95% confidence interval for multiplicative odds factor is 0.86 to 0.99).

Another question that may be asked is "What are the odds of survival for a 55-year-old relative to the odds of survival for a 30-year-old?" In the model fit in Display 20.6, the log odds change by -0.078 for each extra year of age. Therefore, for 25 extra years of age, the log odds of survival change by 25×-0.078 or -1.95. Similarly, a 95% confidence interval for the change in log odds of survival for an additional 25 years of age (based on the results of the preceding paragraph) is 25×-0.1505 to 25×-0.0055, or -3.76 to -0.137. By back-transforming this estimate and confidence interval, the summary statement is this: it is estimated that the odds of survival for a 55-year-old are 0.14 times the odds of survival for a 30-year-old (95% confidence interval for odds ratio is 0.02 to 0.87).

20.4 The Drop-in-Deviance Test

Comparing a Full to a Reduced Model

As in ordinary regression, one often must judge the adequacy of a *reduced* model relative to a *full* model. The reduced model is the special case of the full model obtained by supposing that the hypothesis of interest is true. Typically, the hypothesis is that several of the coefficients in the full model equal zero. For example, if three indicator explanatory variables are used to model the effects of four treatments on the probability of survival of heart disease patients, the hypothesis that no difference exists between treatments is equivalent to the hypothesis that the coefficients of the three indicator variables all equal zero. The reduced model is the logistic regression model excluding those three terms (but including all the other explanatory variables from the full model).

Extra Sums of Squared Residuals

The extra-sum-of-squares F-test in ordinary regression is a formal way to compare the sizes of the residuals from the fit to the reduced model to the sizes of the residuals from the fit to the full model. For each model, the sum of the squared residuals is used as a single summary of the sizes of all residuals. The F-statistic compares the sizes of the two sets of residuals through the extra sum of squares, appropriately scaled for comparison to an F distribution.

The extra-sum-of-squares F-test is motivated by a general procedure, called the *likelihood ratio test*, which is applied to the problem with a normally distributed response. In the context of generalized linear models, the likelihood ratio test is called the *drop in deviance test*. Like the extra-sum-of-squares F-test, the drop in deviance test is based on a sum of squared residuals from the reduced model minus the sum of squared residuals from the full model, but with a new type of residual, called the *deviance residual*.

20.4.1   Deviance Residuals for Generalized Linear Models

If y_i is an observed binary response variable and if $\hat{\pi}_i$ is the estimated mean based on some model, the ordinary residual for observation i is $y_i - \hat{\pi}_i$. This is not particularly useful for models (such as the logistic regression model) for which the response variance is nonconstant, since the residuals themselves may have widely differing variances. For one thing, outlier detection would be complicated with the ordinary residuals. More important for the current discussion, a sum of squares of the ordinary residuals might be influenced unduly by residuals that have large variance. Various other types of residuals have been defined to meet the needs of residual analysis. The deviance residual is one of these, based on the discrepancy in terms of likelihood functions.

Deviance Residuals as Likelihood Discrepancies

Instead of using $y_i - \hat{\pi}_i$ as a measure of the discrepancy between an observation and its estimated mean, one could use $g(y_i) - g(\hat{\pi}_i)$, where g is some function appropriately chosen to meet the needs of residual analysis. Since the likelihood function provides an ordering of likely values of the parameter, it is natural to take the function g to be based on the likelihood. A theoretically appealing choice has g equal to twice the logarithm of the likelihood function, $g(\pi_i) = 2l_i(\pi_i)$, where $l_i(\pi_i)$ is the logarithm of the likelihood function of the ith mean. For binary response variables, the log likelihood is

$$l_i(\pi_i) = y_i \log(\pi_i) + (1 - y_i) \log(1 - \pi_i).$$

The *deviance component* for observation i is defined to be

$$2l_i(y_i) - 2l_i(\hat{\pi}_i).$$

[Notice that $l_i(y_i)$ is $y_i \log(y_i) + (1 - y_i) \log(1 - y_i)$]. This difference is always greater than or equal to zero, and the difference gets larger as $\hat{\pi}_i$ gets farther away from y_i. In this sense it constitutes the likelihood discrepancy between the observation and its estimated mean according to the model. To account for whether y_i is greater than or less than its estimated mean, the deviance residual is defined as the square root of the deviance component as follows.

$$\textit{Deviance component} = 2l_i(y_i) - 2l_i(\hat{\pi}_i)$$

Deviance residual =
Negative square root of deviance component, if $y_i < \hat{\pi}_i$
Positive square root of deviance component, if $y_i > \hat{\pi}_i$.

Example—Donner Party Data

Display 20.7 shows scatterplots of deviance residuals versus estimated means from two different models fit to the Donner Party data.

Display 20.7 Deviance residuals from fits to two logistic regression models for survival—Donner Party data

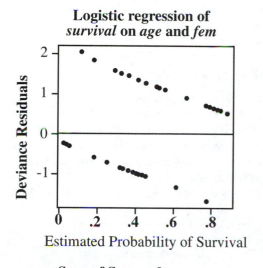

Logistic regression of *survival* **on** *age* **and** *fem*

Deviance Residuals

Estimated Probability of Survival

Sum of Squared Deviance Residuals: 51.26

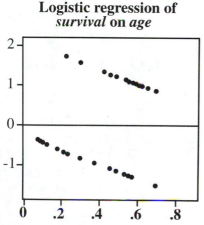

Logistic regression of *survival* **on** *age*

Estimated Probability of Survival

Sum of Squared Deviance Residuals: 56.29

These residual plots are not useful for model evaluation of logistic regression on binary responses. Since there are only two distinct values of y (0 and 1), the residuals tend to be clustered in two groups: those corresponding to observations with y = 1 (which necessarily have positive residuals) and those corresponding to observations with y = 0 (which necessarily have negative residuals). For binary responses, the deviance residual for case i is $+\sqrt{-2\log(\hat{\pi}_i)}$ if y_i is 1 and $-\sqrt{-2\log(1-\hat{\pi}_i)}$ if y_i is 0. The plots in Display 20.7, although not useful in practice, are shown here to illustrate the notion of extra sum of squares—that the residuals from a reduced model will tend to be farther from zero, on average, than those from the full model.

The Deviance Statistic

The sum of squared deviance residuals is called the *deviance* (or the *residual deviance statistic*). It is an overall measure of the discrepancy between the observations of the response variable and the fitted model:

> ### *Deviance = Sum of squared deviance residuals,*
> *with associated degrees of freedom = n − p.*

The associated degrees of freedom here are derived in exactly the same way as those associated with sums of squares residuals in ordinary regression are: they are equal to the number of observations minus the number of parameters used to model the mean.

20.4.2   Drop-in-Deviance Chi-Squared Test

For the Donner Party example the sum of squared deviance residuals is 51.26 for the model with *age* and *fem* as explanatory variables (based on 42 degrees of freedom) and 56.29 for the model with only *age* (based on 43 degrees of freedom). The difference, 5.03, is the amount by which the size of the squared residuals from the model with *age* is reduced when *fem* is included. It is thus the drop in deviance (sum of squared deviance residuals) due to including *fem* in the model that already has *age*.

The extra sum of squared deviance residuals is called the *drop in deviance*:

> ### *Drop in deviance =*
> *Deviance from reduced model − Deviance from the full model.*
> ### *Drop in degrees of freedom = d =*
> *Difference in number of parameters (Full model − Reduced model).*

If the drop in deviance is small, the reduced model does about as good a job as the full model at explaining the responses. If the drop in deviance is large, the reduced model is relatively inadequate.

A *p*-value for the adequacy of the reduced model is the chance that the drop in deviance is as large as its observed value, calculated under the supposition that the reduced model is adequate. Likelihood theory provides an approximate answer:

> *If the reduced model is adequate, the drop in deviance has a χ^2 distribution with d d.f. (approximately).*

The distribution is approximate in the sense that it is theoretically appropriate only for large sample sizes. It turns out to work quite well even for moderately small sample sizes, however, as long as the number of parameters being tested is small relative to the sample size.

Finding the p-Value

To test a hypothesis about some of the parameters in a logistic regression model with the drop-in-deviance test, fit both the full and reduced models, extract the deviance statistics from the fits, and then compare the drop in deviance to a chi-squared distribution with degrees of freedom equal to the drop in degrees of freedom. The p-value is the probability that a variable with a χ^2 sampling distribution on d degrees of freedom will exceed the observed drop in deviance. A small p-value is evidence that the reduced model is inadequate.

Example—Donner Party

Is there a difference between male and female survival probability, after the effect of age has been accounted for? That question is addressed through a test that β_2 is zero in the model

$$\text{logit}(\pi) = \beta_0 + \beta_1 age + \beta_2 fem.$$

The reduced model is formed from this full model by supposing that the hypothesis is true:

$$\text{logit}(\pi) = \beta_0 + \beta_1 age.$$

Display 20.7 shows the sizes of the deviance residuals and of the deviance statistics from these two fits. Is the drop in deviance, 5.03, larger than should be expected if the two models are equivalent? The p-value is the proportion of a chi-squared distribution on 1 degree of freedom that is greater than 5.03. This is .025, as obtained from a calculator, a computer, or Appendix A.3. This represents moderately strong evidence that the reduced model is inadequate; that a difference between males' and females' survival probabilities exists after age is accounted for.

Example—Birdkeeping

The goal of the birdkeeping and lung cancer data was to see whether increased probability of lung cancer is associated with birdkeeping, even after other factors, such as smoking, that may have some bearing on lung cancer are accounted for. In the logistic model shown in connection with step 1 of Display 20.8, the log of the odds of getting lung cancer are explained by regression terms with sex of the individual (FM), age (AG), socioeconomic status (SS), years the individual has smoked (YR), and the indicator variable for birdkeeping

---

Display 20.8   A drop-in-deviance test for the association of the odds of lung cancer with birdkeeping, after accounting for other factors

(1)   *Fit the* ***full model:*** $\text{logit}(\pi) = \beta_0 + \beta_1 FM + \beta_2 AG + \beta_3 SS + \beta_4 YR + \beta_5 BK.$

Variable	Coefficient	Standard Error	z-Statistic	Two-sided p-Value
CONSTANT	−0.2644	1.7649	−0.1498	.8809
FM	−0.7929	0.4393	−1.8049	.0711
AG	−0.0562	0.0361	−1.5568	.1192
SS	−0.0773	0.4614	−0.1675	.8670
YTR	0.0773	0.0236	3.2754	.0010
BK	1.4014	0.4121	3.4006	.0007

Deviance = 153.01   **Degrees of freedom** = 141

(2)   *Fit the* ***reduced model:*** $\text{logit}(\pi) = \beta_0 + \beta_1 FM + \beta_2 AG + \beta_3 SS + \beta_4 YR.$

Variable	Coefficient	Standard Error	z-Statistic	Two-sided p-Value
CONSTANT	1.4560	1.6046	0.9074	.3642
FM	−0.7705	0.4194	−1.8371	.0662
AG	−0.0736	0.0345	−2.1333	.0328
SS	−0.3038	0.4298	−0.7068	.4725
YR	0.0784	0.0236	3.3220	.0009

Deviance = 165.42   **Degrees of freedom** = 142

(3)   *Calculate the* ***drop in deviance*** *and the* ***drop in degrees of freedom.***

$$
\begin{array}{rr}
165.42 & 142 \\
-153.01 & -141 \\
\hline
\text{Drop in Deviance} = \quad 12.41 & \text{Drop in d.f.} = \quad 1
\end{array}
$$

(4)   *Determine the p-value.* $\longrightarrow$ $\Pr(\chi_1^2 > 12.41) = .0004$

Conclusion: There is strong evidence of an association between bird keeping and lung cancer, after accounting for sex, age, status, and smoking years.

(*BK*, which is 1 for birdkeepers). The question of interest here is a hypothesis that the coefficient of *BK* is zero. The reduced model, therefore, is the same model, but without *BK* in the list of explanatory variables. The steps in the drop-in-deviance test are detailed in Display 20.8.

Notes on the Drop-in-Deviance Test

1. As with the extra-sum-of-squares F-test, a few special cases of the drop-in-deviance test deserve mention. A test of whether all coefficients (except the constant term) are zero is obtained by taking the model with no explanatory variables as the reduced model. Similarly, the significance of a single term may be tested by taking the reduced model to be the full model minus the single term (as demonstrated in Display 20.8). This is *not* the same as Wald's test for a single coefficient described in Section 20.3. If the two give different results, the drop-in-deviance test has the more reliable p-value.

2. Confidence intervals can be constructed for a single coefficient from the theory of the likelihood ratio test. This is good practice but is fairly sophisticated and requires a moderate amount of effort. For most purposes, the confidence intervals described in Section 20.3 are adequate.

3. The drop in deviance is not divided by the deviance from the full model to form an F-statistic. This is unnecessary because no extra parameter (like σ in the usual regression model) describes the scale of the extra sum of squared residuals.

20.5 Strategies for Data Analysis Using Logistic Regression

The strategy for data analysis with binary responses and explanatory variables is similar to the general strategy for data analysis with regression, as shown in the chart in Display 9.9 (page 241). Identifying the questions of interest is the important starting point. The data analysis should then revolve around finding models that fit the data and allow the questions to be answered through inference about parameters.

A main difference from ordinary regression is that scatterplots and residual plots are usually of little value, since only two distinct values of the response variables are possible. Fortunately, there is no burden to check for nonconstant variance or outliers, so some functions of usual residual analysis are unnecessary. Model terms and the adequacy of the logistic model, on the other hand, must be checked. Informal testing of extra terms, such as squared terms and interaction terms, plays an important role in this regard.

20.5.1 Exploratory Analysis

The following questions should be addressed during the exploratory stage:

1. Which are the important explanatory variables, and which seem less important?

2. Do the effects of the important explanatory variables appear to be linear on the logit scale?

3. Are there possible interactive effects of important explanatory variables on the logit scale?

The answers to those questions are not easy to come by, but they are important to the strength of the main conclusions.

Graphical Methods

A plot of the binary response variable versus an explanatory variable is not worthwhile, since there are only two distinct values for the response variable. Although no graphical approach can be prescribed for all problems, it is occasionally useful to examine a scatterplot of one of the explanatory variables versus another, with codes to indicate whether the response is 0 or 1. An example is shown in Display 20.11.

Grouping

Some data sets, particularly large ones, may contain *near replicates*—several observations with very similar values of the explanatory variables. As part of an informal analysis to see which explanatory variables are important and to obtain some graphical display, it may be useful to group these together. For example, ages may be grouped into 5-year intervals. A sample proportion is obtained as the proportion of observations with binary response of 1 in the group, and the logit transformation can be applied to this sample proportion so that scatterplots may be drawn. The sample logit can be plotted against the midpoint of the grouped version of the explanatory variable, for instance. An example of this approach is shown in Display 20.12.

Informal Testing of Extra Model Terms

One can also construct preliminary logistic regression models that include polynomial terms, interaction terms, and the like, to judge whether the desired model needs expansion to compensate for very specific problems. This is probably the most useful of the three approaches for exploratory data analysis on binary responses.

Model Selection

As in normal regression, the search for a suitable model may encompass a wide range of possibilities. The Bayesian information criterion (BIC) is a model selection device that emphasizes parsimony by penalizing models for having large numbers of parameters. The BIC for generalized linear models is

$$BIC = \text{Deviance} + [p \times \log(n)],$$

where n is the sample size and p is the number of parameters in the model for the mean, including the constant. The model with the smallest BIC is preferred. Further, the uncertainty involved in model selection can be described by Bayes' posterior probabilities, exactly as in Section 12.4.2.

20.6   Analyses of Case Studies

20.6.1   Analysis of Donner Party Data

Whether women and men had different survival chances can best be answered in the inferential model,

$$\text{logit}(\pi) = \beta_0 + \beta_1 fem + \beta_2 age,$$

which provides control for the effects of age while specifying a simple difference between the logits for women and for men. Is this model defensible? There are two possible problems: first, the logit of survival probability may not be linear in age; and second, even if the logit is linear, the slope may not be the same for males and females.

Since one cannot directly assess the proposed model from sample logits, the best alternative is to embed the preceding model in a larger model that allows for departures from these critical assumptions. Quadratic terms, for example, provide evidence of one kind of departure from a straight line.

Display 20.9 shows the output from the fit of the logistic regression of survival on *age*, *fem*, and three extra terms that assess different aspects of model inadequacy. This model allows for distinct quadratic curves for male and female log odds of survival. One way to pare this down is to use a directed backward elimination technique on the extra terms. *Directed* here refers to examining the least desirable terms first (those that would most complicate the statement of results, for example). It is not necessary to have an exact order for examination, but a suitable starting point is the *fem* $\times$ *age*^{2} term. Since its coefficient has a large *p*-value, it may be dropped; and the resulting model is refit. (Remember that the *p*-values change when a term is dropped.) The *p*-value for the coefficient of *age*^{2} in the model without *fem* $\times$ *age*^{2} is .26, so it is dropped.

Display 20.9   Quadratic logistic regression model to assess fit for Donner Party survival

Variable	Coefficient	Standard Error	z-Statistic	p-Value
Constant	−3.318	3.940	−0.84	.40
fem	0.265	10.430	0.03	.98
age	0.183	0.227	0.81	.42
fem $\times$ *age*	0.300	0.694	0.43	.66
age^2	−0.0028	0.0030	−0.94	.35
fem $\times$ *age*^{2}	−0.0074	0.0107	−0.69	.49
Deviance = 45.361	**Degrees of freedom = 39**			

The resulting model is the one examined in Display 20.5, which provided suggestive evidence of an interactive effect. Since the interactive effect is weak and since it would complicate the statement of the results, one might choose to ignore it (realizing that more data would be needed to assess it carefully) and to report the resulting linear additive model—cautiously—as a general indication of the difference between the males and females. The result is

$$\text{logit}(\hat{\pi}) = 1.633 - 0.078age - 1.597fem.$$
$$\quad\quad\quad (1.105) \quad\quad (0.037) \quad\quad\quad (0.753)$$

A plot of the fitted model is shown in Display 20.10. The conclusions of the analyses appear in Display 20.6, and a statement of the summary is provided in Section 20.1.1.

Display 20.10   Fit of logistic regression model for Donner Party Survival, with *age* and *fem* (1 = Female) as explanatory variables

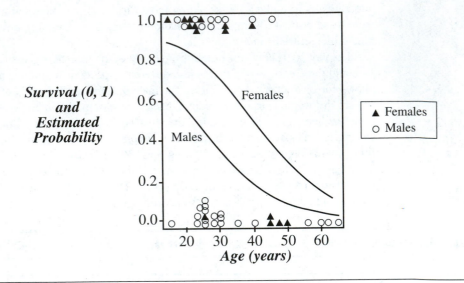

20.6.2 Analysis of Birdkeeping and Lung Cancer Data

The goal is to examine the odds of lung cancer for birdkeepers relative to persons with similar demographic and smoking characteristics who do not keep pet birds.

Effect of Smoking

The first stage of the analysis is to find an adequate model to explain the effect of smoking. Display 20.11 is a coded scatterplot of the number of years that a subject smoked, versus the individual's age. The plotting symbols show whether the individual is a birdkeeper (triangle) or not (circle) and whether the subject is one of the lung cancer cases (filled symbol) or not (unfilled symbol).

Display 20.11   Coded scatterplot of years of smoking versus age of subject: triangles represent birdkeepers, circles are non-birdkeepers, and filled symbols are subjects with lung cancer

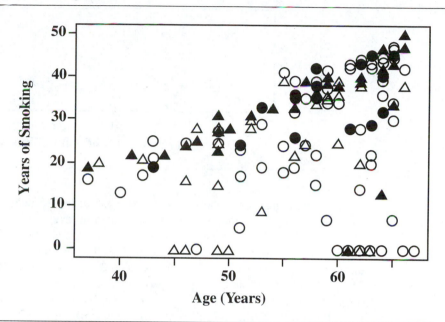

One striking feature of the plot is that, in any vertical column of points in the plot, more dark symbols appear at the top of the column than at the bottom. This indicates that, for similarly aged subjects, the persons with lung cancer are more likely to have been relatively long-time smokers. To investigate the effect of birdkeeping on this plot, one must look at certain regions of the plot—particularly horizontal slices—to compare subjects with similar smoking history and to ask whether a higher proportion of triangles tend to be filled symbols than unfilled symbols. For example, the only lung cancer patient among subjects who did not smoke (for which years of smoking is zero) happened to be a birdkeeper. Similarly, among individuals who smoked for about 10 years, the only lung cancer patient was a birdkeeper. For the upper regions of the scatterplot, it is a bit harder to make a comparison. Slight visual evidence suggests that the proportion of birdkeepers is higher among the lung cancer patients than among the controls.

Display 20.12 is a plot of sample logits, $\log[\overline{\pi}/(1 - \overline{\pi})]$, versus years smoking, where $\overline{\pi}$ is the proportion of lung cancer patients among subjects grouped into similar years of smoking (0, 1–20, 21–30, 31–40, and 41–50). The midpoint of the interval was used to represent the years of smoking for the members of the group.

Evidently, the logit increases with increasing years of smoking. The trend is predominantly linear, but some departure from linearity may be present. This matter can be resolved through informal testing of a model with a quadratic term.

Display 20.12   Scatterplot of sample logits versus years smoking for data grouped by intervals of years of smoking

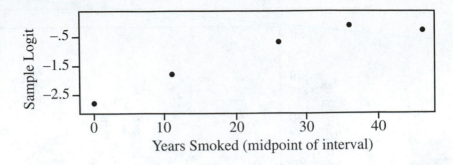

Logistic regression models for the proportion of lung cancer cases as related to the smoking variables, age, and sex may be structured in several ways. Display 20.13 shows the results of one fit, which includes years of smoking, cigarettes per day, and squares and interaction of these. One approach to paring down the model is to use a selective backward elimination of the extra terms involving the squares and interaction of the smoking variables. The interaction term can be dropped (two-sided p-value $= .9093$). Refitting without this term reveals that the squared CD term can be dropped. The squared YR term can be dropped next. Finally, the CD term itself is not needed when YR is included in the model. Thus it appears that the single term YR adequately models the smoking effect, when the matching variables are also considered.

Display 20.13   Logistic regression of lung cancer incidence with rich model; from the birdkeeping and lung cancer case–control study

Variable	Coefficient	Standard Error	z-Statistic	Two-sided p-Value
Constant	0.5620	2.2711	0.2475	.8046
FM	−0.7676	0.4246	−1.8078	.0706
AG	−0.0641	0.0373	−1.7185	.0854
SS	−0.2786	0.4389	−0.6348	.5255
YR	0.0433	0.0961	0.4506	.6523
CD	0.1046	0.1017	1.0285	.3035
YR^2	0.0004	0.0019	0.2105	.8293
CD^2	−0.0018	0.0020	−0.9000	.3637
$YR \times CD$	−0.0003	0.0023	−0.1304	.9093

Deviance $= 163.74$ **Degrees of freedom** $= 138$

Extra Effect of Birdkeeping

The final step adds the indicator variable for birdkeeping into the model with the final list of other explanatory variables already present. The estimated model is summarized in the first table of Display 20.8. The drop in deviance—12.41 with 1 degree of freedom—provides convincing evidence that birdkeeping has a strong association with lung cancer, even after the effects of smoking, sex, age, and socioeconomic status are accounted for (two-sided p-value = .0008).

The estimate of the coefficient of birdkeeping is 1.4014, so the odds of lung cancer for birdkeepers is estimated to be exp(1.34)—or in other words, 4.1—times as large as the odds of lung cancer for nonbirdkeepers, after the other explanatory variables are accounted for. A 95% confidence interval for the coefficient of the birdkeeping indicator variable is 0.5937 to 2.2091. Consequently, a 95% confidence interval for the odds of lung cancer for birdkeepers relative to the odds of lung cancer for nonbirdkeepers is 1.8 to 9.1.

20.7   Related Issues

20.7.1   Matched Case–Control Studies

Matched *case–control* studies are a special type of retrospective sampling in which controls are matched with cases on the basis of equality of certain variables. Only one or two variables are generally used to match subjects, because over-matching can bias important comparisons. The number of controls matched to each case is usually one to three; little is gained from including more. Variables used in the matching process and their possible interactions with other variables should be entered as covariates in logistic regression analyses, much as block effects are included in the analysis of a randomized block experiment. The purpose is to exercise control in the analysis, as well as in the design. The coefficients of these variables cannot be interpreted as odds ratios, and the intercept measures no meaningful feature. Analysis of matched case–control studies is accomplished by using conditional likelihood logistic regression, which differs somewhat from the methods of this chapter.

20.7.2   Probit Analysis

There are several alternatives to logistic regression for binary responses. Any function $F(\pi)$ that has characteristics similar to the logit function—steadily increasing from $-\infty$ to ∞ as π goes from 0 to 1—could easily be chosen as a link. One popular alternative is to choose $F(\pi)$ to be the inverse of the cumulative standard normal probability distribution function; that is, $F(\pi)$ equals the 100πth percentile in the standard normal distribution. This choice leads to *probit regression*. The results from probit regression are similar to those from logistic regression, at least for values of π between .2 and .8.

20.7.3   Discriminant Analysis Using Logistic Regression

Logistic regression may be used to predict future binary responses. On the basis of a training set, a logistic model can be estimated. When applied to a particular set of explanatory variable values for which the binary response is unknown, the estimated probability π may be used as an estimate of the probability that the response is 1.

Logistic regression prediction problems are similar in structure to problems in discriminant analysis, which were discussed in Section 17.5.1 (page 501). In fact, logistic regression is preferable to standard DFA solutions when the explanatory variables have nonnormal distributions (when they are categorical, for example). Logistic regression is nearly as efficient as the standard tools when the explanatory variables do have normal distributions. Therefore, when discrimination between two groups is the goal, logistic regression should always be considered as an appropriate tool.

When estimated from a retrospective sample, logistic regression cannot be used to estimate the probability of group status. If the odds of being a case in the population are known (from some other source) to be ω_p, however, and if the odds based on a subject's explanatory variable values and on the logistic regression (estimated from the retrospective sample) are estimated to be ω_r, then the odds that the subject is a case are $(n_0/n_1)\omega_p\omega_r$, where n_0 and n_1 are the number of controls and the number of cases, respectively, in the sample.

20.8 Summary

The logistic regression model describes a population proportion or probability as a function of explanatory variables. It is a nonlinear regression model, but of the special type called *generalized linear model*, where the logit of the population proportion is a linear function of regression coefficients. If β_3 is the coefficient of an explanatory variable in a logistic regression model, each unit increase of X_3 is associated with an $\exp(\beta_3)$-fold increase in the odds. This interpretation is particularly useful if X_3 is an indicator variable used to distinguish between two groups. Then $\exp(\beta_3)$ is the odds ratio for the two groups, such as the odds of cancer for birdkeepers relative to the odds of cancer for nonbirdkeepers.

Logistic regression analysis for binary responses proceeds in much the same way as regular regression analysis. Since it is difficult to learn much from plots or from residual analysis, model checking is based largely on fitting models that include extra terms (such as squared terms or interaction terms) whose significance would indicate shortcomings of the "target" model. Tests and confidence intervals for single coefficients may be carried out in a familiar way by comparing the coefficient estimate to its standard error. Tests of several coefficients are carried out by the drop-in-deviance chi-squared test. The drop in deviance is the sum of squared deviance residuals from the reduced model minus the sum of squared deviance residuals from the full model.

Donner Party

Given the likely inadequacy of the independence assumption and the inappropriateness of generalizing inferences from these data to any broader population, the analysis is necessarily informal. Nevertheless, it is useful to fit a logistic regression model for survival as a function of sex and age. Initial model fitting requires some examination into the linearity of the effect of age on the logit and of the interaction between sex and age. The coefficient of the sex indicator variable in the additive logistic regression model may be used to make a statement about the relative odds of survival of similarly aged males and females.

Birdkeeping and Lung Cancer

Initial tentative model fitting and model building are used to determine an adequate representation of the smoking variables for explaining lung cancer. Once years of smoking were included in the model, no further smoking variable was found significant when added. Then the variables associated with socioeconomic status and age were included, and interactions were explored. Finally, the effect of birdkeeping was inferred by adding the birdkeeping indicator variable into the model. A drop-in-deviance test established that birdkeeping had a significant effect, even after smoking was accounted for.

Further Reading

An authoritative (and sophisticated) book on generalized linear models, including logistic regression is McCullagh and Nelder (1989). Several books are devoted exclusively to logistic regression, including Hosmer and Lemeshow (1989), which also includes a discussion of discriminant analysis and a trick for performing the correct conditional likelihood analysis for matched case–control studies with standard logistic regression programs (when a single matched case exists for each control). For further reading on case–control studies and in particular on methods for analysis when the matching does not occur according to explicit matching variables, see also Breslow and Day (1980).

20.9 Exercises

Conceptual Exercises

1. **Donner Party.** One assumption underlying the correct use of logistic regression is that observations are independent of one another. (a) Is there some basis for thinking this assumption might be violated in the Donner Party data? (b) Write a logistic regression model for studying whether survival of a party member is associated with whether another adult member of the party had the same surname as that subject, after the effect of age has been accounted for. (c) Prepare a table of grouped responses that may be used for initial exploration of the question in part (b).

2. **Donner Party.** (a) Why should one be reluctant to draw conclusions about the ratio of male and female odds of survival for Donner Party members over 50? (b) What confounding variables might be present that could explain the significance of the sex indicator variable?

3. **Donner Party.** From the Donner Party data, the log odds of survival were estimated to be $1.6 - (0.078 \times age) + (1.60 \times fem)$, based on a binary response that takes the value 1 if an individual survived and with *fem* being an indicator variable that takes the value 1 for females. (a) What would be the estimated equation for the log-odds of survival if the indicator variable for sex were 1 for males and 0 for females? (b) What would be the estimated equation for the log-odds of perishing if the binary response were 1 for a person who perished and 0 for a person who survived?

4. **Birdkeeping.** (a) Describe the retrospective sampling of the birdkeeping data. (b) What are the limitations of logistic regression from this type of sampling?

5. Since the term *regression* refers to the mean of a response variable as a function of one or more explanatory variables, why is it appropriate to use the term *logistic regression* to describe a proportion or probability as a function of explanatory variables?

6. Give two reasons why the simple linear regression model is usually inappropriate for describing the regression of a binary response variable on a single explanatory variable.

7. How can logistic regression be used to test the hypothesis of equal odds in a 2×2 table of counts?

8. How can one obtain from the computer an estimate of the log-odds of $y = 1$ at a chosen configuration of explanatory variables, and its standard error?

Computational Exercises

9. Donner Party. It was estimated that the log odds of survival were $3.2 - (0.078 \times age)$ for females and $1.6 - (0.078 \times age)$ for males in the Donner Party. (a) What are the estimated probabilities of survival for men and women of ages 25 and 50? (b) What is the age at which the estimated probability of survival is 50% (i) for women, and (ii) for men?

10. It was stated in Section 20.2 that, if ω_A and ω_B are odds at A and B, respectively, and if the logistic regression model holds, then $\omega_A/\omega_B = \exp[\beta_1(A - B)]$. Show that this is true, using algebra.

11. Space Shuttle. The data in Display 20.14 are the launch temperatures (degrees Fahrenheit) and an indicator of O-ring failures for 24 space shuttle launches prior to the space shuttle *Challenger* disaster of January 27, 1986. (See the description in Section 4.1.1 on page 82 for more background information).

(a) Fit the logistic regression of *Failure* (1 for failure) on *Temperature*. Report the estimated coefficients and their standard errors. (b) Test whether the coefficient of *Temperature* is 0, using Wald's test. Report a one-sided *p*-value (the alternative hypothesis is that the coefficient is negative; odds of failure decrease with increasing temperature). (c) Test whether the coefficient of *Temperature* is 0, using the drop-in-deviance test. (d) Give a 95% confidence interval for the coefficient of *Temperature*. (e) What is the estimated logit of failure probability at 31°F (the launch temperature on January 27, 1986)? What is the estimated probability of failure? (f) Why must the answer to part (e) be treated cautiously? [*Answer*: It is a prediction outside the range of the available explanatory variable values.]

Display 20.14 Launch temperature (degrees Fahrenheit) and incidence of O-ring failure for 24 space shuttle launches

---

Temperature	Failure	Temperature	Failure	Temperature	Failure
53	Yes	68	No	75	No
56	Yes	69	No	75	Yes
57	Yes	70	No	76	No
63	No	70	Yes	76	No
66	No	70	Yes	78	No
67	No	70	Yes	79	No
67	No	72	No	80	No
67	No	73	No	81	No

---

12. Muscular Dystrophy. Duchenne Muscular Dystrophy (DMD) is a genetically transmitted disease, passed from a mother to her children. Boys with the disease usually die at a young age; but affected girls usually do not suffer symptoms, may unknowingly carry the disease, and may pass it on to their offspring. It is believed that about 1 in 3,300 women are DMD carriers. A woman might suspect she is a carrier when a related male child develops the disease. Doctors must rely on some kind of test to detect the presence of the disease. The data in Display 20.15 are levels of two enzymes in the blood, creatine kinase (CK) and hemopexin (H) for 38 known DMD carriers and 82 women who are not carriers. (Data from D. F. Andrews and A. M. Herzberg, *Data* (New York: Springer-Verlag, 1985.)) It is desired to use these data to obtain an equation for indicating whether a woman is a likely carrier.

Display 20.15 Creatine kinase (CK) and hemopexin (H) for 82 controls ($C = 0$) and 38 muscular dystrophy carriers ($C = 1$)

Controls ($C = 0$)								Muscular Dystrophy Carriers ($C = 1$)			
CK	H	CK	H	CK	H	CK	H	CK	H	CK	H
52	83.5	31	61.5	52	77	21	74.5	167	89	69	111
20	77	62	81	34	75	95	69.8	104	81	48	98
28	86.5	48	79	53	93.2	40	72.7	30	108	109	81
30	104	40	82.5	69	66.7	48	76	65	87	925	81
40	83	55	85.5	24	89.5	39	88.5	440	107	59	93
24	78.8	32	73.8	21	108.5	30	82.7	58	88.2	363	91.3
15	87	26	79.3	51	82	38	85	129	93.1	37	84
22	91	25	91	24	82	27	87.2	265	83.5	101	77.5
42	65.5	30	76	22	102	74	80.4	285	79.5	99	93.2
130	80.3	27	90	32	79.2	34	80.5	124	92	560	106
47	53	72	80.5	24	70.4	28	82.5	53	76	85	94
34	92.7	51	70	20	72	37	93	657	104	197	91.5
40	104.6	67	98	34	91	97.5	34	168	82.5	154	103.5
59	88	50	69	25	92	37	98	286	109.5	80	90.5
42	77	55	78.2	26	109	30	80	73	105.5	28	104
26	84.5	38	82	56	72	37	98	19	100.5	57	88
65	75	27	100	51	91	24	100.5	113	97	326	98
34	86.3	34	84	18	95	41	78.5	57	105	100	101
37	73.3	35	59.4	28	104	30	90.5	78	118	115	79
73	57.4	35	90.3	41	105.5						
35	71	31	75.5	40	81	($n = 82$)				($n = 38$)	

(a) Make a scatterplot of H versus log(CK); use one plotting symbol to represent the controls on the plot and another to represent the carriers. Does it appear from the plot that these enzymes might be useful predictors of whether a woman is a carrier? (b) Fit the logistic regression of carrier on CK and CK-squared. Does the CK-squared term significantly differ from 0? Next fit the logistic regression of carrier on log(CK) and log(CK)-squared. Does the squared term significantly differ from 0? Which scale (untransformed or log-transformed) seems more appropriate for CK? (c) Fit the logistic regression of carrier on log(CK) and H. Report the coefficients and standard errors. (d) Carry out a drop-in-deviance test for the hypothesis that neither log(CK) nor H are useful predictors of whether a woman is a carrier. (e) Typical values of CK and H are 80 and 85. Suppose that a suspected carrier has values of 300 and 100. What are the odds that she is a carrier relative to the odds that a woman with typical values (80 and 85) is a carrier?

13. **Donner Party.** Consider the Donner Party females (only) and the logistic regression model $\beta_0 + \beta_1 age$ for the logit of survival probability. If A^* represents the age at which the probability of survival is 0.5, then $\beta_0 + \beta_1 A^* = 0$ (since the logit is 0 when the probability is one-half). This implies that $\beta_0 = -\beta_1 A^*$. The hypothesis that $A^* = 30$ years may be tested by the drop-in-deviance test with the following reduced and full models for the logit:

$$\text{Reduced:} \quad -\beta_1 30 + \beta_1 age \quad = 0 + \beta_1 (age - 30)$$
$$\text{Full:} \quad \beta_0 + \beta_1 age$$

To fit the reduced model, one must subtract 30 from the ages and drop the intercept term. The drop-in-deviance test statistic is computed in the usual way. Carry out the test that $A^* = 30$ for the Donner Party females. Report the two-sided p-value.

14. **Donner Party.** The estimate in Exercise 13 of A^* is $-b_0/b_1$. A 95% confidence interval for A^* can be obtained by finding numbers A_L^* and A_U^* below and above this estimate such that two-sided p-values for the tests $A^* = A_L^*$ and $A^* = A_U^*$ are both .05. This can be accomplished by trial and error. Choose several possible values for A_L^*, and follow the testing procedure described in the preceding exercise until a value is found for which the two-sided p-value is .05. Then repeat the process to find the upper bound A_U^*. Use this procedure to find a 95% confidence interval for the age at which the Donner Party female survival probability is 0.5.

Data Problems

15. **Spotted Owl Habitat.** A study examined the association between nesting locations of the Northern Spotted Owl and availability of mature forests. Wildlife biologists identified 30 nest sites. (Data from Ripple et al., "Old-growth and Mature Forests Near Spotted Owl Nests in Western Oregon," *Journal of Wildlife Management* 55: 316–18.) The researchers selected 30 other sites at random coordinates in the same forest. On the basis of aerial photographs, the percentage of mature forest (older than 80 years) was measured in various rings around each of the 60 sites, as shown in Display 20.16. (a) Apply two-sample t-tools to these data to see whether the percentage of mature forest is larger at nest sites than at random sites. (b) Construct a binary response variable that indicates whether a site is a nest site. Use logistic regression to investigate how large an area about the site has importance in distinguishing nest sites from random sites on the basis of mature forest. (Notice that this was a case–control study.) You may wish to transform the ring percentages to circle percentages.

16. **Bumpus Natural Selection Data.** Hermon Bumpus analyzed various characteristics of some house sparrows that were found on the ground after a severe winter storm in 1898. Some of the sparrows survived and some perished. The data on male sparrows in Display 20.17 are survival status (1 = survived, 2 = perished), age (1 = adult, 2 = juvenile), the length from tip of beak to tip of tail (in mm), the alar extent (length from tip to tip of the extended wings, in mm), the weight in grams, the length of the head in mm, the length of the humerus (arm bone, in inches), the length of the femur (thigh bones, in inches), the length of the tibio-tarsus (leg bone, in inches), the breadth of the skull in inches, and the length of the sternum in inches. (A subset of this data was discussed in Section 2.1.1 on page 28.)

Analyze the data to see whether the probability of survival is associated with physical characteristics of the birds. This would be consistent, according to Bumpus, with the theory of natural selection: those that survived did so because of some superior physical traits. Realize that (i) the sampling is from a population of grounded sparrows, and (ii) physical measurements and survival are both random. (Thus, either could be treated as the response variable.)

Display 20.16 Spotted Owl study: percentages of mature forest (>80 years) in successive rings around sites in western Oregon forests

| Randomly Chosen Sites | | | | | | | Spotted Owl Nest Sites | | | | | | |
| Outer Radius of Ring (km) | | | | | | | Outer Radius of Ring (km) | | | | | | |
0.91	1.18	1.40	1.60	1.77	2.41	3.38	0.91	1.18	1.40	1.60	1.77	2.41	3.38
26.0	33.3	25.6	19.1	31.4	24.8	17.9	81.0	83.4	88.9	92.9	80.6	72.0	44.4
100.0	92.7	90.1	72.8	51.9	50.6	41.5	80.0	87.3	93.3	81.6	85.0	82.8	63.6
32.0	22.2	38.3	39.9	22.1	20.2	38.2	96.0	74.0	76.7	66.2	69.1	84.5	52.5
43.0	79.7	61.4	81.2	47.7	69.6	54.8	82.0	79.6	91.3	70.7	75.6	73.5	66.8
74.0	61.8	48.3	67.4	74.9	66.0	55.8	83.0	80.6	88.9	79.6	55.8	62.9	52.7
63.0	85.0	85.8	84.9	78.0	78.0	53.6	80.0	80.0	59.3	65.1	61.1	72.2	48.6
60.0	72.2	58.1	54.1	55.6	53.5	67.2	100.0	100.0	89.7	97.0	97.0	90.5	45.1
46.0	68.0	68.8	67.9	71.9	69.5	66.0	67.0	67.0	91.1	87.3	87.9	74.7	66.8
80.0	62.9	42.0	59.6	68.4	59.7	60.0	81.0	71.2	70.1	75.0	75.0	75.0	46.5
76.0	63.8	60.7	59.1	44.2	53.4	29.5	89.0	67.0	86.9	64.2	72.6	83.5	55.6
73.0	60.8	64.6	58.1	59.6	57.5	24.3	72.0	79.3	71.6	82.9	70.6	64.2	47.6
50.0	54.9	65.8	51.6	49.6	54.0	41.8	81.0	56.6	74.4	67.6	65.6	63.5	46.6
77.0	74.6	69.1	74.0	63.1	61.2	56.8	77.0	89.2	82.0	77.6	81.0	70.2	59.7
68.0	63.1	55.7	63.0	79.3	81.1	83.2	100.0	85.3	83.7	82.1	89.0	71.7	60.6
66.0	66.0	62.6	60.6	20.4	66.8	46.7	83.0	68.3	77.0	77.0	82.4	75.8	62.7
88.0	92.9	83.1	88.0	82.6	84.8	63.6	78.0	75.6	66.7	56.2	80.9	67.7	63.9
90.0	94.9	81.7	75.7	80.6	67.7	56.6	58.0	80.0	84.2	80.9	84.9	71.7	65.9
50.0	52.4	61.3	45.1	52.0	62.8	69.2	58.0	38.4	50.0	58.9	62.9	62.6	43.7
70.0	72.4	67.6	74.4	65.6	76.5	77.1	89.0	93.9	91.0	86.6	79.1	72.9	62.7
15.0	39.4	52.6	59.7	49.9	49.6	51.1	81.0	71.2	77.0	68.1	58.7	67.7	65.9
82.0	74.7	65.2	61.7	55.7	58.2	49.7	68.0	53.3	62.0	66.4	63.0	63.0	65.0
68.0	55.8	56.1	56.6	49.1	58.0	43.7	86.0	78.7	79.6	73.1	80.0	69.2	40.4
47.0	44.6	52.9	21.3	52.9	50.5	32.7	51.0	43.7	44.6	38.1	28.7	48.5	34.8
75.0	75.0	78.4	80.4	55.2	49.3	62.0	86.0	90.9	74.2	84.0	78.6	83.0	74.9
62.0	49.8	50.1	37.2	51.0	64.0	59.0	72.0	74.4	93.7	79.0	79.0	74.7	58.7
74.0	42.2	47.2	57.0	62.4	62.3	60.0	73.0	70.6	68.6	79.9	78.4	82.6	71.9
29.0	51.0	48.3	49.9	43.0	38.7	41.0	76.0	76.0	72.6	70.6	74.0	76.2	68.9
73.0	65.7	73.4	79.9	78.4	71.8	73.0	70.0	62.7	63.6	48.2	56.6	56.7	36.6
81.0	76.1	68.7	53.8	60.1	60.4	60.9	87.0	82.1	85.0	71.7	82.0	86.3	57.5
57.0	69.2	65.4	76.3	76.9	74.5	75.1	62.0	47.3	66.3	54.6	52.6	52.7	53.0

17. **Catholic Stance.** The Catholic church has explicitly opposed authoritarian rule in some (but not all) Latin American countries. Although such action could be explained as a desire to counter repression or to increase the quality of life of its parishioners, A. J. Gill supplies evidence that the underlying reason may be competition from evangelical Protestant denominations. He compiled measures of: (1) *Repression* = Average civil rights score for the period of authoritarian rule until 1979; (2) *PQLI* (Physical Quality of Life Index in the mid-1970s) = Average of life expectancy at age one, infant mortality, and literacy at age 15+; and (3) *Competition* = Percentage increase of competitive religious groups during the period 1900–1970. His goal was to determine which of these factors could be responsible for the church being pro-authoritarian in six countries and anti-authoritarian in six others. (Data from A. J. Gill, "Rendering unto Caesar? Religious competition and Catholic political strategy in Latin America, 1962–1979," *American Journal of Political Science* 38(2) (1994): 403–25.)

To what extent do these measurements (see Display 20.18) distinguish between countries where the church took pro- and anti-authoritarian positions? Is it possible to determine the most influential variable(s) for distinguishing the groups?

Display 20.17 Data on 51 male sparrows that survived (*SV* = 1) and 36 that perished (*SV* = 2) during a severe winter storm: Age (*AG*) is 1 for adults, 2 for juveniles; *TL* is total length; *AE* is alar extent; *WT* is weight; *BH* is length of beak and head; *HL* is length of humerus; *FL* is length of femur; *TT* is length of tibio-tarsus; *SK* is width of skull; and *KL* is length of keel of sternum

SV	AG	TL	AE	WT	BH	HL	FL	TT	SK	KL	SV	AG	TL	AE	WT	BH	HL	FL	TT	SK	KL
1	1	154	241	24.5	31.2	.687	.668	1.022	.587	.830	2	1	162	247	27.6	31.8	.731	.719	1.113	.597	.869
1	1	160	252	26.9	30.8	.736	.709	1.180	.602	.841	2	1	163	246	25.8	31.4	.689	.662	1.073	.604	.836
1	1	155	243	26.9	30.6	.733	.704	1.151	.602	.846	2	1	161	246	24.9	30.5	.739	.726	1.138	.580	.803
1	1	154	245	24.3	31.7	.741	.688	1.146	.584	.839	2	1	160	242	26.0	31.0	.745	.713	1.105	.600	.803
1	1	156	247	24.1	31.5	.715	.706	1.129	.575	.821	2	1	162	246	26.5	31.5	.720	.696	1.092	.606	.809
1	1	161	253	26.5	31.8	.780	.743	1.144	.607	.893	2	1	160	249	26.0	31.4	.726	.689	1.097	.602	.850
1	1	157	251	24.6	31.1	.741	.736	1.153	.610	.862	2	1	161	250	27.1	31.6	.737	.711	1.120	.631	.852
1	1	159	247	24.2	31.4	.728	.718	1.126	.609	.793	2	1	162	248	25.1	31.9	.744	.722	1.154	.591	.839
1	1	158	247	23.6	29.8	.703	.673	1.079	.602	.820	2	1	165	252	26.0	32.3	.726	.710	1.145	.609	.887
1	1	158	252	26.2	32.0	.749	.739	1.153	.614	.857	2	1	161	243	25.6	32.5	.709	.707	1.122	.607	.832
1	1	160	252	26.2	32.0	.741	.723	1.129	.624	.892	2	1	161	244	25.0	31.3	.702	.685	1.082	.595	.874
1	1	162	253	24.8	32.3	.766	.752	1.134	.633	.923	2	1	162	248	24.6	31.0	.713	.700	1.086	.590	.837
1	1	161	243	25.4	31.8	.721	.722	1.126	.597	.891	2	1	164	244	25.0	31.2	.703	.690	1.074	.608	.795
1	1	160	250	23.7	29.8	.730	.703	1.103	.590	.820	2	1	158	247	26.0	32.0	.729	.710	1.145	.607	.803
1	1	159	247	25.7	31.4	.729	.717	1.141	.592	.927	2	1	162	253	28.3	31.8	.752	.718	1.152	.600	.857
1	1	158	253	25.7	31.9	.743	.699	1.150	.600	.860	2	1	156	239	24.6	30.5	.659	.658	1.042	.570	.810
1	1	159	247	26.5	31.6	.733	.714	1.155	.611	.923	2	1	166	251	27.5	31.5	.720	.691	1.118	.612	.847
1	1	166	253	26.7	32.5	.767	.765	1.230	.600	.878	2	1	165	253	31.0	32.4	.765	.750	1.183	.613	.905
1	1	159	247	23.9	31.4	.752	.723	1.113	.602	.825	2	1	166	250	28.3	32.4	.754	.718	1.179	.607	.916
1	1	160	248	24.7	31.3	.752	.737	1.176	.603	.803	1	2	156	246	24.6	32.0	.744	.735	1.167	.592	.849
1	1	161	252	28.0	31.8	.770	.731	1.190	.590	.885	1	2	156	245	25.5	32.1	.761	.717	1.147	.620	.816
1	1	163	251	27.9	31.9	.769	.745	1.168	.622	.860	1	2	163	248	24.8	32.2	.742	.733	1.165	.606	.854
1	1	156	242	25.9	32.0	.723	.711	1.116	.609	.886	1	2	163	248	26.3	33.0	.736	.704	1.148	.609	.839
1	1	165	251	25.7	32.2	.751	.742	1.161	.613	.865	1	2	160	250	24.4	31.5	.746	.715	1.173	.604	.893
1	1	160	247	26.6	32.4	.728	.707	1.108	.590	.836	1	2	156	237	23.3	30.6	.692	.664	1.011	.588	.774
1	1	158	244	23.2	31.6	.730	.713	1.142	.585	.888	1	2	162	253	26.7	32.0	.759	.734	1.197	.630	.878
1	1	160	242	25.7	31.6	.709	.705	1.124	.620	.788	1	2	163	254	26.4	32.0	.766	.750	1.165	.605	.886
1	1	157	245	26.3	32.2	.741	.726	1.143	.595	.850	1	2	164	251	26.9	32.0	.755	.742	1.171	.620	.886
1	1	159	244	24.3	31.5	.723	.698	1.107	.615	.847	1	2	163	244	24.3	31.3	.718	.680	1.082	.610	.892
1	1	160	253	26.7	32.1	.739	.714	1.117	.592	.864	1	2	160	247	27.0	31.5	.764	.732	1.177	.617	.846
1	1	158	245	24.9	31.4	.726	.703	1.119	.580	.854	1	2	160	250	26.8	32.5	.764	.729	1.123	.635	.842
1	1	161	247	23.8	31.4	.735	.694	1.101	.602	.780	1	2	158	247	24.9	32.4	.745	.724	1.139	.588	.865
1	1	160	247	25.6	32.3	.756	.745	1.135	.607	.902	1	2	158	249	26.1	32.2	.742	.736	1.148	.602	.817
1	1	160	247	27.0	32.0	.755	.736	1.174	.631	.873	1	2	158	243	26.6	32.4	.747	.711	1.163	.612	.891
1	1	153	241	24.7	32.2	.728	.680	1.092	.592	.884	1	2	155	237	23.3	30.2	.685	.653	1.011	.587	.794
2	1	165	249	26.5	31.0	.738	.704	1.095	.606	.847	2	2	160	249	24.0	30.4	.740	.717	1.130	.620	.840
2	1	160	245	26.1	32.0	.736	.709	1.109	.611	.842	2	2	156	236	26.8	30.2	.690	.671	1.067	.563	.832
2	1	161	249	25.6	32.3	.743	.718	1.128	.602	.828	2	2	158	240	23.5	31.0	.715	.702	1.113	.595	.805
2	1	162	246	25.9	32.3	.738	.709	1.135	.607	.869	2	2	166	245	26.9	31.7	.715	.695	1.107	.601	.847
2	1	163	250	25.5	32.5	.752	.731	1.197	.623	.888	2	2	165	255	28.6	31.5	.766	.744	1.175	.613	.854
2	1	162	247	27.6	31.8	.731	.719	1.113	.597	.869	2	2	157	238	24.7	31.2	.680	.677	1.156	.599	.769
2	1	163	246	25.8	31.4	.689	.662	1.073	.604	.836	2	2	164	250	27.3	31.8	.764	.726	1.171	.588	.860
2	1	161	246	24.9	30.5	.739	.726	1.138	.580	.803	2	2	166	256	25.7	31.7	.752	.751	1.187	.595	.858
2	1	160	242	26.0	31.0	.745	.713	1.105	.600	.803	2	2	167	255	29.0	32.2	.765	.745	1.197	.638	.855
2	1	162	246	26.5	31.5	.720	.696	1.092	.606	.809	2	2	161	246	25.0	31.5	.739	.707	1.123	.587	.850
2	1	163	250	25.5	32.5	.752	.731	1.197	.623	.888	2	2	166	254	27.5	31.4	.760	.742	1.124	.604	.914
											2	2	161	251	26.0	31.5	.731	.707	1.122	.589	.828

Display 20.18 Religious competition, quality of life, and repression in 12 predominantly Catholic Latin American countries

Catholic Church Stance	Country	POLI	Repression	Competition (% increase)
Pro-authoritarian	Argentina	85	5.3	2.7
	Bolivia	39	4.3	4.1
	Guatemala	54	3.5	6.3
	Honduras	53	3.0	3.1
	Paraguay	75	5.2	2.1
	Uruguay	86	4.7	1.2
Anti-authoritarian	Brazil	66	4.8	12.0
	Chile	79	5.0	15.5
	Ecuador	69	3.7	2.9
	El Salvador	64	4.4	5.5
	Nicaragua	55	4.3	5.6
	Panama	79	5.7	4.4

Answers to Conceptual Exercises

1. (a) Yes. Members within the same family may have been likely to share the same fate; and if so, the binary responses for members within the same family would not be independent of one another. (It is difficult to investigate this possibility or to correct it for this small data set. Further discussion of this type of violation is provided in the next chapter.) (b) $\text{logit}(\pi) = \beta_0 + \beta_1 age + \beta_2 sur$, where $sur = 1$ if the individual had a surname that was shared by at least one other adult party member, and 0 if not. (c) Here is one possibility:

Age group	Number in group with shared surname	Proportion surviving	Number in age group without shared surname	Proportion surviving
15–25	13	.625	8	.375
26–45	10	.800	7	.000
>45	6	.160	1	.000

2. (a) There were no females over 50. Any comparisons for older people must be based on an assumption that the model extends to that region. Such an assumption cannot be verified with these data. (b) If males and females had different tasks and if survival was associated with task, the tasks would be a confounding variable.

3. (a) $3.2 - 0.78 age - 1.6 male$. (b) Same equation but with all coefficients negative.

4. A sample of 49 individuals from all those with lung cancer was taken from a population. These were the "cases." Another sample of 98 individuals was taken from a similar population of individuals who did not have lung cancer. (b) Logistic regression may be used, with the 0–1 response representing lung cancer; but the intercept cannot be interpreted.

5. The mean of a binary response is a proportion (or a probability, if the populations are hypothetical).

6. (i) Proportions must fall between 0 and 1, and lines cross these boundaries. (ii) The variance is necessarily nonconstant.

7. Use a binary response to distinguish levels of the response category and an indicator variable to distinguish the levels of the explanatory category. The coefficient of the indicator variable is the log odds of one group minus the log odds of the other group. Carry out inference about that coefficient.

8. Subtract the specified value of each variable from all the variable values to create a new set of variables centered (zeroed) at the desired configuration. Run a logistic regression analysis on the new set, and record the "constant" coefficient and its standard error (see Section 10.2 on page 264).

CHAPTER 21

Logistic Regression for Binomial Counts

This chapter extends logistic regression models to the case where the responses are proportions. These may be grouped binary responses (like the proportion out of a certain number of 50-year-olds in the data set who have lung cancer) or a proportion based on a count for each subject (like the proportion of 100 cells from each subject that have chromosome aberrations). Although the response variable is more general, the logistic regression model is the same as in Chapter 20. The population proportion or probability is modeled, through the logit link, by a linear function of regression coefficients.

One new aspect of the analysis for this type of response is that some care must be taken in checking distributional assumptions. (This was not a problem for ungrouped binary responses.) Some additional tools for data analysis aid in this checking: scatterplots of the sample logits, residual analysis, and a goodness of fit test.

21.1 Case Studies

21.1.1 Island Size and Bird Extinctions—An Observational Study

Scientists agree that preserving certain habitats in their natural states is necessary to slow the accelerating rate of species extinctions. But they are divided on how to construct such reserves. Given a finite amount of available land, is it better to have many small reserves or a few large ones? Central to the debate on this question are observational studies of what has happened in island archipelagos, where nearly the same fauna tries to survive on islands of different sizes.

In a study of the Krunnit Islands archipelago, researchers presented results of extensive bird surveys taken over four decades. They visited each island several times, cataloging species. If a species was found on a specific island in 1949, it was considered to be at risk of extinction for the next survey of the island in 1959. If it was not found in 1959, it was counted as an "extinction," even though it might reappear later. The data on island size, number of species at risk to become extinct, and number of extinctions are shown in Display 21.1. (Data from Vaisanen and Jarvinen, "Dynamics of Protected Bird Communities in a Finnish Archipelago," *Journal of Animal Ecology* 46 (1977): 891–908.)

Display 21.1   Island area, number of bird species present in 1949, and number of these not present in 1959: the Krunnit Islands study

Island	Area (km^2)	Species at Risk	Extinctions
Ulkokrunni	185.80	75	5
Maakrunni	105.80	67	3
Ristikari	30.70	66	10
Isonkivenletto	8.50	51	6
Hietakraasukka	4.80	28	3
Kraasukka	4.50	20	4
Länsiletto	4.30	43	8
Pihlajakari	3.60	31	3
Tyni	2.60	28	5
Tasasenletto	1.70	32	6
Raiska	1.20	30	8
Pohjanletto	0.70	20	2
Toro	0.70	31	9
Luusiletto	0.60	16	5
Vatunginletto	0.40	15	7
Vatunginnokka	0.30	33	8
Tiirakari	0.20	40	13
Ristikarenletto	0.07	6	3

The data are plotted in Display 21.2. Each island is indicated by a single dot. The horizontal measurement shows the island's area on a logarithmic scale. The vertical measurement is the logit of the proportion of the island's at-risk occasions that resulted in an extinction. The scatterplot shows dramatically that larger islands had smaller extinction rates. Another striking feature of the scatterplot is how well a straight line would approximate the apparent relationship.

Display 21.2 Extinctions of bird species versus island size in the Krunnit Island archipelago

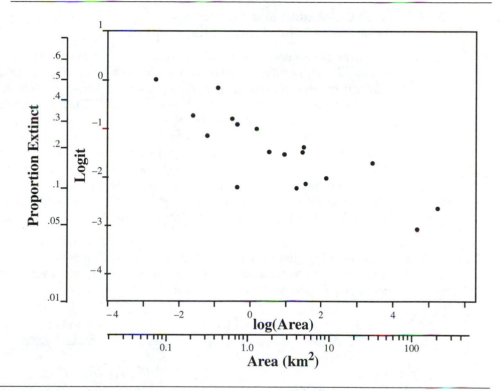

Summary of Statistical Findings

The strong linear component in the relationship is undeniable. Each 50% reduction in island area is associated with a 23% increase in the odds of extinction (95% confidence interval: 13.6% to 31.5%).

Two difficulties with the statistical analysis are evident: it assumes that each species found on an island has the same chance of becoming extinct, and it assumes that extinctions occur independently. Although these assumptions are highly questionable, the linearity in the scatterplot and the conclusiveness of the p-values make any inaccuracy in p-values and standard errors fairly unimportant.

Scope of Inference

Since the data are observational, confounding variables associated with both island size and the event of a species extinction may be responsible for the observed association in this data. In addition, these islands are not a sample of any larger population of islands. Therefore, although the data are consistent with the theory that the odds of extinction decrease with increasing island size, the statistical analysis alone cannot be used to infer that the same relationship would exist in other islands or habitat areas or that island size is a cause of extinction in these islands.

21.1.2   Moth Coloration and Natural Selection—A Randomized Experiment

Population geneticists consider clines particularly favorable situations for investigating evolutionary phenomena. A cline is a region where two color morphs of one species arrange themselves at opposite ends of an environmental gradient, with increasing mixtures occurring between. Such a cline exists near Liverpool, England, where a dark morph of a local moth has flourished in response to the blackening of tree trunks by air pollution from the mills. The moths are nocturnal, resting during the day on tree trunks, where their coloration acts as camouflage against predatory birds. In Liverpool, where tree trunks are blackened by smoke, a high percentage of the moths are of the dark morph. One encounters a higher percentage of the typical (pepper-and-salt) morph as one travels from the city into the Welsh countryside, where tree trunks are lighter. J. A. Bishop used this cline to study the intensity of natural selection. Bishop selected seven locations progressively farther

Display 21.3   Numbers of light (Typicals) morph and dark (Carbonaria) morph moths placed by researchers and numbers of these removed by predators, at each of seven locations of varying distances from Liverpool

Location	Distance from Liverpool (km)	Morph	Number of Moths Placed	Number Removed
Sefton Park	0.0	light	56	17
		dark	56	14
Eastham Ferry	7.2	light	80	28
		dark	80	20
Hawarden	24.1	light	52	18
		dark	52	22
Loggerheads	30.2	light	60	9
		dark	60	16
Llanbedr	36.4	light	60	16
		dark	60	23
Pwyllglas	41.5	light	84	20
		dark	84	40
Clergy Mawr	51.2	light	92	24
		dark	92	39

from Liverpool. At each location, Bishop chose eight trees at random. Equal numbers of dead (frozen) light (*Typicals*) and dark (*Carbonaria*) moths were glued to the trunks in lifelike positions. After 24 hours, a count was taken of the numbers of each morph that had been removed—presumably by predators. The data are shown in Display 21.3. (Data from J. A. Bishop, "An Experimental Study of the Cline of Industrial Melanism in *Biston betularia* [Lepidoptera] Between Urban Liverpool and Rural North Wales," *Journal of Animal Ecology* 41 (1972): 209–43.)

The question of interest is whether the proportion removed differs between the dark morph moths and the light morph moths and, more importantly, whether this difference depends on the distance from Liverpool. If the relative proportion of dark morph removals increases with increasing distance from Liverpool, that would be evidence in support of survival of the fittest, via appropriate camouflage.

Summary of Statistical Findings

The logit of the proportion of moths removed, of each morph at each location, is plotted in Display 21.4 against the distance of the location from Liverpool. From this it appears that

---

Display 21.4   Proportions of moths of two color morphs taken by predators at seven locations near Liverpool, England

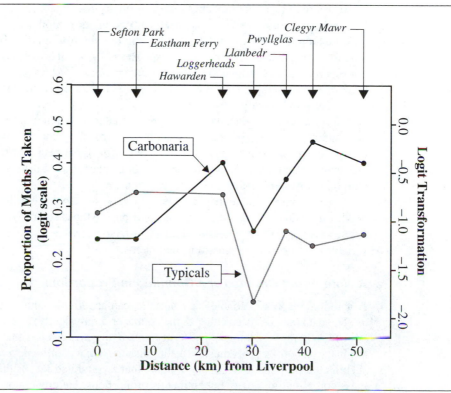

the odds that a moth will be removed by a predator increase with increasing distance from Liverpool for the dark morph and decrease with increasing distance from Liverpool for the light morph. A logistic regression analysis confirms that the odds of removal for the dark morph, relative to the odds of removal for the light morph, increase with increasing distance from Liverpool (one-sided p-value $= .0006$ for interaction of morph and distance). The odds of removal increase by an estimated 2% for each kilometer in distance from Liverpool for the dark morph; and they decrease by an estimated 1% per kilometer, for the light morph.

Scope of Inference

This is not a fully randomized experiment, since the moths were not randomly assigned to trees and to positions on trees. It can be argued, however, that the haphazard assignment of moths to trees is just as good as randomization. The possibility of a confounding variable seems slight.

21.2   Logistic Regression for Binomial Responses

21.2.1   Binomial Responses

As noted in Section 18.2.1 (page 519), a binomial count is the sum of independent binary responses with identical probabilities. If m binary responses constitute a random sample from a population of binary responses with population proportion π, their sum (which must be an integer between 0 and m) has the *binomial*(m, π) distribution. The observed proportion of 1's among the m binary responses is Y/m and is called a *binomial proportion*. For this reason, m is called the *binomial denominator*. (Although it may seem natural to think of m as a sample size, the term *sample size* is reserved for the number of binomial proportions in the study.)

In regression settings, the ith response variable Y_i is binomial(m_i, π_i). The binomial denominators m_i need not be the same, but they must be known. In the first example, Y_i is the number of extinctions on island i, out of m_i bird species at risk for extinction. In the second example, Y_i is the number of moths taken, out of the m_i moths of a particular morph, that were placed at one of the locations.

The response variables here are the binomial counts (or the corresponding proportions), not the individual binary indicators that contribute to those counts. As in the previous chapter, however, a logistic regression model is used to describe π_i as a function of explanatory variables and unknown regression coefficients.

A Note About Counted and Continuous Proportions

A binomial response variable is a *count* that can acquire the integer values between 0 and m. The data may be listed either as the counts or (equivalently) as proportions Y/m, but in both cases m must be known. The term *counted proportion* refers to the proportion of 1's in m binary responses. This must be distinguished from a *continuous proportion* of amounts, such as the proportion of fat (by weight) that is saturated fat, or the proportion of body mass composed of water. In continuous proportions, the numerator and denominator are not integers (except possibly due to rounding), and no m is involved. When they function as response variables, continuous proportions must be handled with ordinary regression methods, possibly after a transformation. The methods of this chapter should never be used with continuous proportions as responses, or with arbitrary and fictitious values of m such as 100.

21.2.2 The Logistic Regression Model for Binomial Responses

As described in Chapter 20, logistic regression is a special case of generalized linear models. The elements of such a model are a distribution for the response variable and a function that links the mean of the distribution to the explanatory variables. A binomial logistic regression with two explanatory variables would be specified as follows:

Binomial Logistic Regression Model
1. Y is binomial(m, π).
2. $\text{logit}(\pi) = \beta_0 + \beta_1 X_1 + \beta_2 X_2$.

The interpretation of the logistic model here is exactly the same as in Chapter 20. The model shows how the population proportion or probability π depends on the explanatory variables through a nonlinear link function. But the nonlinear part is completely captured by the logit of π. Logistic regression on binary responses is a special case of the binomial model in which all m_i's are 1.

Krunnit Islands Example

Let Y_i represent the number of extinctions on island i, out of m_i species at risk, for islands $i = 1, 2, \ldots, 18$. For example, for $i = 1$, the island Ulkokrunni had $Y_1 = 5$ extinctions out of $m_1 = 75$ species at risk. If the logit of the probability of extinctions is assumed to be linear in $X = \log$ of island area (Ulkokrunni has log area $X_1 = \log(185.50\,\text{km}^2) = 5.2231$), the logistic regression model would specify that $\text{logit}(\pi_1) = \beta_0 + \beta_1 5.2231$ for Ulkokrunni, and that a similar expression would describe the proportions of extinctions for other islands. If β_1 in the model is zero, the odds of extinction are not associated with island area.

21.3 Model Assessment

21.3.1 Scatterplot of Empirical Logits Versus an Explanatory Variable

Associated with observation i is a response proportion, $\overline{\pi}_i = Y_i / m_i$. The observed logits, or *empirical logits*, are $\log[\overline{\pi}_i / (1 - \overline{\pi}_i)]$, which may also be written as $\log[Y_i / (m_i - Y_i)]$. Plots of the observed logits versus one or more of the explanatory variables are useful for visual examination, paralleling the role of ordinary scatterplots in linear regression. The scatterplot in Display 21.2, for example, suggests that the logit of extinction probability is linear in log of island area.

If the m_i's are small for some observations, there may be only a few possible values for the observed proportion, making the display less informative. If some of the observed proportions are 0 or 1, a modification is necessary because the logit function is undefined at these values. For purposes of gaining a useful visual display, add a small amount to the numerator and denominator, such as 0.5, so that all the logits, $\log[(Y_i + .5)/(m_i - Y_i + .5)]$, are defined. (No such addition is required for fitting the model, however.)

21.3.2 Examination of Residuals

Two types of residuals are widely used for logistic regression for binomial counts: deviance residuals and Pearson residuals. The formula of a *deviance residual*—the residual that measures likelihood discrepancy, as discussed in Section 20.4.1 (page 578)—is shown in the following box for a binomial response based on a binomial denominator of size m_i. The part of the definition multiplying the square root, $\text{sign}(Y_i - m_i\hat{\pi}_i)$, is either $+1$ (if Y_i is above its estimated mean) or -1 (if below).

$$
\textbf{Deviance Residual}
$$
$$
Dres_i = \text{sign}(Y_i - m_i\hat{\pi}_i)\sqrt{2\left\{Y_i \log\left(\frac{Y_i}{m_i\hat{\pi}_i}\right) + (m_i - Y_i)\log\left(\frac{m_i - Y_i}{m_i - m_i\hat{\pi}_i}\right)\right\}}.
$$

The *Pearson residual* is more transparent. It is defined as an observed binomial response variable minus its estimated mean, divided by its estimated standard deviation. Thus it is constructed to have, at least roughly, mean 0 and variance 1.

$$
\textbf{Pearson Residual}
$$
$$
Pres_i = \frac{Y_i - m_i\hat{\pi}_i}{\sqrt{m_i\hat{\pi}_i(1-\hat{\pi}_i)}}.
$$

Although the two often exhibit little difference, the Pearson residuals are easier to understand and therefore are more useful for casual examination. The deviance residuals, on the other hand, have the advantage of being directly connected to extra-sum-of-squared residual tests. Either set of residuals can be examined for patterns in plots against suspected explanatory variables.

For large values of the binomial denominator, both types of residuals tend to behave as if they came from a standard normal distribution—if the model is correct. Paying close attention to those that exceed 2 in magnitude identifies possible outliers. If the m_i's are small—less than 5, say—the relative sizes of the residuals can be compared, but decisions based on comparison to the standard normal distribution should be avoided.

Example—Krunnit Islands Extinctions

The Pearson and deviance residuals from the fit to the Krunnit Islands extinction data are shown in Display 21.5. The logistic regression model for estimating the proportions of extinctions is: $\text{logit}(\hat{\pi}) = -1.196 - 0.297\log(\text{Area})$. For this example, the two types of residuals differ little, and no evidence of outlying observations is present.

Display 21.5 Residuals from the logistic regression of extinction probability on log of island area: Krunnit Islands data (Raw Residual is $Y/m - \hat{\pi}$)

Island	Observed Proportion	Model Estimate	Raw Residual	Pearson Residual	Deviance Residual
Ulkokrunni	0.067	0.061	0.006	0.24	0.23
Maakrunni	0.045	0.070	−0.026	−0.82	−0.87
Ristikari	0.152	0.099	0.053	1.44	1.35
Isonkivenletto	0.118	0.138	−0.020	−0.42	−0.43
Hietakraasukka	0.107	0.159	−0.052	−0.76	−0.80
Kraasukka	0.200	0.162	0.038	0.46	0.45
Länsiletto	0.186	0.164	0.022	0.39	0.39
Pihlajakari	0.097	0.171	−0.075	−1.10	−1.18
Tyni	0.179	0.185	−0.007	−0.09	−0.09
Tasasenletto	0.188	0.205	−0.018	−0.25	−0.25
Raiska	0.267	0.223	0.044	0.58	0.57
Pohjanletto	0.100	0.252	−0.152	−1.56	−1.72
Toro	0.290	0.252	0.039	0.50	0.49
Luusiletto	0.313	0.260	0.052	0.48	0.47
Vatunginletto	0.467	0.284	0.182	1.57	1.50
Vatunginnokka	0.242	0.302	−0.059	−0.74	−0.76
Tiirakari	0.325	0.328	−0.003	−0.04	−0.04
Ristikarenletto	0.500	0.400	0.100	0.50	0.50

21.3.3 The Deviance Goodness-of-Fit Test

When a linear regression problem, such as the insulating fluid example of Section 8.1.2 on page 200, has replicate observations at each configuration of the explanatory variable(s), one

Display 21.6   Deviance goodness-of-fit test for Krunnit Islands data

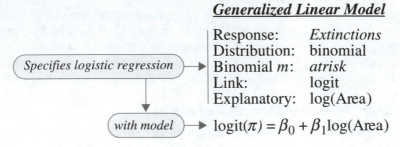

Generalized Linear Model

Response:	*Extinctions*
Distribution:	binomial
Binomial *m*:	*atrisk*
Link:	logit
Explanatory:	log(Area)

Specifies logistic regression

with model → $\text{logit}(\pi) = \beta_0 + \beta_1 \log(\text{Area})$

Computer output:

Variable	Coefficient	Standard Error	z-Statistic
CONSTANT	−1.196	0.118	−10.10
log(Area)	−0.297	0.054	−5.42

Deviance = 12.06   **Degrees of freedom = 16**

Goodness-of-Fit Test: p-value = $\Pr\{\chi_{16}^2 > 12.06\} = \mathbf{0.74}$

Conclusion: There is no evidence that the model is inadequate.

can compare estimates of group means derived from some model to group sample averages in a test of goodness-of-fit for the model. The problems of this chapter are analogous because one compare estimates of proportions derived from a model to the sample proportions. The comparison is like an extra-sum-of-squares test for comparing the model of interest to a general model with a separate population proportion for each observation. The test is only appropriate if the denominators are large: each should be at least 5. Informally, the null hypothesis is that the model of interest is adequate, and the alternative is that more structure is needed to fit the data adequately. More precisely,

Model of interest: $\text{logit}(\pi_i) = \beta_0 + \beta_1 x_{i1} + \cdots$ (*p* parameters);

Saturated model: $\text{logit}(\pi_i) = \alpha_i$ ($i = 1, 2, \ldots, n$, with *n* different parameters).

The extra-sum-of-squares test takes a simple form for this hypothesis, since the deviance residuals from a fit to the full model are all zero. The drop in deviance—due to the $n - p$ additional parameters in the full model that are not included in the reduced model—is the sum of squared deviance residuals from the model of interest (the *deviance statistic*).

If the trial model fits *and the denominators are all large*, the deviance statistic has approximately a chi-squared distribution on $n - p$ degrees of freedom.

$$\text{Approximate } p\text{-value} = \Pr(\chi^2_{n-p} > \text{Deviance Statistic}).$$

Attempts to assess the model with this approach when a substantial proportion of the m_i's are less than 5 can lead to misleading conclusions.

The deviance statistic and its p-value are included in standard output from logistic regression programs. Large p-values indicate that either the model is appropriate or the test is not powerful enough to pick up the inadequacies. Small p-values can mean one of three things: the model for the mean is incorrect (for example, the set of explanatory variables is insufficient); the binomial is an inadequate model for the response distribution; or there are a few severely outlying observations.

In view of the various possibilities just described, it is often difficult to make strong conclusions solely on the basis of the deviance goodness-of-fit test. Further discussion of the case where the binomial model is inadequate for the response distribution is given in Section 21.5 on extra-binomial variation. An example of the deviance goodness-of-fit test is provided in Display 21.6. Notice that the deviance statistic in Display 21.6 is the sum of squared deviance residuals from Display 21.5.

21.4   Inferences About Logistic Regression Coefficients

21.4.1   Wald's Tests and Confidence Intervals for Single Coefficients

Maximum likelihood estimates of the logistic regression coefficients and their standard errors are also included in standard output for logistic regression. Their use, interpretation, and properties are the same as in Chapter 20. The estimated coefficients are approximately normal, so confidence intervals and Wald's tests for single coefficients are accomplished in the same way. The normal approximation is adequate if either the overall sample size n is large or the binomial denominators are all large. It is difficult to state any precise rules about how large a sample size or denominator is large enough, so it is best to label the standard errors and the inferential statements *approximate*.

Example—Krunnit Islands Extinctions

The output in Display 21.7 is from the logistic regression of the proportion of extinctions on the log of island area. This is the fit to the model $\text{logit}(\pi) = \beta_0 + \beta_1 \log(\text{Area})$, where π is the extinction probability.

Is extinction probability associated with island size? If β_1 is zero, the probability of extinction does not depend on area. The question, therefore, is addressed through a test of the hypothesis that $\beta_1 = 0$. The approximate two-sided p-value, as shown in Display 21.7, is extremely small, confirming the visual evidence in the scatterplot that extinction probabilities are associated with island area.

Display 21.7 Wald's test and confidence interval for a single coefficient, from the logistic regression of extinction probability on log of island area: Krunnit Islands data

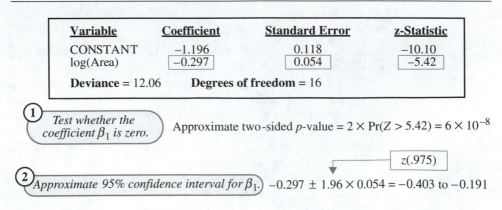

Variable	**Coefficient**	**Standard Error**	**z-Statistic**
CONSTANT	−1.196	0.118	−10.10
log(Area)	−0.297	0.054	−5.42

Deviance = 12.06   **Degrees of freedom** = 16

① *Test whether the coefficient β_1 is zero.* Approximate two-sided p-value = $2 \times \Pr(Z > 5.42) = 6 \times 10^{-8}$

② *Approximate 95% confidence interval for β_1.* −0.297 ± 1.96 × 0.054 = −0.403 to −0.191   [z(.975)]

Interpretation of the Coefficient of a Logged Explanatory Variable

Since the logit is the log of the odds, a unit increase in an explanatory variable is associated with a multiplicative change in the odds by a factor of $\exp(\beta)$, where β is its coefficient in the logistic regression. If the explanatory variable is $\log(X)$, then (as in ordinary regression) the interpretation is most conveniently worded in terms of a twofold, tenfold, or some other k-fold change in X. For each doubling of X, for example, $\log(X)$ increases by $\log(2)$ units, so that the log odds change by $\beta\log(2)$ units. The odds, therefore, change by a factor of $\exp[\beta\log(2)] = 2^{\beta}$. The result is stated generally as follows:

> If $\text{logit}(\pi) = \beta_0 + \beta \log(X)$, then each k-fold increase in X is associated with a change in the odds by a multiplicative factor of k^{β}.

A confidence interval for k^{β} extends from k raised to the lower endpoint to k raised to the upper endpoint of a confidence interval for β. As shown in the accompanying Krunnit Islands example, some customizing of this result may improve the wording of the conclusion.

Example—Krunnit Islands Extinctions

The estimated coefficient of log island area is −0.297. Based on the preceding results with $k = 2$, each doubling of island area is associated with a change in the odds of extinction by an estimated factor of $2^{-0.297}$, or 0.813. It is estimated that the odds of extinction for an at-risk species on an island of area $2A$ is only 81.3% of the odds of extinction for such a species on an island of area A.

Since the effect of reduced habitat area on the odds of extinction is of interest, a researcher might prefer to express the results of the study as the change in odds associated with a *reduction* in island area. With $k = 0.5$, for example: each halving of island area is associated with a multiplicative change in the odds of extinction of $.5^{-0.297}$, or 1.23. The odds of extinction of an island with area $A/2$ are 123% of the odds of extinction of an island of area A. Since an approximate 95% confidence interval for β_1 is -0.403 to -0.191, an approximate 95% confidence interval for the factor associated with a halving of island area is $0.5^{-0.403}$ to $0.5^{-0.191}$, which is to 1.14 to 1.32.

Finally, a 123% change can also be expressed as a 23% increase. Therefore, each halving of area is associated with an increase in the odds of extinction by an estimated 23%. An approximate 95% confidence interval for the percentage increase in odds is 14% to 32%.

21.4.2   The Drop-in-Deviance Test

Drop-in-deviance tests are extra-sum-of-squares tests based on the deviance residuals, as discussed in Section 20.4.2. The issues involved here are the same as those described in Chapter 20 for binary response models. For binomial responses the tests are appropriate if either the overall sample size n is large or the binomial denominators are all large. Studies suggest that the drop-in-deviance test performs well even when the sample size or the denominators are fairly small; again, however, it is appropriate to label the inferences *approximate*.

Example—Krunnit Islands

In the previous section, Wald's test was used to test whether β_1 is zero, resulting in a two-sided p-value of 6×10^{-8}. Alternatively, the drop-in-deviance test can be used to assess the same hypothesis. The reduced model has β_1 set to 0. Its deviance is obtained by fitting a logistic regression model without any explanatory variables (but including a constant term). From computer output (not shown), this deviance is found to be 45.34, with 17 degrees of freedom. In Display 21.7, the deviance statistic from a fit of the model with log(Area) included is 12.06, with 16 degrees of freedom. The deviance dropped by 33.28, with a drop of one degree of freedom. The associated p-value from a chi-square distribution is 8.0×10^{-9}. This is a two-sided p-value, and it provides essentially the same result as that given by Wald's test.

Wald's Test or Drop-in-Deviance Test?

If the hypothesis involves several coefficients, the drop-in-deviance test is the only practical alternative. For a hypothesis that a single coefficient is zero, either test can be used. The approximate p-value from the drop-in-deviance test is generally more accurate than the approximate p-value from the Wald's test. The latter, however, is easier to obtain and is usually satisfactory for conducting tests of secondary importance (such as for informally testing extra terms in an exploratory model-building stage) and when the *p*-value is clearly large or clearly very small.

21.5   Extra-binomial Variation

A binomial count is the sum of independent binary responses with identical π's. If it is the sum of m_i such binary responses, its distribution is known to have mean $m_i\pi_i$ and variance $m_i\pi_i(1 - \pi_i)$. If the binary trials are not independent, or the π's for the binary responses are not the same, or important explanatory variables are not included in the model for π, then the response counts will not have binomial distributions. Typically, the variance of the Y_i's will be greater than expected from a binomial distribution. The terms *extra-binomial variation* and *overdispersion* describe the inadequacy of the binomial model in these situations.

When the response variance is larger than the binomial variance, the regression parameter estimates based on the binomial model will not be seriously biased, but their standard errors will tend to be smaller than they should be; in addition, *p*-values will tend to be too small, and confidence intervals will tend to be too narrow. Thus, if extra-binomial variation exists, alternatives to the usual inference tools must be found.

When in doubt, it is safer to suppose that extra-binomial variation is present than to ignore it. The consequences of assuming the presence of extra-binomial variation when there is actually none are minor. Inferences may be less precise, but they will not be misleading. More severe consequences result from assuming that the responses are binomial when they are not: *p*-values will tend to be too small, and confidence intervals will tend to be too narrow.

21.5.1   Extra-binomial Variation and the Logistic Regression Model

Several procedures allow for extra-binomial variation; but one, the *quasi-likelihood approach*, is particularly easy to use and works in most situations. Unlike the maximum likelihood approach, this method does not require complete specification of the distribution of the response variables. It is based only on the mean and the variance, and these are taken to match the binomial model, except for the presence of a *dispersion parameter*. As shown in the following box, the dispersion parameter, represented by σ^2, accounts for extra-binomial variation. If σ^2 is greater than one, the variance is greater than the binomial variance. However, σ^2 does not represent a variance in this model; it represents a multiplicative departure from binomial variation.

Logistic Regression with Extra-binomial Variation

for quasi-likelihood approach (with two explanatory variables)

$$\mu\{Y_i \mid X_{1i}, X_{2i}\} = m_i\pi_i; \qquad \text{Var}\{Y_i \mid X_{1i}, X_{2i}\} = \sigma^2 m_i\pi_i(1 - \pi_i);$$
$$\text{logit}(\pi_i) = \beta_0 + \beta_1 X_{1i} + \beta_2 X_{2i}.$$

21.5.2   Checking for Extra-binomial Variation

There are three tasks in checking for extra-binomial variation: considering whether extra-binomial variation is likely in the particular responses; examining the deviance goodness-of-fit test after fitting a rich model; and examining residuals to see whether a large deviance statistic may be due to one or two outliers rather than to extra-binomial variation.

Considering Whether Extra-binomial Variation Is Present

In thinking about whether extra-binomial variation is likely, ask the following questions. Are the binary responses included in each count unlikely to be independent? Are observations with identical values of the explanatory variables likely to have different π's? Is the model for π somewhat naive? A yes answer to any of these questions should make a researcher cautious about using the binomial model. Even so, extra-binomial variation is rarely a serious problem when the binomial denominators are small (less than five, say). It is more likely to be a problem for larger denominators.

Extra-binomial variation may be present in the moth coloration responses for a number of reasons:

1.   There may be a lack of independence of binary responses (moth removals) at each site. For example, if predators succeed in finding one moth, their success in finding moths of the same coloration may increase. Thus the odds that a moth of a certain color morph is taken may depend on whether moths of the same morph at the same location are taken or not.

2.   Binary responses at each site may have different probabilities. If the odds that a moth will be removed do indeed depend on the color of the tree where it is placed, and if a site has variation in tree color, moths placed on different trees at the site will have different odds of being taken.

3.   The model may be incorrect. If the odds of removal do depend on tree color, two sites equally distant from Liverpool do not necessarily have the same level of tree coloration, so the model in which the logit of π is a linear function of distance from Liverpool is slightly naive.

Examining the Goodness-of-Fit Test

A large deviance goodness-of-fit test statistic (Section 21.3.3) indicates that the binomial model with specified explanatory variables is not supported by the data. If it can be established that the set of explanatory variables is suitable, however, the large deviance statistic probably indicates extra-binomial variation. This should be examined at an early stage in the analysis by fitting a rich model that includes all the terms that might be important—not excluding polynomial and interaction terms.

Examining Residuals

Since the deviance statistic is the sum of squared deviance residuals, it is useful to examine the deviance residuals themselves to see whether a few of these are responsible for a large deviance statistic. If so, the analysis should focus on the outlier problem rather than on the extra-binomial variation problem.

21.5.3   Quasi-likelihood Inference When Extra-binomial Variation Is Present

The *quasi-likelihood* approach was developed in such a way that the *maximum quasi-likelihood estimates* of parameters are identical to the maximum likelihood estimates, but standard errors are larger and inferences are adjusted to account for the extra-binomial variation.

One possible estimate of the dispersion parameter is the deviance statistic divided by its degrees of freedom:

$$\hat{\sigma}^2 = \frac{\text{Deviance}}{\text{Degrees of freedom}}.$$

The degrees of freedom are $n - p$, the sample size minus the number of parameters in the model for the mean. Apart from the degrees of freedom adjustment, this amounts to a sample variance of the deviance residuals. It should be approximately equal to 1 if the data are binomial, and larger than 1 if the data contain more variation than would be expected from the binomial distribution.

The standard errors of the maximum quasi-likelihood estimates are $\hat{\sigma}$ times the standard errors for the maximum likelihood estimates. The greater the variation in the responses, the greater the uncertainty in the estimates of the parameters.

Inference Based on Asymptotic Normality

Tests and confidence intervals for single coefficients may be accomplished in the usual way with the estimated coefficients and their adjusted standard errors. The inferences are based on the approximate normality of the estimators in situations involving either large sample sizes or large denominators. A test statistic for a hypothesis that β is 7, say, is $t = (\hat{\beta} - 7)/\text{SE}(\hat{\beta})$. The theory suggests that this be compared to a standard normal distribution. Some statisticians, however, prefer to compare it to a t-distribution on $n - p$ degrees of freedom. Similarly, a confidence interval for β may be taken as $\hat{\beta} \pm [t_{\text{df}}(1 - \alpha/2) \times \text{SE}(\hat{\beta})]$, where $t_{\text{df}}(1 - a/2)$ is the appropriate multiplier from the t-distribution on $n - p$ degrees of freedom. Although no theory justifies using t-distributions, in this setting the practice parallels that of ordinary regression models, since the standard error involves an estimate of the unknown σ^2 on $n - p$ degrees of freedom. And although the normal and t-distributions usually yield very similar results, when there is a difference, the approach using the t is more conservative.

Drop-in-Deviance F-Tests

It is also reasonable to consider the drop in the deviance as a way to test whether a single coefficient is zero or whether several coefficients are all zero—as well as to test other hypotheses that can be formulated in terms of a reduced and a full model. Since there is an additional parameter in the response variance, however, the drop in deviance must be standardized by the estimate of σ^2. The resulting statistic is compared to an F-distribution, rather than to a chi-square. This F-test is equivalent to the extra-sum-of-squares F-test for ordinary regression, except that deviance residuals are used rather than ordinary residuals:

$$\boxed{\begin{array}{c} \textbf{Drop-in-Deviance } F\textbf{-Test when Extra-binomial Variation Is Present} \\[1em] F\text{-stat} = \dfrac{\text{Drop in deviance}/d}{\hat{\sigma}^2} \end{array}}$$

where d represents the number of parameters in the full model minus the number of parameters in the reduced model, and $\hat{\sigma}^2$ is the estimate of the dispersion parameter from the full model. The p-value is $\Pr(F_{d,n-p} > F\text{-stat})$ where p is the number of parameters in the full model. No solid theory justifies using this test, but it is a sensible approach, motivated by the analogies to ordinary linear regression.

Notes

1. Many statisticians recommend using the sum of squared Pearson residuals divided by $(n - p)$ rather than the sum of squared deviance residuals divided by $(n - p)$ to estimate σ^2. Although this issue has not been completely resolved, it is always wiser to use one of these approaches than to assume that $\sigma^2 = 1$ when extra-binomial variation is present.
2. Deviance and Pearson residuals from the quasi-likelihood fit are identical to those from the maximum likelihood fit. They may be examined for peculiarities, but they should no longer be compared to a standard normal distribution for outlier detection.
3. Some statistical computer programs allow the user to request quasi-likelihood analysis for binomial-like data. It is evident from the preceding discussion, however, that the quasi-likelihood inferences can be computed with standard output from a binomial logistic regression routine.

21.6   Analysis of Moth Predation Data

In the moth study, two color morphs were placed on trees at various distances from Liverpool. The question is whether the relative rate of removal depends on tree color. Distance is used as a surrogate for tree color. A logistic regression model can isolate a difference that changes steadily with distance by including the product variable *dark* × *dist*, where *dark* is an indicator of dark morph and *dist* is the actual distance. The model

$$\text{logit}(\pi) = \beta_0 + \beta_1 dark + \beta_2 dist + \beta_3 (dark \times dist),$$

enables the analyst to answer the question directly. The constant term β_0 is the log odds for removal of the light morph at Sefton Park (for which $dist = 0$); $\beta_0 + \beta_1$ is the log odds for removal of the dark morph at Sefton Park. So $\beta_1 = (\beta_0 + \beta_1) - \beta_0$ is the log of the odds *ratio* for removal of the dark morph relative to the light morph, at Sefton Park. The log odds for removal of the light morph at a site that is D kilometers distant from Sefton Park is $\beta_0 + \beta_2 D$; and for the dark morph, it is $(\beta_0 + \beta_1) + (\beta_2 + \beta_3)D$. Their difference, $\beta_1 + \beta_3 D$, is the log of the odds ratio for removal of the dark morph relative to the light morph at distance D from Sefton Park. The rate of change in the log odds ratio with distance is

$$[(\beta_1 + \beta_3 D) - \beta_1]/D = \beta_3.$$

Drop-in-Deviance Test for Effect of Distance on Relative Removal Rates

If $\beta_3 = 0$, the relative odds that the light morph will be taken are the same at all locations as the odds that the dark morph will be taken. But if $\beta_3 \neq 0$, the relative odds change with distance. So the question can be addressed by testing the hypothesis that $\beta_3 = 0$, in the context of the preceding model. This may be accomplished by using either Wald's test in the fit of the model or the drop in deviance associated with including the product term in a model that includes only *dark* and *dist*. The latter approach is illustrated in Display 21.8, where the answer to the question seems convincing.

Checking for Extra-binomial Variation

The argument was made (see page 613) that extra-binomial variation may be suspected in this example, based on the study design. Therefore, a close look at the goodness-of-fit deviance statistic from the full model is in order. Overall, the magnitude of the deviance statistic, 13.230, is not large for its degrees of freedom, 10. The p-value for goodness-of-fit, 0.23, shows no substantial problem.

On the other hand, there is no reason to anticipate a linear relationship between logit of removal probability and distance from Liverpool; and the scatterplot of sample logits versus distance (in Display 21.4) suggests departures from a straight line. The issue is clarified by examining the individual deviance residuals in Display 21.9.

One residual—for the light morph at Loggerheads—is somewhat extreme. It might be ignored except that the residual for the dark morph at Loggerheads is also extreme in the same direction. Meanwhile, both residuals at Hawarden are mildly extreme in the opposite direction. These features suggest that each location has unique characteristics—that the differences in removal probabilities at different locations may be due to factors besides the distance from Liverpool.

Adjustment for Extra-binomial Variation

The quasi-likelihood procedure for adjusting the inference based on the Wald's test appears in Display 21.10. The test statistic for the coefficient of the interaction term (which was 3.44 for the binomial model) is 2.99 from the quasi-likelihood fit.

The drop-in-deviance test is calculated from the same deviances that were obtained from the maximum likelihood fit, but the F-version of the statistic is used to allow for extra-binomial variation:

$$\text{F-statistic} \quad = \quad \frac{(11.931)/(1)}{(13.230)/10} \quad = \quad 9.02.$$

Drop in deviance/drop in d.f.

Deviance/d.f.

The p-value is $\Pr(F_{1,10} > 9.02) = .013$.

Display 21.8 The drop-in-deviance chi-squared test for the interaction term in the logistic regression of *removed* (number of moths removed out of the number *placed*) on *dark*, *dist* (distance from Liverpool), and *dark* × *dist*: moth coloration data

$$\text{logit}(\pi) = \beta_0 + \beta_1 dark + \beta_2 dist + \beta_3 dark \times dist$$

1 Fit full model.

Variable	Coefficient	Standard Error	z-Statistic
Constant	−0.718	0.190	−3.77
dark	−0.411	0.275	−1.50
dist	−0.00929	0.00579	−1.60
dark × *dist*	0.0278	0.0081	3.44
Deviance = 13.230	**Degrees of freedom** = 10		

2 Fit reduced model.

$$\text{logit}(\pi) = \beta_0 + \beta_1 dark + \beta_2 dist$$

Variable	Coefficient	Standard Error	z-Statistic
Constant	−1.137	0.157	−7.25
dark	0.404	0.139	2.90
dist	0.0053	0.0040	1.33
Deviance = 25.161	**Degrees of freedom** = 11		

3 Calculate the drop in deviance.

$25.161 - 13.230 = 11.931$
d.f.$= 11 - 10 = 1$

4 Determine the p-value.

p-value $= \Pr(\chi_1^2 > 11.931) = \mathbf{0.0006}$

Conclusion: Substantial evidence indicates that the relative odds of removal of the two color morphs change steadily with distance from Liverpool.

Display 21.9   Deviance residuals from fit of the full model in the moth study

Location	Distance from Liverpool (km)	Morph	Binomial Proportion	Model Estimate	Deviance Residual
Sefton Park	0.0	light	0.304	0.328	− 0.39
		dark	0.250	0.244	0.10
Eastham Ferry	7.2	light	0.350	0.313	0.70
		dark	0.250	0.270	− 0.40
Hawarden	24.1	light	0.346	0.281	1.03
		dark	0.423	0.336	1.31
Loggerheads	30.2	light	0.150	0.269	− 2.21
		dark	0.267	0.361	− 1.56
Llanbedr	36.4	light	0.267	0.258	0.15
		dark	0.383	0.388	− 0.08
Pwyllglas	41.5	light	0.238	0.249	− 0.24
		dark	0.476	0.411	1.21
Clergy Mawr	51.2	light	0.261	0.233	0.63
		dark	0.424	0.455	− 0.59

Location Treated as a Factor with Seven Levels

An alternative, less restrictive model lets the logits for the light morphs be arbitrary at each location, but the difference between the dark morph logit and the light morph logit changes linearly with distance:

$$\text{logit}(\pi) = LOCATION + dark + (dark \times dist).$$

LOCATION contains six indicator variables to represent the seven locations, but the interaction between them is still modeled with the numerical explanatory variable *dist* multiplied by the indicator variable for the dark species.

This model may be a bit more realistic, since it allows the overall log odds of removal to differ for different locations (that is, it does not restrict them to falling on a straight-line function of distance). The interaction of interest, however, continues to be represented by a single term: the difference in log odds of removal between the light and dark morphs is modeled to fall on a straight line function of distance from Liverpool. The output is reported in Display 21.11. Here, cause for concern about extra-binomial variation has greatly diminished, since the *p*-value from the goodness-of-fit test is .72. Furthermore, this approach sharpens the inference about the key issue: the *z*-statistic for the interaction term is 3.43, producing a smaller *p*-value—more convincing evidence—than did the previous approach.

Display 21.10 Quasi-likelihood (QL) inference about distance-by-morph interaction, based on maximum likelihood (ML) output: moth coloration data

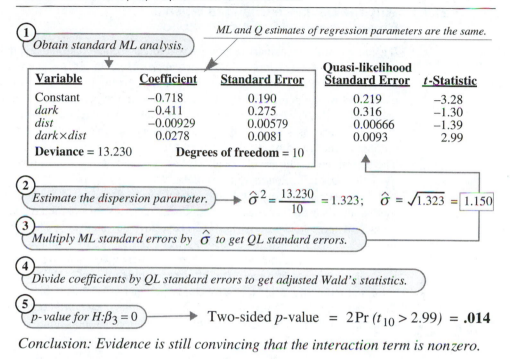

Conclusion: *Evidence is still convincing that the interaction term is nonzero.*

Display 21.11 Logistic regression of proportion of moths removed on location (treated as a factor with seven levels), morph, and a distance × morph interaction

Variable	Coefficient	Standard Error	z-Statistic
Constant	−0.767	0.246	−3.11
EF	0.020	0.274	0.74
HW	0.163	0.305	0.54
LH	−0.798	0.330	−2.42
LB	−0.282	0.322	−0.88
PG	−0.214	0.314	−0.68
CM	−0.434	0.340	−1.28
dark	−0.406	0.275	−1.47
dark × *dist*	0.0277	0.0081	3.43
Deviance = 2.876	**Degrees of freedom** = 5		

21.7   Related Issues

21.7.1   Why the Deviance Changes When Binary Observations Are Grouped

If a number of distinct binary responses exist for each combination of explanatory variables, the data analyst may adopt either of two approaches: fit logistic regression models with the individual binary variables as responses (for example, the 1 or 0 response for each of the 968 moths in the coloration study); or group the observations according to distinct explanatory variable combinations and use as responses the binomial counts—the sums of the binary responses at each distinct explanatory variable combination (for example, the number of moths out of m_i placed, for each of the 14 combinations of type and location). The inferential results are identical; but because the latter approach reduces the data set to a smaller size, it is often more convenient.

As a check, one could use both approaches to fit the same logistic regression model; this would confirm that tests and confidence intervals for coefficients are identical. One feature of the output differs, however: the deviance statistic. The deviance statistic is the *likelihood ratio* test for comparing the fitted model to a saturated model—a model that has a distinct parameter for each observation. The test statistic is different in the two approaches because there are different saturated models. With binary responses, the saturated model has a separate probability parameter for each subject (968 parameters in the moth example). With grouped responses, an *observation* is defined as the binomial count at a distinct explanatory variable combination, so the saturated model has only 14 parameters.

It should not be surprising or alarming, therefore, that the deviance statistic differs in these two approaches, even though all other aspects of estimation, testing, and confidence intervals remain the same. This circumstance is mentioned because the different deviances sometimes cause confusion. The deviance goodness-of-fit test remains inappropriate for binary responses.

21.7.2   Logistic Models for Multilevel Categorical Responses

Multinomial Logits

Suppose that each subject falls into one of three categories of a categorical response variable—for instance, HIV negative, HIV positive but without AIDS symptoms, and HIV positive with AIDS symptoms. Let π_1, π_2, and π_3 be the probabilities of falling into the three categories. It might be desired to see how these probabilities depend on an explanatory variable, such as number of sexual contacts with an infected partner. Since the probabilities sum to 1, any two of them determine the third. Consequently, the analyst need only indicate how any two of them depend on the explanatory variables. Multinomial logits may be defined as follows:

$$\theta_2 = \log(\pi_2/\pi_1)$$
$$\theta_3 = \log(\pi_3/\pi_1).$$

In the example, θ_2 is the log of the odds of being HIV positive relative to being HIV negative, and θ_3 is the log odds of being HIV positive with AIDS symptoms, relative to being HIV negative. Each θ gives the log odds relative to the first category.

If there is no ordering in the response categories, the multinomial logistic regression model has a separate logistic regression for each of the two logits—for example,

$$\log(\pi_{2i}/\pi_{1i}) = \beta_0 + \beta_1 contacts_i$$

and

$$\log(\pi_{3i}/\pi_{1i}) = \alpha_0 + \alpha_1 contacts_i.$$

The different Greek letters for regression coefficients are needed to distinguish the regression coefficients for the two logits.

Some statistical computer packages produce the estimated coefficients and standard errors needed for this model. The packages may differ in their definition of multinomial logits—for example, the choice of the reference level may differ—so some care must be exercised to ascertain what model is being fit by the package. Drop-in-deviance tests are available as before.

Ordered Categorical Responses

The levels of a categorical response commonly have some internal order. Thus the categories just discussed might be considered as ranging essentially from no disease to some disease to severe disease. Because the multinomial logistic regression model has no features that can capture this ordering, it wastes the information from this valuable structure.

Some models do account for such ordering. Their implementation depends on whether they are featured in available statistical computer packages. References with more details are provided at the end of the chapter. The models listed here represent the case involving three response categories; but they extend in a straightforward manner to cases involving more than three categories, and in fact they are most helpful for handling cases involving substantially more categories.

The *common slopes* model stipulates that the coefficients of the explanatory variable (but not the intercepts) in the multinomial model are the same for each multinomial logit:

$$\log(\pi_{2i}/\pi_{1i}) = \beta_0 + \beta_1 contacts_i$$

and

$$\log(\pi_{3i}/\pi_{1i}) = \alpha_0 + \beta_1 contacts_i.$$

This implies that the odds of being HIV positive relative to being HIV negative and the odds of having AIDS symptoms relative to being HIV negative increase at the same rate with increasing numbers of sexual contacts. This model is appropriate if the odds of having

AIDS relative to the odds of being HIV positive do not depend on the number of sexual contacts.

The *linear trends over response categories* model is

$$\log(\pi_{ji}/\pi_{1i}) = \beta_0 + \beta_1(j-1)contacts_i,$$

for j equal to 2 up to the total number of categories (3, in the example). In this model, the odds of having AIDS symptoms relative to being HIV negative increase with the number of sexual contacts at twice the rate of the odds of being HIV positive relative to being HIV negative. This model has the same number of parameters as the common slopes model; but it emphasizes, for example, the greater chance of being in the next highest category with increasing levels of an explanatory variable.

The *proportional odds model* defines logistic regression models for *cumulative probabilities*. If γ_j represents the probability that a subject will respond in category j or less, then $\gamma_1 = \pi_1$, $\gamma_2 = \pi_1 + \pi_2$, and $\gamma_3 = 1$. Models can then be specified for each cumulative probability except the last:

$$\log[\gamma_1/(1 - \gamma_1)] = \beta_0 + \beta_1 contacts_i$$
$$\log[\gamma_2/(1 - \gamma_2)] = \alpha_0 + \beta_1 contacts_i.$$

The first model is for the odds of being HIV negative. The second is for the odds of being AIDS-free. (Here, unlike in the preceding models, the coefficient of *contacts* would be negative.)

21.7.3   The Maximum Likelihood Principle

The *maximum likelihood (ML) principle*, attributed to the British statistician Sir Ronald Fisher, is the approach statisticians predominantly use to select estimation procedures. When they say that a weighted average is preferable to an unweighted one in a specific situation, their justification is often premised on the ML principle. An estimation procedure that has figured prominently in this book—the *least squares principle*—arises directly from the maximum likelihood principle when the response distribution is assumed to be normal.

The ML principle states that the parameters chosen for a model should be those that yield the highest probability of observing precisely what was observed. Application of the ML principle to many statistical problems entails differential calculus and often some formidable algebra. Fortunately, computers handle such problems routinely, which eliminates the need to fathom the mathematics. Such is the case with logistic regression analysis.

This section describes the ML solution to one specific problem (not a logistic regression problem), which was chosen for its simplicity so that the principle itself could be explained.

The Bucket Problem

A bucket contains 12 marbles. Each marble is either red or white, but the numbers of red and white marbles are not known. To gain some information on the bucket's composition, you are allowed to select and view five of them. What inference about the bucket's composition can be drawn from the sample information?

Notation.   The following symbols represent the features of this game:

N = Number of marbles in the bucket (= 12)
θ = Number of red marbles in the bucket (unknown)
n = Sample size = Number of marbles selected for viewing (= 5)
x = Number of red marbles in the sample.

A rule is desired for estimating θ once x is known.

Probability Model.   Imagine that the 12 marbles are numbered consecutively, 1–12, and that the first θ of them are red. A sample of five marbles thus consists of five distinct marble numbers—for example, {2, 5, 6, 9, 12}. If the bucket contained $\theta = 7$ reds, this sample would have $x = 3$ red marbles (numbers 2, 5, and 6).

Assume that the sample selection process ensures that all 792 possible samples of size five from the 12 have an equal chance of selection (1/792). Then for any given number of red marbles in the bucket, one can determine what the chances are of drawing the numbers of red marbles actually obtained in the sample. These chances form the rows of Display 21.12.

For example, if no red marbles were in the bucket to begin with (row 1), the chances are certain (100%) that the sample will have no red marbles. If instead the bucket contained two red marbles (row 3), there is slightly better than a 50-50 chance that the sample will include exactly one red marble; the chances of obtaining no red marbles in this case are double the chances of selecting both red marbles.

The numbers in Display 21.12 come from the hypergeometric probability distribution (previously encountered in the discussion of Fisher's Exact Test). The hypergeometric formula for the probabilities that x will take on each of its possible values when the bucket has θ red marbles is not given here. Each row shows the chances of different outcomes to the sampling experiment when the bucket composition is known.

For the inferential problem—making a statement about the likely values of θ based on the observed sample x—the *columns* of Display 21.12 are important. After the sampling is done and three reds (for example) are observed, one can look at the entries in the column headed by $x = 3$ to see how to use the sample information. Clearly, $\theta = 0, 1$, and 2 are impossible. There must have been at least three reds in the bucket, because three showed up in the sample. Similarly, since two whites appeared in the sample, there must be at least that many whites in the bucket. Thus, the column also rules out $\theta = 11$ and 12.

Can more be said? It is possible that the bucket began with three reds and all were selected; and it is also possible that the bucket began with seven reds and three were selected. But although both are possible, the likelihoods of these two cases are not equal. If only three reds were in the bucket, finding three in the sample would be a relatively *rare* event—an outcome expected to occur in fewer than one in 20 trials. But if seven reds were in the bucket, finding three in the sample would be a *common* occurrence, to be expected in nearly half of all such trials. Since common events are expected and rare events are not, it is natural

Display 21.12　Probability distributions and likelihood functions for the bucket with 12 marbles and five draws

Number (x) of Red Marbles Found in the Sample

	0	1	2	3	4	5
0	1	0	0	0	0	0
1	.5833	.4167	0	0	0	0
2	.3182	.5303	.1515	0	0	0
3	.1591	.4773	.3182	.0454	0	0
4	.0707	.3535	.4243	.1414	.0101	0
5	.0265	.2210	.4419	.2652	.0442	.0012
6	.0076	.1136	.3788	.3788	.1136	.0076
7	.0012	.0442	.2652	.4419	.2210	.0265
8	0	.0101	.1414	.4243	.3535	.0707
9	0	0	.0454	.3182	.4773	.1591
10	0	0	0	.1515	.5303	.3182
11	0	0	0	0	.4167	.5833
12	0	0	0	0	0	1

Number (θ) of Red Marbles Originally in the Bucket (row labels at left)

> When you know how many red marbles are in the *bucket*, look across that row to see how many might be observed in the sample. (Rows are **probability functions**.)

> When you know how many red marbles were in the *sample*, look down that column to infer how many might have been in the bucket. (Columns are **likelihood functions**.)

to look at the possibilities and—having observed three reds in the sample—conclude that it is more likely that the bucket began with seven reds.

The probabilities in Display 21.12 are called *likelihoods* when viewed down a column. The column as a whole is a *likelihood function*, meaning that it is a function of the unknown parameter θ. Each observed number of reds in the sample gives rise to a different likelihood function. Adding probabilities across a row always yields a sum of 1.0; however, the same is *not* true for adding down a column. Therefore, likelihoods must not be confused with probabilities. It is not appropriate to say that, if $x = 3$, the probability that $\theta = 7$ is 0.4419. On the other hand, likelihoods can be expressed in a relative sense. For $x = 3$, the statement "$\theta = 7$ is 10 times more likely than $\theta = 3$" is a legitimate interpretation.

The Maximum Likelihood Estimator for the Bucket Problem

To make a best guess at θ, according to the maximum likelihood principle—as the name implies—one should select the θ for which the likelihood is largest. This amounts to guessing that θ has *the value that gives the highest probability of observing what was, in fact, observed.*

Suppose that a particular sample yielded $x = 3$. In the likelihood column for "$x = 3$," the largest likelihood, .4419, corresponds to $\theta = 7$. Then the best (ML) guess is that the number of reds originally in the bucket was seven.

Examining all columns in Display 21.12 leads to a rule describing which θ to guess for each outcome x. Display 21.13 shows this *maximum likelihood estimator* for the bucket problem.

Display 21.13 The maximum likelihood estimator for the unknown number of red marbles in a bucket of 12, based on a sample of five

Observed number of reds in sample (x):	0	1	2	3	4	5
$\hat{\theta}$ = ML estimate of number of reds in bucket:	0	2	5	7	10	12

Notes About Maximum Likelihood Estimation

1. The probability distribution of the response variables must be specified entirely, except for the values of the unknown parameters. This contrasts to the method of least squares and the method of quasi-likelihood, which only require knowledge of the mean and variance of the response distribution.

2. Display 21.13 also provides the sampling distribution of the ML estimator. The rows of the table show the probabilities for the sample outcomes, and hence for the values of the estimator. In more complex situations, considerable effort is required to determine the sampling distribution of an ML estimator.

3. The ML principle itself does not guarantee that the ML estimator has suitable properties, but theoretical statisticians have shown that the resulting procedure does indeed lead to desirable estimators.

21.8 Summary

Binomial logistic regression is a special type of generalized linear model in which the response variable is binomial and the logit of the associated probability is a linear function of regression coefficients. The case in which all the binomial denominators are 1 was referred to as binary logistic regression and was discussed in Chapter 20.

In analyzing proportions, one should begin, if possible, with scatterplots of the sample logits versus explanatory variables. The model-building stage consists of checking for extra-binomial variation and then finding an appropriate set of explanatory variables. The main inferential statements are based on Wald's z-test or the associated confidence interval for a single coefficient, or on the drop-in-deviance chi-squared test for a hypothesis about one or several coefficients. If extra-binomial variation is suspected, these tests are replaced by t-tests and F-tests, respectively.

Krunnit Islands Data

The analysis comes into focus with a plot of empirical logits versus the log of island area. It is evident from Display 21.2 that the log odds of extinction are linear in the log of island area. This can be confirmed by fitting the model and observing that the deviance statistic is not large (the p-value from the deviance goodness-of-fit test is .74), and also by testing for the significance of a squared term in log island area. The remainder of the analysis only requires that the inferential statement about the association between island area and log odds of extinction be worded on the original scale.

Moth Coloration Data

A plot of the empirical logits versus distance from Liverpool, with different codes for the light and dark moths, revealed initial evidence of an interactive effect—particularly that the odds of light morph removal decrease with increasing distance from Liverpool while the odds of dark morph removal increase with increasing distance from Liverpool. Based on the scatterplot (Display 21.4), the linearity of the logistic regression on distance is questionable, particularly since the logits drop rather dramatically at Loggerheads. There seems to be good reason to believe that distance from Liverpool is not entirely adequate as a surrogate for darkness of trees. Two avenues for dealing with this problem were suggested. One involved using quasi-likelihood analysis, which incorporated the model inadequacy into a dispersion parameter. The other, more appealing model, however, included location as a factor with seven levels (rather than relying on their distance from Liverpool as a single numerical explanatory variable) but retained the product of the morph indicator variable and distance so that a single parameter represented the interactive effect of interest. This model fit well and maximum likelihood analysis was used to draw inferences about the effect.

Further Reading

Logistic regression references were provided in the previous chapter. One useful book on ordered categorical data analysis is by Agresti (1990). The particular models for ordered categorical responses that appear in this chapter are discussed by Aitkin, Anderson, Francis, and Hinde (1989), who also detail the computer trick for fitting two of these with a log-linear regression program.

21.9 Exercises

Conceptual Exercises

1. **Krunnit Islands Extinctions.** Logistic regression analysis assumes that, after the effects of explanatory variables have been accounted for, the responses are independent of each other. Why might this assumption be violated in these data?

2. **Moth Coloration.** (a) Why might one suspect extra-binomial variation in this problem? (b) Why does the question of interest pertain to an interaction effect? (c) How (if at all) can the response variable be redefined so that the question of interest is about a "main effect" rather than an interaction?

3. Which of the following are not *counted proportions*? (a) The proportion of cells with chromosome aberrations, out of 100 examined. (b) The proportion of body fat lost after a special diet. (c) The proportion of 10 grams of offered food eaten by a quail. (d) The proportion of a mouse's four limbs that have malformations.

4. Suppose that Y has a binomial(m, π) distribution. (a) Is the variance of the binomial count Y an increasing or decreasing function of m? (b) Is the variance of the binomial proportion Y/m an increasing or decreasing function of m?

5. To confirm the appropriateness of the logistic regression model logit$(\pi) = \beta_0 + \beta_1 x$, it is sometimes useful to fit logit$(\pi) = \beta_0 + \beta_1 x + \beta_2 x^2$ and test whether β_2 is zero. (a) Does the reliability of this test depend on whether the denominators are large? (b) Is the test more relevant when the denominators are small?

6. What is the quasi-likelihood approach for extending the logistic regression model to account for a case in which more variation appears than was predicted by the binomial distribution?

7. **Malformed Forelimbs.** In a randomized experiment, mouse fetuses were injected with 500, 750, or 1,000 mg/kg of acetazolamide on day 9 of gestation. (Data from L. B. Holmes, H. Kawanishi, and A. Munoz, "Acetazolamide: Maternal Toxicity, Pattern of Malformations, and Litter Effect," *Teratology* 37 (1988): 335–42.) On day 18 of gestation, the fetus was removed and examined for malformations in the front legs. Display 21.14 shows the results. (a) Suppose that the response variable for each mouse fetus is the number of forelimbs with malformations. Explain how logistic regression can be used to study the relationship between malformation probability and dose. (b) What is the sample size? (c) Is it necessary to worry about extra-binomial variation? (d) Is it appropriate to use the deviance goodness-of-fit test for model checking? (e) Is it appropriate to obtain a confidence interval for the effect of dose? (f) How would the resulting confidence interval be used in a sentence summarizing the statistical analysis?

Display 21.14 Dose of acetazolamide and number of malformed forelimbs in mouse fetuses

Dose (mg/kg)	Number of Malformed Forelimbs			Sample Size
	0	1	2	
500	137	40	15	192
750	81	68	42	191
1,000	87	49	69	205

8. If your computer software always produces binary response deviances, how can you determine the lack-of-fit chi-square for the moth data? (See Section 21.7.1.)

Computational Exercises

9.   Moth Coloration. For the moth coloration data in Section 21.1.1, consider the response count to be the number of light moths removed and the binomial denominator to be the total number of moths removed (light and dark) at each location. (a) Plot the logit of the proportion of light moths removed versus distance from Liverpool. (b) Fit the logistic regression model with distance as the explanatory variable; then report the estimates and standard errors. (c) Compute the deviance goodness-of-fit test statistic, and obtain a p-value for testing the adequacy of the model.

10.   Death Penalty and Race of Victim. Reconsider the data of Display 19.2 on page 540 involving death penalty and race of victim. Reanalyze these data using logistic regression. The response variable is the number of convicted murderers in each category who receive the death sentence, out of the m convicted murderers in that category. (a) Plot the logits of the observed proportions versus the level of aggravation. The logit, however, is undefined for the rows where the proportion is 0 or 1, so compute the empirical logit $= \log[(y+.5)/(m-y+.5)]$ and plot this versus aggravation level, using different plotting symbols to distinguish proportions based on white and black victims. (b) Fit the logistic regression of death sentence proportions on aggravation level and an indicator variable for race of victim. (c) Report the p-value from the deviance goodness-of-fit test for this fit. (d) Test whether the coefficient of the indicator variable for race is equal to 0, using the Wald's test. (e) Construct a confidence interval for the same coefficient, and interpret it in a sentence about the odds of death sentence for white-victim murderers relative to black-victim murderers, accounting for aggravation level of the crime.

11.   Death Penalty and Race of Victim. Fit a logistic regression model as in Exercise 10, but treat aggravation level as a factor. Include the products of the race variable with the aggravation level indicators to model interaction, and determine the drop in deviance for including them. This constitutes a test of the equal odds ratio assumption, which could be used to check the assumption for the Mantel–Haenszel test. What can you conclude about the assumption?

12.   Vitamin C and Colds. Reconsider the data in Display 18.2 on page 517 from the randomized experiment on 818 volunteers. Let Y_i represent the number of individuals with colds in group i, out of m_i. Let X_i be an indicator variable that takes on the value 1 for the placebo group and 0 for the vitamin C group. Then the observed values for the two groups are:

Group (i)	Individual with Colds (Y)	Group Size (m)	Placebo Indicator (X)
1	335	411	1
2	302	407	0

Consider the logistic regression model in which Y_i is binomial(π_i, m_i) and logit(π_i) $= \beta_0 + \beta_1 X_i$. (a) Using a computer package, obtain an estimate of and a confidence interval for β_1. (b) Interpret the results in terms of the odds of cold for the placebo group relative to the odds of a cold for the vitamin C group. (c) How does the answer to part (b) compare to the answer in Display 18.9?

13.   Vitamin C. Between December 1972 and February 1973, a large number of volunteers participated in a randomized experiment to assess the effect of large doses of vitamin C on the incidence of colds. (Data from T. W. Anderson, G. Suranyi, and G. H. Beaton, "The Effect on Winter Illness of Large Doses of Vitamin C," *Canadian Medical Association Journal* 111 (1974): 31–36.) The subjects were given tablets to take daily, but neither the subjects nor the doctors who evaluated them were aware of the dose of vitamin C contained in the tablets. Shown in Display 21.15 are the proportion of subjects in each of the four dose categories who did not report any illnesses during the study period.

Display 21.15   Vitamin C and colds

Daily *dose* of Vitamin C (g)	Number of Subjects	Number with No Illnesses	Proportion with No Illnesses
0	1,158	267	0.231
0.25	331	74	0.224
1	552	130	0.236
2	308	65	0.211

(a) For each of the four dose groups, calculate the logit of the estimated proportion. Plot the logit versus the dose of vitamin C. (b) Fit the logistic regression model, $logit(\pi) = \beta_0 + \beta_1\ dose$. Report the estimated coefficients and their standard errors. Report the p-value from the deviance goodness-of-fit test. Report the p-value for a Wald's test that β_1 is 0. Report the p-value for a drop-in-deviance test for the hypothesis that β_1 is 0. (c) What can be concluded about the adequacy of the binomial logistic regression model? What evidence is there that the odds of a cold are associated with the dose of vitamin C?

Data Problems

14.   Belief Accessibility.   Increasingly, politicians look to public opinion surveys to shape their public stances. Does this represent the ultimate in democracy? Or are seemingly scientific polls being rigged by the manner of questioning? Psychologists believe that opinions—expressed as answers to questions—are usually generated at the time the question is asked. Answers are based on a quick sampling of relevant beliefs held by the subject, rather than a systematic canvas of all such beliefs. Furthermore, this sampling of beliefs tends to overrepresent whatever beliefs happen to be most accessible at the time the question is asked. This aspect of delivering opinions can be abused by the pollster. Here, for example, is one sequence of questions: (1) "Do you believe the Bill of Rights protects personal freedom?" (2) "Are you in favor of a ban on handguns?" Here is another: (1) "Do you think something should be done to reduce violent crime?" (2) "Are you in favor of a ban on handguns?" The proportion of yes answers to question 2 may be quite different depending on which question 1 is asked first.

To study the effect of *context questions* prior to a *target question*, researchers conducted a poll involving 1,054 subjects selected randomly from the Chicago phone directory. To include possibly unlisted phones, selected numbers were randomly altered in the last position. (Data from R. Tourangeau, K. A. Rasinski, N. Bradburn, and R. D'Andrade, "Belief Accessibility and Context Effects in Attitude Measurement," *Journal of Experimental Social Psychology* 25 (1989): 401–21.) The data in Display 21.16 show the responses to one of the questions asked concerning continuing U.S. aid to the Nicaraguan Contra rebels. Eight different versions of the interview were given, representing all possible combinations of three factors at each of two levels. The experimental factors were CONTEXT, MODE, and LEVEL. CONTEXT refers to the type of context questions preceding the question about Nicaraguan aid. Some subjects received a context question about Vietnam, designed to elicit reticence about having the U.S. become involved in another foreign war in a third-world country. The other context question was about Cuba, designed to elicit anti-communist sentiments. MODE refers to whether the target question immediately followed the context question or whether there were other questions scattered in between. LEVEL refers to two versions of the context question. In the HIGH LEVEL the question was worded to elicit a higher level of agreement than in the LOW LEVEL wording.

Display 21.16   Opinion polls and belief accessibility: CONTEXT refers to the context of the question preceding the target question about U.S. aid to the Nicaraguan Contra rebels; MODE is "scattered" if the target question was not asked directly after the context question; and LEVEL refers to the wording of the question ("High" designed to elicit a higher favorable response)

GROUP (i):	1	2	3	4	5	6	7	8
CONTEXT:	Vietnam	Cuba	Vietnam	Cuba	Vietnam	Cuba	Vietnam	Cuba
MODE:	scattered	scattered	scattered	scattered	not	not	not	not
LEVEL:	high	high	low	low	high	high	low	low
NUMBER (m):	132	132	132	131	132	131	132	132
PERCENTAGE IN FAVOR OF CONTRA AID:	20.5	31.1	28.0	38.9	34.1	48.9	23.5	45.5

Analyze these data to answer the following questions of interest: (a) Does the proportion of favorable responses to the target question depend on the wording (LEVEL) of the question? If so, by how much? (b) Does the proportion depend on the context question? If so, by how much? (c) Does the proportion depend on the context question to different extents according to whether the target and context questions are scattered? (*Hint:* Use indicator explanatory variables for the factors.)

15.   **Aflatoxicol and Liver Tumors in Trout.** An experiment at the Marine/Freshwater Biomedical Sciences Center at Oregon State University investigated the carcinogenic effects of aflatoxicol, a metabolite of Aflatoxin B1, which is a toxic by-product produced by a mold that infects cottonseed meal, peanuts, and grains. Twenty tanks of rainbow trout embryos were exposed to one of five doses of Aflatoxicol for one hour. The data in Display 21.17 (from George Bailey and Jerry Hendricks) represent the numbers of fish in each tank and the numbers of these that had liver tumors after one year. Describe the relationship between dose of Aflatoxicol and odds of liver tumor. It is also of

---

Display 21.17   Aflatoxicol and liver tumors in trout (4 tanks at each dose)

Dose (ppm)	*Number of Trout with Liver Tumors (Y) / Number in Tank (m)*			
0.010	9/87	5/86	2/89	9/85
0.025	30/86	41/86	27/86	34/88
0.050	54/89	53/86	64/90	55/88
0.100	71/88	73/89	65/88	72/90
0.250	66/86	75/82	72/81	73/89

interest to determine the dose at which 50% of the fish will get liver tumors. (*Note: Tank effects* are to be expected, meaning that tanks given the same dose may have slightly different π's. Thus, one should suspect extra-binomial variation.)

Answers to Conceptual Exercises

1. Neighboring islands are likely to be more similar in the types of species present and in the likelihood of extinction for each species, than islands farther apart. Another possibility is that extinctions of (migratory) shorebirds can occur in another part of the world and simultaneously affect all or several of the Krunnit Islands populations.

2. (a) Sites the same distance from Liverpool may have different probabilities of removal for dark and light morphs because the blackening of the trees may depend on more than simple distance from Liverpool. (b) The question is not whether the probability of removal depends on the covariate distance from Liverpool or on the factor morph, but whether the probability of removal depends on distance from Liverpool to a different extent for light morphs than for dark morphs. (c) Let m be the number of moths removed at a location. Let Y be the number of these that are dark morphs. How does the proportion of removed moths that are dark morphs depend on distance from Liverpool?

3. Choices (b) and (c).

4. (a) Increasing. (b) Decreasing.

5. (a) Partly. The Wald's test and the drop-in-deviance test both can be used if either the sample size is large or the binomial denominators are large. (b) The test is useful as an informal device for assessing the model in each case. It may be more relevant when the denominators are small, however, since few alternatives are available for model checking in this case.

6. The assumed variance is $\sigma^2 m_i \pi_i (1 - \pi_i)$. The dispersion parameter σ^2 is introduced to allow for variance greater than (or possibly less than) that expected from a binomial response.

7. (a) Suppose that Y_i is binomial$(2, \pi_i)$ with logit$(\pi_i) = \beta_0 + \beta_1 dose_i$ for $i = 1, \ldots, 588$. Explore the adequacy of this model. If it fits, make inferences about β_1. (b) 588. (c) Probably not, since the denominators are all so small. (d) No, the denominators are too small. (e) Yes. (f) "The odds of a malformation are estimated to increase by (lower limit) to (upper limit) for each 100 mg/kg of acetazolamide (95% confidence interval)." *Note:* If L and U are the lower and upper endpoints of a confidence interval for β_1, then the interval in this sentence is exp$(100L)$ and exp$(100U)$.

8. Compare the deviance from the model of interest to the deviance from the full model, with a separate mean for each morph at each location.

Log-Linear Regression
for Poisson Counts

The number of traffic accidents at an intersection over a year, the number of shrimp caught in a 1-kilometer tow of a net, the number of outbreaks of an infectious disease in a county during a year, and the number of flaws on a computer chip are examples of responses that consist of integer counts. Such counts differ from binomial counts of successes in a fixed number of trials, where each trial contributes either +1 or 0 to the total. They are counts of the occurrences of some event over time or space, without a definite upper bound. The *Poisson probability distribution* is often useful for describing the population distribution of these types of counts. A related regression model—Poisson log-linear regression—specifies that the logarithms of the means of Poisson responses are linear in regression coefficients.

Poisson log-linear regression, like binomial logistic regression, is a generalized linear model. The tools for analysis are similar to those described in Chapters 20 and 21. Furthermore, the tools extend to account for responses that are Poisson-like but possess excess variation, through addition of a dispersion parameter and use of quasi-likelihood analysis.

22.1 Case Studies

22.1.1 Age and Mating Success of Male Elephants—An Observational Study

Although male elephants are capable of reproducing by 14 to 17 years of age, young adult males are usually unsuccessful in competing with their larger elders for the attention of receptive females. Since male elephants continue to grow throughout their lifetimes, and since larger males tend to be more successful at mating, the males most likely to pass their genes to future generations are those whose characteristics enable them to live long lives. Joyce Poole studied a population of African elephants in Amboseli National Park, Kenya, for 8 years. Display 22.1 shows the number of successful matings and ages (at the study's beginning) of 41 male elephants. (Data from J. H. Poole, "Mate Guarding, Reproductive Success and Female Choice in African Elephants," *Animal Behavior* 37 (1989): 842–49.) What is the relationship between mating success and age? Do males have diminished success after reaching some optimal age?

Display 22.1 Age at beginning of study and number of successful matings for 41 African elephants

Age	Matings	Age	Matings	Age	Matings
27	0	33	3	39	1
28	1	33	3	41	3
28	1	33	3	42	4
28	1	33	2	43	0
28	3	34	1	43	2
29	0	34	1	43	3
29	0	34	2	43	4
29	0	34	3	43	9
29	2	36	5	44	3
29	2	36	6	45	5
29	2	37	1	47	7
30	1	37	1	48	2
32	2	37	6	52	9
33	4	38	2		

Summary of Statistical Findings

The average number of successful matings increased with age, for male elephants between 25 and 50 years old (Display 22.2). There is no evidence that the mean reached some maximum value at any age less than 50 (the one-sided p-value for significance of a quadratic term in the log-linear regression is .33). The estimated mean number of successful matings (in 8 years) for 27-year-old males was 1.31, and the mean increased by a factor of 2.00

for each 10-year increase in age up to about 50 years (95% confidence interval for the multiplicative factor: 1.52 to 2.60).

Display 22.2 Counts of successful matings versus age for 41 male African elephants (jittered to avoid overlapping points): the curve is the estimated log-linear regression model for the mean count

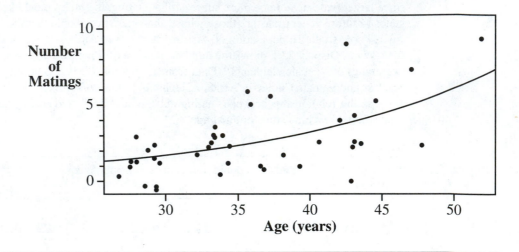

Scope of Inference

There may be some bias in these results, due to measurement error in estimating the age of each elephant and lack of independence of the counts for different males. Bias would also result if successful matings were not correctly attributed. These problems have only minor effects on the relatively informal conclusions obtained. Attempting to apply inferences from these data to a wider elephant population would require careful attention to the method of sampling used.

22.1.2 Treatment for Epileptic Seizures—
A Randomized Experiment

To study the anti-epileptic drug progabide, researchers randomly assigned 59 patients suffering from epileptic seizures to receive either progabide or a placebo, in addition to standard chemotherapy. Display 22.3 shows data from part of the experiment. (Data from Leppik et al., "A Double-blind Crossover Evaluation of Progabide in Partial Seizures," *Neurology* 35 (1985): 285.) Display 22.3 lists the number of epileptic seizures in the 8 weeks prior to administration of the treatment, the number of seizures in the 8 weeks after start of treatment, and the age of each patient.

Display 22.3   Counts of seizures for 59 epileptic patients 8 weeks before and 8 weeks after treatment by placebo (Treatment = 0) or progabide (Treatment = 1), and age

Treatment (0 = Control)	Age (years)	Baseline Count	Posttreatment Count	Treatment (1 = Treated)	Age (years)	Baseline Count	Posttreatment Count
0	31	11	14	1	18	76	42
0	30	11	14	1	32	38	28
0	25	6	11	1	20	19	7
0	36	8	13	1	30	10	13
0	22	66	55	1	18	19	19
0	29	27	22	1	24	24	11
0	31	12	12	1	30	31	74
0	42	52	95	1	35	14	20
0	37	23	22	1	27	11	10
0	28	10	33	1	20	67	24
0	36	52	66	1	22	41	29
0	24	33	30	1	28	7	4
0	23	18	16	1	23	22	6
0	36	42	42	1	40	13	12
0	26	87	59	1	33	46	65
0	26	50	16	1	21	36	26
0	28	18	6	1	35	38	39
0	31	111	123	1	25	7	7
0	32	18	15	1	26	36	32
0	21	20	16	1	25	11	3
0	29	12	14	1	22	151	302
0	21	9	14	1	32	22	13
0	32	17	13	1	25	41	26
0	25	28	30	1	35	32	10
0	30	55	143	1	21	56	70
0	40	9	6	1	41	24	13
0	19	10	10	1	32	16	15
0	22	47	53	1	26	22	51
				1	21	25	6
				1	36	13	0
				1	37	12	10

Summary of Statistical Findings

Baseline seizure counts were transformed to a logarithmic scale for analysis. Even then, the methods of Chapter 11 identify case #49 (with 302 posttreatment seizures) as having high leverage and exerting strong influence. Eliminating this case, one must restrict inferences to patients with baseline counts below 150. In this range, strong associations of posttreatment counts with age and log-baseline counts may be observed (see Display 22.4), but no evidence indicates that the progabide treatment had any effect on the mean posttreatment counts (p-value $= .25$, from an F-test for the effects of treatment and for interaction of treatment with log-baseline counts, with 2 and 53 d.f.; from the log-linear regression of posttreatment counts).

Display 22.4   Numbers of seizures in the 8 weeks following treatment versus number of seizures in the 8 weeks prior to treatment, for 58 epileptic patients (case #49 omitted): curves represent fitted log-linear regression model, at *age* = 30

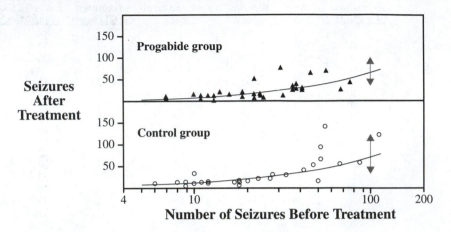

22.2 Log-Linear Regression for Poisson Responses

22.2.1 Poisson Responses

The response variables in the two case studies are counts—the count of successful elephant matings in 8 years of observation, and the count of seizures in an 8-week period. Counts of events or objects occurring in specified time intervals or in specified geographical areas differ from binomial counts.

The Poisson Probability Distribution

Siméon Denis Poisson's treatise on probability theory, "Recherches sur la probabilité des jugements en matière criminelle et en matière civile" (1837), examines the form that the binomial distribution takes for large numbers of trials with small probabilities of success on any given trial. The limiting form of the distribution has no upper limit to the count of successes; the probability of obtaining Y successes is given by the formula

$$\text{Probability}\{Y\} = \exp(-\mu) \times \mu^{Y}/Y!, \quad \text{for any } Y = 0, 1, 2, \dots,$$

where μ is the mean. The distributions described by this mathematical rule have continued to carry Poisson's name.

Since Poisson's time, the distribution has been used to model a wide variety of situations. It is most appropriate for counts of rare events that occur at completely random points in space or time. Its general features make it a reasonable approximation to a wider class of problems involving count data where spread increases with the mean.

The Poisson distribution has several characteristic features:

1. The variance is equal to the mean.
2. The distribution tends to be skewed to the right, and this skewness is most pronounced when the mean is small.
3. Poisson distributions with larger means tend to be well-approximated by a normal distribution.

22.2.2 The Poisson Log-Linear Model

With a count response and explanatory variables, the Poisson log-linear regression model specifies that the distribution of the response is Poisson and that the log of the mean is linear in regression coefficients:

$$\textbf{A Poisson Log-Linear Regression Model}$$
$$Y \text{ is Poisson, with } \mu\{Y|X_1, X_2\} = \mu$$
$$\log(\mu) = \beta_0 + \beta_1 X_1 + \beta_2 X_2.$$

This is a generalized linear model with a Poisson response distribution. The link function, which relates the response mean to the explanatory variables, is the logarithm.

A hypothetical example is illustrated in Display 22.5. The distribution of the response, as a function of a single explanatory variable X, is Poisson; and with μ representing $\mu\{Y|X\}$, the log-linear model is $\log(\mu) = -1.7 + 0.20X$. Poisson probability histograms are displayed at $X = 5$, 14, and 20. Notice that the regression is nonlinear in X, as expected, since $\mu\{Y|X\} = \exp(-1.7 + 0.20X)$. In addition, the variance is greater for distributions with larger means. The skewness of the Poisson distribution is substantial when the mean is small, but it diminishes with increasing mean.

Poisson Regression or Transformation of the Response?

Taking the logarithm of the response helps to straighten out the relationship. For the situation in Display 22.5, $\mu\{\log(Y)|X\}$ is approximately a straight-line function of X. On the other hand, the variance is still nonconstant after this transformation. The transformation that stabilizes the variance is the square root: $\text{Var}\{Y^{1/2}|X\}$ is approximately constant. Both transformations, therefore, are somewhat unsatisfactory. Poisson log-linear regression offers a more suitable approach and does not involve a transformation of the response.

Example—Elephant Matings

The response variable is the number of successful matings, and the explanatory variable is the elephant's age. If the Poisson log-linear model is used, the numbers of successful matings are taken to be Poisson distributed, and the mean number of counts as a function

Display 22.5 A representation of a log-linear model in which the distribution of Y (as a function of X) is Poisson with mean μ and $\log(\mu) = -1.7 + .20X$: the histograms are the Poisson distributions of Y at three values of X

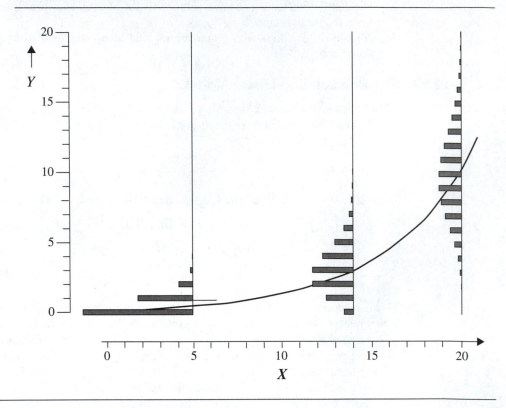

of age is specified by the log-linear model:

$$\log(\mu) = \beta_0 + \beta_1 age \qquad \text{or equivalently,} \qquad \mu = \exp(\beta_0 + \beta_1 age).$$

Thus an increase of one year in age is associated with a K-fold change in the mean number of matings, where $K = \exp(\beta_1)$. The estimated log-linear model is drawn on the scatterplot in Display 22.2. The estimate of β_1 is 0.069, so an increase in 1 year of age is estimated to be associated with a 1.07-fold change in the mean number of successful matings (a 7% increase in the mean for each additional year of age).

22.2.3 Estimation by Maximum Likelihood

Maximum likelihood fitting by a Poisson log-linear regression computer package yields a familiar display of coefficients and standard errors. A fit of a quadratic log-linear model to the elephant mating data appears in Display 22.6. The tools involved in the maximum likelihood analysis, including residuals, the deviance goodness-of-fit test, Wald's tests and

Display 22.6 Poisson log-linear regression of number of successful matings on age and age-squared, from observations on 41 male elephants: $\log(\mu) = \beta_0 + \beta_1 \text{ } age + \beta_2 \text{ } age^2$

Variable	Coefficient	Standard Error	z-Statistic	Two-Sided p-Value
Constant	−2.857	3.036	0.941	.3467
age	0.136	0.158	0.861	.3894
age^2	−0.00086	0.00201	0.427	.6692

Deviance = 50.83   **d.f.**= 38

confidence intervals, and the drop-in-deviance test are used in the same way as for binomial logistic regression. These are discussed in the next few sections and demonstrated in connection with the case studies.

22.3 Model Assessment

22.3.1 Scatterplot of Logged Counts Versus an Explanatory Variable

Initial exploration often involves plotting the logarithms of responses versus one or more explanatory variables, using graphical techniques such as coding and jittering. Display 22.2 shows the number of matings (not the log of the number of matings) versus elephant age. This is useful in the summary, but it does not indicate the appropriateness of the model. A plot of the log of the number of counts versus age would better illustrate the linearity on the log scale. As with empirical logits, however, some small amount like 1/2 needs to be added to the counts if some of them are zero, so that the logarithm can be computed.

22.3.2 Residuals

The issues involved in using residuals are similar to those for binomial logistic regression. The formal definition of a deviance residual—as the maximum possible value of the logarithm of the likelihood at observation i minus the value of the logarithm of the likelihood estimated from the current model—remains unchanged. However, the definition leads to the following somewhat different formula for Poisson regression:

Deviance Residual for Poisson Regression

$$Dres_i = \text{sign}(Y_i - \hat{\mu}_i) \sqrt{2[Y_i \times \log\left(\frac{Y_i}{\hat{\mu}_i}\right) - Y_i + \hat{\mu}_i]} \text{ .}$$

As before, practitioners with access to computer software can disregard the unpleasant appearance of this formula. Subsequent uses of these residuals are straightforward and resemble uses of standard residuals in ordinary regression.

A Pearson residual is an observed response minus its estimated mean, divided by its estimated standard deviation. Since the mean and the variance are both μ for Poisson response variables, the formula for a Pearson residual is

---

Pearson Residual for Poisson Regression

$$Pres_i = \frac{Y_i - \hat{\mu}_i}{\sqrt{\hat{\mu}_i}}.$$

Typically, the analyst examines one set of residuals, not both, depending on what is available in the computer package being used. The deviance residuals are somewhat more reliable for detecting outliers, but the Pearson residuals are easier to interpret. Usually, the two sets of residuals are similar. More important differences occur when they are used for extra-sum-of-squares tests to compare models.

If the Poisson means are large, the distributions of both deviance and Pearson residuals are approximately standard normal. Thus, problems are indicated if more than 5% of residuals exceed 2 in magnitude or if one or two residuals greatly exceed 2. If the Poisson means are small—as in the elephant mating example, where a substantial proportion of the estimated means are less than 5—neither set of residuals has a normal distribution, so comparison to a standard normal distribution is inappropriate. An example of a plot of deviance residuals for the epilepsy days is shown in Display 22.7; this example is discussed further in the next section. The overall pattern of large-magnitude residuals here is noteworthy.

22.3.3   The Deviance Goodness-of-Fit Test

A deviance goodness-of-fit test provides an informal assessment of the adequacy of a fitted model. The term *informal* is used here to emphasize that the *p*-value is sometimes unreliable. The test is not particularly good at detecting model inadequacies, and it needs to be used in conjunction with plots and tests of model terms for assessing the adequacy of a particular fit. A *large* p-value indicates either that the model is adequate or that insufficient data are available to detect inadequacies. A *small* p-value may indicate any of three things: the model for the mean is incorrect (for example, because the set of explanatory variables is insufficient); that the Poisson is an inadequate model for the response distribution; or that a few severely outlying observations are contaminating the data.

The deviance goodness-of-fit test is an extra-sum-of-squares test for comparing the model of interest to the saturated model in which each observation has its own mean pa-

Display 22.7   Deviance residuals versus baseline seizure count from the fit of the log-linear
regression of 8-week seizure count on age, log-baseline count, treatment
indicator, their three interactions, age-squared, and the square of log-baseline

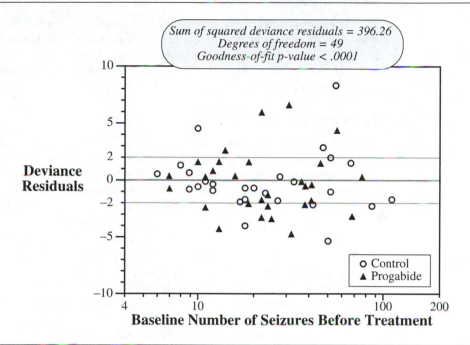

rameter. In the saturated model, the estimated mean for observation i is Y_i, the deviance
residuals are all zero, and the sum of squared deviance residuals is zero. The extra-sum-
of-squares for this test is the sum of squared deviance residuals from the fit to the model
of interest—the deviance statistic. The p-value is obtained by comparing the deviance
statistic to a chi-squared distribution on $n - p$ degrees of freedom, where p is the number
of parameters in the model:

Deviance Statistic = Sum of Squared Deviance Residuals

Approximate p-value $= \Pr(\chi^2_{n-p} >$ Deviance Statistic$)$.

The chi-square approximation to the null distribution of the deviance statistic is valid
when the Poisson means are large. If a substantial proportion of the cases have estimated
means of less than 5, it is potentially misleading to rely on the deviance test as a tool for

model assessment. Other tools, particularly testing of extra terms that represent model inadequacy, should be used instead.

Elephant Matings Example

The deviance statistic, 50.83, is provided in the output of Display 22.6 for the log-linear regression of number of matings on age and age-squared. As Display 22.2 makes clear, however, most of the estimated means are less than 5 and many are less than 2. Thus, it is not prudent to compare 50.83 to a chi-square distribution as a goodness-of-fit test for this model with these data.

Epilepsy Example

The goal of the epilepsy data analysis is to see whether the mean number of counts is smaller for patients who received the progabide treatment than for control patients, after accounting for age and the baseline number of seizures. The baseline count is an important explanatory variable. An initial scatterplot of the log of the number of posttreatment seizures versus the log of the number of pretreatment seizures, with a code to distinguish the progabide patients from controls, is informative (but not shown here).

Initial fitting indicates that case #49 is highly influential. Estimated effects and interaction terms are much more significant when this individual is included than when he is not. He had the largest baseline count (151 seizures in the 8 weeks prior to treatment). When this observation is set aside, attention is restricted to the effect of progabide on patients with baseline counts of 111 or less. Without further discussion of the influence issue, all remaining analyses of this data set exclude case #49.

Further fits of models indicate large deviance statistics. This is true even for fits to models with a substantial number of terms (including interaction terms of the various explanatory variables). For example, the deviance from the fit to the model described in Display 22.7 is 396.26 on 49 degrees of freedom. Since the counts are fairly large, it is appropriate to compare this to a chi-squared distribution (on 49 degrees of freedom) to judge goodness of fit. Since $\Pr(\chi^2_{49} > 396.26)$ is less than .00001, there is substantial evidence that the Poisson log-linear model does not fit. Furthermore, as is evident in Display 22.7, the large deviance statistic is not due to one or two outlying observations; quite a few of the deviance residuals are larger than 2. Evidently the responses show more variability than is predicted by the Poisson distribution. Fortunately, the quasi-likelihood approach (Section 22.5), which allows for extra-Poisson variation, can be used to draw inferences about individual terms in the model.

22.3.4 The Pearson Chi-Squared Goodness-of-Fit Test

Another goodness-of-fit test statistic with the same purpose as the deviance goodness-of-fit test statistic is the sum of squared Pearson residuals. Its p-value, too, is obtained by comparing the observed statistic to a chi-squared distribution on $n - p$ degrees of freedom. Although the deviance goodness-of-fit test is generally preferred, the Pearson test is important for historical reasons and for its current widespread use to analyze tables of counts.

22.4   Inferences About Log-Linear Regression Coefficients

With the Poisson log-linear regression model, tests and confidence intervals for parameters are based on maximum likelihood analysis. The Wald procedure applies to tests and confidence intervals for single coefficients. The drop-in-deviance test compares a full to a reduced model, particularly for hypotheses that several regression coefficients are simultaneously zero.

22.4.1   Wald's Test and Confidence Interval for Single Coefficients

The estimated regression coefficients and their standard errors are calculated by the maximum likelihood method. The sampling distributions of the estimated coefficients are approximately normal if either the sample size is large or the Poisson means are large. Tests and confidence intervals are constructed with the estimates, their standard errors, and tabled values of the standard normal distribution.

Example—Elephant Matings

A main question asked is whether the mean number of successful matings tends to peak at some age and then decrease, or whether the mean continues to increase with age. The inclusion of an age-squared term in the model addresses this question. If a peak occurs at some age, the coefficient of the age-squared term is not zero. Therefore, the test for a peak is accomplished through a test that the coefficient of age-squared is zero. Display 22.5 indicates that the two-sided p-value is .67. There is no evidence of any need for a quadratic term in the log-linear model.

Display 22.8 shows the fit of the data without the quadratic term. A confidence interval for the coefficient of age is constructed in the usual way. The estimate of the coefficient, .0687, is interpreted as follows: a 1-year increase in age is associated with a multiplicative change in the mean number of successful matings by a factor of $\exp(0.0687) = 1.071$. In

---

Display 22.8   Poisson log-linear regression of number of matings on age

Coefficients	Estimate	Standard Error	z-Statistic	p-Value
Intercept	−1.582	0.545	−2.905	.0037
age	0.0687	0.0138	4.996	<.0001

Deviance = 51.01 d.f. = 39

95% Confidence interval for coefficient of age:
$$0.0687 \pm (1.96)(0.0138) = 0.0417 \, to \, .0956$$

Conclusion: The mean number of successful matings increases by 7.1% for each additional year of age; 95% confidence interval is 4.3% to 10.0%.

words: each additional year of age is associated with a 7.1% increase in the mean number of successful matings. A 95% confidence interval for the coefficient of age is 0.0417 to 0.0956. Anti-logarithms produce a 95% confidence interval for the multiplicative factor by which the mean changes with one year of age—1.043 to 1.100.

22.4.2 The Drop-in-Deviance Test

The drop-in-deviance test compares a full model to a reduced version. For the test of a single coefficient, the analyst may use either the drop-in-deviance test or Wald's test. Wald's test is more convenient, but the drop-in-deviance test is more reliable. Its approximate p-value is more accurate, especially if the sample size and the Poisson means are not large. A common practice is to adopt Wald's test for a casual determination of significance or when the test's p-value is very small or very large, and to use the drop-in-deviance test if the test is an especially important aspect of the analysis or when the Wald's test p-value does not provide a conclusive result.

Like other tests based on extra sums of squared residuals, this test is appropriate when a full model is deemed to fit well and when the hypothesis of interest places restrictions on some of the coefficients. The drop-in-deviance test statistic is the deviance statistic from a fit to the reduced model minus the deviance statistic from the full model. This extra sum of squared residuals is the variability explained by the terms that are included in the full model but absent from the reduced model. If the reduced model is adequate, the sampling distribution of the drop in deviance has an approximate chi-squared distribution with degrees of freedom equal to the number of unknown coefficients in the full model minus the number of unknown coefficients in the reduced model.

Example—Elephant Matings

The Wald's test for the significance of the age-squared term was included in the output of the fit to the model with age and age-squared (Display 22.6). To obtain the more accurate p-value from the drop-in-deviance test, fit both models and compute the drop in deviance. The deviance from the fit without age-squared is 51.01. The deviance from the fit with age-squared is 50.83. The drop in deviance due to including age-squared in the model that already has age, therefore, is 0.18. The p-value (two-sided) is the probability that a chi-squared distribution on 1 degree of freedom is greater than 0.18. The answer, .67, is essentially the same as that provided by the two-sided p-value from the Wald's test.

22.5 Extra-Poisson Variation and the Log-Linear Model

22.5.1 Extra-Poisson Variation

The Poisson distribution is one of many distributions that assign probabilities to outcomes for counted responses. It is a good choice, especially for processes that generate events uniformly over time or space. Sometimes, however, unmeasured effects, clustering of events, or other contaminating influences combine to produce more variation in the responses than is predicted by the Poisson model. Using Poisson log-linear regression when such extra-Poisson variation is present has three consequences: parameter estimates are still roughly unbiased; but standard errors are smaller than they should be; and tests give smaller p-values than are truly warranted by the data. The last two consequences are problems that must be addressed.

The quasi-likelihood approach, as shown in Chapter 21, is a method for extending the model to allow for this possibility. It is based on a slight modification of the model for the mean and variance of the response. In particular, the model specifies only the mean and the variance of the responses (not the entire distribution), and it allows an additional parameter to account for more variability than is predicted by the Poisson distribution. For the case of two explanatory variables, the model for extra-Poisson variation is

---

A Log-Linear Model That Permits Extra-Poisson Variation

$$\mu\{Y_i \mid X_{1i}, X_{2i}\} = \mu_i$$
$$\text{Var}\{Y_i \mid X_{1i}, X_{2i}\} = \sigma^2 \mu_i$$
$$\log(\mu_i) = \beta_0 + \beta_1 X_{1i} + \beta_2 X_{2i}.$$

The extra parameter σ^2, which again is referred to as the *dispersion parameter*, captures the extra variation. If the variance of the responses is like the Poisson variance, then $\sigma^2 = 1$. If the variance is larger or smaller than the mean, then σ^2 will be larger or smaller than 1.

22.5.2 Checking for Extra-Poisson Variation

There are four ways to check for extra-Poisson variation:

1. Think about whether extra-Poisson variation is likely.
2. Compare the sample variances to the sample averages computed for groups of responses with identical explanatory variable values.
3. Examine the deviance goodness-of-fit test after fitting a rich model.
4. Examine residuals to see if a large deviance statistic may be due to outliers.

A combination of these usually decides the issue. When the situation is dubious, it is safest to assume that extra-Poisson variation is present.

Consider the Situation

Extra-Poisson variation should be expected when important explanatory variables are not available, when individuals with the same level of explanatory variables may for some reason behave differently, and when the events making up the count are clustered or spaced systematically through time or space, rather than being randomly spaced. However, if the means are all fairly small—less than 5, say—there is seldom reason to worry about extra-Poisson variation.

Plot Variances Against Averages

If responses are grouped by identical values of the explanatory variables, a sample variance and a sample average can be computed for each group. If the responses are Poisson, the sample variances should be approximately equal to the averages. Plotting variances against averages can reveal gross exceptions, and it can also suggest whether the quasi-likelihood model of proportionality is preferable. Plots based on grouping subjects with similar, but not identical, explanatory variable values can be misleading, because the grouping itself induces extra-Poisson variability.

Fit a Rich Model

If the Poisson means are large, the deviance goodness-of-fit test can be examined after a fit to a rich model (one with terms that reflect possible departures from the proposed model). The idea is that including unnecessary terms will not have much effect on the residuals, and a large deviance statistic will indicate inadequacies with the Poisson assumption rather than inadequacies with insufficient explanatory variable terms.

Look for Outliers

The deviance statistic might be large because of one or two outliers. Therefore, it is appropriate to examine the individual deviance residuals in addition to the sum of their squares.

In dubious cases, the quasi-likelihood model provides a more conservative assessment than the maximum likelihood method. On the other hand, it leads to less powerful inferences than the maximum likelihood approach when the Poisson model is correct. For the epileptic seizure data, the deviance statistic and the residuals indicate more variation than would be expected from the Poisson distribution (Display 22.7). Perhaps seizures occur in clusters, rather than being distributed randomly over time; or maybe the mean changes over the 8-week period. The quasi-likelihood model and analysis are strongly indicated for these data.

22.5.3 Inferences When Extra-Poisson Variation Is Present

Tests and confidence intervals are carried out in the same way as they were for extra-binomial variation in Section 21.5.3 (page 614):

1. An estimate of σ^2 is the sum of squared deviance residuals divided by the residual degrees of freedom—the sample size minus the number of parameters in the mean.

2. The quasi-likelihood standard errors are the maximum likelihood standard errors multiplied by the estimate of σ (the square root of the estimate of σ^2). Ideally, the quasi-likelihood approach can be requested from the statistical computing program, so standard errors will be properly adjusted. If not, the user must manually compute the estimate of σ and multiply the given standard errors by this estimate.

3. All t-tests and confidence intervals are based on the coefficient estimates and the adjusted standard errors. It is prudent to use the t-distribution rather than the standard normal distribution as a reference, even though there is no established theory for this practice. The degrees of freedom are $(n - p)$.

4. The drop in deviance per drop in degrees of freedom is divided by the estimate of σ^2 to form an F-statistic. This is analogous to the extra-sum-of-squares F-test in ordinary regression, except that deviance residuals (rather than ordinary residuals) are used.

Example—Epilepsy Experiment

Once the presence of extra-Poisson variation has been established, the next step in the analysis is to use quasi-likelihood methods to build a model. An ideal model for investigating the effect of progabide has

$$\log(\mu) = \beta_0 + \beta_1 age + \beta_2 lbase + \beta_3 prog,$$

where *prog* is an indicator variable that takes on the value 1 for the patients given progabide and 0 for the control patients, and where *lbase* is the logarithm of the baseline count. The effectiveness of the drug could be addressed directly through the parameter β_3, but there might be an interactive effect. This would happen if the effectiveness of the drug depends on the severity of the seizures before treatment. Other possible interaction terms and quadratic terms can be included to find a good working model. The approach demonstrated in Display 22.9 is to add extra terms to the model with *age*, *lbase*, *prog*, and *prog* × *lbase*. A drop-in-deviance F-test shows no evidence that these extra terms are needed.

A t-test for testing the significance of the interaction term in model 2 has a two-sided p-value of .645, so there appears to be no evidence of an interactive effect of treatment and pretreatment seizure severity. Consequently, it is appropriate to fit the model without the interaction. The results of this fit are shown in Display 22.10. The Wald's test (two-sided p-value = .56) shows no evidence of a treatment effect. A confidence interval for the coefficient of progabide (using a t-multiplier rather than a z-multiplier) is $-0.244 \pm (2.00 \times 0.423)$, which is -1.09 to 0.60. In words: after age and initial severity of seizures are accounted for, the mean number of seizures for progabide patients is estimated to be 78% of the mean number of seizures for control patients. A 95% confidence interval is 34% to 183%. In this case the best single estimate corresponds to a (presumably) practically significant difference (a 22% reduction in the mean number of seizures), but there is no statistical significance: the confidence interval is quite large.

Display 22.9   Informal testing of extra terms, with a drop-in-deviance F-test, to find a good fitting model—epilepsy data

(1)   *Fit a rich model, including interactions and quadratic terms.*

Response: *post*
Explanatory variables: *age, lbase, prog, lbase × prog, age × prog, lbase × age, age^2, lbase^2*
Distribution: Quasi-likelihood, with variance = $\sigma^2\mu$
Link: *logarithm*

Variable	Coefficient	Standard Error	t-Statistic	p-Value
Constant	1.663	3.408	0.488	.6277
age	0.032	0.139	0.232	.8174
lbase	−0.335	1.194	0.280	.7805
prog	−0.738	1.291	0.572	.5702
age × prog	−0.0116	0.0267	0.432	.6677
lbase × prog	0.265	0.268	0.990	.3271
lbase × age	0.0358	0.0203	1.766	.0836
age^2	−0.00215	0.00187	1.153	.2546
lbase^2	0.032	0.139	0.233	.8167

Deviance = 396.26   **d.f.**= 49

(2)   *Fit the inferential model, with extra terms deleted.*

Response: ***post***
Explanatory variables: ***age, lbase, prog, lbase×prog***
Distribution: Quasi-likelihood, with variance = $\sigma^2\mu$
Link: *logarithm*

Variable	Coefficient	Standard Error	t-Statistic	p-Value
Constant	−0.457	0.587	0.779	.4392
age	0.0252	0.0117	2.143	.0367
lbase	0.953	0.125	7.614	<.0001
prog	−0.635	0.859	0.739	.4630
lbase × prog	0.112	0.242	0.463	.6450

Deviance = 435.67   **d.f.**= 53

(3)   *Test adequacy of the inferential model.*

$$F\text{-statistic} = \frac{(435.67 - 396.26)/4}{(396.26)/49} = 1.218; \qquad p\text{-value} = 0.3151$$

Conclusion: There is no evidence of a need for the extra terms included in model 1.

Display 22.10   Quasi-likelihood fit to a log-linear regression model for the number of seizures in the 8 weeks after treatment—epilepsy data

Variable	Coefficient	Standard Error	t-Statistic	p-Value
Constant	−0.526	1.61	−0.32	.75
age	0.0237	0.032	0.74	.46
lbase	0.983	0.302	3.25	.002
prog	−0.244	0.423	−0.58	.56

Deviance = 437.44   **d.f.** = 54

22.6   Further Issues

22.6.1   Log-Linear Models for Testing Independence in Tables of Counts

The Poisson distribution also arises in the discussion of tables of counts (Chapter 19). There, Poisson sampling means that a fixed amount of effort is used to sample subjects, who are then categorized into one of the cells in the table. In contrast to the other sampling schemes, neither the marginal totals nor the grand total is known before the data are collected.

It is possible to analyze Poisson counts in tables of counts by treating the cell counts as Poisson response variables and employing a log-linear model with factors representing the different dimensions in the table. The Samoan obesity and heart disease data, for example, involved four cells in a table consisting of two rows and two columns. The means of the counts may be described by a log-linear model, as shown in Display 22.11.

The log-linear model involves obesity as its row factor and CVD death as its column factor. If *obese* and *death* act independently to determine the counts, there is no need for an interaction term. Therefore, a test for the *obese* × *death* interaction in the log-linear model provides a test of independence in the table of counts.

For a table with r rows and c columns, a total of $(r-1)(c-1)$ terms are needed to describe all row × column interactions. These terms may be tested with an extra-sum-of-squared-residuals test. Although some controversy surrounds the question of which residuals are best, common practice is to use Pearson residuals for the analysis of tables of counts. Since the full model—with the interaction terms—is the saturated model (where all residuals are zero), the test statistic for independence equals the sum of the squared Pearson residuals from the log-linear fit to the model without interaction. The test statistic is compared to a chi-square distribution on $(r-1)(c-1)$ degrees of freedom.

Notes About Log-Linear Analysis of Tables of Counts

1.   The Poisson log-linear model describes the mean count in terms of the row and column factors, and these two factors are treated equally. This is unlike ordinary regression, where the distinction between response and explanatory variables is important.

Display 22.11   Log-linear analysis of a 2 × 2 table of counts—Samoan obesity and heart disease (see Section 18.1.1)

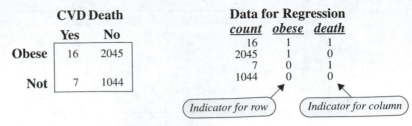

	CVD Death	
	Yes	No
Obese	16	2045
Not	7	1044

Data for Regression

count	*obese*	*death*
16	1	1
2045	1	0
7	0	1
1044	0	0

(Indicator for row)   (Indicator for column)

MODEL: Y_i *is Poisson with mean* μ_i $(i = 1, 2, 3, 4)$
$log(\mu_i) = \beta_0 + \beta_1 obese_i + \beta_2 death_i + (\beta_3 obese_i \times death_i)$

(*A test for independence of row and column factors is a test that the interaction term in the log-linear model is zero.*)

---

2. For 2 × 2 tables of counts, the Poisson analysis with the Pearson statistic is identical to the chi-squared test presented in Section 19.3.2 (page 546). Both forms may also be applied to more complex tables of counts.

3. The log-linear model approach parallels ordinary regression and analysis of variance, with independence between two factors being tested through the significance of their interaction terms.

22.6.2   Poisson Counts from Varying Effort

In the previous examples, the response counts were obtained by observing some process for a fixed amount of time. All epilepsy patients were monitored for 8 weeks; all male elephants were observed for 8 years; and the CVD death counts for the Samoan women were based on a 7-year observation period. That is, each count arises from the same amount of sampling effort. In many situations that arise, however, counts are based on different amounts of effort. It is natural that higher sampling effort yields higher mean counts, and the analysis should concentrate on some measure of average counts per unit of effort, or count rates.

The easiest way to include effort in a Poisson regression equation is to include log(*effort*) as an additional explanatory variable in the log-linear model:

When the Mean Count Is an Occurance Rate Times Effort

$$\mu(Y_i) = \mu_i = effort_i \times rate_i$$

A log-linear model for rate is:

$$log(rate_i) = \beta_0 + \beta_1 X_{1i} + \beta_2 X_{2i} + \cdots .$$

So, a model for the mean count is:

$$log(\mu_i) = log(effort_i) + \beta_0 + \beta_1 X_{1i} + \beta_2 X_{2i} + \cdots .$$

Effort is typically the time of observation during which each count is made, but it may also be an area. If the effort is the same for all observations in the data set, the term involving effort is a constant and can be absorbed into β_0.

The coefficient of log(*effort*) is known to be 1. In generalized linear models, a term with a known coefficient can be inserted into the regression equation. This term is called an *offset*. The mechanism for inserting it depends on the statistical package used. If a Poisson log-linear regression program is available that does not have the capability of including an offset, include log(*effort*) as an explanatory variable with the others having an unknown coefficient. This generally has little effect on the other estimates.

If scatterplots are drawn, it is necessary to plot *rate* (*count/effort*) or the log of rate—instead of raw counts—versus explanatory variables. In the analysis, a check for extra-Poisson variation remains necessary. Its presence can be handled in the usual way, using the quasi-likelihood method.

22.7 Summary

Poisson regression should be considered whenever the response variable is a count of events that occur over some observation period or spatial area. The main feature of the Poisson distribution is that the variance is equal to the mean. Consequently, if ordinary regression is used, the residual plot will exhibit a funnel shape. In the Poisson log-linear regression model, the distribution of the response variables for each combination of the explanatory variables is thought to be Poisson, and the logarithms of the means of these distributions are modeled by formulas that are linear functions of regression coefficients. The method of maximum likelihood is used to estimate the unknown regression coefficients.

Elephant Mating Data

A scatterplot of the number of successful matings versus age for the 41 male elephants indicates the inadequacy of a straight-line regression and of a constant variance assumption for these data. Linear regression after a log transformation of number of matings is possible, but the zeros in the data set are a nuisance in this approach. Poisson log-linear regression provides an attractive alternative. The counts are too small to justify the deviance goodness-of-fit test, but plots and informal testing of extra terms imply that the model with the log of the mean count as a straight-line function of age is adequate. One question of interest is whether the mean count of successful matings peaks at some age. This can be investigated by testing the significance of an age-squared term. The drop-in-deviance test does not provide any evidence of a quadratic effect.

Epilepsy Data

There is some reason to expect extra-Poisson variation in the number of epileptic seizures, since individuals with identical values of the explanatory variables are likely to differ in their seizure frequency and since seizures might come clumped in time rather than uniformly distributed. Examination of the deviance goodness-of-fit test confirms this to be the case. Quasi-likelihood analysis of the log-linear model in which the variance is proportional to the mean may be used. The log of the mean is modeled by a regression structure that includes the log of the baseline count and an indicator variable for treatment group. A drop-in-deviance test provides no evidence of a treatment effect. Another suitable way to analyze these data is with ordinary regression models for the log of the ratio of the post-treatment count to the pretreatment count.

Further Reading

Log-linear regression is discussed in books on generalized linear models; see Dobson (1983) and McCullagh and Nelder (1989). There is also a passage on Poisson regression in the *Encyclopedia of Statistical Science*, vol. 7 (Kotz, Johnson, and Read 1986).

22.8 Exercises

Conceptual Exercises

1.   Elephant Mating. Both the binomial and the Poisson distributions provide probability models for random counts. Is the binomial distribution appropriate for the number of successful matings of the male African elephants?

2.   What is the difference between a log-linear model and a linear model after the log transformation of the response?

3.   If a confidence interval is obtained for the coefficient of an indicator variable in a log-linear regression model, do the anti-logarithms of the endpoints describe a confidence interval for a ratio of means or for a ratio of medians?

4.   Elephant Mating. In Display 22.2 the spread of the responses is larger for larger values of the mean response. Is this something to be concerned about if Poisson log-linear regression is used?

5.   Elephant Mating. From the estimated log-linear regression of elephants' successful matings on age (Display 22.8), what are the mean and the variance of the distribution of counts of successful matings (in 8 years) for elephants who are 25 years at the beginning of the observation period? What are the mean and the variance for elephants who are 45 years?

6.   (a) Why are ordinary residuals—$(Y_i - \hat{\mu}_i)$—not particularly useful for Poisson regression? (b) How are the Pearson residuals designed to deal with this problem?

7.   Epilepsy. Why does Display 22.7 indicate that the Poisson model is inadequate for the epilepsy data?

8.   Consider the deviance goodness-of-fit test. (a) Under what conditions is it valid for Poisson regression? (b) When it is valid, what possibilities are suggested by a small p-value? (c) When it is valid, what possibilities are suggested by a large p-value?

9.   (a) Why is it more difficult to check the adequacy of a Poisson log-linear regression model when the counts are small than when they are large? (b) What tools are available in this situation?

10.   (a) How does the drop-in-deviance test for Poisson log-linear regression resemble the extra-sum-of-squares test in ordinary regression? (b) How does it differ?

11.   (a) How does the drop-in-deviance test for the quasi-likelihood version of the log-linear regression model resemble the extra-sum-of-squares test in ordinary regression? (b) How does it differ?

12.   How does the quasi-likelihood version of the log-linear regression model allow for more variation than would be expected if the responses were Poisson?

13. If responses follow the Poisson log-linear regression model, the Pearson residuals should have variance approximately equal to 1. If, instead, the quasi-likelihood model with dispersion parameter σ^2 is appropriate, what is the approximate variance of the Pearson residuals?

14. Is it acceptable to use the quasi-likelihood model when the data actually follow the Poisson model?

15. Consider a table that categorizes 1000 subjects into 5 rows and 10 columns. (a) If Poisson log-linear regression is used to analyze the data, what is the sample size? (That is, how many Poisson counts are there?) (b) How would one test for independence of row and column factors?

Computational Exercises

16. Elephant Mating and Age. Give an estimated model for describing the number of successful matings as a function of age, using (a) simple linear regression after transforming the number of successful matings to the square root; (b) simple linear regression after a logarithmic transformation (after adding 1); (c) log-linear regression. (d) Do the models used in parts (a) or (b) exhibit obvious inadequacies?

17. Epilepsy. (a) Construct the analyses described in Display 22.9, remembering to exclude case #49. (b) Compare the results of the F-test for quadratic and interaction terms with a drop-in-deviance chi-square test that ignores the extra-Poisson variation. (c) Compute the standard error of the estimated interaction coefficient, first ignoring extra-Poisson variation and then accounting for it.

18. Epilepsy. (a) Complete the analysis of this data set by fitting the model with only *age* and *lbase* as explanatory variables. Compute the F-test for the significance of the two coefficients, and write a summary statement of the conclusions.

19. Epilepsy. Refit the three models (two from Display 22.9 and one from Exercise 18) using quasi-likelihood, but with case #49 included. Compare the conclusions reached here with those obtained earlier case when case #49 was excluded.

20. Suicide–Murders by Deliberate Plane Crash. Some sociologists suspect that highly publicized suicides may trigger additional suicides. In one investigation of this hypothesis, D. P. Phillips collected information about 17 airplane crashes that were known (because of notes left behind) to be murder–suicides. For each of these crashes, Phillips reported an index of the news coverage (circulation in nine newspapers devoting space to the crash multiplied by length of coverage) and the number of multiple-fatality plane crashes during the week following the publicized crash. The data are exhibited in Display 22.12. (Data from D. P. Phillips, "Airplane Accident Fatalities Increase Just

Display 22.12 Multiple-fatality plane crashes in the week following a murder–suicide by plane crash, and the amount of newspaper coverage given the murder–suicide

Index of Coverage	Number of Crashes	Index of Coverage	Number of Crashes
376	8	63	2
347	5	44	7
322	8	40	4
104	4	5	3
103	6	5	2
98	4	0	4
96	8	0	3
85	6	0	2
82	4		

After Newspaper Stories About Murder and Suicide," *Science* 201 (1978): 748–50.) Is there evidence that the mean number of crashes increases with increasing levels of publicity of a murder–suicide?

21.   **Obesity and Heart Disease.** Analyze the table of counts in Display 22.11 as suggested there. What is the *p*-value for testing independence of obesity and CVD death outcome?

22.   **Galapagos Islands.** Reanalyze the data in Display 12.17 (page 360) with number of native species as the response, but using log-linear regression. (a) Fit the model with log area, log elevation, log of distance from nearest island, and log area of nearest island as explanatory variables; and then check for extra-Poisson variation. (b) Use backward elimination to eliminate insignificant explanatory variables. (c) Describe the effects of the remaining explanatory variables.

23.   **Galapagos Islands.** Repeat the previous exercise, but use the number of non-native species as the response variable (total number of species minus the number of native species).

24.   **Cancer Deaths of Atomic Bomb Survivors.** The data in Display 22.13 are the number of cancer deaths among survivors of the atomic bombs dropped on Japan during World War II, categorized by time (years) after the bomb that death occurred and the amount of radiation exposure that the survivors received from the blast. (Data from D. A. Pierce, personal communication). Also listed in each cell is the *person-years at risk*, in 100s. This is the sum total of all years spent by all persons in the category. Suppose that the mean number of cancer deaths in each cell is Poisson with mean $\mu = risk \times rate$, where *risk* is the person-years at risk and *rate* is the rate of cancer deaths per person per year. It is desired to describe this rate in terms of the amount of radiation, adjusting for the effects of time after exposure. (a) Using log(*risk*) as an offset, fit the Poisson log-linear regression model with time after blast treated as a factor (with seven levels) and with *rads* and *rads*-squared treated as covariates. Look at the deviance statistic and the deviance residuals. Does extra-Poisson variation seem to be present? Is the *rads*-squared term necessary? (b) Try the same model as in part (a); but instead of treating time after bomb as a factor with seven levels, compute the midpoint of each interval and include log(*time*) as a numerical explanatory variable. Is the deviance statistic substantially larger in this model, or does

Display 22.13   Cancer deaths among Japanese atomic bomb survivors, categorized by estimated exposure to radiation (in rads) and years after exposure; below the number of cancer deaths are the person-years (in 100s) at risk

					Years After Exposure			
exposure **(rads)**		**0–7**	**8–11**	**12–15**	**16–19**	**20–23**	**24–27**	**28–31**
0	*deaths:*	10	12	19	31	35	48	73
	risk:	262	243	240	237	233	227	220
25	*deaths:*	17	17	17	47	50	65	71
	risk:	313	290	285	280	275	269	262
75	*deaths:*	0	2	1	5	8	7	12
	risk:	38	36	35	34	34	33	32
150	*deaths:*	1	0	4	1	6	12	11
	risk:	28	26	25	25	24	24	23
250	*deaths:*	1	1	0	4	3	7	13
	risk:	13	12	12	12	11	11	10
400	*deaths:*	0	2	5	3	2	3	5
	risk:	15	14	14	14	13	13	13

it appear that time can adequately be represented through this single term? (c) Try fitting a model that includes the interaction of log(*time*) and exposure. Is the interaction significant? (d) Based on a good-fitting model, make a statement about the effect of radiation exposure on the number of cancer deaths per person per year (and include a confidence interval if you supply an estimate of a parameter).

Data Problems

25. **Emulating Jane Austen's Writing Style.** When she died in 1817, the English novelist Jane Austen had not yet finished the novel *Sanditon*, but she did leave notes on how she intended to conclude the book. The novel was completed by a ghost writer, who attempted to emulate Austen's style. In 1978, a researcher reported counts of some words found in chapters of books written by Austen and in chapters written by the emulator. These are reproduced in Display 22.14. (Data from A. Q. Morton, *Literary Detection: How to Prove Authorship and Fraud in Literature and Documents*, New York: Charles Scribner's Sons, 1978.) Was Jane Austen consistent in the three books in her relative uses of these words? Did the emulator do a good job in terms of matching the relative rates of occurrence of these six words? In particular, did the emulator match the relative rates that Austen used the words in the first part of *Sanditon*?

Display 22.14 Occurrences of six words in various chapters of books written by Jane Austen (*Sense and Sensibility*, *Emma*, and the first part of *Sanditon*) and in some chapters written by the writer who completed *Sanditon* (Sanditon II)

Word	Book			
	Sense and Sensibility	*Emma*	*Sanditon* I	*Sanditon* II
a	147	186	101	83
an	25	26	11	29
this	32	39	15	15
that	94	105	37	22
with	59	74	28	43
without	18	10	10	4

26. **Space Shuttle O-ring Failures.** On January 27, 1986, the night before the space shuttle *Challenger* exploded, an engineer recommended to the National Aeronautics and Space Administration (NASA) that the shuttle not be launched in the cold weather. The forecasted temperature for the *Challenger* launch was 31°F—the coldest launch ever. After an intense 3-hour telephone conference, officials decided to proceed with the launch. Shown in Display 22.15 are the launch temperatures and the number of O-ring problems in 24 shuttle launches prior to the *Challenger* (Chapter 4). Do these data offer evidence that the number of incidents increases with decreasing temperature?

27. **Characteristics Associated with Salamander Habitat.** To study the habitat characteristics of the Del Norte Salamander, and particularly these salamanders' tendency to reside in dwindling old-growth forest, researchers selected 47 sites from plausible salamander habitat in national forest and parkland in northern California. (Data from H. H. Welsh and A. J. Lind, "Habitat Correlates of the Del Norte Salamander, *Plethodon elongatus* (Caudata: Plethodontidae), in Northwestern California," *Journal of Herpetology* 29(2) (1995): 198–210.) Various forest characteristics were measured at each site, and the total number of Del Norte Salamanders were counted in a 7-meter by 7-meter search

Display 22.15 Launch temperatures (°F) and number of O-ring incidents in 24 space shuttle flights

Launch Temperature (°F)	Number of Incidents	Launch Temperature (°F)	Number of Incidents
53	3	70	1
56	1	70	1
57	1	72	0
63	0	73	0
66	0	75	0
67	0	75	2
67	0	76	0
67	0	76	0
68	0	78	0
69	0	79	0
70	0	80	0
70	1	81	0

Display 22.16 Salamanders found in a 49 m² area, percentage canopy cover, and forest age in 47 northern California sites

Site	Salamanders	% Cover	Forest Age	Site	Salamanders	% Cover	Forest Age
1	13	85	316	25	1	46	30
2	11	86	88	26	1	80	215
3	11	90	548	27	1	86	586
4	9	88	64	28	1	88	105
5	8	89	43	29	1	92	210
6	7	83	368	30	0	0	0
7	6	83	200	31	0	1	4
8	6	91	71	32	0	3	3
9	5	88	42	33	0	5	2
10	5	90	551	34	0	8	10
11	4	87	675	35	0	9	8
12	3	83	217	36	0	11	6
13	3	87	212	37	0	14	49
14	3	89	398	38	0	17	29
15	3	92	357	39	0	24	57
16	3	93	478	40	0	44	59
17	2	2	5	41	0	52	78
18	2	87	30	42	0	77	50
19	2	93	551	43	0	78	320
20	1	7	3	44	0	80	411
21	1	16	15	45	0	86	133
22	1	19	31	46	0	89	60
23	1	29	10	47	0	91	187
24	1	34	49				

area. Shown in Display 22.16 are the counts of salamanders, the percentage of forest canopy cover in the site, and the age of the forest, in years. Is the size of the salamander count associated with canopy cover and forest age?

28.   Valve Failure in Nuclear Reactors. Display 22.17 shows characteristics and numbers of *failures* observed in 90 valve types from one pressurized water reactor.   There are five explanatory factors: *system* (1 = containment, 2 = nuclear, 3 = power conversion, 4 = safety, 5 = process auxiliary); *operator type* (1 = air, 2 = solenoid, 3 = motor-driven, 4 = manual); *valve* type (1 = ball, 2 = butterfly, 3 = diaphragm, 4 = gate, 5 = globe, 6 = directional control); *head* size (1 = less than 2 inches, 2 = 2–10 inches, 3 = 10–30 inches); and operation *mode* (1 = normally closed,

Display 22.17   Valve characteristics and numbers of failures in a nuclear reactor

system	operator	valve	size	mode	failures	time	system	operator	valve	size	mode	failures	time
1	3	4	3	1	2	4	3	3	4	1	1	0	7
1	3	4	3	2	2	4	3	3	4	1	2	0	4
1	3	5	1	1	1	2	3	3	4	2	1	8	8
2	1	2	2	2	0	2	3	3	4	2	2	0	3
2	1	3	2	1	0	2	3	3	4	3	1	13	2
2	1	3	2	2	0	1	3	3	4	3	2	3	3
2	1	5	1	1	2	4	3	3	5	1	2	0	3
2	1	5	1	2	4	6	3	3	5	2	2	0	5
2	1	5	2	1	1	1	3	4	4	2	2	1	4
2	1	5	2	2	2	1	3	4	4	3	2	1	10
2	2	5	2	2	3	2	3	4	5	2	2	0	4
2	3	4	2	1	0	2	4	3	3	3	2	2	1
2	3	4	2	2	0	4	4	3	4	2	1	2	8
2	3	4	3	1	0	3	4	3	4	2	2	0	4
2	3	4	3	2	0	1	4	3	4	3	2	7	3
2	3	5	1	1	1	2	4	3	5	1	2	0	1
2	3	5	2	2	0	4	5	1	2	2	1	0	3
2	3	5	3	2	0	2	5	1	2	2	2	0	2
2	4	3	1	2	0	1	5	1	2	3	1	0	1
2	4	3	2	1	0	1	5	1	2	3	2	0	5
2	4	4	1	1	2	1	5	1	3	1	1	0	1
2	4	5	2	1	0	2	5	1	3	1	2	0	3
3	1	1	2	1	1	36	5	1	3	2	2	0	2
3	1	1	2	2	2	4	5	1	4	2	1	3	4
3	1	1	3	2	0	2	5	1	4	2	2	0	4
3	1	2	2	1	0	2	5	1	5	1	1	3	1
3	1	2	3	1	3	8	5	1	5	1	2	2	3
3	1	3	2	1	1	15	5	1	5	2	2	0	8
3	1	3	2	2	0	4	5	1	6	1	1	0	1
3	1	4	1	1	0	1	5	1	6	2	2	0	2
3	1	4	1	2	0	2	5	2	3	2	2	0	11
3	1	4	2	1	5	11	5	2	4	1	1	0	1
3	1	4	2	2	23	6	5	3	2	2	1	0	1
3	1	4	3	2	21	4	5	3	2	2	2	0	2
3	1	5	1	1	0	4	5	3	2	3	1	2	4
3	1	5	1	2	0	4	5	3	2	3	2	0	2
3	1	5	2	1	11	31	5	3	4	2	1	2	5
3	1	5	2	2	3	31	5	3	4	2	2	1	14
3	1	5	3	2	2	1	5	3	5	2	2	0	2
3	1	6	2	1	1	2	5	4	3	1	1	1	5
3	1	6	2	2	0	1	5	4	3	1	2	0	2
3	1	6	3	2	0	1	5	4	3	2	1	0	3
3	2	6	2	2	1	2	5	4	4	1	2	0	1
3	3	2	2	1	0	1	5	4	4	2	1	0	1
3	3	2	3	2	0	1	5	4	5	2	2	0	1

2 = normally open). The lengths of observation periods are quite different, as indicated in the last column, *time*. Using an offset for observation time, identify the factors associated with large numbers of valve failures. (Data from L. M. Moore and R. J. Beckman, "Appropriate One-Sided Tolerance Bounds on the Number of Failures using Poisson Regression," *Technometrics* 30 (1988): 283–90.)

Answers to Conceptual Exercises

1. No. The binomial count has a definite upper limit.

2. In a log-linear model, the mean of Y is μ and the model is $\log(\mu) = \beta_0 + \beta_1 X_1$. Y is not transformed. If simple linear regression is used after a log transformation, the model is expressed in terms of the mean of the logarithm of Y.

3. Ratio of means. The model states that the log of the mean—not the median—has the regression form. It is not necessary to introduce the median of the Poisson distribution here.

4. No. The nonconstant variance is anticipated by the Poisson model. The maximum likelihood procedure correctly uses the information in the data to estimate the regression coefficients.

5. For 25-year-old elephants: Mean = 1.15; Variance = 1.15. For 45-year-old elephants: Mean = 4.53; Variance = 4.53.

6. (a) The residuals with larger means will have larger variances. Thus, if an observation has a large residual it is difficult to know whether it is an outlier or an observation from a distribution with larger variance than the others. (b) The residuals are scaled to have the same variance.

7. It is apparent that a good number of observations have residuals larger than 2 or smaller than −2. Since a fairly rich model was fit, extra-Poisson variation is a likely explanation.

8. (a) Large Poisson means. (b) The Poisson distribution is an inadequate model, the regression model terms are inadequate, or there are a few contaminating observations. (c) Either the model is correct, or there is insufficient data to detect any inadequacies.

9. (a) Scatterplots are uninformative, the deviance goodness-of-fit test cannot be used, and the approximate normality of residuals is not guaranteed. (b) One may try to add extra terms, such as a squared term in some explanatory variable, to model a simple departure from linearity and to test its significance with a drop-in-deviance or Wald's test.

10. (a) It is based on a comparison of the magnitudes of residuals from a full to a reduced model. (b) The residuals used are deviance residuals, and a chi-squared statistic is used rather than an F-statistic.

11. (a) Same as 10(a). In this case, an F-statistic is formed just as with the usual extra-sum-of-squares test. (b) It is based on deviance residuals.

12. The variance of the responses is $\sigma^2 \mu$, where σ^2 is unknown. A value of σ^2 greater than 1 allows for more variation than anticipated by a Poisson distribution.

13. σ^2.

14. Yes; the Poisson mean–variance relationship is a special case of the quasi-likelihood model (when σ^2 is 1). (More powerful comparisons result, however, if the Poisson model is used when it definitely applies.)

15. (a) 50. (b) Test for the significance of the 36 interaction terms in the log-linear regression with row, column, and row-by-column effects. This may be accomplished by fitting the model without interaction and comparing the Pearson statistic to a chi-squared distribution on 36 degrees of freedom.

Elements of Research Design

Throughout this book, case studies have highlighted fruitful designs. It is now time to take a closer look at study design, with the objective of setting down the basic principles and illustrating the steps leading to a well-designed study. The discussion of research design takes up the next two chapters. This first chapter lays out the general principles. It provides practical suggestions on the steps to follow, including the important task of determining sample size. The second design chapter (Chapter 24) presents some standard statistical design structures that allow straightforward analysis and interpretation of results.

23.1 Case Study

23.1.1 Biological Control of a Noxious Weed— A Randomized Experiment

With increasing evidence of harmful environmental and health effects of herbicides and pesticides has come an increased interest in the use of biological controls for pests. Researchers investigated a system where two insects—the cinnabar moth and the ragwort flea beetle—were introduced to control tansy ragwort, a noxious weed that is toxic to livestock and aggressively displaces desirable plants. (Data from P. B. McEvoy, N. T. Rudd, C. S. Cox, and M. Huso, "Disturbance, Competition, and Herbivory Effects on Ragwort Senecio Jacobaea Populations," *Ecological Monographs* 63(1) (1993): 55–75).

The researchers designed an experiment to manipulate conditions on square plots, 0.5 meters on a side. The treatments were (1) presence or absence of cinnabar moths, (2) presence or absence of flea beetles, and (3) the degree of competition from perennial grasses. They then observed the reduction of ragwort on each plot.

They were interested in three primary questions: (1) Do the biological control agents (moths and flea beetles) work by reducing weed densities uniformly to low levels or by totally eliminating weeds locally so that weeds survive only in shifting patches? (2) Is the combination of moths and flea beetles more effective in controlling ragwort than the sum of their individual effects? This interaction is quite possible, since moths are active in early summer while flea beetles are active in late summer and autumn. (3) Is the extent of control by moths and flea beetles greater in the presence of competing grasses? This would be the case if the effect of the insect controls were to inhibit ragwort growth to the point where the grasses crowded it out.

The general findings were that flea beetles are more effective control agents than cinnabar moths and that there is weak evidence of an interactive effect of the flea beetles with the degree of competition from perennial grasses. The reader is referred to the original paper for a full discussion. The main concern here is with the experimental design. In particular, question (2) was not resolved by the experiment because of substantial uncertainty in the estimate of the interactive effect of flea beetles and moths. How would a new experiment be designed to sufficiently resolve the question?

23.2 Considerations in Forming Research Objectives

The goal of any research design is to arrive at clear answers to questions of interest while expending a minimum of resources. Some tools to use in meeting this end are discussed in the next section. This section concerns issues relating to the starting point for any study: forming the objectives.

Keep It Simple. Unforeseen complications plague even the most simple studies. Units do not respond, equipment fails to operate as planned, weather does not cooperate, animals die mysteriously, treatments have bizarre interactions with uncontrollable factors, and on and on. To paraphrase R. A. Fisher, what began as an experiment ends up as an experience. Imagine the havoc these problems create in studies that begin with highly complex questions involving large numbers of experimental factors.

Think of the Context.   Scientific inquiry is a community activity. Each study will repeat aspects of and expand upon others' studies. Others, in turn, will repeat, further develop, and challenge what the current study has found. All this frees one from the obligation of discovering all nature's secrets in a single study.

Be Objective.   Researchers must be prepared for evidence against their theories and should strive to conduct studies that provide opportunities for such evidence to arise. They should not become emotionally committed to obtaining desired outcomes. Study design should protect against potential biases due to personal hopes and prejudices. And the results of statistical analyses should not be overstated.

Learn.   Of course there is more to conducting a study than doggedly pursuing the answer to the primary research question. Unexpected relationships appear. New hypotheses emerge. At reporting time, research questions that were posed before the experiment merit strong inferential language, while unforeseen discoveries must be presented as "preliminary findings in need of further study." This wording is necessary to account for the data snooping involved in testing hypotheses that are suggested by the data.

23.3   Research Design Tool Kit

The tools described below are primarily for controlled, randomized experiments, but the ideas are relevant to the design of observational studies—particularly those that involve sampling. In either case, attacking a research question involves making many choices. What will constitute a sampling or experimental unit? What populations of study units are of interest? Which units should be selected for study? How many? How should treatments be allocated to units? In what order should treatments be applied? What confounding factors should be held fixed? What covariates should be measured? Answers to these questions play critical roles in determining the success or failure of a study.

This section lists the basic tools used to create a study design. Each tool is accompanied by a short explanation of its main uses.

23.3.1   Controls and Placebos

A control group provides a baseline for comparison with test groups. Controls that use standard treatments also give critics a means of deciding whether the study's results are consistent with past experience. Using treatments that other researchers have used provides a link to previous work. If the controls behave as others' have, critics will feel more confident that differences reported for the experimental subjects really are related to the test treatment. If the controls behave differently, however, reviewers will take that as a signal that confounding factors may be at work.

A placebo treatment mimics the new treatment in every aspect except the test ingredient. Study units may respond positively to an ineffective treatment simply because they are pleased to receive treatment. If controls are not given a similar experience, a difference may reflect a response to being treated rather than an effect of the treatment. For many studies a placebo cannot be incorporated, but the design can still achieve the more general aim: that all units be treated as similarly as possible except for the critical element of the treatment specified by the question of interest. See the vitamin C and cold study of Section 18.1.2 (page 517) for an experiment that uses a placebo.

23.3.2 Blinding

Persons who administer treatments, subjects who receive them, and researchers who measure responses all have certain hopes and expectations associated with the outcome of a study. Blinding means performing the study in such a way that they do not know which treatment is applied to which units. When successful, blinding eliminates the possibility that the end comparison measures expectations rather than real treatment differences.

A blinded study should always be checked to determine if the blinding was successful. Units will naturally try to break the blind—to figure out what treatment they received. Researchers can often break the blind also, by noting patterns of response. Blinding may be checked by analyzing a table of contents formed by cross-classifying subjects according to actual treatment assigned and suspected treatment assigned (where the suspicious may be the subject's or the researcher's). (See Exercise 16 in Chapter 19 for an example.) When blinding has been broken, the analysis of results should use success-or-failure of the blinding process as a covariate.

23.3.3 Blocking

Blocking means arranging units into homogeneous subgroups so that treatments can be randomly assigned to units within each block. The objectives are

1. To improve precision for treatment comparisons. Blocking effectively removes large block-to-block variations from the estimated treatment differences and from the residual standard deviation used to compute standard errors.
2. To control for confounding variables by grouping experimental units into blocks with similar values of the variable.
3. To expand the scope of inference about treatment differences. Different blocks can be selected to represent a wide range of conditions.

See the seaweed grazers study of Section 13.1.1 (page 363) and the pygmalion study of Section 13.1.2 (page 365) for examples of blocking.

23.3.4 Stratification

Like blocking, stratification means arranging units into homogeneous subgroups for analysis. Stratification, however, refers to the partitioning of the entire population into strata, with random sampling conducted separately within each stratum. The objectives of stratification and blocking are basically the same, but stratification pertains to sampling rather than randomization.

23.3.5 Covariates

A covariate is an auxiliary measurement taken on each unit. It is not one of the variables that directly address a question of interest, but it may be closely related to such responses. The researcher does not directly control a covariate. Either the covariate cannot be controlled (like the weather) or its value is determined in the course of the study. Lacking control over the covariate at the design stage is rectified by asserting control at the analysis stage through its inclusion as an explanatory variable along with the design variables. The objectives of using covariates are

1. To control for potentially confounding factors in the analysis.
2. To improve precision in treatment comparisons. Again, systematic variation associated with the covariates is removed from the residual standard deviation.
3. To assess the model. Introducing covariates into a tentative model can reveal whether confounding variables must be included.
4. To expand the scope of inference. Allowing a confounding variable to vary naturally in the course of a study provides a more realistic frame of reference than when its values are fixed at a few experimental settings.

To be useful in getting other effects out of the way, the measurement of the covariate must not be affected by the treatment given. See the blood–brain barrier study of Section 11.1.2 (page 294) for analyses of experimental data including covariates.

23.3.6 Randomization

Randomization means employing a well-understood random procedure—flipping coins, rolling dice, performing a lottery, or using computer-generated pseudorandom numbers—to assign experimental units to different treatment groups. The objectives of randomization are

1. To control for factors not explicitly controlled for in the design (by blocking) or in the analysis (by covariates). This use of randomization is reserved for the unforeseen and nonmeasurable factors. It is not a substitute for control through design or control through analysis for measurable factors whose influences are known or anticipated. (Block for what you can, and randomize for what you can't.)
2. To permit causal inferences. Randomization may yield one group that has substantially larger responses than the others, and that outcome may be associated with some causal variable. The chance that the randomization—not the treatment—is responsible for the difference is described by the measure of uncertainty attached to the inference (the p-value or confidence level).
3. To provide the probability model—the randomization distribution—for drawing inferences.

23.3.7 Random Sampling

Random sampling means employing a well-understood random procedure to select units from a population. The objectives are

1. To ensure that the sample is representative of the population (in the same way that randomization controls for unforeseen and nonmeasurable factors).

2. To permit an inference that patterns observed in the sample are characteristic of patterns in the population as a whole.

3. To provide the probability model—random sampling from populations—for drawing inferences.

23.3.8 Replication

Replication means conducting copies of a basic study pattern. It refers to assigning one treatment to multiple units within each block. The objectives are

1. Increased precision for treatment effects, through increased sample sizes.

2. Model assessment. Replication can provide a pure error estimate of variance for the purpose of determining whether treatment group differences are adequately represented by a particular statistical model.

Note: The term *replication* is often used to describe a repetition of an entire study at several times or in several locations. The different times or locations, however, should be thought of as blocks. This kind of replication meets objective (1), but it does not provide a pure error estimate of variance unless there are also multiple experimental units with the same treatment combinations *within* each block.

23.3.9 Balance

Balance means having the same number of units assigned to each treatment group. The objectives are

1. To optimize the precision for treatment comparisons.

2. To ensure independence of estimated treatment effects. Whereas balance does not necessarily guarantee that different treatment effects will be statistically independent, a lack of balance virtually guarantees that they will not be independent.

Occasionally there are reasons for designing unbalanced studies. One reason would be the need to gain more information about one specific treatment group than about the others. Another reason would be to sample more heavily a group that is expected to be much more variable than the others.

23.4 Design Choices That Affect Accuracy and Precision

23.4.1 Attaching Desired Precision to Practical Significance

A successful experiment has both statistical and practical significance. But the judgment about what constitutes a practically meaningful result is determined by the researcher's understanding of the science. The role of the experiment is to draw a clear distinction between practically meaningful alternatives.

Often the result of a study may be summarized by a confidence interval on a key parameter. The parameter in question might have some noninteresting value, indicated by a null hypothesis. Or it might have some interesting value that the researcher has specified as practically significant. The researcher would certainly want the experiment to be able to distinguish that interesting value from the null hypothesis value. Given this framework, Display 23.1 shows four possible outcomes to the confidence interval procedure.

Display 23.1   Four possible outcomes to a confidence interval procedure

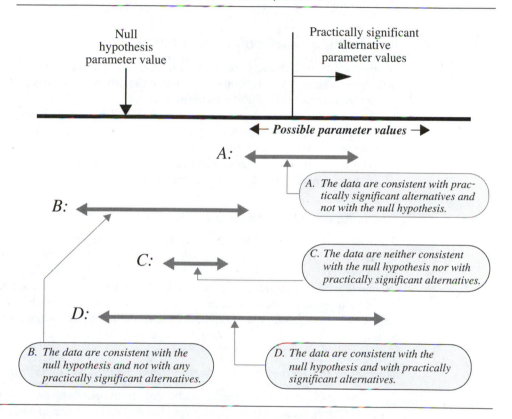

In case *A*, the confidence interval includes a practically meaningful value while ruling out the null value. The researcher may confidently conclude that the parameter does in fact have some interesting value. In case *B*, the researcher cannot rule out the noninteresting null hypothesis, and the evidence strongly suggests that no practically meaningful alternative exists. This is as close as one ever gets to proving the null hypothesis. In case *C*, the researcher can confidently rule out the null hypothesis, but at the same time can confidently rule out any practically meaningful alternative. This happens with very large data sets. In terms of practical significance, this case and case *B* differ little.

These first three outcomes—*A*, *B*, and *C*—are successes, so far as the study is concerned. In each outcome, it is possible to draw an inferential conclusion that distinguishes between the important alternatives one way or the other. The confidence interval in case *D*, however, includes both the null and alternative values of the parameter. This unfortunate outcome signals that the study was not adequate to distinguish between important alternatives. If the analysis utilized the best statistical procedures available, the only way to resolve the issue is to conduct further studies.

The role of the research design is to avoid outcome *D*. This is accomplished by making the confidence interval short enough that it cannot simultaneously include both parameter values.

23.4.2 How to Improve a Confidence Interval

To illustrate how the basic tools of experimental design affect the accuracy and precision of a study, consider the problem of estimating the difference between means of a treated group and a control group. Refer to Display 23.2.

Randomization, blocking, blinding, placebos, and the use of covariates are, as Display 23.2 suggests, important tools for eliminating possible biases in sample averages. The next goal of the research design is to shorten the confidence interval.

The halfwidth of the confidence interval is the product of three factors, any one of which may be made smaller by certain design choices.

Reducing the Halfwidth via the t-Distribution Multiplier

There are two ways to make the multiplier smaller. One way is to lower the level of confidence, which is not particularly satisfying. The other is to increase the sample sizes, by further replication. Reference to a table of the *t*-distributions reveals that the multiplier decreases with increasing degrees of freedom. But the only substantial drop occurs with very low degrees of freedom; after about 10 degrees of freedom, the multiplier declines very slowly. Therefore, one should not look to the multiplier as the best way to shorten an interval.

Reducing the Halfwidth via the Response Standard Deviation

The standard deviation may be a pooled estimate from both samples. It may also be pooled across blocks in a randomized block design. It may also be the residual standard deviation of the responses from a regression model using covariates as explanatory variables. Successful blocking or inclusion of covariates reduces the size of the standard deviation in the halfwidth formula.

Blocking units into groups of homogeneous units and comparing treatments within the blocks takes block-to-block variations out of the standard deviation used for calculating the standard error of treatment differences. If some variable is strongly associated with the response, then blocking on different levels of that variable can substantially decrease the standard deviation and, hence, the halfwidth of a confidence interval.

Display 23.2 The 100 $(1-\alpha)$% confidence interval for the difference between the means of two groups of study units

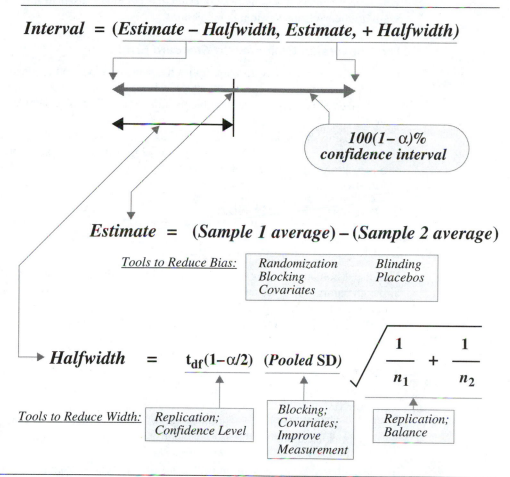

$$Interval \ = \ (Estimate - Halfwidth, \ Estimate, + Halfwidth)$$

$100(1-\alpha)$% confidence interval

$$Estimate \ = \ (Sample \ 1 \ average) - (Sample \ 2 \ average)$$

Tools to Reduce Bias:

Randomization	Blinding
Blocking	Placebos
Covariates	

$$Halfwidth \ = \ t_{df}(1-\alpha/2) \ (Pooled \ SD) \sqrt{\frac{1}{n_1} + \frac{1}{n_2}}$$

Tools to Reduce Width:

Replication; Confidence Level	Blocking; Covariates; Improve Measurement	Replication; Balance

Using covariates reduces the variance of the responses by $100(1 - R_C^2)$%, where R_C^2 is the proportion of variance (the coefficient of determination) explained by the covariates in a multiple linear regression analysis (that also includes the design variables). Such a reduction can improve the precision of the confidence interval dramatically.

If the covariates are not strongly related to the response or if there is virtually no block-to-block variation, introducing covariates and/or blocking could conceivably decrease accuracy slightly. The existence of such situations is easily overemphasized. In practice, one should always be looking for blocking factors and covariates as the best means for improving confidence intervals.

Another way to reduce the standard deviation is to improve the measurement technology. Such improvements were demonstrated in Display 2.12, which documented the reductions in interval widths over time for the parameter used to distinguish Einstein's relativity theory from Newtonian predictions.

The Conversion Factor for the Standard Error

The final factor in Display 23.2 multiplies the standard deviation to make up the standard error of the estimate. In the two-sample problem, this is the square root of the sum of sample size reciprocals. In other problems, this term will involve samples sizes in similar ways.

Increasing the total sample size by replication reduces the size of this factor overall. For fixed total sample size, $n = n_1 + n_2$, choosing equal sample sizes (that is, creating a balanced study), makes this factor as small as possible.

23.5 Choosing a Sample Size

23.5.1 Studies with a Numerical Response

The discussion above demonstrates that selecting a sample size requires an understanding of which parameter values need to be distinguished. The objective of sample size selection is to avoid case D in Display 23.1.

The following ingredients are required for the calculation of an appropriate sample size:

1. An estimate of the standard deviation must be available. If blocking and/or covariate analysis is to be done, then the *residual* standard deviation needs to be known approximately.
2. One specific linear combination of group means should be identified. This linear combination should be the one that addresses the most important question of interest.
3. A confidence level, $100(1 - \alpha)\%$, must be selected.
4. The difference between the null value and the smallest practically significant alternative must be specified.

Suppose that there are K different study groups, and that the study will be balanced with n observations taken from each group. Let $C_1, C_2, \ldots, C_K$ be the coefficients for the linear combination of interest. If s_e is the available estimate of the standard deviation, then the standard error for the linear combination's estimate (Chapter 6) for the balanced case is

$$SE(g) = s_e \frac{\sqrt{C_1^2 + C_2^2 + \cdots + C_k^2}}{\sqrt{n}}.$$

The sample size desired is the one leading to a confidence interval whose width is no larger than the specified practically significant difference. The value of n that makes the interval width—$2[t_{df}(1 - \alpha/2)]SE(g)$—just equal to the practically significant difference is

$$n = 4 \frac{[t_{df}(1-\alpha/2)]^2 \, s_e^2}{(\text{Practically significant difference})^2} (C_1^2 + C_2^2 + \cdots + C_K^2).$$

Because the t-distribution multiplier relies on the sample size, some trial and error may be necessary to get approximate equality. For a 95% confidence interval, calculate the right-hand side of the equation using a t-multiplier of 2.0. The answer can be used to determine what the t-multiplier would be, and the sample size can be recalculated. *Note*: The prescribed sample size does not absolutely guarantee that the resulting confidence interval cannot include both the null and the practically significant values. The residual standard deviation in the study could be larger than the available estimate, s_e.

23.5.2   Studies Comparing Two Proportions

For studies with a binary response, the situation is similar. Two samples of equal size are drawn from separate populations or created by randomization. Suppose that the proportion of successes in the control group is known approximately. A practically significant difference in the treated group can be specified either by an alternative proportion or by an odds ratio. The procedure for determining the sample sizes is summarized in Display 23.3.

The two groups may be the result of a prospective study where units are assigned to treatment groups, or a prospective study where two populations are sampled to compare their response rates, or a retrospective study where affected units and nonaffected units are randomly sampled.

23.5.3   Sample Size for Estimating a Regression Coefficient

The sample size formula for obtaining a confidence interval of a prescribed width for a regression coefficient requires the variance s_X^2 of the explanatory variable of interest. If the researcher can determine the variance by choosing values of the explanatory variable, this requirement is met. If the explanatory variable is not under the researcher's control, some estimate of the anticipated variance must be found.

Simple Linear Regression

The model for a simple linear regression of the response Y on the explanatory variable X is

$$\mu\{Y|X\} = \beta_0 + \beta_1 X.$$

Display 23.3   Choosing sample sizes for comparing two proportions or odds

---

① Specify the expected "control" proportion

      "Control" proportion $=$ π_C.

② Specify a practically significant difference, either with a proportion π_A or an odds ratio R. Calculate the intermediate values below.

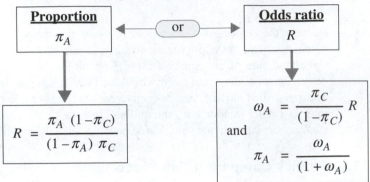

Meaningfully different alternatives

③ Determine the sample size for each group so that the $100(1-\alpha)\%$ confidence interval for the odds ratio will not simultaneously include both 1 and R.

$$n_1 \;\; = \;\; n_2 \;\; = \;\; \frac{4\,[z_{1-\alpha/2}]^2}{[\,\log(R)\,]^2}\left\{ \frac{1}{\pi_C(1-\pi_C)} + \frac{1}{\pi_A(1-\pi_A)} \right\}$$

The objective is to construct a $100(1-\alpha)\%$ confidence interval for the slope parameter β_1 that has width δ. If s_r is an estimate of the standard deviation around the regression line, the required number of (X, Y) pairs is approximately

$$n = \frac{4[t_{1-\alpha/2}]^2 s_r^2}{\delta^2 s_X^2} + 1.$$

Multiple Linear Regression

In a multiple linear regression problem, relationships between explanatory variables can drive up the sample size requirements. With the same objective as in the simple linear regression case, the required number n of cases is approximately

$$n = \frac{4[t_{1-\alpha/2}]^2 s_r^2}{\delta^2 s_X^2 (1 - R_X^2)} + 1,$$

where R_X^2 is the proportion of the variance of X that can be explained by its regression on the other explanatory variables in the model. If X is uncorrelated with the other explanatory variables—as it would be in a balanced factorial experiment—then $R_X^2 = 0$. The s_r in this version is the standard deviation of the response, given all the explanatory variables. Ideally, this can be guessed from related studies or estimated from a pilot study.

23.6   Steps in Designing a Study

Display 23.4 presents a checklist that includes most of the decisions one should consider before beginning a study. The order in which the steps are taken may be different in different studies.

23.6.1   Stating the Objective

What Is the Question of Interest?

Although a study may have many objectives, it is important to concentrate on one specific question, a question whose resolution would make the study successful. Resolution of other issues will be possible, but not necessarily assured by the design.

Example.   In the proposed new ragwort study, the main question is whether there is an interactive effect between cinnabar moths and flea beetles in controlling ragwort.

23.6.2   Determining the Scope of Inference

What Experimental Units Will Be Used?

The scope of inference will be limited by consideration of the sampling procedure (Chapter 1) and by the nature of the units chosen. Whenever treatments are applied to specific objects, the objects become the units. The choice of study units is often obvious, as when a new vaccine is being tested on humans.

Many studies, however, involve several possible choices about study units. In the ragwort study, for example, plots of ground were the obvious units, but their optimum size needed to be determined. When making such decisions, one should ask whether the size of the unit itself affects the responses and whether effects are likely to be clearly measurable on the units selected.

Ragwort Example.   Treated units are square plots with sides 0.5 meters. In large part, this size is determined by plant size and the seed dispersal pattern. A limiting factor on how small a plot can be is whether the plot size itself affects the outcome. It would be unfortunate, for example, if the plot size were so small that insects could not survive on the few hosts present. A factor influencing the upper size limit is the sensitivity of the plot to the treatment. If insects attack only a few plants in a year, plots containing a large number of plants may have many that are not affected even when the insects do considerable damage. Other considerations are physical, e.g., how much effort is required to record the important variables?

Display 23.4   Checklist of tasks involved in the design of a study

☐ 1. State the objective. *What is the question of interest?*

☐ 2. Determine the scope of inference.
   Will this be a randomized experiment or an observational study?
   What experimental or sampling units will be used?
   What are the populations of interest?

☐ 3. Understand the system under study.

☐ 4. Decide how to measure a response.

☐ 5. List factors that can affect the response.
   Design factors
      Factors to vary (treatments & controls)
      Factors to fix
   Confounding factors
      Factors to control by design (blocking)
      Factors to control by analysis (covariates)
      Factors to control by randomization

☐ 6. Plan the conduct of the experiment (time line).

☐ 7. Outline the statistical analysis.

☐ 8. Determine the sample size

What Are the Populations of Interest?

In randomized experiments, the goal is often to show an effect *somewhere*. Experimental units are often chosen to be those that are either most likely to be affected, those that show little natural variability, or those that are most convenient. But when the object of a study is to generalize a statement to a wider population, that population must be carefully described so that units can be randomly selected.

Ragwort Example.   The ragwort experiment is to be conducted in pasture land along the coast of Oregon. Although one may view the outcomes as typical of all such environments, the only random selection occurs when plots are selected within blocks. Blocks were 5-meter by 8-meter regions of a study pasture, so the population to which statistical inference is appropriate is to the set of all such plots in the blocks in that particular pasture. This, of course, is not a serious limitation because the goal of the experiment was to study causal relationships.

23.6.3   Understanding the System

A description of the system amounts to listing the important factors at work and explaining the theory about how they influence each other. Such a description should emphasize the role of the research questions in clarifying the theory.

Ragwort Example.   Ragwort seeds fall within a few meters of the parent plant and build up sizable seed reserves in the soil surrounding where plants have previously grown. Succeeding populations of ragwort are activated by local disturbance of the soil, either in the fall or the spring. The activation is rapid, short-term, and local. The life cycle of new plants is usually 2 years. Approximately 90% of the new plants die before flowering, and most of the 10% that flower do so once in the second year and then die. Stresses of crowding or defoliation, however, can cause plants to live longer and have several flowerings. A plant produces anywhere from 65 to 2,500 flower heads, and flower heads produce an average of 70 seeds.

It is hypothesized that the ragwort populations are inhibited by herbivores and by the competition from perennial grasses. Cinnabar moths emerge in May and June to lay eggs on the plants. The resulting larvae consume the leaves and flower heads, but this rarely kills the plant. The ragwort flea beetles lay eggs singly on the plants or in the surrounding soil. Adults emerge in early summer, hibernate through late summer, then resume activity in the fall. Larvae and adults tunnel into and feed on leaves and stems; larvae also attack roots. The ragwort seed head fly also occurs on these sites, occurring with the moths and destroying a small fraction of the flower heads left by them. Perennial grasses of several species compete for growing space, which is believed to be the principal limiting factor on the spread of ragwort.

23.6.4   Deciding How to Measure a Response

In some studies, the choice of a response variable is obvious. In others there is no clear choice for a single response, or many responses must be measured to answer several questions. If the responses that are used to answer the research questions are decided upon after viewing the available data, this severely restricts the inferential value of the study. So making the decisions in advance is recommended whenever possible.

Determining a sample size will require focusing on a single response. When there are several research questions, different ones will be addressed with different response summaries. The researcher should determine sample size requirements for each question, then make a compromise decision on a final number.

Ragwort Example.   Each plant was located on a plot map, and its progress of growth to maturity, reproduction, and ultimate death was recorded. All measurements for individual plants were summarized to the plot level. These were colonization, survival, reproduction (fecundity, fertility, and establishment rate), and population dynamics. All these measurements are important. They provide insight into different aspects of the control mechanisms. The multivariate nature of "reduction" in ragwort requires a multivariate response and analysis. Nevertheless, the focus question will be addressed by the single response summary, $Y = $ integrated ragwort biomass. At regular times through the lifetimes of the plants, biomass is recorded. Y is the cumulative biomass of all plants on the plot, over their lifetimes.

23.6.5   Listing Factors That Can Affect the Response

Listing factors that affect responses is an exercise in brainstorming. List all factors that might contribute directly. Then list factors that influence the factors already on the list, and so on. The important point is not to overlook a critical factor.

After the list is settled, decide how the factors will be addressed in the design. The *design factors* are those that the researcher manipulates directly and that are directly related to the questions of interest. Other factors will be held fixed. Although possibly influential, these are not related directly to the questions of interest, and it is satisfactory that inferences be restricted to the conditions of their fixed levels. *Potentially confounding factors* should be identified as those that are influential but not directly related to the questions of interest. The researcher controls for measurable confounding factors in the design through blocking, in the analysis by including their effects as covariates in the model, or both.

Some confounding factors are not controlled by blocking because their values cannot be determined prior to the experiment, because they are difficult to categorize, or because they were overlooked at the blocking stage. These can be included as covariates in the analysis. There is nothing wrong, incidentally, with blocking according to a numerical variable like age (*young* being <20 years, *middle* being 20 to 50 years, and *old* being >50 years) and then entering the actual numerical age as a covariate in the analysis.

Ragwort Example
Factors to vary (treatments & controls):
    Cinnabar moth exposure (yes/no)
    Ragwort flea beetle exposure (yes/no)
Factors to hold fixed:
    Climate (coastal Oregon)
    Habitat (pasture land)
    Rodent damage (eliminate using buried wire mesh cages)
    Other plant competitors (roots removed at time of disturbance)
Confounding factors:
    Factors to control by design (blocking)
        Local topography
        Proximity to pasture edge
    Factors to control by analysis (covariates)
        Cinnabar moth density (actual density observed on the plot)
        Flea beetle density (actual density observed on the plot)
        Ragwort seed head fly density (another insect)
    Factors to control by randomization
        Soil fertility
        Moisture
        Light exposure
        Temperature

23.6.6 Planning the Conduct of the Experiment

Establish a time line for the conduct of the experiment. This is particularly important when repeated measures are to be taken or when treatments must be applied in sequence. Have a strict protocol for the conduct of the experiment, with careful thought about how to deal with problems that may arise. Train and evaluate technicians who will administer the protocol and measure the results.

23.6.7   Outlining the Statistical Analysis

Describe how measured variables will be summarized into responses that will be analyzed to answer the questions of interest. List the statistical analysis tools that will be used. Draw up the form of summary tables and graphs that will be incorporated in the final report. This exercise is useful for visualizing how the analysis will proceed.

Ragwort Example.   The response variables will be analyzed with regression models, possibly after transformation. The following will be treated as categorical explanatory variables:

Blocks (with B levels)
Cinnabar moth exposure (2 levels)
Ragwort flea beetle exposure (2 levels).

The interaction of the last two factors will also be included. In addition, the following numerical explanatory covariates will be investigated in a preliminary analysis and retained in the model if they appear statistically significant:

Cinnabar moth density
Ragwort flea beetle density
Ragwort seed head fly density.

23.6.8   Determining the Sample Size

Sample size computations require all four ingredients mentioned at the beginning of Section 23.5.1 on page 668. This is usually challenging. The best situation is one in which the standard deviation can be determined approximately from a similar study. In other situations, one makes a rough guess—usually a conservative one. The specification of a practically significant alternative requires careful thought and judgement on the part of the researcher.

Ragwort Example

The example will focus on finding the number of blocks needed to estimate the interactive effect of the presence of moths and of ragwort flea beetles on the integrated biomass of ragwort.

Ingredient 1: An Estimate of σ.   The proposed study is similar to the parent study described in Section 23.1. That study found it best to analyze integrated biomass on the log scale, and its estimated standard deviation of the transformed response, given the explanatory variables, was about $s_e = 1.38$.

Ingredient 2: The Linear Combination of Interest.   Suppose that *moth* is an indicator variable for the presence of cinnabar moths, with a value of 1 for plots where moths were allowed access and a value of 0 for plots where moths were excluded. Let *beetle* be a similar indicator for the presence of ragwort flea beetles. Ignoring any covariates, the basic model for the mean of the log integrated biomass is

$$\mu\{Y \mid moth, beetle\} = \beta_0 + \beta_1 moth + \beta_2 beetle + \beta_3 moth \times beetle.$$

In terms of the means of the treated groups, this model has

$$\mu\{Y|0, 0\} = \beta_0$$
$$\mu\{Y|1, 0\} = \beta_0 + \beta_1$$
$$\mu\{Y|0, 1\} = \beta_0 + \beta_2$$
$$\mu\{Y|1, 1\} = \beta_0 + \beta_1 + \beta_2 + \beta_3.$$

The question of interest is whether there is an interactive effect between *moth* and *beetle*. The parameter of interest in the regression model is the coefficient β_3, which is a linear combination of the group means, as follows:

$$\beta_3 = \mu\{Y|1, 1\} - \mu\{Y|1, 0\} - \mu\{Y|0, 1\} + \mu\{Y|0, 0\}$$

where the coefficients of the group means are $+1$, -1, -1, and $+1$.

Ingredient 3: A Confidence Level.   This is somewhat arbitrary. The usual choice is 95%, and there is no apparent reason to avoid it here.

Ingredient 4: A Practical Significant Difference.   In the absence of an interactive effect, β_3 is zero, and the effects of the cinnabar moths and of the ragwort flea beetles can be described separately as being multiplicative on the median integrated biomass. The question now is how large a value for β_3 has practical relevance in the study of the control agents. To think about this, it helps to see how the results of the statistical analysis will be worded once the estimates of the parameters are available. The combined effect of cinnabar moths and ragwort flea beetles will be described as producing a median integrated biomass that is $\exp(\beta_3)$ times as large as the median integrated biomass anticipated from the sum of their individual effects. For example, if the effect of cinnabar moths alone was to change median integrated biomass by a factor of 0.8 (a reduction of 20%) and the effect of ragwort flea beetles alone was a factor of 0.6 (a reduction of 40%), and if there were no interaction, then the combined effect of the two agents would be a factor of $(0.8)(0.6) = 0.48$, or a reduction of 52%. The interactive effect would multiply this by $\exp(\beta_3)$.

The situation where β_3 is negative might be described as being *synergistic*, in the sense that the reduction from both exceeds the reduction expected from their individual effects. The situation where β_3 is positive might be described as being *redundant*, since the reduction anticipated from both is not as great as the sum of their effects. If, in the above example, cinnabar moths contributed nothing to what is accomplished by the ragwort flea beetles, then the reduction from both agents would be 0.6—the same as from the flea beetles alone— which would make the interactive effect equal to $\exp(\beta_3) = 1.25$, or $\beta_3 = 0.223$. The same magnitude synergistic effect, from $\beta_3 = -0.223$, changes the median integrated biomass with both control agents by a factor of $(0.8)(0.6)(0.8) = 0.384$, or a 61.4% reduction. In the absence of expert opinion about whether such changes are meaningful, the following discussion will suppose that these do represent values the researcher would want to detect.

Calculation of the Sample Size

It is assumed that there will be B blocks with four plots in each block. The number of observations for each treatment group average will then be $n = B$. (If there are r replicates of each treatment combination in each of B blocks, then each treatment average is based on $n = r \times B$ observations.) Using the formula from Section 23.5.1, and using 2.00 as the t-multiplier, the number of observations per treatment combination required to distinguish that magnitude interaction from no interaction is

$$n = 4\{[2]^2(1.38)^2/(.223)^2\}\{(1)^2 + (-1)^2 + (-1)^2 + (1)^2\} = 2{,}451.$$

Thus, with 2,451 blocks, one could be assured that a 95% confidence interval for the interactive effect of moths and flea beetles could not include both 1 and 0.8 (or 1.25). This is a sobering figure, most likely beyond the resources of the experimenter.

Solutions to the Design Problem

What can be done when a sample size calculation shows that a question cannot be resolved with a manageable sample size? (1) Increasing the size of the practical significant difference lowers the sample size requirement. In the ragwort example, however, the size of interaction was specified to be large enough to completely nullify the effect of the cinnabar moths when plants are exposed to both agents. (2) Decrease the level of confidence. Again, this may work, but it is less than satisfying. (3) Consider a repeated measures crossover design. If all treatments can be applied without carry-over effect to each unit, the analysis would eliminate unit-to-unit variation. (This is not an option in the ragwort example).

The best way to solve the sample size problem is to look to the experimental technique. Blocking and covariate control were both used in the original study, but still the residual SD was 1.38. This means that, typically, in plots treated the same way and having similar co-variate values, responses still differed by multiplicative factors ranging from 1/4 to 4 times the median level. The best way to design a study that could answer the question within reasonable budget levels is to reduce this spread. To accomplish this, the researchers may need to abandon some natural conditions of the study. They allowed regeneration to occur naturally after disturbing existing plots. An alternative would be to establish starts of plots in greenhouse conditions and to move these into the field before the exclusion studies begin. This would allow researchers to exercise control over the numbers and sizes of regenerating ragwort plants and would reduce the size of the response standard deviation.

23.7 Related Issue—A Factor of Four

Readers should be aware that many other texts on experimental design determine sample sizes by asking that the interval *halfwidth* be no larger than the practical significant difference. If such an interval is centered on the practical significant difference, that prescription gives an interval excluding zero. Requiring that the full width be no larger than the practical significant difference results in the sample sizes determined here being four times as large as those prescribed elsewhere. The difference is not one of method but of perspective. The method here provides much stronger guarantees of distinguishing alternatives. *Note*: Other texts determine sample sizes based on specifying the power of hypothesis tests to reject the zero value when there is a difference as large as the practical significant one specified. These result in different sample size requirements, depending on the desired power.

23.8 Summary

In a randomized experiment it is useful to block units according to levels of confounding variables known to be associated with the response. Similar units are grouped together in blocks, then randomly assigned to treatment levels. If there are only two experimental units per block, then blocking is the same as pairing. Other measured confounding variables that were missed at the blocking stage can be accounted for in the analysis by including them in regression models as covariates. The variation of the response after accounting for blocks and covariates is smaller than it would be had these not been included in the analysis. In that respect, blocks and covariates act similarly in improving precision in treatment comparisons.

There are two ways to increase sample size in a randomized block experiment: increase the number of replicate experimental units within each block (if possible) and increase the number of blocks. Both of these improve the precision in treatment comparisons. Only with replication within a block, however, can a pure error estimate of variance be obtained for model checking and examination of block-by-treatment interactions. If there is suspicion of such interactions, it is important to replicate. On the other hand, making blocks larger to include more experimental units may increase the variability of responses within the blocks, thereby decreasing precision.

Some of the elements of experimental design are relevant in random sampling. For example, stratification plays a role similar to blocking. Strata within the population are identified, and random sampling is conducted separately within the different strata. Here, as with experiments, it is not possible to account for all confounding variables simultaneously, so it is important to identify covariates and include them in the analysis.

Further Reading

An excellent discussion of principles of research design, focusing primarily on controlled experiments, is the first chapter of the book by Kuehl (1994). Similar introductions are provided in many other experimental design textbooks. See also the Design of Experiments section in the *Encyclopedia of Statistical Science*, vol. 2 (Johnson, Kotz, and Read 1982), which contains an annotated bibliography. Thompson's *Sampling* (1992), reviews designs for observational studies, including discussions of plot size.

23.9 Exercises

Conceptual Exercises

1. How does a researcher determine whether to control for a potentially confounding factor by blocking or by covariate analysis?

2.   Why is it so important to state the specific research questions in a study beforehand? Why not see what the data suggest first?

3.   What is balance in an experimental design? What is its purpose?

4.   What is *blinding*? What is its purpose? When is it effective? Does blinding always work?

5.   Which elements of a confidence interval offer the most promise for reducing the overall width?

6.   Vitamin C Study.   In one vitamin C study, subjects were randomly assigned to one of two treatment groups. The treated group received daily pills with vitamin C; the control group received daily placebo pills with no active ingredient. It was found later that the taste of the pills allowed many subjects to break the blind and guess their treatment group. Let X be an indicator of those that correctly guessed their group. (a) How could this indicator be used as a covariate? Explain the model and what it can answer. (b) Let V be an indicator of whether subjects believed they were in the treated group. Explain why V might be a better covariate than X.

Study Designs for Discussion

7.   Scientists debate the effects of increased carbon dioxide in the atmosphere, largely the result of burning fossil fuels. But atmospheric carbon dioxide levels have varied widely in the past in response to swings in the earth's general metabolism. If scientists knew what past levels were, they might be better able to predict what current levels portend. Scientists in the Netherlands and in Florida have proposed a method of determining carbon dioxide levels in past atmospheres by examining fossilized plants. They have discovered that leaves of certain plants respond to different levels of carbon dioxide by developing different numbers of stomata (pores) on the undersides of their leaves. Since fossilized leaves dating back to 10 million years ago are available, a technique based on stomata counts may provide strong insights. (Data from R. Monastersky, "The Leaf Annals," *Science News* 144 (August 28, 1993): 140–41).

How could one design a study with live plants in a laboratory setting that would calibrate stomata counts in such a way that carbon dioxide levels could be estimated? Discuss confounding variables, blocking, etc. Describe the analysis. Do not attempt a sample size calculation.

8.   Police searches of the living quarters of accused rapists generally find more pornography than do searches of the living quarters of persons accused of other crimes. Does pornography *cause* such violent behavior toward women? Discuss the issues involved in designing a study that could address this question.

9.   There is considerable controversy among scientists about the potential carcinogenic effects of magnetic fields. Several epidemiology studies have suggested a possible link between electromagnetic fields (EMFs) and a variety of human ailments. How would studies be designed to determine if EMFs are associated with cancers (a) in human tissues and (b) in live humans? What confounding variables can be controlled (c) in the designs and (d) in the analyses?

10.   Although considerable evidence suggests that reward systems have negative effects on productivity (Section 1.1.1), many American companies have elaborate incentive plans. Studies of the effects of such plans usually ask which of several plans works best. How could a study be designed to ask whether they have any benefit at all?

11.   Many studies have been conducted to investigate the ability of cloud seeding to increase rainfall. The detection of even a large effect is difficult because of the large natural variation in rainfall between storms. One study proceeded as follows:

Every day in the study season was evaluated on its suitability for seeding. (Seeding has little chance of being effective if there are no clouds.) On the first suitable day, a random mechanism like the flip of a coin was used to determine whether the clouds would be seeded or not. The rainfall in the target area was measured after treatment. On the next suitable day, the opposite action was taken—the clouds were seeded if they had not been seeded on the previous day and were not seeded otherwise. Again the rainfall in the target area was measured. On the next suitable day, the random mechanism was used again, and on the next the opposite action taken. In this way, consecutive pairs of suitable days contained both actions. (a) Explain why randomization is critically important in this experiment. (b) Explain how blocking was used and under what conditions this blocking might help. (c) Would it be useful to have available, for each suitable day, a measurement of the rainfall in an area adjacent to the target area? Why? How useful might it be, relative to the blocking?

Computational Exercises

12.   Research suggests that hypertension adversely affects the speedy retrieval of newly learned information. One test involves showing subjects two, four, or six digits on a video screen and later asking whether a particular digit was part of the set. It is anticipated that hypertensive individuals will take longer to respond than nonhypertensive individuals. Consider forming two groups of equal size to test this. Suppose that the spread (SD) in response times is about 3 seconds and that researchers consider a mean difference of 5 seconds to be practically important. What sample sizes should be used to make a 95% confidence interval small enough to distinguish a 5-second difference? (Data from "How Hypertension Affects Memory," *Science News* 143 (March 20, 1993): 186.)

13.   The standard treatment of ovarian cancer by the chemicals cytoxan and cisplatin substantially shrinks or eliminates tumors in about 60% of patients. Taxol, derived from the bark of yew trees, is expected to improve upon those response rates. If a clinical trial is conducted to randomly assign patients to the standard chemicals or to taxol, with equal numbers in each group, how large a sample size would be needed to distinguish a 75% response rate for taxol, at 95% confidence? (Data from "Yew Drugs Show Their Mettle," *Science News* 143 (May 29, 1993): 344.)

14.   The 1954 Salk vaccine trials for infantile paralysis show the difficulty faced in prospective trials of preventative treatment for rare diseases. (Data from D. Freedman, R. Pisani, R. Purves, and A. Adhikari, *Statistics*, 2d ed., (New York: W. W. Norton, 1991) chapter 1.) The general rate of infantile paralysis varied highly from year to year, but was somewhere in the vicinity of 1 in 4,000 children. What sample sizes would be required in two groups—a control group and a vaccinated group—in order to determine whether the vaccine reduces the odds of contracting the disease by a factor of three?

15.   Infantile paralysis was linked to hygiene in a strange way that complicated the issue of parental consent. Children from generally high-income families were more likely to obtain consent to participate in the trials, but they were also more likely to contract the disease, because they had been raised in a cleaner environment where they were not exposed to antigens that would build up their immunity. Children from generally low-income families were less likely to obtain consent, and also less likely to contract the disease because of early exposure to antigens. This theory could be tested by a case-control study in which equal numbers of infantile paralysis victims and nonvictims were randomly sampled to determine the proportions in each group that come from low-income families. If, in the general population (the nonvictim group), the proportion of low-income family children was 20%, what sample sizes would be needed to determine that the odds of contracting the disease for high-income family children was 3 times higher than for low-income family children? Compare the sample size requirements in this and the previous problem.

16.   Ragwort Control. Given that the reported study showed that ragwort flea beetles were more effective than cinnabar moths at controlling ragwort, a managerial question is whether there is any reason to introduce the moths in addition to the flea beetles. Suppose that there is no interactive effect, as in Section 23.6.8. (a) In terms of the regression coefficients, what is the effect on median ragwort integrated biomass of including cinnabar moths to plots that may or may not already have exposure to flea beetles? (b) What is the linear combination of the group means that represents this question?

(c) How many blocks would be needed to ensure that a 95% confidence interval for the effect on the median did not simultaneously contain 1 and exp(.8)?

17.   Big Bang (Chapter 7). Suppose that with the best technology one can measure nebulae distances with a standard deviation of 0.25 megaparsecs about their regression on recession velocity. Suppose further that one can observe nebulae whose recession velocities have a standard deviation of about 400 km/sec. How many nebulae should be sampled to obtain a 95% confidence interval on the age of the universe, with a width of 1 billion years? [Recall that 1 billion years is about 0.001021 megaparsec-seconds per kilometer.] How many nebulae are required if one can sample nebulae with recession velocities having a standard deviation of 600 km/sec?

Design Problems

18.   Nibbling and Heart Disease. Some studies by J. Mann at the University of Otago, New Zealand, suggest that nibbling—eating small amounts of food frequently—rather than eating three larger meals each day may be a way to lower cholesterol levels. These were small studies involving hospital patients (Data from J. Mann, "Cholesterol Worries? Nibble More on Less," *Science News* 143 (March 13, 1993): 165.) Design a study that would answer the question, "Does consuming nine small snack-sized meals each day, rather than consuming three meals, reduce LDL cholesterol levels?" Assume that LDL cholesterol will need no transformation, and that the ratio of the standard deviation to the mean cholesterol level is about 0.1 in persons on a three-meal diet. Assume that you wish to determine if nibbling can result in a drop of 6% or more in the mean cholesterol level.

19.   Seasonal Dyslexia. Researchers at the University of Arkansas theorize that exposure to influenza and other viral diseases in the second trimester of pregnancy may be a factor increasing the odds that the child will be dyslexic. (Dyslexia is a reading disability, defined as a reading score on standard tests falling at least 2 years behind the expected level, despite a normal IQ). One piece of evidence that could support this would be a high rate of dyslexia in children born in the summer months, because the second trimester would have fallen in the influenza season. Design an observational study to examine this conclusion. Assume that this will be a case–control study with a random sample of dyslexic boys compared to an equally-sized random sample of nondyslexic boys. Compare the frequency of births in June, July, and August with the frequency of births in December, January, and February. Ignore births in the other months. In the control group, you expect about 50% to be born in the summer months. How large should each sample be if you wish to detect a twofold difference in the odds?

20.   Using the steps in Section 23.6, design a study related to your own field of research. (If you are not yet specialized, select a topic from a data example in a previous chapter.) Make the study specific to one research question. Write a short report in two parts. In part A, discuss briefly the step-by-step decisions involved in the design. List possible confounding variables, and explain how they will be addressed by the design. Explain how treatments will be arranged and how blocking will be done (if necessary). Determine the number of units required to confidently distinguish between practically significant alternatives. In part B, write a brief summary of the results of the study, as you expect them to look. Include figures that would graphically display anticipated results.

Solutions to Conceptual Exercises

1.   Blocking is possible when the levels of the confounding factor are known prior to the performance of the study. If not, the levels must be measured at the time of the study and control must be achieved by covariate analysis.

2.   Whenever the data are consulted to determine the "right" questions, accurate assessment of the strength of statistical evidence must incorporate the consultation process. Standard statistical procedures available in computer software packages do not do this (except in the case of multiple comparisons tests in the separate means analysis of variance). So the standard statistical analysis results do not provide appropriate measures of uncertainty concerning *post hoc* questions.

3. A balanced design is one with equal numbers of units assigned to each treatment group. Balance generally reduces the variability of effect estimates, for a fixed sample size. In most designs, balance also results in statistical independence of different effect estimates.

4. *Blinding* refers to not informing study units and response measurers about treatment group memberships. Its purpose is to eliminate the effects of group membership knowledge on the response—which can be considerable. Blinding is useful when units can affect study conditions and outcomes, and when researchers have particular expectations for certain treatments. But blinding does not always work. In many studies, the units and researchers involved can break the blind.

5. Reducing the standard deviation of the response about its mean, if possible, is a very desirable approach. It is possible through blocking, inclusion of covariates, or improved measurement. The halfwidth can also be reduced by increasing the sample size.

6. Let Y be the binary response indicating having caught a cold. Let π represent the probability of catching a cold given X and T, where T indicates whether the subject actually received the vitamin C pills. Then a logistic regression model can be used to estimate the coefficient of T. This would provide a statement about the reduction in odds of a cold due to vitamin C after accounting for whether the subject correctly guessed their group. (b) Since the placebo effect in this study might lead to reporting fewer colds for individuals who thought they received the active treatment, this covariate is probably more useful than X.

Factorial Treatment Arrangements and Blocking Designs

Good experimental design techniques combine clever approaches to estimating various effects, insightful methods of reducing residual variation, and sensible ways to do all this with a small number of experimental units. Fortunately, each individual researcher does not have to invent a study design for each new study. Some standard design patterns can be applied broadly to cover the vast majority of research situations.

This chapter reviews these basic underlying patterns. It is not intended to provide a full exposition of experimental designs. To do so would require a full textbook (of which there are many excellent examples). Instead, a range of design patterns are discussed in general terms.

Many of the design patterns presented here have appeared in previous chapters. A good exercise for fixing the ideas consists of recalling which case studies from previous chapters fit each design.

24.1 Case Study

24.1.1 Amphibian Crisis Linked to Ultraviolet—A Randomized Experiment

In the early 1990s, scientists became concerned about rapid declines in populations of amphibians worldwide. As is often the case when species decline, the suspects are what Edward O. Wilson describes as "the four horsemen of the environmental apocalypse"— disease, predators, habitat destruction, and hunting. The reasons for amphibian decline, however, are perplexing. Hunting has never been a serious threat to frogs, toads, skinks, and salamanders. Habitat destruction and environmental pollution seem unlikely because some of the declining species inhabit remote, pristine environments. Disease and predators also appear unlikely because species suffering catastrophic losses are found in close proximity to unaffected species.

The puzzle led researchers to consider global atmospheric causes, such as thinning of the ozone layer and the resulting increased exposure to damaging ultraviolet (UV-B) light. There is evidence of ozone loss over temperate regions of the earth where some of the amphibian losses have occurred. Amphibians have no protective hair or feathers, and many lay their eggs in clusters exposed to full sunlight. They could be sensitive indicators of the effects of ozone depletion.

If UV-B is the cause, differential impact on species could be explained two ways: (1) differing exposure of eggs to sunlight, and (2) differences in specific activity of photolyase, a photoreactivating enzyme in many animals that repairs DNA damaged by UV-B, providing protection to eggs. Chromatographic assays and literature review confirmed large photolyase activity differences among 10 amphibians that are found in the northwestern United States (Display 24.1).

Display 24.1 Specific activity of photolyase, mode of egg laying, and relative exposure of eggs to sunlight in 10 amphibian species

Species	Group	Specific Activity of Photolyase (10^{11} CBPDs/hr/µg)	Egg-Laying Mode	Exposure to Sunlight
Ambysoma gracile	Salamander	1.0	open water	some
A. macrodactylum	Salamander	0.8	open water	some
Bufo boreas	Toad	1.3	open shallow water	high
Hyla regilla	Tree frog	7.5	open shallow water	high
Plethodon dunni	Salamander	<0.1	hidden	none
P. vehiculum	Salamander	0.5	hidden	none
Rana cascadae	Frog	2.4	open shallow water	high
Rhyacotriton varieganus	Salamander	0.3	hidden	none
Taricha granulosa	Newt	0.2	hidden	limited
Xenopus laevis	Clawed frog	0.1	under vegetation	limited

The researchers wish to determine if there are UV-B effects on the failure rate of egg hatching and if the failure rate is different for species with different levels of specific activity of photolyase. In theory, higher specific activity of photolyase affords greater protection of eggs from UV-B damage. What animal(s) should they use in their study? Should they study the problem in a laboratory or in the field? What experimental design should be used?

Summary of the Statistical Design

To establish a cause-and-effect relationship between UV-B exposure and egg failure rate, researchers designed a randomized block experiment. Egg clusters from three species that place their eggs in exposed shallow water—the toad (*Bufo boreas*), the tree frog (*Hyla regilla*), and the Cascade frog (*Rana cascadae*)—were exposed to a UV-B filtering treatment having three levels: (1) a UV-B blocking filter; (2) a UV-B transmitting filter; and (3) no filter. Special enclosures were constructed to contain clusters of 150 eggs each. Filtering treatments and egg species were randomly assigned to enclosures. The resulting percentage of egg failures was measured within each enclosure.

Enclosures were located at four sites in the Oregon Cascade mountains. At three sites, only a single species occurs naturally: Sparks Lake (tree frog), Small Lake (Cascade frog), and Lost Lake (toad). All three species occur naturally at the fourth site—Three Creeks. Only eggs of naturally occurring species were assigned to enclosures at each site. Twelve enclosures were placed at each of the single-species sites, four enclosures for each treatment level. Thirty-six enclosures were placed at Three Creeks, with four enclosures assigned to each combination of species and treatment.

Amphibian species represents a blocking factor in this experiment, but it sufficiently resembles a treatment (at Three Creeks) to be considered as one here. At Three Creeks, then, the experiment can be described as a 3^2-factorial experiment, replicated four times. At the other sites (blocks), this was a single treatment factor experiment. The experiment was carried out in the field, where inferences can be drawn to the prevailing natural conditions. (The data, appearing in Display 24.12, are from A. R. Blaustein et al., "UV Repair and Resistance to Solar UV-B in Amphibian Eggs: A Link to Population Declines?" *Proceedings of the National Academy of Science, USA* 91 (March 1994): 1791–95.)

24.2 Treatments

24.2.1 Choosing Treatment Levels

Control

The importance of a control group and tools associated with control—placebo and blinding—were discussed in the previous chapter. In the amphibian study there must be a group of eggs exposed to UV-B and a group unexposed. The researchers left some egg clusters unexposed as a control (to receive ambient UV-B exposure) and a treated group with a filtering apparatus to block UV-B radiation. Since the apparatus itself may have an effect on hatching, they created an additional treatment level by constructing a similar filtering apparatus that had no UV-B blockage (analogous to a placebo).

Number of Treatment Levels When the Levels Are Discrete

Discrete means that the different treatment levels are nominal, such as "no UV-B filtering," "UV-B blockage with a filter," and "no UV-B blocking but the use of a transmitting filter"; rather than numerical, such as filtering at 0%, 50%, and 100%. The choice of the number of discrete treatment levels is usually not a statistical issue, but some statistical issues do have some bearing on the choice: (1) more levels require more experimental units and (2) if there are no planned hypotheses then multiple comparison adjustments are required, making it more difficult to detect differences when they really exist. The main point is that a researcher should be stingy about adding more treatment levels to a study.

Number of Treatment Levels When the Levels Are Numerical

If the effect of a numerical explanatory variable is known to be linear throughout some range, then the design that optimizes the precision in estimating the slope places half of the experimental units at the minimum of the range and half at the maximum. This design, however, prohibits the investigation of any nonlinearity. Therefore, if there is some uncertainty about the linearity, or if the goal is to estimate a response surface (to find some minimizing or maximizing value, for example), then more levels must be used. Three levels allow the researcher to fit a quadratic (parabolic) response curve; four make a cubic response possible; and, generally, $(q + 1)$ levels allow one to fit a polynomial of order q.

The choice is determined by a combination of the degree of uncertainty about a linear relationship and the available number of experimental units. In any case, there is substantial advantage to retaining replicate values at each of the distinct levels.

In the amphibian study the researchers chose three species of amphibians in order to include a variety of values of photolyase activity, which will serve as a numerical covariate in the analysis. The choice permits the modeling of a linear and a quadratic effect of photolyase activity on hatching success. It suggests an inference to other species with intermediate photolyase activity levels, although such an inference requires random assignment of photolyase activity levels.

24.2.2 The Rationale for Several Factors

For cause-and-effect inferences, it seems difficult to improve on a study that deliberately manipulates one factor while holding all other factors fixed. So if there is interest in two factors, why not run two separate experiments? Or to save subjects, why not create three groups—one "control," one with factor A changed, and one with factor B changed—with the control being used as a baseline for both factors? A principal benefit of examing UV-B blockage and photolyase activity simultaneously in the amphibian study is that inferences can be made about whether UV-B affects hatching success more for species with low photolyase activity. Such an inference would strengthen the argument for UV-B as the agent for egg failures.

The reason has to do with interaction and efficiency. Neither separate studies nor the three-group design permits the estimation of interaction between the two factors. If there is an interest in interactions, the effects of each factor must be measured at different levels of the other factor to see if the effects are the same at all levels. In addition, for a fixed number of experimental units, there is greater precision in estimating each of two treatment effects from a single experiment in which both are allowed to vary than from two separate experiments in which each one of the treatments is held fixed while the other varies.

24.3 Factorial Arrangement of Treatment Levels

24.3.1 Definition and Terminology for a Factorial Arrangement

In an experiment with more than one treatment factor, a *treatment combination* specifies the combination of levels for the different factors.

> *In a **factorial** arrangement, treatment combinations are constructed for all possible combinations of factor levels.*

To determine how many treatment combinations are required in a factorial experiment, multiply the numbers of factor levels together. The ragwort study of Chapter 23 had four factors. The *moth*, *beetle*, and *time* factors had two levels each, while the *competition* factor had three levels. So the full factorial arrangement has $2 \times 2 \times 2 \times 3 = 24$ treatment combinations. It is customary to refer to such an arrangement as a "$2 \times 2 \times 2 \times 3$ factorial" or—more simply—a "$2^3 \times 3$ factorial." In 2^3, the 2 refers to the number of levels, and the superscript 3 refers to the number of factors having 2 levels. As a more complex example, a "$2^5 \times 3^4 \times 6$ factorial" arrangement involves 10 factors: there are five factors with two levels, four factors with three levels, and one factor with six levels. The number of treatment combinations is the indicated product—a whopping 15,552.

24.3.2 The 2^2 Factorial Structure

The effect of a certain factor with two levels is the change in the mean response associated with changing that factor from a low to a high level while holding all other factors fixed. The rationale underlying factorial designs is apparent in the most simple example—the 2^2 design involving two factors each at two levels. The factorial design produces *two* estimates of the effect of varying each factor. One estimate of the effect of factor #1 holds factor #2 fixed at a low level; the second estimate of the effect of factor #1 holds factor #2 fixed at a high level. If the two estimates agree, then the effects may be combined into an overall estimate of a meaningful *main effect* for factor #1. But if the two estimates are different, there is an *interaction*, where the effect of changing factor #1 depends on the level of factor #2.

Assume that the factors represent explanatory variables, X_1 and X_2, each of which may be set at either a low or a high level. The mean of the response variable, $\mu\{Y \mid X_1, X_2\}$, at the four treatment combinations is depicted in Display 24.2.

Display 24.2 Treatment effects in a 2^2 factorial experimental design

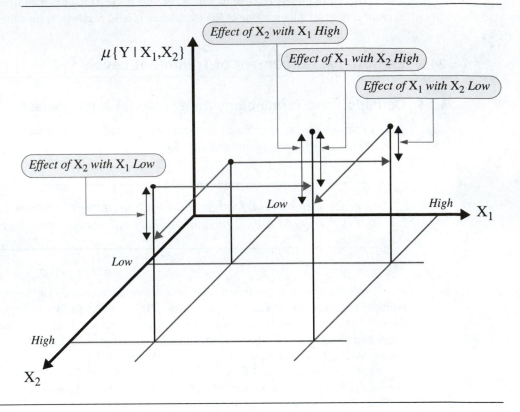

Display 24.2 shows how the 2^2 factorial design measures four different effects:

$$\alpha_{1,H} = \text{Effect of } X_1 \text{ with } X_2 \text{ High}$$
$$= \mu\{Y \mid X_1 = \text{High}, X_2 = \text{High}\} - \mu\{Y \mid X_1 = \text{Low}, X_2 = \text{High}\}$$

$$\alpha_{1,L} = \text{Effect of } X_1 \text{ with } X_2 \text{ Low}$$
$$= \mu\{Y \mid X_1 = \text{High}, X_2 = \text{Low}\} - \mu\{Y \mid X_1 = \text{Low}, X_2 = \text{Low}\}$$

$$\alpha_{2,H} = \text{Effect of } X_2 \text{ with } X_1 \text{ High}$$
$$= \mu\{Y \mid X_1 = \text{High}, X_2 = \text{High}\} - \mu\{Y \mid X_1 = \text{High}, X_2 = \text{Low}\}$$

$$\alpha_{2,L} = \text{Effect of } X_2 \text{ with } X_1 \text{ Low}$$
$$= \mu\{Y \mid X_1 = \text{Low}, X_2 = \text{High}\} - \mu\{Y \mid X_1 = \text{Low}, X_2 = \text{Low}\}.$$

Interpretation of main effects depends on whether the effects of one factor are the same or different at the two levels of the other factor; that is, on whether $\alpha_{1,H} = \alpha_{1,L}$ and $\alpha_{2,H} = \alpha_{2,L}$. The difference between two effects of one variable, measured at different levels of the other, is the *interaction* between the two factors:

$$\begin{aligned}
\beta_{1,2} &= \text{Interaction between } X_1 \text{ and } X_2 \\
&= \alpha_{1,H} - \alpha_{1,L} \\
&= \alpha_{2,H} - \alpha_{2,L} \\
&= \mu\{Y \mid X_1 = \text{High}, X_2 = \text{High}\} - \mu\{Y \mid X_1 = \text{High}, X_2 = \text{Low}\} \\
&\quad - \mu\{Y \mid X_1 = \text{Low}, X_2 = \text{High}\} + \mu\{Y \mid X_1 = \text{Low}, X_2 = \text{Low}\}.
\end{aligned}$$

The second and third lines show that the interaction is the same whether it is called the difference between factor #1 effects at the two levels of factor #2 or is called the difference between factor #2 effects at the two levels of factor #1. The last line emphasizes that the interaction is a linear combination of means.

Ragwort Example. The "silver bullet" theory holds that there is one and only one natural enemy of the pest that will control it. If this pest for the ragwort plant is the ragwort flea beetle (Section 23.1), there would be a reduction of ragwort on plots exposed to flea beetles, and the amount of reduction would not depend on whether the plots were exposed to cinnabar moths. There would be no interaction effect. If, however, control is achieved by a combination of complementary enemies, plots exposed to only one of the insects may have little or no ragwort reduction, while plots exposed to both would have considerable reduction. Measurement of the interaction term in the statistical model allows the researchers to distinguish between these two mechanisms.

Main Effects

If there is no interaction ($\beta_{1,2} = 0$), the effect of X_1 is the same at both levels of X_2. Similarly, the effect of X_2 is the same at both levels of X_1. Then the factorial arrangement provides two estimates of each effect, which may be combined. The average effects are called main effects of the separate factors:

$$\alpha_1 = \text{Main effect of } X_1 = \frac{\alpha_{1,H} + \alpha_{1,L}}{2}$$

$$\alpha_2 = \text{Main effect of } X_2 = \frac{\alpha_{2,H} + \alpha_{2,L}}{2}.$$

Main effects are also linear combinations (contrasts) of treatment means.

Main effects have straightforward interpretation when there is no interaction. When there is an interaction, however, main effects usually make little sense.

Graphical Presentation of Interactions

The best way to present a two-factor interaction is to plot the mean response (or its estimate) against one factor, separately for the levels of the other factor. Display 24.3 shows the format when there is an interaction. (*Note:* Good practice includes drawing in confidence limits for the means.) With no interaction, the two effects—$\alpha_{1,H}$ and $\alpha_{1,L}$—would be equal, and the lines would be parallel.

Display 24.3   Graphical representation of the two-factor interaction in a 2^2 design

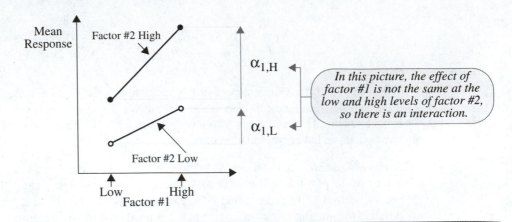

Comparison with Two One-at-a-Time Experiments

Compare this 2^2 arrangement to two "one-at-a-time" experiments, where X_1 and X_2 are changed individually from a low setting to a high setting, but where the treatment of both X_1 and X_2 high is not used. (A one-at-a-time design was encountered in the diet and longevity study of Chapter 5.) Without the both-high treatment combination, it is not possible to estimate interaction. Only one estimate of the effect of each factor is available. Each estimate excludes one-third of the data. With the both-high treatment combination, however, interaction can be estimated, two estimates of each effect are available, and estimates of main effects and interaction use all the data.

24.3.3   The 2^3 Factorial Structure

When there are three factors each at two levels, the eight treatment combinations may be viewed as corners on a cube. Display 24.4 illustrates the layout, with average responses shown as the lengths of arrows.

Definition of Effects

This design provides four estimates of the effect of changing each factor while holding the others fixed. For example, the effects for factor #1 are

$$\alpha_{1,H,H} = \text{Effect of } X_1, \text{ with } X_2 \text{ High and } X_3 \text{ High}$$
$$= \mu\{Y \mid X_1 = H, X_2 = H, X_3 = H\} - \mu\{Y \mid X_1 = L, X_2 = H, X_3 = H\}$$
$$\alpha_{1,H,L} = \text{Effect of } X_1, \text{ with } X_2 \text{ High and } X_3 \text{ Low}$$
$$\alpha_{1,L,H} = \text{Effect of } X_1, \text{ with } X_2 \text{ Low and } X_3 \text{ High}$$
$$\alpha_{1,L,L} = \text{Effect of } X_1, \text{ with } X_2 \text{ Low and } X_3 \text{ Low}.$$

Display 24.4 The representation of a 2^3 factorial design

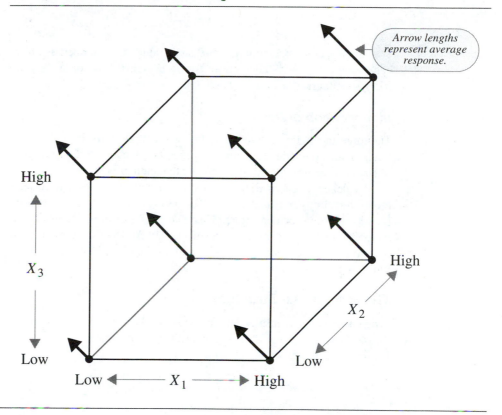

Whether it makes sense to consider an average of estimates of these effects as a main effect estimate depends again on whether there are interactions. There are two estimates for each two-factor interaction. For example, the (factor #1) × (factor #2) interactions are

$$\beta_{1,2H} = \text{Interaction of } X_1 \text{ and } X_2, \text{ with } X_3 \text{ High} = \alpha_{1,H,H} - \alpha_{1,L,H}$$
$$\beta_{1,2L} = \text{Interaction of } X_1 \text{ and } X_2, \text{ with } X_3 \text{ Low} = \alpha_{1,H,L} - \alpha_{1,L,L}.$$

If these two interactions are equal, their separate estimates may be averaged to estimate their common value. But if they are unequal, once again it makes little sense to average their estimates and attempt an explanation. The issue revolves around the *three-factor interaction*,

Two-factor interaction of X_1 and X_2, with X_3 High
− Two-factor interaction of X_1 and X_2, with X_3 Low
= Three-factor interaction of X_1, X_2, and X_3

or

$$\beta_{1,2H} - \beta_{1,2L} = \gamma_{1,2,3}.$$

This three-factor interaction has the same value when defined as above, as the difference between the X_1-by-X_3 interactions for X_2 High and Low, or as the difference between the X_2-by-X_3 interactions for X_1 High and Low.

Interpretation of Results

The three-factor interaction is the first estimate to consider. If there is no evidence of a three-factor interaction, examine the three two-factor interactions. If there is no evidence of any two-factor interactions, consider the main effects.

Evidence of interactions at some level, however, makes it difficult to interpret lower level interactions and main effects. The best approach is to estimate separately those effects for which there is evidence of a difference. If, for example, there is no evidence of a three-factor interaction but there is evidence of an X_2-by-X_3 interaction, the X_2 effect would be estimated separately for X_3 High and for X_3 Low. Plots of averages—with confidence limits—are the best way to display interactions.

24.3.4 The 3^2 Factorial Structure

When a curved response is anticipated for each of two explanatory variables whose interaction is suspected, a 3^2 structure gives the user the wherewithal to estimate all the relevant features and decide which ones are important.

Linear and Quadratic Effects

To understand the nature of the estimable effects, consider a situation involving only one factor at three levels. For illustration, consider the three-level covariate, photolyase activity, from the amphibian study. Display 24.5 shows the proportion of eggs that did not hatch, *fail*, at each of the three levels (averaged over values of the other factors).

Because the factor has three levels, differences between average responses are described by two degrees of freedom. The two features can be expressed in several ways. First, one can report a pair of effects. There is an effect associated with increasing activity from 1.3 (Bufo) to 2.4 (Rana),

$$\alpha_L = \mu\{fail \mid 2.4\} - \mu\{fail \mid 1.3\},$$

estimated to be $45.3\% - 39.2\% = 6.1\%$. And there is an effect associated with increasing activity from 2.4 (Rana) to 7.5 (Hyla),

$$\alpha_H = \mu\{fail \mid 7.5\} - \mu\{fail \mid 2.4\},$$

estimated to be $1.9\% - 45.3\% = -43.4\%$.

An equivalent representation is in terms of linear and quadratic effects. The *linear effect* (λ) is the overall increase in mean response as photolyase changes from 1.3 (Bufo)

Display 24.5 Linear and quadratic effects of one factor set at three levels

to 7.5 (Hyla),

$$\lambda = \mu\{fail \mid 7.5\} - \mu\{fail \mid 1.3\}$$
$$= \alpha_L + \alpha_H,$$

estimated to be $1.9\% - 39.2\% = 6.1\% + (-43.4\%) = -37.3\%$.

The quadratic effect (θ) is the difference between the mean response at the central activity point (2.4 for Rana) and the value predicted by a linear interpolation between the two end points. The interpolated value would be

$$\mu\{fail \mid 1.3\} + \frac{(2.4 - 1.3)}{(7.5 - 1.3)}[\mu\{fail \mid 7.5\} - \mu\{fail \mid 1.3\}],$$

which is estimated to be 32.6%. So the quadratic effect is estimated by 12.7%. The two estimable features can either be expressed in terms of (α_L, α_H) or in terms of (λ, θ), but the interpretation of the latter pair is more natural.

Geometry of the 3^2 Factorial Design

Display 24.6 depicts the results of a 3^2 factorial design, which can be applied to the amphibian study if one considers the factor *FILTER* (with three levels) and the numerical explanatory variable *photolyase* (which has three distinct values). The base plane in Display

Display 24.6   Representation of a 3^2 factorial design

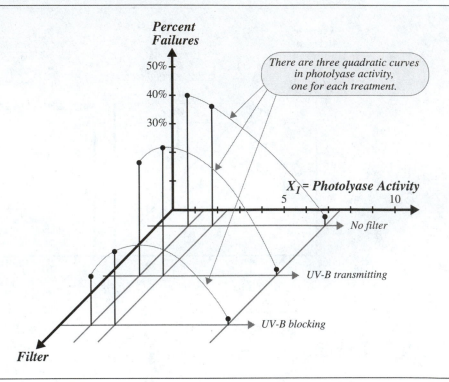

24.6 contains two explanatory factors—photolyase activity and treatment—at three levels each. The vertical axis shows the average response. If both factors were numerical, one could imagine a smooth surface describing the mean response at every pair of explanatory factor values. This display shows three points on each of three slices through the surface. A slice is made by holding *FILTER* fixed and varying photolyase from 1.3 to 2.4 to 7.5.

On any given slice, it is possible to estimate linear and quadratic effects associated with the variations in photolyase activity. Therefore, the kinds of questions that can be answered by such a design involve differences in linear effects and differences in quadratic effects that occur as a result of treatments. The most appealing graphical presentation for the results will be similar to Display 24.3, with three curves displaying the average responses against photolyase on slices of the treatment factor.

Effects in the 3^2 Design

Let the linear and quadratic effects associated with X_1 be as follows:

$$\lambda_{1,L}, \theta_{1,L} = \text{Linear and quadratic effects of } X_1, \text{ with } X_2 \text{ at level 1}$$
$$\lambda_{1,M}, \theta_{1,M} = \text{Linear and quadratic effects of } X_1, \text{ with } X_2 \text{ at level 2}$$
$$\lambda_{1,H}, \theta_{1,H} = \text{Linear and quadratic effects of } X_1, \text{ with } X_2 \text{ at level 3}$$

with similar notation describing effects associated with X_2. These can be summarized by linear and quadratic main effects,

$$\lambda_1 = \frac{\lambda_{1,L} + \lambda_{1,M} + \lambda_{1,H}}{3} \quad \text{and} \quad \theta_1 = \frac{\theta_{1,L} + \theta_{1,M} + \theta_{1,H}}{3},$$

provided that the effects being averaged are all equal; that is, provided that no interactions occur between the effects of X_1 and the levels of X_2.

Interactions in the 3^2 Factorial Arrangement

If the linear or quadratic effects of X_1 differ on different levels of X_2, interaction exists, and main effects have questionable meaning. There are several ways to describe the $2 \times 2 = 4$ interaction possibilities. One method consists of constructing measures of the linear and quadratic changes in the linear and quadratic effects of X_1 with changes in X_2. This produces single estimates for the linear-by-linear interaction, the linear-by-quadratic interaction, the quadratic-by-linear interaction, and the quadratic-by-quadratic interaction. As in all these models, each feature is a linear combination of mean responses, and the methods for drawing inferences appear in Chapter 5.

A more convenient method for the amphibian example, where the treatment factor is not numerical, would be to compare features of the parabolas for the two controls. If they differ, compare the treatment features to the features of the UV-B transmitting control. If they do not, then compare the treatment features to the combined features of the two controls.

For many situations, the set of interactions just described are too complex. Unless there is specific interest in one of the contrasts, a general ANOVA F-test for the combined set may suffice. Evidence of interactions suggests plotting the estimated means (as earlier) with confidence intervals. If the test shows no evidence of interactions, the analysis can be summarized by estimating the main effect parameters.

Amphibian Example.   In the amphibian experiment the factors *FILTER* and *SPECIES* appear in a 3^2 factorial arrangement. If UV-B causes diminished egg hatching success then a treatment main effect should be present. If, in addition, the extent of the UV-B effect depends on the level of photolyase activity, which would indicate a mechanism explaining the UV-B damage, then an interactive effect should be present. A key question is whether the effect of UV-B on hatching failure is greater for species with low levels of photolyase activity than it is for species with high levels.

24.3.5 Higher-order Factorial Arrangements

Factorial arrangements can be expanded to include any number of factors at any number of levels. The advantages to expansion remain: increasing the number of levels allows for a more detailed view of the response curves; and increasing the number of factors allows for exploration of more interactions. The principal disadvantage to expansion lies in the way the number of treatment combinations explodes. Larger numbers of treatment combinations require more subjects, which generally means that subjects will be more diverse. Hence the study loses accuracy that might be available at a smaller scale.

Constructing Error Estimates

There are ways around the problem of too many treatment combinations. Along with the increased number of two-factor interactions and quadratic effects that come with increased levels and factors comes a host of higher-order interactions and higher-order polynomial terms (with numerical explanatory variables). If the researcher is willing to assume that these effects are truly zero, the information normally available for estimating these complex terms may be used to estimate σ. This can reduce the need for replication (See Chapter 14).

An example illustrates the point. In a 2^8 factorial design—eight factors, each at two levels—there are 256 treatment combinations requiring 256 subjects just to use each treatment combination once. In the analysis, there are 8 main effects, 28 two-factor interactions, 56 three-factor interactions, 70 four-factor interactions, 56 five-factor interactions, 28 six-factor interactions, 8 seven-factor interactions, and 1 eight-factor interaction. A researcher willing to assume there are no meaningful interactions above two-factor may use the mean square that would have acted as a numerator mean square in an F-test for the combined set of $56 + 70 + 56 + 28 + 8 + 1 = 219$ higher-order interactions as a 219 degrees of freedom residual mean square. ("Residual" is correct here, because this would be the residual mean square in a regression model containing only main effect and two-factor interaction terms.) Each main effect is a contrast between two averages, each containing 128 observations; each interaction is a linear combination of four averages, each containing 64 observations; and there are 219 degrees of freedom associated with residuals—all this without any replication.

Fractional Replication

As the example suggests, even one replication of a 2^8 factorial arrangement may give more than adequate information about all main effects and two-factor interactions. A researcher can also take advantage of an assumption of no high-order interactions at the design stage to reduce the number of treatment combinations. The method is called *fractional replication*.

Although the mechanics of fractional replication lie outside the scope of this text, the basic structure can be sketched by continuation of the example involving eight factors at two levels each. If only half the 256 treatment combinations are used, it is possible to obtain an analysis as in Display 24.7. That is, the reduced number of treatment combinations appears to affect only the degrees of freedom available for residual.

Display 24.7   Proposed structure for the ANOVA in a half-replicate of a 28 factorial

Source of Variation	Sum of Squares	d.f.	Mean Square	F-Statistic	P-Value
Main effects	xx	8	xx	xx	xx
Two-factor interactions	xx	18			
Residual	xx	101			
Total	xx	127			

Each feature remaining in the model—each main effect and each two-factor interaction—will be completely confounded with a higher-order interaction. Thus the estimate of a main effect will be the same linear combination of observations that would be used to estimate the high-order interaction. To assign the estimates to the low-order features requires the assumption that the higher-order interaction is indeed zero.

Reservations such as these are minor in comparison with the gains that fractional replication achieves in reducing the number of subjects. Specific fractional replication plans are available for a wide range of treatment level combinations.

24.4 Blocking

Complex studies and the desire for high precision necessitate a large number of experimental units, but a common consequence of a large number of experimental units is increased heterogeneity. To counter this, experimental units can be grouped into blocks. Units within each block are homogeneous, so that the heterogeneity is captured by systematic block-to-block variation. By making treatment comparisons within blocks, the block-to-block variation does not affect treatment effect estimates. Including block effects in the models isolates the block-to-block variation so that it also does not affect the residuals. The result is increased precision without loss of bias, the only price being the degrees of freedom necessary to estimate the block mean square (one less than the number of blocks used). This section introduces some of the basic blocking patterns.

24.4.1 Randomized Blocks

In a randomized block experiment treatment combinations are assigned randomly to subjects within a block; and separate randomizations are conducted for each block. It is a randomized *complete* block design (RCBD) if each block contains all treatment combinations. Display 24.8 depicts an example of an RCBD involving eight treatment combinations (e.g., from a 2^3 factorial design).

Suppose that experimental units are plots in a field. Suppose further that plots at the same latitude are expected to have relatively homogeneous responses if treated the same way, but plots at different latitudes are expected to have very different responses. For example, the rows in the display may correspond to plots at the same distance from a river. Then the homogeneous plots should be grouped into blocks; that is, the treatment levels should be assigned to plots separately for each row.

RCBDs may have each treatment combination replicated r times within each block. Whenever $r > 1$, the replicates form a group from which within-group variation can be measured. This allows the researcher to explore possible block-by-treatment interactions (see Section 13.2). Assessing these interactions when $r = 1$ is more difficult, requiring a search for specific interaction patterns that can be modeled using regression tools (Chapter 14).

Even when exploration of block-by-treatment interactions is feasible, the RCBDs are most appropriate when there is an expectation of no interactions—that treatment effects will be the same in all blocks. This is because results about treatment effects are stronger when there is some universality to them; that is, they don't change from block to block.

Display 24.8   A randomized complete block design

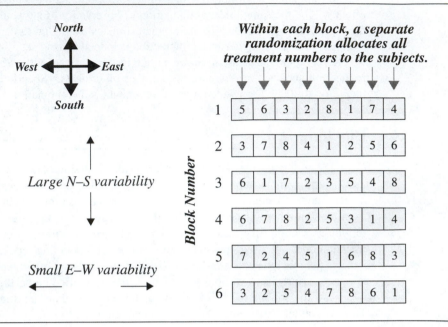

Randomized Incomplete Block Designs

Attaining blocks of homogeneous units often requires very small block sizes. In many studies, block size is as low as two, as in the case study of schizophrenia using twins (Section 2.1.2). When the number of treatments exceeds the number of units in each block, randomized incomplete block designs (RIBD) offer researchers the ability to measure all important effects. The reader should refer to experimental design texts for a systematic presentation of incomplete block techniques. The basic idea is conveyed by the example that follows.

Example. Two treatments, A and B, are proposed to prevent leaf scale on roses. An interesting question concerns the effect of applying the two treatments together—that is, the interaction. Because of large variation in the amount of scale expected to develop on different parts of a rose bush, the researcher wishes to use one leaf as a block with the leaf halves as experimental units. Hence there are only two experimental units in a block, whereas there are $2^2 = 4$ treatments. Display 24.9 illustrates how the study can be arranged.

    The trick is to ensure that each of the basic effects is measured within a block. The effect of treatment A with B absent is estimated by the difference between the response to treatment combination (A, –) and the response to treatment combination (–, –), from Leaf #2. The estimate of treatment A with B present is estimated by the difference between the response to treatment combination (A, B) and the response to treatment combination (–, B), from Leaf #1. The estimate of treatment B with A absent is estimated by the difference between the response to treatment combination (–, B) and the response to treatment

Display 24.9 A randomized incomplete block design for a 2^2 factorial treatment arrangement in blocks of size two

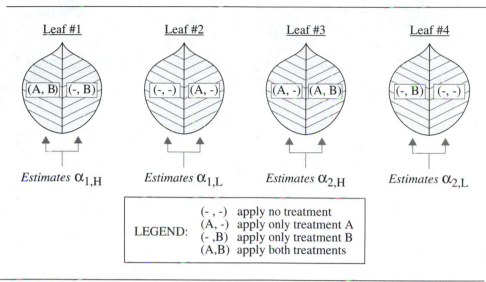

combination $(-, -)$, from Leaf #4. The estimate of treatment B with A present is estimated by the difference between the response to treatment combination (A, B) and the response to treatment combination $(A, -)$, from Leaf #3. Each estimate is made within a leaf, so leaf-to-leaf differences do not influence the estimates.

This design affords two estimates of interaction, from the difference between estimates of $\alpha_{1,H}$ and $\alpha_{1,L}$ and from the difference between estimates of $\alpha_{2,H}$ and $\alpha_{2,L}$. The average difference is the overall estimate of interaction. Similarly, the main effect estimates may be estimated, and their interpretation is clear when there is no evidence of interaction.

This design requires four blocks for two replicates, and multiple replicates are advisable. Separate randomizations assign treatment combination pairs to leaves and then assign treatment combinations to leaf halves within each leaf.

Extensions to More Treatments

Randomized incomplete block designs for more than two treatments (and for more levels) require many more blocks if all effects are to be estimated. Savings are once again possible when the researcher is willing to assume high-order interactions are negligible and need not be estimated. The low-order effects that are estimated will again be confounded with the high-order ones, but this is often a small price in comparison with the advantages to obtaining adequate accuracy from a modestly-sized study.

24.4.2 Latin Square Blocking

More complex blocking patterns account for stronger natural patterns of variation in subjects. One such extension is the Latin square. In Display 24.10, there are eight treatment

Display 24.10 A Latin Square blocking pattern

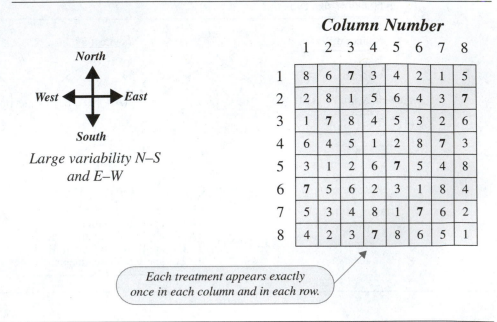

Each treatment appears exactly
once in each column and in each row.

combinations (as in, for example, a 2^3 factorial arrangement), but the experimental units (plots) exhibit variation in both cardinal directions.

Sixty-four plots form a square with sides of eight units. With no treatments applied, it is anticipated that the mean response depends on the row and on the column of the subject, but in an additive way; that is, the mean response can be described by an additive model involving row and column effects.

A Latin Square is one in which all units are numbered from one up to the number of rows (columns) in such a way that each number occurs exactly once in each row and exactly once in each column. There are many different Latin Squares with a given number of rows. The user picks one having the number of sides equal to the number of treatment combinations. The treatment combinations are then randomly assigned to the numbers and applied to the subjects with those numbers assigned.

If there are no position-by-treatment interactions, this method yields an analysis where treatment differences are unaffected by row and column differences. The analysis of variance table for a single replication looks like Display 24.11.

24.4.3 Split Plot Designs

Sometimes different treatments cannot be conveniently applied to the same experimental units. A typical situation: new methods for teaching ecology at the elementary school level involve both the study materials given to the students and the training of the teacher.

Display 24.11 The analysis of variance table for a single replication of a Latin Square design with eight treatments

Source of Variation	Sum of Squares	d.f.	Mean Square	F-Statistic	p-Value
Rows	xx	7	xx		
Columns	xx	7	xx		
Treatments	xx	7	xx	xx	xx
Residual	xx	42	xx		
Total	xx	63			

Different study materials can be randomized to students within a class, but the teachers must be randomized to classes. Consistent sources of variation affect different classes and do not influence student-to-student variation within a class—time of day, for example.

Suppose, to continue the example, that teachers have been trained by either method *A* or method *B*, and suppose that there are two sets of instructional materials, *R* and *S*. The teachers are randomized to classes, which are called *whole plots*. Within each of the classes, instructional materials *R* and *S* are randomized to students. In the end, all students (the *split plots*) are given a standard achievement test.

In terms of the analysis of instructional materials, this design looks like a randomized block design with classes as blocks, and the replication of treatments within the classes allows the researcher to investigate possible teacher training-by-instructional material interaction. When it comes to determining the main effects of teacher training, however, the classes should be viewed as the subjects, with the different student responses viewed as repeated measures. Ordinarily, the average response of all students in a class will summarize the repeated measures for this analysis.

24.5 Summary

In determining treatment structure, two questions arise: (1) How many factors should be examined? and (2) How many levels should each factor have? The number of levels of any given factor is often naturally determined. For numerical treatment levels, the number depends on the complexity of the response curve. Factors should be included with others in the same study when their interactions are of interest.

A study using all combinations of treatment levels is said to have a *factorial* arrangement. Factorial arrangements allow the researcher to investigate very complex interactions. When the researcher is confident that the mean response is not so complex, *fractional replication* can be used to reduce the number of treatment combinations while preserving information about important effects.

In a *randomized complete block design*, all treatment combinations are randomly assigned to experimental units within each of a set of blocks. Blocking experimental units into homogeneous groups reduces the variation and increases the accuracy of treatment comparisons. When the number of treatments exceeds a reasonable block size, it is still possible to estimate effects by using a larger number of blocks in a *balanced incomplete block design*. More complex natural variations in experimental subjects can be controlled by advanced blocking patterns such as the *Latin Square*.

Amphibian Study

The experimental plan called for random assignment of the nine treatment combinations (the factorial combinations of three levels of amphibian species and three levels of UV-B filtering). Treatment combinations were replicated four times at each site. The analysis of the results focuses on the proportion of eggs that fail to hatch. It is possible that these could be analyzed either with binomial logistic regression (with all m_i's being 150 and with careful attention to possible extra-binomial variation) or by ordinary regression and analysis of variance techniques (but with consideration of transformation if the proportions vary widely). In either case, a FILTER effect would be incorporated into the model by, for example, the inclusion of two indicator variables and a SITE effect could be incorporated by the inclusion of five indicator variables. Initial investigation might focus on a SPECIES effect, using two indicator variables, with particular attention to the FILTER-by-SPECIES interaction. Further investigation of the interaction of the filtering treatment and photolyase activity is accomplished by replacing the SPECIES factor by the photolyase numerical explanatory variable. It would be possible to include a linear and a quadratic term (if necessary) for this explanatory variable, and to examine interactions between FILTER and photolyase activity.

Further Reading

The classical references are still worth reading: *The Design of Experiments* (Fisher 1935), and *Design and Analysis of Experiments* (Kempthorne 1952). See also the experimental design references in Chapter 23.

24.6 Exercises

Conceptual Exercises

1. It is believed that forest canopy closure (X, in percent) is an important factor in determining the suitability of a forest for use by flying squirrels. Suppose that the density of flying squirrels is Y, in numbers per hectare. (a) Describe a design situation in which you deliberately alter forest plots to have specified discrete levels of canopy closure and then observe what the resulting flying squirrel density becomes. (b) Describe a study in which you randomly sample forest plots and measure their naturally occurring values (X, Y).

2. Amphibian Decline. Referring to Display 24.1, notice that several species lay their egg clusters in areas protected from the sun. Could the UV-B hypothesis be tested by comparing hatching success of egg clusters laid in areas exposed to the sun with the hatching success of egg clusters laid in, say, in hidden places? Why didn't the researchers include species that lay their eggs in hidden places or in open water?

3. One researcher advocates an "order +2" rule, in which the number of levels of an experimental factor is set at the anticipated order of the polynomial curve, plus two. For example, if the relationship is expected to be a straight line (order = 1), this researcher would use three levels. What could be the justification for this rule?

4. With a $2^4 \times 3^1$ factorial arrangement, how many factors are there in all? How many treatment combinations are involved?

5. What are the differences among these three terms: (a) the interaction of X_1 with X_2; (b) the interaction of X_2 with X_1; and (c) the interaction between X_1 and X_2?

6. Two explanatory factors, temperature and salinity, interact in their association with biodiversity in sea samples. The estimates of main effects are large. Are the main effects meaningful? Why or why not?

7. Draw pictures, similar to Display 24.3, with the mean response versus factor #1 for a 3×2 factorial design depicting: (a) with no interaction, (i) the case with no quadratic effect for factor X_1; and (ii) the case with a quadratic effect; (b) with interaction, (i) the case where only the linear effect of X_1 differs according to the level of X_2; (ii) the case where only the quadratic effect of X_1 differs according to the level of X_2; and (iii) the case where both the linear and quadratic effects of X_1 differ according to the level of X_2.

8. Two researchers examine the results from a 2^6 factorial arrangement with one replication. Researcher A decides in advance to use the residuals from a model with main effects and 2-factor interactions to estimate variability, assuming all 3-factor, 4-factor, 5-factor, and 6-factor interaction terms are zero. Researcher B is willing to assume that the 5-factor and 6-factor interaction terms are zero, but uses a backward stepwise procedure to eliminate 3-factor and 4-factor interactions if they fail to meet the $p = .05$ criterion for significance. Researcher B finds that all 3-factor and 4-factor interactions can be eliminated and therefore ends up with the same model structure as Researcher A. (a) Can the same inferences be drawn from these two approaches? (b) Should the two report their results the same way? (c) Is one approach preferred?

9. (a) What are the residual degrees of freedom in one replication of a Latin Square design with K treatments? (b) How many subjects go into each treatment combination average? (c) Why is this important?

Computational Exercises

10. (*With a computer*) Denoting a treatment combination in a 2^5 factorial as $x_1x_2x_3x_4x_5$ with each x_i being either H or L, suppose that only the following 16 treatment combinations are used: HLLLL, LHLLL, LLHLL, LLLHL, HLLLH, LHLLH, LLHLH, LLLHH, HHHLL, HHLHL, HLHHL, LHHHL, HHHLH, HHLHH, HLHHH, LHHHH. (a) Generate 16 standard normal observations for the responses. (b) Create an indicator variable for the High level of each factor, calling the variables A, B, C, D, and E. (c) Define products of indicators as follows: AB, AC, BC, AE, BE, CE, and DE. (d) Construct a regression equation with the 12 explanatory variables created in parts (b) and (c), determining that those main effects and 2-factor interactions can be estimated. (e) Try to include the 2-factor interaction CD into the equation in (d). The result indicates that CD is completely confounded with one of the variables already in the equation. It is AB. To see this, drop AB and then add CD.

11. (a) Use a randomization procedure to produce three replicates of the Randomized Incomplete Block Design for the rose scale example in Display 24.9. Each replicate has four leaves of two halves each, so there are 24 subjects in all. (b) *With a computer*: (i) Generate 24 standard normal observations for the responses. (ii) Define the appropriate indicator variables for the leaves and for the treatments and construct a multiple linear regression analysis. Use an analysis that assumes there is no treatment-by-leaf interaction but there is an A-by-B treatment interaction. (iii) Select three leaves; add (6, 2, 15) units to each of the responses from selected leaves (1, 2, 3); and recompute the regression analysis, verifying that the treatment conclusions are unchanged.

12. (*With a computer*) (a) Generate 64 standard random normal observations, and assume these are the responses to the 64 subjects in a Latin Square with eight treatments. Define appropriate indicator variables and construct a multiple linear regression analysis resulting in the analysis of variance table in Display 24.11. (b) Add $6 \times I$ to all observations in row I, for each $I = 1, \ldots, 8$. Add $3 \times J$ to all observations in column J, for each $J = 1, \ldots, 8$. Rerun the regression model and verify that the analysis of treatments is unchanged.

Data Problems

13. Amphibian Crisis and UV-B. The percents of unsuccessful hatching from enclosures containing 150 eggs each in the UV-B study appear in Display 24.12. Analyze these data, assessing the evidence supporting the hypothesis that UV-B is responsible for low hatch rates. Fit a rich regression model including interactions of site-by-treatment and species-by-treatment. (There is no possibility of measuring site-by-species interactions.) Examine a residual plot to see if a transformation is appropriate. Determine if site-by-treatment interactions can be ignored. Then assess the evidence for an interaction between UV-B treatment and species. Assess whether the dependency on species can be explained by effects that are linear in photolyase activity. Construct estimates of median percent failure for each species under the blocking filter and under the transmitting filter, at a reference site of Three Creeks. Include Bonferroni 95% confidence limits. Write a paragraph summarizing the statistical findings and explaining what inferences are warranted by the experimental design.

14. Using the residual standard deviation estimate from the previous problem, determine a sample size that would estimate the difference between the UV-B effects on *Rana cascadae* and on *Bufo boreas* to within 10% of the value of the UV-B effect on *Bufo boreas*.

Display 24.12   Percents of frog eggs failing to hatch under different UV-B treatments

| | | *Species:* | *Hyla regilla* | | *Rana cascadae* | | *Bufo boreas* | |
| | | *Photolyase activity:* | 7.5 | | 2.4 | | 1.3 | |
		Locale:	Three Cr.	Sparks Lk.	Three Cr.	Small Lk.	Three Cr.	Lost Lk.
Treatment	**No Filter**		6.0	1.5	38.7	36.7	42.0	54.0
			4.7	0.8	44.0	69.6	50.7	54.7
			0.7	2.9	30.0	39.3	32.7	48.0
			*	3.9	38.7	34.0	44.0	36.7
	UV-B Transmitting		0.9	0.7	28.7	70.0	47.3	46.0
			6.7	2.1	32.7	54.0	22.0	46.7
			2.7	0.0	36.0	48.7	37.3	36.0
			0.7	1.4	40.7	51.3	43.3	35.3
	UV-B Blocking		4.7	4.5	25.3	24.7	18.7	12.7
			0.7	0.0	18.7	25.3	17.3	17.3
			4.7	0.0	21.3	39.3	16.0	31.3
			0.7	0.0	16.7	32.7	4.7	17.3

* Enclosure lost.

Answers to Conceptual Exercises

1. This goes back to Chapter 1. (a) Select a number of plots and randomly assign them to different canopy closure levels. Then observe. This is the approach to be taken when it is desired to show a cause-and-effect relationship between the two variables. (b) By randomly sampling, you learn about the overall structure of the population of (X, Y) pairs that exists naturally within the region sampled, but no cause-and-effect relationship can be inferred.

2. To do so would confound the exposure difference with species differences. Aside from that, they would not be able to randomize exposure conditions to similar egg clusters, which would preclude cause-and-effect inferences. By selecting only species that lay eggs in open shallow water, they are controlling for possible differences between hatching success related to different environments. Remember that cause-and-effect can be established on a subpopulation, and that is what was attempted in this study.

3. The rule includes one more level than is needed to fit the anticipated polynomial. The purpose for the rule is to always provide for one degree of freedom for lack of fit to the polynomial.

4. There are 5 $(= 4 + 1)$ factors. Four have two levels each, while the fifth has three levels. The total number of treatment combinations in a full factorial is the indicated product, 48.

5. None. They are equivalent ways of saying the same thing.

6. No. The main effect of salinity, for example, does not describe the effect of changing salinity on biodiversity. Interaction means that you need to know what the temperature is before determining the effect of salinity. At one temperature the effect of increased salinity may be increased biodiversity, but at another temperature the effect of the same increased salinity could decrease biodiversity.

7. Refer to Display 24.13 for examples. Notice why the interaction has 2 degrees of freedom.

Display 24.13 Examples of response patterns for a 3 × 2 factorial study

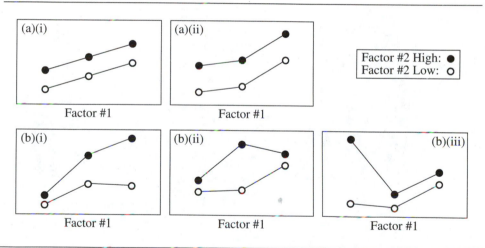

---

8. (a) Technically, no. A may use the p-values directly from a regression or ANOVA output; B uses the data to determine the inferential model, so valid p-values are based on a sequence of trials where the search for a model is incorporated as part of the analysis. There is no easy way to do this, however; so ... (b) No. The best way for them to help their readers evaluate the strength of their evidence is to describe their methodologies—which were different. (c) B addresses one aspect of model inadequacy that A does not.

9. (a) The general formula is: $(K - 1)(K - 2)$. (b) K subjects. (c) These are important for determining the numbers of replications (sample size) necessary to answer the questions of interest.

Tables

Table A.1 Probabilities of the Standard Normal Distribution

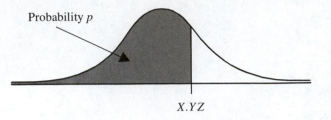

Probability p

Tabled value is p, the probability that a standard normal is less than $X.YZ$

$X.YZ$

					.0Z					
$X.Y$	.00	.01	.02	.03	.04	.05	.06	.07	.08	.09
−3.4	.0003	.0003	.0003	.0003	.0003	.0003	.0003	.0003	.0003	.0002
−3.3	.0005	.0005	.0005	.0004	.0004	.0004	.0004	.0004	.0004	.0003
−3.2	.0007	.0007	.0006	.0006	.0006	.0006	.0006	.0005	.0005	.0005
−3.1	.0010	.0009	.0009	.0009	.0008	.0008	.0008	.0008	.0007	.0007
−3.0	.0013	.0013	.0013	.0012	.0012	.0011	.0011	.0011	.0010	.0010
−2.9	.0019	.0018	.0018	.0017	.0016	.0016	.0015	.0015	.0014	.0014
−2.8	.0026	.0025	.0024	.0023	.0023	.0022	.0021	.0021	.0020	.0019
−2.7	.0035	.0034	.0033	.0032	.0031	.0030	.0029	.0028	.0027	.0026
−2.6	.0047	.0045	.0044	.0043	.0041	.0040	.0039	.0038	.0037	.0036
−2.5	.0062	.0060	.0059	.0057	.0055	.0054	.0052	.0051	.0049	.0048
−2.4	.0082	.0080	.0078	.0075	.0073	.0071	.0069	.0068	.0066	.0064
−2.3	.0107	.0104	.0102	.0099	.0096	.0094	.0091	.0089	.0087	.0084
−2.2	.0139	.0136	.0132	.0129	.0125	.0122	.0119	.0116	.0113	.0110
−2.1	.0179	.0174	.0170	.0166	.0162	.0158	.0154	.0150	.0146	.0143
−2.0	.0228	.0222	.0217	.0212	.0207	.0202	.0197	.0192	.0188	.0183
−1.9	.0287	.0281	.0274	.0268	.0262	.0256	.0250	.0244	.0239	.0233
−1.8	.0359	.0351	.0344	.0336	.0329	.0322	.0314	.0307	.0301	.0294
−1.7	.0446	.0436	.0427	.0418	.0409	.0401	.0392	.0384	.0375	.0367
−1.6	.0548	.0537	.0526	.0516	.0505	.0495	.0485	.0475	.0465	.0455
−1.5	.0668	.0655	.0643	.0630	.0618	.0606	.0594	.0582	.0571	.0559
−1.4	.0808	.0793	.0778	.0764	.0749	.0735	.0721	.0708	.0694	.0681
−1.3	.0968	.0951	.0934	.0918	.0901	.0885	.0869	.0853	.0838	.0823
−1.2	.1151	.1131	.1112	.1093	.1075	.1056	.1038	.1020	.1003	.0985
−1.1	.1357	.1335	.1314	.1292	.1271	.1251	.1230	.1210	.1190	.1170
−1.0	.1587	.1562	.1539	.1515	.1492	.1469	.1446	.1423	.1401	.1379
−0.9	.1841	.1814	.1788	.1762	.1736	.1711	.1685	.1660	.1635	.1611
−0.8	.2119	.2090	.2061	.2033	.2005	.1977	.1949	.1922	.1894	.1867
−0.7	.2420	.2389	.2358	.2327	.2296	.2266	.2236	.2206	.2177	.2148
−0.6	.2743	.2709	.2676	.2643	.2611	.2578	.2546	.2514	.2483	.2451
−0.5	.3085	.3050	.3015	.2981	.2946	.2912	.2877	.2843	.2810	.2776
−0.4	.3446	.3409	.3372	.3336	.3300	.3264	.3228	.3192	.3156	.3121
−0.3	.3821	.3783	.3745	.3707	.3669	.3632	.3594	.3557	.3520	.3483
−0.2	.4207	.4168	.4129	.4090	.4052	.4013	.3974	.3936	.3897	.3859
−0.1	.4602	.4562	.4522	.4483	.4443	.4404	.4364	.4325	.4286	.4247
−0.0	.5000	.4960	.4920	.4880	.4840	.4801	.4761	.4721	.4681	.4641

(Table A.1 continued)

X.Y	.00	.01	.02	.03	.04	.05	.06	.07	.08	.09
0.0	.5000	.5040	.5080	.5120	.5160	.5199	.5239	.5279	.5319	.5359
0.1	.5398	.5438	.5478	.5517	.5557	.5596	.5636	.5675	.5714	.5753
0.2	.5793	.5832	.5871	.5910	.5948	.5987	.6026	.6064	.6103	.6141
0.3	.6179	.6217	.6255	.6293	.6331	.6368	.6406	.6443	.6480	.6517
0.4	.6554	.6591	.6628	.6664	.6700	.6736	.6772	.6808	.6844	.6879
0.5	.6915	.6950	.6985	.7019	.7054	.7088	.7123	.7157	.7190	.7224
0.6	.7257	.7291	.7324	.7357	.7389	.7422	.7454	.7486	.7517	.7549
0.7	.7580	.7611	.7642	.7673	.7704	.7734	.7764	.7794	.7823	.7852
0.8	.7881	.7910	.7939	.7967	.7995	.8023	.8051	.8078	.8106	.8133
0.9	.8159	.8186	.8212	.8238	.8264	.8289	.8315	.8340	.8365	.8389
1.0	.8413	.8438	.8461	.8485	.8508	.8531	.8554	.8577	.8599	.8621
1.1	.8643	.8665	.8686	.8708	.8729	.8749	.8770	.8790	.8810	.8830
1.2	.8849	.8869	.8888	.8907	.8925	.8944	.8962	.8980	.8997	.9015
1.3	.9032	.9049	.9066	.9082	.9099	.9115	.9131	.9147	.9162	.9177
1.4	.9192	.9207	.9222	.9236	.9251	.9265	.9279	.9292	.9306	.9319
1.5	.9332	.9345	.9357	.9370	.9382	.9394	.9406	.9418	.9429	.9441
1.6	.9452	.9463	.9474	.9484	.9495	.9505	.9515	.9525	.9535	.9545
1.7	.9554	.9564	.9573	.9582	.9591	.9599	.9608	.9616	.9625	.9633
1.8	.9641	.9649	.9656	.9664	.9671	.9678	.9686	.9693	.9699	.9706
1.9	.9713	.9719	.9726	.9732	.9738	.9744	.9750	.9756	.9761	.9767
2.0	.9772	.9778	.9783	.9788	.9793	.9798	.9803	.9808	.9812	.9817
2.1	.9821	.9826	.9830	.9834	.9838	.9842	.9846	.9850	.9854	.9857
2.2	.9861	.9864	.9868	.9871	.9875	.9878	.9881	.9884	.9887	.9890
2.3	.9893	.9896	.9898	.9901	.9904	.9906	.9909	.9911	.9913	.9916
2.4	.9918	.9920	.9922	.9925	.9927	.9929	.9931	.9932	.9934	.9936
2.5	.9938	.9940	.9941	.9943	.9945	.9946	.9948	.9949	.9951	.9952
2.6	.9953	.9955	.9956	.9957	.9959	.9960	.9961	.9962	.9963	.9964
2.7	.9965	.9966	.9967	.9968	.9969	.9970	.9971	.9972	.9973	.9974
2.8	.9974	.9975	.9976	.9977	.9977	.9978	.9979	.9979	.9980	.9981
2.9	.9981	.9982	.9982	.9983	.9984	.9984	.9985	.9985	.9986	.9986
3.0	.9987	.9987	.9987	.9988	.9988	.9989	.9989	.9989	.9990	.9990
3.1	.9990	.9991	.9991	.9991	.9992	.9992	.9992	.9992	.9993	.9993
3.2	.9993	.9993	.9994	.9994	.9994	.9994	.9994	.9995	.9995	.9995
3.3	.9995	.9995	.9995	.9996	.9996	.9996	.9996	.9996	.9996	.9997
3.4	.9997	.9997	.9997	.9997	.9997	.9997	.9997	.9997	.9997	.9998

.0Z

Table A.2 Selected Percentiles of t-Distributions

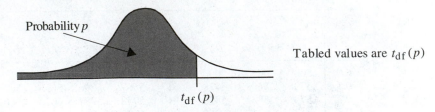

Probability p

Tabled values are $t_{\mathrm{df}}(p)$

$t_{\mathrm{df}}(p)$

d.f.	.75	.80	.85	.90	.95	.975	.98	.99	.995	.9975	.999	.9995
1	1.000	1.376	1.963	3.078	6.314	12.71	15.89	31.82	63.67	127.3	318.3	636.6
2	0.816	1.061	1.386	1.886	2.920	4.303	4.849	6.965	9.925	14.09	22.32	31.60
3	0.765	0.978	1.250	1.638	2.353	3.182	3.482	4.541	5.841	7.453	1.215	12.92
4	0.741	0.941	1.190	1.533	2.132	2.776	2.999	3.747	4.604	5.598	7.173	8.610
5	0.727	0.920	1.156	1.476	2.015	2.571	2.757	3.365	4.032	4.773	5.893	6.869
6	0.718	0.906	1.134	1.440	1.943	2.447	2.612	3.143	3.707	4.317	5.208	5.959
7	0.711	0.896	1.119	1.415	1.895	2.365	2.517	2.998	3.499	4.029	4.785	5.408
8	0.706	0.889	1.108	1.397	1.860	2.306	2.449	2.896	3.355	3.833	4.501	5.041
9	0.703	0.883	1.100	1.383	1.833	2.262	2.398	2.821	3.250	3.690	4.297	4.781
10	0.700	0.879	1.093	1.372	1.812	2.228	2.359	2.764	3.169	3.581	4.144	4.587
11	0.697	0.876	1.088	1.363	1.796	2.201	2.328	2.718	3.106	3.497	4.025	4.437
12	0.695	0.873	1.083	1.356	1.782	2.179	2.303	2.681	3.055	3.428	3.930	4.318
13	0.694	0.870	1.079	1.350	1.771	2.160	2.282	2.650	3.012	3.372	3.852	4.221
14	0.692	0.868	1.076	1.345	1.761	2.145	2.264	2.624	2.977	3.326	3.787	4.140
15	0.691	0.866	1.074	1.341	1.753	2.131	2.249	2.602	2.947	3.286	3.733	4.073
16	0.690	0.865	1.071	1.337	1.746	2.120	2.235	2.583	2.921	3.252	3.686	4.015
17	0.689	0.863	1.069	1.333	1.740	2.110	2.224	2.567	2.898	3.222	3.646	3.965
18	0.688	0.862	1.067	1.330	1.734	2.101	2.214	2.552	2.878	3.197	3.610	3.922
19	0.688	0.861	1.066	1.328	1.729	2.093	2.205	2.539	2.861	3.174	3.579	3.883
20	0.687	0.860	1.064	1.325	1.725	2.086	2.197	2.528	2.845	3.153	3.552	3.850
21	0.686	0.859	1.063	1.323	1.721	2.080	2.189	2.518	2.831	3.135	3.527	3.819
22	0.686	0.858	1.061	1.321	1.717	2.074	2.183	2.508	2.819	3.119	3.505	3.792
23	0.685	0.858	1.060	1.319	1.714	2.069	2.177	2.500	2.807	3.104	3.485	3.768
24	0.685	0.857	1.059	1.318	1.711	2.064	2.172	2.492	2.797	3.091	3.467	3.745
25	0.684	0.856	1.058	1.316	1.708	2.060	2.167	2.485	2.787	3.078	3.450	3.725
26	0.684	0.856	1.058	1.315	1.706	2.056	2.162	2.479	2.779	3.067	3.435	3.707
27	0.684	0.855	1.057	1.314	1.703	2.052	2.158	2.473	2.771	3.057	3.421	3.690
28	0.683	0.855	1.056	1.313	1.701	2.048	2.154	2.467	2.763	3.047	3.408	3.674
29	0.683	0.854	1.055	1.311	1.699	2.045	2.150	2.462	2.756	3.038	3.396	3.659
30	0.683	0.854	1.055	1.310	1.697	2.042	2.147	2.457	2.750	3.030	3.385	3.646
40	0.681	0.851	1.050	1.303	1.684	2.021	2.123	2.423	2.704	2.971	3.307	3.551
50	0.679	0.849	1.047	1.299	1.676	2.009	2.109	2.403	2.678	2.937	3.261	3.496
60	0.679	0.848	1.045	1.296	1.671	2.000	2.099	2.390	2.660	2.915	3.232	3.460
70	0.678	0.847	1.044	1.294	1.667	1.994	2.093	2.381	2.648	2.899	3.211	3.435
80	0.678	0.846	1.043	1.292	1.664	1.990	2.088	2.374	2.639	2.887	3.195	3.416
90	0.677	0.846	1.042	1.291	1.662	1.987	2.084	2.368	2.632	2.878	3.183	3.402
100	0.677	0.845	1.042	1.290	1.660	1.984	2.081	2.364	2.626	2.871	3.174	3.390
500	0.675	0.842	1.038	1.283	1.648	1.965	2.059	2.334	2.586	2.820	3.107	3.310
1000	0.675	0.842	1.037	1.282	1.646	1.962	2.056	2.330	2.581	2.813	3.098	3.300
∞	0.674	0.842	1.036	1.282	1.645	1.960	2.054	2.326	2.576	2.807	3.090	3.291

Table A.3 Selected Percentiles of Chi-Squared Distributions

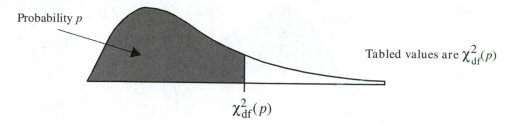

Probability p

Tabled values are $\chi^2_{df}(p)$

$\chi^2_{df}(p)$

d.f.	.75	.80	.85	.90	.95	.975	.98	.99	.995	.9975	.999	.9995
1	1.32	1.64	2.07	2.71	3.84	5.02	5.41	6.63	7.88	9.14	10.83	12.12
2	2.77	3.22	3.79	4.61	5.99	7.38	7.82	9.21	10.60	11.98	13.82	15.20
3	4.11	4.64	5.32	6.25	7.81	9.35	9.84	11.34	12.84	14.32	16.27	17.73
4	5.39	5.99	6.74	7.78	9.49	11.14	11.67	13.28	14.86	16.42	18.47	20.00
5	6.63	7.29	8.12	9.24	11.07	12.83	13.39	15.09	16.75	18.39	20.52	22.11
6	7.84	8.56	9.45	10.64	12.59	14.45	15.03	16.81	18.55	20.25	22.46	24.10
7	9.04	9.80	10.75	12.02	14.07	16.01	16.62	18.48	20.28	22.04	24.32	26.02
8	10.22	11.03	12.03	13.36	15.51	17.53	18.17	20.09	21.95	23.77	26.12	27.87
9	11.39	12.24	13.29	14.68	16.92	19.02	19.68	21.67	23.59	25.46	27.88	29.67
10	12.55	13.44	14.53	15.99	18.31	20.48	21.16	23.21	25.19	27.11	29.59	31.42
11	13.70	14.63	15.77	17.28	19.68	21.92	22.62	24.72	26.76	28.73	31.26	33.14
12	14.85	15.81	16.99	18.55	21.03	23.34	24.05	26.22	28.30	30.32	32.91	34.82
13	15.98	16.98	18.20	19.81	22.36	24.74	25.47	27.69	29.82	31.88	34.53	36.48
14	17.12	18.15	19.41	21.06	23.68	26.12	26.87	29.14	31.32	33.43	36.12	38.11
15	18.25	19.31	20.60	22.31	25.00	27.49	28.26	30.58	32.80	34.95	37.70	39.72
16	19.37	20.47	21.79	23.54	26.30	28.85	29.63	32.00	34.27	36.46	39.25	41.31
17	20.49	21.61	22.98	24.77	27.59	30.19	31.00	33.41	35.72	37.95	40.79	42.88
18	21.60	22.76	24.16	25.99	28.87	31.53	32.35	34.81	37.16	39.42	42.31	44.43
19	22.72	23.90	25.33	27.20	30.14	32.85	33.69	36.19	38.58	40.88	43.82	45.97
20	23.83	25.04	26.50	28.41	31.41	34.17	35.02	37.57	40.00	42.34	45.31	47.50
21	24.93	26.17	27.66	29.62	32.67	35.48	36.34	38.93	41.40	43.78	46.80	49.01
22	26.04	27.30	28.82	30.81	33.92	36.78	37.66	40.29	42.80	45.20	48.27	50.51
23	27.14	28.43	29.98	32.01	35.17	38.08	38.97	41.64	44.18	46.62	49.73	52.00
24	28.24	29.55	31.13	33.20	36.42	39.36	40.27	42.98	45.56	48.03	51.18	53.48
25	29.34	30.68	32.28	34.38	37.65	40.65	41.57	44.31	46.93	49.44	52.62	54.95
26	30.43	31.79	33.43	35.56	38.89	41.92	42.86	45.64	48.29	50.83	54.05	56.41
27	31.53	32.91	34.57	36.74	40.11	43.19	44.14	46.96	49.64	52.22	55.48	57.86
28	32.62	34.03	35.71	37.92	41.34	44.46	45.42	48.28	50.99	53.59	56.89	59.30
29	33.71	35.14	36.85	39.09	42.56	45.72	46.69	49.59	52.34	54.97	58.30	60.73
30	34.80	36.25	37.99	40.26	43.77	46.98	47.96	50.89	53.67	56.33	59.70	62.16
40	45.62	47.27	49.24	51.81	55.76	59.34	60.44	63.69	66.77	69.70	73.40	76.09
50	56.33	58.16	60.35	63.17	67.50	71.42	72.61	76.15	79.49	82.66	86.66	89.56
60	66.98	68.97	71.34	74.40	79.08	83.30	84.58	88.38	91.95	95.34	99.61	102.70
70	77.58	79.71	82.26	85.53	90.53	95.02	96.39	100.40	104.20	107.80	112.30	115.60
80	88.13	90.41	93.11	96.58	101.90	106.60	108.10	112.30	116.30	120.10	124.80	128.30
90	98.65	101.10	103.90	107.60	113.20	118.10	119.70	124.10	128.30	132.30	137.20	140.80
100	109.10	111.70	114.70	118.50	124.30	129.60	131.10	135.80	140.20	144.30	149.50	153.20

Table A.4 Selected Percentiles of F-Distributions

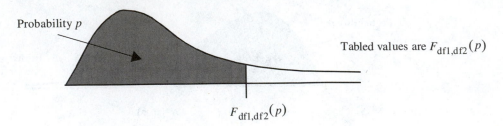

Probability p

Tabled values are $F_{df1,df2}(p)$

$F_{df1,df2}(p)$

		Numerator degrees of freedom (d.f.1)								
d.f.2	p	1	2	3	4	5	6	7	8	9
1	.9	39.86	49.50	53.59	55.83	57.24	58.20	58.91	59.44	59.86
	.95	161.45	199.50	215.71	224.58	230.16	233.99	236.77	238.88	240.54
	.975	647.79	799.50	864.16	899.58	921.85	937.11	948.22	956.66	963.28
	.99	4052.18	4999.50	5403.35	5624.58	5763.65	5858.99	5928.36	5981.07	6022.47
	.999	405284	500000	540379	562500	576405	585937	92873	598144	602284
2	.9	8.53	9.00	9.16	9.24	9.29	9.33	9.35	9.37	9.38
	.95	18.51	19.00	19.16	19.25	19.30	19.33	19.35	19.37	19.38
	.975	38.51	39.00	39.17	39.25	39.30	39.33	39.36	39.37	39.39
	.99	98.50	99.00	99.17	99.25	99.30	99.33	99.36	99.37	99.39
	.999	998.50	999.00	999.17	999.25	999.30	999.33	999.36	999.37	999.39
3	.9	5.54	5.46	5.39	5.34	5.31	5.28	5.27	5.25	5.24
	.95	10.13	9.55	9.28	9.12	9.01	8.94	8.89	8.85	8.81
	.975	17.44	16.04	15.44	15.10	14.88	14.73	14.62	14.54	14.47
	.99	34.12	30.82	29.46	28.71	28.24	27.91	27.67	27.49	27.35
	.999	167.03	148.50	141.11	137.10	134.58	132.85	131.58	130.62	129.86
4	.9	4.54	4.32	4.19	4.11	4.05	4.01	3.98	3.95	3.94
	.95	7.71	6.94	6.59	6.39	6.26	6.16	6.09	6.04	6.00
	.975	12.22	10.65	9.98	9.60	9.36	9.20	9.07	8.98	8.90
	.99	21.20	18.00	16.69	15.98	15.52	15.21	14.98	14.80	14.66
	.999	74.14	61.25	56.18	53.44	51.71	50.53	49.66	49.00	48.47
5	.9	4.06	3.78	3.62	3.52	3.45	3.40	3.37	3.34	3.32
	.95	6.61	5.79	5.41	5.19	5.05	4.95	4.88	4.82	4.77
	.975	10.01	8.43	7.76	7.39	7.15	6.98	6.85	6.76	6.68
	.99	16.26	13.27	12.06	11.39	10.97	10.67	10.46	10.29	10.16
	.999	47.18	37.12	33.20	31.09	29.75	28.83	28.16	27.65	27.24
6	.9	3.78	3.46	3.52	3.18	3.11	3.05	3.01	2.98	2.96
	.95	5.99	5.14	4.76	4.53	4.39	4.28	4.21	4.15	4.10
	.975	8.81	7.26	6.60	6.23	5.99	5.82	5.70	5.60	5.52
	.99	13.75	10.92	9.78	9.15	8.75	8.47	8.26	8.10	7.98
	.999	35.51	27.00	23.70	21.92	20.80	20.03	19.46	19.03	18.69
7	.9	3.59	3.26	3.07	2.96	2.88	2.83	2.78	2.75	2.72
	.95	5.59	4.74	4.35	4.12	3.97	3.87	3.79	3.73	3.68
	.975	8.07	6.54	5.89	5.52	5.29	5.12	4.99	4.90	4.82
	.99	12.25	9.55	8.45	7.85	7.46	7.19	6.99	6.84	6.72
	.999	29.25	21.69	18.77	17.20	16.21	15.52	15.02	14.63	14.33
8	.9	3.46	3.11	2.92	2.81	2.73	2.67	2.62	2.59	2.56
	.95	5.32	4.46	4.07	3.84	3.69	3.58	3.5	3.44	3.39
	.975	7.57	6.06	5.42	5.05	4.82	4.65	4.53	4.43	4.36
	.99	11.26	8.65	7.59	7.01	6.63	6.37	6.18	6.03	5.91
	.999	25.41	18.49	15.83	14.39	13.48	12.86	12.4	12.05	11.77

			Numerator degrees of freedom (d.f.1)						
10	*12*	*15*	*20*	*25*	*30*	*40*	*60*	*120*	*1000*
60.19	60.71	61.22	61.74	62.05	62.26	62.53	62.79	63.06	63.3
241.88	243.91	245.95	248.01	249.26	250.10	251.14	252.2	253.25	254.19
968.63	976.71	984.87	993.10	998.08	1001.41	1005.6	1009.8	1014.02	1017.75
6055.85	6106.32	6157.28	6208.73	6239.83	6260.65	6286.78	6313.03	6339.39	6362.68
605621	610668	615764	620908	624017	626099	628712	631337	633972	636301
9.39	9.41	9.42	9.44	9.45	9.46	9.47	9.47	9.48	9.49
19.40	19.41	19.43	19.45	19.46	19.46	19.47	19.48	19.49	19.49
39.40	39.41	39.43	39.45	39.46	39.46	39.47	39.48	39.49	39.50
99.40	99.42	99.43	99.45	99.46	99.47	99.47	99.48	99.49	99.50
999.40	999.42	999.43	999.45	999.46	999.47	999.47	999.48	999.49	999.50
5.23	5.22	5.20	5.18	5.17	5.17	5.16	5.15	5.14	5.13
8.79	8.74	8.70	8.66	8.63	8.62	8.59	8.57	8.55	8.53
14.42	14.34	14.25	14.17	14.12	14.08	14.04	13.99	13.95	13.91
27.23	27.05	26.87	26.69	26.58	26.50	26.41	26.32	26.22	26.14
129.25	128.32	127.37	126.42	125.84	125.45	124.96	124.47	123.97	123.53
3.92	3.90	3.87	3.84	3.83	3.82	3.80	3.79	3.78	3.76
5.96	5.91	5.86	5.80	5.77	5.75	5.72	5.69	5.66	5.63
8.84	8.75	8.66	8.56	8.50	8.46	8.41	8.36	8.31	8.26
14.55	14.37	14.20	14.02	13.91	13.84	13.75	13.65	13.56	13.47
48.05	47.41	46.76	46.10	45.70	45.43	45.09	44.75	44.40	44.14
3.30	3.27	3.24	3.21	3.19	3.17	3.16	3.14	3.12	3.11
4.74	4.68	4.62	4.56	4.52	4.50	4.46	4.43	4.40	4.37
6.62	6.52	6.43	6.33	6.27	6.23	6.18	6.12	6.07	6.02
10.05	9.89	9.72	9.55	9.45	9.38	9.29	9.20	9.11	9.03
26.92	26.42	25.91	25.39	25.08	24.87	24.60	24.33	24.06	23.86
2.94	2.90	2.87	2.84	2.81	2.80	2.78	2.76	2.74	2.72
4.06	4.00	3.94	3.87	3.83	3.81	3.77	3.74	3.70	3.67
5.46	5.37	5.27	5.17	5.11	5.07	5.01	4.96	4.90	4.86
7.87	7.72	7.56	7.40	7.30	7.23	7.14	7.06	6.97	6.89
18.41	17.99	17.56	17.12	16.85	16.67	16.44	16.21	15.98	15.78
2.70	2.67	2.63	2.59	2.57	2.56	2.54	2.51	2.49	2.47
3.64	3.57	3.51	3.44	3.40	3.38	3.34	3.3	3.27	3.23
4.76	4.67	4.57	4.47	4.40	4.36	4.31	4.25	4.20	4.15
6.62	6.47	6.31	6.16	6.06	5.99	5.91	5.82	5.74	5.66
14.08	13.71	13.32	12.93	12.69	12.53	12.33	12.12	11.91	11.72
2.54	2.5	2.46	2.42	2.40	2.38	2.36	2.34	2.32	3.00
3.35	3.28	3.22	3.15	3.11	3.08	3.04	3.01	0.97	2.93
4.30	4.20	4.10	4.00	3.94	3.89	3.84	0.78	3.73	3.68
5.81	5.67	5.52	5.36	5.26	5.20	3.12	5.03	4.95	4.87
11.54	11.19	10.84	10.48	10.26	10.11	9.92	9.73	9.53	9.36

		Numerator degrees of freedom (d.f.1)								
d.f.2	p	1	2	3	4	5	6	7	8	9
9	.9	3.36	3.01	2.81	2.69	2.61	2.55	2.51	2.47	2.44
	.95	5.12	4.26	3.86	3.63	3.48	3.37	3.29	3.23	3.18
	.975	7.21	5.71	5.08	4.72	4.48	4.32	4.20	4.10	4.03
	.99	10.56	8.02	6.99	6.42	6.06	5.80	5.61	5.47	5.35
	.999	22.86	16.39	13.90	12.56	11.71	11.13	10.70	10.37	10.11
10	.9	3.29	2.92	2.73	2.61	2.52	2.46	2.41	2.38	2.35
	.95	4.96	4.10	3.71	3.48	3.33	3.22	3.14	3.07	3.02
	.975	6.94	5.46	4.83	4.47	4.24	4.07	3.95	3.85	3.78
	.99	10.04	7.56	6.55	5.99	5.64	5.39	5.20	5.06	4.94
	.999	21.04	14.91	12.55	11.28	10.48	9.93	9.52	9.20	8.96
11	.9	3.23	2.86	2.66	2.54	2.45	2.39	2.34	2.30	2.27
	.95	4.84	3.98	3.59	3.36	3.20	3.09	3.01	2.95	2.90
	.975	6.72	5.26	4.63	4.28	4.04	3.88	3.76	3.66	3.59
	.99	9.65	7.21	6.22	5.67	5.32	5.07	4.89	4.74	4.63
	.999	19.69	13.81	11.56	10.35	9.58	9.05	8.66	8.35	8.12
12	.9	3.18	2.81	2.61	2.48	2.39	2.33	2.28	2.24	2.21
	.95	4.75	3.89	3.49	3.26	3.11	3.00	2.91	2.85	2.80
	.975	6.55	5.10	4.47	4.12	3.89	3.73	3.61	3.51	3.44
	.99	9.33	6.93	5.95	5.41	5.06	4.82	4.64	4.5 0	4.39
	.999	18.64	12.97	10.80	9.63	8.89	8.38	8.00	7.71	7.48
13	.9	3.14	2.76	2.56	2.43	2.35	2.28	2.23	2.20	2.16
	.95	4.67	3.81	3.41	3.18	3.03	2.92	2.83	2.77	2.71
	.975	6.41	4.97	4.35	4.00	3.77	3.60	3.48	3.39	3.31
	.99	9.07	6.70	5.74	5.21	4.86	4.62	4.44	4.30	4.19
	.999	17.82	12.31	10.21	9.07	8.35	7.86	7.49	7.21	6.98
14	.9	3.10	2.73	2.52	2.39	2.31	2.24	2.19	2.15	2.12
	.95	4.60	3.74	3.34	3.11	2.96	2.85	2.76	2.70	2.65
	.975	6.30	4.86	4.24	3.89	3.66	3.50	3.38	3.29	3.21
	.99	8.86	6.51	5.56	5.04	4.69	4.46	4.28	4.14	4.03
	.999	17.14	11.78	9.73	8.62	7.92	7.44	7.08	6.80	6.58
15	.9	3.07	2.70	2.49	2.36	2.27	2.21	2.16	2.12	2.09
	.95	4.54	3.68	3.29	3.06	2.90	2.79	2.71	2.64	2.59
	.975	6.20	4.77	4.15	3.80	3.58	3.41	3.29	3.20	3.12
	.99	8.68	6.36	5.42	4.89	4.56	4.32	4.14	4.00	3.89
	.999	16.59	11.34	9.34	8.25	7.57	7.09	6.74	6.47	6.26
16	.9	3.05	2.67	2.46	2.33	2.24	2.18	2.13	2.09	2.06
	.95	4.49	3.63	3.24	3.01	2.85	2.74	2.66	2.59	2.54
	.975	6.12	4.69	4.08	3.73	3.50	3.34	3.22	3.12	3.05
	.99	8.53	6.23	5.29	4.77	4.44	4.20	4.03	3.89	3.78
	.999	16.12	10.97	9.01	7.94	7.27	6.80	6.46	6.19	5.98
17	.9	3.03	2.64	2.44	2.31	2.22	2.15	2.10	2.06	2.03
	.95	4.45	3.59	3.20	2.96	2.81	2.70	2.61	2.55	2.49
	.975	6.04	4.62	4.01	3.66	3.44	3.28	3.16	3.06	2.98
	.99	8.4	6.11	5.18	4.67	4.34	4.10	3.93	3.79	3.68
	.999	15.72	10.66	8.73	7.68	7.02	6.56	6.22	5.96	5.75
18	.9	3.01	2.62	2.42	2.29	2.20	2.13	2.08	2.04	2.00
	.95	4.41	3.55	3.16	2.93	2.77	2.66	2.58	2.51	2.46
	.975	5.98	4.56	3.95	3.61	3.38	3.22	3.10	3.01	2.93
	.99	8.29	6.01	5.09	4.58	4.25	4.01	3.84	3.71	3.60
	.999	15.38	10.39	8.49	7.46	6.81	6.35	6.02	5.76	5.56
19	.9	2.99	2.61	2.40	2.27	2.18	2.11	2.06	2.02	1.98
	.95	4.38	3.52	3.13	2.90	2.74	2.63	2.54	2.48	2.42
	.975	5.92	4.51	3.90	3.56	3.33	3.17	3.05	2.96	2.88
	.99	8.18	5.93	5.01	4.50	4.17	3.94	3.77	3.63	3.52
	.999	15.08	10.16	8.28	7.27	6.62	6.18	5.85	5.59	5.39

			Numerator degrees of freedom (d.f.1)						
10	*12*	*15*	*20*	*25*	*30*	*40*	*60*	*120*	*1000*
2.42	2.38	2.34	2.30	2.27	2.25	2.23	2.21	2.18	2.16
3.14	3.07	3.01	2.94	2.89	2.86	2.83	2.79	2.75	2.71
3.96	3.87	3.77	3.67	3.60	3.56	3.51	3.45	3.39	3.34
5.26	5.11	4.96	4.81	4.71	4.65	4.57	4.48	4.40	4.32
9.89	9.57	9.24	8.90	8.69	8.55	8.37	8.19	8.00	7.84
2.32	2.28	2.24	2.20	2.17	2.16	2.13	2.11	2.08	2.06
2.98	2.91	2.85	2.77	2.73	2.70	2.66	2.62	2.58	2.54
3.72	3.62	3.52	3.42	3.35	3.31	3.26	3.20	3.14	3.09
4.85	4.71	4.56	4.41	4.31	4.25	4.17	4.08	4.00	3.92
8.75	8.45	8.13	7.80	7.60	7.47	7.30	7.12	6.94	6.78
2.25	2.21	2.17	2.12	2.10	2.08	2.05	2.03	2.00	1.98
2.85	2.79	2.72	2.65	2.60	2.57	2.53	2.49	2.45	2.41
3.53	3.43	3.33	3.23	3.16	3.12	3.06	3.00	2.94	2.89
4.54	4.40	4.25	4.10	4.01	3.94	3.86	3.78	3.69	3.61
7.92	7.63	7.32	7.01	6.81	6.68	6.52	6.35	6.18	6.02
2.19	2.15	2.10	2.06	2.03	2.01	1.99	1.96	1.93	1.91
2.75	2.69	2.62	2.54	2.50	2.47	2.43	2.38	2.34	2.30
3.37	3.28	3.18	3.07	3.01	2.96	2.91	2.85	2.79	2.73
4.30	4.16	4.01	3.86	3.76	3.70	3.62	3.54	3.45	3.37
7.29	7.00	6.71	6.40	6.22	6.09	5.93	5.76	5.59	5.44
2.14	2.10	2.05	2.01	1.98	1.96	1.93	1.90	1.88	1.85
2.67	2.6	2.53	2.46	2.41	2.38	2.34	2.30	2.25	2.21
3.25	3.15	3.05	2.95	2.88	2.84	2.78	2.72	2.66	2.60
4.10	3.96	3.82	3.66	3.57	3.51	3.43	3.34	3.25	3.18
6.80	6.52	6.23	5.93	5.75	5.63	5.47	5.30	5.14	4.99
2.10	2.05	2.01	1.96	1.93	1.91	1.89	1.86	1.83	1.80
2.60	2.53	2.46	2.39	2.34	2.31	2.27	2.22	2.18	2.14
3.15	3.05	2.95	2.84	2.78	2.73	2.67	2.61	2.55	2.50
3.94	3.80	3.66	3.51	3.41	3.35	3.27	3.18	3.09	3.02
6.40	6.13	5.85	5.56	5.38	5.25	5.10	4.94	4.77	4.62
2.06	2.02	1.97	1.92	1.89	1.87	1.85	1.82	1.79	1.76
2.54	2.48	2.40	2.33	2.28	2.25	2.20	2.16	2.11	2.07
3.06	2.96	2.86	2.76	2.69	2.64	2.59	2.52	2.46	2.40
3.80	3.67	3.52	3.37	3.28	3.21	3.13	3.05	2.96	2.88
6.08	5.81	5.54	5.25	5.07	4.95	4.80	4.64	4.47	4.33
2.03	1.99	1.94	1.89	1.86	1.84	1.81	1.78	1.75	1.72
2.49	2.42	2.35	2.28	2.23	2.19	2.15	2.11	2.06	2.02
2.99	2.89	2.79	2.68	2.61	2.57	2.51	2.45	2.38	2.32
3.69	3.55	3.41	3.26	3.16	3.10	3.02	2.93	2.84	2.76
5.81	5.55	5.27	4.99	4.82	4.70	4.54	4.39	4.23	4.08
2.00	1.96	1.91	1.86	1.83	1.81	1.78	1.75	1.72	1.69
2.45	2.38	2.31	2.23	2.18	2.15	2.10	2.06	2.01	1.97
2.92	2.82	2.72	2.62	2.55	2.50	2.44	2.38	2.32	2.26
3.59	3.46	3.31	3.16	3.07	3.00	2.92	2.83	2.75	2.66
5.58	5.32	5.05	4.78	4.60	4.48	4.33	4.18	4.02	3.87
1.98	1.93	1.89	1.84	1.80	1.78	1.75	1.72	1.69	1.66
2.41	2.34	2.27	2.19	2.14	2.11	2.06	2.02	1.97	1.92
2.87	2.77	2.67	2.56	2.49	2.44	2.38	2.32	2.26	2.20
3.51	3.37	3.23	3.08	2.98	2.92	2.84	2.75	2.66	2.58
5.39	5.13	4.87	4.59	4.42	4.30	4.15	4.00	3.84	3.69
1.96	1.91	1.86	1.81	1.78	1.76	1.73	1.70	1.67	1.64
2.38	2.31	2.23	2.16	2.11	2.07	2.03	1.98	1.93	1.88
2.82	2.72	2.62	2.51	2.44	2.39	2.33	2.27	2.20	2.14
3.43	3.30	3.15	3.00	2.91	2.84	2.76	2.67	2.58	2.50
5.22	4.97	4.70	4.43	4.26	4.14	3.99	3.84	3.68	3.53

d.f.2	p	Numerator degrees of freedom (d.f.1)								
		1	2	3	4	5	6	7	8	9
20	.9	2.97	2.59	2.38	2.25	2.16	2.09	2.04	2.00	1.96
	.95	4.35	3.49	3.10	2.87	2.71	2.60	2.51	2.45	2.39
	.975	5.87	4.46	3.86	3.51	3.29	3.13	3.01	2.91	2.84
	.99	8.10	5.85	4.94	4.43	4.10	3.87	3.70	3.56	3.46
	.999	14.82	9.95	8.10	7.10	6.46	6.02	5.69	5.44	5.24
21	.9	2.96	2.57	2.36	2.23	2.14	2.08	2.02	1.98	1.95
	.95	4.32	3.47	3.07	2.84	2.68	2.57	2.49	2.42	2.37
	.975	5.83	4.42	3.82	3.48	3.25	3.09	2.97	2.87	2.80
	.99	8.02	5.78	4.87	4.37	4.04	3.81	3.64	3.51	3.40
	.999	14.59	9.77	7.94	6.95	6.32	5.88	5.56	5.31	5.11
22	.9	2.95	2.56	2.35	2.22	2.13	2.06	2.01	1.97	1.93
	.95	4.30	3.44	3.05	2.82	2.66	2.55	2.46	2.40	2.34
	.975	5.79	4.38	3.78	3.44	3.22	3.05	2.93	2.84	2.76
	.99	7.95	5.72	4.82	4.31	3.99	3.76	3.59	3.45	3.35
	.999	14.38	9.61	7.8	6.81	6.19	5.76	5.44	5.19	4.99
23	.9	2.94	2.55	2.34	2.21	2.11	2.05	1.99	1.95	1.92
	.95	4.28	3.42	3.03	2.80	2.64	2.53	2.44	2.37	2.32
	.975	5.75	4.35	3.75	3.41	3.18	3.02	2.90	2.81	2.73
	.99	7.88	5.66	4.76	4.26	3.94	3.71	3.54	3.41	3.30
	.999	14.20	9.47	7.67	6.70	6.08	5.65	5.33	5.09	4.89
24	.9	2.93	2.54	2.33	2.19	2.10	2.04	1.98	1.94	1.91
	.95	4.26	3.40	3.01	2.78	2.62	2.51	2.42	2.36	2.30
	.975	5.72	4.32	3.72	3.38	3.15	2.99	2.87	2.78	2.70
	.99	7.82	5.61	4.72	4.22	3.9	3.67	3.50	3.36	3.26
	.999	14.03	9.34	7.55	6.59	5.98	5.55	5.23	4.99	4.80
25	.9	2.92	2.53	2.32	2.18	2.09	2.02	1.97	1.93	1.89
	.95	4.24	3.39	2.99	2.76	2.60	2.49	2.40	2.34	2.28
	.975	5.69	4.29	3.69	3.35	3.13	2.97	2.85	2.75	2.68
	.99	7.77	5.57	4.68	4.18	3.85	3.63	3.46	3.32	3.22
	.999	13.88	9.22	7.45	6.49	5.89	5.46	5.15	4.91	4.71
26	.9	2.91	2.52	2.31	2.17	2.08	2.01	1.96	1.92	1.88
	.95	4.23	3.37	2.98	2.74	2.59	2.47	2.39	2.32	2.27
	.975	5.66	4.27	3.67	3.33	3.10	2.94	2.82	2.73	2.65
	.99	7.72	5.53	4.64	4.14	3.82	3.59	3.42	3.29	3.18
	.999	13.74	9.12	7.36	6.41	5.80	5.38	5.07	4.83	4.64
27	.9	2.90	2.51	2.30	2.17	2.07	2.00	1.95	1.91	1.87
	.95	4.21	3.35	2.96	2.73	2.57	2.46	2.37	2.31	2.25
	.975	5.63	4.24	3.65	3.31	3.08	2.92	2.80	2.71	2.63
	.99	7.68	5.49	4.60	4.11	3.78	3.56	3.39	3.26	3.15
	.999	13.61	9.02	7.27	6.33	5.73	5.31	5.00	4.76	4.57
28	.9	2.89	2.5	2.29	2.16	2.06	2.00	1.94	1.90	1.87
	.95	4.20	3.34	2.95	2.71	2.56	2.45	2.36	2.29	2.24
	.975	5.61	4.22	3.63	3.29	3.06	2.9	2.78	2.69	2.61
	.99	7.64	5.45	4.57	4.07	3.75	3.53	3.36	3.23	3.12
	.999	13.50	8.93	7.19	6.25	5.66	5.24	4.93	4.69	4.50
29	.9	2.89	2.50	2.28	2.15	2.06	1.99	1.93	1.89	1.86
	.95	4.18	3.33	2.93	2.70	2.55	2.43	2.35	2.28	2.22
	.975	5.59	4.2	3.61	3.27	3.04	2.88	2.76	2.67	2.59
	.99	7.60	5.42	4.54	4.04	3.73	3.50	3.33	3.20	3.09
	.999	13.39	8.85	7.12	6.19	5.59	5.18	4.87	4.64	4.45
30	.9	2.88	2.49	2.28	2.14	2.05	1.98	1.93	1.88	1.85
	.95	4.17	3.32	2.92	2.69	2.53	2.42	2.33	2.27	2.21
	.975	5.57	4.18	3.59	3.25	3.03	2.87	2.75	2.65	2.57
	.99	7.56	5.39	4.51	4.02	3.70	3.47	3.30	3.17	3.07
	.999	13.29	8.77	7.05	6.12	5.53	5.12	4.82	4.58	4.39

	Numerator degrees of freedom (d.f.1)								
10	12	15	20	25	30	40	60	120	1000
1.94	1.89	1.84	1.79	1.76	1.74	1.71	1.68	1.64	1.61
2.35	2.28	2.20	2.12	2.07	2.04	1.99	1.95	1.90	1.85
2.77	2.68	2.57	2.46	2.40	2.35	2.29	2.22	2.16	2.09
3.37	3.23	3.09	2.94	2.84	2.78	2.69	2.61	2.52	2.43
5.08	4.82	4.56	4.29	4.12	4.00	3.86	3.70	3.54	3.40
1.92	1.87	1.83	1.78	1.74	1.72	1.69	1.66	1.62	1.59
2.32	2.25	2.18	2.10	2.05	2.01	1.96	1.92	1.87	1.82
2.73	2.64	2.53	2.42	2.36	2.31	2.25	2.18	2.11	2.05
3.31	3.17	3.03	2.88	2.79	2.72	2.64	2.55	2.46	2.37
4.95	4.70	4.44	4.17	4.00	3.88	3.74	3.58	3.42	3.28
1.90	1.86	1.81	1.76	1.73	1.70	1.67	1.64	1.60	1.57
2.30	2.23	2.15	2.07	2.02	1.98	1.94	1.89	1.84	1.79
2.70	2.60	2.50	2.39	2.32	2.27	2.21	2.14	2.08	2.01
3.26	3.12	2.98	2.83	2.73	2.67	2.58	2.50	2.40	2.32
4.83	4.58	4.33	4.06	3.89	3.78	3.63	3.48	3.32	3.17
1.89	1.84	1.80	1.74	1.71	1.69	1.66	1.62	1.59	1.55
2.27	2.20	2.13	2.05	2.00	1.96	1.91	1.86	1.81	1.76
2.67	2.57	2.47	2.36	2.29	2.24	2.18	2.11	2.04	1.98
3.21	3.07	2.93	2.78	2.69	2.62	2.54	2.45	2.35	2.27
4.73	4.48	4.23	3.96	3.79	3.68	3.53	3.38	3.22	3.08
1.88	1.83	1.78	1.73	1.70	1.67	1.64	1.61	1.57	1.54
2.25	2.18	2.11	2.03	1.97	1.94	1.89	1.84	1.79	1.74
2.64	2.54	2.44	2.33	2.26	2.21	2.15	2.08	2.01	1.94
3.17	3.03	2.89	2.74	2.64	2.58	2.49	2.40	2.31	2.22
4.64	4.39	4.14	3.87	3.71	3.59	3.45	3.29	3.14	2.99
1.87	1.82	1.77	1.72	1.68	1.66	1.63	1.59	1.56	1.52
2.24	2.16	2.09	2.01	1.96	1.92	1.87	1.82	1.77	1.72
2.61	2.51	2.41	2.30	2.23	2.18	2.12	2.05	1.98	1.91
3.13	2.99	2.85	2.70	2.60	2.54	2.45	2.36	2.27	2.18
4.56	4.31	4.06	3.79	3.63	3.52	3.37	3.22	3.06	2.91
1.86	1.81	1.76	1.71	1.67	1.65	1.61	1.58	1.54	1.51
2.22	2.15	2.07	1.99	1.94	1.90	1.85	1.80	1.75	1.70
2.59	2.49	2.39	2.28	2.21	2.16	2.09	2.03	1.95	1.89
3.09	2.96	2.81	2.66	2.57	2.50	2.42	2.33	2.23	2.14
4.48	4.24	3.99	3.72	3.56	3.44	3.30	3.15	2.99	2.84
1.85	1.80	1.75	1.70	1.66	1.64	1.60	1.57	1.53	1.50
2.20	2.13	2.06	1.97	1.92	1.88	1.84	1.79	1.73	1.68
2.57	2.47	2.36	2.25	2.18	2.13	2.07	2.00	1.93	1.86
3.06	2.93	2.78	2.63	2.54	2.47	2.38	2.29	2.20	2.11
4.41	4.17	3.92	3.66	3.49	3.38	3.23	3.08	2.92	2.78
1.84	1.79	1.74	1.69	1.65	1.63	1.59	1.56	1.52	1.48
2.19	2.12	2.04	1.96	1.91	1.87	1.82	1.77	1.71	1.66
2.55	2.45	2.34	2.23	2.16	2.11	2.05	1.98	1.91	1.84
3.03	2.90	2.75	2.60	2.51	2.44	2.35	2.26	2.17	2.08
4.35	4.11	3.86	3.60	3.43	3.32	3.18	3.02	2.86	2.72
1.83	1.78	1.73	1.68	1.64	1.62	1.58	1.55	1.51	1.47
2.18	2.10	2.03	1.94	1.89	1.85	1.81	1.75	1.70	1.65
2.53	2.43	2.32	2.21	2.14	2.09	2.03	1.96	1.89	1.82
3.00	2.87	2.73	2.57	2.48	2.41	2.33	2.23	2.14	2.05
4.29	4.05	3.80	3.54	3.38	3.27	3.12	2.97	2.81	2.66
1.82	1.77	1.72	1.67	1.63	1.61	1.57	1.54	1.50	1.46
2.16	2.09	2.01	1.93	1.88	1.84	1.79	1.74	1.68	1.63
2.51	2.41	2.31	2.20	2.13	2.07	2.01	1.94	1.87	1.80
2.98	2.84	2.70	2.55	2.45	2.39	2.30	2.21	2.11	2.02
4.24	4.00	3.75	3.49	3.33	3.22	3.07	2.92	2.76	2.61

Numerator degrees of freedom (d.f.1)

d.f.2	p	1	2	3	4	5	6	7	8	9
40	.9	2.84	2.44	2.23	2.09	2.00	1.93	1.87	1.83	1.79
	.95	4.08	3.23	2.84	2.61	2.45	2.34	2.25	2.18	2.12
	.975	5.42	4.05	3.46	3.13	2.90	2.74	2.62	2.53	2.45
	.99	7.31	5.18	4.31	3.83	3.51	3.29	3.12	2.99	2.89
	.999	12.61	8.25	6.59	5.70	5.13	4.73	4.44	4.21	4.02
50	.9	2.81	2.41	2.20	2.06	1.97	1.90	1.84	1.80	1.76
	.95	4.03	3.18	2.79	2.56	2.40	2.29	2.20	2.13	2.07
	.975	5.34	3.97	3.39	3.05	2.83	2.67	2.55	2.46	2.38
	.99	7.17	5.06	4.20	3.72	3.41	3.19	3.02	2.89	2.78
	.999	12.22	7.96	6.34	5.46	4.90	4.51	4.22	4.00	3.82
60	.9	2.79	2.39	2.18	2.04	1.95	1.87	1.82	1.77	1.74
	.95	4.00	3.15	2.76	2.53	2.37	2.25	2.17	2.10	2.04
	.975	5.29	3.93	3.34	3.01	2.79	2.63	2.51	2.41	2.33
	.99	7.08	4.98	4.13	3.65	3.34	3.12	2.95	2.82	2.72
	.999	11.97	7.77	6.17	5.31	4.76	4.37	4.09	3.86	3.69
100	.9	2.76	2.36	2.14	2.00	1.91	1.83	1.78	1.73	1.69
	.95	3.94	3.09	2.70	2.46	2.31	2.19	2.10	2.03	1.97
	.975	5.18	3.83	3.25	2.92	2.70	2.54	2.42	2.32	2.24
	.99	6.90	4.82	3.98	3.51	3.21	2.99	2.82	2.69	2.59
	.999	11.50	7.41	5.86	5.02	4.48	4.11	3.83	3.61	3.44
200	.9	2.73	2.33	2.11	1.97	1.88	1.80	1.75	1.70	1.66
	.95	3.89	3.04	2.65	2.42	2.26	2.14	2.06	1.98	1.93
	.975	5.10	3.76	3.18	2.85	2.63	2.47	2.35	2.26	2.18
	.99	6.76	4.71	3.88	3.41	3.11	2.89	2.73	2.60	2.50
	.999	11.15	7.15	5.63	4.81	4.29	3.92	3.65	3.43	3.26
1000	.9	2.71	2.31	2.09	1.95	1.85	1.78	1.72	1.68	1.64
	.95	3.85	3.00	2.61	2.38	2.22	2.11	2.02	1.95	1.89
	.975	5.04	3.70	3.13	2.80	2.58	2.42	2.30	2.20	2.13
	.99	6.66	4.63	3.80	3.34	3.04	2.82	2.66	2.53	2.43
	.999	10.89	6.96	5.46	4.65	4.14	3.78	3.51	3.30	3.13

			Numerator degrees of freedom (d.f.1)						
10	12	15	20	25	30	40	60	120	1000
1.76	1.71	1.66	1.61	1.57	1.54	1.51	1.47	1.42	1.38
2.08	2.00	1.92	1.84	1.78	1.74	1.69	1.64	1.58	1.52
2.39	2.29	2.18	2.07	1.99	1.94	1.88	1.80	1.72	1.65
2.80	2.66	2.52	2.37	2.27	2.20	2.11	2.02	1.92	1.82
3.87	3.64	3.40	3.14	2.98	2.87	2.73	2.57	2.41	2.25
1.73	1.68	1.63	1.57	1.53	1.50	1.46	1.42	1.38	1.33
2.03	1.95	1.87	1.78	1.73	1.69	1.63	1.58	1.51	1.45
2.32	2.22	2.11	1.99	1.92	1.87	1.80	1.72	1.64	1.56
2.70	2.56	2.42	2.27	2.17	2.10	2.01	1.91	1.80	1.70
3.67	3.44	3.20	2.95	2.79	2.68	2.53	2.38	2.21	2.05
1.71	1.66	1.60	1.54	1.50	1.48	1.44	1.40	1.35	1.30
1.99	1.92	1.84	1.75	1.69	1.65	1.59	1.53	1.47	1.40
2.27	2.17	2.06	1.94	1.87	1.82	1.74	1.67	1.58	1.49
2.63	2.50	2.35	2.20	2.10	2.03	1.94	1.84	1.73	1.62
3.54	3.32	3.08	2.83	2.67	2.55	2.41	2.25	2.08	1.92
1.66	1.61	1.56	1.49	1.45	1.42	1.38	1.34	1.28	1.22
1.93	1.85	1.77	1.68	1.62	1.57	1.52	1.45	1.38	1.30
2.18	2.08	1.97	1.85	1.77	1.71	1.64	1.56	1.46	1.36
2.50	2.37	2.22	2.07	1.97	1.89	1.80	1.69	1.57	1.45
3.30	3.07	2.84	2.59	2.43	2.32	2.17	2.01	1.83	1.64
1.63	1.58	1.52	1.46	1.41	1.38	1.34	1.29	1.23	1.16
1.88	1.80	1.72	1.62	1.56	1.52	1.46	1.39	1.30	1.21
2.11	2.01	1.90	1.78	1.70	1.64	1.56	1.47	1.37	1.25
2.41	2.27	2.13	1.97	1.87	1.79	1.69	1.58	1.45	1.30
3.12	2.90	2.67	2.42	2.26	2.15	2.00	1.83	1.64	1.43
1.61	1.55	1.49	1.43	1.38	1.35	1.30	1.25	1.18	1.08
1.84	1.76	1.68	1.58	1.52	1.47	1.41	1.33	1.24	1.11
2.06	1.96	1.85	1.72	1.64	1.58	1.50	1.41	1.29	1.13
2.34	2.20	2.06	1.90	1.79	1.72	1.61	1.50	1.35	1.16
2.99	2.77	2.54	2.30	2.14	2.02	1.87	1.69	1.49	1.22

Table A.5 Selected Percentiles of Studentized Range Distributions

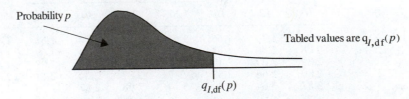

Probability p

Tabled values are $q_{I,\mathrm{df}}(p)$

$q_{I,\mathrm{df}}(p)$

					Number of groups, I				
d.f.	p	2	3	4	5	6	7	8	9	10
1	.9	8.93	13.44	16.36	18.49	20.15	21.51	22.64	23.62	24.48
	.95	17.97	26.98	32.82	37.08	40.41	43.12	45.40	47.36	49.07
	.99	90.03	135.0	164.3	185.6	202.2	215.8	227.2	237.0	245.6
2	.9	4.13	5.73	6.77	7.54	8.14	8.63	9.05	9.41	9.72
	.95	6.08	8.33	9.80	10.88	11.74	12.44	13.03	13.54	13.99
	.99	14.04	19.02	22.29	24.72	26.63	28.20	29.53	30.68	31.69
3	.9	3.33	4.47	5.20	5.74	6.16	6.51	6.81	7.06	7.29
	.95	4.50	5.91	6.82	7.50	8.04	8.48	8.85	9.18	9.46
	.99	8.26	10.62	12.17	13.33	14.24	15.00	15.64	16.20	16.69
4	.9	3.01	3.98	4.59	5.03	5.39	5.68	5.93	6.14	6.33
	.95	3.93	5.04	5.76	6.29	6.71	7.05	7.35	7.60	7.83
	.99	6.51	8.12	9.17	9.96	10.58	11.10	11.55	11.93	12.27
5	.9	2.85	3.72	4.26	4.66	4.98	5.24	5.46	5.65	5.82
	.95	3.64	4.60	5.22	5.67	6.03	6.33	6.58	6.80	6.99
	.99	5.70	6.98	7.80	8.42	8.91	9.32	9.67	9.97	10.24
6	.9	2.75	3.56	4.07	4.44	4.73	4.97	5.17	5.34	5.50
	.95	3.46	4.34	4.90	5.30	5.63	5.90	6.12	6.32	6.49
	.99	5.24	6.33	7.03	7.56	7.97	8.32	8.61	8.87	9.10
7	.9	2.68	3.45	3.93	4.28	4.55	4.78	4.97	5.14	5.28
	.95	3.34	4.16	4.68	5.06	5.36	5.61	5.82	6.00	6.16
	.99	4.95	5.92	6.54	7.01	7.37	7.68	7.94	8.17	8.37
8	.9	2.63	3.37	3.83	4.17	4.43	4.65	4.83	4.99	5.13
	.95	3.26	4.04	4.53	4.89	5.17	5.40	5.60	5.77	5.92
	.99	4.75	5.64	6.20	6.62	6.96	7.24	7.47	7.68	7.86
9	.9	2.59	3.32	3.76	4.08	4.34	4.54	4.72	4.87	5.01
	.95	3.20	3.95	4.41	4.76	5.02	5.24	5.43	5.59	5.74
	.99	4.60	5.43	5.96	6.35	6.66	6.91	7.13	7.33	7.49
10	.9	2.56	3.27	3.70	4.02	4.26	4.47	4.64	4.78	4.91
	.95	3.15	3.88	4.33	4.65	4.91	5.12	5.30	5.46	5.60
	.99	4.48	5.27	5.77	6.14	6.43	6.67	6.87	7.05	7.21
11	.9	2.54	3.23	3.66	3.96	4.20	4.40	4.57	4.71	4.84
	.95	3.11	3.82	4.26	4.57	4.82	5.03	5.20	5.35	5.49
	.99	4.39	5.15	5.62	5.97	6.25	6.48	6.67	6.84	6.99
12	.9	2.52	3.20	3.62	3.92	4.16	4.35	4.51	4.65	4.78
	.95	3.08	3.77	4.20	4.51	4.75	4.95	5.12	5.27	5.39
	.99	4.32	5.05	5.50	5.84	6.10	6.32	6.51	6.67	6.81
13	.9	2.50	3.18	3.59	3.88	4.12	4.30	4.46	4.60	4.72
	.95	3.06	3.73	4.15	4.45	4.69	4.88	5.05	5.19	5.32
	.99	4.26	4.96	5.40	5.73	5.98	6.19	6.37	6.53	6.67
14	.9	2.49	3.16	3.56	3.85	4.08	4.27	4.42	4.56	4.68
	.95	3.03	3.70	4.11	4.41	4.64	4.83	4.99	5.13	5.25
	.99	4.21	4.89	5.32	5.63	5.88	6.08	6.26	6.41	6.54
15	.9	2.48	3.14	3.54	3.83	4.05	4.23	4.39	4.52	4.64
	.95	3.01	3.67	4.08	4.37	4.59	4.78	4.94	5.08	5.20
	.99	4.17	4.84	5.25	5.56	5.80	5.99	6.16	6.31	6.44

				Number of groups, I					
11	*12*	*13*	*14*	*15*	*16*	*17*	*18*	*19*	*20*
25.24	25.92	26.54	27.10	27.62	28.10	28.54	28.96	29.35	29.71
50.59	51.96	53.20	54.33	55.36	56.32	57.22	58.04	58.83	59.56
253.2	260.0	266.2	271.8	277.0	281.8	286.3	290.4	294.3	298.0
10.01	10.26	10.49	10.70	10.89	11.07	11.24	11.39	11.54	11.68
14.39	14.75	15.08	15.38	15.65	15.91	16.14	16.37	16.57	16.77
32.59	33.40	34.13	34.81	35.43	36.00	36.53	37.03	37.50	37.95
7.49	7.67	7.83	7.98	8.12	8.25	8.37	8.48	8.58	8.68
9.72	9.95	10.15	10.35	10.52	10.69	10.84	10.98	11.11	11.24
17.13	17.53	17.89	18.22	18.52	18.81	19.07	19.32	19.55	19.77
6.49	6.65	6.78	6.91	7.02	7.13	7.23	7.33	7.41	7.50
8.03	8.21	8.37	8.52	8.66	8.79	8.91	9.03	9.13	9.23
12.57	12.84	13.09	13.32	13.53	13.73	13.91	14.08	14.24	14.40
5.97	6.10	6.22	6.34	6.44	6.54	6.63	6.71	6.79	6.86
7.17	7.32	7.47	7.60	7.72	7.83	7.93	8.03	8.12	8.21
10.48	10.70	10.89	11.08	11.24	11.40	11.55	11.68	11.81	11.93
5.64	5.76	5.87	5.98	6.07	6.16	6.25	6.32	6.40	6.47
6.65	6.79	6.92	7.03	7.14	7.24	7.34	7.43	7.51	7.59
9.30	9.48	9.65	9.81	9.95	10.08	10.21	10.32	10.43	10.54
5.41	5.53	5.64	5.74	5.83	5.91	5.99	6.06	6.13	6.19
6.30	6.43	6.55	6.66	6.76	6.85	6.94	7.02	7.10	7.17
8.55	8.71	8.86	9.00	9.12	9.24	9.35	9.46	9.55	9.65
5.25	5.36	5.46	5.56	5.64	5.72	5.80	5.87	5.93	6.00
6.05	6.18	6.29	6.39	6.48	6.57	6.65	6.73	6.80	6.87
8.03	8.18	8.31	8.44	8.55	8.66	8.76	8.85	8.94	9.03
5.13	5.23	5.33	5.42	5.51	5.58	5.66	5.72	5.79	5.85
5.87	5.98	6.09	6.19	6.28	6.36	6.44	6.51	6.58	6.64
7.65	7.78	7.91	8.03	8.13	8.23	8.33	8.41	8.49	8.57
5.03	5.13	5.23	5.32	5.40	5.47	5.54	5.61	5.67	5.73
5.72	5.83	5.93	6.03	6.11	6.19	6.27	6.34	6.40	6.47
7.36	7.49	7.60	7.71	7.81	7.91	7.99	8.08	8.15	8.23
4.95	5.05	5.15	5.23	5.31	5.38	5.45	5.51	5.57	5.63
5.61	5.71	5.81	5.90	5.98	6.06	6.13	6.20	6.27	6.33
7.13	7.25	7.36	7.46	7.56	7.65	7.73	7.81	7.88	7.95
4.89	4.99	5.08	5.16	5.24	5.31	5.37	5.44	5.49	5.55
5.51	5.61	5.71	5.80	5.88	5.95	6.02	6.09	6.15	6.21
6.94	7.06	7.17	7.26	7.36	7.44	7.52	7.59	7.66	7.73
4.83	4.93	5.02	5.10	5.18	5.25	5.31	5.37	5.43	5.48
5.43	5.53	5.63	5.71	5.79	5.86	5.93	5.99	6.05	6.11
6.79	6.90	7.01	7.10	7.19	7.27	7.35	7.42	7.48	7.55
4.79	4.88	4.97	5.05	5.12	5.19	5.26	5.32	5.37	5.43
5.36	5.46	5.55	5.64	5.71	5.79	5.85	5.91	5.97	6.03
6.66	6.77	6.87	6.96	7.05	7.13	7.20	7.27	7.33	7.39
4.75	4.84	4.93	5.01	5.08	5.15	5.21	5.27	5.32	5.38
5.31	5.40	5.49	5.57	5.65	5.72	5.78	5.85	5.90	5.96
6.55	6.66	6.76	6.84	6.93	7.00	7.07	7.14	7.20	7.26

		Number of groups, I								
d.f.	p	2	3	4	5	6	7	8	9	10
16	.9	2.47	3.12	3.52	3.80	4.03	4.21	4.36	4.49	4.61
	.95	3.00	3.65	4.05	4.33	4.56	4.74	4.90	5.03	5.15
	.99	4.13	4.79	5.19	5.49	5.72	5.92	6.08	6.22	6.35
17	.9	2.46	3.11	3.50	3.78	4.00	4.18	4.33	4.46	4.58
	.95	2.98	3.63	4.02	4.30	4.52	4.70	4.86	4.99	5.11
	.99	4.10	4.74	5.14	5.43	5.66	5.85	6.01	6.15	6.27
18	.9	2.45	3.10	3.49	3.77	3.98	4.16	4.31	4.44	4.55
	.95	2.97	3.61	4.00	4.28	4.49	4.67	4.82	4.96	5.07
	.99	4.07	4.70	5.09	5.38	5.60	5.79	5.94	6.08	6.20
19	.9	2.45	3.09	3.47	3.75	3.97	4.14	4.29	4.42	4.53
	.95	2.96	3.59	3.98	4.25	4.47	4.65	4.79	4.92	5.04
	.99	4.05	4.67	5.05	5.33	5.55	5.73	5.89	6.02	6.14
20	.9	2.44	3.08	3.46	3.74	3.95	4.12	4.27	4.40	4.51
	.95	2.95	3.58	3.96	4.23	4.45	4.62	4.77	4.90	5.01
	.99	4.02	4.64	5.02	5.29	5.51	5.69	5.84	5.97	6.09
24	.9	2.42	3.05	3.42	3.69	3.90	4.07	4.21	4.34	4.44
	.95	2.92	3.53	3.90	4.17	4.37	4.54	4.68	4.81	4.92
	.99	3.96	4.55	4.91	5.17	5.37	5.54	5.69	5.81	5.92
30	.9	2.40	3.02	3.39	3.65	3.85	4.02	4.16	4.28	4.38
	.95	2.89	3.49	3.85	4.10	4.30	4.46	4.60	4.72	4.82
	.99	3.89	4.45	4.80	5.05	5.24	5.40	5.54	5.65	5.76
40	.9	2.38	2.99	3.35	3.60	3.80	3.96	4.10	4.21	4.32
	.95	2.86	3.44	3.79	4.04	4.23	4.39	4.52	4.63	4.73
	.99	3.82	4.37	4.70	4.93	5.11	5.26	5.39	5.50	5.60
60	.9	2.36	2.96	3.31	3.56	3.75	3.91	4.04	4.16	4.25
	.95	2.83	3.40	3.74	3.98	4.16	4.31	4.44	4.55	4.65
	.99	3.76	4.28	4.59	4.82	4.99	5.13	5.25	5.36	5.45
120	.9	2.34	2.93	3.28	3.52	3.71	3.86	3.99	4.10	4.19
	.95	2.80	3.36	3.68	3.92	4.10	4.24	4.36	4.47	4.56
	.99	3.70	4.20	4.50	4.71	4.87	5.01	5.12	5.21	5.30
∞	.9	2.33	2.90	3.24	3.48	3.66	3.81	3.93	4.04	4.13
	.95	2.77	3.31	3.63	3.86	4.03	4.17	4.29	4.39	4.47
	.99	3.64	4.12	4.40	4.60	4.76	4.88	4.99	5.08	5.16

Number of groups, I

11	12	13	14	15	16	17	18	19	20
4.71	4.81	4.89	4.97	5.04	5.11	5.17	5.23	5.28	5.33
5.26	5.35	5.44	5.52	5.59	5.66	5.73	5.79	5.84	5.90
6.46	6.56	6.66	6.74	6.82	6.90	6.97	7.03	7.09	7.15
4.68	4.77	4.86	4.93	5.01	5.07	5.13	5.19	5.24	5.30
5.21	5.31	5.39	5.47	5.54	5.61	5.67	5.73	5.79	5.84
6.38	6.48	6.57	6.66	6.73	6.81	6.87	6.94	7.00	7.05
4.65	4.75	4.83	4.90	4.98	5.04	5.10	5.16	5.21	5.26
5.17	5.27	5.35	5.43	5.50	5.57	5.63	5.69	5.74	5.79
6.31	6.41	6.50	6.58	6.65	6.73	6.79	6.85	6.91	6.97
4.63	4.72	4.80	4.88	4.95	5.01	5.07	5.13	5.18	5.23
5.14	5.23	5.31	5.39	5.46	5.53	5.59	5.65	5.70	5.75
6.25	6.34	6.43	6.51	6.58	6.65	6.72	6.78	6.84	6.89
4.61	4.70	4.78	4.85	4.92	4.99	5.05	5.10	5.16	5.20
5.11	5.20	5.28	5.36	5.43	5.49	5.55	5.61	5.66	5.71
6.19	6.28	6.37	6.45	6.52	6.59	6.65	6.71	6.77	6.82
4.54	4.63	4.71	4.78	4.85	4.91	4.97	5.02	5.07	5.12
5.01	5.10	5.18	5.25	5.32	5.38	5.44	5.49	5.55	5.59
6.02	6.11	6.19	6.26	6.33	6.39	6.45	6.51	6.56	6.61
4.47	4.56	4.64	4.71	4.77	4.83	4.89	4.94	4.99	5.03
4.92	5.00	5.08	5.15	5.21	5.27	5.33	5.38	5.43	5.47
5.85	5.93	6.01	6.08	6.14	6.20	6.26	6.31	6.36	6.41
4.41	4.49	4.56	4.63	4.69	4.75	4.81	4.86	4.90	4.95
4.82	4.90	4.98	5.04	5.11	5.16	5.22	5.27	5.31	5.36
5.69	5.76	5.83	5.90	5.96	6.02	6.07	6.12	6.16	6.21
4.34	4.42	4.49	4.56	4.62	4.67	4.73	4.78	4.82	4.86
4.73	4.81	4.88	4.94	5.00	5.06	5.11	5.15	5.20	5.24
5.53	5.60	5.67	5.73	5.78	5.84	5.89	5.93	5.97	6.01
4.28	4.35	4.42	4.48	4.54	4.60	4.65	4.69	4.74	4.78
4.64	4.71	4.78	4.84	4.90	4.95	5.00	5.04	5.09	5.13
5.37	5.44	5.50	5.56	5.61	5.66	5.71	5.75	5.79	5.83
4.21	4.28	4.35	4.41	4.47	4.52	4.57	4.61	4.65	4.69
4.55	4.62	4.68	4.74	4.80	4.85	4.89	4.93	4.97	5.01
5.23	5.29	5.35	5.40	5.45	5.49	5.54	5.57	5.61	5.65

References

Aitkin, M., Anderson, D., Francis, B., and Hinde, J. 1989. *Statistical Modelling in GLIM*. Oxford: Clarendon Press.

Agresti, A. 1990. *Categorical Data Analysis*. New York: John Wiley & Sons.

Bowerman, B. L., and O'Connell, R. T. 1993. *Forecasting and Time Series: An Applied Approach*, 3d ed. Belmont, Calif.: Duxbury Press.

Box, G. E. P., Hunter, W. G., and Hunter, J. S. 1978. *Statistics for Experiments: An Introduction to Design, Data Analysis, and Model Building*. New York: John Wiley & Sons.

Breslow, N. E., and Day, N. E. 1980. *Statistical Methods in Cancer Research, Vol. 1: The Analysis of Case-Control Studies*. Lyon: World Health Organization.

Chatfield, C. 1984. *Analysis of Time Series: An Introduction*, 3d ed. London: Chapman & Hall.

Cleveland, W. S. 1993. *Visualizing Data*. Murray Hill, N.J.: AT&T Bell Laboratories.

Daniel, C., and Wood, F. S. 1990. *Fitting Equations to Data: Computer Analysis of Multifactor Data*, 2d ed. New York: John Wiley & Sons.

Devore, J. L., and Peck, R. 1993. *Statistics: The Exploration and Analysis of Data*, 2d ed. Belmont, Calif.: Duxbury Press.

Dobson, A. J. 1983. *An Introduction to Statistical Modelling*. London: Chapman & Hall.

Draper, N. R., and Smith, H. 1981. *Applied Regression Analysis*. New York: John Wiley & Sons.

Durbin, J., and Watson, G. S. 1951. "Testing for Serial Correlation in Least Squares Regression II." *Biometrika* 38: 159–78.

Fisher, L. D., and van Belle, G. 1993. *Biostatistics: A Methodology for the Health Sciences*. New York: John Wiley & Sons.

Fisher, R. A. 1935. *The Design of Experiments*. New York: Hafner Library.

Freedman, D., Pisani, R., Purves, R., and Adhikari, A. 1991. *Statistics*. New York: W. W. Norton.

Fuller, W. A. 1987. *Measurement Error Models*. New York: John Wiley & Sons.

Good, P. 1994. *Permutation Tests: A Practical Guide to Resampling Methods for Testing Hypotheses*. New York: Springer-Verlag.

Hand, D. J., and Taylor, C. C. 1987. *Multivariate Analysis of Variance and Repeated Measures: A Practical Approach for Behavioural Scientists*. London: Chapman & Hall.

Hill, M. O. 1974. "Correspondence Analysis: A Neglected Multivariate Method." *Applied Statistics* 23: 340–354.

Hoaglin, D. C. 1988. "Transformations in Everyday Experience." *Chance* 1: 40–45.

Hosmer, D. W., Jr., and Lemeshow, S. 1989. *Applied Logistic Regression*. New York: John Wiley & Sons.

Johnson, A. R., and Wichern, D. W. 1992. *Applied Multivariate Statistical Analysis*, 3d ed. Englewood Cliffs, N.J.: Prentice Hall.

Kempthorne, O. 1952. *The Design and Analysis of Experiments*. New York: John Wiley & Sons.

Kotz, N. L., Johnson, S., and Read, C. B. 1988. *Encyclopedia of Statistical Sciences*, vols. 8 and 9. New York: John Wiley & Sons.

————. 1986. *Encyclopedia of Statistical Sciences*, vol. 7. New York: John Wiley & Sons.

————. 1985. *Encyclopedia of Statistical Sciences*, vols. 5 and 6. New York: John Wiley & Sons.

————. 1983. *Encyclopedia of Statistical Sciences*, vols. 3 and 4. New York: John Wiley & Sons.

————. 1982. *Encyclopedia of Statistical Sciences*, vols. 1 and 2. New York: John Wiley & Sons.

Kuehl, R. O. 1994. *Statistical Principles of Research Design and Analysis*. Belmont, Calif.: Duxbury Press.

Mallows, C. L. 1986. "Augmented Partial Residuals." *Technometrics* 28: 313–19.

Mardia, K. V., Kent, J. T., and Bibby, J. M. 1979. *Multivariate Analysis*. New York: Academic Press.

McCullagh, P., and Nelder, J. A. 1989. *Generalized Linear Models*, 2d ed. London: Chapman & Hall.

Mendenhall, W. 1993. *Beginning Statistics A to Z*. Belmont, Calif.: Duxbury Press.

Miller, R. G., Jr. 1986. *Beyond ANOVA: Basics of Applied Statistics*. New York: John Wiley & Sons.

Moore, D. S., and McCabe, G. P. 1993. *Introduction to the Practice of Statistics*, 2d ed. New York: Freeman.

Morrison, D. F. 1990. *Multivariate Statistical Methods*, 3d ed. New York: McGraw Hill.

Mosteller, F., and Tukey, J. 1968. "Data Analsyis—Including Statistics." In G. Lindzey and E. Aronson, eds., *Handbook of Social Psychology, Vol. 2*, 2d ed. (Cambridge, Mass: Addison-Wesley), pp. 80–203.

————. 1977. *Data Analysis and Regression: A Second Course in Statistics*. Reading, Mass.: Addison-Wesley.

Neter, J., Wasserman, W., and Kutner, M. 1990. *Applied Linear Statistical Models*, 3d ed. Homewood, Ill.: Irwin.

Pearson, E. S., and Hartley, H. O. 1972. *Biometrika Tables for Statisticians, Vol. II*, 3d ed. Cambridge, U.K.: University Press.

Picard, R. R., and Berk, K. N. 1990. "Data Splitting." *American Statistician* 44(2): 140–47.

Rice, J. A. 1995. *Mathematical Statistics and Data Analysis*, 2d ed. Belmont, Calif.: Duxbury Press.

Sprent, P. 1993. *Applied Nonparametric Statistical Methods*. London: Chapman & Hall.

Swed, F. S., and Eisenhart, C. 1943. "Tables for Testing Randomness of Groupings in a Sequence of Alternatives." *Annals of Mathematical Statistics* 14: 66–87.

Tanur, J. 1972. *Statistics: A Guide to the Unknown*. San Francisco: Holden-Day.

Thompson, S. K. 1992. *Sampling*. New York: John Wiley & Sons.

Van Ryzin, J., ed. 1977. *Classification and Clustering*. New York: Academic Press.

Weisberg. S. 1985. *Applied Linear Regression*. New York: John Wiley & Sons.

Wetherill, G. B. 1981. *Intermediate Statistical Methods*. London: Chapman & Hall.

Yates, F. 1984. "Tests of Significance for 2×2 Tables." *Journal of the Royal Statistical Society, A*, 147: 426–63.

Answer Section

Chapter 1: Drawing Statistical Conclusions

1.17 The 2-sided p-value is $3/35 = 0.0857$.

1.19 Coin flips will not divide the subjects in such a way that there is an exact age balance. However, it is impossible to tell prior to the flips which group will have a higher average age.

1.21 There is no computation involved. This is, however, a sobering exercise.

1.25
  a. No radiation median is 0; radiation median is 1.
  b. Both distributions are positively skewed. The radiation group has a larger spread.
  c. Over half the numbers in the set are $= 0$.
  d. It is observational data, so a strict interpretation would say that causation cannot be inferred. But what else could it be?

Chapter 2: Inference Using *t*-Distributions

2.13
  a. (Fish, Regular): Averages are $(6.571, -1.143)$; SDs are $(5.855, 3.185)$
  b. Pooled SD $= 4.713$
  c. SE for difference $= 2.519$
  d. d.f. $= 12$; $t_{12}(.975) = 2.179$
  e. 95% CI from 2.225 to 13.203 mm
  f. t-stat $= 3.062$
  g. 1-sided p-value $= .005$.

2.15 t-statistic $= 9.32$, with 174 d.f. Very convincing, indeed.

2.19
  a. Average $= -1.14$; SD $= 3.18$; d.f. $= 6$.
  b. SE $= 1.20$
  c. 95% CI: from -4.09 to 1.80
  d. t-statistic $= -0.95$; 2-sided p-value $= .38$.

Chapter 3:   A Closer Look at Assumptions

3.21 **a.** Yes. One should expect the rates to follow a time series where serial correlation is present.

  b. There is a problem: there is a steady increase, or 'trend', in the series. There is also a somewhat cyclic behavior.

3.23 Use the computer. Refer to Display 3.6.

3.25 **a.** (i) Oil exporters are positively skewed. Industrialized are reasonable symmetric. (ii) The log makes the oil exporters look OK, but the industrialized group gets squashed and has its outlier made more prominent. (iii) estimate $= 1.703$, SE $= 0.280$. (iv) The median per-capita income in the industrialized countries is 5.5 times that in the oil exporting companies. (v) From 1.128 to 2.278. (vi) The industrialized-to-oil exporting ratio of median per-capita incomes is estimated to be between 3.1 and 9.75 (95% confidence interval).

  b. (i) Oil exporters group is a bit long-tailed, and it has greater spread than industrialized group. (ii) Uniform health standards; technology that insulates health conditions from environmental factors. (iii) The log, square root, and reciprocal will fail because the group with the higher average has the smaller spread. The square might work. (iv) These are the conditions where the t-tools fail - the group with the higher spread has the smaller sample size.

3.27 With all the data, the 1-sided p-value is .0405; without the 0.659 value, the 1-sided p-value is .0900. This is a fair swing; the evidence goes from suggestive to faint.

Chapter 4:   Alternatives to the t-Tools

4.15 One-sided p-value $= 2/10 = .20$.

4.17 .0396.

4.19 **a.** 0.1718
  b. Normal approximation
  c. Continuity correction
  d. t-test gives $p = .081$; t-test with removal gives $p = .180$; rank sum gives $p = .1718$.
  e. The rank sum test is valid AND it uses all the data.

4.21 2-sided p-value $= .0314$

4.23 2-sided p-value $= .00643$, compared to .00537.

4.25 CI: (39.59, 165.81), based on Welch's t with 97 d.f. 2-sided p-value $= .0016$. No. It looks like something else is involved.

4.27 2-sided p-value $= .00452$, from signed rank test on the log(ratio) values. On the straight difference scale, the signed rank gives .00208 ... close. It is not particularly apparent.

Chapter 5:   Comparisons Among Several Samples

5.15 **a.** 6.914, with 39 d.f.
  b. t-statistic $= 5.056$, with 39 d.f., giving 1-sided p-value $< .0001$.

5.19 **a.**

CPFA50	CPFA150	CPFA300	CPFA450	CPFA600	Control
168.3	171.7	146.7	151.0	152.3	185.6

There is no suggestion that the size of residuals depends on the average protein level. There is a suggestion that the mean level may change from one day to another.

b. There is convincing evidence that the means are different (p-value $= .0001$). There is ample evidence to suggest that the means under the control treatment are different on different days (p-value $= .0021$).

5.21 See Display 5.23.

Chapter 6: Linear Combinations and Multiple Comparisons of Means

6.13 The 95% confidence intervals are: (Amputee − Crutches) between −3.009 and 0.025; (Amputee − Wheelchair) between −2.431 and 0.603; (Crutches − Wheelchair) between −0.939 and 2.095.

6.15 **a.** 4.484
  b. 1/3, −1/2, −1/2, 1/3, 1/3
  c. $g = 3.000$, SE$(g) = 1.3645$; $t_{40}(.975) = 2.021$, HW $= 2.758$, giving 95% confidence interval: (0.24, 5.76).

6.17 $g = 0.803$

Chapter 7: Simple Linear Regression: A Model for the Mean

7.13 **a.** 15.53
  b. 3.94
  c. 4

7.15 **a.** Intercept estimate $= 0.7648$; slope estimate $= 0.5369$.
  d. 0.1357, with 8 d.f.

7.17 **b.** Estimate of $\mu\{$life $|$ income$\} = 68.87 + 0.00077$ income. 1-sided p-value $= 0.0053$
  c. 0.572.

7.19 The left hand side expressions require a second pass through the data, because one full pass must be made to get the averages of X and Y. Therefore, either the data must be entered twice or it must be stored the first time through.

7.21 **a.** SE$\{$pred$\} = 0.0875$
  b. $5.6139 <$ mean pH < 6.0175.

7.23 About 109.

7.25 **a.**

	H. nudus	*L. bellus*	*C. productus*
slope	0.4083	2.9737	2.0685
SE(slope)	0.5426	0.6125	0.4275

  b. *C. productus* vs. *L. bellus*: t-stat $= 1.212$. 2-sided p-value $= 0.24$
  C. productus vs. *H. nudus*: t-stat $= 2.403$. 2-sided p-value $= 0.025$.

Chapter 8: A Closer Look at Assumptions for Simple Linear Regression

8.15 **b.** Estimated mean number of species $= 24.04928 + 0.00211 \times$ Area.
  c. The regression line does not come near hitting the center of the distribution of species numbers from islands with similar area. There is a pronounced curvature in the residual plot.

8.17   **b.**   Estimated mean log(Mass) $= 3.797 - 0.262 \times$ sqrt(Load)

   c.   The residual plot looks satisfactory.

8.19   **a.**   A residual plot from a fit of all data shows possible curvature and possible outliers.

   b.   Transformation of the scales does not accomplish much in clarifying the situation.

   c.   Fits with and without the bees with duration over 30 seconds give quite different results for the regression. Examination of a residual plot from the fit without durations over 30 seconds shows no further problems.

   d.   Conclusion: For visits under 30 seconds, a straight line regression appears to give a reasonable summary. That description does not extend to visits over 30 seconds.

Chapter 9:   Multiple Regression

9.13   **a.**   The 2-sided p-value is .0132.

   b.   The 2-sided p-value is .50.

   c.   Yes, the scale on which the squared term is insignificant might be more appropriate.

9.15   **b.**   Estimate of $\mu\{yield \mid rain\} = -5.0 + \quad 6.0rain - \quad .229rain^2$

   SEs: (11.4)   (2.0)   (.089)

   d.   Estimate of $\mu\{yield \mid rain, year\} = -263 + \quad 5.7rain - \quad .216rain^2 + \quad .136year$

   SEs: (98)   (1.900)   (0.082)   (0.052)

   e.   Estimate of $\mu\{yield \mid rain, year\} =$

           $-1909 + \quad 159rain - 0.186rain^2 + \quad 1.00year - 0.081rain \times year$

   SEs: (486)   (45)   (.072)   .26   .023

   The 2-sided p-value for significance of the interaction term is .0016. This indicates that the effect of rainfall on yield is smaller for years closer to 1927 than for years closer to 1890. One possible explanation is that in later years the yield became less dependent on rainfall.

9.17   **b.**   Let *datej* represent the indicator variable for date j, for $j = 2, \ldots, 8$).

   Estimate of $\mu\{interval \mid dur, DATE\} =$

   $32.9 + \quad 10.9dur + 1.3date2 + 0.8date3 + 0.2date4 + 0.2date5 + 2.0date6 - 0.2date7 - 0.7date8$

   (3.1)   (0.66)   (2.7)   (2.7)   (2.6)   (2.6)   (2.6)   (2.7)   (2.7)

Chapter 10:   Inferential Tools for Multiple Regression

10.9   **a.**   32.

   b.   .0014

   c.   0.053 to 3.267

10.11   **a.**   The 2-sided p-value $= .2443$; the 1-sided p-value is .1222; yes. Failure to account for sampling effort could lead to concluding there was a relationship between species numbers and reserve size, when the full data would suggest that different effort may be a better explanation.

   b.   The p-value is $< .0001$.

   c. -0.1634 to $+0.3252$.

10.13   **b.**   The slope estimate is 0.8150 for all three groups. The intercept estimates are: (i) -1.5764 for non-echolocating bats; (ii) -1.4741 for birds; and (iii) -1.4977 for echolocating bats.

   d.   Same as b.

   e.   2-sided p-value $= .8828$.

10.15 The F-statistic for inclusion of the day indicators is $F = 0.209$. The p-value is .98.

10.17
  a. $R^2 = 92.64\%$; adjR$^2 = 91.16\%$
  b. $R^2 = 99.03\%$; adjR$^2 = 98.55\%$
  c. $R^2 = 99.94\%$; adjR$^2 = 99.87\%$
  d. $R^2 = 99.98\%$; adjR$^2 = 99.95\%$
  e. $R^2 = 99.996\%$; adjR$^2 = 99.977\%$
  f. $R^2 = 100\%$; adjR$^2 =$ undefined

10.19
  c. The p-value $= .89$.

Chapter 11: Model Checking and Refinement

11.11
  a. Cook's distance identifies sample #17 as being highly influential. This is the result of a high leverage (0.755) in combination with a large Studentized residual (3.468).
  b. Sample #1 now becomes a slight problem.

11.13 We do, naturally.

11.15
  c. The externally Studentized residual for case 22 is 3.327.
  d. They differ because including case 22 increases the estimate of the residual standard deviation, which is a divisor of the raw residual.
  e. The long-tailed aspect of these data is apparent in a normal plot of the residuals.
  f. The externally Studentized residual for case $22 = 3.327$. This is very large for a normal deviate. However, recall that predictions are reliant on the normal distribution assumption, which is in question from other cases besides 22 in the normal plot. Because the normal plot shows that case 22 belongs to the same long-tailed pattern exhibited by the remainder of the data, there is not convincing evidence that this election was fraudulent.

11.17
  d. The strongest relationship is between the response and Sacrifice Time.

11.19
  a. Estimates for means of log tumor-to-liver ratio are (SEs in parentheses):

	Sacrifice Time (hours)			
	0.5	3.0	24	72
BD	−3.505 (0.195)	−2.371 (0.205)	+0.752 (0.209)	+1.649 (0.209)
NS	−4.302 (0.205)	−3.168 (0.195)	−0.044 (0.209)	+0.852 (0.209)

  b. Same results!

Chapter 12: Model Selection with Large Numbers of Explanatory Variables

12.11 Forward selection settles on the model B.

12.13 Here are the results of one simulation. Each solution will differ, however, in detail.
  a. 11.9%.
  b. One variable entered, using a 4.0 cutoff for the F-statistic.
  c. The Cp statistic chose a model with two variables.
  d. The BIC (correctly) chose a model with the constant term only.

12.15   **a.**   The model with all four design variables has the smallest Cp statistic, at 3.50.

b.   Forward selection proceeds directly to the model with the four design variables, adding Sac Time 72 ($F = 24.83$), then Sac Time 24 ($F = 126.28$), then Sac Time 3 ($F = 10.86$), and then BD ($F = 18.89$).

c.   Backward elimination gets to the same model.

d.   Stepwise does not alter the selection outcome.

e.   The smallest BIC ($= -25.18$) is for the model with only the four design variables.

12.17   **a.**   The p-value $= .0083$.

b.   The p-value $= .0308$. The conclusion is about the same, although there is some disagreement about the strength of the evidence.

Chapter 13: The Analysis of Variance for Two-way Classification

13.11   **a.** p-value for interaction $= .72$.

b. p-value for treatment effect in the additive model, $= .012$.

c.   Yes.

13.15   **b.** p-value $= .92$.

c. p-value for effect of treatment $= .00016$. p-value for effect of sacrifice time $< .0001$.

Chapter 14: Multifactor Studies

14.9   **b.**   The difference between slopes at $H_2O = -.05$ and $-.40$ is -1.44713. A 95% confidence interval for difference between slopes is $(-4.24109, +1.34683)$. Exponentiate to get an estimate of 0.235, with a 95% CI of (0.014, 3.845).

14.11   **c.**   There is a much more obvious treatment effect when the sex is accounted for. Although the estimate of the effect is the same in both cases, the standard error is much smaller when sex is accounted for (in the randomized block version of the experiment in part **b**). It is smaller because SD{*score* | *treat, sex*} $<$ SD{*score* | *treat*}.

14.13   **a.**   Two-sided p-value $= .50$.

Chapter 15: Adjustment for Serial Correlation

15.7   **b.**   2-sided p-value, .0201.

c.   The first serial correlation coefficient is $r_1 = -0.3610$; two-sided p-value $= .0002$.

d.   two-sided p-value $= .9856$

15.9   **c.**   For this segment of the sunspot series, an AR(2) model for the square root transformed numbers appears satisfactory.

Chapter 16: Repeated Measures

16.5   **a.** $r = 0.1310$

b.   (i) R-squared $= 11.08\%$ (ii) p-value $= .7379$ (iii) R-square $= 0.49\%$ (iv) p-value $= .0153$

16.7   **a.**   Neither 95% confidence interval includes zero.

b. p-value $= .0643$.

   c.   The resulting 95% confidence interval includes $(0, 0)$, because the p-value exceeds .05.

   d.   This may also seem to be contradictory, but only to those who cling doggedly to the .05 level cutoff. The evidence registered against zero values—.0409, .0366, .0643—is relatively consistent.

16.9   Here are four examples representing different regions around the confidence ellipse.

Short term	Long term	T-square	F-statistic	p-value
8	8	13.76	6.45	.0095
11	–7	7.80	3.65	.0509
25	0	5.61	2.63	.1049
0	0	26.30	12.53	.0007

16.11   **d.**   (i) A multivariate regression is the best way to view the situation, with the *P2P ratio* and the *%Indigenous* as responses to *%Catholic* as the explanatory variable. (ii) The inference from multivariate regression can be approximated by separate inferences from univariate regression. (iii) Countries with high population percents being Catholic generally have low numbers of priests per parishioner; these countries with low numbers of priests per parishioners generally have higher percents of the priests being non-indigenous.

Chapter 17: Exploratory Tools for Summarizing Multivariate Responses

17.9   The first three principal components account for about 95% of the total variation. (1) (M2 + M5 + M6 + M7 + M11)/5; M2; and (M5 + M11)/2 − (M6 + M7)/2; or (2) (M2 + M5 + M7)/3; M2; and M5 − M7.

17.11   **a.**   The largest canonical correlation between passionate responses and compassionate responses is 0.5433. The test that all four canonical correlations are zero has a p-value of .4375, indicating no evidence of any correlation between the two sets of responses.

   b.   The largest canonical correlation between husbands' responses and wives' responses is .5717, with associated p-value = .4572. Again, there is no evidence that the two sets of responses are correlated in any way.

Chapter 18: Comparisons of Proportions or Odds

18.9   **a.**   (i) 0.01832 in obese group; 0.01537 in not obese group. (ii) 0.00505. (iii) −0.00696 to +0.01285.

   b. z-statistic = 0.5825; one-sided p-value = .2801.

   c.   (i) 0.01866 and 0.01561 (ii) 1.195 (iii) 0.304 (iv) 0.66 to 2.17.

   d.   The odds of CVD death in the obese group are estimated to be between 0.66 and 2.17 times the odds of CVD death in the not obese group (95% confidence interval).

18.11   **a.**   (i) 0.00050 (ii) 0.000250 (iii) 0.6429

   b.   (i) 0.00025 (ii) 0.00025 (iii) 0.4737

   c.   (i) 0.00025 (ii) 0.00025 (iii) 0.1692

         Retrospective samples (of equal size) do not estimate the population proportions.

18.13   **a.**   If, for example, $\pi_u = .0010$ and $\pi_v = .0002$, the relative risk is $\rho = 5.0$, while the odds ratio is $\omega = 5.004$.

b.   3.69

c.   2.403

Chapter 19: More Tools for Tables of Counts

19.11 $C_{503,0} \times C_{317,5}/C_{820,5} = (317 \times 316 \times 315 \times 314 \times 313)/(820 \times 819 \times 818 \times 817 \times 816) = .0085.$

19.13   **a.**   Excess = 5.5; SE(Excess) = 1.963; z-statistic = 2.80; one-sided p-value = .0026.

        b.   Fisher's Exact Test: one-sided p-value = .0044.

19.15   **a.**   No. The expected number of used and not used houses of old wood are both smaller than 5.

        b.   1-sided p-value = .0057.

Chapter 20: Logistic Regression for Binary Response Variables

20.9   **a.**   Estimated survival probabilities for males are .413 at age 25 and .091 at age 50. For females, the estimates are .777 at age 25 and .332 at age 50.

        b.   For females, the age of 50% survival is 41.0 years; for males it is 20.5 years.

20.11   **a.**
$$\text{logit}(\hat{\pi}) = 10.8753 - 0.1713 \text{ temperature}$$
$$(5.7031) \quad (0.0834)$$

        b.   One-sided p-value = .0200, from z-statistic = −2.054.

        c.   Drop in deviance = 5.9441 with 1 d.f. gives p-value = .0148. This is a two-sided p-value for the coefficient, so the approximate one-sided p-value based on the deviance would be .0074.

        d.   The 95% confidence interval extends from −0.3348 to −0.0078.

        e.   logit = 5.5650; estimated probability of failure = .9962.

        f.   It represents an extrapolation beyond the range of the available data.

20.13 p-value = .3008.

Chapter 21: Logistic Regression for Binomial Counts

21.9   **b.**   The estimate of the intercept is 0.2805, with standard error = 0.2309. The slope estimate is −0.0187, with standard error 0.0068.

        c.   The p-value for goodness of fit is .8388.

21.11   The drop in deviance for interaction terms is 2.2391, with 5 d.f. (p-value = .8152). There is no evidence to suggest that the equal odds ratio assumption is inadequate.

21.13   **b.**   The fit:
$$\text{logit}(\hat{\pi}) = -1.2003 - 0.0346 \text{ dose}$$
$$(0.0617) \quad (0.0711)$$

        The goodness of fit p-value = .7674, from the deviance = 0.5296 with 2 d.f.. Wald's test for the coefficient of dose is z = −0.4866, giving a 2-sided p-value = .6265. The drop in deviance test has a p-value = .6253, based on the drop = 0.2385 with 1 d.f.

        c.   There is not evidence that the model is inadequate. Nor is there much evidence that the odds of a cold are associated with the daily dose of vitamin C.

Chapter 22: Log-Linear Regression for Poisson Counts

22.19   There are substantial differences when case 49 is included. A main one is that the effect of treatment effect is significant and negative. Apparently patient 49, who was in the treated group, had an extremely large post-treatment count (substantially larger than the pre-treatment count), thus making it appear that number of seizures were greater for those receiving the treatment than for those who did not.

22.21   The two-sided p-value from the deviance test is .73.

22.23   **a.**   The deviance goodness of fit p-value is less than .0001, providing overwhelming evidence of lack of fit of the Poisson model.

   b.   Using quasi-likelihood analysis and backward elimination—discarding a variable at each step if its t-statistic is smaller in magnitude than 2—leads to the model with log area and log of area of the nearest neighbor for describing the mean number of non-native species.

   c.   For each doubling of island area the mean number of nonnative species increases by 35%. For a given island area, the mean number of nonnative species decreases by 9% for each doubling of area of the nearest island.

Chapter 23: Elements of Research Design

23.13   Here the response is binary: $Y = 1$ for responding to the drug and $= 0$ for non-response. The control proportion is 0.60 for the standard treatment, and the desired alternative is 0.75 for taxol. The desired odds ratio is 2.0. The sample size in each group should be 304.

23.15   The non-victim proportion is 0.20, corresponding to odds of 0.25. An odds ratio of 3.0 translates to odds of 0.75, or a victim proportion of 0.4286. Distinguishing the two can be done with a sample of 132 from each group.

Chapter 24: Factorial Treatment Arrangements and Blocking Designs

24.11   **b.**   There are indicator variables for A, for B, for the interaction of A and B, and for 11 of the 12 leaves (blocks). The resulting regression fit is: $0.715 + 0.616A - 0.272B + 0.119AB - 0.433L2 - 0.841L3 + 0.009L4 - 1.039L5 - 1.200L6 - 2.013L7 - 1.144L8 - 1.803L9 - 1.294L10 + 0.013L11 - 1.211L12$. Adding 6 to both responses in leaf 8, 2 to both responses in leaf 11, and 15 to both responses in leaf 3 does not change coefficients of A, B, and AB. The coefficients of L3, L8, and L11 increase by 15, 6, and 2.

Index